Biophotonics
Visions for Better Health Care

Edited by
Jürgen Popp and Marion Strehle

Related Titles

Popp, J., Liedtke, S.

Laser, Licht und Leben

Techniken in der Medizin

202 pages
2006
Hardcover
ISBN 3-527-40636-0

Prasad, P. N.

Introduction to Biophotonics

593 pages
2003
Hardcover
ISBN 0-471-28770-9

Heath, J. P.

Dictionary of Microscopy

358 pages
2005
Paperback
ISBN 0-470-01199-8

Saliterman, S.

Fundamentals of BioMEMS and Medical Microdevices

576 pages
2005
Hardcover
ISBN 0-8194-5977-1

Hao, L., Lawrence, J.

Laser Surface Treatment of Bio-Implant Materials

232 pages
2006
Hardcover
ISBN 0-470-01687-6

Biophotonics

Visions for Better Health Care

Edited by
Jürgen Popp and Marion Strehle

WILEY-VCH

WILEY-VCH Verlag GmbH & Co. KGaA

The Editors

Prof. Dr. Jürgen Popp
Institute of Physical Chemistry
Friedrich-Schiller-University Jena, Germany
juergen.popp@uni-jena.de

Dr. Marion Strehle
Institute of Physical Chemistry
Friedrich-Schiller-University Jena, Germany
marion.strehle@uni-jena.de

Forschungsschwerpunkt
Biophotonik

GEFÖRDERT VOM

Bundesministerium
für Bildung
und Forschung

Co-ordinated by the VDI-Technologiezentrum.

Library of Congress Card No.: applied for.

British Library Cataloging-in-Publication Data:
A catalogue record for this book is available from the British Library.

Bibliographic information published by Die Deutsche Bibliothek
Die Deutsche Bibliothek lists this publication in the Deutsche Nationalbibliografie; detailed bibliographic data is available in the Internet at <http://dnb.ddb.de>.

Typesetting Uwe Krieg, Berlin
Printing betz-Druck GmbH, Darmstadt
Binding J. Schäffer GmbH, Grünstadt
Cover Design Peter Hesse, Berlin

Printed in the Federal Republic of Germany
Printed on acid-free paper

ISBN-13: 978-3-527-40622-7
ISBN-10: 3-527-40622-0

Contents

Biophotonics: Visions for Better Health Care. Jürgen Popp and Marion Strehle (Eds.)

ISBN: 3-527-40622-0

List of Contributors

Volker Andresen *Ch. 7*
LaVision BioTec GmbH
Meisenstr. 65
33607 Bielefeld
Germany
andresen@lavisionbiotec.com

Dario Anselmetti *Ch. 7*
Experimental Biophysics & Applied Nanosciences
Faculty of Physics
University of Bielefeld
Universitätsstraße 25
33615 Bielefeld
Germany
dario.anselmetti@physik.uni-bielefeld.de

Mathias Arnold *Ch. 9*
DASGIP AG
Rudolf Schulten Straße 5
52428 Jülich
Germany
m.arnold@dasgip.de

Gert von Bally *Ch. 6*
Universitätsklinikum Münster
Labor für Biophysik
Robert-Koch-Str. 45
48129 Münster
Germany
LBiophys@uni-muenster.de

Jürgen Beuthan *Ch. 6*
Charité - Universitätsmedizin Berlin
Institut für Medizinische Physik und Lasermedizin
Fabeckstr. 60–62
14195 Berlin
Germany
juergen.beuthan@charite.de

Uwe Bindig *Ch. 5*
Laser- und Medizin-Technologie GmbH, Berlin
Fabeckstraße 60–62
14195 Berlin
Germany
u.bindig@lmtb.de

Sergey Borisov *Ch. 9*
Institut für Analytische Chemie, Chemo- und Biosensorik
Universität Regensburg
Universitätsstraße 31
93040 Regensburg
Germany
sergey.borisov@chemie.uni-regensburg.de

Albrecht Brandenburg *Ch. 2*
Fraunhofer Institute for Physical Measurement Techniques
Department of Optical Spectroscopy and Systems
Heidenhofstr. 8
79110 Freiburg
Germany
albrecht.brandenburg@ipm.fraunhofer.de

Biophotonics: Visions for Better Health Care. Jürgen Popp and Marion Strehle (Eds.)

ISBN: 3-527-40622-0

Gernot Breitfuss *Ch. 2*
Breitfuss Messtechnik GmbH
Danziger Str. 20
27243 Harpstedt
Germany
g.breitfuss@breitfuss.de

Andrea Buchstaller *Ch. 6*
Ludwig-Maximilians-Universität München
Pathologisches Institut
Thalkirchner Str. 36
80337 München
Germany
andrea.buchstaller@med.uni-muenchen.de

Mario Bürger *Ch. 5*
GeSiM – Gesellschaft für Silizium-Mikrosysteme mbH
Rossendorfer Technologiezentrum
Bautzner Landstraße 45
01454 Grosserkmannsdorf
Germany
buerger@gesim.de

Hans Burkhardt *Ch. 2, 3*
Lehrstuhl für Mustererkennung und Bildverarbeitung
Institut für Informatik
Albert-Ludwigs-Universität Freiburg
Georges-Koehler-Allee Geb. 052
79110 Freiburg
Germany
hans.burkhardt@informatik.uni-freiburg.de

Daniel Carl *Ch. 6*
Universitätsklinikum Münster
Labor für Biophysik
Robert-Koch-Str. 45
48129 Münster
Germany
LBiophys@uni-muenster.de

Julien Colombelli *Ch. 10*
Cell Biology and Biophysics Unit
European Molecular Biology Laboratory
Meyerhofstraße 1
69117 Heidelberg
Germany
julien.colombelli@embl-heidelberg.de

Christoph Cremer *Ch. 4*
Kirchhoff Institut für Physik
Universität Heidelberg
Im Neuenheimer Feld 227
69120 Heidelberg
Germany
cremer@kip.uni-heidelberg.de

Dalia Denapaite *Ch. 4*
Fachbereich Biologie
Abt. Mikrobiologie
Technische Universität Kaiserslautern
Paul-Ehrlich-Straße
Gebäude 23
6766 Kaiserslautern
Germany
denapait@rhrk.uni-kl.de

Philip Denner *Ch. 10*
Schering AG
Enabling Technologies / AD-HTS
Muellerstr. 170-178
13342 Berlin
Germany
philip.denner@schering.de

Joachim Diebold *Ch. 10*
Pathologisches Institut der Universität München
Thalkirchner Str. 36
80337 München
Germany
joachim.diebold@lrz.uni-muenchen.de

Christian Dietrich *Ch. 6, 10*
Carl Zeiss Jena GmbH
Research Center
Carl-Zeiss-Promenade 10
07745 Jena
Germany
c.dietrich@zeiss.de

Johannes Donauer *Ch. 8*
University Hospital Freiburg
Hugstetter Straße 55
79106 Freiburg
Germany
donauer@medizin.ukl.uni-freiburg.de

Wilhelm Dunkhorst *Ch. 2*
Fraunhofer Institute for Toxicology and Experimental Medicine
Department of Aerosol Technology
Nikolai-Fuchs-Str. 1
30625 Hanover
Germany
wilhelm.dunkhorst@item.fraunhofer.de

Markus von Ehr *Ch. 2*
Fraunhofer Institute for Physical Measurement Techniques
Department of Optical Spectroscopy and Systems
Heidenhofstr. 8
79110 Freiburg
Germany
vonEhr@ipm.fraunhofer.de

Bernd M. Fischer *Ch. 8*
Freiburger Materialforschungszentrum
Stefan-Meier-Straße 21
79104 Freiburg
Germany
bernd.fischer@physik.uni-freiburg.de

Rüdiger Frank *Ch. 8*
Eberhard-Karls University
Institute of Physical and Theoretical Chemistry
Auf der Morgenstelle 8
72076 Tübingen
Germany
ruediger.frank@ipc.uni-tuebingen.de

Marcus Fratz *Ch. 2*
Fraunhofer Institute for Physical Measurement Techniques
Department of Optical Spectroscopy and Systems
Heidenhofstr. 8
79110 Freiburg
Germany
markus.fratz@ipm.fraunhofer.de

Andreas Frey *Ch. 5*
Laborgruppe Mukosaimmunologie
Abteilung Klinische Medizin
Forschungszentrum Borstel
Parkallee 22
23845 Borstel
Germany
afrey@fz-borstel.de

Dirk Gansert *Ch. 9*
Institut für Ökologische Pflanzenphysiologie und Geobotanik
Universität Düsseldorf
Universitätstraße 1
40225 Düsseldorf
Germany
gansert@uni-duesseldorf.de

Günter Gauglitz *Ch. 8*
Eberhard-Karls University
Institute of Physical and Theoretical Chemistry
Auf der Morgenstelle 8
72076 Tübingen
Germany
guenter.gauglitz@ipc.uni-tuebingen.de

Ingo Gersonde *Ch. 6*
LMTB GmbH Berlin
Fabeckstr. 60–62
14195 Berlin
Germany
i.gersonde@lmtb.de

Dominik Giel *Ch. 2*
Fraunhofer Institute for Physical Measurement Techniques
Department of Optical Spectroscopy and Systems
Heidenhofstr. 8
79110 Freiburg
Germany
dominik.giel@ipm.fraunhofer.de

Carsten Glasenapp *Ch. 8*
Laboratory for Micro-optics
Department of Microsystems Engineering – IMTEK
University of Freiburg
Georges-Koehler-Allee 102
79110 Freiburg
Germany
glasenap@imtek.uni-freiburg.de

Steffi Grohmann *Ch. 7*
Institute for Bioprocessing and Analytical Measurement Techniques
Department of Biomaterials
Rosenhof
37308 Heiligenstadt
Germany
steffi.grohmann@iba-heiligenstadt.de

Bernhard Häfner *Ch. 4*
Physikalisch-Chemisches Institut
Universität Heidelberg
Im Neuenheimer Feld 253
69120 Heidelberg
Germany
bernhard.haefner@gmx.de

Regine Hakenbeck *Ch. 4*
Fachbereich Biologie
Abt. Mikrobiologie
Technische Universität Kaiserslautern
Paul-Ehrlich-Straße
Gebäude 23
6766 Kaiserslautern
Germany
regine.hakenbeck@t-online.de

Michaela Harz *Ch. 3*
Institut für Physikalische Chemie
Friedrich-Schiller-Universität Jena
Helmholtzweg 4
07743 Jena
Germany
michaela.harz@uni-jena.de

Michael Hausmann *Ch. 4*
Kirchhoff Institut für Physik
Universität Heidelberg
Im Neuenheimer Feld 227
69120 Heidelberg
Germany
hausmann@kip.uni-heidelberg.de

Ulrich Heimann *Ch. 2*
German Weather Service
Department of Human Biometeorology
79104 Freiburg
Germany
ulrich.heimann@dwd.de

Jürgen Helfmann *Ch. 5*
Laser- und Medizin-Technologie GmbH, Berlin
Fabeckstraße 60–62
14195 Berlin
Germany
j.helfmann@lmtb.de

Hanspeter Helm *Ch. 8*
Department of Molecular and Optical Physics
University of Freiburg
Stefan-Meier-Str. 19
79104 Freiburg
Germany
hanspeter.helm@physik.uni-freiburg.de

Dirk-Peter Herten *Ch. 4*
Physikalisch-Chemisches Institut
Universität Heidelberg
Im Neuenheimer Feld 253
69120 Heidelberg
Germany
dirk-peter.herten@urz.uni-hd.de

Gerhard Hildebrand *Ch. 7*
Institute for Bioprocessing and Analytical Measurement Techniques
Department of Biomaterials
Rosenhof
37308 Heiligenstadt
Germany
gerhard.hildebrand@iba-heiligenstadt.de

Paul Hing *Ch. 8*
Sensovation AG
Ludwigshafener Straße 29
78333 Stockach
Germany
paul.hing@sensovation.com

Stefan Hofer *Ch. 3*
Kayser-Threde GmbH
Wolfratshauser Str. 48
81379 München
Germany
stefan.hofer@kayser-threde.com

Matthias Hoffmann *Ch. 8*
Department of Molecular and Optical Physics
University of Freiburg
Stefan-Meier-Str. 19
79104 Freiburg
Germany
matthias.hoffmann@physik.uni-freiburg.de

Lutz Höring *Ch. 10*
Carl Zeiss AG
Forschungszentrum Oberkochen
Carl Zeiss-Str. 22
73446 Oberkochen
Germany
l.hoering@zeiss.de

Klaus Irion *Ch. 6*
Karl Storz GmbH & Co. KG
Mittelstr. 8
78532 Tuttlingen
Germany
k.irion@karlstorz.de

Peter Uhd Jepsen *Ch. 8*
Research Center COM
Technical University of Denmark
2800 Kgs. Lyngby
Denmark
jepsen@com.dtu.dk

Michael Kempe *Ch. 6*
Carl Zeiss Jena GmbH
Research Center
Carl-Zeiss-Promenade 10
07745 Jena
Germany
m.kempe@zeiss.de

Björn Kemper *Ch. 6*
Universitätsklinikum Münster
Labor für Biophysik
Robert-Koch-Str. 45
48129 Münster
Germany
LBiophys@uni-muenster.de

Thomas Kleine-Ostmann *Ch. 8*
Institut für Hochfrequenztechnik
TU Braunschweig
Schleinitzstraße 22
38106 Braunschweig
Germany
t.kleine-ostmann@tu-bs.de

Jens-Peter Knemeyer *Ch. 4*
Physikalisch-Chemisches Institut
Universität Heidelberg
Im Neuenheimer Feld 253
69120 Heidelberg
Germany
j.knemeyer@gmx.de

Sabine Knoche *Ch. 6*
Universitätsklinikum Münster
Labor für Biophysik
Robert-Koch-Str. 45
48129 Münster
Germany
LBiophys@uni-muenster.de

Martin Koch *Ch. 8*
Institut für Hochfrequenztechnik
TU Braunschweig
Schleinitzstraße 22
38106 Braunschweig
Germany
martin.koch@tu-bs.de

Wolfgang Koch *Ch. 2*
Fraunhofer Institute for Toxicology and Experimental Medicine
30625 Hanover
Germany
koch@item.fhg.de

Christian Krause *Ch. 9*
PreSens Precision Sensing GmbH
Josef Engert Straße 9
93053 Regensburg
Germany
presens@t-online.de

Holger Krause *Ch. 8*
Laboratory for Micro-optics
Department of Microsystems Engineering – IMTEK
University of Freiburg
Georges-Koehler-Allee 102
79110 Freiburg
Germany
krause_ho@web.de

Mario Krause *Ch. 3*
Institut für Physikalische Chemie
Friedrich-Schiller-Universität Jena
Helmholtzweg 4
07743 Jena
Germany
mario.krause@uni-jena.de

Gernot Langer *Ch. 10*
Schering AG
Enabling Technologies / AD-HTS
Muellerstr. 170-178
13342 Berlin,
Germany
gernot.langer@schering.de

Markus Lankers *Ch. 3*
rap.ID Particle Systems GmbH
Köpenicker Str. 325/ Haus 40
12555 Berlin
Germany
markus.lankers@rap-id.com

Nicolae Leopold *Ch. 8*
Eberhard-Karls University
Institute of Physical and Theoretical Chemistry
Auf der Morgenstelle 8
72076 Tübingen
Germany
nicolae.leopold@ipc.uni-tuebingen.de

Susanne Liedtke *Ch. 1, 11*
Institut für Physikalische Chemie
Friedrich-Schiller-Universität Jena
Helmholtzweg 4
07743 Jena
Germany
susanne.liedtke@uni-jena.de

Klaus Liefeith *Ch. 7*
Institute for Bioprocessing and Analytical Measurement Techniques
Department of Biomaterials
Rosenhof
37308 Heiligenstadt
Germany
klaus.liefeith@iba-heiligenstadt.de

Hubert Lödding *Ch. 2*
Fraunhofer Institute for Toxicology and Experimental Medicine
Medicine Department of Aerosol
Nikolai-Fuchs-Str. 1
30625 Hanover
Germany
hubert.loedding@item.fraunhofer.de

Udo Löhrs *Ch. 10*
Pathologisches Institut der Universität München
Thalkirchner Str. 36
80337 München
Germany
udo.loehrs@med.uni-muenchen.de

Jörg Martini *Ch. 7*
Experimental Biophysics & Applied Nanosciences
Faculty of Physics
University of Bielefeld
Universitätsstraße 25
33615 Bielefeld
Germany
jmartini@physik.uni-bielefeld.de

Barbara Meckelein *Ch. 5*
Forschungszentrum Borstel
Laborgruppe Mukosaimmunologie
Abteilung Klinische Medizin
Forschungszentrum Borstel
Parkallee 22
23845 Borstel
Germany
bfrey@fz-borstel.de

Wolfgang Mönch *Ch. 8*
Laboratory for Micro-optics
Department of Microsystems Engineering – IMTEK
University of Freiburg
Georges-Koehler-Allee 102
79110 Freiburg
Germany
moench@imtek.uni-freiburg.de

Volker Mordstein *Ch. 10*
Pathologisches Institut der Universität München
Thalkirchner Str. 36
80337 München
Germany
volker.mordstein@med.uni-muenchen.de

Hans-Walter Motzkus *Ch. 3*
Schering AG
Müllerstr. 178
13353 Berlin
Germany
HansWalter.Motzkus@schering.de

Andrea Müller *Ch. 9*
DASGIP AG
Rudolf Schulten Straße 5
52428 Jülich
Germany
m.arnold@dasgip.de

Matthias Müller *Ch. 4*
Physikalisch-Chemisches Institut
Universität Heidelberg
Im Neuenheimer Feld 253
69120 Heidelberg
Germany
tiasmueller@web.de

Werner Müller *Ch. 2*
Helmut Hund GmbH
Wilhelm-Will-Str. 7
35580 Wetzlar
Germany
w.mueller@hund.de

Tina Mutschler *Ch. 8*
Eberhard-Karls University
Institute of Physical and Theoretical Chemistry
Auf der Morgenstelle 8
72076 Tübingen
Germany
tina.mutschler@ipc.uni-tuebingen.de

Andreas Nolte *Ch. 10*
Carl Zeiss AG
Geschäftsbereich Lichtmikroskopie
Königsallee 9–21
38081 Göttingen
Germany
a.nolte@zeiss.de

Oliver Nolte *Ch. 4*
Physikalisch-Chemisches Institut
Universität Heidelberg
Im Neuenheimer Feld 253
69120 Heidelberg
Germany
webmaster@coleopterologe.de

Karsten Parczyk *Ch. 10*
Schering AG
Enabling Technologies / AD-HTS
Muellerstr. 170-178
13342 Berlin,
Germany
karsten.parczyk@schering.de

Klaus-Dieter Peschke *Ch. 3*
Lehrstuhl für Mustererkennung und Bildverarbeitung
Institut für Informatik
Albert-Ludwigs-Universität Freiburg
Georges-Koehler-Allee Geb. 052
79110 Freiburg
Germany
peschke@informatik.uni-freiburg.de

Renate Petry *Ch. 3*
Institut für Physikalische Chemie
Friedrich-Schiller-Universität Jena
Helmholtzweg 4
07743 Jena
Germany
renate.petry@uni-jena.de

Jürgen Popp *Ch. 1, 3, 11*
Institut für Physikalische Chemie
Friedrich-Schiller-Universität Jena
Helmholtzweg 4
07743 Jena
Germany
and
Institut für Physikalische Hochtechnologie
Albert-Einstein-Str. 9
07745 Jena
Germany
juergen.popp@uni-jena.de

Alfons Riedinger *Ch. 10*
EMBL Workshops
European Molecular Biology Laboratory
Meyerhofstraße 1
69117 Heidelberg
Germany
alfons.riedinger@embl.de

Stefan Prechtl *Ch. 10*
Schering AG
Enabling Technologies / AD-HTS
Muellerstr. 170-178
13342 Berlin
Germany
stefan.prechtl@schering.de

Rainer Riesenberg *Ch. 3*
Institut für Physikalische Hochtechnologie
Albert-Einstein-Str. 9
07745 Jena
Germany
rainer.riesenberg@ipht-jena.de

Niels Röckendorf *Ch. 5*
Laborgruppe Mukosaimmunologie
Abteilung Klinische Medizin
Forschungszentrum Borstel
Parkallee 22
23845 Borstel
Germany
nroeckendorf@fz-borstel.de

Olaf Ronneberger *Ch. 2, 3*
Lehrstuhl für Mustererkennung und Bildverarbeitung
Institut für Informatik
Albert-Ludwigs-Universität Freiburg
Georges-Koehler-Allee Geb. 052
79110 Freiburg
Germany
ronneber@informatik.uni-freiburg.de

Petra Rösch *Ch. 3*
Institut für Physikalische Chemie
Friedrich-Schiller-Universität Jena
Helmholtzweg 4
07743 Jena
Germany
petra.roesch@uni-jena.de

Frank Rutz *Ch. 8*
Institut für Hochfrequenztechnik
TU Braunschweig
Schleinitzstraße 22
38106 Braunschweig
Germany
f.rutz@tu-bs.de

Nicholas Salmon *Ch. 10*
SLS Software Technologies GmbH
Meyerhofstraße 1
D-69117 Heidelberg
Germany
Salmon@SLS-Software.de

Markus Sauer *Ch. 4*
Angewandte Laserphysik und Laserspektroskopie
Universität Bielefeld
Universitätsstr. 25
33615 Bielefeld
Germany
sauer@physik.uni-bielefeld.de

Daniela Schädel *Ch. 5*
Laser- und Medizin-Technologie GmbH, Berlin
Fabeckstraße 60–62
14195 Berlin
Germany
d.schaedel@lmtb.de

Ronald Schade *Ch. 7*
Institute for Bioprocessing and Analytical Measurement Techniques
Department of Biomaterials
Rosenhof
37308 Heiligenstadt
Germany
ronald.schade@iba-heiligenstadt.de

Stefan Scharring *Ch. 2*
German Weather Service
Department of Human Biometeorology
79104 Freiburg
Germany
stefan.scharring@dwd.de

Günther Schmauz *Ch. 3*
Fraunhofer Institut für Produktionstechnik und Automatisierung
Nobelstr. 12
70569 Stuttgart
Germany
schmauz@ipa.fraunhofer.de

Marcus Alexander Schmidt *Ch. 5*
Institut für Infektiologie
Zentrum für Molekularbiologie der Entzündung
Universitätsklinikum Münster
von-Esmarch-Straße 56
48149 Münster
infekt@uni-muenster.de

Eberhard Schmitt *Ch. 4*
Fritz-Lipmann Institut (FLI)
Beutenbergstr. 11
07745 Jena
Germany
eschmitt@imb-jena.de

Michael Schmitt *Ch. 1, 11*
Institut für Physikalische Chemie
Friedrich-Schiller-Universität Jena
Helmholtzweg 4
07743 Jena
Germany
m.schmitt@uni-jena.de

Jürgen Schnekenburger *Ch. 6*
Universitätsklinikum Münster
Medizinische Klinik B
Domagkstr. 3A
48129 Münster
Germany
schnekenburger@uni-muenster.de

Andreas Schüle *Ch. 3*
Fraunhofer Institut für Produktionstechnik und Automatisierung
Nobelstr. 12
70569 Stuttgart
Germany
schuele@ipa.fraunhofer.de

Eckart Schultz *Ch. 2*
German Weather Service
Department of Human Biometeorology
79104 Freiburg
Germany
eckart.schultz@dwd.de

Karin Schütze *Ch. 6*
P.A.L.M. Microlaser Technologies GmbH
Am Neuland 9
82347 Bernried
Germany
karin.schuetze@palm-microlaser.com

Jutta Schwarz-Finsterle *Ch. 4*
Kirchhoff Institut für Physik
Universität Heidelberg
Im Neuenheimer Feld 227
69120 Heidelberg
Germany
jfin@kip.uni-heidelberg.de

Titus Sparna *Ch. 8*
University Hospital Freiburg
Core Facility Genomics
Stefan Meier Straße 19
79106 Freiburg
Germany
titus.sparna@pharmazie.uni-freiburg.de

Heinrich Spiecker *Ch. 7*
LaVision BioTec GmbH
Meisenstr. 65
33607 Bielefeld
Germany
spiecker@lavisionbiotec.com

Achim Stangelmayer *Ch. 9*
PreSens Precision Sensing GmbH
Josef Engert Straße 9
93053 Regensburg
Germany
presens@t-online.de

Stefan Stein *Ch. 4*
Kirchhoff Institut für Physik
Universität Heidelberg
Im Neuenheimer Feld 227
69120 Heidelberg
Germany
stein@kip.uni-heidelberg.de

Ernst H. K. Stelzer *Ch. 10*
Cell Biology and Biophysics Unit
European Molecular Biology Laboratory
Meyerhofstraße 1
69117 Heidelberg
Germany
stelzer@embl-heidelberg.de

Monika Stich *Ch. 6*
P.A.L.M. Microlaser Technologies GmbH
Am Neuland 9
82347 Bernried
Germany
monika.stich@palm-microlaser.com

Katharina Stöhr *Ch. 4*
Physikalisch-Chemisches Institut
Universität Heidelberg
Im Neuenheimer Feld 253
69120 Heidelberg
Germany
katharina.stoehr@pci.uni-heidelberg.de

Marion Strehle *Ch. 1, 11*
Institut für Physikalische Chemie
Friedrich-Schiller-Universität Jena
Helmholtzweg 4
07743 Jena
Germany
marion.strehle@uni-jena.de

Michel Stutz *Ch. 6*
Carl Zeiss Jena GmbH
Research Center
Carl-Zeiss-Promenade 10
07745 Jena
Germany
m.stutz@zeiss.de

Gerd Sulz *Ch. 2*
Fraunhofer Institute for Physical Measurement Techniques
79110 Freiburg
Germany
gerd.sulz@ipm.fraunhofer.de

Hans Thiele *Ch. 3*
Kayser-Threde GmbH
Wolfratshauser Str. 48
81379 München
Germany
hans.thiele@kayser-threde.com

Katja Tönsing *Ch. 7*
Experimental Biophysics & Applied Nanosciences
Faculty of Physics
University of Bielefeld
Universitätsstraße 25
33615 Bielefeld
Germany
katja.toensing@physik.uni-bielefeld.de

Claude-Dietrich Voigt *Ch. 10*
Schering AG
Technical Development Laboratory
Muellerstr. 170-178
13342 Berlin
Germany
ClaudeDietrich.Voigt@schering.de

Qing Wang *Ch. 2*
Institute of Computer Sciences
University of Freiburg
79110 Freiburg
Germany
QWang@informatik.uni-freiburg.de

Katrin Wehry *Ch. 5*
Laborgruppe Mukosaimmunologie
Abteilung Klinische Medizin
Forschungszentrum Borstel
Parkallee 22
23845 Borstel
Germany
kwehry@fz-borstel.de

Otto Wolfbeis *Ch. 9*
Institut für Analytische Chemie, Chemo- und Biosensorik
Universität Regensburg
Universitätsstraße 31
93040 Regensburg
Germany
otto.wolfbeis@chemie.uni-regensburg.de

Jürgen Wolfrum *Ch. 4*
Physikalisch-Chemisches Institut
Universität Heidelberg
Im Neuenheimer Feld 253
69120 Heidelberg
Germany
wolfrum@urz.uni-heidelberg.de

Ralf Wolleschensky *Ch. 6*
Carl Zeiss Jena GmbH
R & D Advanced Imaging
Carl-Zeiss-Promenade 10
07745 Jena
Germany
wolleschensky@zeiss.de

Andreas Wuttig *Ch. 3*
Institut für Physikalische Hochtechnologie
Albert-Einstein-Str. 9
07745 Jena
Germany
andreas.wuttig@ipht-jena.de

Hans Zappe *Ch. 8*
Laboratory for Micro-optics
Department of Microsystems Engineering – IMTEK
University of Freiburg
Georges-Koehler-Allee 102
79110 Freiburg
Germany
zappe@imtek.uni-freiburg.de

1 Introduction: Biophotonics – Visions for Better Health Care

Susanne Liedtke, Michael Schmitt, Marion Strehle, Jürgen Popp[1]

Since the late 1980s, natural scientists have introduced a multiplicity of new terms and definitions. We had to learn the difference between genes and proteins, we have been taken into the miniature world of viruses and prions, and the newspapers report on a great many new technologies, such as biotechnology, information technology and nanotechnology, and nowadays even on bionanotechnology. The development of those new technologies combines an increase of both the scientific and also the technological understanding and knowledge of, for example, life processes and has already led to economic profit as well as to enormous stock-exchange quotations. However, apart from undreamed-of possibilities for mankind, the new technologies also entail ethical risks and problems as we can conclude from the discussion on the first cloned sheep "Dolly" or the application of embryonic stem cells. Rather than entering into a detailed discussion, we will stop here, and instead introduce a fairly new discipline of natural science: Biophotonics.

Biophotonics deals with the interaction between light and biological systems. The word itself is a combination of the Greek syllables *bios* standing for life and *phos* standing for light. Photonics is the technical term for all procedures, technologies, devices, etc. utilizing light in interaction with any matter. Before we discuss Biophotonics in more detail, we focus first and foremost on the achievements of Photonic Technologies, which are often used synonymously with Photonics. The advanced control and manipulation of light now available make Photonics as powerful as Electronics. One major goal is to incorporate even more photonically driven processes into our daily lives. Therefore, Photonics is considered as a key technology of the twenty-first century.

Photonics encompasses the entire physical, chemical and biological laws of nature, together with technologies for the generation, amplification, control, manipulation, propagation, measurement, harnessing and any other type of utilization of light. This rather broad definition of Photonics emphasizes the huge importance of light for our modern human society. Photonics is a key technology for solving momentous problems in the domains of health

1) Corresponding author

Biophotonics: Visions for Better Health Care. Jürgen Popp and Marion Strehle (Eds.)

ISBN: 3-527-40622-0

care, food production and technology, environmental protection, transportation, mobility, etc. It is a pacemaking technology for other developments such as communication and production technology, biotechnology and nanotechnology. Just one brief example of what the near future holds: conventional lighting by light bulbs, neon lamps or fluorescent tubes will be a thing of the past. The new wonder is organic light-emitting diodes (OLED). Those innovative diodes provide more efficient and long-lasting illumination, with energy consumption reduced to a minimum. Thus, the wide-spread application of these diodes will allow our economy to save billion of Euros both on the cost of energy and also on consequential costs, such as, for example, the cost of environmental protection. Furthermore, the use of these diodes for displays will revolutionize graphic screen technology. In comparison to TFT (Thin Film Transistor) screens, an OLED display does not need background illumination. The display has the thickness of a plastic transparency, and its flexibility. Completely new applications, such as the use as "information wallpaper", are in the minds of scientists and advertising experts. However, in other fields of Photonics the future has already started. The information carried via the World Wide Web is mainly carried by light. From these few examples, we can recognize that photonic technologies are already pervading our entire life.

Now let us come back to another field in which Photonics is seen as a key technology for future scientific and economic progress – the field called "Biophotonics". What is special about Biophotonics? In order to answer this question we have to shed light on life as well as on light itself because, as already mentioned, Biophotonics deals with both light and life. Both phenomena will be important issues in this book. While the "life" aspect will be covered in the various chapters, a short summary of the properties of light, as well as on the interactions of light and matter, will be given later on in this introductory chapter. From time immemorial both phenomena have fascinated people. Despite life and light have being ubiquitous and self-evident, an understanding of the scientific bases of light and life was and is still a special challenge. As we can see from history, there was a long and roundabout journey from the first description of the nature of light in early antiquity to the photon theory of Albert Einstein. In a similar but no less complicated way, the phenomenon of life has occupied scientists and philosophers from Aristotle to the modern students of genomics in the twenty-first century.

It makes sense to put both life and light together. We can, indeed we must, learn from nature. Nature demonstrates how valuable and fruitful the interaction between light and biological systems, and thus Biophotonics as defined above, can be for life. Consider the "harnessing" of photons. Plants utilize light via photosynthesis as an energy source. In the process of vision light generates pictures of our environment in our brains by a quite complicated but very efficient pathway. The scientist dealing with Biophotonics attempts to understand nature by mimicking the basic principles and taking advantage of the same tools. Using this approach, highly innovative future-oriented technologies can be brought into the real world. This makes the business quite thrilling. Biophotonics is not a science as an end in itself. In fact it opens undreamed-of possibilities for fundamental research, the pharmaceutical and food industries, biotechnology, medicine, etc.

The investigation of biological materials by means of innovative optical techniques has led to totally new optical technologies, including techniques capable of, for example, yielding snapshots of cellular conditions and monitoring dynamic processes. The ultimate goal of Biophotonics is to unravel life processes within cells, cell colonies, tissues, or even whole organs. Biophotonics seeks to provide a comprehensive multidimensional understanding of the various processes occurring in an organism. Therefore, Biophotonics combines all optical methods to investigate the structural, functional, mechanical, biological and chemical properties of biological material and systems. The optical phenomena exploited to gain all this information include all the interactions of electromagnetic radiation with living organisms or organic material, such as absorption, emission, reflection, scattering, etc. Other areas of Biophotonics use light as a miniaturized tool, e.g., optical tweezers or a laser scalpel.

Why do we use light? What are the advantages of applying light to the study of biological matter? The three major advantages are: (1) Light measures without contact, i.e., light allows processes taking place within a living cell to be studied without disturbing or affecting the biological activity. (2) Light measures more quickly and yields instantaneous information, i.e., the complex preparation needed in conventional methods to obtain, for example, a reliable diagnosis, which may take days, or even weeks, are no longer required or can be performed in a much shorter time. (3) Light measures more precisely, i.e., optical methods allow ultrasensitive detection, down to the single-molecule detection necessary for the life science sector.

As already indicated by the title "Biophotonics – Visions for better health care" the main focus of this contribution is on the topic "health care". However, the reader should bear in mind that Biophotonics is not limited to dealing with "health care". The final chapter of this book will mention other topics and challenges of Biophotonics. With the help of this new discipline we hope to get a precise understanding of the origin of diseases so that in the future we can prevent diseases or at least diagnose them more precisely and at an earlier stage so that we can treat them more efficiently. In principle, this looks quite simple. However, to be successful in achieving these ultimate goals scientists have to look beyond their own noses. This means that the developers of photonic technologies, mainly physicists, physical chemists, chemists and engineers, have to be in close contact with the possible users in biology, medicine, the pharmaceutical and food industries and environmental research. Otherwise the great potential of Biophotonics cannot be applied. In various contributions to this book we shall see that to some extent the innovations of Biophotonics are based on precise observation of a natural phenomenon by a biologist or physician, which is then transformed with the help of photonic technologies into new diagnostic methods and technologies. Other progress is based on optical and spectroscopic innovation made by physicists and afterwards adapted to appropriate biological and medical problems. However, there is still a large gap between those scientists developing optical technologies and those scientists who like to use optical technologies. Thus, bringing together the various disciplines is one of the greatest challenges of Biophotonics.

1.1 The Situation of Biophotonics in Germany and Other Countries

The German government recognized the great potential of Biophotonics, not only as a key technology but also as a bridging technology, and installed a multidisciplinary operating research network. The main research framework Biophotonics, funded by the German Federal Ministry of Education and Research (BMBF), has two major common goals.

- Scientific goal: Shedding light on biological processes, i.e. understanding life processes on a functional as well as a molecular level. This will allow scientists to understand the origin of diseases and to invent new strategies to defeat them more effectively and will be the base for real prevention and effective therapies.
- Technological goal: Developing innovative light-based technologies to achieve the above-mentioned goal.

A prerequisite for reaching these two goals is that scientists from the various disciplines mentioned above should work closely together. This means that the potential user, e.g. the physician, has to be involved in the scientific and technical development from the very beginning. Another quite important requirement is that the industry, with all its technological knowledge and experience, cooperates right from the start. The involvement of both parties is seen as one important key to a successful conclusion.

This multi- and interdisciplinary research should make a strong impact on society. First of all, Biophotonics will lead to better health. Diseases, whether cancer or infectious diseases, will be defeated very effectively. Thus, the quality of life will be dramatically improved and the cost of health care considerably reduced. Areas in which Biophotonics is already operating successfully include pathology, oncology, dermatology, cardiology, urology, ophthalmology, gastroenterology and dentistry. Secondly, in addition to scientific and technical progress, the research framework strives for the important goal of strengthening the position of Germany as a centre for technology by creating new employment and protecting highly qualified workers and sustainable jobs. As a consequence, Biophotonics will become as important as other leading technologies, such as nanotechnology, genomics and proteomics.

Germany is not the only country to have recognized the importance of Biophotonics. All over the world scientists from academia and industry are working on this topic. A short overview on the different activities of Germany, USA, France and England is given in **Table 1.1** which has been published in report on Biophotonics in Germany by Deloitte Consulting and Kraus Technology Consulting 2005.

Biophotonics is also very important in other countries, e.g., Canada, Japan, China, Australia and all over Europe. There is almost no country in the world where scientists are not becoming aware of the possibilities and potential achievement of Biophotonics. The major focus of research activities lies on topics from medicine and biotechnology. To summarize all these activities is way beyond the scope of this book.

Table 1.1: Biophotonics in four major countries.

	Germany	USA	France	UK
Organized partnership and cooperation in network	yes	yes	Le Club Biophotonique [1]	not known
Annual statistics	no	yes	yes [2]	not known
Kick-off meeting and process of formation of opinion	Deutsche Agenda "Optische Technologien"[3]; Studie zur Biophotonik	Harnessing Light: Optical Science and Engineering for the 21st Century [4]	La Biophotonique française – Perspectives de développement 2003 [5]	not known
Coordinated Project executing organization	yes [6]	Center for Biophotonics – Science and Technology (CBST)	cooperation between various research centers in France	concentrated in Glasgow
Decided research center	no	CBST, founded in 2002	concentration of various partners around Paris	Institute of Photonics and Center for Biophotonics, University of Strathclyde
Decided research program	yes	yes	yes, but within existing research programs	not known

Table 1.1: (continued)

	Germany	USA	France	UK
Other scientific panels, organization, etc.	Deutsche Gesellschaft für angewandte Optik (DGaO), network of competence, scientific councils at universities	in collaboration with CBST	some collaboration between different groups from different disciplines from various universities	Work group Scottish Optoelectronics Association
Amount of public financial support p.a. (without military projects)	5 Million EUR p.a.	Million EUR p.a. for 10 Years	not known	not known
Programs for public relations, education and advanced training	general reports on the demand on qualified employees in the field of optical technology	public relation using the keyword "Biophotonics"	support by Opticis valley, annual symposia in Paris	not known

©2005 Deloitte Consulting GmbH Kraus Technology Consulting— "Biophotonik in Deutschland: Wohin geht die Reise?"
1) www.paris-biophotonique.org
2) Annuaire de la Biophotonique en Ile-de-France
3) bmbf, VDI-TZ
4) National Research Council
5) Opticsvalley et Grenoble® Evry En collaboration avec l'ADIT
6) VDI-TZ

1.2 The Interplay Between Light and Matter: Interactions Allowing Us to Understand Our Environment

Before we introduce the marvelous world of Biophotonics we shall first give a brief introduction into the basics of light–matter interactions. The interaction of light with matter, in particular with biological material, i.e., what happens if a light wave or photon hits matter, is an exciting topic. The various light–matter interaction phenomena enable us to observe our environment and are responsible for the existence of life on earth. What follows is a short and very general introduction into the manifold interplay (e.g. absorption, emission, scattering, reflection, refraction, diffraction, dispersion, polarization, etc.) between light and matter. The advanced reader may therefore skip the following pages.

Light propagates as electromagnetic waves. Electromagnetic waves are wavelike perturbations (invisible perturbations of the so-called force field) recurring periodically over a certain distance called the wavelength. Light waves are described as oscillating electric (E) and magnetic fields (H), which are perpendicular to each other. Time-dependent changes of the electric field are always combined with spatial changes of the magnetic field. Similarly, time-dependent changes of the magnetic field are connected with spatial changes of the corresponding electric field. Electromagnetic waves can propagate in a vacuum at a speed of $c = 299\,792.458\ \mathrm{km\,s^{-1}}$. The time-dependent electric field E can be described as: $E(z,t) = E_0 \cos 2\pi\nu(t - z/c)$ where $E(z)$ = electric field at position z; E_0 = maximum electric field; ν = light frequency; t = time; c = speed of light. It is just this oscillating electric field that can interact with matter and transfer energy to or from matter. The connection between frequency and wavelength is given by $\lambda\nu = c$. The spectroscopically more common unit wavenumber is defined by $\tilde{\nu} = 1/\lambda$.

The interaction of electromagnetic radiation with matter can polarize matter and induce a dipole moment. Matter, e.g., molecules, consists of atoms held together by electrons. If the binding electrons are distributed between the atoms, a covalent bond exists. If the binding electrons are located on an atom, ionic bonding is present. These two main models of bonding are extreme cases. The only covalent bonds which have no ionic character are those between identical atoms. However, molecules formed between different atoms may exhibit partial charges due to an asymmetric distribution of the binding electrons between the atoms. If no side of the molecule has more negative or positive partial charge than the other side, the molecule is nonpolar. However, molecules exhibiting an asymmetric partial charge distribution are polar. An example of a polar molecule is the water molecule, consisting of two hydrogen atoms and one oxygen atom held together by two common electron pairs. Since the oxygen atoms attract electrons more than the hydrogen atoms the binding electrons in a water molecule are asymmetrically distributed, leading to a negative partial charge on the oxygen atom and positive partial charges on the hydrogen atoms. These partial charges make water act as a dipole. The interaction of polar molecules like water with electromagnetic waves of a certain frequency leads to an orientation of the water molecules.

To learn more about electromagnetic waves and their influence on matter, we look first at a plate-type capacitor whose polarity is changing with a certain frequency. Water molecules are aligned within a plate-type capacitor in such a way that the negative pole of the water molecules points towards the positively charged plate while the positive pole orients towards the negative plate. The water dipoles orient themselves according to the applied electric field of the capacitor, i.e., the water molecules become oriented. This phenomenon is called orientation polarization. Changing the polarity of the capacitor plates leads to a reorientation of the water molecules.

What happens if nonpolar molecules like nitrogen N_2, oxygen O_2 or carbon dioxide CO_2 are brought between the plates of a capacitor? Applying an external electric field leads to a distortion of the electrons compared to the positively charged atomic nuclei. This distortion creates partial charges, i.e., the electric field induces a dipole moment. This polarization effect is called distortion polarization. The more easily the electrons can be displaced compared to the positively charged nuclei the bigger the induced dipole moment. The induced dipole moment μ_{ind} is direct proportional to the applied electric field E where the proportionality constant α is called the molecule's polarizability: $\mu_{\text{ind}} = \alpha E$. α is a measure of how easily electrons can be moved or displaced within a molecule. Thus nonpolar molecules within a plate-type capacitor also become oriented owing to the induced dipole moments. Changing the polarity of the electric field changes the induced dipole moment accordingly. However, the nonpolar molecules are not reoriented; only the electrons will be moved in another direction.

The general form of the polarization depends on the frequency of the applied field. Radio frequency radiation (100 MHz) leads to an alignment of polar molecules according to the external electric field, i.e., a reorientation of the complete molecules (orientation polarization) takes place. The same is true for the electrons, which can follow the changing electric field much more easily owing to their low mass. However, higher frequencies lead to a distortion polarization of the electrons ("induced electric dipole moment") since molecules exhibiting a permanent dipole can no longer follow such rapidly oscillating fields. The induced dipole moment μ_{ind} oscillates with the same frequency as the exciting oscillating electric field E: $\mu_{\text{ind}} = \alpha E_0 \cos 2\pi\nu(t - z/c)$ where α is the molecule's polarizability. Since the oscillating induced dipole moment is simply an oscillating charge, matter radiates light of the same frequency as the initial exciting radiation field (secondary radiation). This is in total analogy to a Hertzian dipole acting as a broadcasting and receiving antenna, which is based on the radiation of electromagnetic waves from a dipole. In the antenna, electrons are driven by a generator to the top or the bottom. This generates a charge distribution similar to a dipole.

The interplay of molecules with light, in particular with visible light, which polarizes molecules and leads to the emission of a secondary radiation of the same frequency as the polarizing field in all directions, is called elastic light scattering or Rayleigh scattering. It is precisely this undirected scattered radiation which enables us to observe our environment in the presence of sunlight or light from a lamp. The difference between the earth and outer space is that the earth has an atmosphere consisting of molecules which can be polarized. The darkness in outer space is due to the lack of an atmosphere. Rayleigh scattering scales with the fourth power of the light frequency (ν^4), i.e., short wavelengths or high-frequency blue light is scattered significantly more strongly than low-frequency red or infrared light. Hence the midday sky is blue and the sun appears more yellow or red than it really is. The ν^4 dependence becomes especially obvious when the path of the sunlight through the atmosphere is longest. Thus the rising or setting sun appears especially red, since then the less-scattered red light can better reach our eyes.

For a description of microscopic systems like atoms or molecules, quantum effects need to be considered. The term quantum (Latin, "how much") refers to discrete units assigned to certain physical quantities, such as the energy of an atom or molecule. This means that a physical quantity that appears macroscopically continuous appears in the microscopic world only in well defined values that cannot be further divided. The energy of a microscopic system is quantized, i.e., divided into well defined energy portions. An illustration of this phenomenon might be a staircase, in which each step marks an energy portion. Molecules exhibit certain movement patterns, i.e., the molecular degrees of freedom can be classified into translation, rotation and vibration. The energy of all these degrees of freedom is quantized. Similarly, the electron energy in an atom or molecule is quantized. What happens if light interacts with such a quantized rotating and vibrating molecule? The energy of the molecule's degrees of freedom are quantized, and the energy of light is quantized too. Light can exhibit properties of both waves and particles. This phenomenon is known as wave–particle dualism. Einstein postulated the existence of photons, which are quanta of light energy with particle character. By postulating these photons, Einstein was able to explain the photoelectric effect which cannot be explained by the wave theory of light. The photoelectric effect describes the emission of electrons from matter after the absorption of high-energy ultraviolet light due to a collision with the particle-like photon. Each photon possesses the energy $E = h\nu$ where h is Planck's constant ($6.626 \times 10^{-}3$ J sec) and ν the frequency of the light, i.e., electromagnetic radiation of the frequency ν can only carry energy of $0, h\nu, 2h\nu, \ldots$. Only photons possessing energy above a certain threshold lead to an ejection of electrons. According to the wave theory, the electromagnetic field E exerts an oscillating force on the electrons within the matter. This, however, would mean that electrons are ejected with increasing amplitude and not frequency, which is contrary to experiment. Since the energy of matter, i.e., molecules, and electromagnetic radiation cannot be varied continuously, light can only promote matter from one discrete energy level to an energetically higher one if the photon possesses an energy matching the energy difference between two quantum states of the system. This process is called absorption.

Depending on the light wavelength, rotations, vibrations or electrons can be excited within a molecule. Microwaves can excite a transition between two rotational energy states, infrared light is necessary to promote a molecule into a higher vibrational state and visible or ultraviolet light transfers electrons from one electronic state in an energetically higher electronic quantum state. An excited system, however, can relax into the lowest energy state by releasing the additional energy in terms of heat via collisions with the surrounding environment or even by the emission of light. This special form of light–matter interaction is the basis for modern molecular diagnostic procedures in life sciences and medicine. In the following we will concentrate on the vibrational and electronic excitation of molecules, since the excitation of rotations in condensed matter plays no significant role for diagnostic purposes.

A polyatomic molecule exhibits a multitude of vibrations, but of interest are the so-called normal mode vibrations; an N-atomic molecule has $3N - 6$ normal modes. How can one derive this little relationship? The molecular degrees of freedom can be divided into translation, rotation and vibration. Atomic motion through space can be described by the three directions in space x, y and z. As already mentioned, these three coordinates are therefore enough to describe the translation motion of an atom, i.e., an atom has three translational degrees of freedom. Rotations and vibrations do not exist for an atom. If we consider a molecule consisting of three atoms, e.g., water H_2O, the three atoms could move independently through space if they were not connected via a chemical bond. This independent motion would result in nine translational degrees of freedom. However, we know of course that the atoms of a molecule cannot move independently through space but only the whole molecule as an entity. Therefore we need to subtract the three degrees of freedom describing the collective motion of the whole molecule in space from the totality of motions, i.e., of the nine degrees of freedom six remain. By taking into account that a molecule can rotate along the three axes of the coordinate system we need to subtract three more degrees of freedom for the rotational motion. From the original nine degrees of freedom only three remain. These remaining degrees of freedom can be assigned to the vibrational degrees of

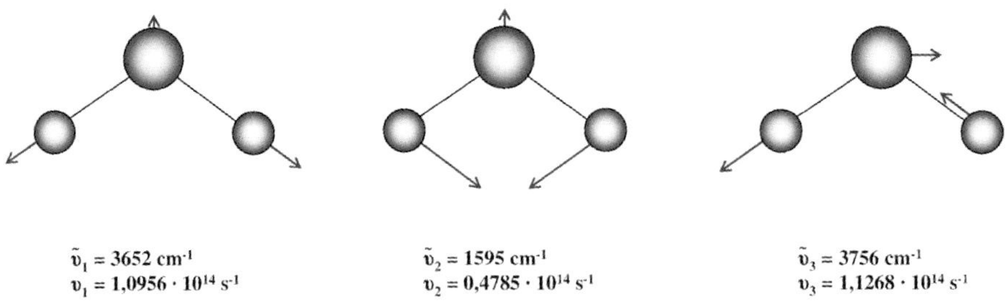

Figure 1.1 Three normal modes of water. The atoms move along the arrows. These three vibrational motions exhibit different vibrational frequencies and can be excited independently.

freedom, i.e., the normal modes of water. The three normal modes of water are shown in **Figure 1.1**. These vibrational motions differ from each other by different atomic displacements. One can differentiate between pure stretching (chemical bonds are stretched or compressed) or pure bending (bond angles are changed) vibrations and mixed forms exhibiting both stretching and compression of chemical bonds and extension or reduction of the bond angles. The easiest way to describe such a vibration is by approximating or describing the chemical bond as atoms held together by a spring. The vibrational frequency then depends on the atomic mass and the spring force constant. The atoms move during a vibrational period towards or away from each other. This movement is repeated periodically and can be easily described by a harmonic oscillator. A harmonic oscillator is a system performing periodic vibrations about its equilibrium position where the restoring force F is directly proportional to the displacement x. **Figure 1.2** shows such a vibrational motion in a harmonic parabolic potential. The minimum of the potential corresponds to the equilibrium distance, i.e., the nuclear distance exhibiting the lowest energy. According to quantum theory, the vibrational energy cannot be continuously but quantized. This quantum harmonic oscillator is symbolized by the horizontal lines inside the parabolic potential. The vibrating quantum mechanical system possesses a zero point energy, i.e., molecules are always vibrating and are never at rest. If the displacement from the equilibrium position is minimal, which is true for small vibrational quantum numbers, the harmonic oscillator model is well suited to describe a vibrational absorption process. However, for large displacements, i.e., for high vibrational states, the model of the harmonic oscillator model is problematic: Firstly, it is impossible to put an arbitrary amount of energy into the system without destroying the molecule. However, it is known that chemical bonds can break, i.e., dissociate, or molecules can fall apart if they are heated too much. To allow for dissociation to take place the harmonic oscillator model has been refined by the anharmonic oscillator model. The anharmonic oscillator model considers

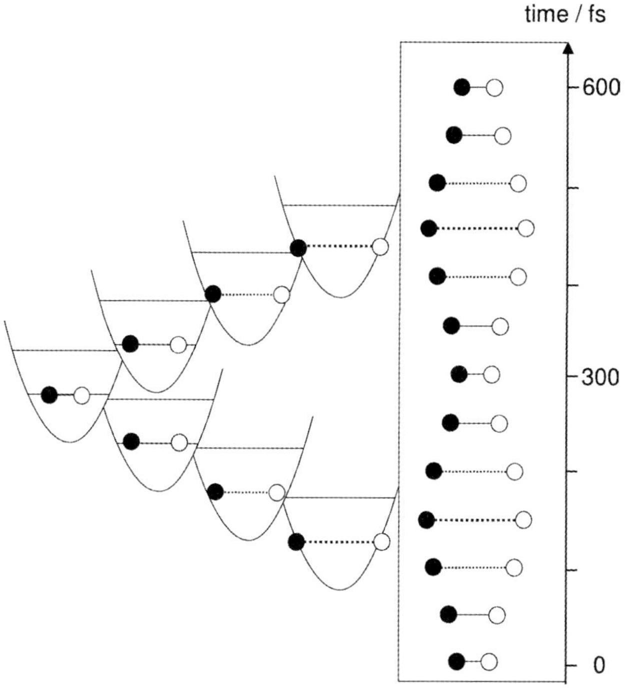

Figure 1.2 Harmonic Oscillator Model. The molecule performs a harmonic vibration about its equilibrium position, i.e., the restoring force is direct proportional to the displacement of the atoms from their equilibrium position (right side). The molecule is modelled as balls connected by a spring. The vibrational frequency depends on the mass of the balls and the spring force constant. The left side depicts the vibrational motion in a harmonic parabolic potential. According to quantum theory the vibrational energy is quantized which is symbolized by the horizontal lines within the harmonic potential.

the fact that molecules dissociate at high vibrational energies, i.e., the spring between the atoms breaks. While for the harmonic oscillator the energy difference between the quantized vibrational states is always the same, this difference for the anharmonic oscillator decreases with increasing energy until a continuum is reached where all vibrational states have almost the same energy. Within this continuum dissociation occurs, i.e., the atoms of the molecule can leave the molecule's force field (anharmonic potential see **Figure 1.3**).

After all this theory the question remains, "What can we learn from vibrating molecules?" Most normal modes if they are not degenerated (degeneracy = same energy) exhibit different vibrational frequencies. The vibrational frequencies of the three normal modes of water are: $\nu_1 = 3652\ \text{cm}^{-1}$ (symmetrical stretch motion), $\nu_2 = 1595\ \text{cm}^{-1}$ (symmetrical bending motion) and $\nu_3 = 3756\ \text{cm}^{-1}$ (asymmetrical stretch vibration). Light with the appropriate frequency can be absorbed by the ensemble of vibrating molecules, promoting them from the vibrational ground state into the first excited vibrational state.

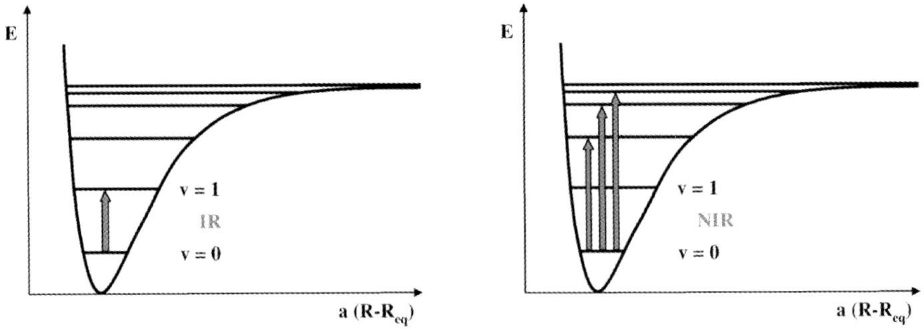

Figure 1.3 Comparison between harmonic (left) and anharmonic oscillator (right). The anharmonic oscillator considers the fact that molecules can dissociate for high vibrational energies. For the harmonic oscillator the energy distance between the vibrational levels is constant. This is in contrast to the anharmonic oscillator, where the energy difference decreases for increasing energy until a continuum is reached where dissociation takes place.

During the course of the absorption process only the amplitude of the vibrational motion out of the rest position changes, while the vibrational frequency stays constant. The light frequencies required to directly excite vibrational absorptions covers the spectral range from 2.5 μm to 1 mm or in spectroscopic wavenumber units 400–4000 cm^{-1}. This frequency range is also called the far-infrared region. The spectral area below 400 cm^{-1} called terahertz radiation has been opened recently for vibrational spectroscopy. Terahertz radiation is extremely promising for Biophotonics, since the penetration depth of this radiation into biological tissue is extremely large.

We shall now consider IR absorption, i.e., the direct absorption of IR radiation by vibrating molecules. What can vibrating molecules tell us about the molecules themselves or the surroundings in which they are embedded? The answer is a lot, since both the number and the type of vibrations depend directly on the atoms present in the molecule and, in particular, how these atoms are chemically bonded to each other. Absorption of IR radiation with the appropriate frequency promotes the molecule from its vibrational ground state into the first excited vibrational molecular state. This absorption process decreases the transmitted intensity with respect to the incident intensity. A plot of the transmitted intensity versus the radiation frequency yields an IR spectrum. For this reason the molecule provides detailed information about itself via interaction (absorption) with an appropriate electromagnetic field (e.g. IR radiation). Moreover, the energy of the vibrational transition depends on the chemical environment the molecules are embedded in. Hence, vibrational spectroscopy provides a key to the molecular environment of the molecules. In addition to the transition from the vibrational ground state to the first excited vibrational state (the fundamental transition), direct absorption processes can also promote the molecule into the second, third or even higher vibrational state, although with much less probability than the fundamental transition. These higher transitions are called overtones. The energy required to excite overtones moves from the IR region into the mid- (2.5–50 μm) or even the near-IR (800 nm to 2.5 μm) spectral region. Overtone vibrational spectroscopy is an important well established method in quality control, but plays only a minor role in the field of health care.

Vibrational transitions can also take place via an inelastic light-scattering process. We have shown already that light can polarize molecules. For the visible wavelength region, the main contribution to the induced polarization comes from electrons, whose distribution relative to the atomic nuclei is distorted by the interacting electromagnetic field. Thus this type of light–matter interaction induces an electric dipole moment within the molecules. The polarizability α is a measure of how easily the electron distribution can be distorted within a molecule. The induced dipole, oscillating with the frequency of the electromagnetic field, emits an electromagnetic wave in all directions. If the polarizability does not change with time, the frequency of the emitted secondary wave corresponds to the frequency of the oscillating induced dipole, i.e., the frequency of the external electromagnetic wave inducing the dipole (= Rayleigh scattering). However, since molecules are always vibrating, the polarizability α is not constant over time, but changes according to the different vibrational frequencies of the molecule's normal modes. Therefore, the induced dipole moment and thus the emitted secondary radiation are also modulated by the different vibrational frequencies. Consequently, the secondary radiation emitted by the molecule is a superposition of the exciting frequency and the various vibrational frequencies of the molecule. Dispersing this secondary radiation into its frequency components yields beside the strong Rayleigh scattering also weak sidebands. The distance between the Rayleigh wavelength and the wavelength of the sidebands corresponds to the vibrational frequencies of the molecule. The appearance of these sidebands arising from an inelastic light-scattering process was first discovered in 1928 by C.V. Raman. This so-called Raman Effect marks an indirect approach to the excitation of molecular vibrations. The Raman Effect can be interpreted quantum mechanically as an inelastic collision between photons and vibrating molecules. Photons can be scattered from molecules. This scattering process corresponds to a molecular transition into an extremely short-lived transition state, the so-called virtual level (= collective quantum energy state of the entity molecule and photon). The molecule can subsequently relax from this virtual level into the original molecular state or to an energetically excited molecular state. If the scattering process starts from the vibrational ground state and ends up in a vibrationally excited state via a transition into the virtual state it is called Stokes–Raman scattering. If the molecules are initially already in a vibrationally excited state and are transferred by the scattering process into the vibrational ground state one refers to as anti-Stokes Raman scattering. Since at room temperature the vibrational ground state is significantly more populated than the vibrationally excited states, the Stokes–Raman spectrum of a sample is more intense than the anti-Stokes Raman spectrum. For Rayleigh scattering, the state before and after the scattering process is the same. These scattering processes are classified as two-photon processes since two photons are involved.

The two vibrational spectroscopic methods Raman and IR absorption spectroscopy are complementary methods based on two different light–matter interaction phenomena and thus exhibit different selection rules. Selection rules determine which vibration of a molecule can be excited by what method. In the case of IR absorption, one photon directly promotes the molecule into a higher vibrational state while the Raman scattering process involves two photons. In order for a molecular vibration to absorb an IR photon, the dipole moment of the molecule has to change during the course of the vibration, i.e., only those vibrations which give rise to an oscillating dipole are IR active. The polarizability has to change during the vibration so that a molecular vibration can be promoted via an inelastic scattering process into a higher vibrational state.

Both Raman and IR absorption spectra can be considered as molecular fingerprints of the molecules existing in a biological sample. The important role of vibrational spectroscopy, and in particular Raman spectroscopy, in Biophotonics can be found in Chapter 3, where representative examples of IR absorption, Raman spectra and a more detailed introduction to Raman spectroscopy (theory, instrumentation, etc.) is given.

Light scattering can take place over the complete electromagnetic spectrum, although the scattering power scales with the fourth power of the exciting frequency, i.e., short-wavelength radiation is scattered strongly while long-wavelength radiation is only scattered poorly. Direct absorption of microwaves or IR radiation can excite rotations or vibrations, respectively. However, the interaction of matter with visible or UV light can also lead to an absorption of this radiation. In this case, light–matter interaction leads to electronic excitation. We can differentiate between two spectral regions: the region between 200 and 380 nm is called the ultraviolet (UV) region, while the area between 380 and 700 nm spans the visible (VIS) wavelengths. What happens if UV-VIS radiation is absorbed by a molecular system? So far, the description of IR absorption and Raman scattering has been limited to the electronic ground state, which was described as a harmonic or anharmonic oscillator (parabolic potential curves, see also **Figures 1.2** and **1.3**). **Figure 1.4** displays the electronic ground state, denoted S_0, as well as the first electronically excited state S_1. In order to simplify the presentation, the harmonic or anharmonic poten-

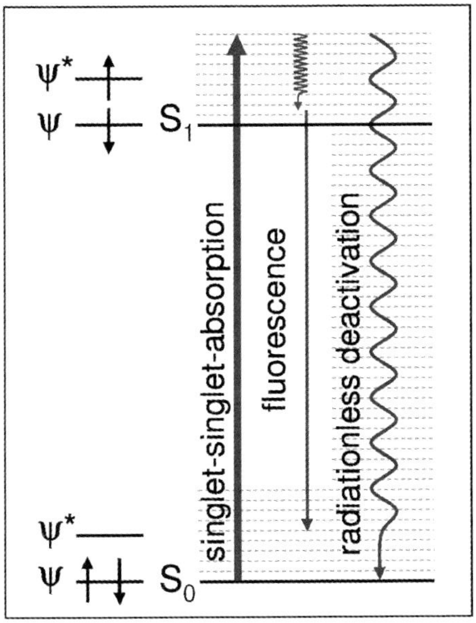

Figure 1.4 Electronic energy diagram. The horizontal lines represent the electronic energy at the equilibrium position. The S_0 state is the electronic ground state of the molecule. A UV or a VIS photon can promote the molecule from its electronic ground state S_0 into the first electronic excited state S_1. The excitation takes place from the vibrational ground state of the electronic ground state into a vibrationally excited state of the electronic excited state. The molecule can lose its additional vibrational energy within the S_1 state via collisions with, for example, solvent molecules to relax into the vibrational ground state of the S_1 state. From there the molecules can relax to the electronic ground state by emission of fluorescence light. Besides the radiative decay processes, radiationless decay processes from the S_1 into the S_0 state also exist.

tial curves are not shown and the electronic states are depicted as horizontal lines, reflecting the electronic energy at the equilibrium geometry of the relevant molecule. S stands for a singlet state. Two electrons are involved in a chemical bond. As mentioned before in a covalent bond, the two binding electrons are mainly located between the two atomic nuclei. If the electrons are mainly located more or less on one side the chemical bond is classified as polar or an ionic bond. Without going into too much detail we need to consider that electrons possess a negative charge as well as a spin. This means in the figurative sense that the electrons spin on their own axes. Depending on the rotating direction we differentiate between electrons having α-spin (clockwise spin) and β-spin (counter-clockwise spin). Why is this quantum mechanical detail of importance? So far electrons have been considered as particles but wave–particle dualism also applies to electrons, i.e., an electron can also be described as a wave. If electrons are moving around atomic nuclei in the form of a wave the wave must reproduce itself on successive circuits. Thus a waveform in which after one or more circuits a wave peak meets a wave valley is not allowed since then it would interfere destructively with itself and would not survive. This simple picture reveals that only special, discrete waves are acceptable to describe the electronic motion. The spatial distributions of these special waves (in the strict sense the square modulus of theses discrete waves) are called orbitals. The discrete solutions lead to quantization, and physicists denote the possible single-valued electron waves with quantum numbers. Wolfgang Pauli found that electrons occupying a single orbital are not allowed to be identical, i.e., the existence of two electrons exhibiting exactly the same quantum numbers is not allowed. Thus an orbital can be occupied by maximally two electrons differing in their spin state, i.e., their intrinsic angular momentum. In the case of chemical bonds, the quantized electron waves occupied by the bonding electrons are called molecular orbitals. According to the Pauli principle, electrons need to be paired in a chemical bond, i.e., one electron has α-spin and the other β-spin. In **Figure 1.4** the electron having α-spin corresponds to arrow up ($\uparrow$) and the electron exhibiting β-spin is denoted by a down arrow ($\downarrow$). Paired electrons within the S_0 state are denoted by ($\uparrow\downarrow$). The absorption process promotes an electronic excitation, i.e., an electron from the S_0 state is transferred into an energetically higher lying orbital, the S_1 state, while the spin state is conserved. This excitation proceeds in analogy to the aforementioned IR absorption process via direct absorption of an appropriate photon. Thus, electronic excitation takes place from the vibrational ground state ($v = 0$) of the electronic ground state into a vibrational state v' of the first excited electronic state. Which vibrational states within the electronic excited states are populated depends on the geometrical rearrangement taking place upon electronic excitation. If we measure, as described above for IR absorption spectroscopy, the ratio between the initial light intensity I_0 and the transmitted intensity I vs. the light wavelength it is possible to determine the concentration of absorbing molecules within a sample via the so-called Lambert–Beer law: $E(\nu) = \epsilon(\nu)cd$. The attenuation of light due to an electronic absorption is described by $I(\nu) = I_0 \times 10^{-E}$. $\epsilon(\nu)$ is the molar decade absorption coefficient, E corresponds to the absorption, c is the desired concentration and d represents the thickness of the sample cell.

Now the questions arise, "What happens to electronically excited molecules? What is the residence time of the molecules in the excited state?" An excited molecule tends to release its additional energy in any form to get back into the lowest energy state, i.e., the most stable ground state. What happens in detail? If the electronic excitation generated vibrationally excited molecules within the excited electronic state, the molecule rapidly releases the additional vibrational energy via collisions with the surroundings (e.g. solvent molecules) to pass into the vibrational ground state of the excited electronic state. The time-scale of this ultrafast vibrational relaxation is 10^{-14}–10^{-12} s. In polyatomic molecules, vibrational relaxation can also take place without the presence of solvent molecules via a redistribution of the additional vibrational energy from one specific mode populated upon electronic absorption to other vibrational modes.

From the vibrational ground state of the excited electronic state the molecule can spontaneously relax to the electronic ground state by emission of light. This radiational transition is called fluorescence. The time-scale for fluorescence to take place is 10^{-9}–10^{-8} s. Since the time-scale for vibrational relaxation is orders of magnitudes shorter than that for the emission of fluorescence light, the emission of molecules in condensed phases or solid state always starts from the vibrational ground state of the first excited electronic state S_1. However, the emission of fluorescence light following an electronic absorption is not the most common electronic relaxation mechanism but rather an exception. In nature, radiationless transitions are the dominant form of electronic relaxation. Collisional deactivation processes lead to a decay of the excited electronic state S_1 directly back into the S_0 state. The color of plants, fruits, etc. are not due to the emission of fluorescence following the electronic absorption of light but rather a consequence of light absorption and reflection (vide infra). White light results from a superposition of all wavelengths in the UV-VIS electromagnetic spectrum. A tomato appears red under irradiation with white light because it absorbs all colors from the white light spectrum except red. Thus the color red is reflected from a tomato.

Besides fluorescence and vibrational relaxation, several other electronic relaxation mechanisms exist. However, a detailed description of these is beyond the scope of this general introduction.

So far this short and general introduction of light–matter interactions has mainly concentrated on single molecules. However, matter is generally not present in the form of single molecules but rather as molecular aggregates. The aggregates can exhibit different spatial dimensions and might range from a few associated molecules called clusters via nano- and microparticles to large molecular aggregates, e.g., crystals, visible to the naked eye. Nanoparticles are aggregates of a few hundreds of molecules or atoms forming discrete units with a size in the nanometer range. When light interacts with matter whose particles are larger than the light wavelength, i.e. larger than 300 nm, beside absorption and scattering new light–matter-interaction phenomena occur. Such new light–matter interactions are, for example, reflection and diffraction of light. These mechanisms play an important role in the interplay of light with biological matter, e.g., united cell structures or tissues. For these phenomena the aforementioned molecular polarization is of particular importance. In the case of extended matter, with spatial dimensions greater than the light wavelength, the polarizability is described as the sum of the molecular properties, i.e., as an averaged value. In total analogy to the molecular polarizability, the bulk polarizability describes the dipole moment

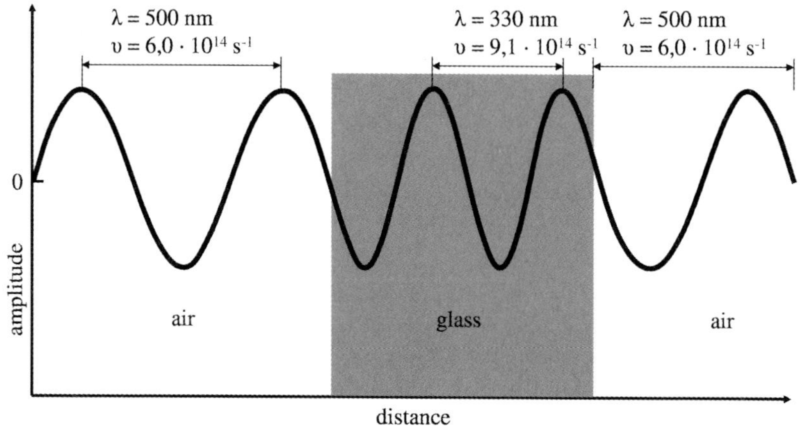

Figure 1.5 Light wave of a wavelength of 500 nm ($\nu = 6.0 \times 10^{14}$ s^{-1}) travelling through air hits a glass medium ($n = 1.5$). Because of the refractive index of glass, light no longer travels at almost 300 000 km s^{-1} but only with a reduced speed of 198 000 km s^{-1}. Since the frequency of the light wave remains constant while travelling through matter the wavelength is reduced from 500 nm to about 330 nm.

induced into bulk matter by an external electric field. Put simply, a value can be derived from the bulk polarizability indicating how fast light travels through matter. This value is called the refractive index. If no matter is present, i.e., in a vacuum, the speed of light is $c_0 = 3 \times 10^8$ m s^{-1}. If, however, light travels through matter the speed of light c is reduced. The ratio of the two speed values determines the refractive index: $n(\nu) = n(\lambda) = c_0/c$. Like the polarizability the refractive index depends on the light frequency ν or the wavelength λ. If a light beam hits matter, e.g., a piece of glass, the light beam experiences a refractive index difference from n_1 to n_2. The speed of light reduces from around 300 000 km s^{-1} to about 198 000 km s^{-1} if the glass has a refractive index of n(500 nm) = 1.515. The refractive index of the vacuum is by definition 1. Since frequency (ν), wavelength (λ) and speed of light (c) are related by the equation $\nu = c/\lambda$ and the speed of light c is a function of the refractive index, the following equation results: $\lambda = c_0/(\nu n)$. This equation raises the question, "Which value stays constant when light passes through matter: the wavelength or the frequency?" This question can be easily answered by taking into account the effect a visible-frequency electromagnetic wave causes within matter. The oscillating electric field polarizes the molecule and induces a dipole moment oscillating with the same frequency as the external field. The evolving secondary wave has the same frequency as the exciting one. Thus the frequency remains constant, while the wavelength decreases (see **Figure 1.5**).

What other effects can be seen if a light beam hits a glass plate? As is well known, the light beam is reflected or refracted at the glass surface. These

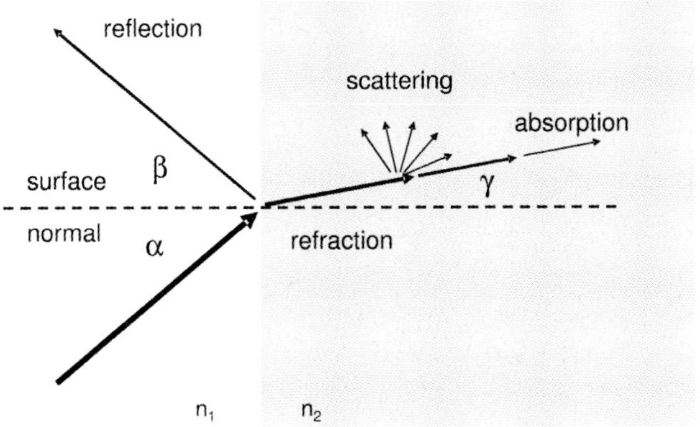

Figure 1.6 Illustrative description of the optical processes: reflection, refraction, scattering and absorption taking place if a light beam passes from one material to an optically different material.

phenomena can be easily explained by geometrical optics. Depending on the nature of the material, a certain fraction of the incident light is reflected, and the angles from the normal of the incident and reflected waves are identical. Refraction of the wave into the medium takes place if the medium does not absorb the radiation. **Figure 1.6** summarizes the light–matter interactions of reflection, refraction, scattering and absorption. All these processes are of special importance while investigating biological cells or tissue by means of optical methods.

The change of the propagation angle of the light beam depends on the refractive angle and can be described by Snell's law: $n_1 \sin \alpha = n_2 \sin \gamma$; $n_1 < n_2$. In case $n_1 > n_2$, i.e., n_1 is the optically more dense medium, the light is refracted away from the normal and not towards the normal. Before dealing with the interaction of light with biological matter in more detail another important phenomenon, total reflection, needs to be explained. The phenomenon of total reflection is depicted in **Figure 1.7**. Total reflection means the complete reflection of the light beam. A prerequisite for total reflection to take place is the light beam needs to travel from an optically more dense to an optically less dense medium. If the angle of incidence is larger than the threshold angle θ_c, total reflection takes place. The threshold angle θ_c is defined as $\theta_c = \arcsin(n_2/n_1)$ where $n_2 < n_1$.

If light hits biological tissue it may be partly reflected from the surface or it will be in parts refracted into the tissue. Within the tissue light can be absorbed or scattered. The proportion of light refracted increases as the angle between the incident light and the surface increases (see **Figure 1.8**), i.e., as the angle of incidence α (**Figure 1.6**) decreases. If as much light as possible is to penetrate into the tissue the light must hit the tissue at a right angle. Since biological tissue is rather inhomogeneous, the various light–matter effects occur

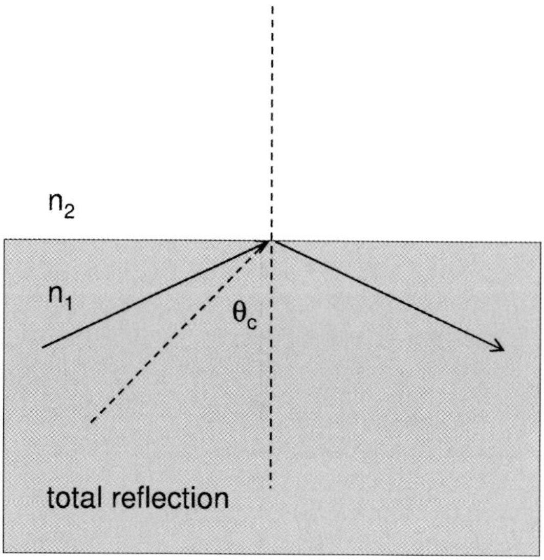

Figure 1.7 If a light beam coming from an optical thicker medium hits an interface to an optical thinner medium ($n_1 > n_2$) the light beam can only pass over into the optically thinner medium, if the incident normal angle is smaller than a critical angle θ_c. In case the incident normal angle is bigger than θ_c total reflection occurs.

in different proportions. Biological tissue is usually a strongly light-scattering material. Depending on the tissue constituents, we can differentiate between elastic light (Rayleigh) scattering and Mie scattering. Rayleigh scattering occurs predominantly from cell constituents smaller than the light wavelength. If the size of the tissue constituents is the same as the light wavelength, a new scattering phenomenon occurs, the so-called Mie scattering discovered by Gustav Mie 1908. In contrast to Rayleigh scattering, Mie scattering shows a less pronounced wavelength dependence and the light is mainly scattered in the forward direction. Both light scattering and absorption lead to an attenuation of the light beam. The absorption of light originates from the multitude of molecules present in biological tissue. The attenuation can be described by the Lambert–Beer law $I(z) = I_0 \exp[-(\alpha(\nu) + \alpha_s)z]$. $I(z)$ characterizes the light intensity at the position z within the biological tissue, z equals the penetration depth, $\alpha(\nu)$ is the absorption coefficient and α_s denotes the scattering coefficient. Both coefficients describe the loss of light intensity within the tissue.

Because of the multitude of molecular constituents which can be found in biological tissue, e.g., proteins, peptides, Desoxyribonucleic Acid (DNA) and Ribonucleic Acid (RNA), haemoglobin, melanin and water, biological tissue absorbs over a wide spectral range. **Figure 1.9** shows various wavelength regions as well as absorption data of typical components of biological tissue,

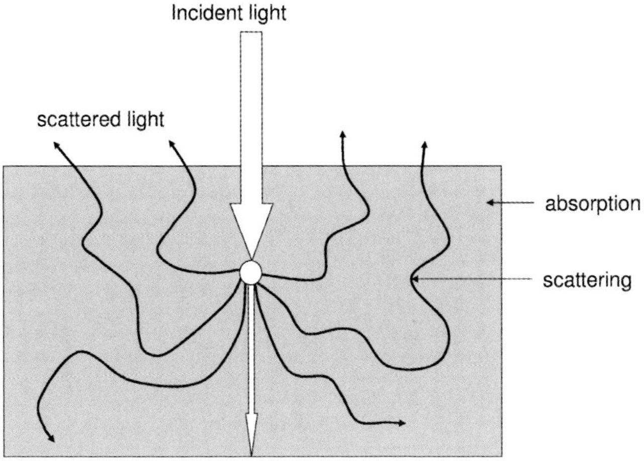

Figure 1.8 If a light beam strikes perpendicularly on biological tissue the proportion directly reflected from the surface becomes minimal. The light penetrating into the tissue is absorbed more or less depending on the wavelength. Besides absorption, the straight-forward propagation of light is hindered within biological tissue by Rayleigh and Mie scattering. Depending on the biological tissue, only a minimal intensity of the originally applied electromagnetic radiation reaches through the tissue.

such as blood, melanosome and epidermis. Owing to the high absorption of biological tissue over a relatively broad spectral range, the penetration depth of light strongly depends on the light wavelength used. To some extent, light can only penetrate by a fraction of a millimeter before it is totally absorbed or scattered. Blue and green light will be absorbed strongly while red light is almost not absorbed by biological tissue. Furthermore, short-wavelength light is scattered strongly (ν^4 dependency). Overall this means that only red light can travel through biological tissue without experiencing too many losses. Near-IR light exhibits the highest penetration depth of 2 to 5 mm. Green light will be totally absorbed or scattered after a penetration depth of 0.5 to 2 mm.

1.3 A Fascinating Tour Across Biophotonics

After this very brief introduction into light–matter interactions we will come back to the question of the scope of this book. Our aim is to take you on a fascinating tour across Biophotonics. We will tell you nine different stories, which result from nine projects funded within the Biophotonics research framework of the German Federal Ministry of Education and Research (BMBF) since the year 2001. Each chapter describes the basic principles, methods and results of a single network project, showing the long journey from a scientific idea, via constant improvements for various scientific and business applications, to a

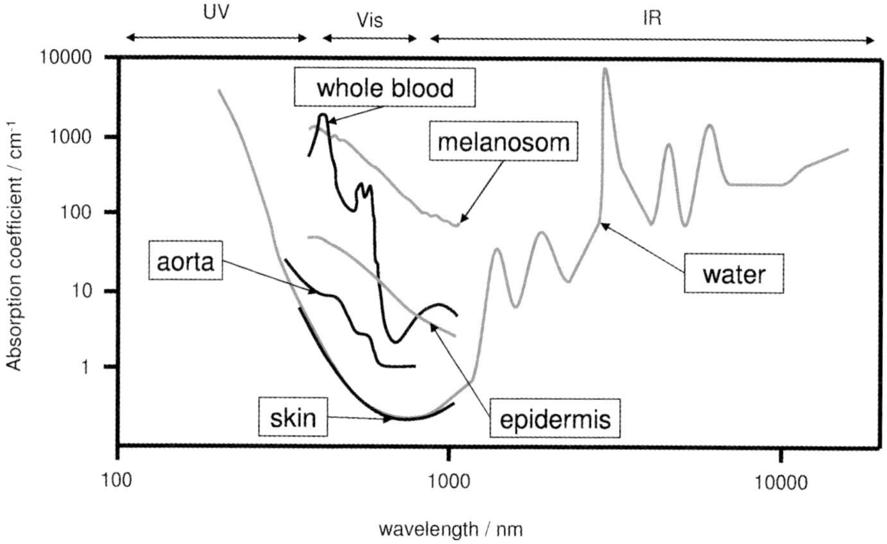

Figure 1.9 The main constituents of the absorption spectrum of biological tissue: (1) Absorption in the UV is increased by proteins, DNA and other molecules. (2) In the IR the absorption increases for longer wavelengths because of the water (75%) present in biological tissue. (3) In the red spectral region as well as the near-IR the absorption of all molecular constituents of biological tissue is minimal. This frequency region is therefore called the diagnostic or therapeutic window since using these wavelengths allows one to penetrate deeply into biological tissue. (4) Blood is a strong absorber in the red/NIR region. However, since blood is only present to a small percentage within biological tissue the average absorption coefficient is not influenced to a great extent. However, if a photon hits a blood vessel it will be absorbed, i.e., the spatially varying absorption properties of biological tissue determine the light–tissue interactions and the average absorption properties determine the light transport through biological tissue. (5) Melanosomes are strong absorbers. However, they are only present in a small percentage in the epidermis, i.e., the local interaction of light with melanosomes is great, but the contribution of melanosomes to the average absorption coefficient is rather low, i.e., the light transport is only slightly influenced by these molecules.

product which will very soon be ready to be brought to the market.

Themed by the key phrase "Light for better physical health", the projects deal with bio-processes, cell–cell communication and biological interactions in the whole organism. As mentioned before, the intention is to gain a deeper understanding of the processes that lead to the outbreak of common diseases. To achieve this end will take time, and as in an old saying so also in Biophotonics, even the longest journey begins with a single small step. So our first purpose has to be to learn more about the correlation between the various influences from outside and inside the human body shown in **Figure 1.10**, which draw the distinction between health and illness. Internal influences are based on the genetic profile of the individual, and are responsible for changes in protein production and metabolism, which can have severe effects no matter

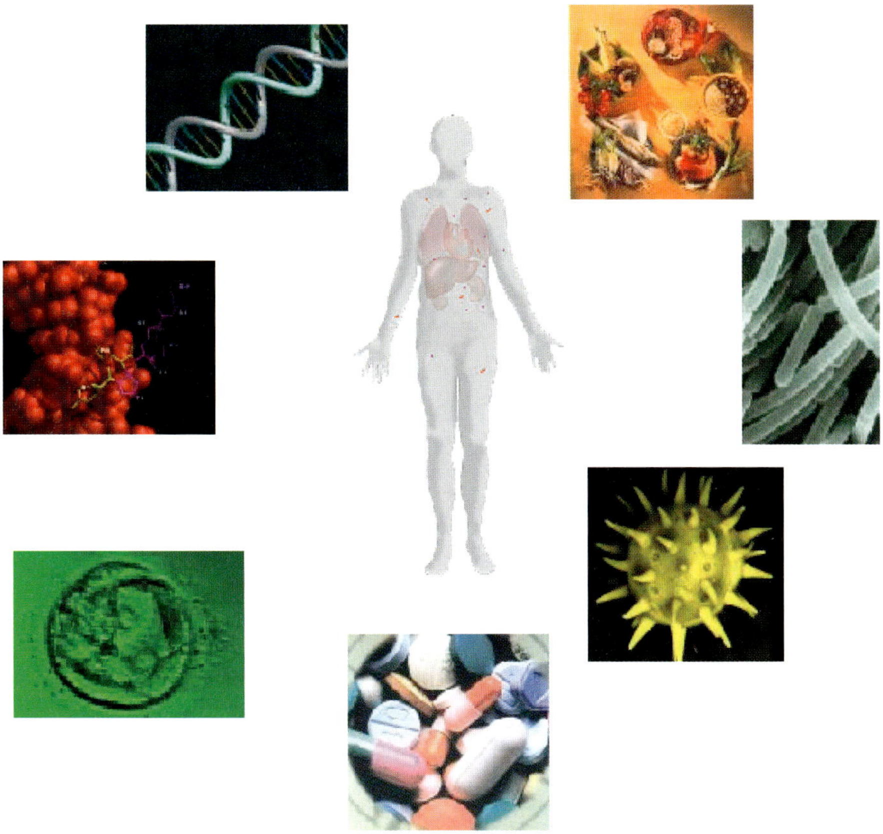

Figure 1.10 The individual projects of the German main research topic Biophotonics work on different biological levels. Some deal with the genomic level, i.e., DNA mutations, others with the interaction of proteins or changes in the metabolism of a single cell. Other projects concentrate on influences from outside, such as airborne microorganisms or pollen grains or the effects of food and drug ingredients. However, always at center stage stands the human being and its physical health.

how marginal they might be. External parameters can also be found inside the body, such as microorganisms living in our gut. But despite that, and because our organism is an open system, there is a continuous intake and excretion of substances that also influence our state of health, e.g., food substances, drugs and airborne biotic and abiotic particles. You will find this variety of influencing agents reflected in the topics of Chapters 2 to 10, which highlight the so far undreamed-of possibilities being provided by the fast-emerging field of Biophotonics. In the following we will give a short overview of the content of the various Biophotonics projects.

A better quality of life for the millions of allergic persons is the ambitious goal of the network project "Online Monitoring of Airborne Allergenic Particles" (OMNIBUSS). To allow people suffering from hay fever or asthma a nearly normal way of life and to avoid unnecessary intake of pharmaceuticals, a continuously and routinely updated knowledge of pollen concentration in the air is essential. Until now, pollen forecasting has employed manual microscopy techniques, which are not only time-consuming and labor-intensive but also provide results of undefined and unsatisfactory quality. As described in Chapter 2, p. 31ff., OMNIBUSS has developed a new microscope-based fully automated monitoring method, which is characterized by high temporal resolution, excellent reproducibility, detection limit, recall and precision, and which therefore meets the strong public demand for absolutely reliable of pollen concentration data. The device the project has led to combine continuous sampling and automatic preparation of aerosol, automated particle imaging and automated identification of pollen grains based on mathematical fingerprints.

Just as in allergy prevention, the presence of airborne biotic particles plays a very important role in connection with clean-room processes. In industrial food or pharmaceutical production, such bioaerosol may lead to fatal consequences. To drastically reduce the time needed for quality assurance of life science products, traces of contamination need to be tracked down reliable and rapidly. Chapter 3, p. 89ff., "Online Monitoring and Identification of Bioaerosols", shows how the new approach of the research network OMIB can make a decisive contribution to the monitoring process for a rapid detection of aerosol and an identification of airborne microorganisms without loss of time. For standard microbiological tests, the microorganisms are collected on a growth medium, bred and eventually counted. Under certain conditions even the combination of several microbiological tests leads only to an ambiguous identification. The authors of Chapter 3 describe how they developed totally new equipment, which combines differentiation of biological from non-biological particles by means of fluorescence detection with an identification step by vibrational spectroscopy, in particular Raman spectroscopy. This method offers an enormous time gain compared to the conventional methods applied until now.

Infectious agents are gaining ground again – and they can not only cause industry a loss of time and money but each one of us a loss of life. Germs such as the tuberculosis-causing *Mycobacteria* are horrifying the world, not only by their rapid spread around the globe but also by their ability to cope with antimicrobial drugs. One new powerful weapon is presented in Chapter 4 called "Novel Singly Labeled Probes for Identification of Microorganisms, Detection of Antibiotic-resistant Genes and Mutations, and Tumor Diagnosis". "Smart Probes" (Chapter 4, p. 167ff.) are fluorescently labelled DNA-hairpin structures, which have the potential to open a new avenue in molecular diagnostics by their ability to discriminate between wild type and resistant bacteria. Besides a detailed description of Smart Probes applications in the detection of antibiotic-resistant genes and mutations as well as in tumor diagnosis, the chapter gives a competent general survey of single-molecule fluorescence spectroscopy and 3D fluorescence nanoscopy.

The abbreviation PLOMS (Chapter 5, p. 231ff.) conceals an innovative way to detect cancer of the colon in its very early stages, thus dramatically increasing healing rates simultaneously cutting the expenses in the health system. The authors of Chapter 5 captioned their contribution "Early Diagnosis of Cancer". They give a very detailed introduction to the epidemiological and biological background of cancer, the mechanisms of carcinogenesis and the impact of early cancer diagnosis before outlining their new approach for very early detection of the first stages of tumor growth in the colon. This new approach is based on the fact that the structure of the glycocalyx of normal mucosa cells changes as they degenerate into cancer cells. Like an old blanket becoming threadbare, the glycocalyx forms small holes, which allow the scientists of the network project PLOMS to discriminate between healthy and degenerated cells in the colon via sophisticated labelling strategies.

Chapter 6, p. 301ff., entitled "New Ways for Marker-free Live Cell and Tumor Analysis" also deals with early tumor diagnosis. MIKROSO is a network project developing digital holographic microscopy as a very new approach for label-free quantitative imaging of living cells and therefore as a useful tool for seeing the signs that reveal even marginal pathological changes in cells and tissue, and for watching very closely the behavior of healthy and diseased cells. The authors point out the many advantages of digital holographic microscopy in comparison with standard methods of cell microscopy, and describe the various application possibilities of combinations of digital holographic microscopy with other techniques, such as phase-contrast and fluorescence imaging or laser micromanipulation. In addition, the article provides an introduction to optical coherence tomography as well as to minimally invasive holographic endoscopy.

The aim of "Regenerative Surgery" is to heal diseased tissue by full or partial reconstruction and to support the regeneration of organs if not actually to substitute them. Chapter 7, p. 361ff., gives an account of the medical and biological background not only of regenerative surgery but also of tissue engineering, with a strong emphasis on cell and tissue culture technologies. The network project MeMo designs novel techniques to improve such cell and tissues cultures by laser optical on-line monitoring. Innovative biophotonic technologies have been developed for a more efficient evaluation of tissues. Taking as an example the replacement of human cartilage tissue and chondrocytes, the authors outline the advantages of recent methods like three-dimensional laser-scanning microscopy, fluorescence lifetime measurements and parallelized two-photon measurement systems for rapid high-resolution tissue imaging.

Chapter 8, p. 405ff., brings us from the surgery back to the laboratory, and in particular to the "lab on a chip". Microarrays have been one of the enabling technologies of the 1990s, and have greatly increased the possibilities not only for basic research in molecular biology but also for the identification and validation of drug targets. But as a technique which can perform "thousands of reactions on a small chip" as indicated by the chapter title, microarrays demand efficient, reliable and comprehensive analysis methods. In this regard, optical systems provide a wide choice of techniques. The scientists of the MOBA network project concentrate on terahertz spectroscopy to obtain very rapid high-quality results, previously unknown. The binding of DNA and other biomolecules can be analyzed to learn more about the genetic profile of the scanned samples. As well as the specific adjustment of drugs for individual gene profiles, further applications lie in the areas of bioweapon analysis and telemedicine. An overview of fluorescence and label-free techniques completes this chapter.

A deeper understanding of the processes of life from the molecular structure to the whole organism demands methods that provide very high resolution in time and space. Therefore, analytical tools are required which cope with the need for minimally invasive measurement techniques featuring not only high precision and selectivity but also the ability to process a large number of samples at the same time. Chapter 9, p. 477ff., called "Hybrid Optodes" describes the work of the research network HYBOP aiming for the development of a novel fluorescence-based hybrid technology which will be employed in spatiotemporal high-resolution bioprocess analysis. This new technology is based on indicator-specific polymer surface coatings which supply two kinds of information at the same time by means of optoelectronic measurements. Thus, in each case two parameters, e.g., temperature and oxygen or carbon dioxide and pH, can be measured with high precision simultaneously. The chapter gives an overview of the principles of hybrid optodes and their applications and perspectives in biotechnology.

With the vision of an early detection of diseases and tailor-made therapies the ODMS network project is consistent with the aims of the main research topic "Biophotonics" in general. As described in Chapter 10, p. 519ff., entitled "Digital Microscopy", ODMS has developed an "Ocularless Digital Microscope System" for the on-line *in vivo* measurement of biological or biomedical parameters. There are three main areas of application to which the device will be adapted. One is the target assay development on the cell-culture level, providing a comprehensive visualization of the pharmaceutical effects of a new drug. Telepathology, that is the transfer of histological medical findings in the form of digitalized data ("virtual slides"), is another application for ODMS. This technology reduces the time-consuming dispatch of samples to different consultants. Thirdly, ODMS will aid cell and developmental biological studies. The aims are in particular to minimize stress for the biological probe and to use light to its full capacity.

We hope that with these short summaries we have aroused your curiosity so that you continue to read. But first we should like to thank the Ministry of Education and Research and the Association of German Engineers (VDI) for their financial support of the work of all the network projects and their excellent cooperation over the years. In addition, the editors would like to thank all the people who have contributed to this book. Above all, we should like to thank Andreas Thoss and the team at Wiley for providing us with the idea for this project. We appreciate the confidence and patience they have shown us very much indeed. To cope with the task of editing a book like this cannot be done without many helping hands. So our gratitude goes to the authors of the single chapters of this book for all their efforts to make this volume a valuable one for scientists all over the world. A truly special word of thanks goes to our colleagues at the Institute of Physical Chemistry in Jena, who provided us with invaluable help in the matter of technical support and were able to restore us to a good mood at any time. We must mention by name Kathrin Strehle, Ute Uhlemann, Dana Cialla and Reinhold Gade, but our thanks go also to all the other members of our working group. On a very personal note we should like to thank our families for being so understanding during all the days and nights we were constrained by the work on this book.

1.4 Links and Literature about Biophotonics

The following list is far from complete and is only intended to provide the interested reader with further insight into the rapidly growing subject of biophotonics:

Journal:

- "Biophotonics International", Laurin Publishing Company (www.photonics.com)

Selected "Special Issues":

- Special Issue on Biophotonics, Journal of Physics D: Applied Physics, Volume 36, Number 14, 21 July 2003
- Special Issue on Biomedical Optics, Journal of Physics D: Applied Physics, Volume 38, Number 15, 7 August 2005

- "Biophotonics Micro- and Nano-Imaging", Progress in Biomedical Optics and Imaging, Vol. 5, No. 33, in: Proceedings of SPIE, Volume 5462, 2004.

Books:

- Faupel, M., Brandenburg, A., Smigielski, P. and Fontaine, J. (Eds): Biophotonics for Life Sciences and Medicine, FontisMedia, Lausanne/Formatis, Basel, in press.
- Marriott, Gerard (Ed): Biophotonics, Academic Press, San Diego, 2003.
- Prasad, Paras N.: Introduction to biophotonics, Wiley-Interscience, Hoboken, New Jersey, 2003.
- Shen, X. and van Wijk, R. (Eds): Biophotonics: Optical Science and Engineering for the 21st Century, Springer, Berlin, 2006.
- Wilson, B.C., Tuchin, V.V. and Tanev, S. (Eds): Advances in Biophotonics, NATO Science Series: Life and Behavioural Sciences, Volume 369, IOS Press, Amsterdam, 2005.

Selected Initiatives in Germany:

- Main research topic "Biophotonics" funded by the German Federal Ministry of Education and Research: http://www.biophotonik.org
- OptecNet: Network consisting of several local initiatives dealing with photonics and biophotonics, respectively: http://www.optecnet.de
- Another local network, the "Biotech/Life Science Portal Baden Württemberg" offers a number of articles in the field of biophotonics: http://www.bio-pro.de

2
Online Monitoring of Airborne Allergenic Particles (OMNIBUSS)

Stefan Scharring[1], Albrecht Brandenburg, Gernot Breitfuss[1], Hans Burkhardt[1], Wilhelm Dunkhorst, Markus von Ehr, Marcus Fratz, Dominik Giel, Ulrich Heimann, Wolfgang Koch[1], Hubert Lödding, Werner Müller[1], Olaf Ronneberger, Eckart Schultz, Gerd Sulz[1], Qing Wang

2.1
Introduction

2.1.1
Health-related Impacts of Aerosols

Since the middle of the twentieth century there has been growing concern about air quality. Since the mid-1990s, new European framework directives relating to air quality gave rise to intense discussions about limit values for ambient exposure and their practical application. With respect to the impact on human health, particulate matter has to be differentiated by its aerodynamic diameter [1]. Particles with an aerodynamic diameter less than 10 μm, denoted as PM_{10}, can enter the lower respiratory tract, i.e., the bronchial tubes in the lung. However, the pulmonary alveoli can only be reached by the finer particle fraction with an aerodynamic diameter less than 2.5 μm, denoted as $PM_{2.5}$. A study by the World Health Organization (WHO) notes a correlation of ambient exposure of $PM_{2.5}$ with mortality and cardiovascular diseases [2]. An increased risk of respiratory diseases correlated with ambient exposure to PM_{10} was also noted [3].

Beyond the determination of the total mass concentration of particulate matter, it is reasonable to focus on aerosol components owing to their broad spectrum of health impacts. While soot particles are alleged to be carcinogenic, the beneficial effect of airborne sea salt on respiratory diseases is clearly evident. Microscopy is a powerful tool for the differentiation of coarse particulate matter and allows a detailed analysis of air quality.

1) Corresponding authors

Biophotonics: Visions for Better Health Care. Jürgen Popp and Marion Strehle (Eds.)

ISBN: 3-527-40622-0

2.1.2 Allergies

Among the atopic diseases, hay fever (pollinosis) and asthma affect the human respiratory tract. The prevalence of asthma, allergic rhinitis and conjunctivitis has considerably increased during the second half of the twentieth century. In parallel, sensitization to pollen allergens has increased in most areas of Europe, e.g., sensitization to birch pollen in central and northern Europe, to olive pollen in Mediterranean areas, to ragweed pollen in Hungary or to plane tree and *Cupressaceae*, particularly *Taxaceae* pollen in urban areas, typically. At present, their concentration is mostly unclear, owing to the lack of data suitably resolved over time and space. On the other hand, a knowledge of pollen concentration in the air is an important factor for estimating disease development and outcome. Furthermore, pollens are not only allergen carriers but may also release proinflammatory substances.

According to the so-called hygiene hypothesis, the increased prevalence of allergies is due to reduced exposure to infectious agents [4]. The hygiene hypothesis suggests that an inverse relationship exists between the incidence of infectious diseases in early life and the development of allergies. However, not all infections are associated with a lower prevalence of allergic diseases, but the time and route of infections are crucial factors in conditioning a protective effect on sensitization and allergic diseases [5]. These results show no association between allergic diseases and airborne infections, but a strong negative association with food-borne infections.

According to the European Allergy White Paper, the overall prevalence of seasonal allergic rhinitis in Europe is approximately 15% [6]. The current asthma prevalence rates range between 2.5 and 10%. The prevalence of allergic diseases and asthma has steadily increased in recent years, and this trend seems to continue in most countries. The consequences are an increased morbidity, the loss of productivity, a decreasing quality of life and growing costs in health care in western societies due to pollen and other aeroallergens, e.g., fungal spores. The increase was found to be most pronounced in children, with the prevalence up by about 200% since the mid-1970s [7]. As well as encouraging research into the cause and therapy of allergies, this development has over the years also given rise to various efforts to improve the measurement and forecasting of pollen concentrations in ambient air. The aim of these efforts is to provide information to enable pollen allergen sufferers to adjust doses of medication, personal activities and precaution strategies to current or expected pollen concentration.

Allergy symptoms, especially in phases of high pollen concentration, might be weakened or in some cases even completely suppressed by specific medication. However, many allergic people avoid medical treatment owing to suspected adverse effects or even health risks. Allergic people, therefore, often try to manage without their medication. Instead, they arrange their personal habits, lifestyle and activities to avoid peak pollen concentrations. This requires information that is as detailed as possible about present or expected pollen concentrations. The information should also provide differentiated data that distinguishes between pollen species, because of the different allergenic potentials of pollens and the specific sensitization of the allergic individual.

There is growing evidence that climate change might facilitate the geographical spread of particular plant species to new areas that become climatically suitable. But changes in land use and socio-cultural changes as well as international transport and tourism also obviously promote the spread of these plant species. The occurrence of some invasive species means particular risks for health and requires control, especially of those pollens characterized by high allergenic potentials, e.g., ragweed.

2.1.3 Pollen Counting – State-of-the-Art

For conventional analysis, pollens are sampled from ambient air with instruments that provide samples of pollen deposited on a sticky carrier for subsequent microscopic analysis. Pollen counting under the microscope is done by eye. This work is demanding and time-consuming, even for experienced microscopists. From the samples, a daily average pollen concentration is derived in routine measuring networks. The daily data are not available earlier than the day after sampling. The quality and reliability of routine data vary considerably according to the qualification and commitment of the pollen counters. Late availability of the data and high costs arising from the time-consuming work are other consequences of this procedure.

However, until now pollen measurement has still been based on this technique world-wide. In most European countries pollen networks were set up in the second half of the last century. They may be operated by private groups, universities or national institutions. A European pollen information network was founded in Vienna [8] to standardize and coordinate collection networks, the technique of visual pollen counting and the setting up of networks. However, considerable deviations still exist on the European scale in respect of various aspects, e.g., adequate spatial coverage, frequency and continuation of data transfer, time resolution of pollen data, quality control and assurance and site representativeness.

In Germany, a pollen network is operated and maintained by the Foundation of German Pollen Information Service [9]. Pollen data from this network are the basis for the pollen forecast of the German Weather Service. Since about 1980, daily forecasts have been distributed by the Human Biometeorology department of the German Weather Service at Freiburg [10]. The forecasts are given for the most allergy-relevant pollen in Germany, hazel, alder, birch, sweet grass, rye and mugwort in the pollen season from late winter to early autumn.

A specific disadvantage of the existing pollen data is their low temporal resolution. As an example, illustrated in **Figure 2.36** in Section 2.3.2 below, a moderate daily average concentration may arise on the day from a temporal course with high fluctuations [11, 12]. Thus, a detailed analysis of data might reveal that, for example, only two hours of the day exhibit a peak concentration with high stress for allergy sufferers, while there is a low pollen load from the evening until the following morning. Moreover, the different allergenic potential of the individual pollen species as shown in **Table 2.1** requires detailed and sensitive analysis.

Table 2.1 Allergenic potential of different pollen species in Germany.

	Concentration, $1/m^3$		
Allergic stress	low	moderate	high
Hazel, alder	1–10	11–100	> 100
Birch	1–10	11–50	> 50
Gras	1–5	6–30	> 30
Rye, mugwort	1–2	3–6	> 6

2.1.4
A New Approach to Pollen Information and Forecasting

The pollen forecast is still increasingly requested by allergy sufferers in Germany. A public opinion poll revealed that about 75% of pollen allergy sufferers use pollen forecasts as health-related information and a guide for planning individual activities. However, consumer satisfaction with the quality of forecasts is only moderate owing to a lack of real time data, missing information about daily variations, and insufficiently resolved data relating to the area covered by the forecasts.

More than 20% of the participants are highly interested in an improvement of pollen forecasting in order to overcome the perceived deficiencies in pollen information. There is a strong public demand to provide pollen concentration data that exhibit the following features:

- real time information
- higher temporal resolution
- known quality in relation to reproducibility, measurement uncertainty, detection limit, recall and precision.

Therefore, an online pollen information service will be set up and an objective pollen forecast with a higher temporal and spatial resolution will be established including meteorological, orographical and phenological data [13].

To achieve these goals the following approach was chosen:

- continuous sampling of pollen in ambient air and automatic preparation of the aerosol sample for microscopic analysis at the sampling site
- automated particle imaging by different microscopic techniques in parallel
- automated identification of the pollen grains by a computer-based classifier based on the extraction of a mathematical fingerprint from all pollen grains
- integration of pollen sampling, sample handling, microscopic imaging and pollen recognition in an autonomous measuring system.

This approach will be described in detail in the following sections.

2.2 Monitoring Bioaerosols: State-of-the-Art

2.2.1 Existing Instrumentation for Sampling Environmental Allergens

The size of atmospheric bioaerosol particles varies over three orders of magnitude. Viruses are small, about 100 nm, whereas plant pollen can be as large as 80–100 μm. Within this project the focus is directed to particles ranging in size from approximately 2–3 μm up to 60–80 μm since this regime covers the major environmental allergens: fungal spores and plant pollen. Detection and recognition of bioaerosols is performed by sampling, which means removing particles from the air and subsequently depositing them on a suitable substrate. The desired properties of the collection medium depend strongly on the analytical method used for bioaerosol detection, specification and counting. For spores and pollen one method of choice is aerosol deposition onto an optically transparent substrate, e.g., a foil or a plate of glass or plastic, followed by a preparation step that may include a specific staining procedure and subsequent microscopic analysis for visual determination or pattern recognition.

For environmental allergens two sampling and deposition methods that differ in principle are used nowadays:

1. sampling through a sampling orifice by an active air intake system (suction device) and subsequent deposition on a surface

2. direct deposition of the particles of interest on surfaces rotating in the air at sufficiently high speed.

The Burkard® pollen trap [14] and the Rotorod® sampler [15] (**Figure 2.1**) are instruments based on these two principles that are in operation for atmospheric pollen monitoring.

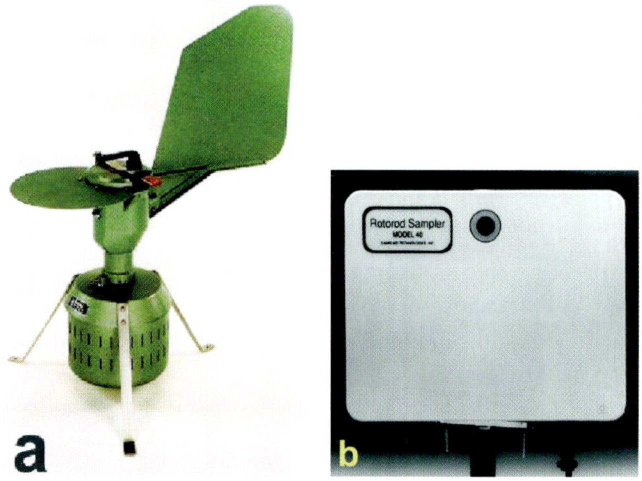

Figure 2.1 The Burkard® pollen trap (a) and the Rotorod® sampler (b) are widely used as aerosol samplers in pollen measurement networks.

In the widely used Burkard® trap, an air stream of 10 $\mathrm{l\,min^{-1}}$ is sucked in through a rectangular inlet of 2 × 14 mm. The air jet is directed onto a sticky, transparent foil inside the instrument. Particles are deposited on the foil by inertial impaction in the stagnation point flow developing above the foil. The inlet system is allowed to align itself to the wind direction by means of a weather vane. The Rotorod® sampler collects the particles on the sticky surface of small plastic rods rotating at high speed in the air. Obviously this sampling method by itself causes the instrument's sampling characteristic to be independent of the wind direction. In both instruments, the foil and the rods, respectively, are removed from the instrument after sampling and are analyzed off-line in the laboratory.

The rotating rod principle was identified as unsuitable for use in an automatic instrument for obvious technical reasons. A one-to-one translation of the sampling design of the Burkard® sampler was also considered to be impractical because of the desired flexibility in the sampling flow rate for a possible adaptation to field conditions and because of the intention to make the instrument mechanically as simple as possible. Therefore, alternative sampling strategies and sampling inlets were investigated with a view to obtaining representative samples of bioaerosols from the outside air.

In this context, previous knowledge in the field of general atmospheric dust sampling as well as dust measurement at workplaces is very helpful. Whereas today atmospheric dust sampling by active sampling systems focuses mainly on the fine PM_{10} particles, the particles of interest in industrial hygiene range up to 100 μm.

Various types of sampling inlets for aerosol sampling devices have been reviewed by, among others, Vincent [16]. Burton and Lundgren [17] and later Holländer et al. [18] have reported on a wide-range aerosol classifier developed as a reference instrument for sampling and size classification of particles up to 100 μm in the atmosphere, even under conditions of high wind speeds. This instrument, with an inlet diameter larger than 50 cm, has very large dimensions and operates at very high intake flow rates $>1000\ m^3\,h^{-1}$. The reason for this is the aim of reaching a 100% sampling efficiency in the entire range of particle sizes and wind speeds. This design cannot directly be used for the new automatic pollen monitor. However, the strict conditions that are met by the large sampler described above are not necessarily appropriate for health-related particle sampling. For this purpose one can refer to the so-called inhalability curve, which has been defined in international standards for workplace dust measurements [19]. This curve mimics, on average, the particle intake by humans, i.e., it characterizes for each particle size the fraction of the suspended airborne particles inspired into the human respiratory system. The curve starts out with 100% at small particle sizes and levels off at 50% for large particles. It is much easier to match this inhalability criterion than to comply with the criterion of 100% sampling efficiency. A few so-called inhalable samplers exist that comply with the inhalability criterion. Many of them, however, have directional inlets, like the Burkard® trap. The performance of some of the samplers has been tested under controlled conditions in wind tunnel experiments by Li et al. [20]. This study shows that omnidirectional slit inlets have good performance in sampling inhalable dust.

2.2.2
Microscopic Techniques

An appropriate preparation of the deposited particles delivers aerosol samples suitable for microscopic analysis. Visual determination of pollens includes their observation with an optical microscope and comparison of grain sizes, shapes, and surface structures with those of reference pollens.

Optical resolution considerations limit microscopic analysis to the supermicron particle range. However, a large number of health-effective particles occur in this range [21]. A typical example of aerosol particles, taken by a Burkard® trap, is shown in **Figure 2.2**.

In general, relevant features for the discrimination and identification of bioaerosol particles are morphological properties, such as size, shape, surface

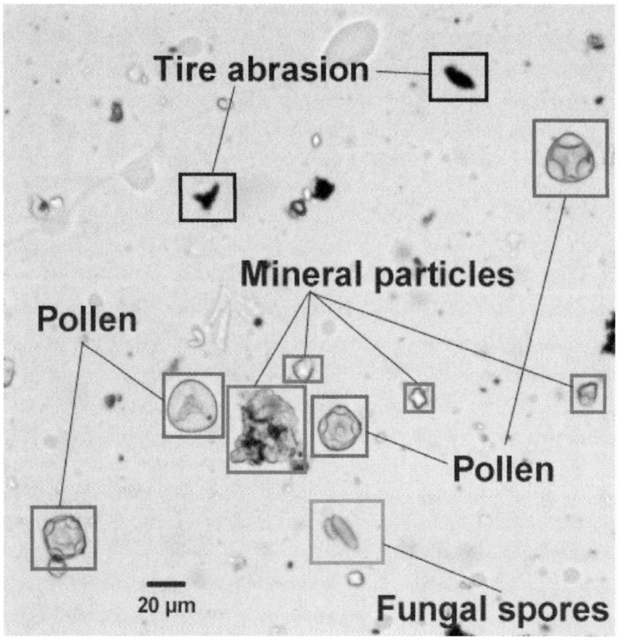

Figure 2.2 Microscopy reveals the content of the air that we breathe. Its variety can be found on such a typical sample (10× magnification, brightfield illumination) from a Burkard® trap with various aerosol particles, e.g. mineral particles, tire abrasion, fungal spores and pollen.

structure and structures inside biological particles [22–24]. Particularly, in the case of allergens such as airborne pollen, these morphological characteristics are conventionally used to identify the pollen classes. The variety of pollen structures, made visible by brightfield microscopy, can be seen in **Figure 2.3**, while the morphologic diversity of airborne fungal spores is depicted in **Figure 2.4** [25, 26].

Besides these structural quantities, other specific optical characteristics of bioaerosols can also be used to distinguish between different classes of particles. Biological particles such as pollen show a strong primary fluorescence in the UV and blue spectral range, allowing reliable differentiation from the background and from other simultaneously occurring particles within the process of image analysis. Colour, birefringence, etc. are other parameters which are sometimes useful to identify particles.

For a reliable and reproducible automatic recognition process, the microscopic images have to contain all the relevant features of the particles that are necessary for an unambiguous identification [27]. There is a range of different microscopic illumination and contrast methods [28] for the imaging of different structural and optical particle properties [29–35]. Regarding illumi-

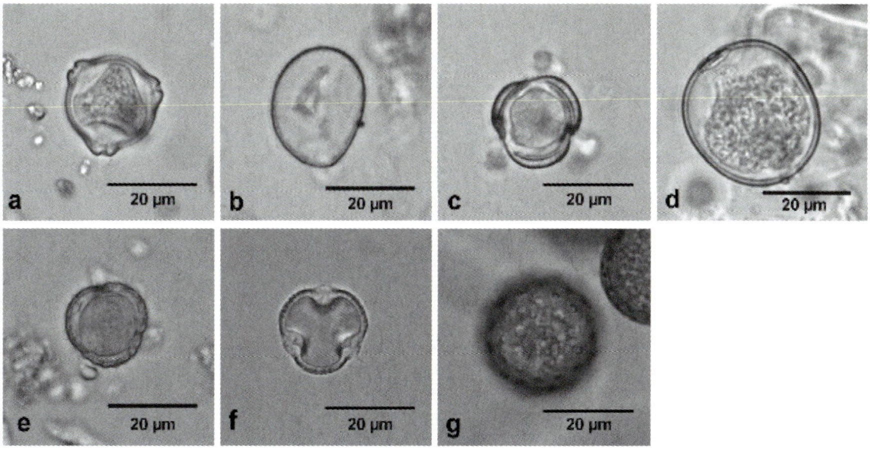

Figure 2.3 Visual pollen counting is based on knowledge and experience of the various morphological forms of pollen grains. These microscopic images (40× magnification, brightfield illumination) show examples of (a) birch (three porate apertures), (b) yew (inaperturate), (c) mugwort (three colporate apertures), (d) grass (monoporate), (e) common ragweed (three colporate apertures), (f) willow (three colporate apertures) and (g) goosefoot (polyperiporate).

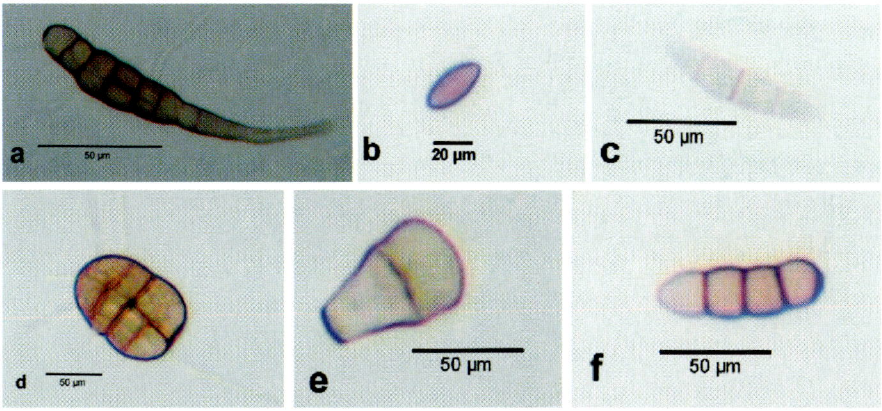

Figure 2.4 Many different shapes can be found among fungal spores. These images taken in brightfield illumination (20× magnification) show examples of (a) *Alternaria* sp., (b) *Chaetomium* sp., (c) *Fusarium* sp., (d) *Pleospora* sp., (e) *Polythrincium* sp. and (f) *Torula* sp.

nation of the sample, transmitted and incident light microscopy have to be considered. Additionally, there are several contrast methods for the imaging of a sample, for example brightfield illumination, darkfield illumination, phase contrast, differential interference contrast (DIC), polarization contrast and the fluorescence method.

For conventional pollen counting, transmitted light microscopy with brightfield illumination (**Figures 2.5** and **2.6**) is used. This is the classical mi-

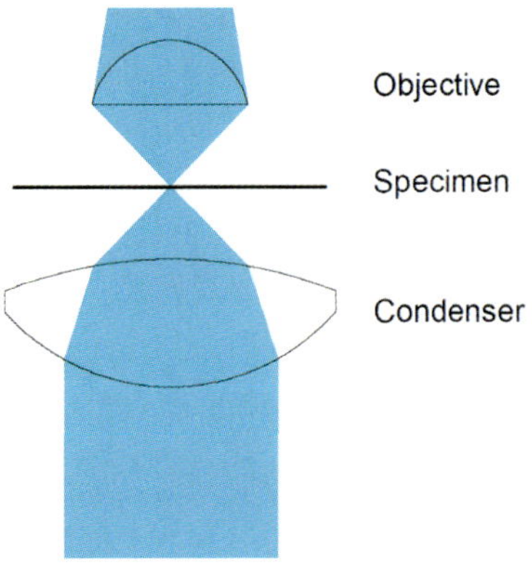

Figure 2.5 Illumination principle of transmitted light brightfield microscopy.

croscopy method mostly used for microscopic imaging of samples. With brightfield illumination, structural information and additional transparency and color characteristics of the samples can be investigated. On the other hand, many unstained biological samples, such as cells, generate images with very little contrast and are sometimes invisible in ordinary brightfield microscopy.

In general an incident wavefront of an illuminating beam on a sample becomes divided into two components: an undeviated wave that passes through

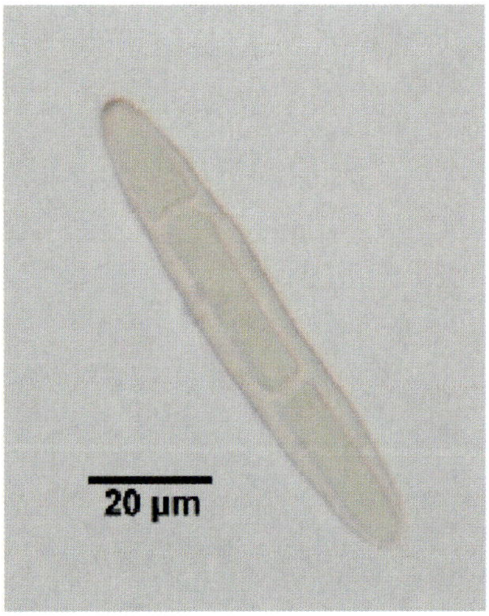

Figure 2.6 *Bipolaris* sp. is a large airborne fungal spore. Though it is rather translucent, three cells can be detected with brightfield illumination ($20\times$ magnification).

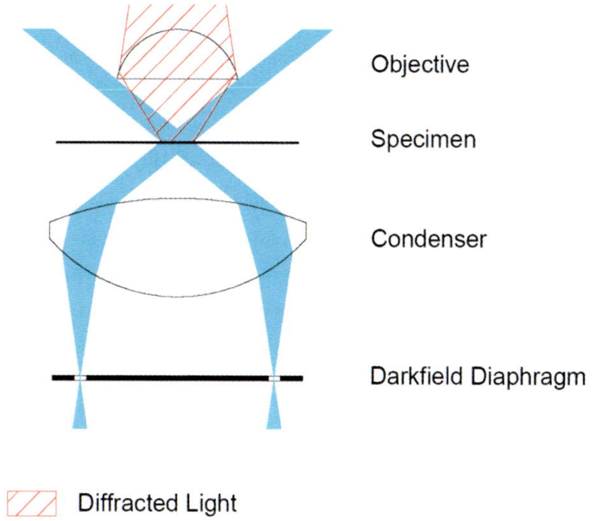

Figure 2.7 Illumination principle of transmitted light darkfield microscopy.

the sample and a deviated or diffracted wave that becomes scattered in many directions. In contrast to brightfield microscopy, where both waves are collected by the microscope objective and interfere in the image plane, in darkfield illumination only the scattered light contributes to the image. In transmitted light darkfield illumination (**Figures 2.7** and **2.8**), the specimen is illuminated at such an angle that no direct light can pass through the objective. If the refractive index in the specimen changes abruptly, diffraction occurs and the light is scattered in various directions. The angle of scattering is wider the greater the change in the refractive index. A part of this scattered light reaches the objective, whereas no scattered light is emitted by the surroundings, which remain invisible.

Darkfield illumination is therefore very powerful for contrasting structures that are mainly based on changes of the refractive index. So the following objects are suitable for this kind of illumination:

- extensive objects of regular structure: diatoms, radiolarians, etc.
- linear structures: flagellates, fibres, crystals, bacteria, etc.

These linear structures will become visible even when their thickness is below the resolving power of the objective.

Point-shaped objects at or below the limit of resolution are represented by a diffraction disc. In this case the darkfield image gives no information about their true shape but only indicates their occurrence.

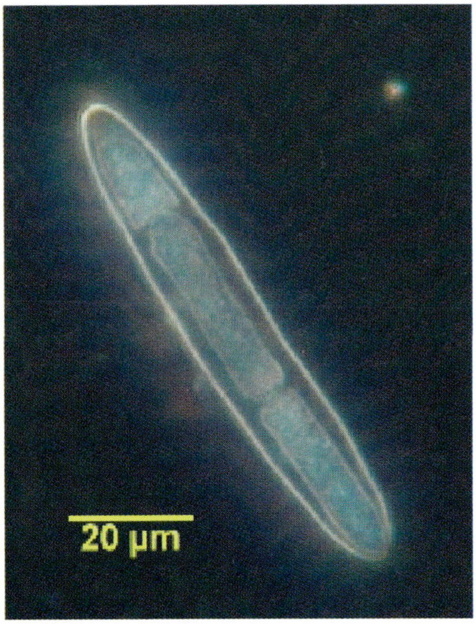

Figure 2.8 Diffraction and scattering of light make fine structures, such as these boundary layers of a *Bipolaris* sp. visible (20× magnification).

Phase contrast microscopy (**Figures 2.9** and **2.10**) permits high-contrast rendering of transparent objects that do not significantly affect either the brightness or the color of the light as it passes through. Differences in the relative phase of object-diffracted waves are transformed into amplitude differences in the image. Structural elements with differential refractive indices will therefore be rendered visible. Hence, this technique is particularly suitable for the visualization of very fine structural features and provides high-contrast images of living or fixed unstained material such as tissue sections, cells, etc. preferably in very thin layers <5 µm.

For the technical realization of phase contrast in a microscope two components are required:

- a condenser with a ring diaphragm in the plane of the aperture diaphragm (annular condenser aperture)
- an objective with a phase ring in the back focal plane.

When working in phase contrast, the image of the light annulus in the condenser and the phase ring have to be superimposed in the back focal plane.

Whereas ordinary brightfield microscopy renders the structures of cells and tissues visible through the absorption and diffraction of light, fluorescence microscopy, cf. **Figures 2.11** and **2.12**, is based on a completely different principle [36]. If fluorescent substances are present in a specimen they will become

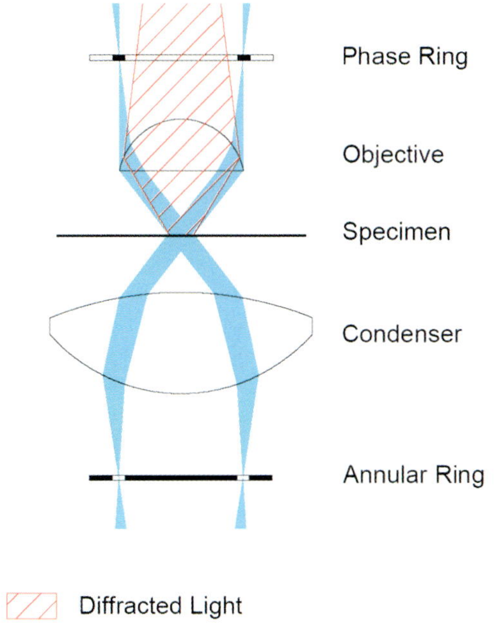

Figure 2.9 Illumination principle of phase contrast microscopy

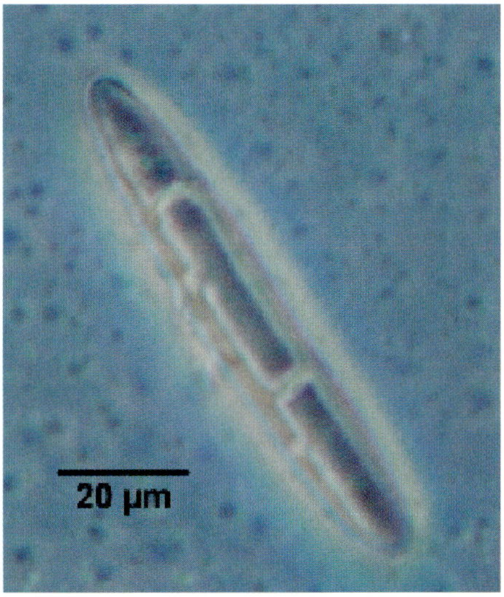

Figure 2.10 Structural features can also be displayed by phase contrast microscopy, exemplified with a *Bipolaris* sp. (20× magnification).

independent light sources as long as they are irradiated with exciting light, whereas the nonfluorescent surroundings remain dark.

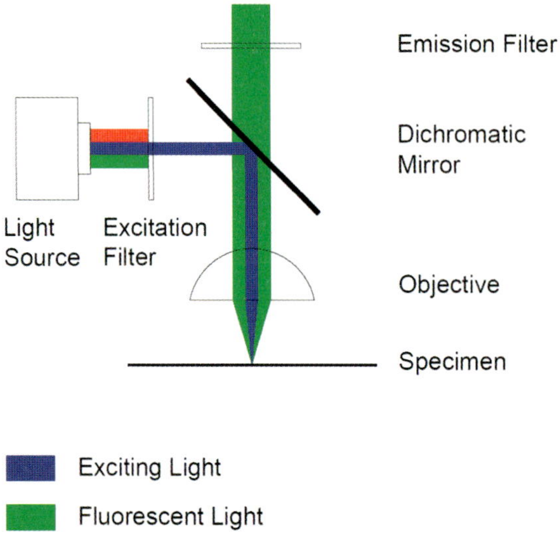

Figure 2.11 Illumination principle of fluorescence microscopy.

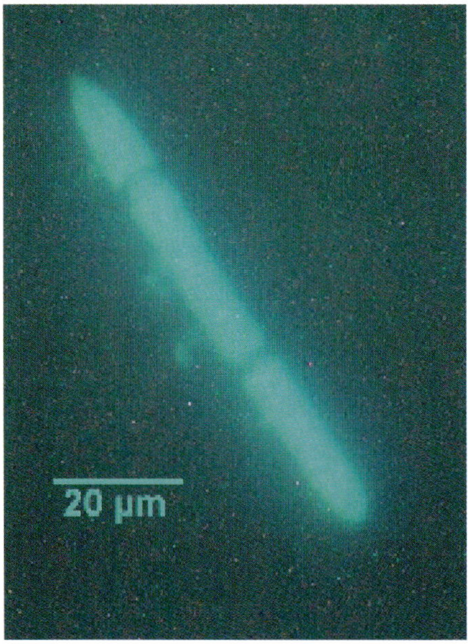

Figure 2.12 Fluorescence excitation highlights the cells of a *Bipolaris* sp. while other parts of the spore remain in the background (20× magnification).

Many biological, medical and petrographic specimens have the ability to emit fluorescence light when excited with appropriate light, without being specially treated, stained or labelled. This phenomenon is called primary fluorescence or autofluorescence.

In most cases, in order to obtain specific and meaningful fluorescence, it is necessary to stain the specimens with fluorescent dyes called fluorochromes or fluorophores. Different components within the specimen will take up these fluorochromes in different concentrations, resulting in different fluorescence colours and brightness, thus allowing qualitative identification and evaluation. Fluorescence resulting from staining the specimen with fluorochromes is called secondary fluorescence.

Pollens show a strong primary fluorescence in the blue and green spectral range if they are excited in the violet spectral range around 380 nm. This provides an easily accessible tool for a reliable separation from the background and from other nonfluorescent particles, e.g., minerals.

Among the contrast-enhancing techniques we use to visualize low-contrast transparent specimens, Normarski's differential interference contrast (DIC), the VAREL contrast method (Zeiss AG) and others are available for high resolution light microscopy and are widely used in biological applications.

In contrast to the above mentioned methods, interferometric techniques have the advantage of yielding quantitative measurements of parameters including the phase distribution produced by transparent specimens. Optical digital holography as an example provides an accurate 3D imaging of biological materials, down to the microscopic scale. Digital recording and numerical reconstruction of holograms by numerical processing of the complex wave front allows simultaneous computation of the intensity and the phase distribution of the propagated wave [37].

2.2.3 Pattern Recognition

Even though pattern recognition of images is widely used in several biological applications, very few papers in the literature deal with pollen recognition and most of them focus on fossil pollen and 2D data [38–42].

Pollen recognition in air samples incorporating 3D data was independently started at the end of 1998 by a French group around Monique Thonnat (INRIA Sophia-Antipolis) and our group around Hans Burkhardt (University of Freiburg, Germany) in cooperation with the German and Swiss meteorological services with two different approaches: Bonton [43] and Boucher [44] proceed in the same way as human microscopists, i.e., they use multiple translucent images of the pollen from different focus planes and extract the same "high-level" features as humans do (e.g., the number of pores) by use of many highly pollen-taxa-specific or even pollen-grain-orientation-specific algorithms for feature extraction and they employ (to the authors' knowledge) a simple hard-coded classifier. Using a set of 350 pollen grains from 30 different taxa the corresponding recognition rate is about 77%.

The more general approach by our group without pollen-specific algorithms using gray-scale invariants on volumetric data from confocal fluorescence microscopy was presented in Refs. [45] and [46], where we reached recognition rates of 92% for 26 species to 97.4% when combining non-allergic relevant pollen into one class. Meanwhile we improved the recognition rate on this dataset to 98.7% for all 26 species [not yet published].

For fossil pollen recognition (which uses extensive chemical preprocessing of the pollen samples), a good summary of recent and older research of the group around J. R. Flenley can be found in Refs. [47, 48] and [49]. Even though the results presented were only based on manually cropped texture parts of 2D translucent images, they also show the clear advantage of gray-value/texture-based feature extraction in combination with a self-learning classifier compared to the use of "human motivated" features.

The main "shortcoming" of all published work on automatic pollen recognition is the missing step to real-world data. All of them work satisfactorily with a few manually prepared clean pollen samples without dust particles on or beside the pollen. For a real-world application, robustness to contamination and a very low number of "false positives" (which are the pollen-like dust particles that are falsely classified as pollen) are the main criteria that separate a good system from an unusable one.

2.3
Online Monitoring of Airborne Allergenic Particles by Microscopic Techniques and Pattern Recognition

2.3.1
Instrumentation

In contrast to conventional methods, where the different steps for the identification of airborne allergenic particles, e.g., sampling, preparation, microscopic inspection and analysis, are performed in different places, an automatic online monitoring and identification system requires an integration of all these functions into one system laid out for correct autonomous operation without any manual intervention.

The principle modular structure of an online monitoring system is shown in **Figure 2.13**. The instrument combines

1. a high-volume sampling unit for the sampling of coarse particles >2.5 μm in combination with electrostatic precipitation of this fraction onto a surface suitable for optical analysis

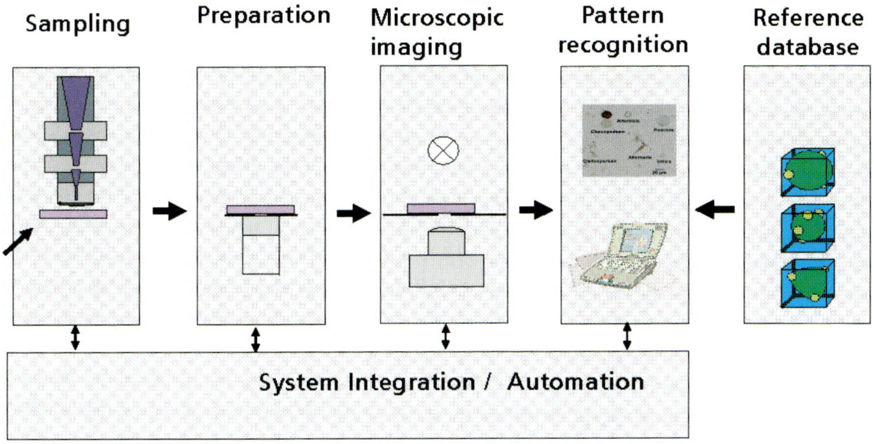

Figure 2.13 Principles of the online monitoring system (for details see text).

2. an automatic preparation unit for microscopic single-particle analysis
3. a digital microscopic imaging unit with different illumination and contrasting methods, e.g., brightfield, fluorescence and darkfield microscopy
4. a pattern recognition unit, which provides feature extraction by gray-scale invariants and classification by self-learning Support Vector Machines
5. a data telecommunication unit for long-distance transmission of the analysis results.

An hourly output of number concentration of airborne pollen, fungal spores and other particles of interest is specified while the focus within this project is laid on bioaerosols, especially pollens.

Several technical and economic requirements have been considered for the construction of such a system. Since the instrument has to be placed outdoors and has to work under very varied weather conditions a very robust and low maintenance construction is necessary. This includes the integration of an air-conditioning system to allow reproducible temperature-dependent preparation steps and to avoid water condensation inside the instrument in the cold season. Special isolation measures against mechanical vibrations caused by internal components, e.g., the air-conditioning system, and outside influences, such as wind, mechanical impact, etc., have been adopted to allow stable operation of the microscope unit under outdoor conditions. Other aspects are the size of the instrument and the data-communication protocols necessary

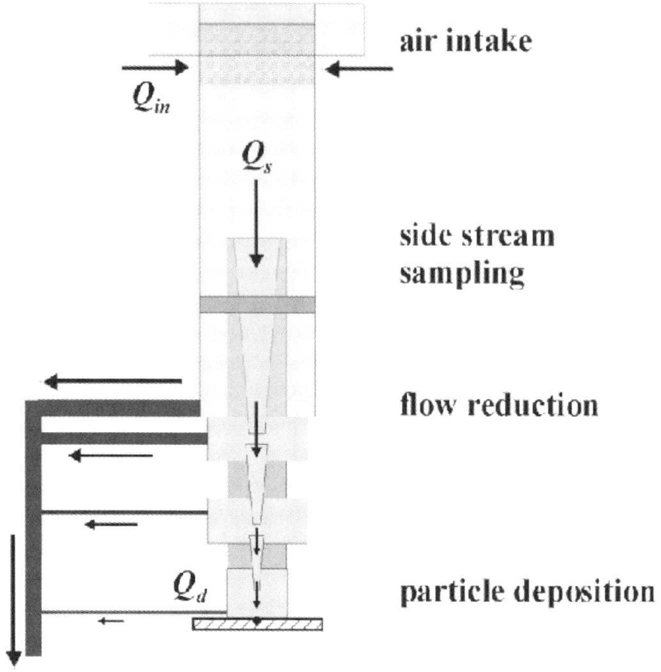

Figure 2.14 Schematic of the design of the sampling and deposition method used in the new instrument. The overall sample flow is maintained by a fan. The internal flows are controlled by orifices. From the overall air stream $Q_{in} = 1167\ l\,min^{-1}$ a representative side stream of $Q_s = 100\ l\,min^{-1}$ is taken isokinetically and subsequently concentrated up resulting in a final flow rate of $Q_d = 2\ l\,min^{-1}$.

for integration into present meteorological networks. Last but not least, the economic requirements for such an instrument are strongly affected by the application fields, cf. Section 2.4.

2.3.1.1 **Sampling**

Sampling and Deposition Scheme of the New Instrument

The sampling unit has to be able to meet both, the criteria for correct health-related particle sampling from the outside air and the needs of the subsequent microscopic detection method. Hence, the various steps of sampling, sample conditioning and particle deposition were combined in such a way that the optimal design parameters for each step could be met independently. The layout is shown schematically in **Figure 2.14**.

A relatively large stream Q_{in} of outside air is sucked in through an inlet system designed for representative sampling including large particles. Only a side stream Q_S, taken isokinetically from the overall sampling flow, is further processed for particle identification and counting. This is the relevant flow rate for determination of the pollen concentration. The reason for the use of this side stream can easily be explained: While Q_{in} is determined by the outside wind speed and the maximum particle size to be analyzed, cf. Section 2.2.1, the smaller flow Q_s can be selected independently according to internal specifications of the instrument, for example the required time resolution, and the area of the particle-loaded spot being analyzed microscopically. The flow Q_s , however, can still be much larger than the maximum flow rate Q_d of the final step of particle deposition onto a surface by filtration, impaction or electrostatic deposition. This requires an additional conditioning unit in which the flow rate is sufficiently reduced without simultaneously decreasing the flux of the relevant particles to be analyzed.

In the instrument, the following values were chosen for the three flow-rates:

$$Q_{in} = 1167\,\mathrm{l\,min^{-1}} = 70\,\mathrm{m^3\,h^{-1}}, Q_s = 100\,\mathrm{l\,min^{-1}}, \text{ and } Q_d = 2\,\mathrm{l\,min^{-1}}.$$

Sampling Atmospheric Particles from Moving Air by Active Intake Systems

As mentioned above, the representativeness of an aerosol sample depends on the sampling characteristics of the corresponding sampler. This section reports on our investigations into the aspiration efficiency of various sampling inlets with respect to different particle size fractions and variable wind speeds. The aspiration efficiency A is defined as the ratio of the number of particles passing the sampling inlet in a certain time period and the number of particles incorporated in an undisturbed atmospheric air volume equivalent to the volume sampled by the instrument in the same time period. Special designs of sampling inlet are required to achieve acceptable aspiration efficiencies, because the relevant particle size range of allergenic pollen reaches up to several tens of micrometers and these particles have to be sampled even at high wind speeds of up to several meters per second.

In this context, particle inertia and gravitational settling are the two mechanisms affecting the aspiration efficiency of sampling inlets. The viscosity of air causes a deceleration of moving particles, which can be characterized in two typical cases. The stop distance s_p is the distance a particle travels when decelerated due to the viscosity of calm air from an initial velocity u to rest. The settling velocity u_{set} is the asymptotic vertical velocity the particle suspended in calm air achieves under the combined influence of gravitational, frictional and buoyant forces. The physical quantity determining both mechanisms is the particle relaxation time τ_p, from which s_p and u_{set} can be calculated as follows:

$$s_p = u \cdot \tau_p \tag{2.1}$$

and

$$u_{set} = g \cdot \tau_p. \tag{2.2}$$

The particle relaxation time increases with the square of the particle diameter. For particles exhibiting a diameter of 50 μm, $\tau_p = 0.008$ s and $u_{set} = 0.076\ \mathrm{m\,s^{-1}}$.

There are different layouts for particle intake systems suitable for collecting large atmospheric particles: omni-directional inlets such as vertical tube inlets and annular slit inlets, and directional inlets such as horizontal tubes facing the wind. The latter are technically more complicated, since they have to rotate, for example using a weather vane, in order to always align with the wind direction, like the Burkard® sampler, shown in **Figure 2.1a**.

Parameters determining the sampling efficiency of inlet systems are: the wind velocity u_W, the ratio R between wind velocity and sampling velocity u_S, $R = u_W/u_S$, the dimension D_s of the sampling inlet and the angle θ between wind direction and inlet flow direction, e.g., $\theta = 90°$ for a tube oriented upwards, $\theta = 0°$ for an inlet facing the wind. The influence of the particle's properties on the aspiration efficiency can be quantified via the Stokes number $\mathrm{Stk} = s_p/D_s$ and the ratio W between settling velocity and sampling velocity, $W = u_{set}/u_S$.

Correct sampling can be assured if the sampling inlet is much larger than the particle's stop distance and if the particle's settling velocity is much lower than the sampling velocity, i.e., if $\mathrm{Stk} \ll 1$ and $W \ll 1$. The second requirement can easily be fulfilled for particles up to 50 μm by choosing a sampling velocity $u_S \geq 1\ \mathrm{m\,s^{-1}}$. However, the first condition will require quite large dimensions of inlets when the aim is to sample large particles at atmospheric wind velocities up to several meters per second. For example the stop distance of particles with a diameter of 50 μm is 4 cm for $u_W = 5\ \mathrm{m\,s^{-1}}$. This would require a sampling inlet diameter of the order of 0.4 m and a minimum flow rate of 450 $\mathrm{m^3\,h^{-1}}$. For directional inlets, the condition $\mathrm{Stk} \ll 1$ can be less strict when the sampling flow rate is close to the wind velocity, i.e., $R \approx 1$.

For practical reasons, a tractable size of the instrument, availability of suction pumps, we restricted our system to a flow rate of the order of $Q_{in} = 70\ \mathrm{m^3\,h^{-1}}$ and a tube dimension of 4 cm inner diameter. Three different configurations were evaluated theoretically for these design parameters:

1. the upward directed tube where the air enters through the top circular plane, **Figure 2.15a**
2. the same tube with a ring orifice of height 3 cm, **Figure 2.15c**
3. the tube directed horizontally against the wind, **Figure 2.15b**.

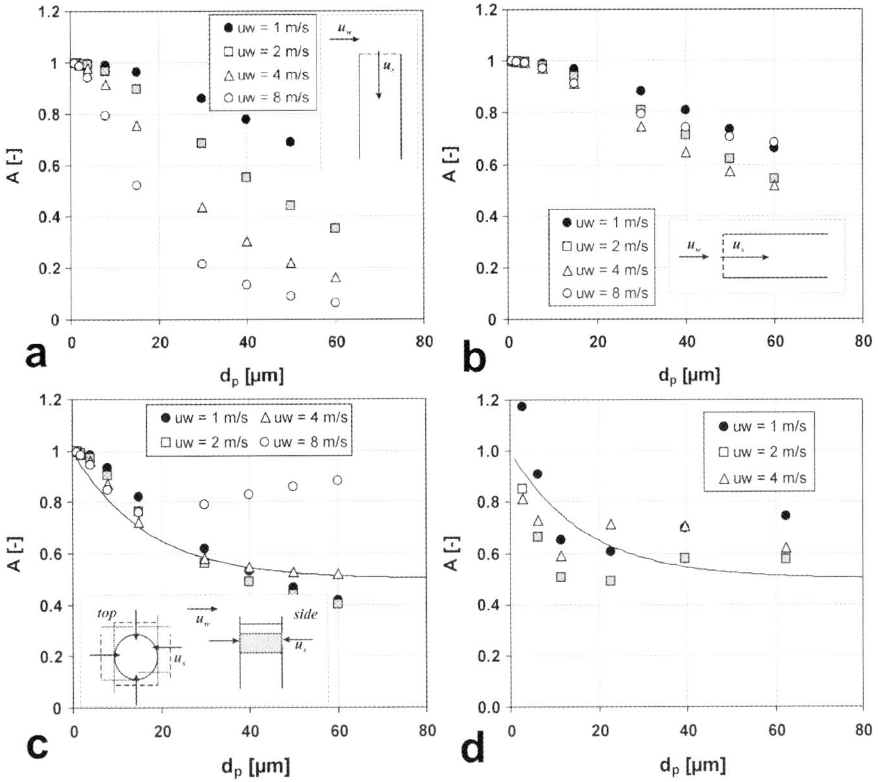

Figure 2.15 Model calculations on the aspiration efficiency of various inlet systems (a–c) and experimental validation of the annular ring inlet system designed for the pollen monitor (d). Full line: definition curve for inhalable particle sampling (d_p = aerodynamic diameter).

At the selected flow rate of $Q_{in} = 70\ m^3\ h^{-1}$ the corresponding inlet velocity is $u_s = 13\ m\ s^{-1}$ for the circular plane inlets (**Figures 2.15a** and **2.15b**) and $u_s = 4\ m\ s^{-1}$ for the ring inlet (**Figure 2.15c**). The aspiration efficiencies of the circular plane inlets were calculated using the correlations of a thin-walled sampling inlet as given by Vincent [16] and by Tsai et al. [50]. In the theoretical calculations, the ring inlet was represented by four identical tube inlets: one facing the wind, one directed opposite and two oriented at $\theta = 90°$ to the wind direction. **Figures 2.15a–c** show the calculated aspiration efficiencies for the three inlet types.

The aspiration efficiency of the upward directed inlet has a strong dependence on both the particle size and the wind velocity u_W. As an example, for particles of 50 μm diameter the Stokes number Stk ranges from 0.2 at $u_W = 1\ m\ s^{-1}$ to 1.6 for $u_W = 8\ m\ s^{-1}$ while correct sampling is subject to the condition Stk $\ll 1$. For the vertical orientation of the inlet tube facing the wind, the aspiration efficiency depends only weakly on the wind velocity and particle size. For the range of wind velocities indicated in the figure the probe samples sub-isokinetically. The efficiency would be one for all particle sizes only for isokinetic sampling conditions, i.e., $u_W = 13\ \mathrm{m\,s^{-1}}$. For wind velocities up to 4 $\mathrm{m\,s^{-1}}$ the ring inlet behaves very similarly to the horizontally directed inlet: low sensitivity of the aspiration efficiency to wind speed but decreasing efficiency with increasing particle size, around 50% for particles with 50-μm diameter. This efficiency curve matches the inhalability curve [19] quite well shown as the full line in **Figure 2.15c**.

From this analysis we concluded that the ring inlet is a good compromise in terms of simplicity, reasonably low sensitivity to wind velocity and compliance with standards of inhalable aerosol sampling, provided the oversampling effect at very large wind speeds can be reduced. This can be achieved by using a wind shroud surrounding the annular air inlet as shown schematically in **Figure 2.14**.

A corresponding inlet system with a wind shroud of 15-cm diameter was built. Its aspiration efficiency was tested in a wind tunnel using polydisperse glass beads as a test aerosol. The aspiration efficiency of the inlet was determined by measuring the number–size distribution in an aerosol stream sampled by the test inlet and in a stream sampled by a reference inlet placed next to the test inlet. The experimental results are given up to a wind speed of 4 $\mathrm{m\,s^{-1}}$, a wind speed of 8 $\mathrm{m\,s^{-1}}$ could not be achieved in our wind tunnel. The experimental data on the aspiration efficiency are shown in **Figure 2.15d**. They confirm the trends predicted by the theoretical analysis. For wind speeds up to at least 4 $\mathrm{m\,s^{-1}}$ there is no significant dependence of the aspiration efficiency on the external wind speed and there is reasonable matching with the convention for inhalable particle sampling.

Sample Conditioning and Deposition

Only one-tenth of the total air intake is used for bioaerosol collection. This side stream of 100 $\mathrm{l\,min^{-1}}$ is taken isokinetically from the main airflow and is subsequently processed by a two stage virtual impactor [51]. In this device, the flow rate is reduced to 10 $\mathrm{l\,min^{-1}}$ in a first stage and, finally to 2 $\mathrm{l\,min^{-1}}$ in the second stage, cf. **Figure 2.14**. The flow rate reduction in each stage is achieved by separating the main flow entering the stage into a minor flow (10% of the main flow) and a major flow (90% of the main flow) in a set-up of two concentric nozzles. The minor flow proceeds straight through the receiving nozzle opposing the inlet nozzle, whereas the major flow is sucked off perpendicularly. All aerosol particles with a Stokes number larger than 0.40 suspended in the main flow of the respective stage will pass straight through the gap between the nozzles and will enter the receiving nozzle. Thus, all large particles will be incorporated in the 10% minor flow. The number flux of the smaller particles will split up according to the flows and is therefore reduced by a factor of ten in the minor flow channel. The Stokes number of the virtual impactor stage is defined in terms of the particle relaxation time, the air velocity in the nozzle, and the nozzle diameter: $\text{Stk} = \tau_p\, u_n / D_n$. **Table 2.2** shows the design criteria of the virtual impactor stages and the diameter d_{crit} of the aerosol particles with a Stokes number of 0.4. For the selected values, all particles larger than ca. 3 μm incorporated in the inlet flow will be transferred to the final deposition stage. This is the particle fraction of interest for monitoring airborne spores and pollen. The number flux of the complementary smaller particles such as the strongly light-absorbing atmospheric soot, and other environmental pollutants with high overall scattering efficiency, is reduced by 98% and, thus, will not contaminate the sample for microscopic analysis.

Table 2.2 Design criteria for the two-stage virtual impactor.

	Inlet dimension	**Flow rate**	**Air velocity at the inlet**	**Cutoff diameter**
	D_n [mm]	Q [l min^{-1}]	u_n [m s^{-1}]	d_{crit} [μm]
Stage 1	6.2	100	55.2	2.7
Stage 2	3.0	10	23.6	2.9
Deposition stage	5.0	2	1.7	11

The final particle removal from the airstream takes place in the deposition stage. In principle, several methods are available for particle collection: filtration, impaction and electrostatic deposition. For reasons related to the requirements that the sample preparation procedure should lead to optimum image quality, filtration was excluded from the above list. Instead, a combination of impaction and electrostatic deposition was employed.

The particles leaving the second stage of the virtual impactor were deposited directly onto the collection medium. This is a transparent foil covered with a thin layer of specially prepared glycerine gelatine. This layer serves as an adhesive for fixing the particles, and is suitable for subsequent preparation, cf. Section 2.3.1.2. In order to avoid particles bouncing from the surface, the particle velocity has to be relatively small so that deposition by impaction can be achieved for the large pollen. The design criteria are given in the third row of **Table 2.2**. The final nozzle is part of an electrostatic precipitator shown in **Figure 2.16**. In this device, particles are first charged in a corona discharge region and subsequently deposited in an electric field established between the outlet of the nozzle and the electrically conducting sample surface.

The design of the deposition stage was based on particle trajectory calculations as well as on extensive experimental pre-tests to characterize particle charging and particle transport in the electric field of the precipitation section.

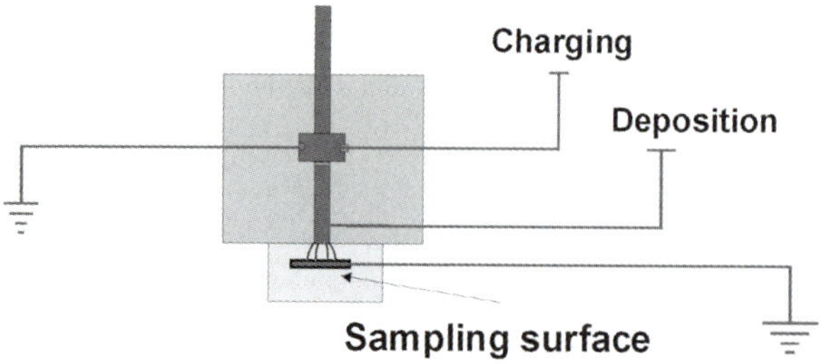

Figure 2.16 Combined electrostatic precipitator and impactor for particle deposition onto the glycerine gelatine substrate for microscopic analysis.

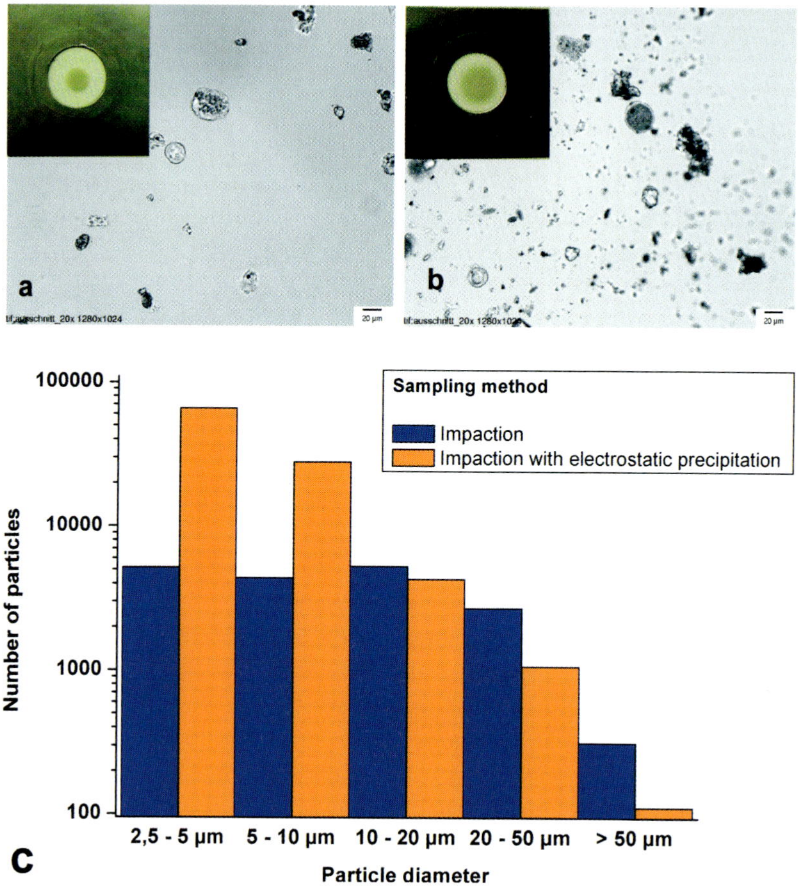

Figure 2.17 Deposition spot for two different operating modes of the deposition stage. The sample was taken from the outside environment. (a): impaction mode, only large particles; (b): electrostatic mode, large and small particles deposited on the surface; (c): histogram of the deposited particles. The inserts are a macroscopic view of a model substrate and the deposition spot.

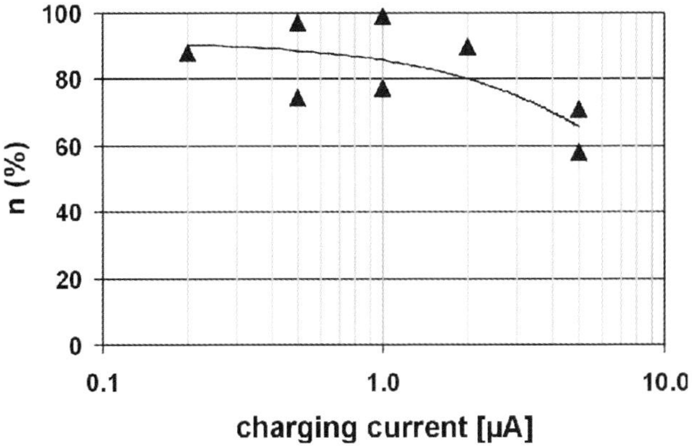

Figure 2.18 The deposition efficiency of pollen as a function of the charging current.

Parameters controlling the overall performance of the precipitator are: charging current, precipitation voltage, nozzle diameter and nozzle exit velocity. This gives flexibility in adjusting the width of the deposition spot as well as the size fraction deposited on the surface, see **Figure 2.17**. Either only pollen or pollen and spores are deposited. The background of the microscopic images is very clear owing to the absence of smaller particles, e.g. soot, which are present in large numbers in the outside air but are not deposited on the substrate. In any case the collection efficiency is sufficiently high over a faily wide range of charging currents used in the corona section of the electrostatic precipitator (**Figure 2.18**). At high charging currents the pollen efficiency starts to decrease because pollens are lost by electrostatic deposition inside the nozzle.

In conclusion, the particle sampling and deposition method developed for the automatic bioaerosol monitor has the following properties:

- representative, sampling of health-related aero-allergens from the outside atmosphere
- *in situ* sample conditioning by removal of the fine particle fraction
- smooth deposition of aero-allergens onto the surface of the sample carrier laid out for microscopy
- adjustable surface coverage and particle size fraction deposited.

2.3.1.2 **Preparation**

For the purpose of a completely automated online analysis of bioaerosols a special focus has to be laid on sample preparation. Components of the deposition unit and preparation of the particle sample must fulfil certain requirements for microscopic analysis.

At the start of our investigations into preparation, we considered two possibilities for particle collection: impaction surfaces and fiber filters. While impaction and electrostatic precipitation are carried out on a flat surface for particle deposition, common fiber filters exhibit a meshwork with a thickness of about 100 to 400 μm.

The meshwork of a fiber filter results in an inhomogeneous optical background that severely interferes with microscopic analysis of the relevant aerosol particles. As a counter measure, a suitable embedding medium can be employed that both embeds the particles within the filter mesh and renders the fibers invisible by matching their refractive index and consequently suppressing light scattering. Nevertheless, mechanical application of an embedding liquid to the filter within the automated process would possibly be error-prone with respect to long-term use under field conditions, apart from the difficulty of assuring a thorough wetting of the filter meshwork without air bubbles.

Impaction and electrostation deposition have therefore been chosen for particle removal in the deposition stage. As a sample carrier, a metal plate with circular holes is used, with the back covered by a transparent window as specified below. The resulting shallow cavities are filled to a set level with a mixture of glycerine, gelatine, water and a surfactant. As already mentioned, this mixture serves as an adhesive for particle sampling. Instead of adding an additional embedding liquid after sampling, the sampled particles are immersed in the same medium that serves for deposition in the first step. The immersion is achieved by heating the sample to about 90 °C. At this temperature the medium is of lower viscosity so that the particles become immersed, see **Figure 2.19**.

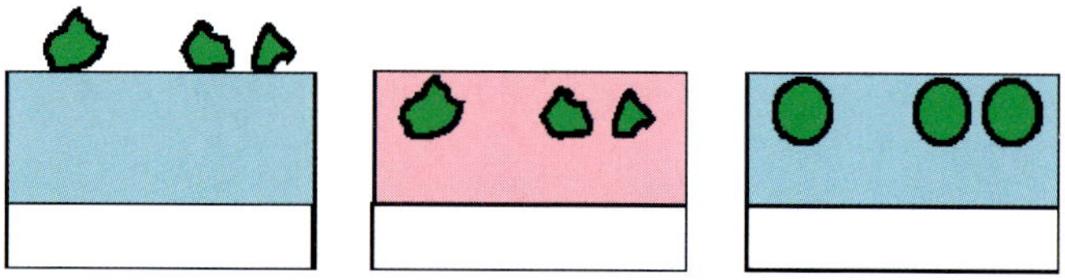

Figure 2.19 Schematic steps of embedding: A mixture of glycerine, gelatine, water and a surfactant serves as an impaction medium for dried out bioaerosols (left) at room temperature. The mixture becomes liquid on heating and the particles are incorporated (middle). After cooling, the bioaerosols have swollen and reached their well known state (right).

Preparation of Bioaerosols for Microscopic Analysis

The fixation of the sampled particles and their embedding in an optically suitable medium are general requirements for a precise, quantitative and reproducible analysis. The embedding has to provide a low background signal with respect to the microscopic techniques employed and it should suppress light scattering at the particle's boundaries. Therefore, the refractive index of the embedding medium is chosen relatively close to the refractive index of the occurring particles, decreasing the intensity of reflected light.

Special attention has to be given to the detection of bioaerosols. Since pollens are mostly transparent, they have to be stained for visual counting, e.g., with safranin or fuchsin. In the case of automatic recognition, the primary fluorescence of pollen is used for detection instead.

(a)

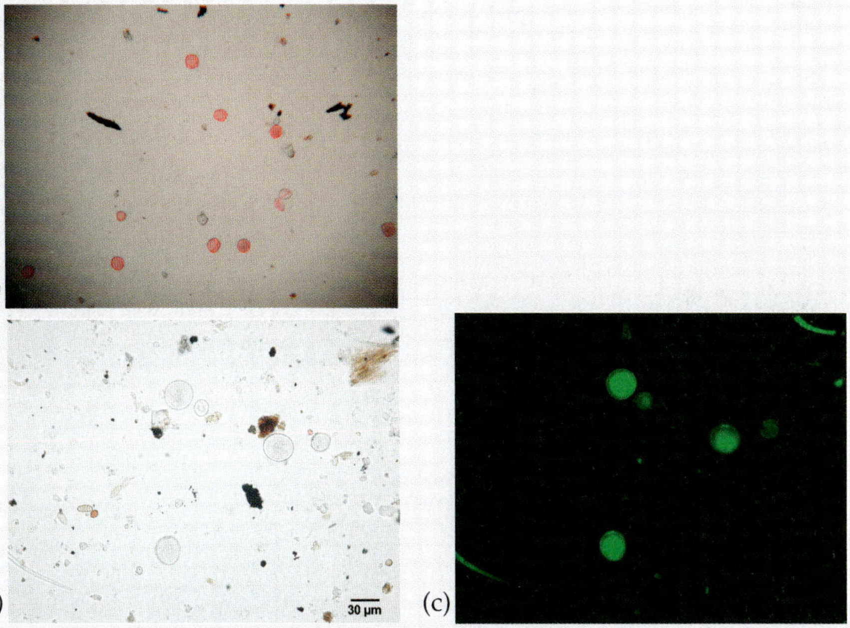

(b) (c)

Figure 2.20 Visual pollen detection requires a special staining (a) for the transparent pollen (b). This step is not necessary in the case of automatic pollen recognition where primary fluorescence of pollen is used for detection (c).

While airborne, pollen dry out and change their shape and size owing to the water loss. Pollen recognition requires the re-establishment of their original shape. This is achieved by the uptake of a polar liquid like water or glycerine, which the pollen grains absorb from the embedding liquid, see **Figure 2.21**. This process typically takes only a few seconds.

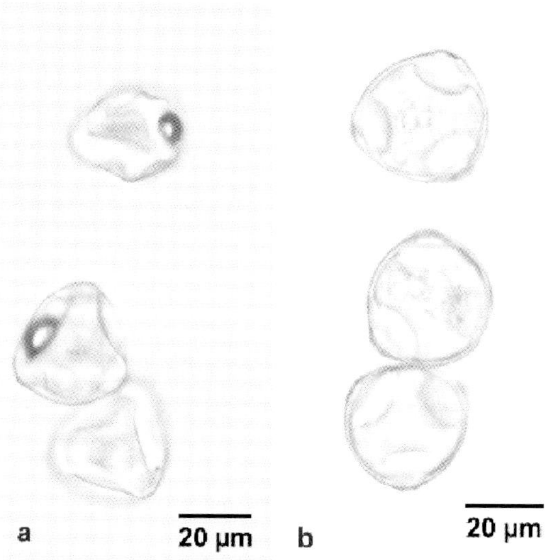

Figure 2.21 Microscopic image ($40\times$ magnification) of dried-out hazel pollen (a) and hazel pollen after rehydration in a polar liquid (b).

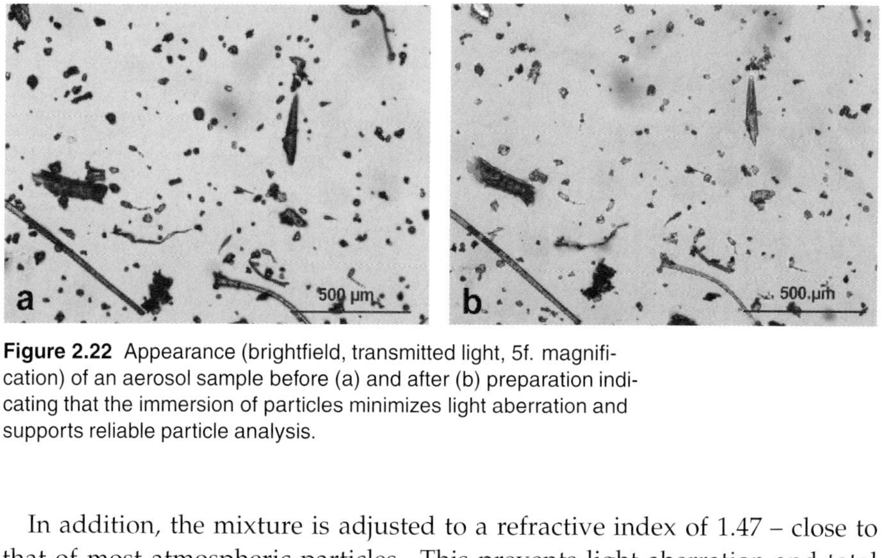

Figure 2.22 Appearance (brightfield, transmitted light, 5f. magnification) of an aerosol sample before (a) and after (b) preparation indicating that the immersion of particles minimizes light aberration and supports reliable particle analysis.

In addition, the mixture is adjusted to a refractive index of 1.47 – close to that of most atmospheric particles. This prevents light aberration and total reflectance at the particle's boundaries, see **Figure 2.22**.

The water content of the embedding medium also provides a conductive surface, which is necessary for electrostatic precipitation.

The strong primary fluorescence of pollen is an important component for the subsequent segmentation as a part of the image analysis, see Section 2.3.1.4. At the same time, background fluorescence may arise from the carrier or the immersion liquid, interfering with the recognition of pollen. For this reason, and to obtain mechanical stability during impaction and preparation, thin cover slides of glass are used as carrier instead of a plastic foil. For the embedding medium, a special gelatine exhibiting only low primary fluorescence has been selected [52].

Samples loaded with grass pollen and mineral dust by the above described sampling technique showed a circular, widely homogeneous coverage of particles sticking to the surface layer. After heating, the particles are immersed in the solution owing to the considerably lowered viscosity and are embedded, in this manner, for microscopic analysis. Microscopic inspection by confocal laser-scanning microscopy showed that particles were not submerged in the solution deeper than 100 μm. Thus the layer thickness was limited to a value of 100 μm as an appropriate fill level in order to reduce the remaining background fluorescence resulting from the embedding medium. A thicker layer was found to affect image quality, a thinner layer tended to dry out and to limit particle incorporation.

For quality assurance of the microscopic analysis, the optical behavior of samples was studied as a function of time. The experiments indicated that even long-term storage for several months did not lead to changes in pollen morphology due to water losses if environmental conditions were specifically controlled. On the other hand, there is evidence that optical properties may change. Fluorescence can individually weaken or change in activity within the pollen. Such effects have to be considered when analyzing aged samples.

2.3.1.3 Microscopic Imaging and System Integration

Optical microscope-based imaging techniques are the basis of the pattern recognition process and the identification of deposited biological particles. Both transmitted light recordings with brightfield illumination and reflected fluorescence light images are used for contrast purposes. Transmitted light recordings provide an overview of the huge variety of deposited particles: mineral dust particles, abrasion from tires, pollen, spores, etc. They also allow a fast automatic focusing procedure, which is necessary for fully automated image acquisition of a relatively large sample area (about 40 mm^2) with sufficient resolution ($<$0.6 μm) in a short time (0.5–1 h). A part of the microscopy unit is shown in **Figure 2.23**.

Because bioaerosols of interest typically exist in small concentrations against a dominant background, a small number of bioaerosol particles must be detected in the presence of a very large concentration of nonbioaerosol particles. The strong primary fluorescence of pollen in the UV and blue spectral range is used to localize organic particles in digital images and permits fast and reliable isolation of these particles from the background and from many other inorganic particles (**Figure 2.24**). Therefore, with a combination of fluorescent and transmitted light images an optical screening for bioaerosols can be achieved that speeds up the time-consuming pattern recognition process.

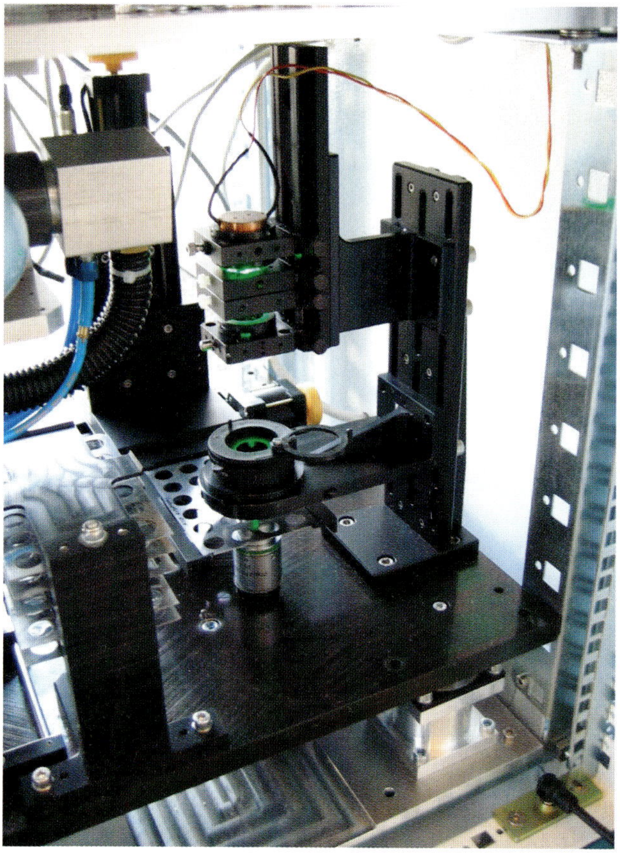

Figure 2.23 Microscopy unit within the online pollen monitor

Furthermore, the bulk of particle information usable for the method of pattern recognition employed can be found in the fluorescence image in the case of pollen. For fungal spores, it is also useful to employ darkfield illumination in parallel.

After the automatic preparation process, particles are located in quite different levels of the embedding medium. To obtain a full 3D volumetric dataset of all particle image stacks for each contrasting technique, steps of 2.5 μm are taken at each position of the sample area.

High-power light-emitting diodes (LEDs) are well suited for microscopic illumination purposes in the pollen monitor. They are used for both transmitted light as well as fluorescence imaging. There are several advantages to using light-emitting diodes. First of all, the lifetime of an LED is about 50 000–100 000 h, which is especially important for an automatic unsupervised system like the pollen monitor. In contrast to this, conventional halogen lamps for transmitted light illumination and high-pressure arc lamps for fluorescence excitation, respectively, have lifetimes in the range of several hundred hours.

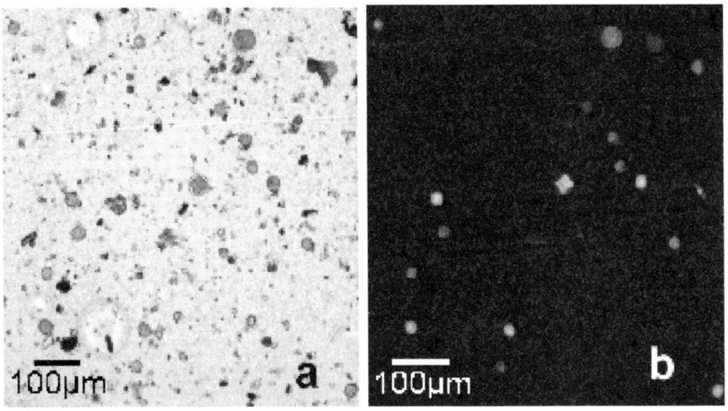

Figure 2.24 Fluorescence microscopy is a powerful tool to separate bioaerosols from the background. An aerosol sample from a Burkard® trap is shown ($10\times$ magnification) in brightfield illumination (a) and fluorescence excitation (b).

Second, the power consumption in comparison to conventional bulbs is very low – therefore heat generation is negligible. Light-emitting diodes have very good spectral and power stability thus providing a good reproducibility of image data. They thus fulfil all the requirements for an outdoor system installation, where long-term stability and low power consumption are important factors in a low-maintenance operation.

In the case of the pollen monitor, a green LED (wavelength $\lambda_{\text{trans}} \sim 533$ nm; power 35 mW) is used for transmitted light recordings. At this wavelength good optical resolution can be obtained and filter sets required for fluorescence illumination do not influence the image. Fluorescence illumination is obtained with a violet LED (wavelength $\lambda_{\text{exc}} \sim 380$ nm; 85 mW), whereas the fluorescence emission is recorded above $\lambda_{\text{em}} \sim 455$ nm.

For an automated microscope-based single-particle analysis in ambient air samples, a system has been developed combining the techniques described above for automatic particle sampling, preparation, sample handling for subsequent microscopic analysis and particle recognition. The principal structure of the system is shown in **Figure 2.25**. Sample carriers for a one-week automatic operation are stored in a stacker. A fully computer controlled three-axis motorized stage transfers carriers between the stacker and the sampling unit and the preparation unit and the specifically adapted microscopy unit. Although using only one motorized stage, the system allows simultaneous sampling of particles onto one carrier and microscopic analysis on another, so that a nearly continuous collection and microscopic imaging of aerosol particles is possible, with only short interruptions during transfer of carriers. Transfer time of the carriers is around half a minute between the stacker and the other units of the system.

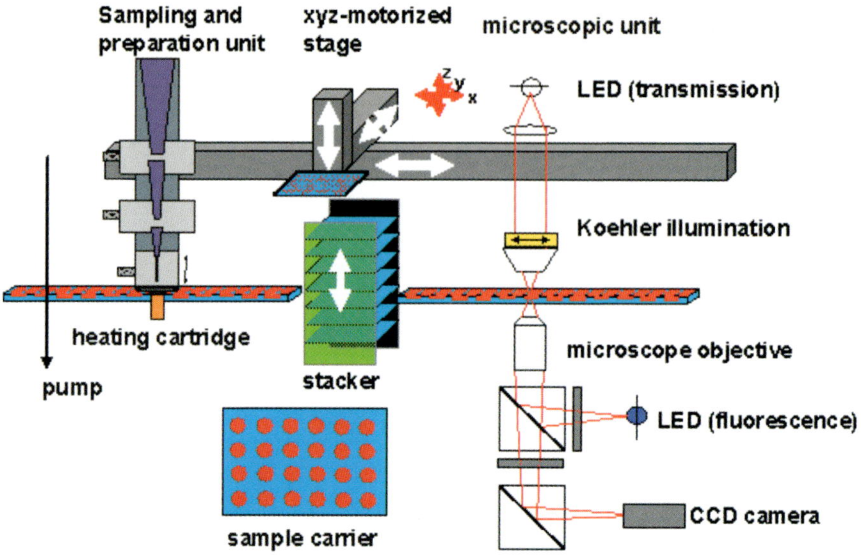

Figure 2.25 Schematic drawing of the sampling, preparation and detection units.

2.3.1.4 **Pattern Recognition**

Segmentation

The first step in pattern recognition is to locate and crop the individual particles ("segmentation"). Biological particles, such as pollen, show a strong primary fluorescence in the UV and blue spectral range allowing reliable separation from the background and from other simultaneously occurring particles. Due to the automatic sampling and preparation process the particles are located in different levels. The segmentation is realized by dividing the shading-corrected fluorescence volume ($1380 \times 1040 \times 40$ voxels) into small blocks of $10 \times 10 \times 40$ voxels. For each of these blocks the optimal focus plane is found by calculating the Laplacian energy for each 10×10 pixel image as the criterion for sharpness. The layer numbers compose a layer image which is smoothed and interpolated to get a layer image of the original size. A patch image is created from the stack according to the layer image (**Figure 2.26**)

This image is used for segmentation. The resulting binary mask for each particle is applied to each plane of the fluorescence and translucent stack to crop each 3D object from the volume.

Gray-scale-based Invariant Features

The general method of object recognition is first to extract appropriate features from the unknown object and then to start a classification based on the

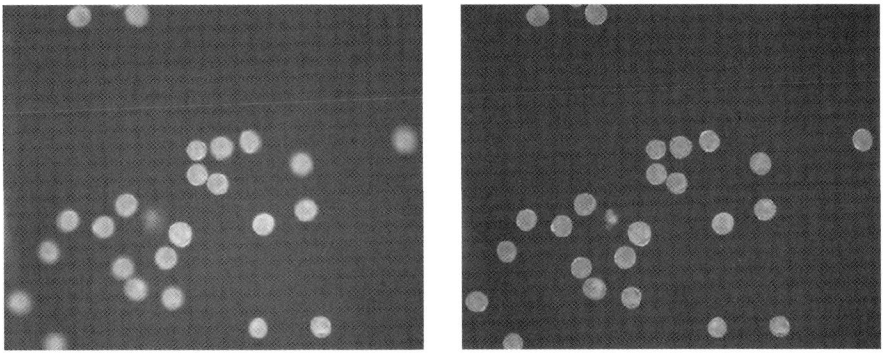

Figure 2.26 Creation of a patch image (right) from an image stack with different focal planes. An example of an image from the middle layer is shown on the left.

resulting set of parameters, with the aim of finding the best match between the parameters of the unknown object and the parameters of the labelled objects in a database.

There are two principal ways to develop an automatic pattern recognition system. The first is to extract highly abstract features from the images, such as the "number of pores", which means to put all the "intelligence" of the software into the feature extraction part. This keeps the classifier very simple, but is usually very demanding in the development of a large collection of highly object-specific functions, e.g., the pore finder, which must be adapted to every pollen species, or even to each possible orientation of a pore.

The alternative is to keep the feature extraction as general as possible and to use a more sophisticated classifier, which can be trained with given samples from a reference database. The advantage of such an approach is the easy adaption to different environmental conditions or even to other objects just by exchanging the reference database and rerunning the training program.

A quite simple but very powerful method for general feature extraction is the calculation of so-called "gray-scale invariants" [53, 54]. The gray-scale invariants do not need any segmentation within the object, but operate directly on the gray-values of the image. Furthermore, they are not limited to 2D image data and can be straightforwardly extended to 3D volumetric data [55].

The aim of such an invariant feature is the following: The scanned 3D volume dataset represents one individual pollen grain, independent of its position and orientation in space. This means that the 3D volume dataset of one individual pollen grain, in all possible positions and orientations (Euclidean motion), represents exactly one equivalence class. An invariant transformation is able to map all representations in the vector space of the equivalence class into one point of the feature space and there represents the intrinsic information of the structure, independent of its position and orientation (see textbox **Pattern Recognition**).

Pattern Recognition

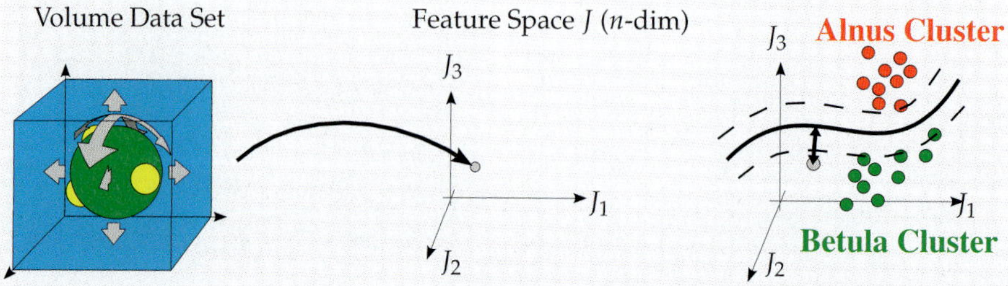

Figure 2.27 Pattern Recognition.

Pattern recognition is usually done in two steps:

1. Previous knowledge is used to map the objects into a low-dimensional feature space. This transformation should map objects of the same class to the same point. For pollen recognition we only use the previous knowledge that our object will appear in an unknown orientation at an unknown position, and therefore the transformation must be invariant to rotations and translations of the object. For this transformation we use gray-scale invariants.

2. A classifier is used to divide the feature space into different clusters to describe the remaining variations within one class. The required information can be extracted from labelled training examples. For pollen recognition we use support vector machines.

The basic idea for the calculation of these invariants is to take a small nonlinear kernel function $f(\mathbf{X})$ for combining some neighboring pixels or voxels and to integrate the results of this function over all possible representations in the equivalence class.

$$T[f](\mathbf{X}) \quad := \quad \int_G f(g\mathbf{X})\,\mathrm{d}g \tag{2.3}$$

G : transformation group
g : one element of the transformation group
f : nonlinear kernel function
$\mathbf{X}$: n-dim, multichannel dataset
$g\mathbf{X}$: the transformed n-dim dataset

For each nonlinear kernel function f this integral returns a scalar value that describes a certain feature of the n-dimensional dataset invariant under the given transformations, as long as the integral exists and is finite.

Reduction to Kernel Functions with Sparse Support

The general formulation of the gray-scale invariants in Eq. (2.3) cannot be computed in an acceptable time. But if we select kernel functions with sparse support we can find fast algorithms for the evaluation. If the kernel function f only depends on a few points of the image or volume, i.e., if we can rewrite $f(\mathbf{X})$ as $f\left(\mathbf{X}(x_1), \mathbf{X}(x_2), \mathbf{X}(x_3), \ldots\right)$, where $\mathbf{X}(x_i)$ is the gray value at position x_i we only need to transform the kernel points $x_1, x_2, x_3, \ldots$ accordingly, instead of the whole dataset $\mathbf{X}$. (We use the term "gray value" even for color or other multichannel data. In that case one "gray value" has multiple components.) This transformation of the kernel points is denoted as $s_g(x_i)$ such that

$$(g\mathbf{X})(x_i) := \mathbf{X}(s_g(x_i)) \quad \forall g, x_i \tag{2.4}$$

With this we can rewrite (2.3) as

$$T[f](\mathbf{X}) := \int_G f\left(\mathbf{X}(s_g(x_1)),\ \mathbf{X}(s_g(x_2)),\ \mathbf{X}(s_g(x_3)), \ldots\right) \mathrm{d}g\,. \tag{2.5}$$

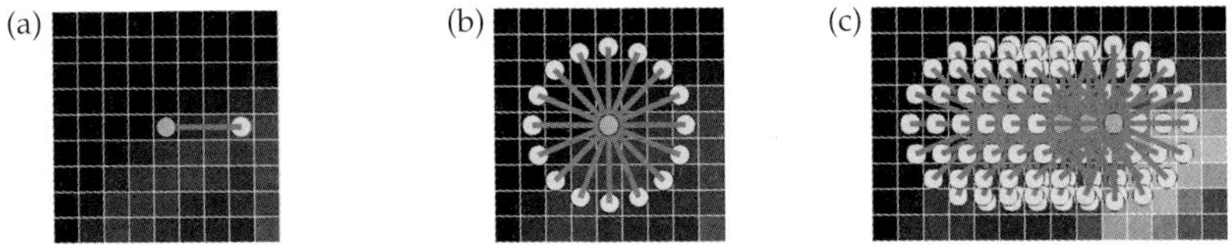

$f(\mathbf{X}) := \mathbf{X}(0,0) \cdot \mathbf{X}(3,0)$

$$T_\varphi = \mathbf{X}(0,0)\cdot\mathbf{X}(3,0) + \mathbf{X}(0,0)\cdot\mathbf{X}(2.7,1.1) + \mathbf{X}(0,0)\cdot\mathbf{X}(2.1,2.1) + \ldots$$

$$T = \mathbf{X}(0,0)\cdot\mathbf{X}(3,0) + \mathbf{X}(0,0)\cdot\mathbf{X}(2.7,1.1) + \mathbf{X}(0,0)\cdot\mathbf{X}(2.1,2.1) + \ldots + \mathbf{X}(1,0)\cdot\mathbf{X}(4,0) + \mathbf{X}(1,0)\cdot\mathbf{X}(3.7,1.1) + \mathbf{X}(1,0)\cdot\mathbf{X}(3.1,2.1) + \ldots$$

$\mathbf{X}$: gray-value image
$\mathbf{X}(x, y)$: interpolated gray value at position x, y

Figure 2.28 Calculation of a 2D gray-scale invariant. (a): Selection of a nonlinear kernel function for combining some neighboring pixels: In this example the kernel function $f(\mathbf{X})$ is defined as the multiplication of two gray values of distance 3. (b): This kernel function is evaluated for all angles and the results are summed up, to become invariant to rotations of the object. (c): This set of rotated kernel functions is evaluated at all possible positions of the image and the results are summed up, to become invariant to translations of the object. As a result, identical values for T are obtained independent of the angle and position of the object in the image.

This considerably speeds up the computation and results for a given kernel in linear complexity $O(N)$ of the algorithm, where N is the number of pixels/voxels in the dataset. An example for the direct evaluation of a two-point-kernel on a 2D image and the transformation group of Euclidian motion is given in **Figure 2.28**.

Multiscale Approach

In the general formulation of the gray-scale invariants, Eq. (2.3), appropriate kernel functions can be used in order to sense any features of the structure at any scale. Computable kernel functions, Eq. (2.5), depend only on a few gray values at certain points. To use them for sensing large-scale information, a multiscale approach is applied [46]. In the continuous domain this is equivalent to applying a certain low-pass filter (e.g., convolution with a Gaussian) to the dataset before evaluating the kernel functions (see **Figure 2.29**).

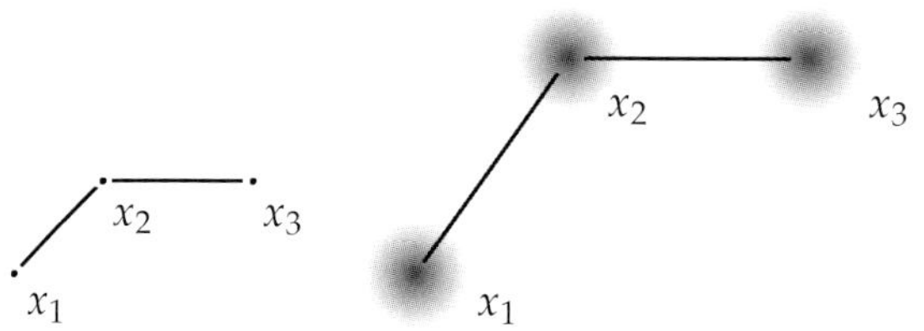

Figure 2.29 Computable kernels rely on a small number of sampling points. To sense information at multiple scales, the sampling points are "enlarged" with Gaussians of multiple size.

Several of those gray-scale-based invariants with different kernel functions and at different scales are computed for each object and their combination builds a kind of "mathematical fingerprint" for this object.

The main advantages of these features, compared to several traditional feature extractors, is their direct operation on the gray values of the voxels without any complicated and, usually, highly application-specific pre-processing, such as edge detection, searching for pores of pollen grains, etc. The gray-scale invariants are therefore applicable to any type of particles and do not require any change to the programs.

Training and Classification

The third step in this recognition pipe is the classification of the extracted "fingerprints". As classifier, so-called Support Vector Machines [56] are employed. A support vector machine is able to "learn" how to distinguish between two given classes, e.g., between the fingerprints of hazel pollen and the fingerprints of miscellaneous particles, just by the machine's being introduced to examples labelled by experienced personnel. The output of this training is a model allowing the classification of fingerprints of new particles (see **Figures 2.30** and **2.31**).

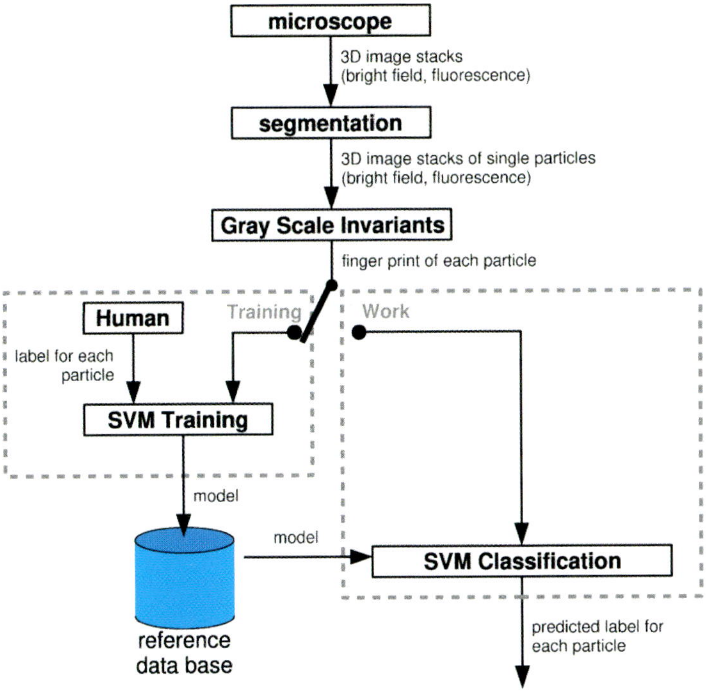

Figure 2.30 Schematic representation of the training and classification process.

2.3.1.5 Integration of the Pollen Monitor in an Online Environmental Monitoring Network

Many countries operate networks for full-coverage monitoring of environmental data. Monitoring stations that are placed countrywide are connected to a monitoring network. Depending on the specific task, the stations are equipped with measuring devices, analyzers and sensors for monitoring meteorological data, air quality, radioactivity or water quality.

Each monitoring station is equipped with a station PC that acquires online data from all monitoring devices, conducts plausibility checks, compresses and saves the data and transfers the information periodically to the monitoring headquarters, see **Figure 2.32**. The transfer cycle can be defined according to the urgency of the information. In addition to the measurements, operating and error status information of the monitoring devices and the station are transferred.

The infrastructure of existing monitoring networks is suitable for pollen monitoring. The mechanical, electrical and communication interfaces of the pollen monitor satisfy all the preconditions for easy integration into an online monitoring network. The pollen monitor can either be used in stand-alone operation mode or it can be integrated into an existing monitoring station.

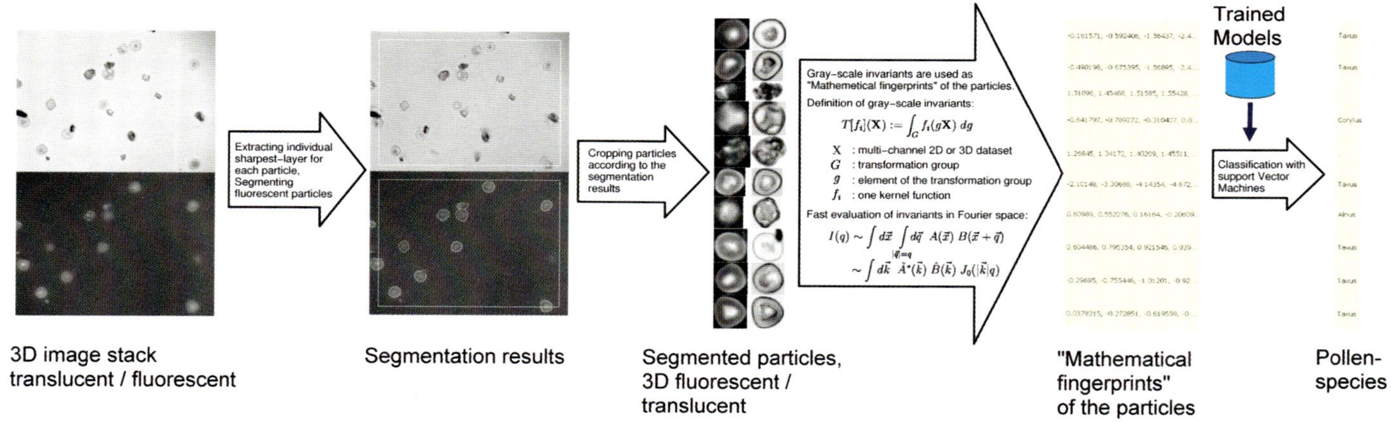

Figure 2.31 Automatic recognition of pollen grains in an ambient air sample.

The Pollen Monitor as Measuring Device in an Ambient Air Monitoring Station

In general ambient air monitoring stations are shelters with standardized 19″ racks for mechanical integration of rack-mounted monitoring devices. They are air conditioned and equipped with infrastructure such as a 230-V power supply, illumination, lightning protection, purified compressed air and a telephone line or mobile telephone. The monitoring stations are equipped with analyzers and sensors connected to the station computer via a multiserial interface for data acquisition, visualization, communication and remote control.

For integration of the pollen monitor 20 height units = 1000 mm of a 19″ rack are required. The upright sampling system dictates that the pollen monitor is mounted on the top of the rack. A roof mounting flange for the sampling tube with an outer diameter of 50 mm needs to be integrated into the shelter roof. For power supply, six power sockets with 16 A fuses are required.

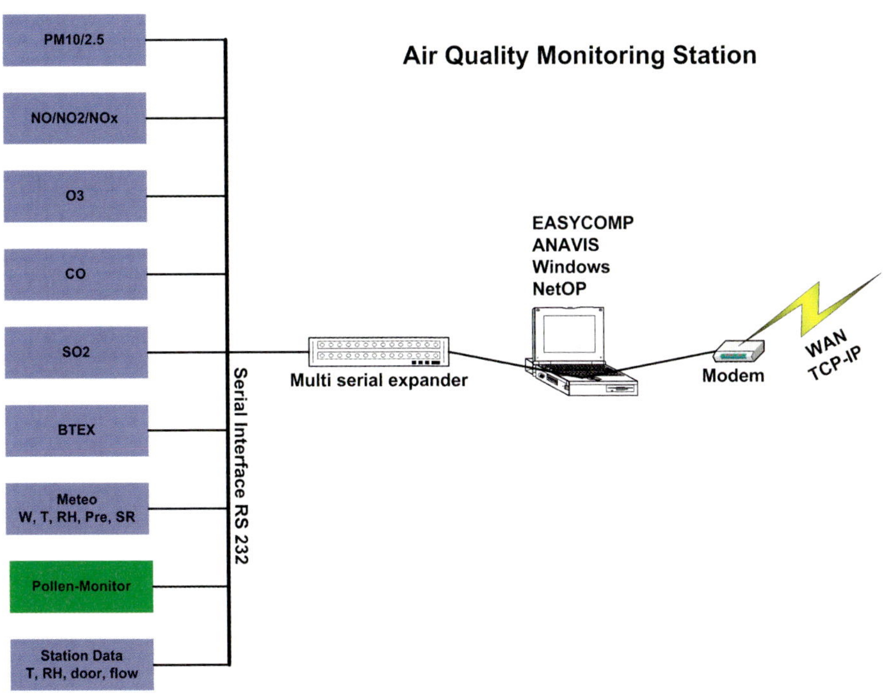

Figure 2.32 In an air quality monitoring station different analyzers are controlled by the station PC collecting data, e.g., about particulate matter (PM_{10}, $PM_{2.5}$), gases (nitrogen oxides, sulfur dioxide, benzene, toluene...) and bioaerosols as well as meteorological data.

Stand-alone Operation of the Pollen Monitor

If pollen concentrations are to be monitored at sites not yet equipped with environmental monitoring stations, the pollen monitor can be placed inside an air-conditioned outdoor cabinet, see **Figure 2.33**. The cabinet consists of a weather-resistant insulated double-wall aluminium shell. Inside the shelter all elements of the monitor are mounted in a 19″ rack. An air conditioner integrated into the side wall of the shelter ensures an inside temperature of 20 °C ± 4 °C and low humidity.

For setting up the outdoor cabinet a planar surface of approximately $1 \times 1\,m^2$ is sufficient. The respective data acquisition and communication software can be installed on the PC of the pollen monitor, which also acts as the station PC and is responsible for data acquisition and transmission.

Monitoring Network

The user programs are laid out

- for the acquisition of measurement data, data visualization, control, data storage and for communication at the monitoring station and
- at the central stations for communicating with the different monitoring stations, control of monitoring stations, remote maintenance, the visualization and evaluation of measurement data and status notifications as well as report and alarm generation.

Communication between monitoring stations and the central station is achieved by modems using GSM/GPRS, analog or digital (ISDN) telephone lines. Data transmission is performed through TCP/IP connections established by the operating system according to the application.

All PCs of the monitoring network are equipped with a remote service program as such NetOP or pcAnywhere. This program enables selected users to directly log into one of the PCs for remote maintenance, alignment of parameters, or installation of software updates, see **Figure 2.34**.

General Features of Monitoring Networks

Notification Periods The system has got two averaging periods. The short averaging period is fixed to one minute and serves, for example, for data compression for local storage. The long average period may be freely set.

Archive Items The period of storage of these average values at the station depends on the number of data channels and the size of the hard disk. Though in the case of the pollen monitor the temporary image data of all particles scanned in one hourly measurement may comprise 40 to 70 GB, the corresponding datasets of gray-scale invariants used for classification sum up to

Figure 2.33 General view of the pollen monitor

less than 1 MB. Thus, feature extraction by the calculation of invariants results in a large data compression.

Presentation Current and stored measured values may be displayed in various layouts and tables. The system offers a high universality with regard to measurement channels, periods, colors, additional information such as percentile, etc. Since these data govern the maintenance work of the service technician at the surveying station, great attention is paid to the provision of a simple and easily understandable display of additional information, such as operational and error status as well as the validity of measured values, in addition to the actual measured values.

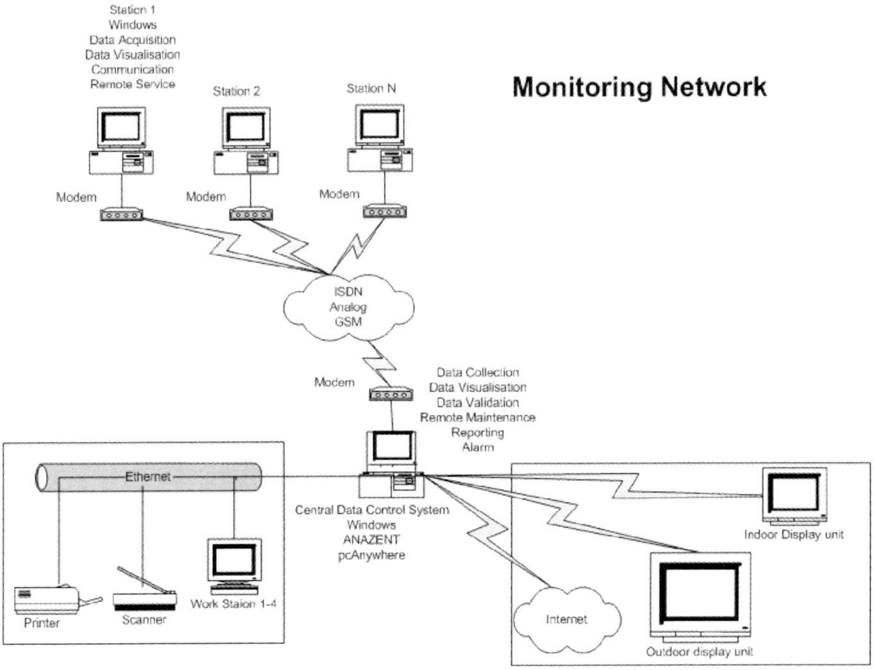

Figure 2.34 Scheme of the communication in an air quality monitoring network.

Central Station Commands to the Surveying Station

- release of a calibration from the central station
- setting up a maintenance program in response to an resp. invalidity status report from the central station
- manual or automatic time synchronization between the central station and a surveying station
- checking and altering software parameters of the surveying station
- file transfer from and to the central station
- release of the station software's restart
- general remote control of the surveying station from the central station.

Communication All station computers are TCP/IP RAS servers. In the case of central station data creation, the central station will establish a TCP/IP connection to the RAS server. Once this connection has been established, it works like a local transparent LAN connection. The central and surveying stations

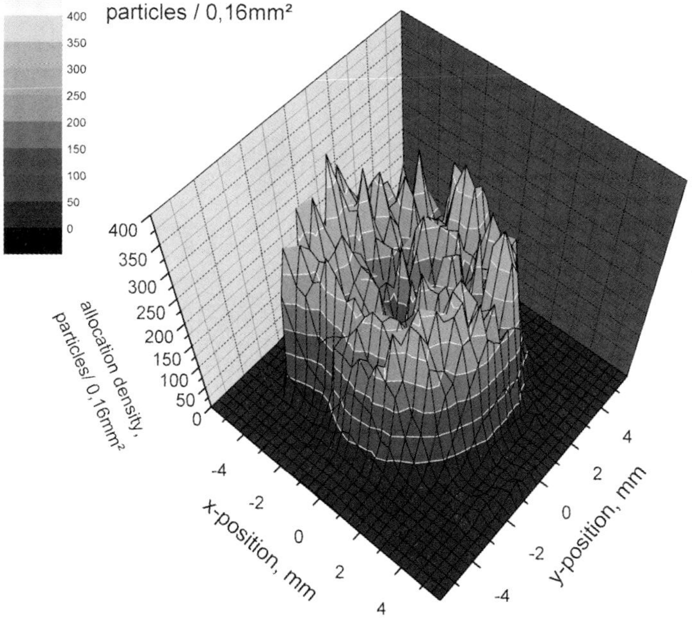

Figure 2.35 Particle allocation on the sampling spot of an outdoor air sample after electrostatic precipitation. The following parameters have been chosen: charging current for aerosol particles: 1 μA, field voltage above the impaction surface: 5 kV

will process their communication via a specific TCP/IP Port, and the central station will interrupt the RAS connection subsequently.

2.3.2 First Results

2.3.2.1 Automated Sampling

The sampling system was found to be suitable to create a circular and homogeneous allocation of particles, see **Figure 2.35**. This feature is necessary to provide high statistical precision for the calculation of pollen concentration after the analysis of only parts of the sample. Hourly measurements are made, yet one-fifth of the whole sampling area corresponds to the same air volume, 1 m^3, that is now usually evaluated by visual pollen counting distributed over the whole day. Thus, this kind of automation has the potential to enhance the power of pollen counting by a factor of almost 120, with temporal resolution better by a factor of 24 and a higher statistical precision due to the possible analysis of 5 m^3 air volume, apart from the benefits of reliable recognition.

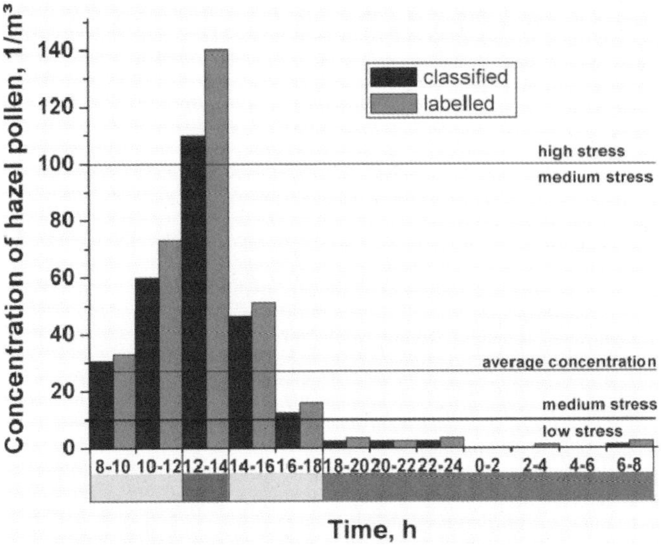

Figure 2.36 Diurnal cycle of pollen concentration, species hazel (*Corylus*), recorded on Feb 5, 2004 in Zurich.

2.3.2.2 **Automated Sample Scanning**

Automated sample scanning and particle recognition itself constitutes a great improvement in pollen forecasting. Even conventional pollen sampling with common Burkard® traps can benefit from this novel technique. **Figure 2.36** shows the diurnal cycle of an allergenic pollen species, taken from a conventional Burkard® sample from MetroSwiss in Zurich. In routine measurements, only daily means of pollen concentration are determined by eye. Automated sample scanning accelerates and simplifies the detailed analysis of different sampling intervals. After checking the recognition results, marked as "classified", by subsequent visual analysis, marked as "labelled", a high precision can be found. The apparently low recall (80.2%) can be ascribed to the impaction method in a Burkard® sampler. The sampling of particles down to 2.5 μm often leads to severe agglomeration of pollen with fine particles. This effect can be reduced in samples from the online monitor owing to the increased sampling cutoff with the operation of impaction without electrostatic precipitation.

2.3.2.3 **Continuous Sampling and Online Analysis**

The whole process of automatic sampling and measurement in the monitor can be described as follows. Two different sampling plates are processed alternately during continuous operation. In the case of an hourly measurement cycle, an aerosol sample from around 5 m^3 sampled air volume is deposited on a certain position of a sample carrier during a period of about 50 minutes. Meanwhile, another position of a different sample carrier from the preceding sampling interval is prepared for microscopic analysis by embedding within five minutes and a subsequent microscope scan in the remaining 45 minutes. The recorded image stacks are processed by segmentation and pattern recognition. The variable intensities of the fluorescence signals require the recording of images with different exposure times. The weak primary fluorescence in some cases requires a large overall time for the recording of all image stacks, restricting the scanning capacity of the monitor within a measurement cycle. Nevertheless, all particles from a sampling volume of around 1 m^3 can be scanned, detected and determined within the hourly cycle. A significant improvement is expected from the insertion of an appropriate illumination unit, enhancing the intensity of fluorescence light and allowing a nearly complete analysis of all deposited particles in the further course of this project.

2.3.2.4 **Pattern Recognition – Pollen**

As a first step, the feasibility and capacity of automated bioaerosol recognition had to be shown. Therefore, investigations under laboratory conditions were the point of origin.

Pollen sources from many different kinds of trees, bushes and grasses were collected, dried and stored as a basic raw material. Mechanically treated, pollens can easily be sampled manually on microscope slides and serve as reference objects. In a test of the pattern recognition scheme and as a predictor of its performance, 1751 pollen grains from manually prepared samples were recorded. A so-called leave-one-out test was carried out by taking one particle out of the dataset, training the classifier with the remaining 1750 particles and classifying the left-out particle with the resulting model. This procedure was repeated for each particle. The resulting recognition rate allows us to estimate the expected recall with unknown data of the same image quality. The results are shown in **Figure 2.37**. The recognition rates for manually prepared samples are very promising and are notably higher than 90%. The low recognition rate for sycamore tree (*Plantago*) of only 69% is deemed to result from the low number of only 53 samples present in the labelled stream. Low sample size may short-change the system by not presenting all possible variations of objects in size, shape and morphology to have maximum learning.

2.3.2.5 **Pattern Recognition – Fungal Spores**

A similar test was carried out for fungal spores. Microscopy images from deposition samples (**Figure 2.38**) provided an image database of 1609 particles. Among them, four morphologically different spore genera, two classes of small morphologically similar spore genera and miscellaneous aerosol par-

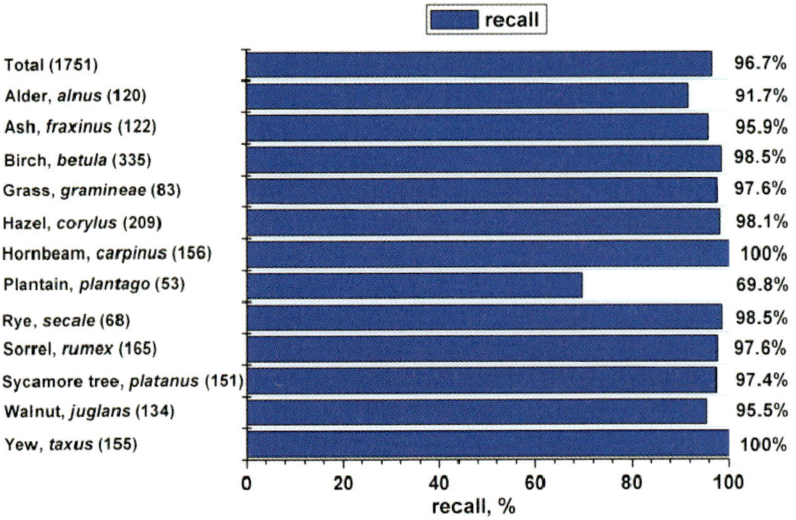

Figure 2.37 Recall from a leave-one-out-test with pollen from microscopic images of high quality.

ticles have been labelled. To permit better recognition of structural features, darkfield illumination was employed in a addition to brightfield and fluorescence. A leave-one-out test was carried out, resulting in recall and precision as shown in **Figure 2.39**. In contrast to recall, which indicates the proportion of particles of one species that is recognized correctly, precision means the proportion of particles that are correctly attributed to a single label. As mentioned above, lower recognition rates for single classes can be ascribed to short datasets (*Epicoccum* sp.) or to poor image quality of small fungal spores (elliptical or spherical type).

Nevertheless, the high recognition rates of 96.7% (pollen) and 93.3% (fungal spores) give an impression of the feasibility, in principle, of the automated recognition of bioaerosols and its large capacity.

2.3.2.6 **Real-world Samples**

The next step was the evaluation of the recall and precision of the online monitor under laboratory conditions. During the winter period, outdoor aerosol samples were taken with typical seasonal pollens (yew, alder, hazel) added. The samples were prepared and microscopically analyzed in the monitor. The automated recognition took place with an already existing and independent database, constituting the real-world scenario as distinct from predictive leave-one-out tests, simply to indicate the possible capacity of an existing reference database. 2286 particles, among them pollen species for the period January to March (yew, alder, hazel), had been used for training the classi-

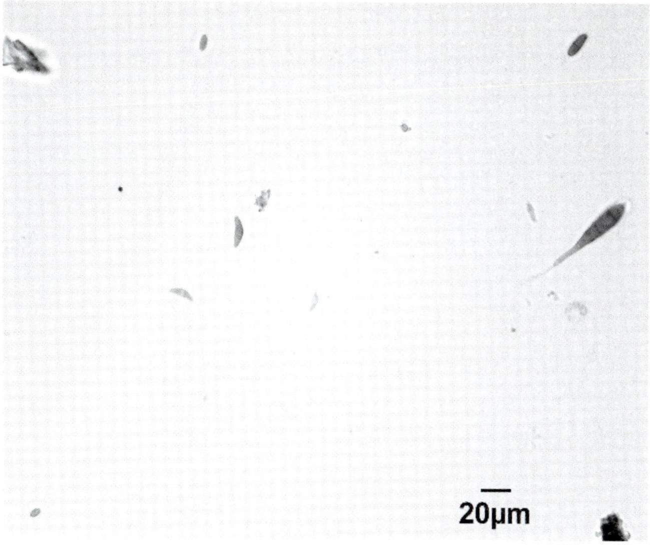

Figure 2.38 Microscopic image of an aerosol sample, different spores (*Alternaria*, *Cladosporium*, *Didymella*) and other aerosols collected by deposition sampling (VDI 2119, part 4).

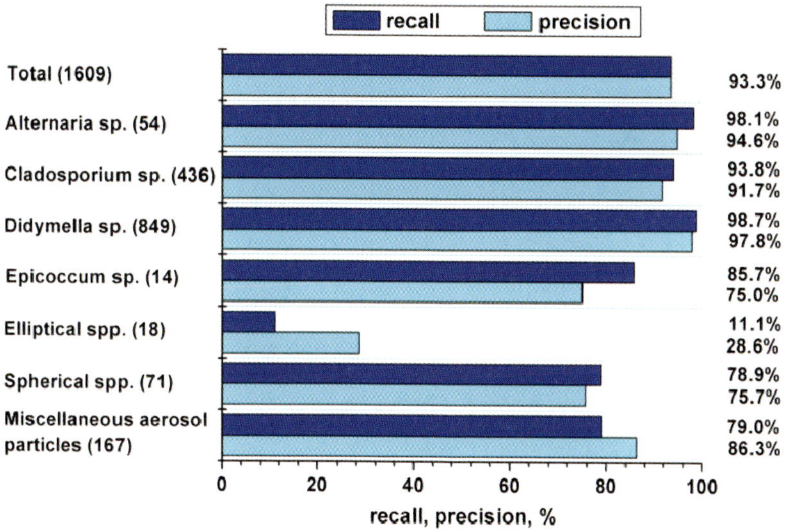

Figure 2.39 Recall and precision from a leave-one-out test with fungal spores from microscopic images of high quality.

fier. The corresponding recognition rates, (**Figure 2.40**), are higher than 90% and prove the success of automatic pollen recognition. Obviously, the recognition rate depends on successful segmentation, and thus can be enhanced, from 91% to 96.8% in this case, by discarding agglomerates. Algorithms for

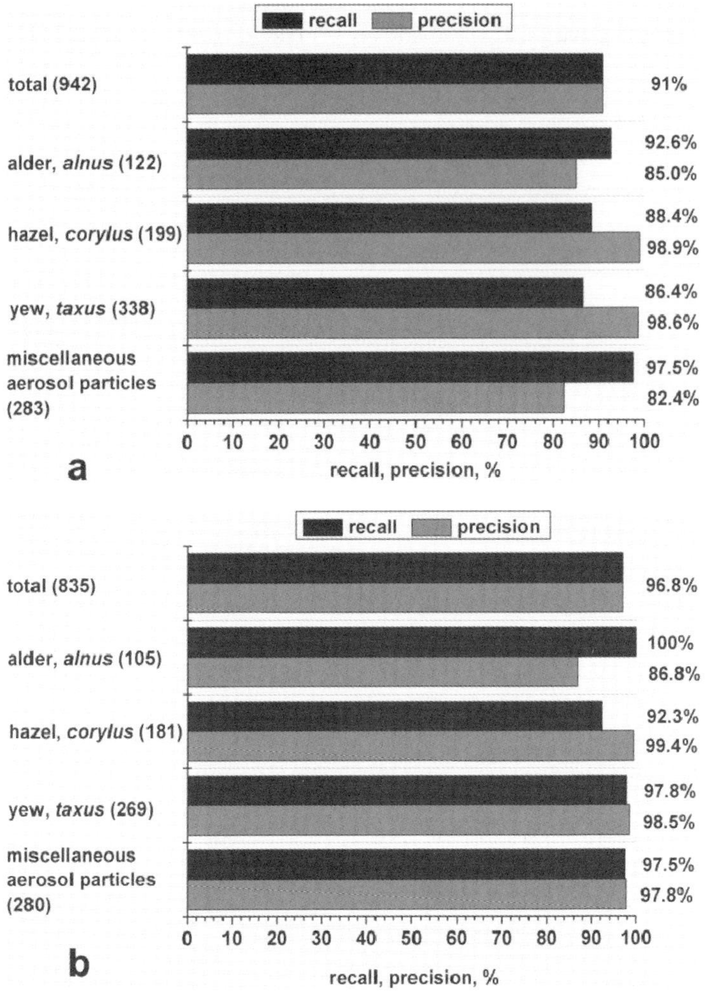

Figure 2.40 Recall and precision for pollen from microscopy images from the pollen monitor using an external reference database including (a) and excluding (b) agglomerates of pollen on the sample.

segmentation are still subject to optimization and new recognition methods are being developed.

2.3.2.7 **Field Experiments**

At present our project focuses on field experiments with the pollen monitor. System components and procedures are being adapted to outdoor requirements. Samples from the 2005 pollen season have been collected and labelled by skilled microscopists. It turnes out that for unambiguous pollen labelling a finer vertical resolution had to be chosen for the image stacks, resulting in

70 images with 1.5-μm layer distance instead of a layer distance of 2.5 μm using 40 images. The derived invariant features of several tens of thousands of particles will build the foundation for the classifier for automatic pollen recognition in the second phase of the field experiments in 2006.

2.4 Summary / Outlook

We have presented and discussed an online monitor for the detection of pollen and other bioaerosols. The system is capable of the automatic collection, preparation, detection and identification of pollen. The instrument's robustness allows it to operate continuously in the field. The monitor yields actual concentration data every hour.

The sampling unit fulfils the specifications with respect to the desired size range of scavenging from about 2.5 μm for fungal spores to even large airborne pollen up to about 60 μm, e.g., maize. A uniform deposition of particles suitable for automated microscopic analysis is effected onto the prepared surface.

The unique impaction surface simultaneously acts as both an adhesive and an embedding liquid owing to its thermoplastic properties. Subsequent microscopic analysis is possible without any additional preparative efforts. The content of polar liquids in the embedding medium affords the restoration of the collected pollen's original size and shape, and its water content provides the conducting surface that is needed for electrostatic precipitation.

The combination of brightfield microscopy and fluorescence microscopy is an effective approach to the automated particle recognition of bioaerosols. High-power LEDs are well suited as a light source for the microscopy techniques employed. Additionally, LEDs satisfy the requirements for outdoor operation, i.e., low power consumption, low maintenance and long-term stability in regard to power, wavelength, and lifetime.

The recognition rates from samples from the first experiments are very promising and notably higher than 90%. Extension of the reference database to cover the most commonly occurring airborne pollen in Middle Europe and adaptation of the monitor for operation under field conditions are subject to actual development.

The final validation in 2006 will provide reliable information on data quality, e.g., the measurement uncertainty derived from duplicate determinations, systematic errors relating to segmentation and agglomerates and the recall and precision of the classifier. At the end of the field validation process a commercial limited-lot production and the trial of a completely automatic network is planned.

Compared with conventional methods, such as the visual counting of Burkard® samples, which deliver daily mean concentration data, hourly measurements with the online monitor provide a higher temporal resolution. Furthermore, fast recognition algorithms provide the capacity for analysis of a larger sampling area and air volume respiration. These features greatly enlarge the database of pollen forecasting and modelling.

The first users of this technique will be the operators of pollutant monitoring stations in meteorological networks. Other possible applications cover the field of occupational health and safety in composting plants and in waste separation companies. Hospitals and pharmacies in Europe as well as waste treatment plants are other potential users.

The benefits of automated pollen monitoring will mainly consist of better pollen information for people who suffer from pollen allergy. A network of automatic pollen monitors will be the basis for future pollen-distribution forecasts. The information will make it easier to react to changes in pollen concentration. They will help to trigger reasonable strategies to avoid exposure and to adapt medical treatment to real needs. These improvements should protect allergic persons from any aggravation of their suffering, probably connected with asthma, but in any case will lead to an improvement in their quality of life.

But beyond that future improvement there is also much potential in using single modules of the monitor for pollen recognition. Automated recognition can assist as a powerful tool for pollen determination in food inspection of honey, in forestry and in the surveillance of genetically modified plants.

In the present work we have developed a completely automatic particle monitor, that can be used for a broad spectrum of applications. This is due to its adaptive and flexible software for pattern recognition. We have already shown that apart from pollen recognition the determination of fungal spores is possible. High recognition rates have been achieved without changes in the software. While the recognition of small, morphologically similar spore genera still remains a challenging task for further research, current monitoring systems are only able to deliver an unspecific detection of bioaerosols. Thus, in the case of fungal spores, manual sampling and visual counting under the microscope or cell-culture technology is the usual determination method. In contrast, with the online monitor there is no significant delay in the detection of health-affecting aerosols.

There is a concrete need for online measurement of airborne fungal spores in several fields of public life, for example in civil defense or the safety of work and trade. But pest management also requires an easy way to monitor airborne fungal spores as well as to measure the indoor climate in the case of formation of mould. In the latter case, other aerosol biogenic particles, such as epithelial layers or plant abrasion, can also be detected by fluorescence excitation.

The modular concept of the instrument means that its application is not restricted to the identification of allergens such as airborne pollen and fungal spores. By specific adaptation of the reference database and the chosen invariants, where necessary, the system is open for analysis of any component accessible by light microscopy, e.g., the analysis of other health-relevant aerosol particles, such as pathogenic microorganisms, tire abrasion particles, etc. Thus many other applications are possible.

Parts of the instrument, especially the microscopy unit combined with the pattern recognition software, can also be used with conventional particle sampling units. In the control of ambient air, such a combination could deliver highly time-resolved data for the atmospheric load of special pollutants, which are detectable by morphological characteristics.

The monitoring of dust can be optimized by the detection of biogenic aerosol with fluorescence microscopy and the employment of polarization methods for sea salt. The concentration of these components varies seasonally and has to be considered when limit values are exceeded. Furthermore, a detailed real-time analysis of coarse particles, such as tire abrasion, fly ash, mineral particles etc., leads to a more precise detection of the individual sources of dust particles, and prepares the ground for the prevention of exceedances of limit values.

Acknowledgements

The contribution of the project partners, the project executing organization by the VDI technology centre and the financial support by the BMBF in the Biophotonik framework under grant No. 13N8368, 13N8367, 13N8487, 13N8437 and 13N8372 are gratefully acknowledged.

Glossary

Aerodynamic diameter The aerodynamic diameter of a particle refers to its aerodynamic properties. As a definition standard, the aerodynamic diameter of a spherical particle with a unit density of $1\ \mathrm{g\,cm^{-3}}$ is identical to its geometric diameter. Particles with a different density and/or a different shape usually exhibit an aerodynamic diameter different from their geometric diameter [1].

Aerosol Aerosols are components of atmospheric pollution. They take part in chemical reactions, play an important role in radiation processes (absorption, scattering, extinction) and act as condensation nuclei in the atmosphere.

Biometeorology Human biometeorology comprises interdisciplinary investigations about weather, climate and human health. The meteorological impacts can be ascribed to three categories: human heat exchange, non-thermal impacts of solar radiation (ultraviolet, visible, infrared) and air pollution [57].

Colpate A pollen grain is called colpate if its germination apertures exhibit a terete shape [24].

Colporate If the germination apertures of a pollen grain exhibit both colpate and porate morphologic features it is called colporate [24].

Cupressaceae The *Cupressaceae* or cypress family is a conifer family of cosmopolitan distribution. The family includes 27 to 30 genera with about 130–140 species. They are trees and shrubs from 1–112 m tall [58].

Cutoff diameter The cutoff diameter is a parameter describing a step-function efficiency curve of an impactor, in which all particles greater than a certain aerodynamic size (= cutoff diameter) are collected and all particles less than that pass through [1].

GSM / GPRS The Global System for Mobile Communications (GSM) is the most popular standard for mobile phones in the world. The General packet radio service (GPRS) is a mobile data service for the users of GSM mobile phones [59].

Inaperturate A pollen grain is called inaperturate if it does not exhibit any germination apertures [24].

Isokinetic Isokinetic sampling is a procedure to ensure that a representative sample of aerosol enters the inlet of a sampling tube when sampling from a moving stream. Sampling is isokinetic when the inlet of the sampler, which may be a thin-walled tube or probe, is aligned parallel to the gas streamlines and the gas velocity entering the probe is identical to the free stream velocity approaching the inlet [1].

Mugwort Common mugwort (*Artemisia vulgaris*) belongs to the family of composites (*Asteraceae/Compositae*). It is a common herbaceous plant on screes and roadsides, growing to 2 m [24].

19″ rack A 19-inch rack is a standardized system for mounting various modules in a "stack", or rack, 19 inches (482.6 mm) wide. Equipment designed to be placed in a rack is typically described as rack-mount, a rack mounted system, a rack mount chassis, subrack, or occasionally, simply shelf [60].

Orographical describing the average height of the earth's surface

Oversampling Oversampling can occur as a consequence of anisokinetic sampling conditions. Depending on particle size, certain fractions may be enriched in the sampled volume, so that their concentration in the ambient air will be overestimated.

Petrographic Petrographic investigations focus on detailed descriptions of rocks and their mineral content. Detailed microscopic analysis of thin sections aids in understanding the origin of the rock [61].

Phenology Phenology comprises the seasonal dependent phenomena of animals and plants, e.g., typical growing and flowering seasons [62].

Pixel Picture element

Polyperiporate A pollen grain is called polyperiporate if it exhibits many germination apertures shaped like a pore and distributed equally over the pollen grain [24].

Porate A pollen grain is called porate if its germination apertures are shaped like a pore [24].

Prevalence frequency of occurrence of a certain disease in a population at the point in time of an examination [63].

Proinflammatory Proinflammatory substances tend to cause inflammations.

Ragweed Ragweed (*Ambrosia artemisiifolia*) belongs to the family of composites (*Asteraceae/Compositae*). It is a common herbaceous plant of high allergenic potential occurring in Switzerland, Hungary and the south of France, but also spreading into Austria and Germany [24].

RAS Remote Access Services (RAS) is a service provided by Windows NT which allows most of the services which would be available on a network to be accessed over a modem link [64].

Taxa generic term of the names of classes in plant or animal kingdom referring to the categorization of plants or other creatures in a systematic way [65].

Taxaceae The family *Taxaceae*, commonly called the yew family, includes three genera and about 7 to 12 species of coniferous plants. They are much-branched, small trees and shrubs. The male cones are 2–5 mm long, and shed pollen in the early spring [66].

Voxel Volume element

Key References

W. C. HINDS, *Aerosol Technology, properties, behavior and measurement of airborne particles,* John Wiley & Sons, New York, 1982.

K. AAS, N. ABERG, C. BACHERT, K. BERGMANN, R. BERGMANN, S. BONINI, J. BOUSQUET, A. DE WECK, I. FARKAS, K. HEJDENBERG. European allergy white paper: Allergic diseases as a public health problem. The UCB Institute of Allergy, Brussels, 1997.

J.H. VINCENT, *Aerosol Sampling,* John Wiley&Sons, Chichester, 1989.

J. BEUG, *Leitfaden der Pollenbestimmung für Mitteleuropa und angrenzende Gebiete,* Verlag Dr. Friedrich Pfeil, Munich, 2004.

PH. GREGORY, *The Microbiology of the Atmosphere,* Leonard Hill Ltd., London, 1961.

D. GERLACH, *Das Lichtmikroskop. Eine Einführung in Funktion und Anwendung in Biologie und Medizin,* Thieme Verlag, Stuttgart, 2nd edition, 1985.

A. EHRINGHAUS, *Das Mikroskop – seine wissenschaftlichen Grundlagen und seine Anwendung,* Teubner Verlag, Stuttgart, 6th edition, 1967.

O. RONNEBERGER, E. SCHULTZ, H. BURKHARDT, Automated Pollen Recognition using 3D Volume Images from Fluorescence Microscopy. *Aerobiologia,* 18:107–115, 2002.

P.J. TSAI, J.H. VINCENT, D. MARK, G. MALDONADO, *Aerosol Science Technol.,* 22:271, 1995.

H. SCHULZ-MIRBACH, Invariant features for gray scale images. In G. Sagerer, S. Posch, and F. Kummert, editors, *17. DAGM-Symposium "Mustererkennung",* Informatik aktuell, pages 1–14, Springer, 1995.

H. BURKHARDT, S. SIGGELKOW. Invariant features in pattern recognition – fundamentals and applications, in: C. Kotropoulos and I. Pitas, editors, *Nonlinear Model-Based Image/Video Processing and Analysis,* pages 269–307, John Wiley & Sons, 2001.

V.N. VAPNIK. *The nature of statistical learning theory,* Springer, Berlin, 1995.

References

1 W. C. HINDS, *Aerosol Technology, properties, behaviour and measurement of airborne particles,* John Wiley & Sons, New York, 1982.

2 WHO, Meta-analysis of time-series studies and panel studies of particulate matter (PM) and ozone (O_3), WHO (E82792), 2004.

3 U. KRÄMER, T. KOCH, U. RANFT, J. RING, H. BEHRENDT, Traffic-related air pollution is associated with atopy in children living in urban areas, *Epidemiology,* 11 (**2000**), pp. 64–70.

4 D. STRACHAN et al., Hayfever, hygiene and household size, *BMJ,* 299 (**1989**), pp. 1259–1260.

5 P. MATRICARDI et al., Hayfever and asthma in relation to markers of infection in the United States, *J. Allergy Clin. Immunol.,* 110 (**2002**), pp. 381–387.

6 K. AAS, N. ABERG, C. BACHERT, K. BERGMANN, R. BERGMANN, S. BONINI, J. BOUSQUET, A. DE WECK, I. FARKAS, K. HEJDENBERG, *European allergy white paper: Allergic diseases as a public health problem,* The UCB Institute of Allergy, Brussels, 1997.

7 D. STRACHAN, B. SIBBALD, S. WEILAND, et al., Worldwide variations in prevalence of symptoms of allergic rhinoconjunctivitis in children: The international study of asthma and allergies in childhood (ISAAC), *Pediatr. Allergy Immunol.,* 8 (**1997**), pp. 161–176.

8 http//www.polleninfo.org.

9 http://www.pollenstiftung.de.

10 http://www.dwd.de/de/WundK/W_aktuell/index.htm.

11 C.-M. LIAO et al., Temporal/seasonal variations of size-dependent airborne fungi indoor/outdoor relationships for a wind-induced naturally ventilated airspace, *Atmospheric Environment*, 38 (**2004**), pp. 4415–4419.

12 L. Makra, M. Juhàsz, E. Borsos, R. Béczi, Meteorological variables connected with airborne ragweed pollen in southern Hungary, *Int. J. Biometeorol.*, 49 (**2004**), pp. 37–47.

13 N. Helbig, B. Vogel, H. Vogel, F. Fiedler, Numerical modelling of pollen dispersion on the regional scale, *Aerobiologia*, 20 (**2004**), pp. 3–19.

14 J.M. Hirst, An automatic volumetric spore trap, *Ann. Appl. Biol.*, 36 (**1952**), pp. 257–265.

15 P.L. Magill, E.D. Lumpkins, J.S. Arveson, A system for appraising airborne populations of pollens and spores, *Am. Ind. Hyg. Assoc.*, 29 (**1968**), p. 293.

16 J.H. Vincent, *Aerosol Sampling*, John Wiley&Sons, Chichester, 1989.

17 R.M. Burton, D.A. Lundgren, Wide range aerosol classifier: A size selective sampler for large particles, *Aerosol Science Technol.*, 6 (**1987**), p. 289.

18 W. Holländer, W. Dunkhorst, G. Pohlmann, A sampler for total suspended particulates with size resolution and high sampling efficiency for large, *Particle & Particle System Characterization*, 6 (**1989**), p. 74.

19 Comité Européen de Normalisation (CEN), *Workplace atmospheres: Size fraction definitions for measurement of airborne particles in the workplace*, CEN Standard EN 481, 1992.

20 S.N. Li, D.A. Lundgren, D. Rovell-Rixx, Evaluation of six inhalable aerosol samplers, *Am. Ind. Hyg. J.*, 61 (**2000**), p. 506.

21 W.C. McCrone, J.G. Delly, *The Particle Atlas*, volume IV, Ann Arbor Science Publishers, Ann Arbor, Michigan, 1973.

22 J. Beug, *Leitfaden der Pollenbestimmung für Mitteleuropa und angrenzende Gebiete*, Verlag Dr. Friedrich Pfeil, Munich, 2004.

23 Institute of Respiratory Medicine, *Airborne allergens, pollen grains and fungal spores*, Institute of Respiratory Medicine, Sydney, 1999.

24 H. Winkler, R. Ostrowski, M. Wilhelm, *Pollenbestimmungsbuch der Stiftung Deutscher Polleninformationsdienst*, TAKT-Verlag, Paderborn, 2001.

25 K. Wilken-Jensen, S. Gravesen, *Atlas of Moulds in Europe causing respiratory Allergy*, ASK Publishing, Copenhagen, 1984.

26 Ph. Gregory, *The Microbiology of the Atmosphere*, Leonard Hill Ltd., London, 1961.

27 F.H.L. Benyon et al. Differentiation of allergenic fungal spores by image analysis, with application to aerobiological counts, *Aerobiologia*, 15 (**1999**), pp. 211–223.

28 http://www.uni-giessen.de/~gi38/publica/mikros.

29 A. Periaswami, *Method in Cellular Imaging*, Oxford University Press, Oxford, 2001.

30 H. Haferkorn, *Optik, Physikalisch-technische Grundlagen und Anwendungen*, 4th edition, Wiley-VCH, Weinheim, 2002.

31 D. Gerlach, *Das Lichtmikroskop. Eine Einführung in Funktion und Anwendung in Biologie und Medizin*, 2nd edition, Thieme Verlag, Stuttgart, 1985.

32 G. Göke, *Moderne Methoden der Lichtmikroskopie*, Francksche Verlagshandlung, Stuttgart, 1988.

33 H. Robenek, *Mikroskopie in Forschung und Praxis*, GIT Verlag, Darmstadt, 1995.

34 A. Ehringhaus, *Das Mikroskop – seine wissenschaftlichen Grundlagen und seine Anwendung*, 6th edition, Teubner Verlag, Stuttgart, 1967.

35 S. Bradbury, *An introduction to the optical microscope*, Oxford University Press, Oxford, 1984.

36 D. Lansing Taylor, Yu-li Wang, *Fluorescence Microscopy of Living Cells in Culture*, Academic Press, Inc., San Diego, 1989.

37 T. Kreis, *Handbook of Holographic Interferometry*, Wiley-VCH, Weinheim, 2005.

38 M. Langford, G. Taylor, J. Flenley, The application of texture analysis for automated pollen identification, in: *Proc. Conf. on Identification and Pattern Recognition*, Univ. Paul Sabatini, Toulouse, 1986, pp. 729–739.

39 M. LANGFORD, G.E. TAYLOR, J.R. FLENLEY, Computerized identification of pollen grains by texture analysis, *Review of Palaeobotany and Palynology*, 64 (**1990**), pp. 197–203.

40 I. FRANCE, A.W.G. DULLER, H.F. LAMB, G.A.T. DULLER, A comparative study of model based and neural network based approaches to automatic pollen identification, in: *British Machine Vision Conference*, **1997**, pp. 340–349.

41 PING LI, J.R. FLENLEY, Pollen texture identification using neural network, *Grana*, 38 (**1999**), pp. 59–64.

42 I. FRANCE, A. W. G. DULLER, G. A. T. GULLER, H. F LAMB, A new approach to automated pollen analysis. *Quaternary Science Reviews*, 19(6, February) (**2000**), pp. 537–546.

43 P. BONTON, A. BOUCHER, M. THONNAT et al., Colour image in 2d and 3d microscopy for the automation of pollen rate measurement, *Image Anal. Stereol.*, 21 (**2002**), pp. 25–30.

44 A. BOUCHER, P. HIDALGO, M. THONNAT et al., Development of a semi automatic system for pollen recognition, *Aerobiologia*, 18 (**2002**), pp. 195–201.

45 O. RONNEBERGER, E. SCHULTZ, H. BURKHARDT, Automated Pollen Recognition using 3D Volume Images from Fluorescence Microscopy, *Aerobiologia*, 18 (**2002**), pp. 107–115.

46 O. RONNEBERGER, H. BURKHARDT, E. SCHULTZ, General-purpose Object Recognition in 3D Volume Data Sets using Gray-Scale Invariants – Classification of Airborne Pollen-Grains Recorded with a Confocal Laser Scanning Microscope, in: *Proceedings of the International Conference on Pattern Recognition*, Quebec, Canada, September 2002.

47 W. J. TRELOAR, G. E. TAYLOR, J. R. FLENLEY, Towards automation of palynology 1: analysis of pollen shape and ornamentation using simple geometric measures, derived from scanning electron microscope images, *Journal of quaternary science*, 19(8) (**2004**), pp. 745–754.

48 P. LI, W. J. TRELOAR, J. R. FLENLEY, L. EMPSON, Towards automation of palynology 2: the use of texture measures and neural network analysis for automated identification of optical images of pollen grains, *Journal of quaternary science*, 19(8) (**2004**), pp. 755–762.

49 Y. ZHANG, D. W. FOUNTAIN, R. M. HODGSON, J. R. FLENLEY, S. GUNETILEKE, Towards automation of palynology 3: pollen pattern recognition using gabor transforms and digital moments, *Journal of quaternary science*, 19(8) (**2004**), pp. 763–768.

50 P.J. TSAI, J.H. VINCENT, D. MARK, G. MALDONADO, Impaction model for the aspiration efficiencies of aerosol samplers in moving air under orientation-averaged conditions, *Aerosol Science Technol.*, 22 (**1995**), p. 271.

51 W. KOCH, W. DUNKHORST, H. LÖDDING, Design and performance of a new personal aerosol monitor, *Aerosol Sci. Technol.*, 31 (**1999**), pp. 231.

52 C.G.B. COLE et al., Gelatine fluorescence and its relationship to animal age and gelatine colour, *The SA Journal of Food Science and Nutrition*, 8(4) (**1996**), pp. 139–143.

53 H. SCHULZ-MIRBACH, Invariant features for gray scale images, in: G. SAGERER, S. POSCH, F. KUMMERT (eds.), *17. DAGM-Symposium "Mustererkennung"*, Informatik aktuell, Springer, 1995, pp. 1–14.

54 H. BURKHARDT, S. SIGGELKOW, Invariant features in pattern recognition – fundamentals and applications, in: C. KOTROPOULOS, I. PITAS (eds.), *Nonlinear Model-Based Image/Video Processing and Analysis*, John Wiley & Sons, New York/Chichester, 2001, pp. 269–307.

55 M. SCHAEL, S. SIGGELKOW, Invariant grey-scale features for 3d sensor-data, in: *Proceedings of the International Conference on Pattern Recognition (ICPR2000)*, Barcelona, Spain, September 2000, pp. 531–535.

56 V. N. VAPNIK, *The nature of statistical learning theory*, Springer, Berlin/Heidelberg/New York, 1995.

57 http://www.dwd.de/de/wir/Geschaeftsfelder/Medizin/Medizin.html.

58 http://en.wikipedia.org/wiki/Cupressaceae.

59 http://en.wikipedia.org/wiki/GSM.

60 http://en.wikipedia.org/wiki/19%22_rack.

61 http://en.wikipedia.org/wiki/Petrographic.

62 www.dwds.de.

63 Roche Lexikon Medizin.

64 http://en.wikipedia.org/wiki/Remote_Access_Service.

65 http://www.rp-kassel.de/static/themen/naturschutz/lrp2000/hilfen/glossar/glossar.htm.

66 http://en.wikipedia.org/wiki/Taxaceae.

3
Online Monitoring and Identification of Bioaerosol (OMIB)

Petra Rösch, Michaela Harz, Mario Krause, Renate Petry, Klaus-Dieter Peschke, Hans Burkhardt, Olaf Ronneberger, Andreas Schüle, Günther Schmauz, Rainer Riesenberg, Andreas Wuttig, Markus Lankers, Stefan Hofer, Hans Thiele, Hans-Walter Motzkus, Jürgen Popp[1]

3.1
Bioaerosol and the Relevance of Microorganisms

The term "aerosol" is used for any solid and liquid particles in the size range from a few nanometers up to some 100 μm which are suspended in air. The origin of the particles may be inorganic, organic or biological. Their chemical composition may vary as widely as their potentially far-reaching and long-lasting effects on flora and fauna. Aerosols influence not only the climate and atmospheric chemistry but also human health. Beside organic and inorganic microparticles, biological microparticles are currently gaining significance in research. One major interest lies in the identification of the constituents as well as in the characterization of the physical and chemical behavior of the bioaerosol [1].

Any airborne particles of biological origin are called bioaerosol. More specifically, this includes all living and dead microorganisms such as viruses, bacteria, spores, yeast, mold, pollen, algae, etc. In addition, larger or smaller fragments of microorganisms, as well as traces from skin, hair and body fluids can be constituents of a bioaerosol. Components of biological origin (proteins, enzymes) bound to abiotic particles (e.g., dust) are also included [2].

The sources of such bioaerosol are various and complex. Composting depots produce huge amounts of bioaerosol. In particular, the process of turning releases a huge number of microorganisms in combination with biomolecules attached to dust. Spume from surge or waterfalls that are loaded with microbes are another source of bioaerosol. In addition, humankind might be the source of hazardous bioaerosol, since certain diseases are transmitted via airborne infections. Thus, bioaerosol plays an important role in human life due to its potentially infectious, allergic and toxic effects.

1) Corresponding author

Biophotonics: Visions for Better Health Care. Jürgen Popp and Marion Strehle (Eds.)

ISBN: 3-527-40622-0

Routine aerosol monitoring is based on particle counting, which provides information on the number and size but not on the type and origin of the particles. In addition, abiotic and biotic particles should be differentiated. Abiotic particles can be classified into organic and inorganic particles. Inorganic particles might derive from metal, metal oxide, building material, glass, ceramics, inorganic excipients, etc., whereas organic particles might occur from plastics, organic fibers, keratin, etc. The composition of organic and inorganic particles can easily be analyzed by spectroscopic methods such as energy-dispersive X-ray analysis (EDX) or Raman spectroscopy.

On the other hand, biotic particles are conventionally analyzed by time-consuming isolation and cultivation of microorganisms. Modern microbial detection methods need to be fast and highly sensitive to give immediate detection of even a single pathogenic organism, which might be enough to cause an infection.

The analysis of bioaerosol has to take into account a huge number of particles containing a great variety of biotic particles, e.g., bacteria, yeast, mold, spores, pollen, algae, viruses, etc. A broad spectrum of investigation techniques is necessary to detect all the above mentioned different kinds of biotic particles.

One example where intensive monitoring of bioaerosol is required is the clean-room production of pharmaceutical drugs. Owing to the artificial environment in this field, the bioaerosol exhibits special characteristics: On the one hand, the total number of particles is limited, which also leads to a small number of biotic particles in the samples. On the other hand, the number of microbial species commonly found in a clean room is well defined. Therefore, only a certain number of species of a few genera need to be identified. A fast and reliable method is required for the identification of microbes in clean air, on instrument surfaces or in pharmaceutical drugs. This chapter reports the introduction of a new method based on a combination of fluorescence imaging and micro-Raman spectroscopy with elaborate statistical methods for a fast, label-free and nondestructive identification of microorganisms.

3.2
Diagnosis of Microorganisms: State-of-the-Art

Microbiological laboratories are able to isolate and identify most of the common microorganisms within 48 h after sampling. The latest diagnostic methods allow the identification of some bacteria within minutes. Immunological and molecular biological methods even allow the identification of many microbes without previous cultivation [3].

In the following sections, different methods for the identification of microorganisms are presented. In principle, two approaches are possible: to detect the microorganisms directly, or to identify the immune reaction and, thus, to detect the pathogen indirectly [4].

Most of the methods presented are used to examine medical samples such as body fluids (blood, liquor, urine, etc.), tissue probes or smears, but they are also suitable for environmental samples, such as soil or aerosol.

3.2.1 Microbiological Diagnosis

The common approach for microbiological diagnosis consists of the isolation and cultivation of the bacteria in order to gain sufficient cell mass, followed by the identification. For most identification methods a pure colony is required, in order to avoid cross-reactions from contaminating microbes [3].

For preselection of microorganisms, different macroscopic and microscopic methods are commonly applied. Morphological parameters such as size and shape in combination with the existence of, for example, flagellates, capsules or spores, allow a first differentiation between microbial species. Staining (e.g., Gram stain) delivers additional information. Using the results of these preliminary tests, appropriate further investigation methods can be chosen [3].

Growth-dependent tests are based on the ability of microbes to metabolize different cultivation media. So-called colored series allow the observation of bacterial reaction on selective and differential media. Selective media contain substances which selectively inhibit the growth of certain bacteria. Differential media contain an indicator to differentiate between various reactions during the growth phase of the bacteria. Here the presence or absence of enzymes, the fermentation of carbohydrates, the change of pH, etc. can be observed. For routine investigations with a colored series, 20 standardized tests are normally performed. For identification of the species the results are then compared with a database [5].

Besides exploiting different metabolic abilities, direct cell attributes can also be used for identification. In the following, immunological reagents as a means to identify bacteria will be discussed.

Two attributes of an immunological test are a measure of its applicability: specificity and sensitivity. Specificity is the ability of an antibody to recognize one single antigen while undergoing no antibody cross-reactions and therefore showing no false positive reactions. Sensitivity defines the minimum amount of antigen necessary for successful detection. A high sensitivity prevents false negative reactions, since even low concentrations of antibodies can be detected [3].

Several methods can be applied for the identification of microbial antigens: agglutination tests, fluorescence labelling or the detection of specific molecules by gas chromatography, mass spectrometry or gel electrophoresis to detect phenotypic differences between microorganisms. By contrast, nucleic acid or DNA probes differentiate directly between the genomes of bacterial strains.

Agglutination tests are based on the bonding of particular antibodies to the antigens of the cell. The antibodies or antigens are bound to latex particles. Positive reactions are detected by agglutination of the antigen- or antibody-functionalized latex particles.

For pathogens which can only be found temporarily, are difficult to cultivate or cannot be cultivated at all, an indirect detection method is used. In this case the antibodies against the pathogen are detected [3].

Another immunological method of identifying microorganisms is the labelling of antigens or antibodies with fluorescence markers. In principle, this allows the detection of the antibody reaction of a single cell. Chemical modification of the antibodies has to be performed in such a way that the specificity of the antibody does not change. This method is used, for example, in cell sorters (fluorescence-activated cell sorter, FACS). These fluorescence labelling methods exhibit a high specificity and a rather lower sensitivity. To achieve an increased sensitivity, the ELISA (enzyme-linked immunosorbent assay) or the RIA (radioimmunoassay) methods are used. ELISA and RIA employ enzymes or radioisotopes, respectively, in order to mark the antibody molecules. The detection limit of radioactivity and specific enzyme reaction products is extremely low. Therefore, the amount of antibody necessary for detection can be minimized [3].

Membrane lipids, peptidoglycans from cell walls or RNA polymerase are also used to differentiate between microbes. These molecules can be identified by means of gas chromatography, mass spectrometry or gel electrophoresis [3].

The methods described so far are phenotypic identification methods. However, genotypic methods such as nucleic acid probes are also used, for instance the classical PCR (polymerase chain reaction) method in combination with sequencing techniques, which use the rRNA (ribosomal ribonucleic acid) molecules 16S RNA or 18S RNA to probe for bacteria or eukaryotes, respectively. These rRNA molecules are particularly suitable for use as molecular fingerprints of microorganisms on a strain level. A disadvantage of the method is the need for a pure colony [3].

An alternative approach is the application of DNA (deoxyribonucleic acid) probes, where one specific DNA sequence is identified instead of the whole organism. For the detection of this specific DNA sequence, a complementary single-stranded DNA is labelled with a reporter molecule, e.g., a radioisotope, an enzyme or a fluorescence marker. Depending on the reporter molecule, up to 25 μg of DNA per sample can be identified. In the case of fluorescence labels, the number of bacteria are counted, for example by flow cytometry. The application of selective DNA markers is only available for certain bacterial species or strains [6]. Other methods for the identification of single microbial cells are the fluorescence imaging staining techniques FRET (fluorescence resonance energy transfer) or FISH (fluorescence *in situ* hybridization). These staining techniques require the complementary DNA marker for every microorganism to be identified. Therefore, it is necessary either to use fluorescence molecules with different absorption/excitation characteristics, or to concentrate the search onto a single distinct specimen, neglecting all others [3].

An alternative approach for the identification of microorganisms is the application of various spectroscopic techniques. Garg et al. [7] performed ^{1}H-NMR spectroscopic investigations in order to identify bacteria from abscesses by their metabolic pattern. The application of matrix-assisted laser deposition/ionization mass spectrometry allows the identification of tryptic digested bacteria from various *Bacillus* species [8].

3.2.2
Vibrational Spectroscopic Methods

The application of vibrational spectroscopic techniques like IR absorption and Raman spectroscopy to investigate and identify microorganisms allows a non-invasive investigation of the cell.

Vibrational spectroscopy yields spectral fingerprints of the bacterial cells [9]. However, vibrational spectra cannot provide information about specific components, since they result from a superposition of the individual spectra of the cell constituents [10]. Since the IR or Raman spectra of different bacterial strains and species are very similar, it is necessary to use pattern recognition methods differentiating between the taxa [11].

Depending on the applied spectroscopic method, it can be shown that different cultivation media influence the vibrational spectra [12]. Therefore, an IR spectroscopic identification is normally performed with standardized cultivation parameters, e.g., the culture medium, cultivation time and temperature should be controlled rigidly [10]. Since water disturbs the IR spectra, sample preparation requires a drying step [13]. The dried bacterial films can, for example, be prepared on a ZnSe window [14]. Alternatively, the bacteria can be lyophilized to generate KBr tablets [15]. For the investigation of biofilms, an ATR device can be used [16]. Goodacre et al. [17] applied IR spectroscopy to discriminate between antibiotic resistant and non-resistant *Staphylococcus aureus* strains.

The number of cells that are necessary for an IR spectroscopic identification of bacteria and yeasts can be reduced to a few hundred cells by the application of an IR microscope [18–20].

In contrast to IR spectroscopy, water is not troublesome for Raman spectroscopy, so biological samples can be investigated without, or with only minor, sample preparation. Since most biomarkers absorb in the region between 280 and 190 nm, UV-resonance Raman spectroscopy is a promising tool for direct monitoring of these molecules inside the cells. Resonance Raman spectroscopy allows one to obtain selective information about macromolecules, e.g., proteins, cytochromes and nucleic acids, which are widely used to identify microbes [21]. In addition, Raman excitation wavelengths below 260 nm lead to fluorescence-free Raman spectra [22].

Wu et al. [23] applied UV-resonance Raman spectroscopy to investigate bacteria and bacterial components at different Raman excitation wavelengths. The influence of different cultivation conditions on Raman spectra of bacteria was studied by Manoharan et al. [24]. Besides bacteria, cyanobacteria were also investigated at different UV Raman excitation wavelengths [25]. Chadha et al. [26] combined UV-resonance Raman spectroscopy with a cryostage and were therefore able to investigate small numbers of bacteria (20 to 50 cells). First attempts at UV-resonance Raman spectroscopical identification of bacteria on a genera level or of the bacillus group was reported by Jarvis et al. [27] or López-Díez et al. [28], respectively.

Calcium dipicolinate (CaDPA) is a characteristic substance that can be found in bacterial spores. Bacterial endospores can be identified via the UV-resonance Raman spectra of the marker signals of CaDPA [29, 30].

Investigation of biological samples by Raman spectroscopy with visible excitation wavelengths is often hampered by the appearance of fluorescence. One possibility to overcome this effect is the use of surface-enhanced Raman spectroscopy (SERS). The SERS effect leads to an enhancement (by a factor of up to 10^{11}–10^{14}) of the Raman scattering of a molecule situated in the vicinity of nanosized metallic structures.

For the investigation of bacteria, different SERS substrates, such as electrochemically roughened silver and gold foils [31], gold island films [32], gold and silver colloids [33, 34], localized gold nanoparticles [35], as well as silver colloids prepared *in situ* inside the bacteria were used [36–38]. For the detection of bacterial endospores, SERS spectra of pure CaDPA were compared to the corresponding SERS spectra of spores [39–41]. Furthermore, Grow et al. [42] used a roughened metal film as a biochip surface for the detection of pathogens.

The first investigations on microorganisms were resonance Raman spectroscopic studies on chromophores such as carotenoids and chlorophylls of chromobacteria [43–45] and algae [46, 47], with an excitation wavelength of 488 or 514 nm.

However, since the resonant excitation of bacterial chromophores in the visible wavelength range leads to rather non-specific Raman spectra, non-resonant Raman spectroscopical investigations of bacteria were performed with excitation wavelengths of 1064 or 830 nm [48]. Several studies report a non-resonant Raman spectroscopic identification of bacterial bulk material [49, 50]. In order to minimize the cultivation time, micro-Raman spectroscopic investigations were performed on microcolonies [49, 50]. Micro-Raman spectroscopy was also applied to the investigation of spores [51] and yeasts [52–54].

One possible method of achieving single-cell information is the combination of Raman spectroscopy with optical tweezers. With this method, bacteria and yeasts [55–58] as well as spores [59] can be investigated on a single-cell level. Admittedly, this technique cannot be used as a screening method.

An alternative approach, which reduces the number of bacteria necessary for identification, is the use of micro-Raman spectroscopy. Schuster et al. [60, 61] investigated single cells of *Clostridium acetobutylicum* and identified different chemical components, e.g., starch, inside the cells. Huang et al. [62] investigated the dependence of cultivation medium and culture age on the Raman spectra of different bacteria. Single spores [63] and yeast cells [64, 65] were also investigated by means of micro-Raman spectroscopy.

For the investigation of fossil microorganisms, which are often incompletely preserved and easily mistaken for non-biological mineral structures, a non-destructive identification method for these very rare and expensive biological structures, like micro-Raman spectroscopy, is required. Brasier et al. [66] used micro-Raman spectroscopy to identify graphitic material in microfossils. Schopf et al. [67] performed Raman imaging experiments to discriminate microbial fossils from microscopic pseudofossils ("lookalikes").

In this chapter, we describe the development of a totally automated device for the identification of microorganisms. For this purpose, a monitoring step first discriminates between abiotic and biotic particles. Then the biotic particles are characterized by micro-Raman spectroscopy and identified with a support vector machine (SVM). The database is built in such a way that new microorganism strains can be incorporated at any stage. Therefore, the whole device can easily be adapted to different requirements and environments.

Raman Spectroscopy

Raman spectroscopy and infrared (IR) spectroscopy are vibrational spectroscopic methods. **Figure 3.1A** shows the IR absorption mechanism. An IR-photon can be absorbed by a molecule leading to an excitation from the vibrational ground to the first excited state. If higher vibrational states are excited, so-called vibrational overtones are detected in the IR spectrum.

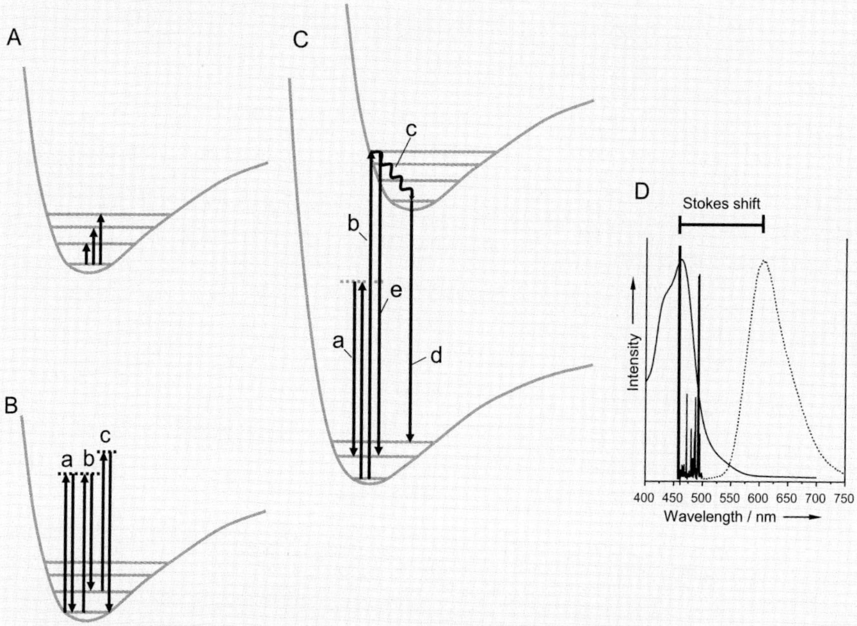

Figure 3.1 A: IR absorption; B: different scattering processes: Rayleigh scattering (a), Stokes–Raman scattering (b), and anti-Stokes–Raman scattering (c); C: comparison between normal Raman scattering (a), absorption (b), vibrational relaxation (c), fluorescence (d), and resonance Raman spectroscopy (e); D: Absorption and fluorescence spectra in comparison to a resonance Raman spectrum.

While in IR absorption spectroscopy the light is adsorbed directly, it is inelastically scattered by the molecule in Raman spectroscopy. **Figure 3.1B** shows the various scattering processes of light in a molecule. A molecule in the vibrational ground state interacts with a photon leading to the promotion of the system to a 'virtual state'. Subsequently, most photons are elastically scattered. This means that the system relaxes into the original vibrational state by emitting a photon of the same energy (wavelength) as the incident photon. The elastic scattering process is called Rayleigh scattering (see **Figure 3.1B** process a).

However, a small fraction of the light (approximately 1 out of 10^8 photons) is scattered at optical frequencies different from the frequency of the incident photons. The process leading to inelastic scattering is called the Raman effect (see **Figure 3.1B**, b and c). As a result of the coupling between the incident radiation and the quantized states of the scattering system, an energy transfer occurs (schematically depicted in **Figure 3.1B**, b and c). Depending on the nature of the coupling, the incident photons either lose or gain energy. The scattered light with lower energy compared to the incident laser light is called Stokes–Raman scattering (process b) and the radiation with higher energy is referred to as anti-Stokes–Raman scattering (process c). The absolute value of the difference between the incident photon and the scattered photon equals the difference between the two vibrational states (e.g., vibrational ground state and first excited vibrational state). The intensity of Stokes–Raman scattering is usually higher than the intensity of anti-Stokes–Raman scattering, since the excited state from which the anti-Stokes–Raman scattering occurs is less populated than the ground state under normal conditions at room temperature. Therefore, in most cases Stokes–Raman scattering is recorded.

The observation of Raman scattering is in some cases limited by the excitation of fluorescence, which typically exhibits an intensity several orders of magnitude stronger than normal Raman scattering. Biomolecules or biological systems often contain fluorophores and, therefore, often generate fluorescence when investigated with visible excitation wavelengths. Hence, the Raman spectrum is often obscured when excited with light in the visible range.

An alternative approach is resonance Raman spectroscopy. In **Figure 3.1C**, the Raman scattering process (a) is displayed in comparison to an electronic absorption process (b). After vibrational relaxation (c) fluorescence (d) can take place. Alternatively, direct emission of light can occur after absorption. This is called the resonance Raman effect (e).

In **Figure 3.1D** an absorption spectrum (solid line) is shown, together with the corresponding fluorescence spectrum (dotted line). If the Stokes shift (wavelength difference between the absorption and the fluorescence emission, see **Figure 3.1D**) is large enough, a resonance Raman spectrum, can be observed. To measure a resonance Raman spectrum the excitation wavelength of the laser needs to be chosen near to, or in the same energy range as, an electronically excited state. The preresonant or resonant excitation results in significant enhancement of the intensity of certain signals within the Raman spectrum.

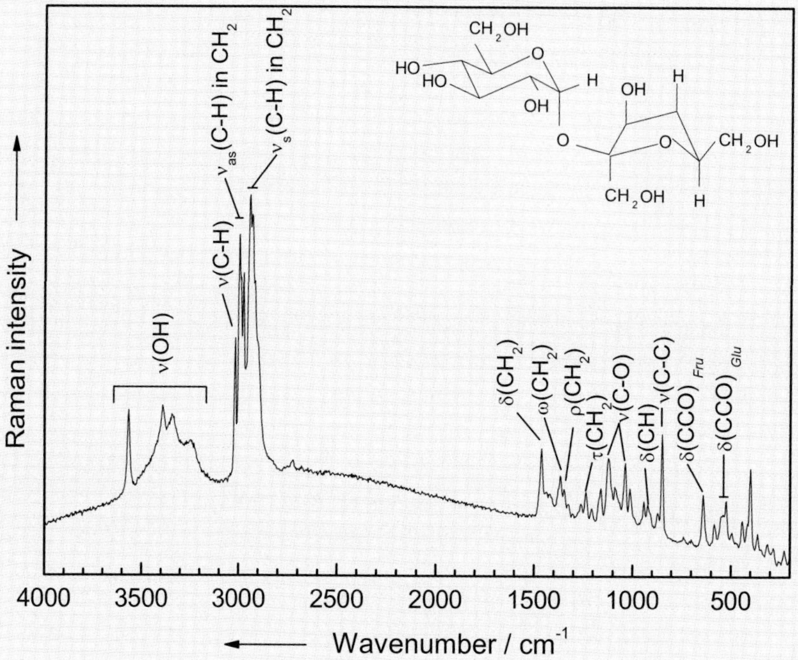

Figure 3.2 Raman spectrum of sucrose with an excitation wavelength of $\lambda = 532$ nm. The chemical structure of sucrose is also shown.

Figure 3.2 shows the Raman spectrum of sucrose, a common biomolecule. Sucrose (β-D-fructofuranosyl-α-D-glucopyranoside) is a disaccharide composed of α-glucose and β-fructose. Since each molecular group exhibits its own characteristic Raman signals, these signals can be used either for the characterization of the whole molecule or as marker signals for the identification of the component. Vibrations, or more precisely normal modes, can be divided into stretching vibrations (ν) and deformation vibrations, where stretching vibrations change the bond length and deformation vibrations the bond angle. Deformation vibrations can be further divided into τ twisting, ρ rocking, ω wagging, and δ bending modes.

The Raman bands of the O-H groups of sucrose and crystal water can be seen between 3451 and 3271 cm^{-1} (ν(O-H)). The CH_x groups exhibit Raman modes at 2982 (ν(C-H)), 2944 (ν_{as}(C-H) in CH_2), 2912 (ν_s(C-H) in CH_2), 1456 (δ(CH_2)), 1366 (ω(CH_2)), 1340 (ρ(CH_2)), 1266 (τ(CH_2)), and 920 cm^{-1} (δ(C-H)). The C-C bond can be found at 836 cm^{-1} (ν(C-C)). In addition, there are two carbon–oxygen bands at 1110 (ν(C-O) endocyclic) and 1064 cm^{-1} (ν(C-O) exocyclic) as well as the corresponding deformation modes at 640 (δ(C-C-O) fructose), 600 (δ(O-C-O)), 548 (δ(C-C-O) endocyclic of glucose), and 528 cm^{-1} (δ(C-C-O) exocyclic of glucose) [68].

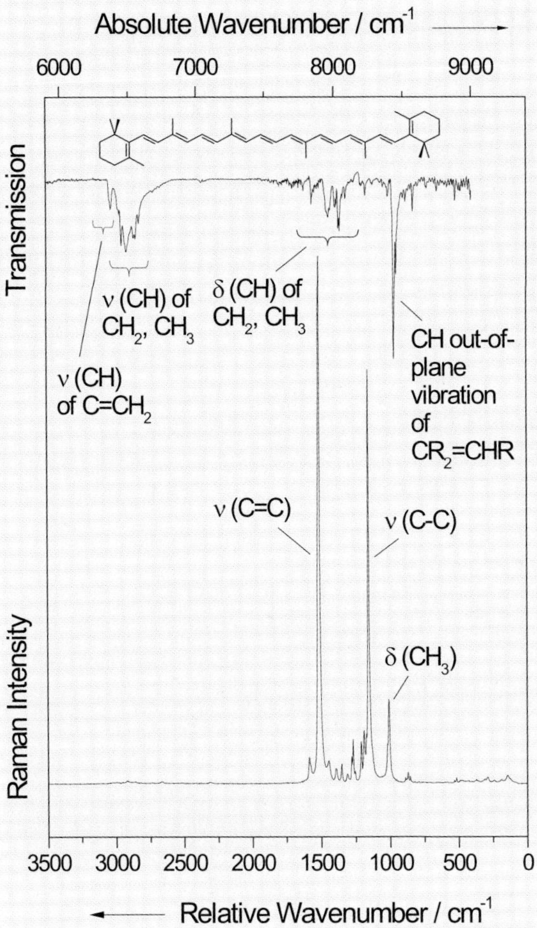

Figure 3.3 IR spectrum of β-carotene in comparison to a Raman spectrum with an excitation wavelength of $\lambda = 1064$ nm.

Figure 3.3 compares the IR and Raman spectra of β-carotene, a common biomolecule found in various biological cells. The structure of β-carotene is displayed as an inset. The chemical groups of the molecule exhibit characteristic IR signals (upper spectrum). The IR spectrum is dominated by the C-H stretching and deformation vibrations. In the Raman spectrum (lower spectrum), the most prominent vibrations are the C=C stretching vibrations of β-carotene at 1526 cm^{-1}, the C-C stretching modes at 1164 cm^{-1} and the CH_3 bending vibrations at 1005 cm^{-1}. IR and Raman spectra usually contain complementary information and permit the complete characterization of the molecule.

3.3
Innovative Optical Technologies to Identify Bioaerosol

Bioaerosol, which is defined as all biological particles suspended in air, might be dandruff, grit, pollen, spores, mold, yeast or bacteria. A few of them cause allergies or diseases or might lead to food spoilage. Therefore, rapid and unambiguous identification of biological contamination (bacteria, spores, mold, yeast) is necessary for the monitoring of aerosols.

Conventional investigations of bioaerosol are performed in two different ways. One approach is to record the total number and size of particles. However, this method provides no information about the type and therefore the origin of the particles. In particular, it is not possible to differentiate between biotic and abiotic particles. The second approach is to apply conventional microbiological techniques. After the isolation and cultivation of microorganisms from the aerosol, pure colonies are used for microbiological differentiation. In this chapter, we describe an alternative approach, which allows us to determine the overall number of particles and the ratio of biological contamination, and also to identify the microorganisms found.

Bacteria play a large role in almost all fields of life. Most problematic are appearances of bacteria, for example, in hospitals (formation of resistance), in the food-processing industry (accelerated food spoilage and toxic byproducts) and in pharmaceutical clean-room production (possible contamination of pharmacological products).

In various industrial production processes (e.g., medical technology, pharmaceutical production, food technology, biotechnology) biotic contamination can affect the quality of the products. Several guidelines and standards demand the routine control of the production environment [69]. According to the European guidelines on "Good manufacturing practices" for medical products [70] routine hygiene control is essential. From January 2006, the European Order 853/2004 [71] requires that every process and all equipment for the production of food must be evaluated for hygiene. Thus, there is a huge industrial demand for a method to do online monitoring and identification of microbial contamination.

In this contribution we focus on the investigation of clean-room environments. Because of the artificial environment in this field the bioaerosol exhibits special characteristics. On the one hand, the total number of particles is limited, leading to a small number of biotic particles in the samples. On the other hand, the number of microbial species is well defined. Therefore, only a certain number of species in a few genera need to be identified (see **Table 3.1**).

Since each substance exhibits its own typical Raman spectrum, Raman spectroscopy allows an unambiguous identification of single microparticles. However, it would be time-consuming to identify every microparticle on a given sample. Even in typical clean-room samples, a large number of particles

Table 3.1 Microorganisms which need to be identified for clean-room detection.

Name	Number of strains
Bacteria	
Bacillus pumilus	10
Bacillus sphaericus	10
Bacillus subtilis	19
Escherichia coli	10
Micrococcus luteus	8
Micrococcus lylae	3
Staphylococcus aureus	3
Staphylococcus cohnii	6
Staphylococcus epidermidis	7
Staphylococcus hominis	3
Staphylococcus warneri	3
Yeast	
Candida albicans	3
Candida glabrata	3
Candida krusei	3
Saccharomyces cerevisiae	3
Mold	
Aspergillus niger	3

would need to be analyzed. For microbiological investigations, only the biotic particles are of interest. They form between 1% and 0.001% of the total number of particles depending on the type of hygienic production (pharmacy, food, cosmetics). Analyzing all particles is therefore not appropriate. In addition, many particles agglomerate and biotic particles can often be found on the surface of larger abiotic particles. In order to localize and differentiate between these particles and in order to minimize the analysis time, pre-sorting is required. This section starts with a description of the concept of monitoring with techniques.

Subsequently the application of different Raman techniques in combination with SVMs to identify microorganisms is described. By using different Raman excitation wavelengths, it is possible to achieve various goals. Excitation in the UV region allows direct investigation of macromolecules such as proteins and DNA/RNA. Therefore direct correlation with genotaxonomic information is possible. Excitation in the visible range reveals information about the chemical composition of the whole cell. This allows a phenotypic characterization of microorganisms.

3.3.1
Monitoring of Biocontamination by Fluorescence Spectroscopy

The following specific requirements for an online monitoring method are essential for the assessment of biocontamination in a hygienic production environment:

- It must be possible to detect biocontamination immediately and online. Online detection is the basis for rapid intervention and for improved quality assurance in hygienic production.
- The detection and recognition of biocontamination must be reliable. Borderline cases, which do not permit firm conclusions as to whether a particle is biotic or not, must be rated as biocontamination from the point of view of quality assurance. Multistage recognition increases recognition certainty.
- Biocontamination must be detected and all kinds of colony-forming units must be recognized in order to obtain a safe basis for industrial use. The method must also be capable of detecting biocontamination which is difficult to detect using established methods.
- Differentiation between biocontamination and abiotic particles must be possible. The total particle contamination is an indication of the degree of cleanliness of the production environment. The differentiated recognition of biocontamination in the total contamination permits conclusions and correction to be made regarding contamination causes.
- A qualitative and quantitative assessment of biocontamination is an important criterion for applying the procedure in automated monitoring. This forms the basis for detecting deviations from given limits for maximum tolerated total numbers of bio-contaminants.
- It must be possible to automate the procedure to detect and identify microorganisms. The ability of the procedure used for in-line monitoring allows the easy automated monitoring of automated process cycles. Automated online monitoring of biocontamination opens up new aspects for quality assurance in the manufacture of hygienic products.

The necessary online aspect of the detection requires an optical detection system. There is no optical test method currently available which completely fulfills the above mentioned requirements. However, by combining the operating principles of glancing illumination and selective fluorescence detection, it is possible to develop a test set-up for the recognition of biocontamination on sampling surfaces. The test method is based on the fluorescent-optical detection of specific metabolites and cellular components (for example

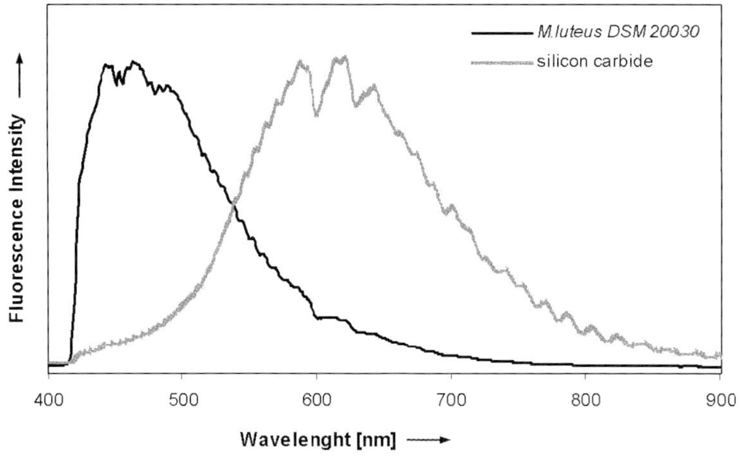

Figure 3.4 Fluorescence spectra of a biotic (*M. luteus* DSM 20030) and an abiotic (silica carbide) particle after excitation with UV light of 365 nm.

NAD(P)H and riboflavine, see **Figure 3.4**) [72–74]. Fluorescent stimulation of biocontamination is carried out in such a way that a representative section of the (rough) technical surface is illuminated in steps with light with wavelengths of 340 nm and/or 370 nm at specific glancing illumination angles. The NAD(P)H-based and/or riboflavine-based fluorescence emitted at 470 nm and/or 520 nm is detected by a CCD camera system [75–77]. The camera system is combined with an image-processing and evaluation system. Depending on surface roughness, the detection of cell fluorescence is increased when the glancing illumination is done above 10°. **Figure 3.5** shows the relative fluorescence intensity depending on the surface roughness and the illumination angle.

Validation of any method developed has to prove that the test set-up can be used as a basis for an industrial in-line procedure. Specific experiments are developed to verify the following requirements:

- reliable recognition of biocontamination
- detection of all biocontamination forming colonies.

The verification is done with a mixture of biocontamination and abiotic contamination. Test organisms are selected from the species of biocontaminants relevant for hygienic production. *S. cerevisiae*, *M. luteus* and *E. coli* were selected for the necessary experiments. Stainless steel particles and polystyrene particles were selected as abiotic test particles.

Reliable recognition of biocontamination could be achieved using the methodical multilevel test procedure (see **Figure 3.6**).

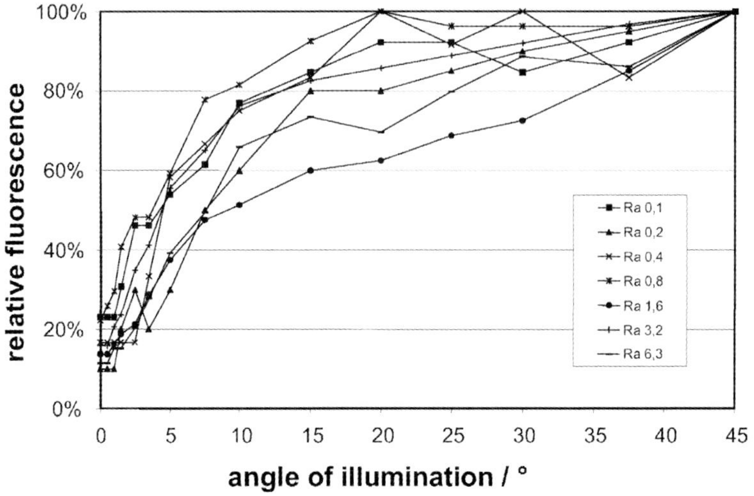

Figure 3.5 Intensity of detectable NAD(P)H-based fluorescence of biocontamination on surfaces of different roughness of sampling surfaces depending on the angle of illumination.

Recognition of the total contamination based on scattered light allows all selected test particles on the test surfaces to be recognized. Contamination by stainless steel particles can already be differentiated from biocontamination by *S. cerevisiae* in this first step. Polystyrene particles emit fluorescence in the same range as NAD(P)H. Using the stimulation of riboflavin-based fluorescence, the polystyrene particles are differentiated from the biocontamination in a second step.

The experiments with *S. cerevisiae*, *M. luteus* and *E. coli* demonstrate that a correct quantitative detection of biocontamination on technical surfaces, independent of category and species, is possible using the developed test set-up. The selected bio-contaminants can be adequately recognized based on NAD(P)H-based and riboflavin-based fluorescence (see **Figure 3.7**).

Evaluation of the results obtained from the experiments and their comparison with the required criteria showed that *in situ* recognition of biocontamination on relevant technical surfaces is possible under the applied conditions with glancing illumination. The developed test method provides a basis for further development of the test set-up for industrial application.

For processing the data, two fluorescence images (one to capture the NAD(P)H fluorescence signal and one to capture the riboflavin fluorescence signal) and a darkfield image are taken. In addition, the particle images are segmented in such a manner that for each particle three images are available, offering information in different spectral ranges. **Figure 3.8** shows the darkfield image of an air sample and a magnification of a region with a biotic

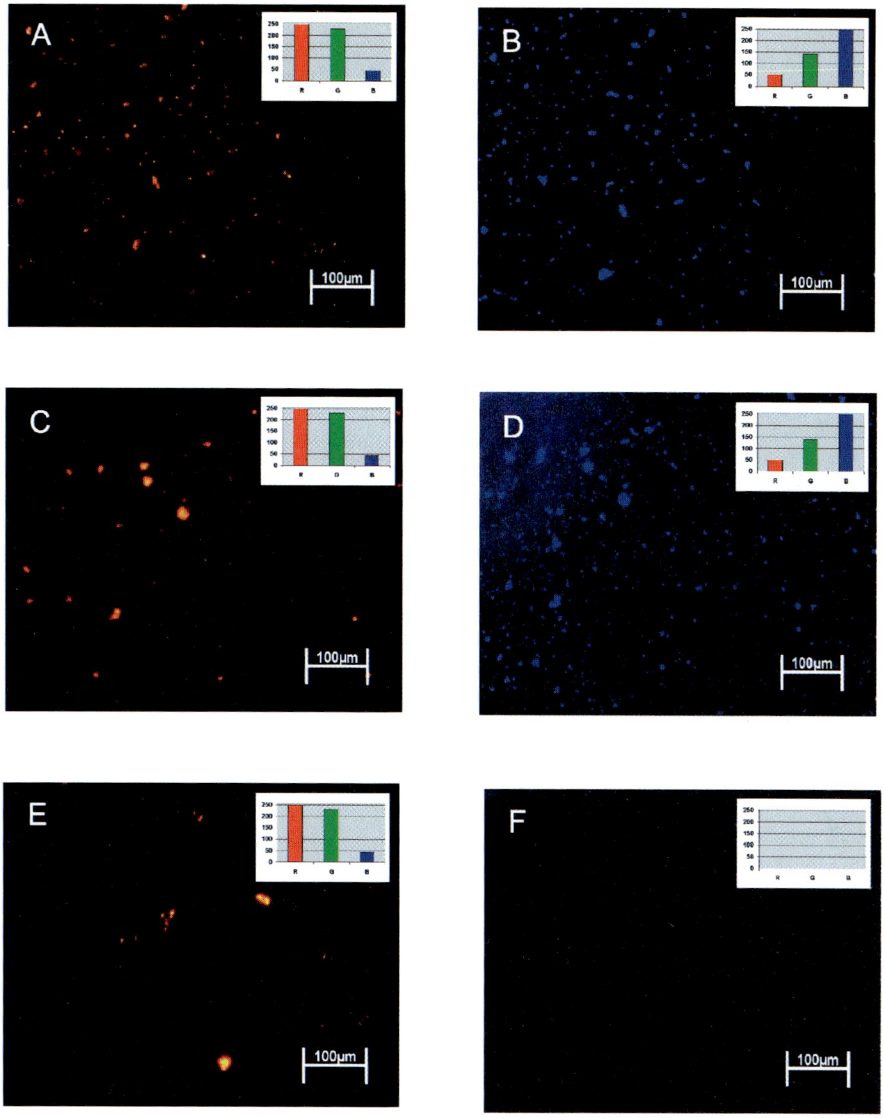

Figure 3.6 Tests with contaminated stainless steel surfaces (roughness $R_a = 0.8$ µm): *S. cerevisiae* scattered light (A) and NAD(P)H-based fluorescence (B); stainless steel particles with scattered light (C) and NAD(P)H-based fluorescence (D); polystyrene particles with scattered light (E) and NAD(P)H-based fluorescence (F).

particle (*B. sphaericus*) and an abiotic particle. The biotic and abiotic particles have different fluorescence characteristics, which can be seen in the fluorescence images filtered with two different emission filters. Here the biotic and abiotic particles can easily be separated. However, in several other cases differentiation is not possible based on the fluorescence information alone. This

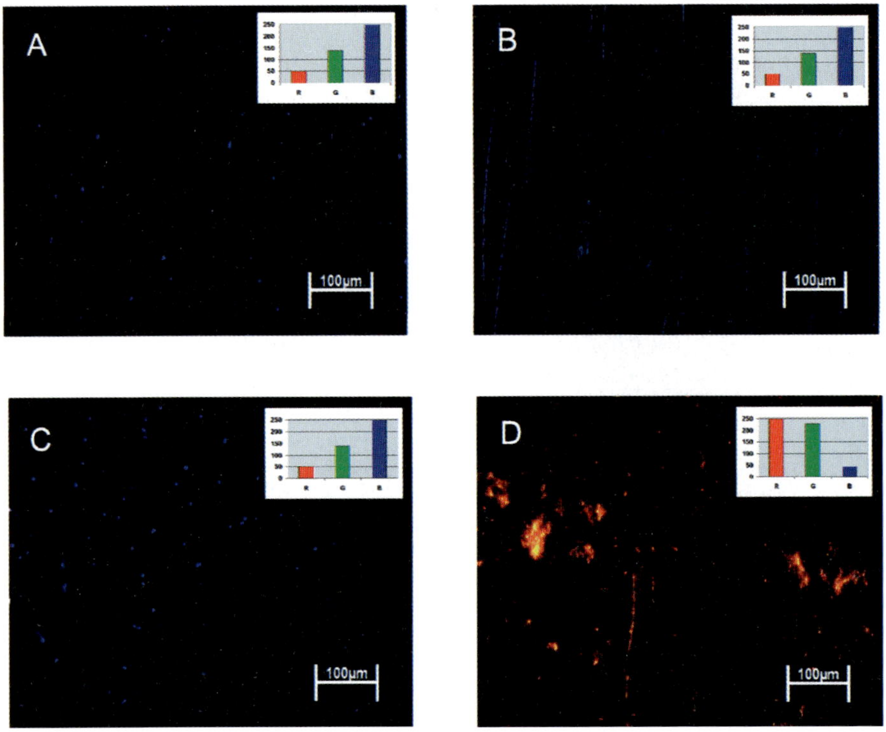

Figure 3.7 Tests with contaminated stainless steel surfaces (roughness $R_a = 0.8$ µm) A: *M. luteus*, NAD(P)H-based fluorescence, B: *E. coli*, NAD(P)H-based fluorescence, C: *S. cerevisiae*, *E. coli*, *M. luteus* and metal particles, NAD(P)H-based fluorescence; D: *S. cerevisiae*, *E. coli*, *M. luteus* and metal particles, scattered light.

can be best seen in **Figures 3.9** and **3.10**, where intensity histograms for 338 biotic particles (*S. cohnii*, *B. sphaericus*, *M. luteus*) and 301 abiotic particles (titanium dioxide) are shown after subtracting the background gray value. Based on those histograms, a threshold to differentiate between biotic and abiotic particles cannot be found using just the fluorescence information.

Therefore, the object images are processed by two successive steps, to improve the discriminatory capabilities. The first step takes into account structural information of each object and the second step uses an SVM as a more advanced classifier (see Section 3.3.2.2). For exploiting the object structure, gray-scale invariants [78] are calculated. Those invariants represent the particles based on their morphological structure, independently of their position and orientation in the image, and have been successfully applied to other biological samples [79]. Furthermore, by using invariants, it is also possible to link the three images of each particle, so that the spectral and the darkfield information are combined. Here two-point invariants are used with a multiplicative combination of the image points. The structural details of the parti-

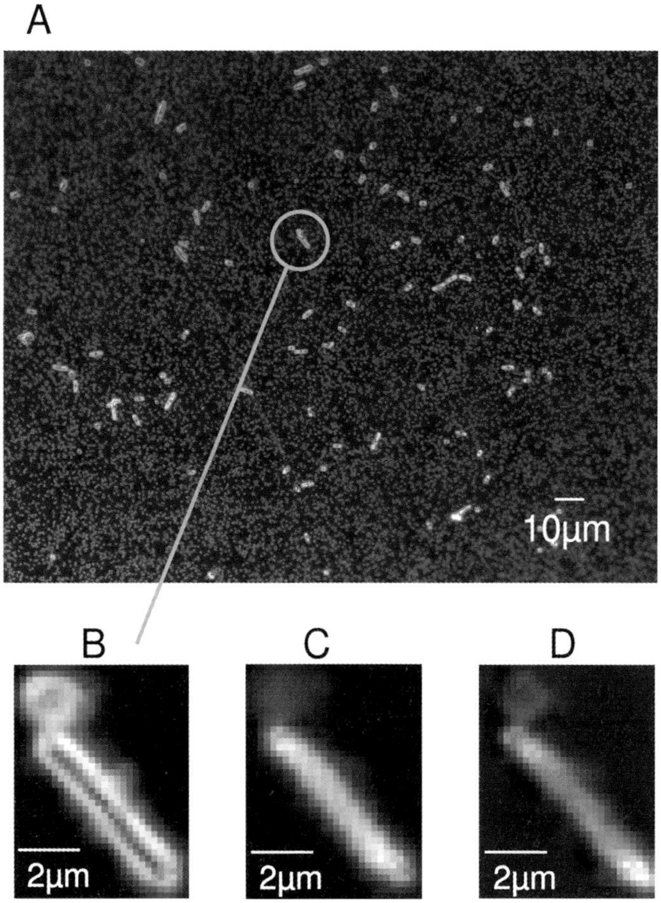

Figure 3.8 Darkfield image of an air sample (A) and a biotic and an abiotic particle extracted (B). Fluorescence images of the particles filtered at 470 nm (C) and 550 nm (D).

cles are taken into account by using a multiscale approach. Afterwards, the invariants are used as features for the discrimination. This step is based on an SVM as a classifier. To control the number of falsely classified particles, a weight is given to the SVM. In that way the SVM can be adapted to the problem and the number of wrongly classified biotic particles (false negatives) can be reduced. However, at the same time the number of wrongly classified abiotic particles (false positives) increases. As described before the classification by SVMs is based on the decision function that yields the decision values and a threshold. In **Figure 3.11** the decision values for the same fluorescence data described above are plotted.

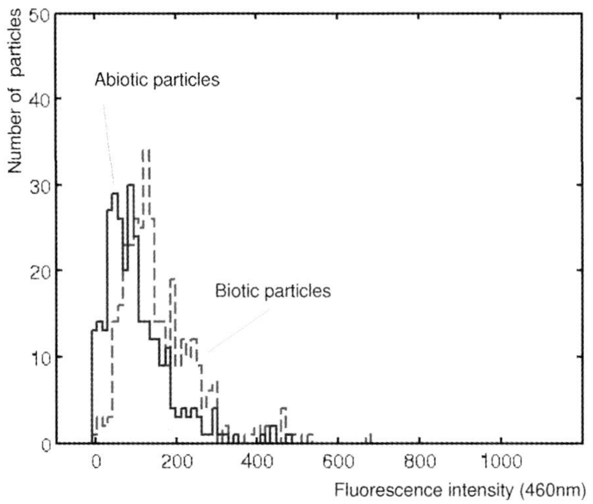

Figure 3.9 Intensity histograms of particles filtered with a fluorescence filter centered at 460 nm.

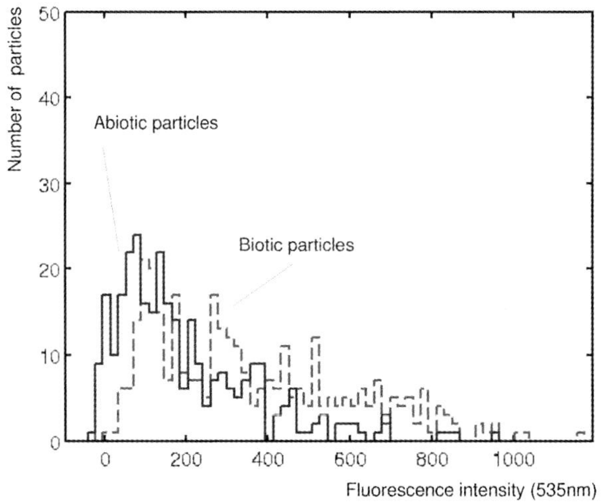

Figure 3.10 Intensity histograms of particles measured with a fluorescence filter centered at 535 nm.

Even when the histograms of the decision values overlap, two distinct peaks can be seen in the plot. In the case shown, the number of false negatives was 1.5% and the number of false positives was 34.5% while the weight was 0.3. A reduction in the number of false negatives will lead to a further increase in the false positive number, i.e., the result is always a tradeoff between the number of false positives and the number of false negatives. However, the

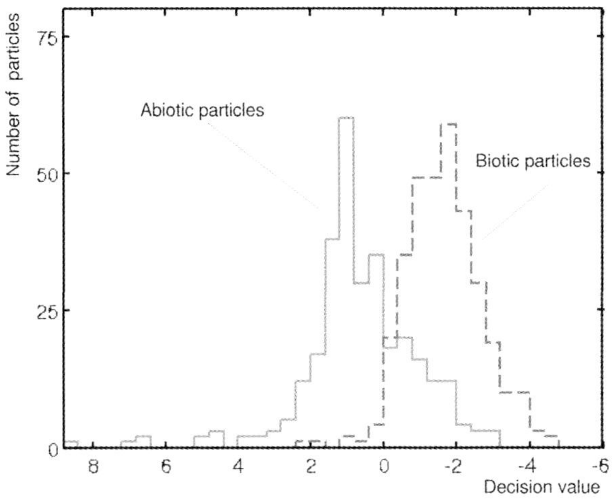

Figure 3.11 Histogram of the decision values for a support vector machine.

reduction in the total number of particles that need further treatment, means that less time is needed for the spectroscopic measurements and hence the overall processing time is reduced.

3.3.2
Identification of Single Microorganisms by Raman Spectroscopy Combined with Statistical Data Evaluation

Since each molecule exhibits its own characteristic Raman spectrum ("spectroscopic fingerprint") it is possible to identify molecules using their Raman spectra. Microbial cells from different species or strains differ in their chemical composition, e.g., the concentration and type of proteins, carbohydrates, DNA/RNA or lipids. These variations can be monitored by means of Raman spectroscopy [80] leading to an identification of the microorganisms.

Different excitation wavelengths can be used for Raman spectroscopy. Excitation in the UV region (below 400 nm) leads to resonance enhancement of macromolecules such as proteins and/or DNA/RNA. The UV-resonance Raman spectra therefore correlate with the composition of the DNA, which leads to a genotypic differentiation. Excitation in the visible range allows investigation of the whole chemical composition of the cell, e.g., proteins, DNA/RNA, lipids and carbohydrates. This overall description of the cell leads to a phenotypic characterization.

3.3.2.1 Genotype Identification of Microorganisms by Means of UV-resonance Raman Spectroscopy

The majority of applicable molecules for taxonomic investigations are macromolecules, in which DNA is of particular interest. As already mentioned signals of proteins and DNA/RNA are selectively enhanced using Raman excitation wavelengths in the UV region. Therefore, it is possible to obtain a genotaxonomic classification by using UV-resonance Raman spectroscopy. The GC-value of bacterial cells can be probed by UV-resonance Raman spectroscopy, since this method yields direct information of the DNA bases guanine (G), adenine (A), cytosine (C) and thymine (T) as well as the RNA base uracil (U). The GC-value is defined as the ratio of the sum of the bases guanine and cytosine to all DNA bases. The GC-value of bacteria varies from 30% for *Streptococcus* species to above 70% for some *Micrococcus* species [3, 81]. Since the GC-value is species-specific it can be used as a taxonomic criterion.

However, because UV laser radiation can cause photo damage of the sample, light exposure of the sample should be kept to a minimum. In order to minimize sample degradation, dried bacteria layers on fused silica plates are rotated and simultaneously moved in the x, y direction. Approximately 10^5 cells contribute to one Raman spectrum. This number of cells requires a microcolony, which can be obtained after about 5 h of cultivation.

In **Figure 3.12** the UV-resonance Raman spectra of *S. warneri* DSM 20316 are displayed for two excitation wavelengths (λ = 244 and 257 nm). The most obvious difference between the two spectra is the increased background of the spectrum excited at 257 nm. This background can be attributed to a higher fluorescence intensity of fused silica slides and the investigated microorganisms at 257 nm.

The UV-resonance Raman signals of both spectra can be assigned mainly to proteins and DNA: the vibrations for the aromatic amino acids phenylalanine, tyrosine and tryptophan can be found at 1122, 1324 and 1607 cm^{-1}. The signals of the DNA bases can be observed at 1247 (G, A, U), 1324 (A, G), 1359 (T), 1475 (G, A), 1524 (C), 1567 (G, A), and 1638 cm^{-1} (T).

The UV-resonance Raman spectra recorded for an excitation wavelength of λ = 244 nm of one strain of nine species are displayed in **Figure 3.13A**. The spectra are very similar, making a chemometric pattern recognition method necessary for identification (see below). The UV-resonance Raman spectra with an excitation wavelength of λ = 257 nm of the same strains are displayed in **Figure 3.13B**.

The classification of UV-resonance Raman spectra can be performed with a variety of different chemometrical methods. Owing to the large number of bacterial cells for each measurement, even unsupervised methods are possible. Since the identification of a single bacterium (see Section 3.3.2.4) is only possible with a SVM, this method is presented here for better comparison.

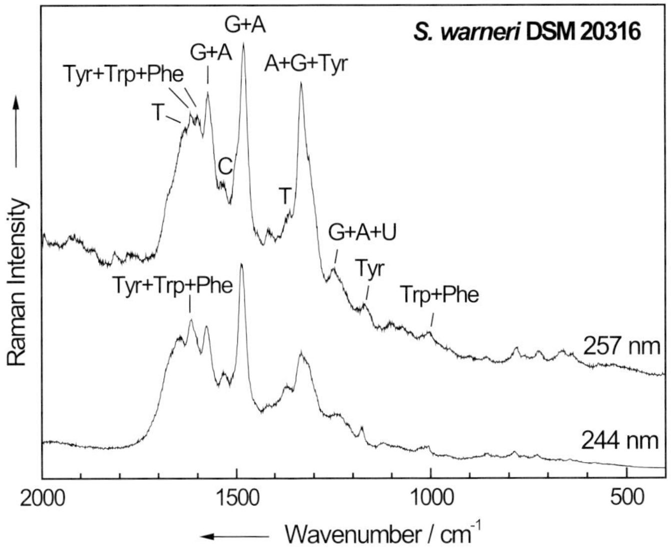

Figure 3.12 Comparison of UV-resonance Raman spectra from *S. warneri* DSM 20316 with excitation wavelengths of 244 and 257 nm.

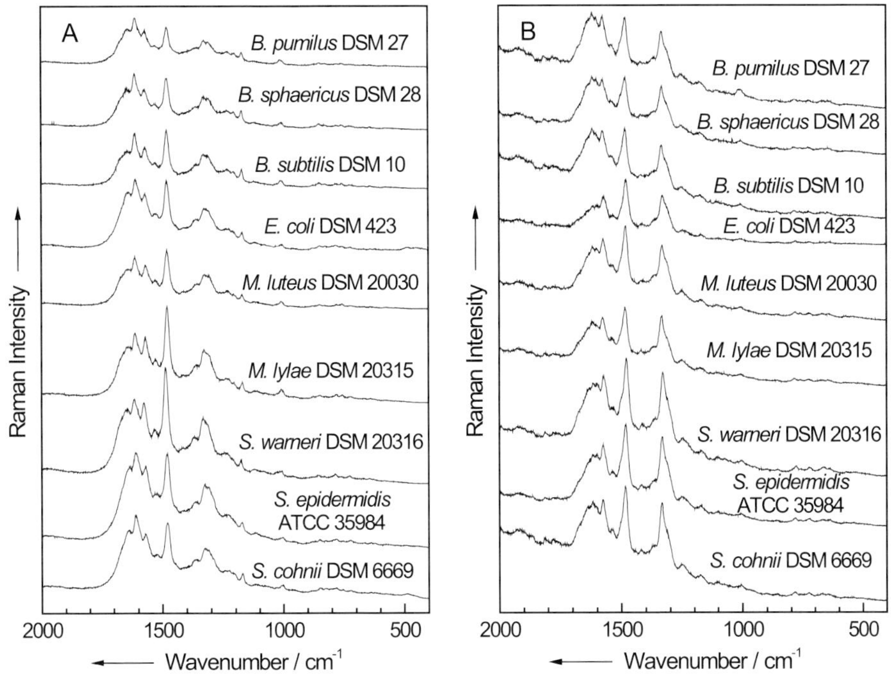

Figure 3.13 UV-resonance Raman spectra of different bacterial strains with an excitation wavelength of $\lambda = 244$ (A) and 257 nm (B).

Chemometrics

The term chemometrics was coined in 1972 by Svante Wold and Bruce R. Kowalski. Chemometrics describes the application of statistical and mathematical methods to the analysis of chemical data in order to properly characterize the chemical system. Chemometric methods can be applied for the planning, development, choice or evaluation of chemical procedures and experiments. While the analysis of a small amount of data or highly distinct spectra is manageable without using chemometric methods, they are indispensable for the classification of capacious datasets such as the spectra of different bacterial strains.

Chemometric methods can be divided into univariate and multivariate methods. In univariate approaches, only one unique feature (e.g., temperature, pH value) is important, whereas in multivariate techniques many features and measuring points (e.g., combination of univariate features or spectra) are considered.

Furthermore, chemometric techniques can be divided into supervised and unsupervised methods for classification. When applying unsupervised techniques, no information about class affiliation (features) of the objects (spectra) is included in the calculation of similarities. These methods are used if the affiliation of the objects is unknown or if the dataset is investigated for native similarities. Hierarchical cluster analysis (HCA) or principal component analysis (PCA) are frequently applied methods of this type. On the other hand, when the features of the samples are considered in addition to the spectral information, so-called supervised methods are used. These supervised methods are applied if the classification is followed by the identification of unknown samples. To do so, a classification model with known samples is generated and the quality of the model is subsequently tested with samples with known features (validation). Common methods are the *k*-nearest neighbor (K-NN) method or the soft independent modelling of class analogies (SIMCA) method.

More recent methods are artificial neural networks (ANNs) and support vector machines (SVMs). Their major goal is to use spectra to build a model for estimating or predicting the properties of a system.

For chemometric investigations the spectra have to be comparable, generating a need for preprocessing with the intention of reducing or even eliminating irrelevant data arising from changes or systematic variations such as fluctuating laser power, noise, cosmic spikes or fluorescence background. The preprocessing of the measurements (e.g., normalization, smoothing, baseline correction) is a crucial point, since it can influence the quality of the model enormously.

Unsupervised Methods

For classification with unsupervised methods, class affiliation is not included in the process. There are different HCA methods that arrange the measured samples in clusters by means of their spectral distances, which are a measure of similarity. The smaller the spectral distances the more similar are the spectra; identical spectra have a spectral distance of zero. The result of the clustering process is represented in a dendrogram that is adopted from taxonomic dendrograms. To accomplish an HCA, the spectral distances of all spectra are calculated with the help of different methods. Afterwards, the two most similar spectra are aggregated into one cluster. This is followed by the calculation of the distances between this cluster and all other spectra or clusters to combine the next two most similar objects. This procedure is repeated until all spectra are merged into one cluster. It is thus of decisive relevance how a third object is tied to an already existing cluster.

The aim of a PCA is to reduce the dimension of the dataset while maintaining the maximum information. To realize that, a coherence between the data must exist. The data are transferred into principal components so that the first principal component (PC) describes the biggest portion of variance, while higher principal components contain mainly noise. Every sample has a set of coordinates compared to the new axes (PCs). The coordinates of the samples with respect to these new PCs are called scores and the contributions of the original variables to the PCs are denoted as loadings. For applying a PCA, a covariance or correlation matrix needs to be computed. To solve this eigenvalue problem, different numerical methods were developed. The calculated eigenvectors of the covariance or correlation matrix correspond to the weighting of the original variable at the generation of the principal components and therefore to the loadings. If the loadings are multiplied with the centered data matrix, the coordinates of the samples in the new coordinate system of the PC (scores) are obtained. With the help of these scores, 2D or 3D figures can be generated that visualize the grouping of similar samples. For a classification of spectra with a PCA, the number of principal components to describe the system is of great importance. If too few principal components are included, the dataset will be described incompletely. In contrast, if too many PCs are involved, the model will be overfitted because many random variations such as noise are included in the description.

Supervised Methods

Applying supervised classification methods means to recognize that samples belonging to one class exhibit similarities to each other. Therefore, class affiliation is considered during model evaluation. The intention of a supervised method is the identification of unknown samples by grouping the unknown object into existing classes. First of all, a model is trained with known samples. For estimation of the model quality a validation is necessary. If the model is too detailed – random changes in the dataset (e.g., noise) are considered for the segregation of classes – many unknown objects will be classified into the wrong class – this phenomenon is called overfitting. The opposite, underfitting, means that the model is too imprecise to separate the objects. The development of a model is the tightrope walk between overfitting and underfitting.

The aim of validation is to decide if the model is well chosen by determining the classification error rate. When applying the resubstitution method, the objects taken for model generation are used for the classification to estimate the error rate. Normally, the determined error rates are too low because the objects are known. Therefore, no prediction on the classification quality of unknown objects can be made. In contrast, with a hold-out method (test-set method) the dataset is divided into two parts, one serving for the model calculation, and the other one being used for validation. It should be mentioned that a large dataset is necessary for test-set methods because only a part of the data is used for modelling and also that the choice of objects for training and testing should occur randomly. A better method, exhibiting a realistic classification error rate, is the leave-one-out method. Here, every object is removed once from the dataset. The model calculation is carried out with the remaining data. Afterwards, the isolated object is classified and the result is saved. Because a new model calculation has to be performed for every sample, this method is very time-consuming. Therefore, mixed techniques have been developed from the leave-one-out method and the hold-out method by randomly removing some objects of the training set.

One supervised classification method is the k-nearest-neighbors technique. This is a model-free approach. An object will be classified into that group where the majority of the closest objects can be found. The Euclidean distance between the unknown sample and all other objects of a known class is calculated as a measure of the spectral distance. The unknown object will be classified to that class that most of its k nearest neighbors belong to. The choice of the dimension of the parameter k is important. If the value is too small, outliers dominate the classification. If the value is too great, the group with the largest number of objects gains too much weight.

Another important supervised method is called the soft independent modelling of class analogies (SIMCA). This technique uses PCA. For every class the principal component model will be calculated separately to obtain the shape and position of the classes that represent the objects. The result are multidimensional geometrical structures that enclose the particular classes. Unknown objects are classified on the basis of the Euclidean distances between the unknown sample and the other samples with known class affiliation, depending on which geometrical structure they are situated in.

Artificial neural networks (ANN) are supervised methods that mimic biological neuronal nets. They consist of neuron layers. From the different input values one output value is calculated in the neural net. All neurons of one layer are connected to all neurons in the previous and the subsequent layer. Each neuron adds weighted input signals. The result is transformed by a nonlinear function and generates the output. This function is a threshold function that produces a high or low output value depending on the amplitude of the sum. Neural networks are suitable for the classification of nonlinear systems. Since they also model random information they are inclined to overfitting.

Support vector machines (SVMs) are supervised methods which are also applied for the classification of big datasets. This method is widely applied in different areas of pattern recognition and will be described in detail in Section 3.3.2.2.

3.3.2.2 Support Vector Machine (SVM)

For the classification of the Raman spectra a support vector machine is used. This classifier, based on the theory of statistical learning [82], is already widely applied in different areas of pattern recognition. However, it is not widespread in the field of chemometrics and only a few articles have been published yet (see for example Refs [83,84].).

SVM Linear Case

Since all multiclass classification tasks can be decomposed into two-class problems, the classification approach based on SVMs can be boiled down to finding a decision function to separate two classes. In the linear case, the decision boundary for a 2D problem is a line. As an example, the data samples $\mathbf{x}$ of a two class problem in a 2D space are shown in **Figure 3.14** while the centered line shows the decision boundary, where the decision function becomes zero and the enveloping two lines define the margin. When dealing with the classification of spectra, the data samples normally have a certain number of spectral values d which leads to a classification problem in a d-dimensional space. This space, where the classification is carried out, is normally referred to as the so-called feature space. For these high-dimensional problems, the decision boundary can be seen as a hyperplane. Generally speaking, the decision function in any dimension is defined as

$$f(\mathbf{x}) = \mathbf{w}^T\mathbf{x} + b \tag{3.1}$$

where the column vector $\mathbf{x}$ is given as a sample and the vector $\mathbf{w}$ is the normal of the separating plane. The value b is a bias value, which can be seen as the

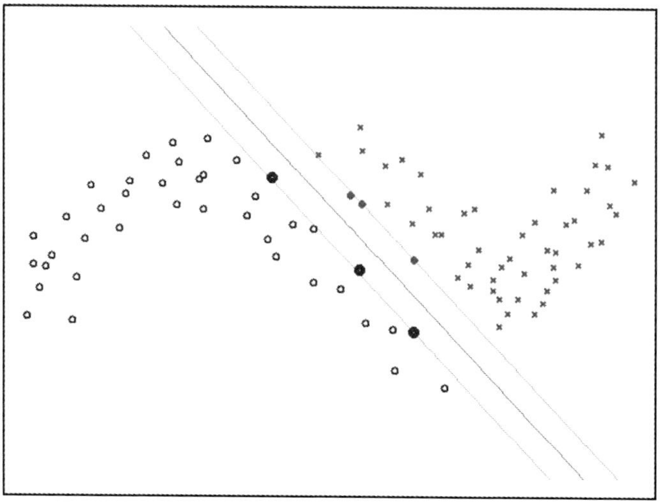

Figure 3.14 Classification of a two-class problem by a linear SVM in a two-dimensional sample space.

offset from the origin. The value $f(\mathbf{x})$ for a certain vector $\mathbf{x}$ is also called the decision value. During the classification step labels are given to the unknown test spectra. In cases where for a certain test spectrum $\mathbf{x}$ the condition $f(\mathbf{x}) > 0$ is fulfilled, the label $y = +1$ is given and the spectrum is assigned to the corresponding class. In all other cases the label $y = -1$ is given. Compared to other approaches, SVM does not try to estimate or model a distribution of a class, but focuses on the margin between two classes. Between all possible hyperplanes that can be found to separate the two classes, the boundary is chosen that defines the largest margin. To find this decision boundary, a quadratic optimization problem with linear constraints must be solved by applying Lagrangian theory. Afterwards, the decision function can be written as

$$f(\mathbf{x}) = \sum_{i=1}^{N_{SV}} \alpha_i y_i \mathbf{x_i}^T \mathbf{x} + b. \tag{3.2}$$

The Lagrange multipliers α_i are given as a solution of the optimization problem and y_i are the labels corresponding to the given training vectors $\mathbf{x_i}$. A remarkable point in Eq. (3.2) is, that the sum need not to be calculated for all training samples, but only for the N_{SV} samples that have a corresponding Lagrange multiplier that is above 0. Samples that fulfil this requirement are called support vectors.

The expression $\mathbf{x_i}^T\mathbf{x}$ in Eq. (3.2) is nothing but a scalar product of two vectors. Encapsulating this scalar product into a so-called kernel function

$K(\mathbf{x_i}, \mathbf{x}) = \mathbf{x_i}^T\mathbf{x}$ leads to an other way of writing Eq. (3.2):

$$f(\mathbf{x}) = \sum_{i=1}^{N_{SV}} \alpha_i y_i K(\mathbf{x_i}, \mathbf{x}) + b. \tag{3.3}$$

As mentioned before, the aim of SVM training is to find a function that discriminates two classes, while looking for a maximum margin between the classes. In cases where the classes overlap, a cost value C is used during the training to punish samples of a class that lie on the wrong side of the boundary. The boundary is completely described by the support vectors, which are plotted in bold symbols in **Figure 3.14**. This leads to the advantage, that for the classification only the support vectors are necessary, and all other training data can be neglected and removed. Moreover, classification based on SVMs offers further important advantages. First, the solution of the quadratic optimization problem yields a global optimum. This is an enormous advantage compared to other classifiers, such as artificial neural networks, where usually only a local optimum is reached. Second, the number of parameters that need to be estimated for an SVM does not depend on the dimensionality of the data, as is the case for classifiers based on estimating a covariance matrix. When using a linear SVM, only one parameter (a cost value) needs to be chosen. Finally, an extension of the SVM to a nonlinear case can easily be achieved by changing the kernel function K. This fact is addressed below.

In a multiclass problem, i.e., when not yet classified samples must be assigned to one out of several different classes, the classification step is split up into several two-class problems. During the experiments within the OMIB project, the so-called one-versus-one approach was used. In that approach the classes are trained successively against each other and it is noted how often the unknown sample is assigned to each class. Afterwards the sample is classified as belonging to the class with the highest count.

When using a linear SVM, the classification output can also be interpreted geometrically. This is shown in **Figure 3.15** where spectra of two classes and the weighting curve from the corresponding two-class SVM are plotted. The spectra in the upper part of the figure belong to class *M. luteus* DSM 348 while the spectra in the lower part belong to the class *B. subtilis* DSM 10. The curve with the classification weights is plotted in the center of the figure. Peaks in the weighting curve show regions that are important for classification. The orientation of the peaks shows which class the important regions belong to. It can be seen that carotenoid peaks of *M. luteus* at about 1531 and 1156 cm^{-1} are quite important for differentiation of the classes. Nevertheless, bacterial cells that do not contain a significant amount of a pigment, or where the pigment is bleached, are also assigned correctly since additional bands (e. g. 1455 cm^{-1}) are also characteristic for *M. luteus* but do not belong to carotenoids.

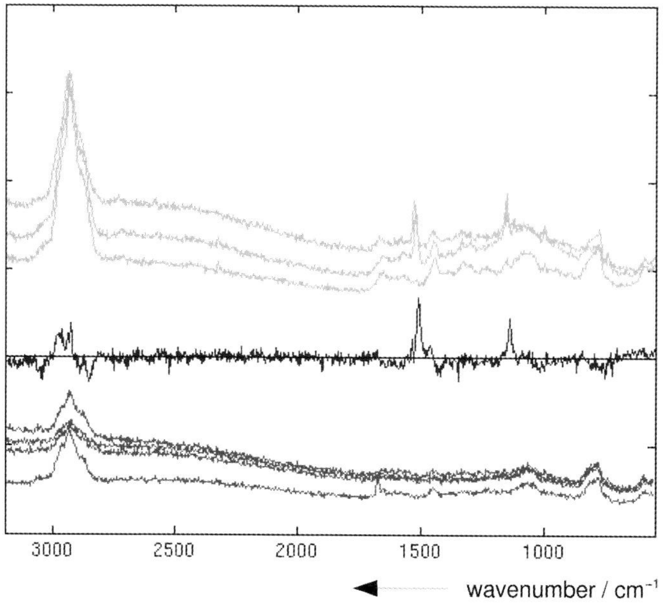

Figure 3.15 Class weights of a linear SVM for spectra of the class *B. subtilis* and *M. luteus*. The spectra of *B. subtilis* are plotted in the lower part of the figure; the spectra of *M. luteus* in the upper part of the figure. The curve representing the SVM weights is plotted in the center.

SVM Nonlinear Case

By using a nonlinear transformation of the data, mapping into a high (or even infinite) dimensional space can be carried out. While the separating boundary in the transformed space is a hyperplane, the corresponding separating boundary in the original space is a hypersphere. This allows the classification of rather complex problems. The mapping and calculation of a distance measure in the feature space is achieved implicitly by using a nonlinear kernel function in Eq. (3.3). This is an advantage, because no explicit transformation needs to be calculated. The conditions that must be fulfilled by the kernel function are stated in Mercer's theorem [82]. A frequently used kernel is given by the so-called Gaussian radial basis function (rbf) kernel, that is defined as:

$$K(\mathbf{x}_i, \mathbf{x}_j) = \exp(-\gamma \|\mathbf{x}_i - \mathbf{x}_j\|^2). \tag{3.4}$$

Compared to the linear kernel described above, the Gaussian rbf kernel has a further parameter γ, that needs to be adjusted.

Using *a priori* Knowledge with SVM

In general, the classification results can be improved by exploiting *a priori* knowledge about the data. For example, knowledge about transformations that the samples might show without changing the meaning of the sample can be incorporated into the classification step. This transformation can be described as a manifold in a parameter space. In spectroscopy, the amount of measured sample area or the measuring time can vary. This leads to different intensity values for different spectra of the same species, which can result in a misclassification of the spectra. However, by including knowledge about scaling transformations, this fact can be considered. Between different kinds of methods for exploiting knowledge, the tangent distance approach has shown to be useful for classification tasks with SVMs [85]. The standard Euclidean distance of two different vectors in the SVM kernel is replaced by the difference of the tangent spaces (linear approximations) of the set of transformation manifolds. In the area of pattern recognition, these distances are called tangent distances. All kinds of different transformation can be thought of, but for spectral data two transformations are reasonable [86]. First, an approximation of the scaling, as mentioned before. Second, combined with the scaling, transformations referring to the baseline shift are used. The baseline variation is normally due to a fluorescence background during the measurement process. Since the variation of the baseline can be approximated by some nonconstant low-frequency curves [87], polynomials of low degree are chosen to model the baseline shift. Lagrangian polynomials up to the degree of 3 are used as a basis for the polynomials. This can be seen in **Figure 3.16**, where a spectrum and its transformations are plotted. The first column shows only the scaling and no baseline transformations. The figures in the following columns give the transformation according to the four Lagrange baseline shifts. The images in the middle row show the original sample.

Experimental results show that the classification rate can be improved by up to 5% when using tangent distances. Moreover, it turns out that the improvement depends on the choice of the base line polynomials, so the selection of the base polynomials is a crucial point in this approach.

Novelty Detection with SVMs

Since SVMs are not based on approximating distributions (or parameters of distributions) of different classes, there is no straightforward way to reject new data that do not belong to known classes. However, a SVM can be trained for the known data, such that the decision boundary encloses the data. Geometrically speaking this can be seen as a sphere that encapsulates the known samples. During the training process the optimal sphere, i.e., one with a minimum volume, is found. New data samples that are not included in this sphere are detected as unknown.

Similar to the classification approach, a nonlinear kernel can be applied, i.e., the data are mapped into a feature space via a kernel transformation. In that way the sphere can be adapted more precisely to the data distribution. **Fig-**

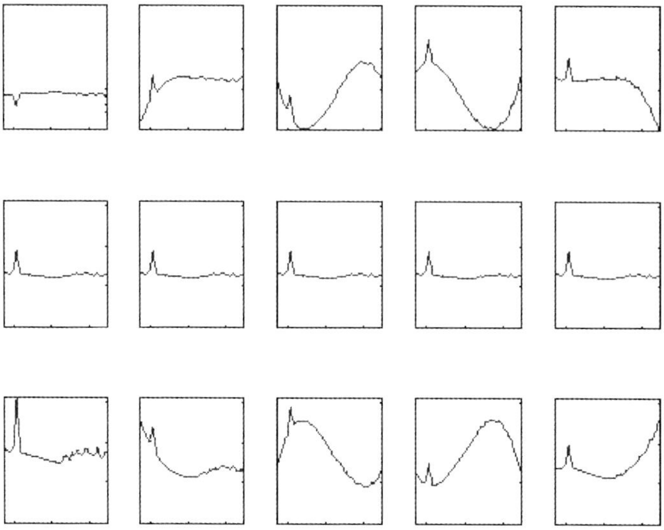

Figure 3.16 Scaling transformation (left column) as well as Lagrangian baseline shifts (center columns and right column) applied to a spectrum. Figures in the middle row show the original spectrum.

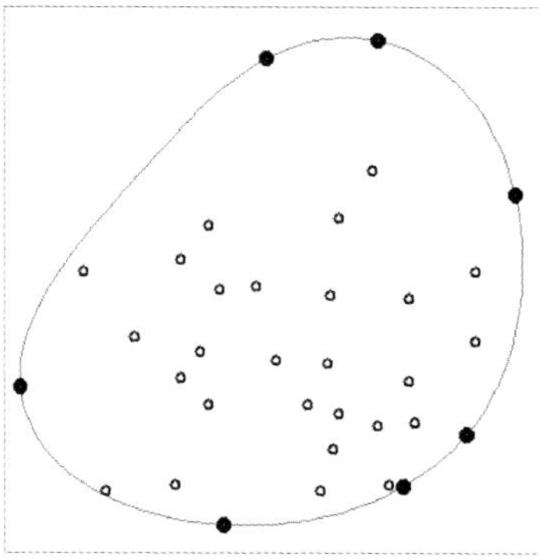

Figure 3.17 Dataset and its enclosing sphere, while using a SVM with a nonlinear (rbf) kernel.

ure 3.17 shows a training set and the enclosing sphere, where the spheres are adapted differently, due to the use of different kernels. Experiments showed that novelty detection is mostly disturbed by variations of the baseline of different spectra. To deal with this variance a baseline correction was calculated as a preprocessing step.

Spectra that were measured with a laser excitation wavelength at 257 nm gave good classification results without any preprocessing. The classification rate for 18 different bacterial strains can be seen in **Table 3.2**. Altogether, the dataset consists of 705 spectra. SVMs are used for the classification as de-

Table 3.2 Results of SVM classification for spectra measured with an excitation wavelength of 257 nm.

Name (DSM number)	Total number of spectra	Number of wrongly classified strain spectra	Recognition rate for strains (%)	Number of wrongly classified species spectra	Recognition rate for species (%)
B. pumilus DSM 27	45	0	100.00	0	100.00
B. pumilus DSM 361	35	0	100.00	0	100.00
B. sphaericus DSM 28	35	2	94.29	0	100.00
B. sphaericus DSM 396	58	0	100.00	0	100.00
E. coli DSM 499	45	1	97.78	0	100.00
E. coli DSM 498	53	1	98.11	1	98.11
M. luteus DSM 348	40	0	100.00	0	100.00
M. luteus DSM 20030	31	0	100.00	0	100.00
M. lylae DSM 20315	39	1	97.44	1	97.44
M. lylae DSM 20318	45	0	100.00	0	100.00
S. cohnii DSM 6669	39	0	100.00	0	100.00
S. cohnii DSM 20260	43	0	100.00	0	100.00
S. cohnii DSM 6718	33	0	100.00	0	100.00
S. cohnii DSM 6719	32	0	100.00	0	100.00
S. epidermidis DSM 1798	41	0	100.00	0	100.00
S. epidermidis ATCC 35984	28	1	96.43	1	96.43
S. warneri DSM 20316	29	1	96.55	1	96.55
S. warneri DSM 20036	34	0	100.00	0	100.00
Average recognition rate			98.92		99.36

scribed above. Here the nonlinear Gaussian rbf kernel with a cost value of 10^6 and a γ of $1.5 \cdot 10^{-8}$ was used. Moreover, the whole training set was normalized by subtracting the mean value and dividing by the standard deviation of all spectra for each grid point. This step results in a homogenization of the training data. The classifier was then tested with a leave-one-out test. During this test, one spectrum is left out of the dataset. The remaining spectra are used for the training and the left-out spectrum is then classified. Afterwards another spectrum is left out and the training and classification is repeated. In that way the largest amount of data (all data except one) is available for training, which leads to a robust characterization of the dataset.

First attempts to classify bacteria with UV-resonance Raman spectra were made by Jarvis et al. [27]. Here, five microorganism strains can be discriminated on the genera level. For the discrimination, both hierarchical cluster analysis (HCA) and principle component analysis (PCA) are used. López-Díez et al. [28] performed a classification of closely related endospore forming bacteria of the genera *Bacillus* and *Brevibacillus* with UV-resonance Raman spectroscopy in combination with HCA and PCA on a species level. In this study, it was not possible to discriminate between certain species. The study presented here reports on a dataset including 18 strains of eight species that can be identified with an average recognition rate of about 99% on the strain and species level.

Raman Set-up and Confocal Raman Spectroscopy

Nowadays, the micro-Raman technique is a well established method for both the investigation of samples in the order of picograms (10^{-9} g) or even less, and the spectroscopic imaging of sample surfaces. A typical micro-Raman set-up is displayed in **Figure 3.18**.

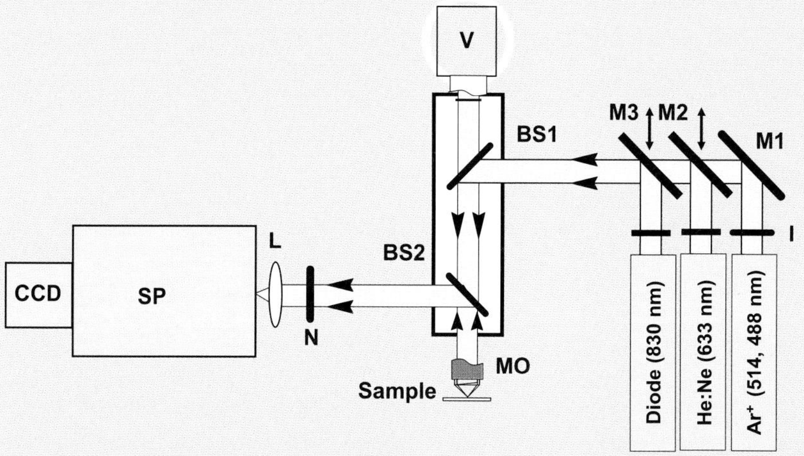

Figure 3.18 Typical micro-Raman set-up: I: Interference filter, M1–M3: mirrors for different laser excitation: BS1, BS2: beam splitter, V: video control, MO: microscope objective, N: Notch filter, L: lens, SP: spectrometer, CCD: charge-coupled device.

A microscope objective (MO) serves to focus the laser beam on the sample (S) and to collect the scattered light. For Raman excitation, different laser lines can be used depending on the investigated substances. An interference filter (I) is used to cut off plasma lines or side bands from the laser light. Microscope objectives with a high numerical aperture and high magnification are used to focus the light down to the diffraction limit ($<1\ \mu m^2$). This allows the investigation of small sample volumes or heterogenous samples such as biological cells or tissues. In addition to laser excitation, an optical survey of the sample is obtained with the help of a camera (V). The scattered light from the sample is collected by the same microscope objective, the elastically scattered light is cut off with a Notch filter (N) and the residual light is focused via a lens (L) onto the entrance slit of a spectrometer (SP) and is then detected with a CCD (charge-coupled device) camera.

In order to obtain spatially resolved information from the sample, Raman mapping or imaging can be applied. With these techniques, 2D spectral information from the sample can be received. It is possible either to scan the sample point by point and to calculate the distribution of a marker band over the scanned area, or to illuminate the whole sample and to measure at only one spectral position.

A major problem with many micro-Raman studies is obtaining a good spatial resolution perpendicular to the optical axis (lateral) as well as along the optical axis of the microscope. In a conventional microscope, the entire field of view is uniformly illuminated and observed. With a confocal arrangement the resolution along the optical axis can be increased. Confocal microscopy uses a pinhole, which isolates the light originating from a small region of the sample coincident with the illuminated spot. This efficiently eliminates the contributions from out-of-focus zones. Thus, the advantage of spatial filtering by an optically conjugated pinhole diaphragm is achieved (**Figure 3.19**). Usually, a physical aperture such as an adjustable pinhole is applied.

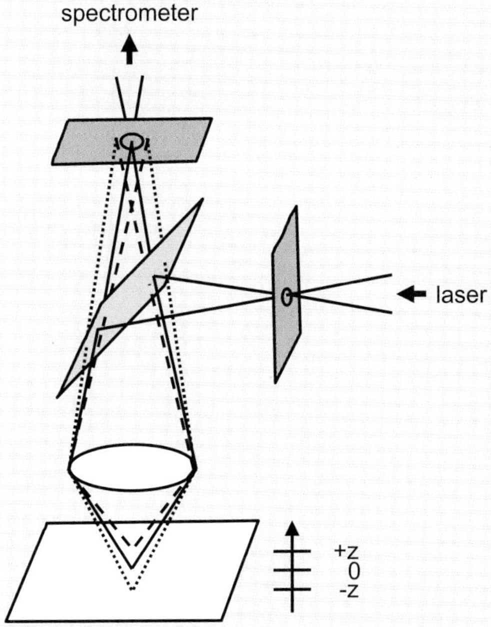

Figure 3.19 Confocal micro-Raman set-up (focus plane: solid line, focus above focal plane: dotted line, focus below focal plane: dashed line).

3.3.2.3 Analysis of Bulk Microbial Samples by Means of Micro-Raman Spectroscopy with Excitation in the Visible (Phenotype Identification)

An alternative approach to UV-resonance Raman spectroscopy on bacteria described in Section 3.3.2.1 is the utilization of visible Raman excitation wavelengths, from which information about the chemical composition of the whole cell can be obtained, allowing a phenotypic characterization of microorganisms.

In order to minimize the cultivation time, single-cell Raman microbial analysis is the method of choice. Micro-Raman spectroscopy excited in the visible spectral region allows the investigation of particles of approximately 1 μm in size. Since bacteria are not destroyed using visible Raman excitation wavelengths, it is possible to investigate a single spatial spot on a prepared multilayer. This lowers the number of investigated bacteria from 10^5 cells for UV-resonance Raman spectroscopy to approximately 30 cells for the visible Raman excitation wavelength.

In this study, excitation wavelengths of 532, 633, 785 and 830 nm were used. Wavelengths in the far red or even NIR region (785 and 830 nm) have the advantage of minimal fluorescence excitation, improving the Raman signal. In the following the results for an excitation wavelength of 532 nm are shown. The results of other wavelengths are quite similar.

Before investigating single bacteria, bulk samples were measured in order to prove the overall discrimination ability of this method. In **Figure 3.20** micro-Raman spectra of two strains from nine different species, respectively, are shown. For an excitation wavelength of 532 nm all chemical components of the cell, such as DNA/RNA, proteins, lipids and carbohydrates, were measured leading to a phenotypic classification of bacteria.

The micro-Raman spectra show a huge fluorescence background with less pronounced signals than the UV-resonance Raman spectra. In addition, this fluorescence background is highly variable from strain to strain. Beside the C-H vibrations at approx. 3000 cm^{-1} most of the strains, e.g., *S. warneri*, exhibit signals due to protein (amide-I at 1665 cm^{-1}) and DNA components (1588 cm^{-1}). In addition to these signals also bands of carbohydrates (1124 cm^{-1}) and aromatic amino acids (phenylalanine 1005 cm^{-1} and tryptophan 747 cm^{-1}) can be observed. The colored strains such as *M. luteus* exhibit additional signals at 1524, 1159 and 1003 cm^{-1} which can be assigned to the carotenoid sarcinaxanthin.

The bulk data spectra were preprocessed by a median filtering and normalization. It was found experimentally that using the C-H peak for normalization gave the best results. The classification was achieved by using a linear SVM. The reported results were acquired by a leave-one-out test (described above). The recognition rate at strain level (98%) was already very good and improved only marginally for recognition at species level, as can be seen in **Table 3.3**.

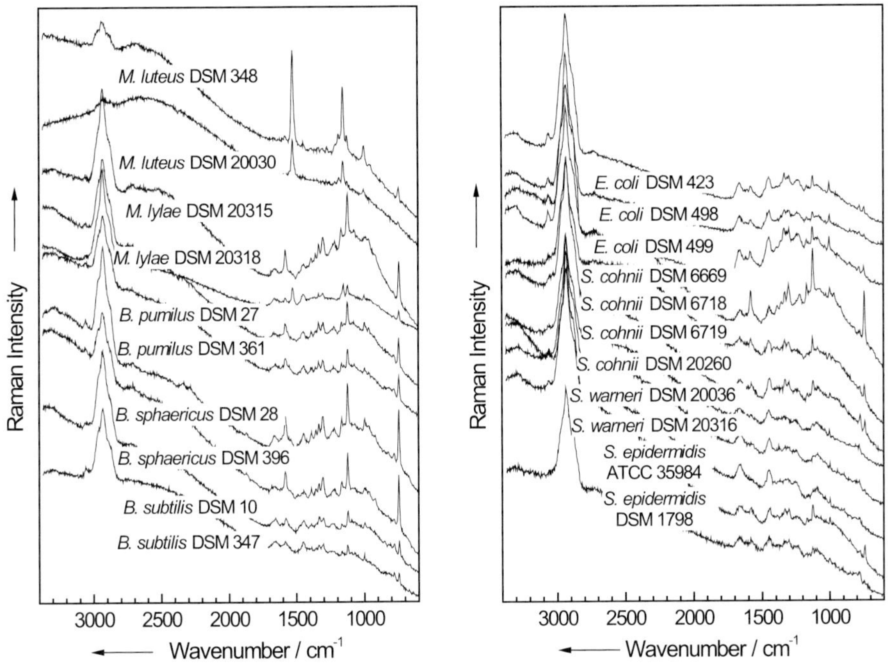

Figure 3.20 Micro-Raman spectra of bulk samples of different bacterial strains with an excitation wavelength of $\lambda = 532$ nm.

Several approaches for a Raman spectroscopic identification of bulk bacterial material have been reported. Microcolonies from blood cultures have been investigated by means of IR and micro-Raman spectroscopy in combination with HCA, linear discriminant analysis (LDA) and artificial neural networks (ANN) in order to differentiate between bacteria and yeasts on the species level [48, 54]. Hutsebaut et al. [50] used smears of bacteria on CaF_2 for micro-Raman spectroscopy. Here, different strains of the genus *Bacillus* were classified with HCA and LDA with respect to different culture media, cultivation times and temperatures, with an average recognition rate of 70 to 100%, depending on the cultivation condition. Dried and grinded bacteria were used for a FT-Raman spectroscopic study of different species with HCA [9]. Food contamination was simulated by Yang et al. [57] by analyzing bacterial smears on apples. The analysis was performed by FT-Raman spectroscopy in combination with PCA at a species level. Jarvis et al. [34] used SERS spectroscopy of various strains to differentiate with PCA and HCA at a genus level. The recognition rate of the bulk database presented within this study is in the range of the cited results.

Table 3.3 Results of SVM classification of bulk spectra (532 nm).

Name (DSM number)	Total number of spectra	Number of wrongly classified strain spectra	Recognition rate for strains (%)	Number of wrongly classified species spectra	Recognition rate for species (%)
B. pumilus DSM 27	12	0	100.00	0	100.00
B. pumilus DSM 361	12	0	100.00	0	100.00
B. sphaericus DSM 28	14	0	100.00	0	100.00
B. sphaericus DSM 396	16	1	93.80	0	100.00
B. subtilis DSM 10	16	0	100.00	0	100.00
B. subtilis DSM 347	10	0	100.00	0	100.00
E. coli DSM 423	12	0	100.00	0	100.00
E. coli DSM 498	20	1	95.00	0	100.00
E. coli DSM 499	20	0	100.00	0	100.00
M. luteus DSM 348	12	0	100.00	0	100.00
M. luteus DSM 20030	20	0	100.00	0	100.00
M. lylae DSM 20315	21	0	100.00	0	100.00
M. lylae DSM 20318	22	0	100.00	0	100.00
S. cohnii DSM 6669	20	2	90.00	2	90.00
S. cohnii DSM 20260	13	1	92.30	1	92.30
S. cohnii DSM 6718	16	0	100.00	0	100.00
S. cohnii DSM 6719	18	1	94.40	0	100.00
S. epidermidis ATCC 35984	25	0	100.00	0	100.00
S. warneri DSM 20036	20	1	95.00	1	95.00
S. warneri DSM 20316	20	0	100.00	0	100.00
Average recognition rate			98.00		98.9

3.3.2.4 **Single Bacterium Analysis with Micro-Raman Spectroscopy**

For the investigation of bacterial bulk samples, grown microcolonies are necessary. However, for really rapid identification, analysis without cultivation would be advantageous, i.e., a method for the identification of a single bacterium is required. As already mentioned, micro-Raman spectroscopy is able to investigate structures in the sub-micrometer range. Therefore, single microbial cells are accessible with this method. Before performing such a single-cell Raman study, certain preliminary tests are necessary. Since the origin of the analyzed single microbial cell is not known, the influence of several environmental effects, such as temperature, nutrition, age, different states of the cell cycle, etc. on the Raman spectra of the microorganisms need to be investigated. First of all it is of interest to elicit if one Raman spectrum is representative of one bacterial strain.

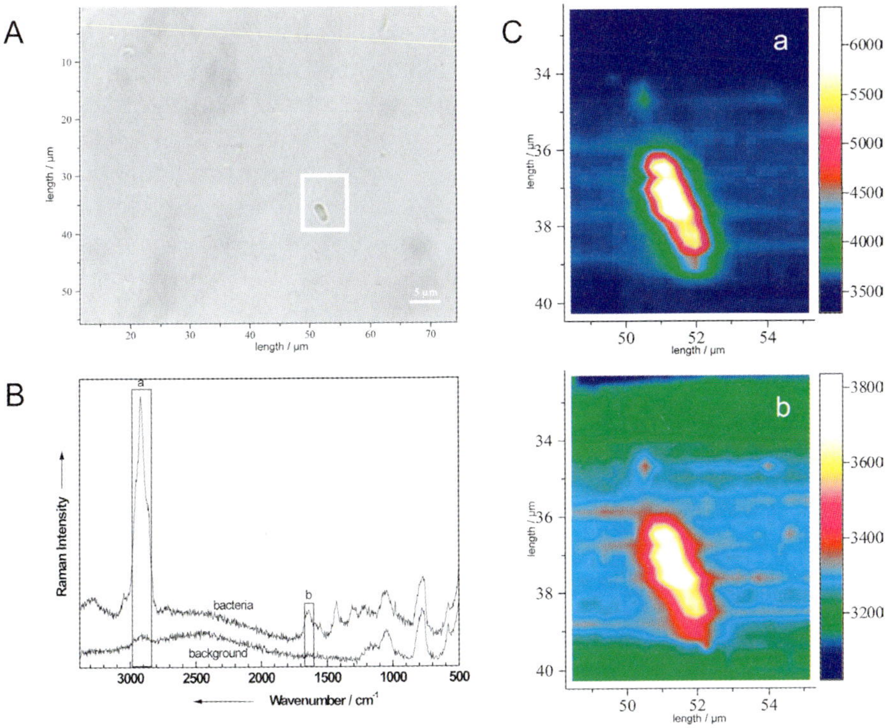

Figure 3.21 Raman mapping of *B. sphaericus* DSM 28; A: microphotograph of a single cell; B: Raman spectra of the cell and the background; C: false color plots of the C-H vibration (a) and the amide-I band (b).

Heterogeneity Testing

Since certain bacterial cells, e.g., *Bacillus*, exhibit a dimension of 0.8×3 µm or even larger, the first step is to check the spatial homogeneity of the sample. **Figure 3.21** shows the result of a Raman mapping experiment over a single bacterial cell. **Figure 3.21A** shows a microphotograph of a single *B. sphaericus* DSM 28 cell on a fused silica plate. The white box indicates the area where the Raman mapping was performed.

The Raman mapping experiment was performed using a step size of 0.3×0.3 µm^2 (total 20×28 points). These parameters are smaller than the spatial resolution of the Raman microscope (0.7 µm) but were chosen to increase the spatial overlap of the Raman mapping experiments. Each spectrum was measured with an integration time of 120 s, which leads to a total measuring time of 20 h. In order to minimize the background of fused silica a pinhole of 500 µm has been used.

The representative Raman spectra of the bacterium and the substrate, which was recorded beside the bacterium, are displayed in **Figure 3.21B**.

For the false color plots two different spectral regions are chosen: (I) the C-H stretching vibration (a) and (II) the amide-I band (b). The false color plots for these signals over the mapping area are shown in **Figure 3.21C** (a) and (b), respectively. The intensity of the signals are displayed as colors – the higher the intensity of a Raman signal at one position, the brighter the color. Both false color plots show a homogeneous distribution of the Raman signal over the bacterial cell [88]. Therefore it is possible to use one Raman spectrum independent of the measuring position as a representative characterization for the whole cell.

In the case of dividing cells, such cells can be observed at a particular time as two separate cells. A Raman spectrum for each cell can be measured following the microscopic image.

The spatial homogeneity of bacteria can be explained by the fact that bacteria normally exhibit no compartments. Some bacteria might exhibit vesicles where, for example, sulfur or PHB (poly-β-hydroxybutyric acid) is stored. As was shown by Schuster and coworkers [60, 61] Raman spectra of single *Clostridium* cells differ, owing to different amounts of starch-like granulose. Using line scans over the cell axis, no variations within the measuring position could be observed.

Beside forming vesicles, some bacteria e.g., *Bacillus*, are able to build dominant bodies (spores). These structures are known to be more complex than vegetative cells as they exhibit several layers. In **Figure 3.22**, panel A shows a microphotograph of both vegetative cells and spores of *B. sphaericus* DSM 28.

The Raman mapping experiment was performed over the marked area in three different layers (see **Figure 3.22C**). In panel B four Raman spectra from the center and rim of a spore, a vegetative cell and the background are displayed, respectively, which were taken from different spatial positions inside the marked area.

For calculating the false color plots two Raman signals were chosen as typical: the C-H vibration (a) which is representative for all biological compounds and the pyridine ring vibration of calcium dipicolinate (CaDPA, b) which is a marker substance for spores.

For the 3D Raman mapping, a lateral and axial spatial resolution of 0.5 μm was used, which leads to a total volume of $22 \times 16 \times 3$ points. Each spectrum was measured with an integration time of 300 s resulting in a total time of approx. 90 h. For the confocal measurements a hole of 200 μm and a slit of 50 μm were used. The laser power at the sample was around 2.5 mW.

The images a1 to a3 show the false color plots of the C-H stretching vibration (2871–2991 cm^{-1}) recorded for the three different depths positions 1, 2 and 3 as shown in **Figure 3.22C**. The C-H signal can be found in both vegetative cells and spores, therefore this signal monitors the distribution of cells. However,

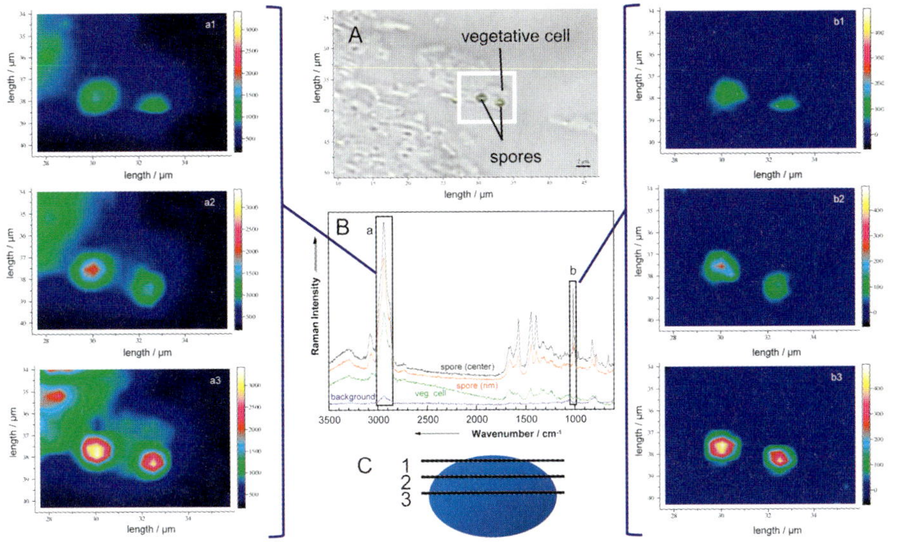

Figure 3.22 Raman mapping of *B. sphaericus* DSM 28 spores and vegetative cells: A: Microphotograph of both vegetative cells and spores on a fused silica substrate; B: Micro-Raman spectra of different positions inside the marked area; C: Depth scheme of the different mapping layers; False color plots of the C-H stretching vibration (a) and the pyridine ring breathing vibration of CaDPA (b) at different depths.

from the three false color plots a1 to a3 it can be seen that spores have a greater depth than vegetative cells. In image a1 Raman signals can mainly be found at positions where spores are located. For images a2 and a3 C-H stretching vibrations can also be found at regions where vegetative cells are located.

On the other hand the images b1 to b3 display the distribution of the marker band for CaDPA (993–1034 cm^{-1}), which can be found exclusively in spores. These false color plots show no information about the vegetative cells. The spatial distribution of the two spores in the C-H and CaDPA false color plots display different spatial depth profiles. Since CaDPA is only located in the cortex layer of the spore, i.e., in an inlaying layer, the dimension of the spores is smaller in the CaDPA plot than in the CH plot.

These results prove that it is possible to identify vegetative bacterial cells from one Raman spectrum. Investigating bacterial spores one spectrum might not be sufficient for the identification of one single spore. Here, at least three to five different measuring positions need to be analyzed.

Cultivation Conditions

As well as confirming the homogeneity of a bacterial cell, it is also necessary to examine the influence of different cultivation conditions on single bacterial cell Raman spectra. Since the origin of a bacterial cell is not known it is necessary to evaluate the influence of different cultivation conditions on the Raman

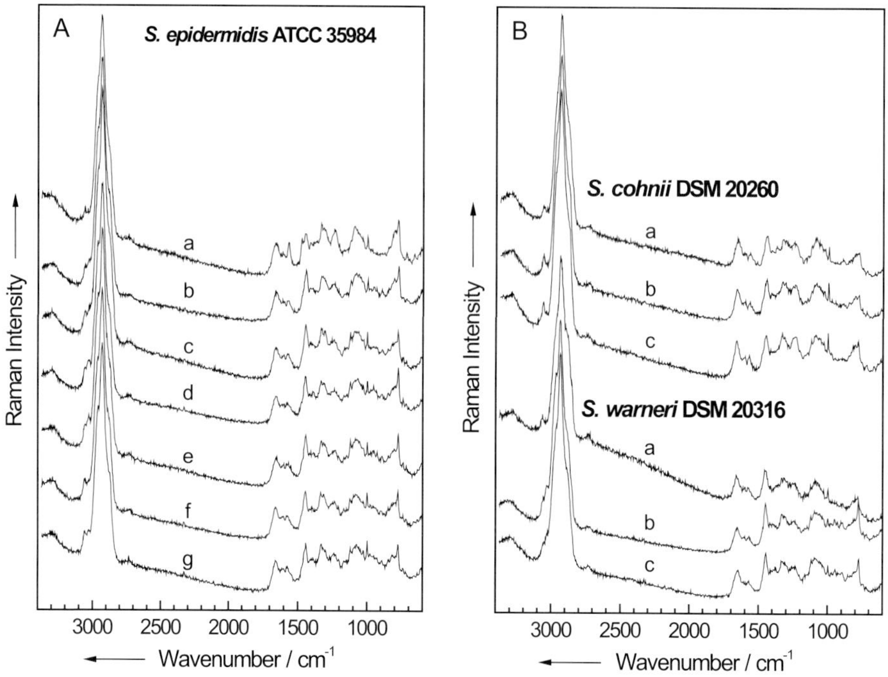

Figure 3.23 Micro-Raman spectra of A: *S. epidermidis* ATCC 35984 with different cultivation ages (a: 6 h, b: 18 h, c: 30 h, d: 42 h, e: 54 h, f: 66 h, g: 72 h) and B: *S. cohnii* DSM 20260 and *S. warneri* DSM 20316 cultivated under different conditions: a: CA at 37 °C, b: CASO at 37 °C, c: CA at 30 °C.

spectra of bacteria. In addition to different cultivation conditions, the age of a culture also influences the chemical composition of the cells and therefore the Raman spectrum [89].

In **Figure 3.23A** micro-Raman spectra of *S. epidermidis* ATCC 35948 grown on CASO-agar at 30 °C are displayed. The single bacteria spectra are measured at different cultivation ages from 6 to 72 h. The first Raman spectra were recorded after only 6 h incubation time. New samples were prepared and measured at 6 h intervals to a total incubation time of 72 h.

Micro-Raman spectra recorded from single cells cultured under these conditions exhibit nearly the same Raman bands, but there is one band in the region around 1575 cm^{-1} that decreases with increasing cultivation time. Every spectrum of *S. epidermidis* ATCC 35984 exhibits a band at 778 cm^{-1} independent of culturing conditions in contrast to the other strains. It can be seen that the Raman spectra of single bacterial cells taken from older cultures exhibit a better signal-to-noise ratio than those from younger cultures. For an unambiguous analysis on a single-cell level, these variations need to be taken into account by including these variations in the database applied for chemometric identification (see next section).

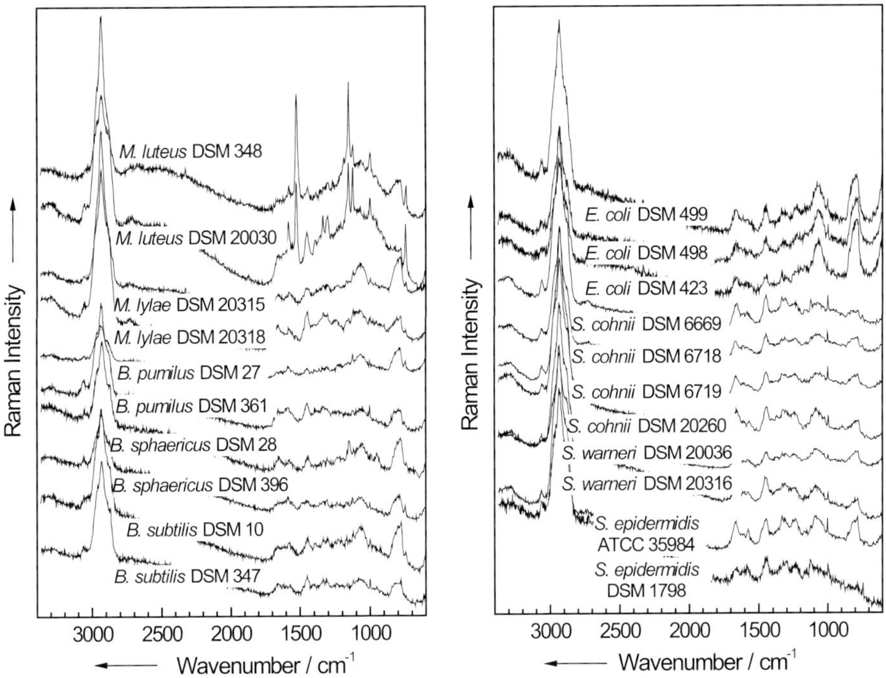

Figure 3.24 Micro-Raman spectra of different single bacterial cells from various strains.

Figure 3.23B shows Raman spectra from single cells of the two *Staphylococcus* species (*S. cohnii* DSM 20260 and *S. warneri* DSM 20316) obtained after cultivation under different growing conditions. The spectra of the single cells from cultures grown at a lower temperature (30 °C instead of 37 °C) exhibit nearly the same background signal, with broader bands than the bacterial spectra grown under 37 °C. The same effect can be observed in the Raman spectra of bacteria grown on CASO-agar instead of a CA-agar and vice versa. In general Raman spectra recorded from single cells cultured under different conditions exhibit nearly equivalent Raman bands and a characteristic peak at 778 cm^{-1}.

Single-cell Identification

In **Figure 3.24** micro-Raman spectra from single cells of two different strains from nine different species are shown as examples. Each spectrum was measured with an integration time of 60 s. Compared to the bulk Raman spectra in **Figure 3.20**, the Raman spectra of single cells show several differences.

The single bacterium spectra exhibit a lower background and a lower signal-to-noise ratio. In addition, signals due to the fused silica plate which all occur because of the very low sample volume of 0.5 (*Micrococcus* and

Staphylococcus) to 2.5 μm^3 (*Bacillus* and *E. coli*) of a single bacterium can be observed [88]. The poor signal-to-noise ratio for each Raman spectrum of the various single cells is a result of the low integration time. However, the quality of the single-cell spectra shown in **Figure 3.24** is sufficient for an identification of the bacterium by means of an SVM (see below and **Table 3.4**). As time is a critical issue for the analysis of clean-room samples, the overall investigation time should be kept as short as possible. Furthermore, some other features appear in the Raman spectra of single cells, e.g., protein signals at 1655 and 1452 cm^{-1} in the single-cell Raman spectra of *M. luteus* DSM 20030. The differences between the single-cell and bulk Raman spectra of *Bacillus* strains as well as of *E. coli* strains are less pronounced compared to the corresponding bulk spectra plotted in **Figure 3.20**. The three single-cell *staphylococcus* Raman spectra and the *M. lylae* spectra are of much better quality than the bulk spectra in **Figure 3.20**.

Experimental Results

The preprocessing of spectra of single cells is carried out by a running median filter to remove cosmic spikes. Moreover, since the spectra were measured at different grid points, the grid points were reinterpolated with a bilinear interpolation. This step was necessary because some of the points varied owing to different systematic conditions. The whole dataset consists of 3768 spectra of 29 different bacterial strains. Each spectrum was measured independently from a single cell, well isolated from other cells on the sample holder. The dataset of each strain comprises Raman spectra from different cultivation conditions (e.g., nutrition, temperature, etc.) and age. The spectra are limited to the range of 550 to 3200 cm^{-1}. SVMs are used for the classification as described above. Here the nonlinear Gaussian rbf kernel with a γ of $1.4 \cdot 10^{-5}$ and a cost value of 10^6 gives the best results.

The results can be seen in **Table 3.4**, where the recognition rates of the different bacterial strains are displayed. The table shows the recognition rate at strain level as well as at species level. The average recognition rate of 82.51 % at strain level and 96.43 % at the species level is quite good. Nevertheless, there are two different groups of strains in the database. Some strains show rather good recognition rates (e.g., *B. subtilis* or *M. luteus*) at strain and species level. The highest recognition rates at strain level are achieved from *M. luteus* DSM 348 with 99.52 %, *S. epidermidis* ATCC 35984 with 99.01 % and *B. subtilis* DSM 10 with 97.06 %. However, some strains like *S. epidermidis* DSM 20042 with 44.34 % or *E. coli* DSM 499 with 51.81 % exhibit a rather limited recognition rate for single strains. The recognition rate of these strains improves remarkably at the species level and shows a recognition rate of 100 %. This shows, that most of the strains were assigned to the correct species.

Table 3.4 Results of SVM classification of Raman spectra from single bacterial cells (532 nm).

Name	Total number of spectra	Number of wrongly classified strain spectra	Recognition rate for strains (%)	Number of wrongly classified species spectra	Recognition rate for species (%)
B. pumilus DSM 27	57	9	84.21	3	94.74
B. pumilus DSM 361	69	11	84.06	4	94.20
B. sphaericus DSM 28	53	9	83.02	6	88.68
B. sphaericus DSM 396	42	10	76.19	7	83.33
B. subtilis DSM 347	42	2	95.24	2	95.24
B. subtilis DSM 10	306	9	97.06	7	97.71
E. coli DSM 613	94	22	76.60	0	100.00
E. coli DSM 429	90	31	65.56	0	100.00
E. coli DSM 423	134	25	81.34	5	96.27
E. coli DSM 499	83	40	51.81	0	100.00
E. coli DSM 498	86	22	74.42	0	100.00
E. coli DSM 1058	71	16	77.46	0	100.00
E. coli DSM 2769	108	22	79.63	0	100.00
M. luteus DSM 348	619	3	99.52	3	99.52
M. luteus DSM 20030	48	5	89.58	2	95.83
M. lylae DSM 20315	45	3	93.33	3	93.33
M. lylae DSM 20318	20	1	95.00	1	95.00
S. cohnii DSM 6669	67	1	98.51	1	98.51
S. cohnii DSM 20260	65	2	96.92	1	98.46
S. cohnii DSM 6718	65	8	87.69	6	90.77
S. cohnii DSM 6719	63	11	82.54	6	90.48
S. epidermidis 195	74	3	95.95	3	95.95
S. epidermidis DSM 20042	106	59	44.34	0	100.00
S. epidermidis DSM 3269	93	39	58.06	1	98.92
S. epidermidis DSM 3270	110	47	57.27	1	99.09
S. epidermidis DSM 1798	215	28	86.98	0	100.00
S. epidermidis ATCC 35984	805	8	99.01	8	99.01
S. warneri DSM 20316	71	10	85.92	5	92.96
S. warneri DSM 20036	67	3	95.52	1	98.51
Average recognition rate			82.51		96.43

To the best of our knowledge only one contribution has been published dealing with the identification of bacteria on the single-cell level [62]. Here, seven bacterial strains were discriminated on the genus level by means of a PCA.

Table 3.5 Results of SVM classification of unknown independent Raman spectra from single bacterial cells (532 nm).

Strain	Number of spectra	Number of correctly classified strain spectra	identified as
B. sphaericus DSM 28	8	8	
B. sphaericus DSM 396	7	7	
B. subtilis DSM 347	8	8	
E. coli DSM 423	7	7	
E. coli DSM 498	7	7	
E. coli DSM 1058	20	17	*E. coli* DSM 423, *E. coli* DSM 499, *E. coli* DSM 2769
M. luteus DSM 20030	6	6	
M. lylae DSM 20315	5	5	
M. lylae DSM 20318	5	5	
S. cohnii DSM 6669	8	8	
S. cohnii DSM 20260	7	7	
S. cohnii DSM 6718	5	5	
S. cohnii DSM 6719	5	5	
S. epidermidis 195	5	5	
S. epidermidis ATCC 35984	20	18	*S. warneri*, *E. coli*
S. warneri DSM 20036	5	5	
Identification	130	125	

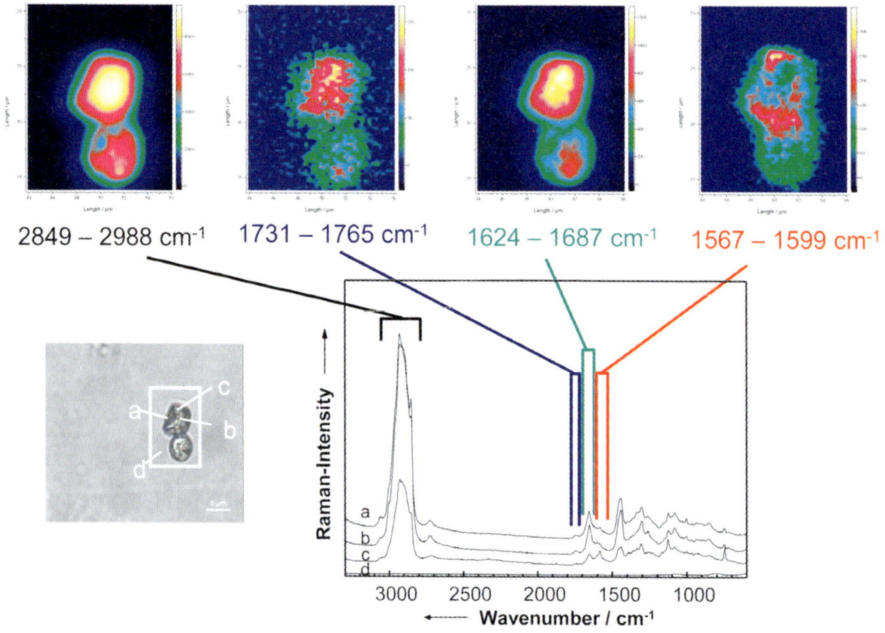

Figure 3.25 Raman mapping over two yeast cells. The Raman spectra (bottom right) are characteristic for different regions in the cells (see microphotograph bottom left). The false color plots display these distributions for different marker bands in the cells.

For identification of an unknown single bacterium, an independent set of different single bacteria from 16 strains, already included in the database, were measured [90]. These spectra were recorded anonymously and then identified with the corresponding dataset. **Table 3.5** shows the result for the unknown sample spectra. From 130 Raman spectra, 125 were correctly identified. Three spectra that were known to be *Escherichia coli* DSM 1058 were identified as different *E. coli* strains. Two spectra of *Staphylococcus epidermidis* 195 were identified as *S. warneri* and *E. coli*, respectively, which is the wrong species or even the wrong genus. This test demonstrates, that from the 130 independent Raman spectra 96.2 % were identified correctly at the strain level and 98.5 % at the species level.

Investigation of Eukaryotic Cells

Alongside bacterial contamination, eukaryotic cells can also be found in bioaerosol. Here, yeast cells and mold spores are particularly relevant. Mold spores can be measured in the same way as bacterial endospores. A classification with, for example, HCA achieves phenotypic relationships comparable to a phylogenetic classification.

As already mentioned, bacteria are prokaryotic cells and show no spatial heterogeneity. On the other hand eukaryotic cells such as yeasts contain compartments like the nucleus or mitochondria. This fact can be visualized by performing a Raman mapping experiment over yeast cells. In **Figure 3.25** (bottom left) a microphotograph of two single *S. cerevisiae* DSM 70449 cells is shown. The marked area is used for Raman mapping experiments. **Figure 3.25** (bottom right) displays four Raman spectra at different spatial positions on the yeast cell as indicated in the microphotograph.

Raman mapping experiments were performed over the area enclosed by the white rectangle in **Figure 3.25** (bottom left) to determine the spatial distribution of the various components within the yeast cells [91]. The upper panels of **Figure 3.25** show four Raman images of four representative Raman bands. The first false color plot (top left) is a result of integrating over the C-H stretch vibration (2849–2988 cm^{-1}), which is a marker band for organic matter. The intensity distribution of this band mirrors the different thickness of the yeast cell. The next Raman false color plot displays the intensity distribution of the C=O stretch vibration (1731–1765 cm^{-1}) marking the lipid fraction. Mapping over the amide-I and the C=C lipid band (1624–1687 cm^{-1}) yields the Raman image shown in **Figure 3.25** (top middle). These Raman bands exhibit considerable intensities only in confined areas. The phenylic C=C Raman band (1567–1607 cm^{-1}) can only be seen in the periphery of the cells (**Figure 3.25** (top right)).

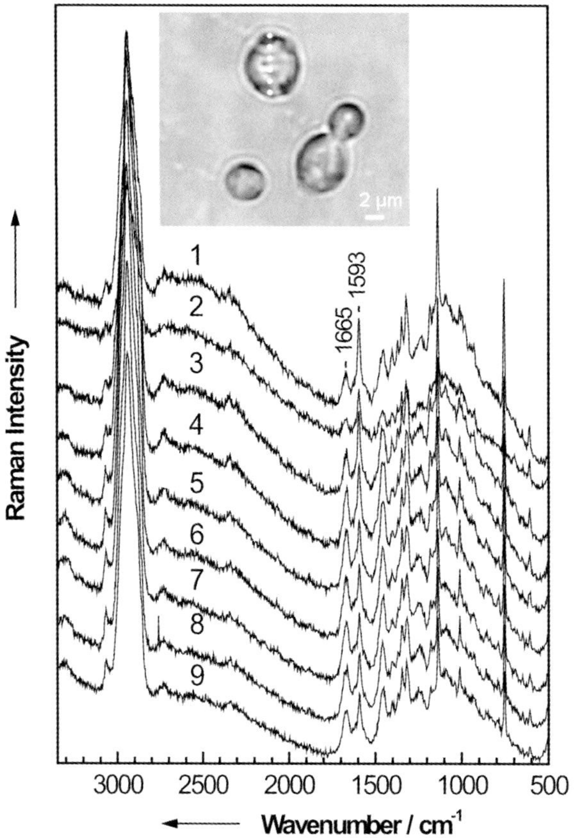

Figure 3.26 Line scan over a dry yeast cell (inset); Micro-Raman spectra from the separate measuring positions.

In order to enable a single-cell analysis of yeast cells it is necessary to incorporate the spatially varying information from the whole cell. Therefore, line scans were performed along the main axis of the cell. The inset of **Figure 3.26** displays a microphotograph of different yeast cells from dry yeast. The dotted line indicates the line scan performed over the main axis of one cell.

The micro-Raman spectra in **Figure 3.26** 1 to 9 are the different point spectra from a line scan. Here, the dependence of the Raman spectra on the position inside the yeast cell can be most clearly seen. The relative intensities of the signals at 1655 cm^{-1} and the peak at 1582 cm^{-1} show the greatest variation.

In order to overcome this spatial heterogeneity, a database for yeast cells is established by creating an average spectrum from a line scan over the main axis of a yeast cell, as shown in **Figure 3.26**. With this method it is possible to include the variations inside a yeast cell and obtain a Raman spectrum which represents the whole cell.

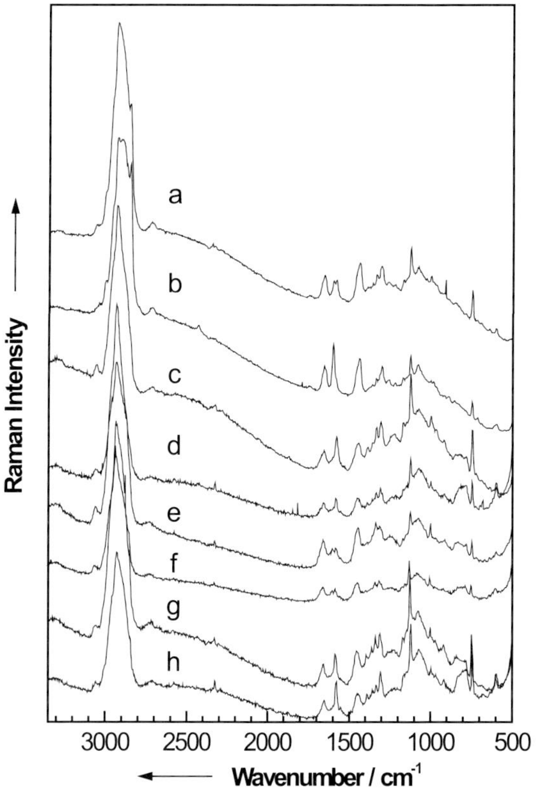

Figure 3.27 Average Raman spectra of different yeasts: (a): dry yeast, (b): *S. cerevisiae* DSM 70449, (c): *S. cerevisiae* DSM 1334, (d): *C. glabrata* DSM 11226, (e): *C. glabrata* DSM 70614, (f): *C. glabrata* DSM 70615, (g): *C. krusei* DSM 70075, (h): *C. krusei* DSM 70086.

Figure 3.27 shows representative average spectra of nine different yeast strains. The Raman spectra of commercial dry yeast (a), *S. cerevisiae* DSM 70499 (b) and *S. cerevisiae* DSM 1334 (c) are compared. Five strains from the genus *Candida* were measured for the dataset: *C. glabrata* DSM 11226 (d), *C. glabrata* DSM 70614 (e), *C. glabrata* DSM 70615 (f), *C. krusei* DSM 70075 (g) and *C. krusei* DSM 70086 (h).

Table 3.6 contains the results for the identification of nine different yeast strains from average spectra. The lowest recognition rate was obtained for *C. glabrata* DSM 70615 with 60 % at strain level. On the other hand all average spectra from *C. glabrata* DSM 70614 were identified correctly. At the strain level an average recognition rate of 86.2 % was obtained. The lowest recognition rate of 81.8 % is obtained for *S. cerevisiae* DSM 1334. Here an overall recognition rate of 94.8 % was achieved at species level.

Table 3.6 Results of SVM classification of single yeasts from average Raman spectra.

Name	Total number of average spectra	Number of wrongly classified strain spectra	Recognition rate for strains (%)	Number of wrongly classified species spectra	Recognition rate for species (%)
C. glabrata DSM 11226	10	1	90.0	0	100.0
C. glabrata DSM 70614	10	0	100.0	0	100.0
C. glabrata DSM 70615	10	4	60.0	1	90.0
C. krusei DSM 70075	10	1	90.0	0	100.0
C. krusei DSM 70086	10	1	90.0	0	100.0
S. cerevisiae DSM 1334	11	3	72.7	2	81.8
S. cerevisiae DSM 70449	13	1	92.3	1	92.3
dry yeast	18	1	94.4	1	94.4
Average recognition rate			86.2		94.8

Two approaches have been reported for Raman spectroscopic identification of bulk samples from various yeast strains. Different yeasts of the genus *Candida* were characterized at the species level using HCA, PCA or linear discriminant analysis (LDA) [52, 53].

3.4
Instrumentation of a Fully Automated Microorganism Fingerprint Sensor

The major objective of the OMIB research initiative was the design and development of a fully automated device for the investigation of bioaerosol contaminants in clean-room environments. Due to the artificial environment in clean rooms the bioaerosol exhibits special characteristics: First of all, the total number of particles is limited which also leads to a small number of biotic particles in the samples. In addition the number of microbial species is well defined. Therefore, only a certain amount of species within a few genera needs to be identified.

Since each substance exhibits its own typical Raman spectrum, the method allows an unambiguous identification of single microparticles, but it is a time-consuming process. Even in typical clean-room samples, a large number of particles must be analyzed. In relation to microbiological contamination, only biotic particles are of interest, comprising only a minor part of the total amount. Analyzing all particles is therefore not appropriate. In order to localize and differentiate biotic particles from abiotic particles and in order to minimize the analysis time, pre-sorting is required. Therefore, a first monitoring concept based on stray light microscopy and fluorescence imaging, is implemented. In a second step the potential biotic particles are analyzed by micro-Raman spectroscopy and identified by SVM.

3.4.1 The Concept of Online Monitoring and Identification of Microbial Contamination

Micro-Raman spectroscopy with a spatial resolution below 1 μm allows the investigation of single particles on a substrate in this size range. **Figure 3.28A** shows a microphotograph of four different particles similar in shape and size. Conventional light microscopy cannot distinguish between these particles. Some of these particles were investigated using micro-Raman spectroscopy as indicated by the white line.

The micro-Raman spectra of the four different types of particles are displayed in **Figure 3.28B**: the bacteria *M. luteus* (a) exhibits a completely different Raman spectra from particles of organic (melamine resin (b) and PMMA (c)) or inorganic origin (TiO_2 (d)). The respective micro-Raman spectra can be used to differentiate between biological, organic and inorganic particles.

Micro-Raman spectroscopy can distinguish between different types of particles which are indistinguishable on the basis of their microscopic image. Since hundreds to thousands of particles can be found even in clean-room samples, it is necessary to minimize the number of particles that need to be analyzed. Therefore, a preselection step is inserted into the OMIB concept [90].

Figure 3.29 schematically displays this concept. First the bioaerosol is impacted on a sample carrier surface. A microscopic image is taken in order to determine the size and position of each particle on the surface. In a second step biotic and abiotic particles are distinguished by the different fluorescence characteristics of the two particle classes. Thus the number of particles to be analyzed is minimized *via* fluorescence images. In a third step the biotic particles are characterized by means of micro-Raman spectroscopy. The measured Raman spectra allow the identification of single microbes with the help of SVM. This leads to a much more rapid analysis [90].

The monitoring step is based on the different fluorescence characteristics of biotic and abiotic particles. **Figure 3.6A** shows fluorescence spectra with an excitation wavelength of 365 nm from different particles.

The fluorescence spectra of the chosen biotic particles differ strongly from abiotic particles such as silicon carbide. This permits discrimination between biotic and abiotic particles. Therefore, the total number of particles is reduced and only the biotic particles need to be analyzed.

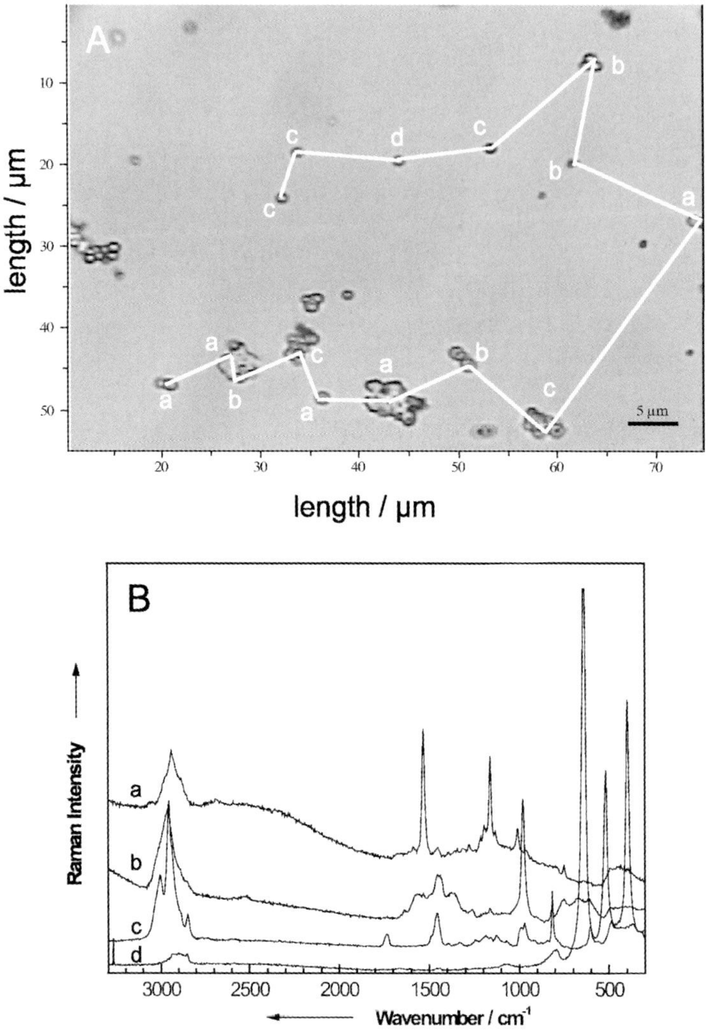

Figure 3.28 (A): Microphotograph of different microparticles on a fused silica plate, the scan is represented by the white line; (B): Raman spectra of different particles: (a): *M. luteus* DSM 348, (b): melamine resin, (c): PMMA, (d): TiO_2.

3.4.2
Instrumentation of Advanced Compact Spectral Sensors

The application of Raman spectroscopy in industrial environments makes great demands on the compactness of the measuring instruments and particularly of the spectral sensors, which are usually the main contributors to the overall size. This imposes the need for miniaturization of the spectral sensors. At the same time, compared to laboratory spectrometers, an extension of

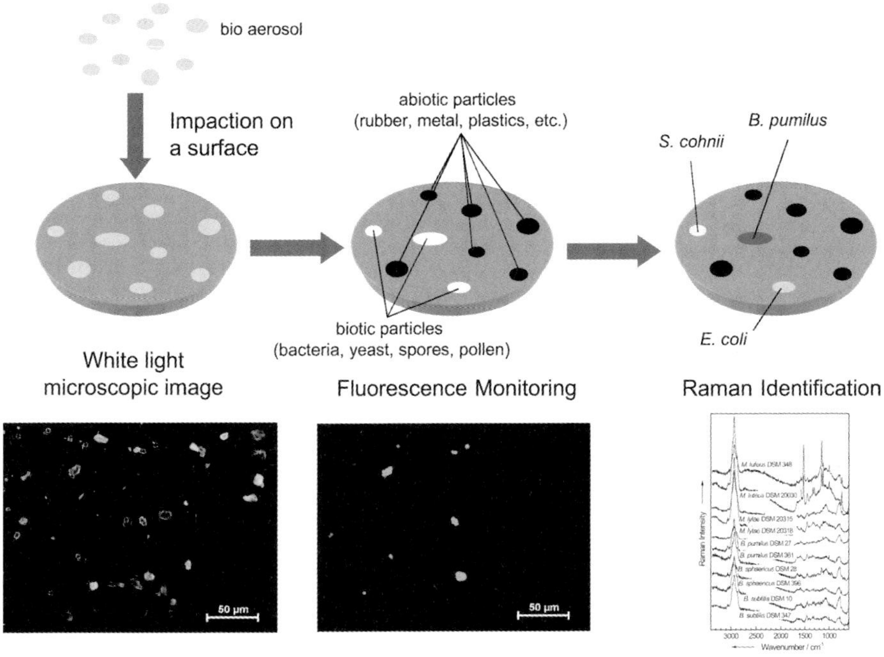

Figure 3.29 Concept of the OMIB principle: First the bioaerosol is impacted on a surface, where a microscopic image is obtained. The same sample is then investigated using fluorescence imaging in order to exclude abiotic particles. These are then identified by micro-Raman spectroscopy in combination with a SVM.

the bandwidth–resolution ratio is desirable to allow the capture of full Raman spectra of about 4000 cm^{-1} bandwidth at sufficient resolution within a single measurement. This needs to be possible for different excitation wavelengths and without any moving mechanical parts such as movable or exchangeable gratings.

However, increasing the bandwidth–resolution ratio and miniaturizing the dispersive spectrometers poses a number of difficulties:

- Pixel resolution: To achieve a spectral resolution $\Delta\lambda$ over a spectral range of width Λ, $\alpha \cdot \Lambda/\Delta\lambda$ spectral support points are needed, where $\alpha = 2$–3, depending on whether the spectrum will be Nyquist sampled or undersampled [92]. As an example, for the UV spectrometer for OMIB, which has a bandwidth of $\Lambda = 116$ nm and a resolution goal of $\Delta\lambda = 0.03$ nm, this means that, on the on hand, roughly 11.600 spectral support points (not including additional points due to nonlinearities) are needed. On the other hand, CCD detector arrays which are sensitive enough for Raman detection have pixel sizes of at least 13 μm which would lead to a spectrum length of 151 mm. Thus, the overall size of a Raman spectrometer that was only a shrunken classical spectrometer would be mainly determined by the detector size. Furthermore, without expensive staggering, such arrays are limited to approximately 4000 pixels in one direction.

- Linear dispersion: As with the detector length, the necessary linear dispersion of a spectrometer is determined by the required spectral resolution, the pixel pitch p and α. In particular, it computes as $\mathrm{d}x/\mathrm{d}\lambda = p \cdot \alpha / \Delta\lambda$. From the spectrometer developer's perspective, the linear dispersion is roughly the product of angular dispersion and focal length. Thus, to miniaturize the spectrometer, i.e., to reduce the focal length, the angular dispersion has to be increased if the pixel size is fixed. The latter becomes increasingly inconvenient for an angular dispersion that significantly exceeds the usual angular dispersions found in high-resolution spectrometers, i.e., this is not really an option.

- Aberrations: As the spectrum length of a spectrometer grows in relation to the overall size of the spectrometer, imaging aberrations become increasingly severe. The main influence is caused by a curvature of the image field, which tends to resemble a deformed sphere and whose cross-section with the optical plane is known as the Rowland circle [93]. Modern holographic gratings reduce this effect [94].

- Light throughput: If miniaturization is achieved by decreasing the real or virtual (see below) pixel size, the slit size also has to be decreased. A key parameter here is the product between light beam or entry slit area A and the square of the aperture ratio $1/f$, which we will refer to as "throughput", if it represents the light acceptance of an optical system, and as "beam volume", if it represents a light beam. A reduction of the throughput below the beam volume (given by the spot diameter at the sample and the collecting aperture of the Raman microscope) results in a light loss.

  Thus smaller slits will also reduce the usable spot size and collecting aperture of the Raman microscope, if no counter-measures are taken.

3.4.2.1 **The Basic Principle of the Double Array Grating Spectrometer**

As illustrated above, the large bandwidth–resolution ratios needed for OMIB are incompatible with a simple miniaturization of a classical spectrometer. A large detector would still be needed, and as a consequence of having a large detector in a small spectrometer strong imaging aberrations and difficulties in achieving sufficiently high linear dispersion would be expected. To overcome this kind of discrepancy, different extended spectrometer concepts have been developed in the past. The most widely known concept is that of the Echelle spectrometer. Here, the spectral range of the spectrometer is distributed over multiple diffraction orders of a blazed grating. Each of these diffraction orders contains only a small portion of the whole spectrum and different diffraction orders are arranged side by side on a two-dimensional detector by cross-dispersing optics. However, Echelle spectrometers still tend to be rather large, partly due to the need for a cross-disperser and partly because they need the same linear dispersion as ordinary spectrometers.

The spectrometer concept used for OMIB is that of the so-called double-array spectrometer [95–98]. In contrast to the Echelle spectrometer, the spectrum is not split into different pieces, but each real detector pixel is subdivided into several virtual sub-pixels, thus breaking the spectral support point density barrier imposed by CCD pixel pitch and linear dispersion. To generate sub-pixel information, the spectrum is measured several times with different sampling rasters, shifted by fractions of the raster pitch. The latter is achieved by changing the position of the entry slit with respect to the spectrometer, leading to an equivalent shift of the center wavelength of each CCD pixel. The resulting raw data or "sub-spectra" are then re-combined into a single, high-resolution spectrum by means of a so-called superresolution algorithm. The primary advantage of this strategy is that the resolution is no longer limited by the CCD pixel pitch but by a fraction of it, which can be lower than 30 % of the physical pixel pitch. This, in turn, allows a relatively small linear dispersion to be used, which can be translated into convenient grating angular dispersions, low aberrations or a short focal length, leading to a miniaturization of the spectrometer.

Instead of the cross-disperser of an Echelle system, the double-array spectrometer uses an entrance slit array which is either one-dimensional and switchable or two-dimensional [98, 99] as the key component responsible for the performance improvement.

The first established double-array spectrometers were based on switchable entrance slit arrays in conjunction with a one-dimensional (1D) detector array. The slits, which had fixed positions, were normally opened and closed in different combinations which made up so-called Hadamard patterns [100, 101]. In this way, the throughput of the spectrometer could be increased to values larger than for single-slit systems.

More recent, 2D double-array spectrometers, such as those implemented for OMIB, generate the different sampling rasters through spatial multiplexing instead of temporal multiplexing. The sub-spectra generated by the different entry slits are arranged side by side on a 2D detector array (see **Figures 3.30** and **3.31**). As opposed to the 1D spectrometer type with switchable entry slits, and similar to a normal diffractive spectrometer, such a spectrometer is capable of measuring spectra in a single shot, which is often advantageous.

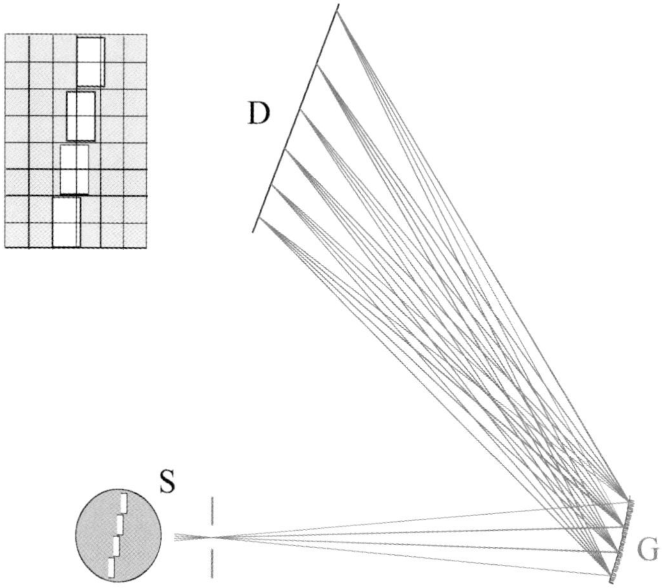

Figure 3.30 Principle of the two-dimensional double-array spectrometer: For every single entry slit of the two-dimensional entrance slit array S a different sub-spectrum image is generated on the detector array D by the imaging grating G. For every single wavelength, an image of the slit array is generated on the detector array, as depicted in the figure. Changing the wavelength leads to a horizontal shift of this image.

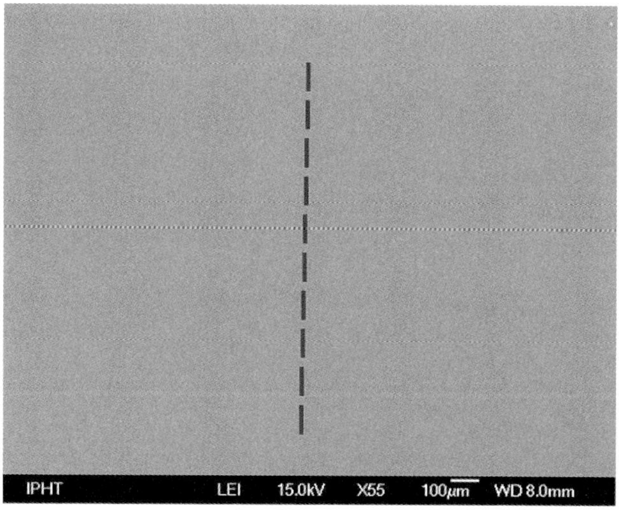

Figure 3.31 Sample SEM image of an entrance slit array with ten single entry slits, which differ in their horizontal position by several micrometers.

3.4.2.2 Resolution, Aberration Variability and the Multifocus Advantage

Apparently, as the different entry slits are located at different positions within the spectrometer, they will be subject to different optical aberrations. This holds for 1D double-array spectrometers, where the entrance field is normally not flattened, as well as for 2D double-array spectrometers, where the image field and the entrance field suffer from curvature along the slit direction. This leads to aberrations, which change with wavelength and from slit to slit. For a successful superresolution reconstruction, these varying aberrations have to be modelled and included in the reconstruction by the superresolution algorithm. From a signal theoretical point of view, the reconstruction of the high-resolution spectrum is possible, if all Fourier components of its sampled representation can be found in the set of measured sub-spectra in a unique way, i.e., if the system of equations, describing the sub-spectra as a function of the high-resolution spectrum, is invertible. For this condition to be satisfied it is not necessary for every sub-spectrum to contain all frequency components and that these can be separated using the different sampling rasters of the different sub-spectra. In fact it is already sufficient that every signal component is contained in at least one of the sub-spectra, as long as their separation is possible using the different sampling rasters or the varying strength of each signal component in different sub-spectra. As a result, the achievable resolution is determined by the resolution at the best-resolved slit and not by the average resolution or the resolution at the worst-resolved slit. The corresponding way to achieve a well focused spectrum for all wavelengths despite a strongly curved image field is shown in **Figure 3.32** for a 1D double-array spectrometer with a Rowland-circle-shaped image field. In the example, different slits cause different Rowland circles, which cut the detector plane for different wavelengths. As a result, for every wavelength there is a slit that is well focused, and reconstruction of a spectrum with high definition at all wavelengths is possible. This effect is called the multifocus effect and can also be well exploited in 2D double-array spectrometers. For the latter, the distribution of well focused wavelengths can be influenced by tilting the entrance slit plane or the image plane. **Figure 3.33** shows a simulation of the effect for a line pair under conditions that match those relevant for OMIB. The simulation clearly shows that a well resolved spectrum can be recovered even for a large, varying defocus, as long as at least one slit is well focused. If the defocus is constant, i.e., no slit is well focused, defocus values that are only a fraction of the maximum defocus of the previous case can already lead to significant loss in resolution of the reconstructed spectrum.

In summary, the multifocus effect can be used to overcome miniaturization limitations related to the curvature of the image field.

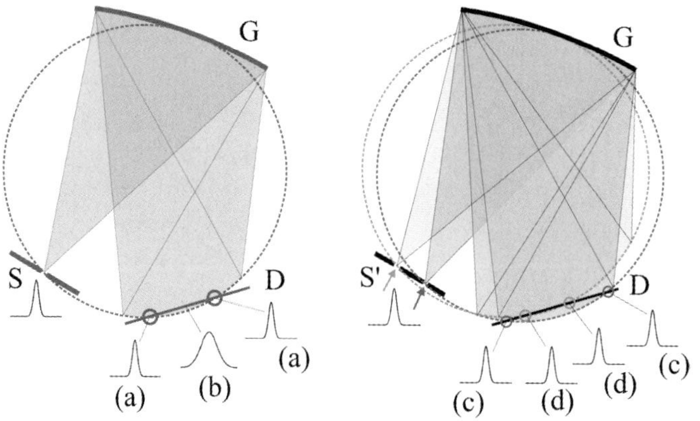

Figure 3.32 Principle of the multifocus effect in the example of a 1D double-array spectrometer: On the left side the situation for a classical spectrometer is shown. The single entrance slit S generates a corresponding Rowland circle, where the slit is imaged by the grating G. This circle is cut by the detector plane D for only two different wavelengths (a). For other wavelengths a defocus with a reduction of resolution results (b). The right side shows the situation for a double-array spectrometer. For each slit S′ of the spectrometer a different Rowland circle is generated. These different Rowland circles cut the detector plane at different positions (c and d) or for different wavelengths, respectively. By arranging the Rowland circles in a proper manner it is possible to have at least one well focused slit image for every wavelength, which will enable the superresolution algorithm to return a perfectly focused spectrum.

3.4.2.3 **Enhancing the Throughput with Two-dimensional Entrance Slit Arrays**

So far we have shown how problems related to pixel size, overall spectrometer size, aberrations and linear dispersion can be overcome by means of the 2D double-array spectrometer architecture. The remaining problem to be addressed is that of throughput limitation due to slit width and aperture ratio restrictions.

For typical Raman applications the effective beam volume at the source, which is given by the source spot diameter and the numerical aperture of the collecting lens, is in the range of about 3 μm^2 in the UV region to about 15 μm^2 in the NIR region. Optical effects related to the fiber coupling can cause these beam volumes to grow to more than 30 μm^2 before the light reaches the spectral sensor entrance slit. Although the throughput of the entrance slit can be significantly larger than this value, only low coupling efficiencies will be achieved with regular conversion optics, since the incoming light has a circular aperture while the slit has a large height but a small width.

One way to overcome this limitation is to use non-regular imaging optics which rearrange portions of the light beam to resemble the slit shape. However, such optics – for instance beam twister arrays [102] or image slicers [103] – are highly complex on the shelf components which can significantly increase the spectral sensor's complexity and its production costs.

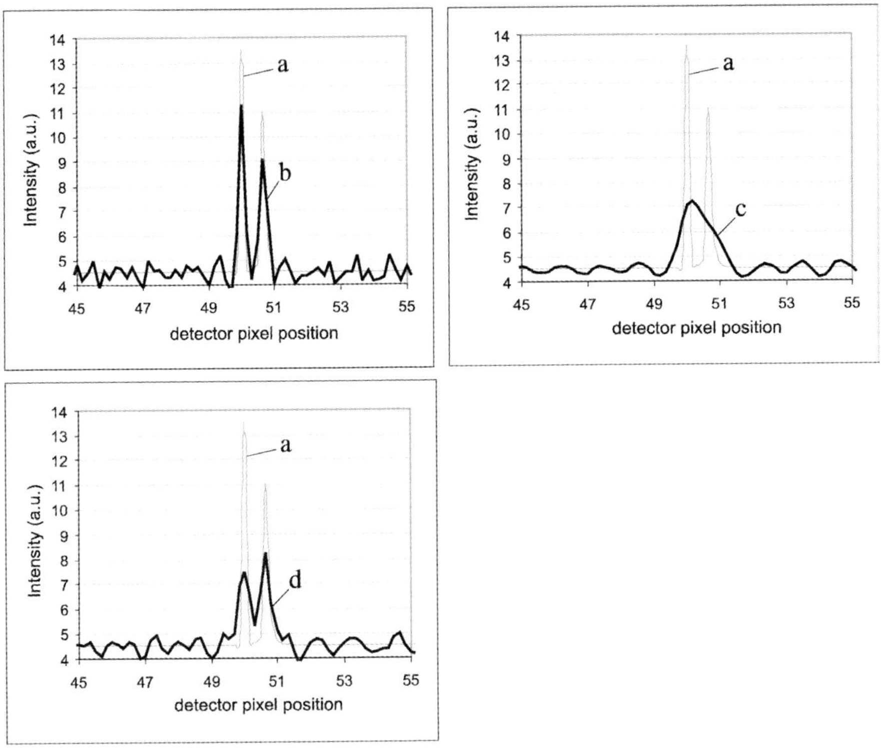

Figure 3.33 Simulation of the multifocus advantage for two spectral lines with $\Delta\lambda = 0.041$ nm at a physical pixel dispersion of 0.057 nm/pixel (actual OMIB UV spectral sensor: 0.060 nm/pixel at $\lambda = 244$ nm to 0.054 nm/pixel at $\lambda = 360$ nm) and an aperture ratio of $f/6$: **top left:** original signal (a) and simulated measurement result (including a certain amount of detector noise) for a perfectly focused spectrometer (b); **top right:** result for strong defocus of 50 μm for all entry slits (c) and **bottom:** result for defocus growing linearly from 0 to 150 μm (d). In case (c) the two spectral lines are no longer resolved. Although for case (d) the defocus is up to three times as large as for case (c), the spectral lines are resolved again with only minor resolution losses and losses in signal-to-noise ratio as compared to case (b). If the image surface of the spectrometer is curved, as is the case with most spectrometers, a situation as in case (d) can be achieved simultaneously for all wavelengths by tilting the detector plane around the axis of the dispersion.

Within the OMIB project, an alternative approach – the use of coded, 2D aperture arrays – was investigated for the first time. Instead of a single, small slit for each sub-spectrum, a certain slit pattern with a much larger overall width was used (see **Figure 3.34**). This is similar to the Hadamard multiplexing done for 1D double-array spectrometers (see above).

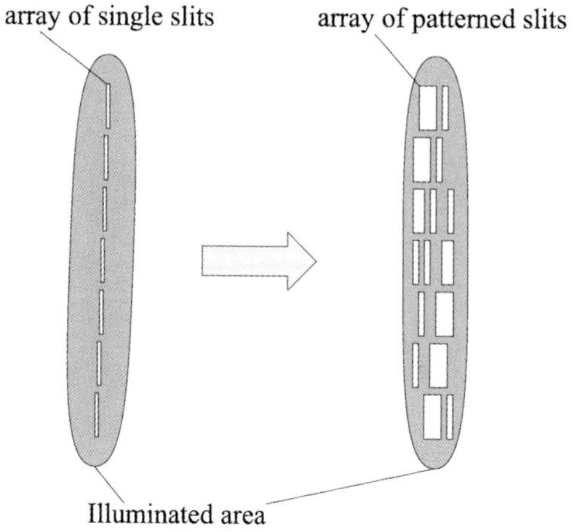

Figure 3.34 Throughput enhancement by introduction of patterned slits. For an array of single slits (left) considerable light loss results if the width of the illuminated area is significantly larger than the slit width. Patterned slits as on the right side achieve the same resolution as the single slits at a significantly increased throughput.

3.4.2.4 **The OMIB Raman Spectral Sensors**

Two compact spectral sensors have been developed for OMIB, one dedicated to UV measurements (see **Figure 3.35**) and the other dedicated to NIR measurements (see also Section 3.5). For reasons of excitation wavelength variability and cost requirements, which rule out expensive mechanics, the OMIB-UV spectrometer has to provide a resolution of $\Delta\nu = 7\ \text{cm}^{-1}$ as a requirement and $\Delta\nu = 5\ \text{cm}^{-1}$ as a goal over a spectral range from 244 to 360 nm (i.e., width of wavelength range $\Lambda = 116$ nm) without mechanical switching. For the second, IR spectrometer $\Delta\nu = 5\ \text{cm}^{-1}$ has to be achieved for a spectral range from 785 to 1100 nm, where silicon detectors finally become insensitive. For both devices, the length has to be less than about 100 to 150 mm. In contrast, commercial state-of-the-art dispersive Raman laboratory spectrometers that achieve a resolution of $5\ \text{cm}^{-1}$ have focal lengths typically in the range of 300 to 500 mm and the wavenumber range of such a spectrometer is about $3000\ \text{cm}^{-1}$, which, for instance, would correspond to a covered wavelength bandwidth of $\Lambda \approx 20$ nm at an excitation wavelength of 244 nm. To change the wavelength range, the gratings have to be moved or exchanged. Compared to such commercial spectrometers, particularly for the OMIB UV spectral sensor, the simultaneously covered bandwidth Λ has to be extended by roughly a factor of 6. Since for a dispersive spectrometer the linear dispersion is constant or even increases towards longer wavelengths, the wavelength resolution requirement of the spectrometer is driven solely by the wavenumber resolution requirement at the shortest wavelength. This means, to achieve a

Table 3.7 Parameters of the OMIB Raman spectral sensors.

Parameter	OMIB-UV	OMIB-IR
	Requirements	
Excitation wavelength	244 or 325 nm	785–830 nm
Wavenumber resolution	7 cm^{-1}	5 cm^{-1}
Required sensor wavenumber range for single excitation wavelength	0–4000 cm^{-1}	0–< 4000 cm^{-1}
Resulting physical parameters and implemented values		
Wavelength resolution	< 0.041 nm	< 0.3 nm
Resulting wavelength range	244–360 nm*	785–1100 nm
CCD model	e2v CCD42-10	e2v CCD40-11
CCD size (pixels, pitch)	2048 × 512, 13.5 μm	1024 × 128, 26 μm
Image distance (size)	137 mm	137 mm

*) corresponds to ≈ 13000 cm^{-1} wavenumber range

wavenumber resolution of 7 cm^{-1} or 5 cm^{-1} starting from a wavelength of 244 nm, a spectral resolution of $\Delta\lambda = 0.041$ nm (0.03 nm) is necessary. The technical parameters of the two spectral sensors are given in **Table 3.7**.

For the OMIB set-up the UV spectral sensor is already in use. The IR spectral sensor is being developed for further applications where small high-performance spectrometers are necessary.

Figure 3.35 Fully assembled UV-Raman spectral sensor. The left part of the system belongs to the spectrometer optics including the coupling optics. This part is approximately 145 mm long; The larger, right part is the commercial CCD detector including a TEC cooler and parts of the CCD electronics.

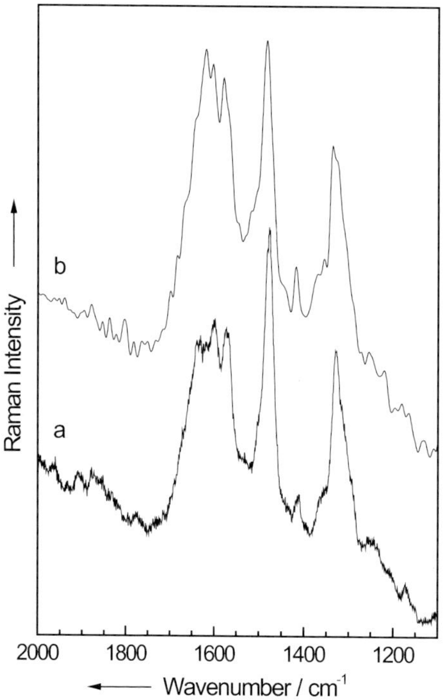

Figure 3.36 UV-resonance Raman spectra of *B. pumilus* DSM 361 measured with a conventional spectrometer (a) and the specially designed OMIB spectral sensor (b).

The optical design was developed in close cooperation with the Zeiss company in Jena, the gratings were fabricated by Zeiss and the design and fabrication of the spectral sensor mechanical frames were carried out by the Fraunhofer Institute for Applied Optics and Precision Engineering in Jena.

The coupling optics of the spectrometer is a separate, small unit, which can be exchanged and adjusted with respect to the main spectral sensor. This allows us to investigate the effect of different slit geometries and the multifocus effect in a real device. For the first time, the application of patterned entry slits to a double-array arrangement and their impact on the multifocus effect were investigated.

In **Figure 3.36** the UV-resonance Raman spectrum of *B. pumilus* DSM 361 obtained with a conventional laboratory spectrometer (a) is compared with a spectrum from the specially designed OMIB spectral sensor. An excitation wavelength of 257 nm was used with an integration time of 100 s and three accumulations each. Both spectra are in principle highly comparable.

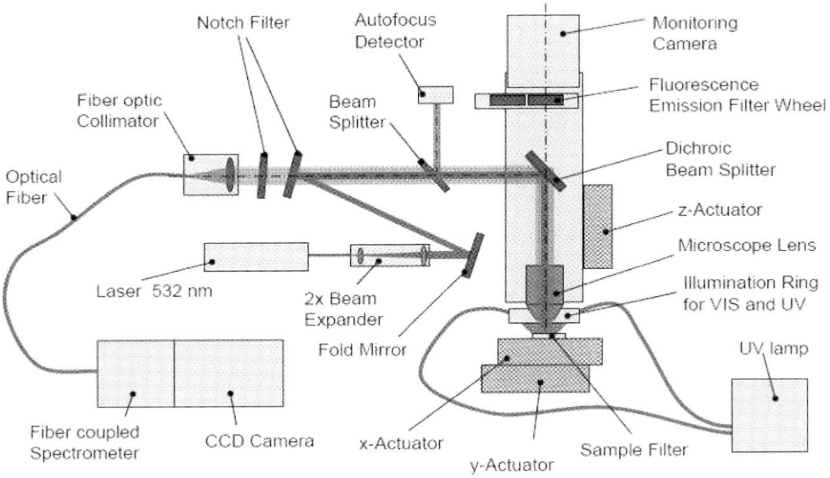

Figure 3.37 Optical configuration of the OMIB Instrument.

3.4.3 Realization of OMIB Instrumentation

So far the detection methods Raman spectroscopy and fluorescence microscopy described above have to be performed on separate instruments and, to a large extent manually. In addition, the derived individual data-sets have to be transferred manually for statistical data evaluation. To overcome these restrictions, the three detection methods microscopy, fluorescence microscopy and Raman spectroscopy and the postdata processing, are integrated in a fully automated laboratory instrument. This allows all measurements and the statistical data evaluation to be performed automatically, even for a large number of particles, and relevant critical bioparticles to be identified very rapidly. **Figure 3.37** indicates the optical layout of the instrument.

The optical microscope observes the sample in the visual range and subsequently in specific fluorescence bands, which can be selected by the filter wheel with dedicated fluorescence emission filters. The illumination ring directly above the sample contains a number of bright LEDs for illumination for the visual monitoring. Additionally, three UV fiber collimators are integrated in the illumination ring to focus the UV light in the field of view of the microscope to excite fluorescence in the relevant particles. These images are evaluated with regard to number, position, size and shape of the particles. With the help of the fluorescence images, additional information can be derived to preselect between biotic and abiotic particles. The dichroic beam splitter within the optical path of the microscope allows the optical path of the Raman set-up to be co-aligned onto the optical axis of the microscope. Thus the focus spot of the Raman system is fixed in relation to the field of view of the microscope. This is essential for being able to focus the Raman measurements on the individual, detected particles with a high spatial resolution of a few micrometers. When a particle is detected as a potentially relevant bioparticle by visual and fluorescence monitoring, its position within the microscope field of view allows the derivation of x, y movement coordinates to center this individual particle directly under the Raman focus spot. The whole sample filter is repositioned with high accuracy by means of a linear actuators in the x, y direction. Once centered, an auto-focus detector measures the reflected Rayleigh scattered light intensity. By fine adjustment of the z-position of the microscope/Raman objective lens, this signal can be maximized, to obtain the best focus position for the Raman measurements. After acquisition of the Raman spectrum of the particle, the next relevant particle within the microscope's field of view is centered under the Raman focus. By subsequent repositioning, followed by Raman measurements, all preselected particles in the field of view are characterized automatically. The resulting dataset, which now contains the position, size, shape, fluorescence behavior and Raman spectrum for each individual particle is fed to the SVM, which identifies the particles by comparison with its database support vectors. The instrument finally automatically displays the identified particles, their statistical distribution and their position, in case they need to be further investigated or cultivated. **Figure 3.38** summarizes all process steps, which are implemented in the automated instrument.

Figure 3.39 shows the basic version of the OMIB instrument [90]. All the required subsystems and components indicated in **Figure 3.37** and all the necessary control and power supply electronics are integrated within a volume of 80 cm × 50 cm × 48 cm. An additional standard laboratory PC runs the control software, performs the data evaluation and displays the measurement results. All the operator has to do is to insert the sample filter, which contains the sampled particles, in the filter transport wheel and start the measurement. Monitoring and fluorescence monitoring then takes a few minutes per field of view (500 μm × 500 μm). Several of these microscope fields of view can be stacked to result in an image mosaic of maximum 5 × 5 mm size. Once a particle is detected within the field of view it takes between several seconds to about one minute to acquire its individual Raman spectrum. Depending on the number of particles on the sample (a highly contaminated sample filter surface might contain several hundreds of relevant bio-particles!) the overall measurement cycle can last between several minutes and several hours. The identification of a single bioparticle, including monitoring, Raman measurement, data processing and identification with the SVM takes just a few minutes.

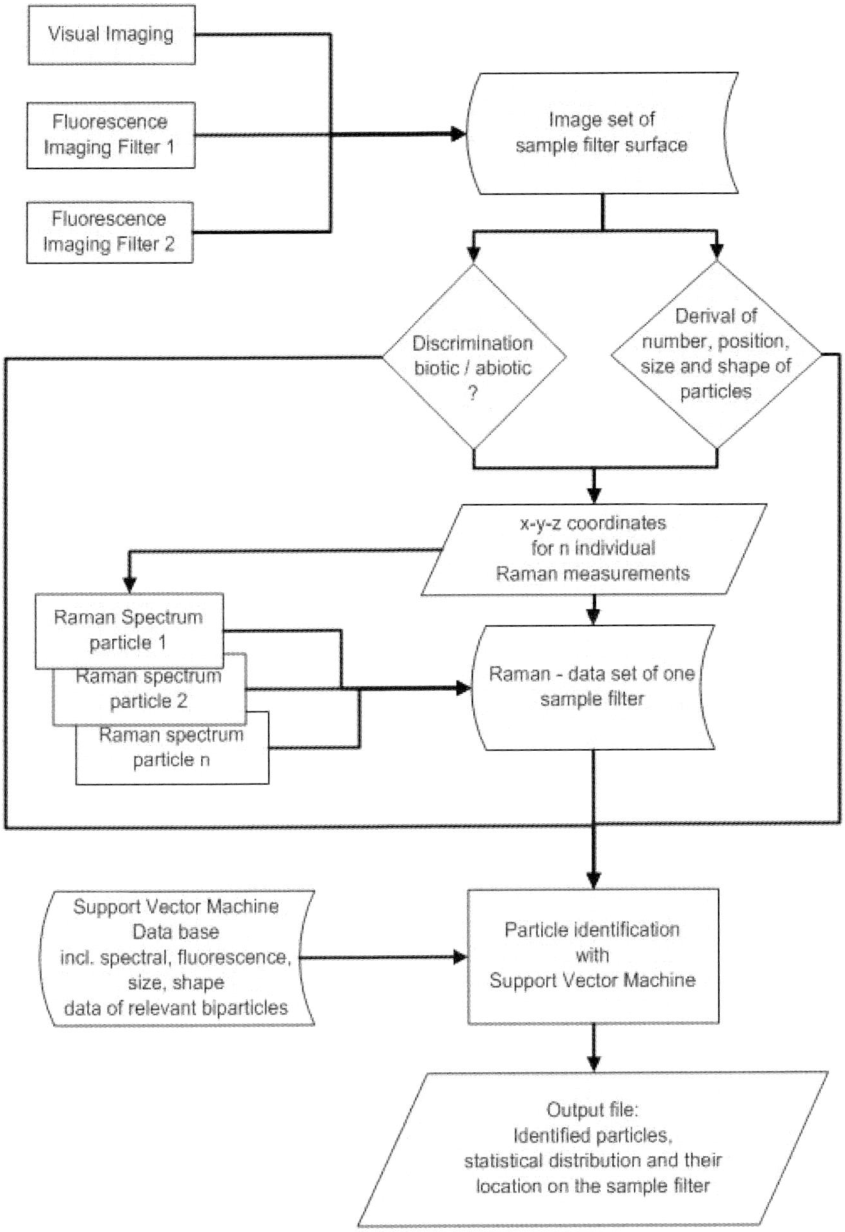

Figure 3.38 Flow chart of the automated measurement, evaluation and identification process.

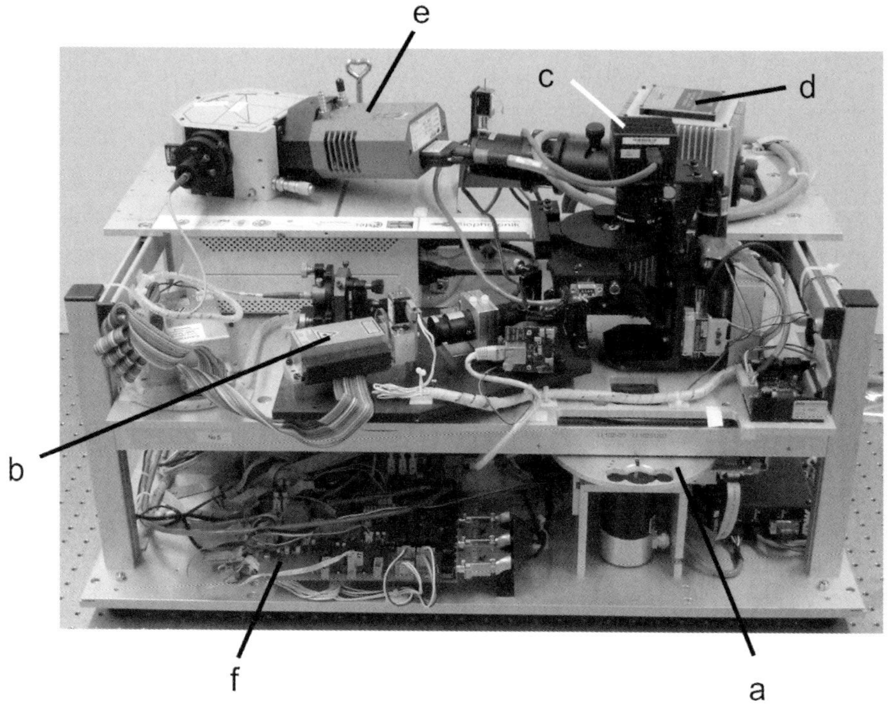

Figure 3.39 OMIB laboratory demonstrator without housing: sample transport wheel (a), excitation laser (b), fluorescence microscope (c), fluorescence excitation source (d), compact spectrometer with CCD detection (e), system control hardware (f).

Figure 3.40 shows Raman spectra of single microorganisms measured with the OMIB demonstrator. The Raman spectrum of the pigmented yeast *Rhodotorula mucilaginosa* DSM 70403 is dominated by the signals of β-carotene. The second yeast *C. glabrata* DSM 70614 exhibits characteristic Raman bands that can be assigned to protein vibrations. The intensity of the Raman spectra of the two bacteria (*M. luteus* and *B. pumilus*) is lower than those of the two yeasts. This can be explained by the lower sample volume of the bacteria. The Raman spectrum of *M. luteus* DSM 348 as a pigmented bacteria exhibit the characteristic signals of sarcinaxanthin, while the spectrum of *B. pumilus* DSM 361 can be assigned to a typical protein spectrum.

These first tests show the feasability of an automated detection of single microorganisms. An extensive database specially designed for the application of the OMIB device in clean-room monitoring is currently under development.

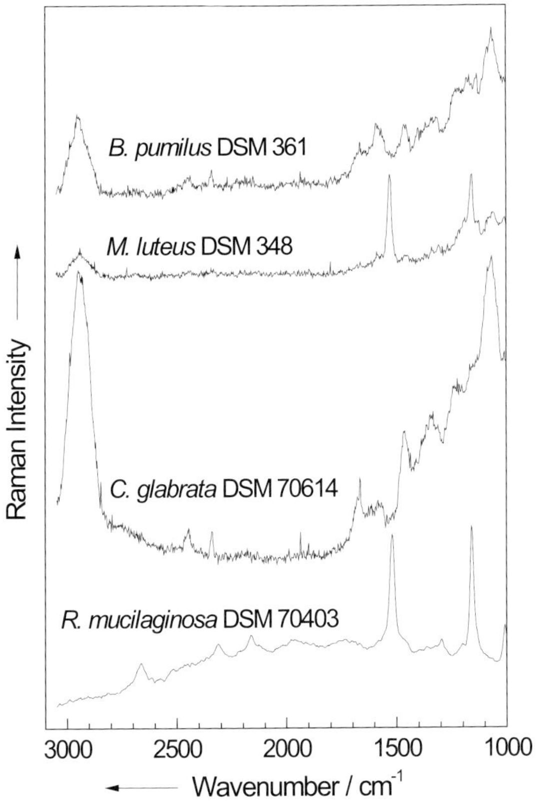

Figure 3.40 First Raman spectra of different bacteria and yeast cells measured with the automated OMIB laboratory demonstrator.

3.4.4 Applications

In clean-room manufacturing, the quality of products originating from the fields of microelectronics, pharmaceutical, life sciences, medical technology, the food and cosmetics industries and also of MEMS (Micro-Electro-Mechanical Systems) and optical instruments depends on a number of factors. One important factor is the degree of cleanliness of the production environment. Contamination is a potential risk in the production environment which not only influences product quality, but also endangers the end-user of contaminated products and the staff involved in their manufacture. The use of clean-room technology helps to reduce relevant contaminants and keep them below critical levels. As a result, production environments can be adapted to meet required cleanliness standards. The concentrations of particles allowed into the hygienic production environment are laid down and defined in several different guidelines as clean-room classes and surface-cleanliness classes.

Parallel to these, there are also guidelines that recommend contamination limits for biocontamination in hygienic production as well as for products. Acceptable concentration levels of particles and microorganisms are related to the potential risk they represent for the product. A differentiation is made between total amount of biocontamination and pathogenic biocontamination for hygienic products. A method for the online detection and identification of particle contamination with regard to their biotic or abiotic origin would drastically reduce the costs and time involved in quality assurance in all hygienic life sciences such as cosmetics, food or pharmaceutical production. The contamination detected could be directly correlated with the respective production conditions. This would enable a fast and systematic reaction to changes detected in particle concentrations and biocontamination levels in the production environment. This would result in the avoidance of yield losses due to punctual adaptation of manufacturing processes. Costs associated with quality assurance could be reduced and a safer production environment created both for personnel and for products.

3.5 Outlook

The OMIB set-up described above is specially designed for the identification of bioaerosol in clean-room environments. Using modified sample holders would make additional application fields possible. Particulate and biological contamination of liquids can be investigated by a filter with a modified smooth surface with a gold or aluminum layer. Possible applications for the analysis of liquids would be the investigation of biological contamination in, for example, juices or water.

In addition, the dataset will be expanded to normal environmental conditions. With this application it should be possible to detect possible contamination arising from bioterrorism.

Currently we are developing miniaturized devices based on the OMIB apparatus described above, with Raman excitation wavelengths in the UV and NIR regions. These devices use the spectral sensors described on pp. 148.

Acknowledgement

Funding of the research project FKZ 13N8365, 13N8366, 13N8369 and 13N8379 within the framework "Biophotonik" from the Federal Ministry of Education and Research, Germany (BMBF) is gratefully acknowledged.

Glossary

Bioaerosol An aerosol is a collection of particles suspended in a gas. The term refers collectively to both the particles and the gas in which the particles are suspended. When the point of interest is biotic particles in air, the aerosol is called bioaerosol.

Calcium dipicolinate The core of a spore has normal cell structures but is metabolically inactive. Up to 15% of the spore's dry weight may consist of dipicolinic acid complexed with calcium ions. Dipicolinic acid could be responsible for the heat resistance of the spore. Calcium may aid in resistance to heat as well as to oxidizing agents. The combination of calcium ions and dipicolinic acid may stabilize spore nucleic acids.

CCD Charge-coupled device: a light sensitive sensor for recording information on place-resolution level (e.g., images). An integrated circuit contains an array of linked, or coupled, capacitors. A CCD is one kind of semiconductor detector.

Chemometrics Chemometric research spans a wide area of different methods which can be applied in chemistry. There are techniques for collecting good data (optimization of experimental parameters, design of experiments, calibration, signal processing) and for getting information from these data (statistics, pattern recognition, modelling, structure–property relationship estimates).

Confocal A hole or slit is used to hide light out of the focus plane. With reduction of scattered light the effective resolution in confocal microscopy can be increased.

Cultivation media Differential media includes an indicator that causes visible, easily detectable changes in the appearance of the agar gel or bacterial colonies in a specific group of bacteria. For example, EMB (Eosin Methylene Blue) agar causes *E. coli* colonies to have a metallic green sheen, and MSA (Mannitol Salt Agar) turns yellow in the presence of mannitol-fermenting bacteria.

Eukaryote All organisms are referred to as eukaryotes which possess a nucleus and a cytoskeleton. Eukaryotes are normally greater as prokaryotes and there is a division of the cell in compartments. Eukaryotic cells are structured with cell organelles which are structures enclosed in the cytoplasm, such as mitochondria and the endoplasm reticulum. In contrast to prokaryotes, they contain a nucleus, which is separated from the cell plasma by the nuclear membrane.

Fingerprint region In IR and Raman spectroscopy, every substance analyzed shows a fingerprint region. By means of fingerprint analysis, similar substances can be differentiated.

Fluorescence Light emission from an electronically excited state into the electronic ground state.

Normalization of spectra Spectra are normalized in order to make different samples comparable. In chemometrics, common normalization methods are normalizing with respect to a certain band or normalizing such that all measured sample points have unit variance.

Microorganism taxonomy A lot of organizations deal with microorganisms (ATCC, DSM, ECACC, etc). To identify and differentiate every kind, a nomenclature is needed. Genus (e.g. *Bacillus*, *Staphylococcus*), species (e.g. *Bacillus subtilis*, *Staphylococcus cohnii*), subspecies (e.g. *S. cohnii* subsp. *urealyticum*) and a once-only number for the strain (e.g. *S. cohnii* subsp. *urealyticum* DSM 6669) are the elements of correct notation.

Micro-Raman set-up Raman spectrometer in combination with a microscope.

NAD(P)H Reduced form of NADP (nicotinamide adenine dinucleotide phosphate); significant coenzyme in anabolism (constructive metabolism) in cells.

Novelty detection Detection of data that does not belong to one of the known classes that the classifier was trained with. A **support vector machine** can be used for novelty detection.

Pathogen A pathogen is a biological agent that can cause disease to its host. A synonym of pathogen is "infectious agent". The term "pathogen" is most often used for agents that disrupt the normal physiology of a multicellular animal or plant. Pathogens can infect unicellular organisms from all the biological kingdoms.

Prokaryote Their cell type is labelled as a protocyte. Prokaryotes combine the domain bacteria and archaea and they are round or bacillary cells from organisms that do not contain a nucleus enclosed by a membrane. Normally they feature only one ring-shaped DNA molecule as a chromosome (nucleoid), which is located free in the cytoplasm. Prokaryotic cells contain ribosomes, cytoplasm and also repository material. They are not compartmented and do not enclose organelles, unlike eukaryotes.

Raman imaging The intensity of a certain spectral region is measured over a defined area and then plotted in a false color plot.

Raman mapping An area is scanned using defined steps between the measured points. A motorized x, y-table moves along the defined area and one spectrum is measured at every chosen position. With the help of a false color plot differences in the intensities of various marker bands can be visualized over the mapped area.

Raman spectroscopy Vibrational spectroscopic method based on the inelastic scattering of light in vibrating molecules.

Recognition rate For an estimate of the classification error probability of the final system, $N - 1$ samples are chosen as a training set and the isolated sample is classified. The average recognition rate is the arithmetic mean of the recognition rates for each strain and species and therefore equalizes the varying number of samples per strain and species in the database.

Riboflavin Riboflavin, also known as vitamin B_2 or vitamin G, is an easily absorbed, water-soluble micronutrient with a key role in maintaining human health. Vitamin B_2 is also required for red blood cell formation and respiration, antibody production and for regulating human growth and reproduction.

Running median filter A running median filter is a median filter that is shifted over a data sample (e.g., a measured spectrum). At each data position, the median is calculated over a certain filter size and the current data point is replaced by the median value.

Spore Specific bacteria produce special structures in the cell – endospores or spores. These are differentiated cells that are very resistant to dehydration, toxic and aggressive substances, aging and heat. Endospores can cease their metabolism and thereby need neither water, nutrients or oxygen. Additionally spores exhibit cell walls that inhibit water evaporation.

SVM (Support vector machine) A classifier that is used to distinguish between several classes (species) of bacterial spectra. Compared to other classifiers, a support vector machines does not try to estimate densities or model classes, but solves classification problems by looking for a separating curve. This is achieved under the constraint that the classes are separated with the largest possible margin. The solution given by the support vector machine for a certain set of parameters is a global optimum.

Key References

Raman Spectroscopy

B. Schrader, *Infrared and Raman spectroscopy*, VCH, Weinheim, **1995**.

G. Turell, J. Corset, *Raman Microscopy: Developments and Applications*, Elsevier, Amsterdam, **1996**.

Chemometric Methods

K. R. Beebe, R. J. Pell and M. B. Seasholtz, *Chemometrics: a Practical Guide*, John Wiley and Sons, New York, **1998**.

H. Martens, T. Naes, *Multivariate Calibration*, John Wiley and Sons, Chichester, **1993**.

Fluorescence Spectroscopy

H. J. Tanke, *Fluorescence Microscopy*, in Microscopy Handbooks Series, Vol. 35, Bios Scientific Publishers, **1997**.

T. Vo-Dinh, *Biomedical Photonics Handbook*, CRC Press, **2003**.

References

1 E. J. Davis, G. Schweiger, *The Airborn Microparticle: Its Physics, Chemistry, Optics and Transport Phenomena*, Springer-Verlag, Berlin, 2002.

2 J. Popp, M. Lankers, Mit optischer Spektroskopie auf der Spur von Bioaerosolen, *Nachr. Chem.*, 51(9) (**2003**), pp. 995–998.

3 M. T. Madigan, J. M. Martinko, J. Parker, *Brock Mikrobiologie*, Spektrum Akademischer Verlag, Heidelberg, 2002.

4 H. Hahn, D. Falke, S. H. E. Kaufmann, U. Ullmann, *Medizinische Mikrobiologie und Infektiologie*, 5th edition, Springer, Berlin, 2005.

5 S. F. Al-Khaldi, M. M. Mossoba, Gene and bacterial identification using high-throughput technologies: genomics, proteomics, and phenomics, *Nutrition*, 20(1) (**2004**), pp. 32–38.

6 H. M. Shapiro, Microbial analysis at the single-cell level: tasks and techniques, *J. Microbiol. Meth.*, 42(1) (**2000**), pp. 3–16.

7 M. Garg, M. K. Misra, S. Chawla, K. N. Prasad, R. Roy, R. K. Gupta, Broad identification of bacterial type from pus by ^{1}H NMR spectroscopy, *European J. Clin. Invest.*, 33(6) (**2003**), pp. 518–524.

8 B. Warscheid, C. Fenselau, A targeted proteomics approach to the rapid identification of bacterial cell mixtures by matrix-assisted laser desorption/ionization mass spectrometry, *Proteomics*, 4(10) (**2004**), pp. 2877–2892.

9 D. Naumann, S. Keller, D. Helm, C. Schultz, B. Schrader, FT-IR spectroscopy and FT-Raman spectroscopy are powerful analytical tools for the noninvasive characterization of intact microbial-cells. *J. Mol. Struct.*, 347 (**1995**), pp. 399–405.

10 D. Naumann, Infrared spectroscopy in microbiology, in: R. A. Meyers (ed.), *Encyclopedia of Analytical Chemistry*, Biomedical Spectroscopy, pp. 102–131, John Wiley and Sons, Chichester, 2000.

11 D. Naumann, D. Helm, H. Labischinski, P. Giesbrecht, The characterization of microorganisms by fouier-transform infrared spectroscopy (FT-IR), in: W. H. Nelson (ed.), *Modern techniques for rapid microbiological analysis*, pp. 43–96, VCH Publishers, New York, 1991.

12 Z. Filip, S. Herrmann, J. Kubat, FT-IR spectroscopic characteristics of differently cultivated bacillus subtilis. *Microbiol. Res.*, 159(3) (**2004**), pp. 257–262.

13 D. Naumann, FT-infrared and FT-Raman spectroscopy in biomedical research, in: H.-U. Gremlich, B. Yan

(eds.), *Infrared and Raman Spectroscopy of Biological Materials*, volume 24 of *Practical Spectroscopy Series*, pages 323–378, Marcel Dekker, New York, 2001.

14 T. Udelhoven, D. Naumann, J. Schmitt, Development of a hierarchical classification system with artificial neural networks and FT-IR spectra for the identification of bacteria, *Appl. Spectrosc.*, 54(10) (**2000**), pp. 1471–1479.

15 M. A. Miguel Gomez, M. A. Bratos Perez, F. J. Martin Gil, A. Duenas Diez, J. F. Martin Rodriguez, P. Gutierrez Rodriguez, A. Orduna Domingo, A. Rodriguez Torres, Identification of species of brucella using Fourier transform infrared spectroscopy, *J. Microbiol. Meth.*, 55(1) (**2003**), pp. 121–131.

16 J. Schmitt, H.-C. Flemming, FT-IR spectroscopy in microbial and material analysis, *Intern. Biodeter. Biodegrad.*, 41(1) (**1998**), pp. 1–11.

17 R. Goodacre, P. J. Rooney, D. B. Kell, Rapid analysis of microbial systems using vibrational spectroscopy and supervised learning methods: Application to the discrimination between methicillin-resistant and methicillin-susceptible staphylococcus aureus, *Proc. SPIE – Int. Soc. Opt. Eng.*, 3257 (Infrared Spectroscopy: New Tool in Medicine) (**1998**), pp. 220–229.

18 A. Fehrmann, M. Franz, A. Hoffmann, L. Rudzik, E. Wust, Dairy product analysis: Identification of microorganisms by mid-infrared spectroscopy and determination of constituents by Raman spectroscopy, *J. AOAC Int.*, 78(6) (**1995**), pp. 1537–1542.

19 M. Wenning, H. Seiler, S. Scherer, Fourier-transform infrared microspectroscopy, a novel and rapid tool for identification of yeasts, *Appl. Environm. Microbiol.*, 68(10) (**2002**), pp. 4717–4721.

20 N. A. Ngo-Thi, C. Kirschner, D. Naumann, Characterization and identification of microorganisms by FT-IR microspectrometry, *J. Mol. Struct.*, 661–662 (**2003**), pp. 371–380.

21 W. H. Nelson, J. F. Sperry, UV resonance Raman spectroscopic detection and identification of bacteria and other microorganisms, in: W. H. Nelson (ed.), *Modern techniques for rapid microbiological analysis*, pp. 97–143, VCH Publischers, New York, 1991.

22 W. H. Nelson, R. Manoharan, J. F. Sperry, UV resonance Raman studies of bacteria, *Appl. Spectrosc. Rev.*, 27(1) (**1992**), pp. 67–124.

23 Q. Wu, T. Hamilton, W. H. Nelson, S. Elliott, J. F. Sperry, M. Wu, UV Raman spectral intensities of E. coli and other bacteria excited at 228.9, 244.0, and 248.2 nm, *Anal. Chem.*, 73(14) (**2001**), pp. 3432–3440.

24 R. Manoharan, E. Ghiamati, S. Chadha, W. H. Nelson, J. F. Sperry, Effect of cultural conditions of deep UV resonance Raman spectra of bacteria, *Appl. Spectrosc.*, 47(12) (**1993**), pp. 2145–50.

25 M. Baek, W. H. Nelson, P. E. Hargraves, Ultra-violet resonance Raman spectra of live cyanobacteria with 222.5–251.0 nm pulsed laser excitation, *Appl. Spectrosc.*, 43 (**1989**), pp. 159–162.

26 S. Chadha, W. H. Nelson, J. F. Sperry Ultraviolet micro-Raman spectrograph for the detection of small numbers of bacterial cells, *Rev. Sci. Instrum.*, 64 (**1993**), pp. 3088–3093.

27 R. M. Jarvis, R. Goodacre, Ultraviolet resonance Raman spectroscopy for the rapid discrimination of urinary tract infection bacteria, *FEMS Microbiol. Lett.*, 232(2) (**2004**), pp. 127–132.

28 E. C. López-Díez, R. Goodacre, Characterization of microorganisms using UV resonance Raman spectroscopy and chemometrics, *Anal. Chem.*, 76(3) (), pp. 585–591, 2004.

29 R. Manoharan, E. Ghiamati, R. A. Dalterio, K. A. Britton, W. H. Nelson, J. F. Sperry, UV resonance Raman spectra of bacteria, bacterial spores, protoplasts, and calcium dipicolinate, *J. Microbiol. Meth.*, 11 (**1990**), pp. 1–15.

30 E. Ghiamati, R. Manoharan, W. H. Nelson, J. F. Sperry, UV resonance Raman spectra of bacillus spores, *Appl. Sepctrosc.*, 46(2) (**1992**), pp. 357–364.

31 K. M. Spencer, J. M. Sylvia, S. L. Clauson, J. A. Janni, Surface-enhanced Raman as a water monitor for warfare agents, *Proc. SPIE – Int. Soc. Opt. Eng.*, 4577 (Vibrational Spectroscopy-Based Sensor Systems) (**2002**), pp. 158–165.

32 J. Pendell Jones, N. F. Fell Jr., T. A. Alexander, K. Dorschner, C. Tombrello, B. Ritz Reis, A. W. Fountain III., Surface-enhanced Raman substrate optimization for bacterial identification, *Proc. SPIE – Int. Soc. Opt. Eng.*, 5071 (Sensors, and Command, Control, Communications, and Intelligence (C3I) Technologies for Homeland Defense and Law Enforcement II) (**2003**), pp. 205–211.

33 R. M. Jarvis, A. Brooker, R. Goodacre, Surface-enhanced Raman spectroscopy for bacterial discrimination utilizing a scanning electron microscope with a Raman spectroscopy interface, *Anal. Chem.*, 76(17) (**2004**), pp. 5198–5202.

34 R. M. Jarvis, R. Goodacre, Discrimination of bacteria using surface-enhanced Raman spectroscopy, *Anal. Chem.*, 76(1) (**2004**), pp. 40–47.

35 W. R. Premasiri, D. T. Moir, M. S. Klempner, N. Krieger, G. Jones II., L. D. Ziegler, Characterization of the surface enhanced Raman scattering (SERS) of bacteria, *J. Phys. Chem. B*, 109(1) (**2005**), pp. 312–320.

36 S. Efrima, B. V. Bronk, Silver colloids impregnating or coating bacteria, *J. Phys. Chem. B*, 102(31) (**1998**), pp. 5947–5950.

37 L. Zeiri, B. V. Bronk, Y. Shabtai, J. Czege, S. Efrima, Silver metal induced surface enhanced Raman of bacteria, *Colloid Surf. A: Physicochem. Eng. Asp.*, 208(1–3) (**2002**), pp. 357–362.

38 L. Zeiri, B. V. Bronk, Y. Shabtai, J. Eichler, S. Efrima, Surface-enhanced Raman spectroscopy as a tool for probing specific biochemical components in bacteria, *Appl. Spectrosc.*, 58(1) (**2004**), pp. 33–40.

39 S. Farquharson, A. D. Gift, P. Maksymiuk, F. E. Inscore, Rapid dipicolinic acid extraction from bacillus spores detected by surface-enhanced Raman spectroscopy, *Appl. Spectrosc.*, 58(3) (**2004**), pp. 351–354.

40 X. Zhang, M. A. Young, O. Lyandres, R. P. Van Duyne, Rapid detection of an anthrax biomarker by surface-enhanced Raman spectroscopy, *J. Am. Chem. Soc.*, 127(12) (**2005**), pp. 4484–4489.

41 C. L. Haynes, C. R. Yonzon, X. Zhang, and R. P. Van Duyne, Surface-enhanced Raman sensors: Early history and the development of sensors for quantitative biowarfare agent and glucose detection, *J. Raman Spectrosc.*, 36(6/7) (**2005**), pp. 471–484.

42 A. E. Grow, L. L. Wood, J. L. Claycomb, P. A. Thompson, New biochip technology for label-free detection of pathogens and their toxins, *J. Microbial Meth.*, 53(2) (**2003**), pp. 221–33.

43 W. F. Howard Jr., W. H. Nelson, J. F. Sperry, A resonance Raman method for the rapid detection and identification of bacteria in water. *Appl. Spectrosc.*, 34(1) (), pp. 72–5, 1980.

44 R. A. Dalterio, W. H. Nelson, D. Britt, J. F. Sperry, F. J. Purcell, A resonance Raman microprobe study of chromobacteria in water, *Appl. Spectrosc.*, 40 (**1986**), pp. 271–272.

45 R. A. Dalterio, M. Baek, W. H. Nelson, D. Britt, J. F. Sperry, F. J. Purcell, The resonance Raman microprobe detection of single bacterial cells from a chromobacterial mixture, *Appl. Spectrosc.*, 41 (**1987**), pp. 241–244.

46 S. K. Brahma, P. E. Hargraves, W. F. Howard, W. H. Nelson, A resonance Raman method for the rapid detection and identification of algae in water, *Appl. Spectrosc.*, 37 (**1983**), pp. 55–58.

47 W.-D. Wagner, W. Waidelich, Selective observation of chlorophyll c in whole cells of diatoms by resonant Raman spectroscpoy, *Appl. Spectrosc.*, 40 (**1986**), pp. 191–195.

48 K. Maquelin, C. Kirschner, L. P. Choo-Smith, N. van den Braak, H. P. Endtz, D. Naumann, G. J. Puppels, Identification of medically relevant microorganisms by vibrational spectroscopy, *J. Microbial Meth.*, 51(3) (**2002**), pp. 255–71.

49 C. Kirschner, K. Maquelin, P. Pina, N. A. Ngo Thi, L. P. Choo-Smith, G. D. Sockalingum, C. Sandt,

D. AMI, F. ORSINI, S. M. DOGLIA, P. ALLOUCH, M. MAINFAIT, G. J. PUPPELS, D. NAUMANN, Classification and identification of enterococci: A comparative phenotypic, genotypic, and vibrational spectroscopic study, *J. Clinical Microbiol.*, 39(5) (**2001**), pp. 1763–1770.

50 D. HUTSEBAUT, K. MAQUELIN, P. DE VOS, P. VANDENABEELE, L. MOENS, G. J. PUPPELS, Effect of culture conditions on the achievable taxonomic resolution of Raman spectroscopy disclosed by three bacillus species, *Anal. Chem.*, 76(21) (**2004**), pp. 6274–6281.

51 S. FARQUHARSON, L. GRIGELY, V. KHITROV, W. SMITH, J. F. SPERRY, G. FENERTY, Detecting bacillus cereus spores on a mail sorting system using Raman spectroscopy, *J. Raman Spectrosc.*, 35(1) (**2004**), pp. 82–86.

52 M. S. IBELINGS, K. MAQUELIN, H. PH ENDTZ, H. A. BRUINING, G. J. PUPPELS, Rapid identification of Candida spp. in peritonitis patients by Raman spectroscopy, *Clinical Microbiol. Infection*, 11(5) (**2005**), pp. 353–358.

53 K. MAQUELIN, L. P. CHOO-SMITH, H. P. ENDTZ, H. A. BRUINING, G. J. PUPPELS, Rapid identification of Candida species by confocal Raman microspectroscopy, *J. Clinical Microbiol.*, 40(2) (**2002**), pp. 594–600.

54 K. MAQUELIN, C. KIRSCHNER, L. P. CHOO-SMITH, N. A. NGO-THI, T. VAN VREESWIJK, M. STAMMLER, H. P. ENDTZ, H. A. BRUINING, D. NAUMANN, G. J. PUPPELS, Prospective study of the performance of vibrational spectroscopies for rapid identification of bacterial and fungal pathogens recovered from blood cultures, *J. Clin. Microbiol.*, 41(1) (**2003**), pp. 324–329.

55 C. XIE, Y.-Q. LI, Confocal micro-Raman spectroscopy of single biological cells using optical trapping and shifted excitation difference techniques, *J. Appl. Phys.*, 93(5) (**2003**), pp. 2982–2986.

56 C. XIE, Y.-Q. LI, W. TANG, R. J. NEWTON, Study of dynamical process of heat denaturation in optically trapped single microorganisms by near-infrared Raman spectroscopy, *J. Appl. Phys.*, 94(9) (**2003**), pp. 6138–6142.

57 H. YANG, J. IRUDAYARAJ, Rapid detection of foodborne microorganisms on food surface using Fourier transform Raman spectroscopy, *J. Mol. Struct.*, 646(1–3) (**2003**), pp. 35–43.

58 R. GESSNER, C. WINTER, P. RÖSCH, M. SCHMITT, R. PETRY, W. KIEFER, M. LANKERS, J. POPP, Identification of biotic and abiotic particles by using a combinationn of optical tweezers and in situ Raman spectroscopy, *ChemPhysChem*, 5 (**2004**), pp. 1159–1170.

59 J. W. CHAN, A. P. ESPOSITO, C. E. TALLEY, C. W. HOLLARS, S. M. LANE, T. HUSER, Reagentless identification of single bacterial spores in aqueous solution by confocal laser tweezers Raman spectroscopy, *Anal. Chem.*, 76(3) (**2004**), pp. 599–603.

60 K. C. SCHUSTER, I. REESE, E. URLAUB, J. R. GAPES, B. LENDL, Multidimensional information on the chemical composition of single bacterial cells by confocal Raman microspectroscopy, *Anal. Chem.*, 72(22) (**2000**), pp. 5529–5534.

61 K. C. SCHUSTER, E. URLAUB, J. R. GAPES, Single-cell analysis of bacteria by Raman microscopy: Spectral information on the chemical composition of cells and on the heterogeneity in a culture, *J. Microbiol. Meth.*, 42(1) (**2000**), pp. 29–38.

62 W. E. HUANG, R. I. GRIFFITHS, I. P. THOMPSON, M. J. BAILEY, A. S. WHITELEY, Raman microscopic analysis of single microbial cells, *Anal. Chem.*, 76(15) (**2004**), pp. 4452–4458.

63 A. P. ESPOSITO, C. E. TALLEY, T. HUSER, C. W. HOLLARS, C. M. SCHALDACH, S. M. LANE, Analysis of single bacterial spores by micro-Raman spectroscopy. *Appl. Spectrosc.*, 57(7) (**2003**), pp. 868–871.

64 Y.-S. HUANG, T. KARASHIMA, M. YAMAMOTO, H. HAMAGUCHI, Molecular-level pursuit of yeast mitosis by time- and space-resolved Raman spectroscopy. *J. Raman Spectrosc.*, 34(1) (**2003**), pp. 1–3.

65 Y.-S. HUANG, T. KARASHIMA, M. YAMAMOTO, T. OGURA, H. HAMAGUCHI, Raman spectroscopic signature of life in a living yeast cell, *J. Raman Spectrosc.*, 35(7) (**2004**), pp. 525–526.

66 M. D. BRASIER, O. R. GREEN, A. P. JEPHCOAT, A. K. KLEPPE, M. J.

VAN KRANENDONK, J. F. LINDSAY, A. STEELE, N. V. GRASSINEAU, Questioning the evidence for earth's oldest fossils, *Nature*, 416(6876) (**2002**), pp. 76–81.

67 J. W. SCHOPF, A. B. KUDRYAVTSEV, D. G. AGRESTI, T. J. WDOWIAK, A. D. CZAJA, Laser-Raman imagery of earth's earliest fossils. *Nature*, 416(6876) (**2002**), pp. 73–76.

68 M. MATHLOUTHI, J. L. KOENIG, Vibrational spectra of carbohydrates, *Adv. Carbohydr. Chem. Biochem.*, 44 (**1986**), pp. 7–86.

69 BPI, *Pharma Kodex-Richtlinien, Gesetze, Empfehlungen*, Volumes 1 and 2, 1st edition, Bundesverband der Pharmazeutischen Industrie e. V., Frankfurt am Main, 1994.

70 The rules governing medical products in the European Union, *Pharma Kodex-Richtlinien, Gesetze, Empfehlungen*, Volume 4, European Commission of the European Union, 1997.

71 Verordnung (EG) Nr. 853/20 4 des Europäischen Parlaments und des Rates mit spezifischen Hygienevorschriften für Lebensmittel tierischen Ursprungs, Frankfurt am Main, 29 April 2004.

72 J. ENG, R. S. LYNCH, R. M. BALABAN, Nicotinamide adenine dinucleotide fluorecence spectroscopy and imaging of isolated cardiac myocytes, *Biophys. J.*, 55(4) (**1989**), pp. 621–630.

73 I. TAKEBE, K. KITAHARA, Levels of nicotinamide nucleotide coenzymes in lactic acid bacteria, *J. Gen. Appl. Microbiol*, 9(4) (**1963**), pp. 31–41.

74 J. LONDON, M. KNIGHT, Concentrations of nicotinamide nucleotide coenzymes in micro-organisms, *J. General Microbiol.*, 44 (**1966**), pp. 241–254.

75 T. G. SCOTT, R. D. SPENCER, N. L. LEONARD et al., Emission properties of NADH-studies of fluorescence lifetimes and quantum efficiencies of NADH, AcPyADH and simplified synthetic models, *J. Am. Chem. Soc.*, 92(3) (**1970**), pp. 687–695.

76 J. R. LAKOWICZ, H. SZMACINSKI, K. NOWACZYK et al., Fluorescence lifetime imaging of free and protein-bound NADH, *Proc. Natl. Acad. Sci.* (**1992**), pp. 1271–1275.

77 A. GALLHUBER, *Pikosekunden-Fluoreszenzspektroskopie am Coenzym NADH*, Carl von Ossietzky Univ., Oldenburg, 1992.

78 H. SCHULZ-MIRBACH, Invariant features for gray scale images, in: G. SAGERER, S. POSCH, F. KUMMERT (eds.), *17. DAGM - Symposium "Mustererkennung", Bielefeld*, pp. 1–14, Reihe Informatik aktuell, Springer, Berlin/Heidelberg, 1995.

79 O. RONNEBERGER, E. SCHULTZ, H. BURKHARDT, Automated pollen recognition using 3D volume images from fluorescence microscopy, *Aerobiologia*, 18 (**2002**), pp. 107–115.

80 P. RÖSCH, M. SCHMITT, W. KIEFER, J. POPP, The identification of microorganisms by micro-Raman spectroscopy, *J. Mol. Struct.*, 661–662 (**2003**), pp. 363–369.

81 H. G. SCHLEGEL, *Allgemeine Mikrobiologie*, 7th edition, Georg Thieme Verlag, Stuttgart, 1992.

82 V. N. VAPNIK *The Nature of Statistical Learning Theory*, Springer, New York, 1995.

83 A.I. BELOUSOV, S.A. VERZAKOV, J. VON FRESE, A flexible classification approach with optimal generalization performance: Support vector machines, *Chemometrics and Intelligent Laboratory Systems*, 64 (**2002**), pp. 15–25.

84 A.I. BELOUSOV, S.A. VERZAKOV, J. VON FRESE, Applicational aspects of support vector machines, *Journal of Chemometrics*, 16 (**2002**), pp. 482–489.

85 B. HAASDONK, D. KEYSERS, Tangent distance kernels for support vector machines, in: *Proceedings of the 16th International Conference on Pattern Recognition*, Volume 2, pp. 864–868, 2002.

86 B. HAASDONK, *Transformation Knowledge in Pattern Analysis with Kernel Methods - Distance and Integration Kernels*, PhD thesis, Computer Science Department, University of Freiburg, Germany, 2005.

87 K. R. BEEBE, R. J. PELL, M. B. SEASHOLTZ, *Chemometrics: a Practical Guide*, Wiley, New York, 1998.

88 P. Rösch, M. Harz, K.-D. Peschke, O. Ronneberger, H. Burkhardt, H.-W. Motzkus, M. Lankers, S. Hofer, H. Thiele, J. Popp, Chemotaxonomic identification of single bacteria by micro-Raman spectroscopy: Application to clean-room-relevant biological contaminations, *Appl. Environm. Microbiol.*, 71 (**2005**), pp. 1626–1637.

89 M. Harz, P. Rösch, K.-D. Peschke, O. Ronneberger, H. Burkhardt, J. Popp, Micro-Raman spectroscopical identification of bacterial cells of the genus staphylococcus in dependence on their cultivation conditions, *Analyst*, 130(11) (**2005**), pp. 1543–1550.

90 P. Rösch, M. Harz, K.-D. Peschke, O. Ronneberger, H. Burkhardt, A. Schüle, G. Schmauz, M. Lankers, S. Hofer, H. Thiele, H.-W. Motzkus, J. Popp, Online monitoring and identification of bio aerosols, *Anal. Chem.*, 78(7) (**2006**), pp. 2163–2170.

91 P. Rösch, M. Harz, M. Schmitt, J. Popp, Raman spectroscopic identification of single yeast cells, *J. Raman Spectrosc.* 36 (**2005**), pp. 377–379.

92 M. Meyer, *Grundlagen der Informationstechnik*, Vieweg, Braunschweig/Wiesbaden, 2002.

93 T. Namioka, Theory of the concave grating. I., *JOSA*, 49(5, May) (**1959**), pp. 446–460.

94 W. R. McKinney, Ch. Palmer, Numerical design method for aberration-reduced concave grating spectrometers, *Appl. Opt.*, 26(15) (**1987**), pp. 3108–3118.

95 R. Riesenberg, W. Voigt, J. Schöneich, Compact spectrometers made by micro system technology, in: *Sensor 97, Nürnberg, 13–15 May 1997, Proceedings*, Volume 2, pp. 145–150, 1997.

96 A. Wuttig, R. Riesenberg, G. Nitzsche, Subpixel analysis of a double array grating spectrometer, *Proc. SPIE*, 4480 (**2002**), pp. 334–344.

97 R. Riesenberg, G. Nitzsche, A. Wuttig, B. Harnisch, Micro spectrometer and MEMS for space, *6th ISU Annual International Symposium, "Smaller Satellites: Bigger Business?"*, Strassbourg, 21–23 May 2001.

98 R. Riesenberg, A. Wuttig, G. Nitzsche, B. Harnisch, Optical MEMS for high-end microspectrometers, in: *Proc. Deutsche Raumfahrtkonferenz, 2002*, Volume 4928, pp. 6–14, 2002.

99 R. Riesenberg, Micro-mechanical slit positioning system as a transmissive spatial light modulator, *Proc. SPIE*, 4457 (**2001**), pages 197–203. invited paper.

100 M. Harwit, N. J. A. Sloane, *Hadamard Transform Optics*, Academic Press, New York, 1979.

101 A. Wuttig, Optimal transformations for optical multiplex measurements in the presence of photon noise, *Appl. Opt.*, 44(14) (**2005**), pp. 2710–2719.

102 P. Schreiber, B. Hoefer, P. Dannberg, U. D. Zeitner, High-brightness fiber-coupling schemes for diode laser bars, *Proc. SPIE*, 5876 (**2005**), pp. 1–10.

103 E. Prieto, Ch. Bonneville, P. Ferruit, J. R. Allington-Smith, R. Bacon, R. Content, F. Henault, O. LeFevre, P. E. Blanc, Great opportunity for NGST-NIRSPEC: a high-resolution integral field unit, *Proc. SPIE*, 4850 (**2003**), pp. 486–492.

4
Novel Singly Labelled Probes for Identification of Microorganisms, Detection of Antibiotic Resistance Genes and Mutations, and Tumor Diagnosis (SMART PROBES)

Oliver Nolte[1], *Matthias Müller, Bernhard Häfner, Jens-Peter Knemeyer, Katharina Stöhr, Jürgen Wolfrum, Regine Hakenbeck, Dalia Denapaite, Jutta Schwarz-Finsterle, Stefan Stein, Eberhard Schmitt, Christoph Cremer, Dirk-Peter Herten, Michael Hausmann*[1], *Markus Sauer*[1]

4.1
Introduction: The Requirement for Novel Probes

4.1.1
A Brief Historical Overview

Robert Koch founded the new science of medical microbiology in the mid-1880s, with his report that the common disease "tuberculosis" (killing every seventh adult in the late nineteenth century) is not a constitutive disease but a disease caused by a microorganism, the acid-fast rod *Mycobacterium tuberculosis*. The findings of Koch and his co-workers – at that time and during the subsequent decades – provided opportunities for the detection and identification of microbes and of targeting infectious disease-causing microorganisms in humans for the first time. The first attempts at specific treatment regimes were proposed by Sir Almroth Wright, a British physician who developed so-called vaccine therapy for the treatment of infectious diseases [1]. It was later the most famous pupil of Wright, Sir Alexander Fleming, who found that certain substances from filamentous fungi (e.g., *Penicillium notatum*) are able to inhibit the growth of microorganisms *in vitro* [2]. The isolation and purification of this substance, penicillin, can be seen as a landmark in medicine, marking the beginning of the modern antibiotic era.

Following the introduction of more and more antibiotic drugs into medicine, the end of the era of infectious diseases was anticipated. This is highlighted for instance by the quote of the Nobel award winner Sir Frank MacFarlane Burnet, who wrote in 1962: "One can think of the middle of the twentieth century as the end of one of the most important social revolutions in history, the virtual elimination of the infectious disease as a significant factor in social life" [3]. This view, however, turned out to be one of the most erroneous

1) Corresponding authors

Biophotonics: Visions for Better Health Care. Jürgen Popp and Marion Strehle (Eds.)

ISBN: 3-527-40622-0

Table 4.1: Overview of the most relevant antimicrobial (antibiotic and antiviral) resistant microorganisms. To show the state-of-the-art in molecular diagnostics (last column) only examples of available test systems are given. The list of available assays is by no means complete.

Resistant microorganism	Resistant to	Abbreviation	Description	Situation	Current state-of-the-art molecular diagnostics[2]
Staphylococcus aureus	methicillin	MRSA	Resistance against all penicillin and cephalosporin drugs, first strains identified which have decreased susceptibility for the last remaining antibiotic class, the glycopeptides (e.g., vancomycin)	Germany: 18.2% in 2003; 19.5% in 2004[1]; increasing in many European countries, e.g., 44.3% of all invasive *S. aureus* isolates in Greece in 2004 (EARSS-website)	Real time PCR [C] [6]/ [H] [7]
Group D streptococci	vancomycin	VRE	Emerging population of group D streptococci (*Enterococcus faecalis*, *Enterococcus faecium* and others). The resistance genes are often located on plasmids providing the possibility of transfer to other strains/species.	VRE outbreaks are increasing in the US and Europe; also co-carriage with MRSA reported [8]. Also aminoglycoside (high level) resistance reaches high rates in many European countries, e.g., Germany: 2001 31.2%; 2002 41.9%, 2003 46.9%, 2004 41.6%) [1]	LightCycler® VRE Detection Kit/Roche Diagnostics [C]/ [H] [9]
Streptococcus pneumoniae	penicillin		Emerging population of pneumococci, resistant to penicillin. Drug resistance is frequently adopted by horizontal genetic transfer.	Significant rates in European countries (source), e.g., Spain 9.6% in 2002, 7.3% in 2003, 8.8% in 2004 (*S. pneumoniae* non-susceptible to Oxacillin and/or Penicillin)[1].	Real time PCR [H] [10]microarray [11] [H]
Klebsiella pneumoniae	penicillins and cephalosporins	ESBL	Also other gram negative bacteria, e.g., *Escherichia coli*, *Proteus mirabilis*, *Salmonella* spp., producing enzymes that specifically hydrolyze beta lactam antibiotics (i.e., monobactams, cephalosporins)	Latin America 45%, Pacific Region 25%, Europe 23%, USA 8%, Canada 5% [12]	Real time PCR [H] [13]; microarray [H] [14]

Table 4.1: (continued)

Resistant microorganism	Resistant to	Abbreviation	Description	Situation	Current state-of-the-art molecular diagnostics[2]
Mycobacterium tuberculosis	divers	MDR-TB	Characterized by resistance to at least two first line drugs, rifampicin and isoniazid, but frequently being resistant to other drugs as well	Decrease in developed countries (e.g., Germany: 2.1% in 2003 [y -RKI-2005]) but increase for instance in Eastern Europe (e.g., Estonia, 36.9% resistant to at least one drug and 14.1% MDR prevalence [15]	Real time PCR [H] [16]; probe-based assays GenoType® MTBDR Hain Lifescience, INNO-LiPA Rif.TB Innogenetics [C]/[H], microarray [17][H], direct sequencing [H] [18]
Neisseria gonorrhoeae	tetracycline	tetR-GO	*Neisseria gonorrohoea*, resistant to tetracycline and other antimicrobials, mostly from the inadequate use of these drugs in Asia; resistance gene may be located on a plasmid.	Resistant rates close to 100% in some asian countries [19, 20], sporadic occurrence in Europe [21, 22]	PCR [H] [20]
HIV			Resistance occurs against reverse transcriptase inhibitors and protease inhibitors. Resistance against one or both inhibitor classes may occur.	Primary resistance rate reaches about 10% in Europe and about 14% in the US.	Combined phenotypic and genotypic test system, (e.g., PhenoSense GT assay/ViroLogic ; virco® TYPE HIV-1/virco [C] Array [23]; phenotypic test [24]

1) data obtained from the EARSS Website (http://www.earss.rivm.nl/) database (last visited/searched: July 25th 2005).
2) **[C]** = commercially available assay, **[H]** = method described as in-house assay, currently not available from a commercial provider.

perspectives ever held. Today, infectious diseases are the most frequent cause of death, accounting for over one-third of all casualties worldwide (**Table 4.1**). The frequent occurrence of completely new infectious diseases and/or microorganisms such as "bird flu", West Nile virus disease or the SARS-causing Corona virus provide a continuous challenge not only in terms of treatment but also for diagnostics.

Moreover, microorganisms possess the notorious tendency of developing mechanisms for escaping the action of antibiotic drugs (resistance). The situation is complicated because different microorganisms may share their acquired resistance with each other (i.e., mobility of resistance genes through horizontal genetic transfer). We know that some have pessimistically announced the end of the antibiotic era. In fact, medical microbiology, whether in academic institutions, clinics or industry, faces an increasing number of microorganisms resistant to anti-infective drugs. Following decades during which new antibiotics were developed by pharmaceutical companies [4] the pipeline is now virtually running dry. The development of new antibiotic drugs has become less attractive for industry owing to the high cost of research and development. Investments are at risk – among other reasons – because of the rapid development of resistance [5]. As a consequence, rapid identification of resistant microorganisms is required in order to control antimicrobial resistance development, to provide the best possible therapy for the affected patient, to reduce cost of treatment and to prevent the spread of diseases.

Unfortunately, the erroneous perspective on infectious diseases was not the only error of "modern" medicine of the 1960s. At that time all over the western world, cancer research centers were founded in order to investigate tumors, to find *the* reason for cancer and search for *the* cancer therapy. So many people at that time thought that within 20 years the problems of cancer would be more or less solved and an appropriate cancer therapy would routinely be administered to nearly all types of tumor.

Today, some 40–50 years later, cancer appears to be more complex than ever and neoplastic differentiations of cells and tissues into tumor cells and cancer tissues are multi-rationale developments. New insights into the molecular functioning of cells have elucidated many functional pathways in cancer development on the genomic and proteomic level. Novel methods and techniques use the results of molecular investigations for tumor marker definitions in cancer diagnostics, and although some highly effective therapeutics are on the market (e.g., Tyrosinkinase-Inhibitors for the treatment of chronic myeloid leukemia), improved diagnostic tools are required for a more detailed classification of tumor cells and improved, tumor-specific treatment.

4.1.2
Molecular Microbiology

The diagnosis of infectious diseases normally requires a disease-causing microorganism to be cultured from a swab or from tissue specimens that have been sent to a routine diagnostic laboratory. In the majority of cases, cultured microorganisms are identified by determining the biochemical profile, testing for specific metabolic reactions or verifying the absence/presence of certain virulence factors. Routine diagnostic service is usually completed once the antimicrobial susceptibility test is finished and a resistance profile of the microorganism is available. This traditional approach clearly has some serious limitations. Culture-based diagnosis is time-consuming, cumbersome and in some instances not possible. Viruses, for instance, need complex cell culture systems in order to be propagated, or, in many cases, are not replicated in cell culture at all. Even some bacterial pathogens fail to grow in the laboratory (e.g., the syphilis-causing spirochete *Treponema pallidum*) or need prolonged incubation times to grow (e.g., *Mycobacterium tuberculosis*, the etiologic agent of one of the most frequent infectious diseases in humans, tuberculosis).

In 1985, the enzymatic amplification of a part of the β-globin gene of humans was reported in *Science* [25]. This method, known as the polymerase chain reaction (PCR) has changed the situation in routine diagnosis dramatically. Many derivatives of classical PCR have been developed and reported as being used for the specific amplification of microbial DNA. The amplification of microbial DNA by PCR constitutes an indirect indication that the particular microorganism was present in the examined specimen. If no amplification is achieved (and all necessary controls are in place) this indicates that the particular microorganism was not present, at least not in detectable amounts. One important characteristic of the available PCR-based methods is that they are highly sensitive. Sensitivity is enhanced by the addition of specific fluorescently labelled probes. The probes should enhance the specificity of the amplification process as well, as they only bind to their specific target but not to a non-specific target that might also be present in a completed PCR reaction.

PCR (Polymerase Chain Reaction)

The polymerase chain reaction (PCR) is a process for the selective amplification of nucleic acid fragments. In principle, a PCR takes advantage of the mechanisms active during DNA duplication in the living bacterial cell. In living cells, duplication of DNA needs a starting point for DNA synthesis. This starting point is provided by the enzyme "primase" which adds a few nucleotides to the partially opened DNA double strand. These nucleotides provide the essential free hydroxy-end without which a DNA polymerase is not able to operate (i.e., synthesizing a new, complementary DNA strand). In the laboratory, target DNA by means of chromosomal DNA (viral, bacterial, fungal or any other origin) in aqueous solution serves as template for amplification. In order to provide the starting points for the DNA polymerase, two short (18 to 20 nucleotides ±5) oligonucleotides are added. These so-called "primers" not only determine the start point for a heat-stable DNA polymerase but define the actual DNA fragment to be amplified. By careful selection of primers, a PCR is therefore specific for a particular microorganism and specific for a defined region of that microorganisms genome. Amplification of that DNA fragment is achieved by applying a cyclic temperature profile (**Figure 4.1**).

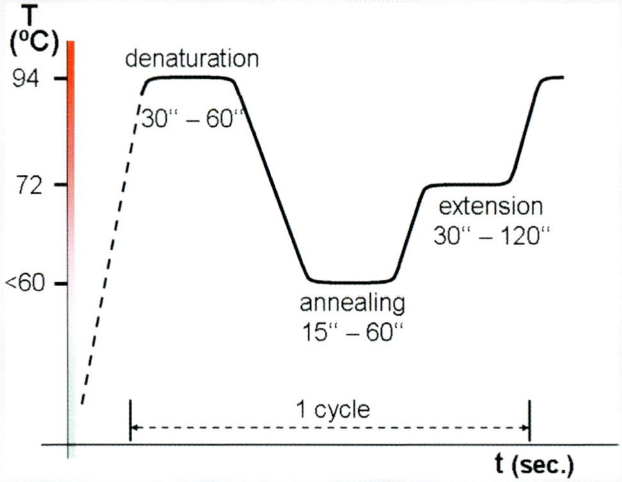

Figure 4.1 Temperature profile for a typical PCR. One cycle of denaturation, annealing and extension is repeated 30 to 40 times. The temperatures and times given are only typical examples.

The DNA double strand is "opened" by heat denaturation, making the hybridization sites for the primers accessible (**Figure 4.2**).

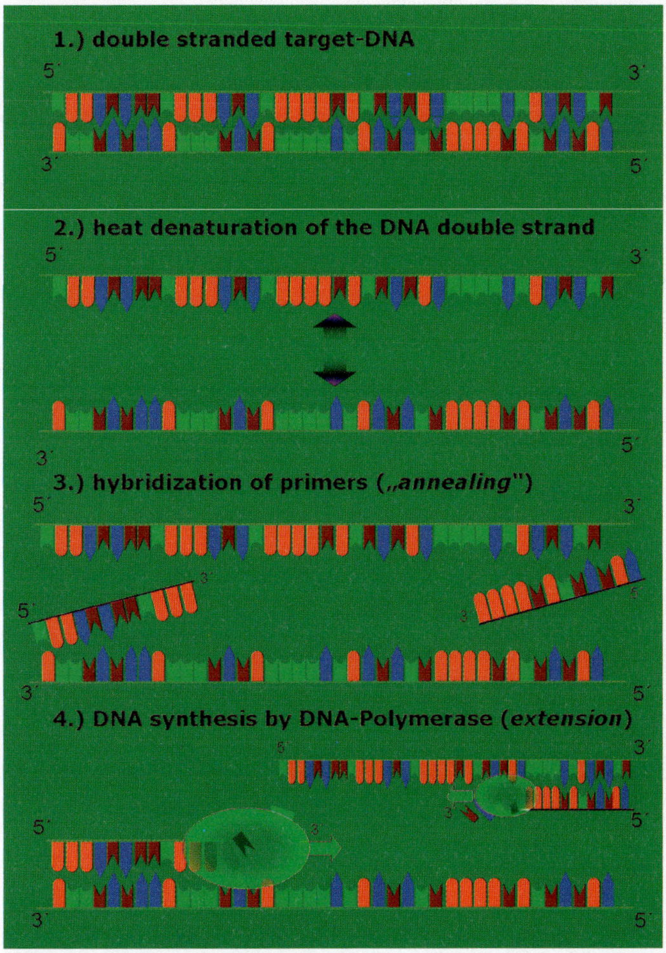

Figure 4.2 Basic process of PCR. The double-stranded template DNA needs to be denatured before the primers anneal. The hybridized primers provide the start (i.e., free hydroxy-end) for the polymerase and define the length and specificity of the amplification process. The newly synthesized double strand serves as a template for the next amplification round.

Once the double strand is opened completely the primers are able to "anneal". Effective annealing depends on the specific melting temperature of the primers (where the melting temperature is the temperature at which 50% of the primers have already hybridized to their target). Typically, annealing temperatures range between 40 °C and 60 °C but variations in each direction can be made if the melting temperature is higher or lower. Upon hybridization of the primers the temperature is shifted to 70–72 °C when the thermo-resistant polymerases commonly used exhibits their optimal performance. The time required to "copy" the desired DNA fragment flanked by the primers depends on the length of the fragment to be amplified and on the processivity of the polymerase used. However, in the classical PCR application, extension times typically range from half a minute to one or two minutes, and the length of the amplified fragments ranges from about 100 bp to 2000–10000 bp. As a result of this temperature profile, two new hetero-double strands exist. One strand of each is the original template strand while the second strand is the newly synthesized one, starting with the primer. Following heat denaturation, four template strands are now available. By repeating this cycle 30 to 40 times, a nearly exponential increase in amplified DNA fragments occurs. This amplified amount of short-length DNA is easy to detect (for instance by gel electrophoresis), compared with the genomic DNA of a microorganism in a clinical sample. Owing to this sensitivity, PCR has become a highly valuable tool in molecular diagnostics. Other applications are the selective amplification of genes or gene fragments in molecular biology for subsequent cloning, sequencing or other analysis. The process shown is only the basic approach and dozens of different modifications of this basic process have been developed. Modern state-of-the-art real time PCR, for instance, is performed very rapidly in thin capillaries or tiny wells of microplates, allowing completion of the amplification process within an hour or less. Amplification itself is monitored in "real time" through the release of fluorescent dyes. (A more detailed explanation of the PCR process can be found in the online dictionary "Wikipedia" by using the following URL: http://en.wikipedia.org/wiki/Polymerase_Chain_Reaction (checked October 5th 2005))

Since the introduction of PCR into routine microbiology in the early 1990s, a plethora of different methods and derivatives of methods have been described. Adaptations made from the original PCR are direct sequencing of PCR products (cycle sequencing), detection of conformational differences in single-stranded DNA molecules, differing by one or more nucleotide bases in sequence using, for example, single-strand conformation polymorphism SSCP [26], or sequence-specific oligonucleotide probe hybridization (in homogeneous test settings as well as chip based) [27, 28]. Nowadays, so-called real time PCR allows the completion of PCR reactions within less than one hour and without the need for gel electrophoresis. Instead, amplification is detected using specific probes and/or specific fluorescent dyes, which label newly synthesized double-stranded DNA. A wide variety of specific protocols for the detection of many currently known microorganisms have been published in the literature. The same is true for the molecular detection of resistance-determining genes and mutations. Whereas resistance-conferring plasmids or resistance-conferring genes are comparatively easy to detect, the specific detection of resistance-conferring single nucleotide polymorphisms (SNP, **Figure 4.3**) is more complicated.

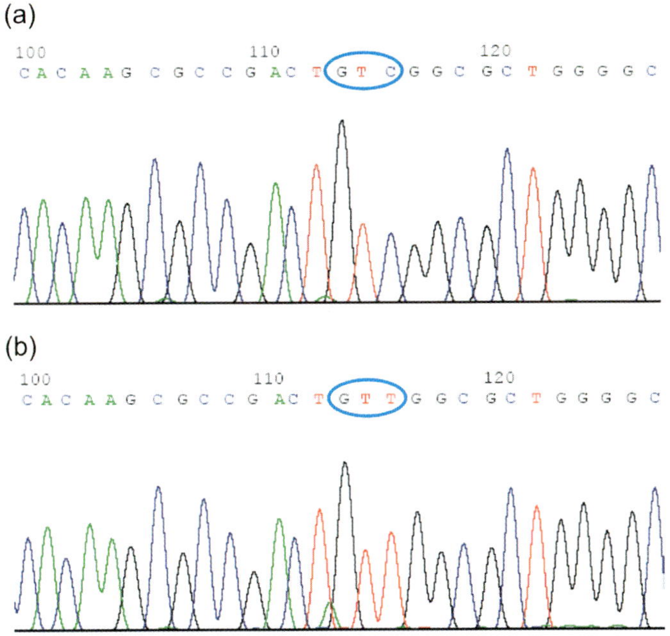

Figure 4.3 Two sequence files, showing a small portion of the *rpoB* gene of *M. tuberculosis* of a rifampicin-susceptible strain (top sequence) and a rifampicin-resistant strain (bottom sequence). The resistance conferring mutation is in position 116 of the sequence files (encircled). Both nucleotide sequences were obtained with an ABI310Prism automatic sequencer and cycle sequencing of specific PCR fragments. The numbering of the sequence corresponds to the number of nucleotide positions sequenced in that particular PCR-fragment and does not reflect the numbering of nucleotide positions in the entire gene. The mutation shown corresponds to the mutated amino acid position 456 in the *rpoB* gene product of *M. tuberculosis*.

However, modern genomics has led to novel methods for the detection of SNPs. Among these technologies are, for instance, microarray-based assays allowing the detection of a set of different disease-causing microorganisms and their major antibiotic-resistance determinants [29], the screening for SNPs conferring resistance (detection of extended spectrum β-lactamase-producing (ESBL) gram negative bacteria) [30], the detection of SNPs conferring rifampicin resistance in *M. tuberculosis* isolates using unlabelled hairpin-shaped oligonuleotides in real time PCR [31], or by pyrosequencing of amplicons obtained from sputum [32]. However, although a multitude of approaches have been developed and described, the majority of these assays are in-house assays, which until now have not been commercialized as validated diagnostic assays. Even microarrays, which have been under development for more than a decade now, are still in-house procedures and are estimated to become a diagnostic feature in specialized settings perhaps within the next ten years [33].

Besides in-house protocols (which are usually of limited distribution) a number of molecular assays are commercially available, e.g., for the detection of the most prominent pathogens like *M. tuberculosis*, *Clamydia pneumoniae* or HIV in swabs or tissue samples. A commercial assay also allows the detection of resistance-conferring mutations (InnoLiPa Rif-TB probe assay to screen for rifampicin-resistant *M. tuberculosis* isolates) evaluated for example by Herrera et al. [34]. These commercial assays have the advantage that they are highly reproducible and usually comparatively easy to perform. The sensitivity and specificity of these assays are assessed rigorously, first by the manufacturer before they are introduced to the market and finally by users after their introduction. The common assays have also been proven to be useful tools in diagnostics by nation-wide proficiency testing in many countries. Despite these positive characteristics, the existing commercial PCR methods, however, display some serious limitations: The majority of commercial assays are adapted to specific and complex instruments. Owing to the high development cost and the complexity of the instrumentation used, each individual assay is very expensive, and therefore not available to laboratories operating in developing countries. The distribution of commercial assays is thus limited to laboratories in the developed countries whilst laboratories in developing countries are still limited to using the much cheaper but less standardized in-house protocols. While in the developed countries a bunch of different high tech methods for diagnostics is available, some of the poorer countries cannot even afford basic amplification approaches in their diagnostic settings.

4.1.3
Needs and Expectations of Molecular Diagnostics in the Twenty-first Century

Infectious diseases, disease-causing microorganisms and antibiotic resistance issues are global challenges. None of these problems can be seen as a geographically restricted problem. In the slipstream of economic development in developed countries, the incidence of infectious diseases has declined remarkably. In contrast, developing countries are still prone to high incidence rates and high mortality rates due to poor living standards, causing reduced life expectancy and accounting for social problems not only in those countries affected but also in the so-called developed world. Diseases and disease-causing microorganisms are transmitted from one country to another, from one continent to another, within hours or days. Globalization forces people more and more to move from one region of a continent to another, from one country to another. Globally operating companies send their employees to developing countries to provide the basis for later economic benefit. Bioterrorism and biological warfare have come into public focus and have caused novel challenges for public health authorities. Thus, the needs and requirements for molecular diagnostics (in particular for the diagnosis of infectious diseases) will need to change from regional or federal needs (influenced by and adapted to the epidemiology of infectious diseases in a particular country) to needs and expectations on a global scale (taking into account trends and novel developments, as well as global epidemiology). Future diagnostics should contribute to the effective management of infectious diseases and affected patients in order to provide the best available treatment for affected patients as fast as possible and in order to minimize cost (to the individual, to the insurance companies and thereby to the public). Effective management of infectious diseases on a global scale is necessary in order to prevent widespread diseases in a population, thereby contributing to an increase in living standards and economic power.

Taking into account global perspectives and future trends, the following needs and expectations can be defined:

1. Fast, efficient and reproducible diagnostic methods are required, allowing the identification of a disease-causing microorganism as well as the detection of resistance genes or resistance-conferring mutations in the genome of that particular microorganism, preferably within a very short time and within a single test setting.

2. Cost-effective diagnostic test settings are required, which are affordable for laboratories all over the world.

3. Test settings need to be further standardized, replacing the various in-house methods established all around the world.

4. Future test settings must be easy to perform, ensuring wide distribution and use, and minimizing the risk of erroneous results.

Besides the dramatic developments of infectious diseases and the requirements for their diagnosis and control, another type of life risk is no less dramatic and also global, the increase in tumor diseases. Moreover, recent findings have indicated an often close coincidence of virus infections with tumor induction, suggesting a causal correlation of infectious disease and tumor development.

As with infectious diseases, molecular techniques derived from PCR have begun a triumphant success in modern tumor diagnosis and therapy control. Fluorescent DNA probes have contributed considerably to the detection of various structural and numerical tumor-specific aberrations in the genomes of single cells. Although this progress in medical routine is still not exhausted, the need for higher specificity and sensitivity in the probing techniques accompanied by demands for cost-effectiveness are continuously growing.

The need for new probes is therefore part of the need for a completely new avenue of diagnostic methods. New probes providing higher sensitivity than those currently in use are absolutely essential. These new probes need to be available in simplified, miniaturized test settings, matching the concepts of a "lab on a chip" [35]. Looking to the future, the next generation of test settings should work without amplification of target DNA and on the sensitivity level of an individual aberrant cell, requiring highly sensitive probes and even more sensitive instrumentation. This instrumentation, in turn, must be robust, affordable and easy to use. Molecular diagnostics thus faces its next challenge, a new era of rapid point-of-care diagnosis.

4.1.4
Development of Novel Probes and Assays

To determine the presence of a specific DNA sequence, it is generally necessary to hybridize fluorescently labelled oligonucleotide probes and immobilize the hybrids on a solid surface. After removal of the non-hybridized probes with a washing step, the remaining fluorescence intensity is measured. Unfortunately, the requirement that non-hybridized probe molecules must be removed precludes the use of this technique for online monitoring of hybridization. In addition, the physical separation of non-hybridized probe molecules in a heterogeneous assay decreases the sensitivity owing to non-specific binding of probe molecules to the surface. Therefore, homogeneous assays using fluorescently labelled oligonucleotides represents a desirable alternative. During the last decades several homogeneous assays based on fluorescently labelled probe molecules have been developed. Among these are

1. double-strand probes [36]
2. single-strand probes [37]
3. cleavable TaqMan probes [38].

However, these probes do not directly indicate the presence of the target DNA sequence but indirectly its amplification by PCR. In other words, the sensitivity of such assays is based on PCR amplification rather than on the probe or the detection method used. Furthermore, they are limited by their moderate specificity and sensitivity.

On the other hand, higher detection sensitivity can be achieved by application of recently developed techniques based on the detection of single molecules [39]. For example, fluorescence correlation spectroscopy (FCS) originally developed in 1972 has experienced a renaissance since 1992 for detecting and investigating minute amounts of proteins and nucleic acids [40]. FCS has successfully been used to measure the hybridization kinetics of fluorescently labelled oligonucleotides to target DNA [41] and the cleavage of M13DNA by several restriction enzymes [42].

The difficulty in detecting the hybridization of labelled oligonucleotide probes to their target DNA sequence arises from the law of mass action. In order to shift the equilibrium towards the bound probe–target complex at low concentrations of target DNA ($<10^{-10}$ M) an excess of fluorescent probe is required, which leads to a highly unspecific fluorescent background of free probe molecules. One way to circumvent this problem is two-color fluorescence cross-correlation spectroscopy (FCCS) which is able to discriminate the bound state from the unbound state by simultaneous detection of two spectrally different dye molecules in the bound state only. The background fluorescence of the unbound state differs from that of the bound state by the fluorescence emission of only one of the two dyes used for labelling [43]. Using two-color detection in combination with two differently labelled nucleic acid probes complementary to different sites on the target DNA, a homogeneous assay can be performed in a concentration range even below 10^{-10} M. This approach has been successfully used to detect specific nucleic acid sequences in non-amplified genomic DNA [44]. Although, it has been shown that this principle can be used widely in the field of life sciences, its applicability for the development of fast and cheap assays suffers from the more complex demands of the experimental methods as well as of the design and synthesis of the employed probes. In order to reach the requirements of novel diagnostic tools as outlined above, the probes and the detection methods have to be simplified instead. Here, fluorescently quenched DNA hairpin probes such as MOLECULAR BEACONS and SMART PROBES offer an elegant alternative.

4.1.5
Novel Probes for *In Vitro* and *In Vivo* Diagnostics

Fluorescent probes are nowadays essential tools in many fields of biological and biomedical research and molecular diagnostics, e.g., molecular pathology. They offer attractive possibilities to label biomolecules in cells with specificity and sensitivity previously unknown so that spectroscopic and microscopic visualization of cellular sub-compartments down to the single molecule level have become possible. In particular, fluorescent DNA probes have become versatile tools routinely applied in medical research, diagnosis and therapy. Fluorescence *in situ* hybridization (FISH) is a routine method by which specific genomic DNA targets and DNA sequence alterations, e.g., numerical or structural gene alterations, can be visualized in cell nuclei and on metaphase spreads. FISH can be performed on isolated cells in suspension, blood or bone marrow cell smears and tissue sections (fresh, frozen, fixed and paraffin-wax embedded). The method has been known for more than 20 years and is an integrated part of medical diagnosis, e.g., in pathology, haematology, genetics, oncology and cardiology [45]. In these fields FISH can also be used as a prognostic and predictive tool in the management of the treatment of patients.

The intensive use of FISH in tumor diagnostics has induced several companies to produce ready-to-use kits for many chromosomal genome loci and disease-specific genomic aberrations. In addition ready-to-use kits are available for whole chromosome painting of each chromosome/chromosome arm, and specific labelling of each centromere and telomere. These kits usually contain a mixture of single-stranded DNA sequences of a given genome region either carrying fluorophores, or hapten or steroid molecules (e.g., biotin or digoxigenin) for specific labelling; the DNA is dissolved in an appropriate buffer supporting cell penetration, target DNA denaturation and specific probe binding. Usually the DNA strands of probe kits are derived from BAC-, Cosmid-, or YAC-clones which are amplified by the standard techniques of molecular biology. Most of these clones are available in so-called DNA libraries established all over the world (see for instance www.ensembl.org). They have been constructed by means of molecular biology, e.g., isolating, enzymatic cutting and cloning of DNA from sorted metaphase chromosomes.

Although the sophisticated techniques of molecular biology have made nearly any genome target accessible for DNA probes, some shortcomings exist in specificity, sensitivity and cell treatment:

1. Many probes, especially BAC probes with a typical length in the order of 100 kb (kilobases DNA), do not exactly map the genome target regions of diagnostic interest. In particular, small genes or translocation breakpoint regions only a few kilobases long cannot be efficiently labelled, so that in many cases several neighboring target sites are probed simultaneously. Moreover, deletions and mutations of a few hundred bases only, as well as the exact breakpoints, are not accessible by standard FISH probes.

2. Probe design so far needs a complex preparation in molecular biology and has several limitations in versatility.

3. Owing to the base composition of the probes the labelling efficiency varies (different binding energy), which requires complex labelling strategies in cases where multiple DNA probes with different spectral signatures are required for diagnosis.

4. Most FISH protocols are work-loaded and time-consuming. Moreover, all current protocols require a denaturation step of the native DNA double strand in the cell nucleus in order to bind the single-stranded DNA probe to the complementary single-stranded target sequence. This denaturation step is performed by heat treatment (typically over 70 °C) accompanied by an extensive use of chaotropic agents like formamide [46].

5. Because target denaturation is necessary, FISH cannot be performed in living cells [47] and destructive effects on the nanostructure of chromatin cannot be excluded [48].

These limitations severely affect FISH diagnosis and therapy control, where gentle specimen treatment is required, as well as cytogenetic and biophysical research, where specific fluorescence labelling of chromatin in nuclei of living cells is required. Recently, COMBO-FISH (COMBinatorial Oligo FISH) has been developed [49] to specifically label genome targets defined precisely by a beginning and end nucleotide in the human DNA sequence database, so that labelling is strongly focused on the regions of interest. COMBO-FISH takes advantage of homopurine/homopyrimidine oligonucleotides that, depending on their 3′- to 5′- strand orientation, form double strands (via so-called Watson–Crick bonding) as well as triple helices (via so-called Hoogsteen bonding) with complementary homopurine/homopyrimidine sequences in intact genomic DNA.

COMBO-FISH in general works without the use of chaotropic agents, and is less destructive of chromatin morphology especially in cases of sensitive cell specimens. Moreover, Hoogsteen bonding can be performed without prior thermal denaturation of the DNA target sequence. This very gentle specimen treatment offers the principal possibility for DNA labelling in living cells with high specificity and without any modification of gene activity and expression. For COMBO-FISH of a given genome target, a set of distinct, singularly co-localizing oligonucleotide probes of about 15–30 bases each, are computer selected and configured from the human genome database (for details see Section 4.4.5) and synthesized as PNA or DNA homopurine/homopyrimidine probes using widely automated oligonucleotide synthesis procedures.

Analysis of human genome databases has shown that homopurine/homopyrimidine sequences longer than 10 DNA bases are nearly homogeneously distributed over the genome and that they represent about 1–2% of the entire genome. Realizing that, for instance, the observation volume in a confocal laser-scanning microscope (equipped with a high numerical aperture lens) contains only a few femtoliters (10^{-15}), and corresponds on average to a $\sim$ 250 kb chromatin domain in a normal mammalian cell nucleus, this volume should typically contain 150–200 homopurine/homopyrimidine stretches. Hence, also smaller genes in the order of a few kilobases can also be labelled using singularly co-localizing oligonucleotide stretches simultaneously.

Owing to the diffraction of a microscope lens, the fluorescence signals of the configured oligonucleotide probe set merge into a typical, nearly homogeneous FISH "spot". Nevertheless, some of the oligonucleotide probes may have additional binding sites somewhere else in the genome. Although clusters in the genome with a few of these oligonucleotide probes are excluded during computer configuration of the probe set, a severe fluorescence background may be expected from causes other than small minor binding sites. In particular, fluorescent probes nonspecifically attached to chromatin, and free probe material, contribute to the fluorescence background. This problem increases in those cases where stringent washing has to be omitted, i.e., in cases of fragile specimen and live-cell labelling.

Despite the advantages of COMBO-FISH, which overcome all the problems of standard FISH described above, the low number of oligonucleotide stretches labelling a small genome target results in a low number of available fluorophores in the target region only. If this would be insufficient for an unequivocal microscopic interpretation, the fluorescence signal-to-background ratio must be improved as much as possible. One possibility would be an amplification of the signal, which is severely restricted by the limited number of fluorophores that can be attached to individual oligonucleotides.

Another possibility to improve the signal-to-background ratio in fluorescence measurements is to reduce the background. Since washing procedures have to be omitted, especially for live-cell labelling, novel generations of "intelligent" probes are required that only fluoresce when they specifically bind to their DNA target site. With the development of SMART PROBES such types of probe have become available for the routine user.

4.2
The Principle of SMART PROBES

The idea underlying the elimination of background fluorescence is to switch light on only upon recognition and subsequent binding of the probe to the target. In terms of fluorescence spectroscopy the process of target recognition has to be linked to a photophysical process that activates fluorescence of an otherwise nonfluorescent or quenched probe. This can be achieved by relating target recognition to a conformational change of the probe that switches the state of a fluorescent dye between quenched and unquenched forms. Several mechanisms of fluorescence quenching, such as FLUORESCENCE RESONANCE ENERGY TRANSFER (FRET) or PHOTO-INDUCED ELECTRON TRANSFER (PET) can be found in the literature, together with numerous substances that can be employed as quenchers [50]. FRET or proximity quenching is used by the MOLECULAR BEACON principle [51,52] in which donor/acceptor- or dye/quencher-labelled molecules change their conformation (and hence their donor/acceptor or dye/quencher distance) upon binding to the target. MOLECULAR BEACONS are single-stranded nucleic acid molecules that adopt a stem–loop structure and are ideally suited for highly sensitive homogeneous DNA binding assays. The loop consists of a probe sequence that is complementary to a portion of the target sequence, whereas the stem is formed by annealing of two complementary arm sequences which are unrelated to the target sequence (**Figure 4.4a**).

The hairpin oligonucleotide is labelled with a fluorescent dye at the 5′-position and another fluorophore or external quencher at the 3′-terminus. Upon close contact of the two fluorophores, quenching occurs, i.e., the closed hairpin oligonucleotide shows only weak fluorescence. When the hairpin probe encounters a target molecule, it forms a longer and more stable hybrid than can be formed by the arm sequences. Consequently, the MOLECULAR BEACON undergoes a conformational change that forces the FRET-donor and -acceptor or quencher apart thereby increasing the measured fluorescence intensity of the donor dye. In addition, hairpin-shaped oligonucleotides are superior for distinguishing single-point mutations (SPMs).

However, the specific labelling of both termini of MOLECULAR BEACONS with different fluorescent dyes is cumbersome and inefficient. Incomplete labelling of the probes with acceptor dye or quencher results in the unquenched donor-only probes contributing to the unwanted fluorescence background. Furthermore, their application in inhomogeneous assays requires further modifications to allow immobilization of the MOLECULAR BEACONS, e.g., on microspheres or solid supports [53]. Therefore hairpin-shaped oligonucleotide probes that carry only a single fluorescent dye constitute a highly desirable alternative.

Instead of using interactions between two extrinsic probes, interactions of fluorophores with DNA bases or amino acids can be used for specific detection of DNA or RNA sequences and antibodies at the single-molecule level [54]. Nucleobase-specific interactions have been reported for several commercially available organic dyes, including rhodamine, oxazine, fluorescein, stilbene, and coumarin derivatives [55]. Recently, we introduced a novel method to synthesize DNA hairpins – so-called SMART PROBES – that takes advantage of the fact that several fluorophores are selectively quenched by guanosine residues (**Figure 4.4b**) [56]. The method uses the differences in specific properties of naturally occurring nucleotides, in particular the low oxidation potential of the DNA base guanosine [55] and the tendency of many fluorophores to aggregate in aqueous environments to decrease their water accessible area. Thus, depending on the reduction potential of the fluorophore used to label the DNA hairpin, efficient photo-induced intramolecular electron transfer (PET) occurs in the excited state on contact with guanosine. In contrast to electronic energy transfer-based systems, where long-range dipole–dipole interactions occur, these probes require close contact between the fluorophore and guanosine residue for efficient PET. With careful design of these conformational flexible probes and the use of appropriate fluorophores (rhodamine and oxazine dyes are very suitable candidates), efficient single-molecule-sensitive DNA hairpins can be produced [56]. If quenching interactions between the

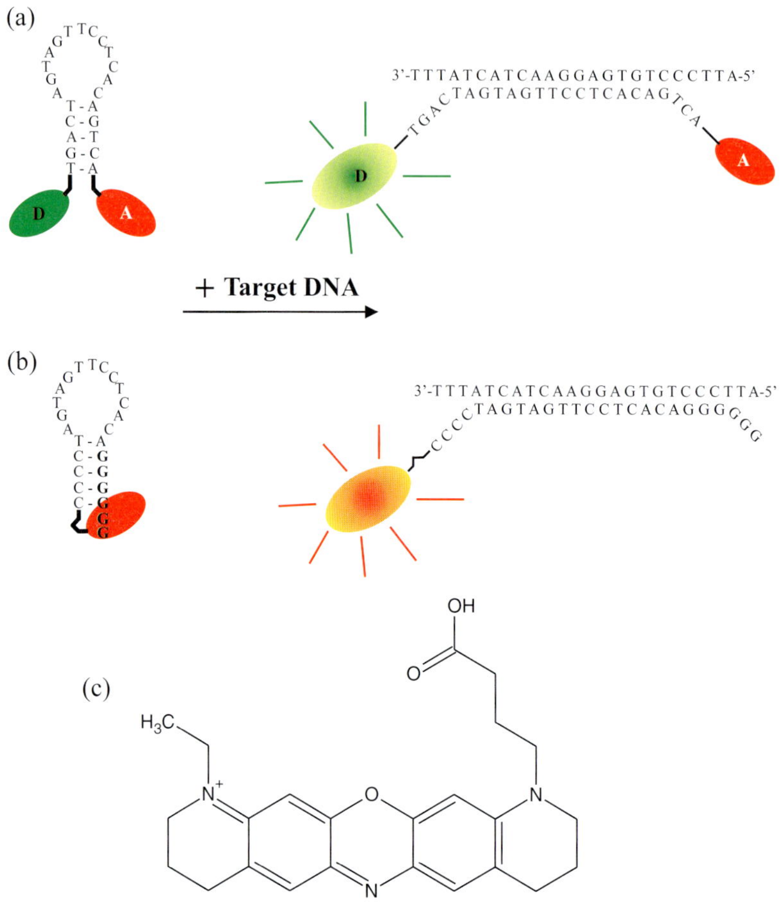

Figure 4.4 Sketch showing the principles of (a) MOLECULAR BEACONS and (b) SMART PROBES. In both systems the loop consists of a probe sequence complementary to a portion of the target DNA, whereas the stem is formed by annealing two complementary arm sequences which are partly unrelated to the target sequence. In MOLECULAR BEACONS the hairpin oligonucleotide is labelled with a fluorescent dye at the 5′-end and another chromophore or quencher at the 3′-end. Close contact of the two chromophores quenches the fluorescence by FLUORESCENCE RESONANCE ENERGY TRANSFER (FRET) or proximity quenching. When the hairpin probe encounters a target molecule, it forms a longer and more stable hybrid than formed by the arm sequences. Consequently, the MOLECULAR BEACON undergoes a conformational change that forces the FRET-donor and -acceptor apart thereby increasing the measured fluorescence intensity of the donor. In the case of SMART PROBES, 4–5 guanosine residues terminating the 3′-end are responsible for the strong fluorescence quenching of the dye labelled to the 5′-end. Upon addition of an excess of target sequence and subsequent hybridization to the target the stem is forced apart and the quenching interaction between the dye and the guanosine residues is interrupted, resulting in a strong increase in fluorescence intensity. (c) Molecular structure of the oxazine dye MR121 used to label the SMART PROBES.

fluorophore and the guanosine residue are diminished upon specific binding to the target, for example because of binding of a complementary DNA sequence or because of cleavage by an endonuclease enzyme, fluorescence of the DNA hairpin is restored (**Figure 4.4b**). DNA hairpins labelled with a single oxazine dye at the 5′-end increase fluorescence upon hybridization six-fold [56], which may provide a basis for a cost-effective and highly sensitive DNA/RNA detection method. Very recently [57], immobilized DNA hairpins based on nucleobase-specific fluorescence quenching of an oxazine dye have been used successfully to detect the presence of a sub-picomolar target DNA sequence. Therefore, SMART PROBES offer an elegant alternative to conventional MOLECULAR BEACONS based on electronic energy transfer processes.

The dyes used for synthesis of SMART PROBES have to be carefully selected. Useful fluorescent dyes are specifically influenced by their environment, i.e., changes in fluorescence emission exhibit information about neighboring groups and differences in the polarity of the microenvironment. In more detail, the principle of SMART PROBES makes use of an interaction between oxazine and rhodamine dyes with guanosine residues, one of the four naturally occurring bases in DNA, which results in more or less pronounced static quenching, depending on the dye used [58]. Covalent linking of these dyes to oligonucleotides terminated by four or five guanosine residues results in a diminished fluorescence quantum yield and a shorter fluorescence lifetime, depending on the distance between the guanosine residue and the fluorescent dye. In addition, it was shown that the greatest effect on the fluorescence kinetics of the dyes occurs when the guanosine residues are located in close proximity to the dye, i.e., upon contact formation [57, 58]. Excitation of the fluorophore leads to an electron-transfer reaction from the ground-state guanosine to the first excited singlet state of the fluorophore, which is reflected in strong fluorescence quenching. To reduce the background due to auto-fluorescence of the sample, we used an oxazine derivative (MR121) or alternatively the dye ATTO 655 (commercially available at ATTO-TEC GmbH, Siegen, Germany) as the fluorescent labels in most experiments, which both absorb and emit in the red wavelength range (absorption maximum ~660 nm, emission maximum ~680 nm). In addition, the use of red-absorbing dyes enables efficient excitation of the sample with cost-effective and robust diode lasers emitting, for example, at 635 nm.

Covalent attachment of the oxazine derivative MR121 (**Figure 4.4c**) to the 5′-end of an appropriately designed hairpin oligonucleotide decreases the fluorescence intensity and lifetime. As shown in **Figure 4.5**, upon addition of a 100-fold excess of target sequence to a solution of SMART PROBES, a longer and more stable hybrid is formed with the target sequence. Thus, the measured fluorescence intensity increases. Using standard fluorescence spectrometers, synthetic target sequences can be easily detected

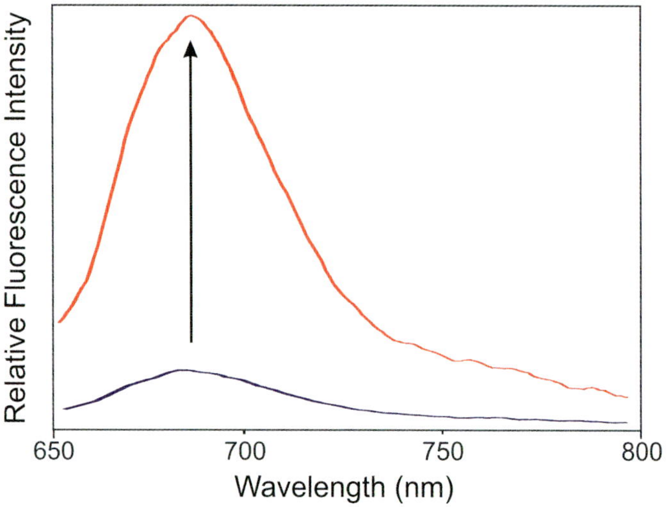

Figure 4.5 Relative fluorescence emission spectra of SMART PROBE MR121-C_6-5′-CCCCC T_{15} GGGGG-3′ in the absence (lower curve) and presence (upper curve) of a 100 fold excess of complementary DNA sequence (5′-CCCCC A_{15}-3′) measured in 100 mM Tris-borate buffer (pH 7.4) containing 140 mM NaCl and 5 mM $MgCl_2$ (excitation at 630 nm).

down to the nanomolar range [56]. The discrimination between closed and hybridized SMART PROBE can be further improved by the substitution of guanosine by 7-deazaguanosine residues. The lower oxidation potential of 7-deazaguanosine of $E_{ox} = 1.00$ V [58] produces higher quenching efficiencies. Using 7-deazaguanosine residues as quencher and additional guanosine residues as a single-stranded overhang at the 3′-end can yield enhancement factors between closed and hybridized SMART PROBES of up to 25 [58].

4.2.1 SMART PROBE Design

At first glance, the design of SMART PROBES seems to be fairly straightforward. One simply has to identify the desired target sequence, order the complementary amino-modified oligonucleotide ($\sim$ 20 bp) extended by four or five G/C base pairs forming the stem of the hairpin, and label a suitable fluorescent dye to the cytosine side of the stem, which is selectively quenched by guanosine. Although experiments with synthetic target sequences (short complementary oligonucleotides) show the expected increase in fluorescence intensity due to spontaneous hybridization at room temperature, preliminary hybridization experiments with genomic DNA of mycobacteria amplified by PCR showed only a small increase in fluorescence intensity due either to in-

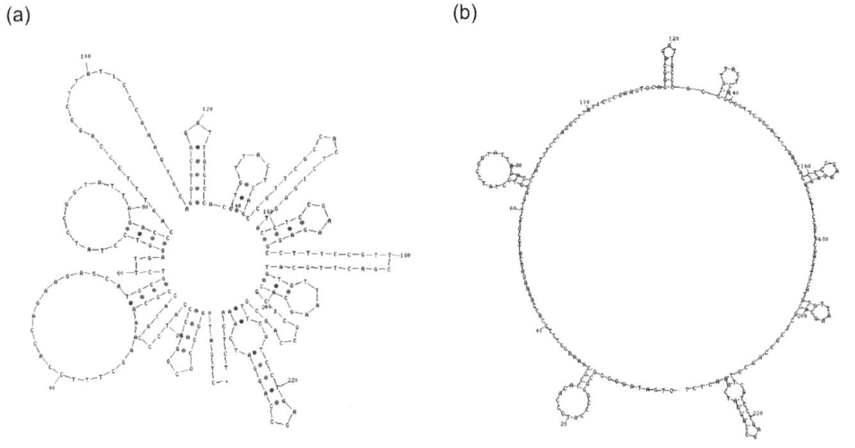

Figure 4.6 Secondary structures of the antisense strand of the 240-bp amplicon of *Mycobacterium avium* modeled with MFOLD at 45 °C (a) and 50 °C (b) for a Na^{+} concentration of 300 mM without Mg^{2+}.

efficient quenching within the SMART PROBE or to a low molar fraction of SMART PROBES forming the double-stranded hybrid. Temperature-dependent and time-resolved fluorescence studies indicate that the kinetics of DNA hairpin formation as well as the formation of suboptimal secondary structures are mainly responsible for the observed inefficient performance. To improve the design of SMART PROBES for the specific identification of target DNA sequences, experimental approaches, i.e., variation of target DNA sequences, stem length and dangling bases, as well as hybridization conditions (salt concentration, temperature and length of the PCR amplicons), were combined with theoretical approaches by modelling the secondary structures of the PCR amplicons and the DNA hairpins using the program MFOLD [59].

As shown in **Figure 4.6**, the target sequence (e.g., the DNA antisense strand of the 240-bp amplicon of *M. avium*) is not necessarily accessible for hybridization of a SMART PROBE at a given temperature. That is, hybridization of the SMART PROBE is thermodynamically favored over its quenched hairpin conformation only when the complementary target sequence is not folded into secondary structures. Therefore, modelling of SMART PROBES as well as hybridization experiments with complementary PCR amplicons was performed at higher temperature (50 °C) and at suitable salt concentrations to ensure the accessibility of the target sequence. Furthermore, the secondary structures of the SMART PROBE itself have to be taken into account. Not only does the thermodynamic stability of the stem have to be balanced with respect to the probe/target hybrid, but intramolecular folding (hybridization) within the SMART PROBE sequence has to be carefully considered for proper design.

While the experimental data is governed by the ensemble of all possible secondary structures, the Gibbs free energy of a certain SMART PROBE reflects the stability of the most stable secondary structure only. Modelling of secondary structures was performed using MFOLD version 3.1.2 [59]. The program uses nearest neighbor energy rules (also called loop-dependent energy rules) to estimate the free energies of possible secondary structures of a given DNA sequence. The secondary structures of the 240-bp PCR amplicon of *M. avium* at 45 °C and 50 °C (**Figure 4.6**) represent the predicted secondary structures that exhibit minimum free energies. In addition, a superposition of all possible folding possibilities within a given energy range of the minimum folding energy can be computed by MFOLD. The multitude of optimal and suboptimal secondary structures is represented in so-called energy dot plots. A dot plot is a triangular plot that depicts base pairs as dots. A dot in column i and j of a triangular array represents the base pair between the i-th and j-th base of the given nucleic acid sequence. For every possible base pair (i, j) MOLECULAR BEACON computes the minimum free energy of any secondary structure that contains the (i, j) base pair. The minimum free energy of a base pair is represented by its color within the energy dots plots shown in **Figure 4.7** [59].

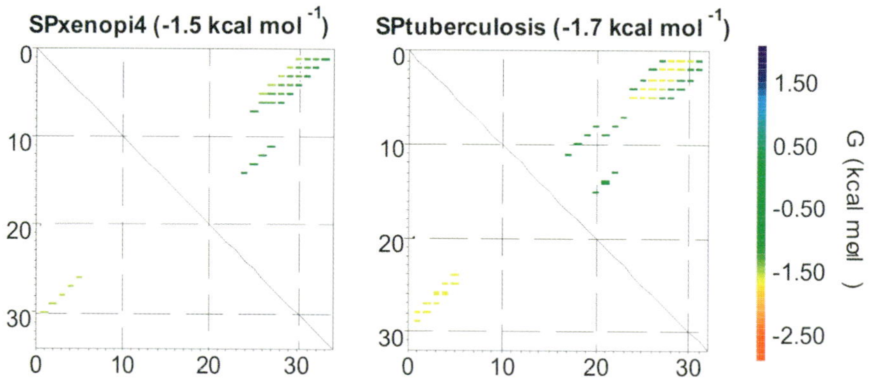

Figure 4.7 Dot plots of two SMART PROBES SPxenopi4 (MR121-C6-5′-CCCCCTAGGACCATTCTGCGC ATGTGGGGGGGG-3′) and SPtuberculosis (MR121-C6-5′-CCCCCGTGGTG GAAAGCGCTTTAGGGGGGGG) with a 3 G's single-stranded overhang at the 3′-terminus directed against *M. xenopi* and *M. tuberculosis*, respectively, modelled with MFOLD at 50 °C and a Na^+ concentration of 300 mM without Mg^{2+}. The axes represent the bases of the oligonucleotides by their sequence number. Intramolecular hybridization is indicated by colored dots, where the color represents the Gibbs free energy of the most stable secondary structure containing the helix. The most stable secondary structure is shown in the lower left triangle of the matrix, while the right upper triangle displays a superposition of all possible secondary structures. The modelled data indicates that the SMART PROBES have only very few and unstable foldings (high ΔG) in the loop, while a multitude of secondary structures is possible within the stem. Taking into account that the stem is formed by G/C base pairs, the dot plots indicate that the fluorescence of the dye attached to the 5′-end, i.e., base no. 1, is quenched in multiple secondary structures.

Figure 4.7 gives dot plots of two SMART PROBES, SPxenopi4 MR121-C6-5′-CCCCCTAGGACCATTCTGCGCATGTGGGGGGG-3′) and SPtuberculosis (MR121-C6-5′-CCCCCGTGGTGGAAAGCGCTTTAGGGGGGG) designed for the detection of *M. xenopi* and *M. tuberculosis*, respectively. The two hairpins exhibit a three- to four-fold increase in fluorescence intensity upon addition of complementary target sequence at 50 °C, i.e., both probes are in principle suitable for the identification of PCR amplicons from different mycobacteria. Comparison of the experimental results with DNA hairpin models, i.e., the dot plots (**Figure 4.7**), suggests that a useful SMART PROBE should form no or only unstable secondary structures in the loop, while the possibility of secondary structures in the stem should preferably be high. The reason for this becomes clear when taking into account that the dye labelled to the 5′-end of the DNA hairpin will always be quenched as long as the 5′-end is part of a C/G base pair. Conformational flexibility in this sense results in a low quantum yield as long as the complementary oligonucleotide is not present and a likewise eased opening for hybridization to the complementary target sequence.

Fluorescence Spectroscopy of Single Molecules

The dream of manipulating individual molecules arose together with the confirmation of their existence. To understand the technological breakthroughs achieved in single molecule fluorescence spectroscopy (SMFS) we have to focus on the central player, usually a fluorophore under laser irradiation. Using a laser source of appropriate wavelength, the fluorophore is excited from the singlet ground state, S_0, to the first excited singlet state, S_1. Subsequently, the fluorophore emits a fluorescence photon spontaneously to relax to the electronic ground state, S_0. Owing to the loss of vibronic excitation energy during the excitation/emission cycle, the fluorescence photon is of lower energy than the excitation photon, i.e., spectrally red-shifted (the so-called Stokes shift), and can be detected efficiently using appropriate optical filtering. The Stokes shift represents the basis for the high sensitivity of fluorescence spectroscopy compared to other spectroscopic methods. Assuming a fluorescence quantum yield of 100%, and a typical fluorescence lifetime of 4 ns, a fluorophore would emit on average 2.5×10^6 fluorescence photons per second if it were always excited immediately after emission of a photon, i.e., under optical saturation conditions. However, owing to the presence of competing depopulation pathways, of which intersystem crossing into long-lived triplet states is the most prominent, the fluorescence quantum yield never approaches 100%. Furthermore, it must be remembered that fluorophores undergo irreversible photobleaching when irradiated with intense laser light because of the increased reaction capability of molecules in electronically excited states. That is, the time available to detect the presence of a single fluorophore is limited, usually to the range of milliseconds to seconds, depending on the excitation intensity. Nevertheless, if only a few percent of the photons emitted finally reach the detector (mainly limited by the collection efficiency of the applied optics and the transmission of the filters), the presence of a single fluorophore would be indicated by the detection of a few tens thousands of photons per second. Of course, the quantum efficiency of common high-sensitivity CCD cameras or avalanche photodiodes is less than 100%, but even the human eye exhibits sufficient sensitivity, especially in the green wavelength range, to detect the fluorescence photons emitted by a single fluorophore. This demonstrates that the detection efficiency does not constitute the crucial parameter in SMFS. It is the background signal that sets the detection limit. Background stems mainly from elastic (Rayleigh) and inelastic (Raman) scattering from solvent molecules as well as from auto-fluorescent impurities. While Rayleigh scattering can be efficiently suppressed by the use of suitable optical filters, the complete suppression of Raman scattering, which is directly proportional to the number of irradiated solvent molecules, is challenging because it occurs, at least partly, in the same spectral range as the fluorescence signal. On the other hand, autofluorescence from impurities strongly depends on the excitation and detection wavelength. Especially in biological samples, luminescent impurities can decrease the sensitivity to, or even prevent the distinct detection of, individual fluorophores. Since the contribution of the background signal is directly proportional to the number of molecules in the excitation volume, the reduction of the excitation/detection volume is crucial for SMFS. Efficient reduction of the excitation volume can be achieved using laser excitation in different configurations, i.e., confocal, evanescent or near-field arrangements. Although, all configurations offer distinct advantages, it was confocal fluorescence microscopy that established single-molecule-sensitive optical techniques as a complementary standard tool in various disciplines ranging from material science to cell biology.

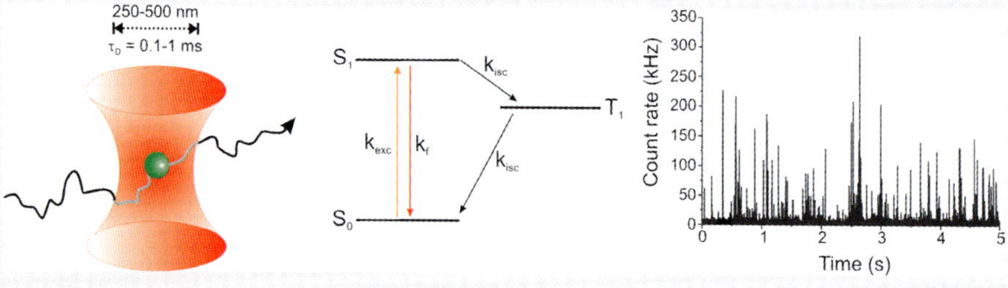

Figure 4.8

The confocal arrangement defines a small, cylindrical volume with a diameter of a few hundred nanometers in the x, y-direction, and ~1 μm in the

z-direction, i.e., a detection volume of 0.5–1.0 femtoliters. For such small volumes, Poisson statistics predict that for concentrations $<10^{-10}$ M, the number of fluorophores present in the detection volume fluctuates predominately between zero and one. Hence, by working with these concentrations we can be sure to observe individual fluorophores diffusing through the detection volume by Brownian motion. While diffusing through the detection volume (typical diffusion times, τ_D, for fluorophores lie in the range of a few hundred microseconds for femtoliter volumes) the fluorophore is repeatedly excited (k_{exc}) and emits fluorescence photons (k_f) dependent on its fluorescence quantum yield. Depending on the intersystem crossing yield, the fluorophore visits the triplet state T_1 before the singlet ground state S_0 is repopulated again via k_{isc}.

4.2.2
SMART PROBES in Heterogeneous Assays

Homogeneous assays are in general restricted with respect to detection sensitivity. Although more sensitive methods, e.g., single-molecule fluorescence spectroscopy (SMFS), are about to break the current barriers of sensitivity, alternative approaches such as heterogeneous assays can be used to increase the sensitivity using less-demanding techniques [60]. The approach presented here for microsphere-based assays combines the increase in fluorescence intensity of SMART PROBES upon specific binding with the accumulation of fluorescence on the surface of appropriately modified microspheres. **Figure 4.9** depicts three different possibilities for using SMART PROBES in a microsphere-based heterogeneous assay: (a) The PCR amplicons can be immobilized on streptavidin-coated microspheres if the PCR is performed with biotinylated primers. In the presence of specific SMART PROBES, fluorescence accumulates on the bead surface and the fluorescence of the microspheres can be easily detected at the bottom of the reaction chamber. Because unbound SMART PROBES diffuse freely in solution with reduced fluorescence intensity they can be applied at relatively high concentrations to shift the thermodynamic equilibrium to the hybridized fluorescent state. Thus, in contrast to other heterogeneous assays, the method can be used without washing steps. (b) SMART PROBES also offer the opportunity for direct immobilization of probe molecules on modified (streptavidin-coated) microspheres by functionalizing the free end of the oligonucleotide e.g., with biotin. The subsequently added target DNA is then hybridized to the immobilized SMART PROBES, inducing conformational reorganization accompanied by an increase in fluorescence intensity. Here, background fluorescence might result from unquenched DNA hairpins on the microspheres. (c) By using an additional biotinylated oligonucleotide complementary to a second region on the target DNA, the assay can even be designed for sensing the presence of two different sequences within

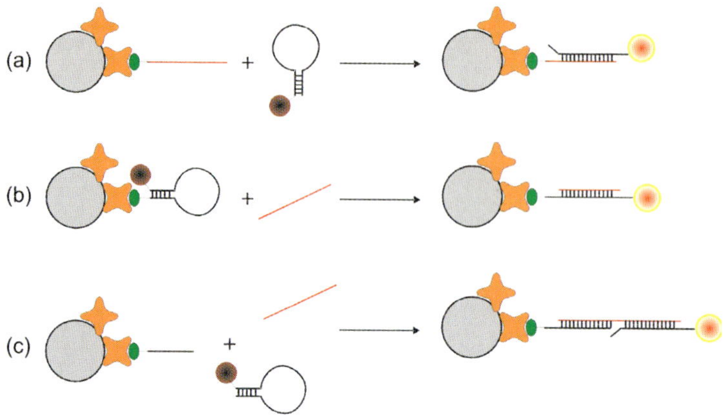

Figure 4.9 Schematic drawing of three different possibilities for the application of SMART PROBES in heterogeneous microsphere-based fluorescence assays. (a): The target DNA is amplified by PCR using a biotinylated primer and subsequently coupled to streptavidin-coated microspheres. The complementary SMART PROBE is then hybridized to the target sequence (red) yielding highly fluorescent microspheres. (b): The SMART PROBE is biotinylated at its free 3′-end and immobilized on the microsphere. Upon hybridization of the complementary PCR amplicon (red) the local fluorescence intensity of the microsphere increases. (c): By additional use of a second biotinylated oligonucleotide bound to the microsphere (green) two specific sequences within a single PCR amplicon can be probed simultaneously. The PCR amplicon is immobilized by hybridization to the first oligonucleotide, and the signal on the microsphere is generated only if in addition the SMART PROBE hybridizes to its specific target sequence on the immobilized PCR amplicon.

the DNA by formation of a sandwich-like double-stranded DNA conjugate. Upon specific hybridization of the target DNA to the immobilized oligonucleotide, the first sequence is recognized and only upon hybridization of the SMART PROBE to the second sequence does the fluorescence on the microspheres increase. The latter two schemes have the advantage over the first one that in principle PCR could easily be avoided if the sensitivity of the assay can be sufficiently high tuned, since no modification of the target DNA is necessary.

In our first experiments, we immobilized short complementary DNA oligonucleotides on 5 μm microspheres via biotin/streptavidin binding (shown in **Figure 4.9a**). The sensitivity of the assay is demonstrated in **Figure 4.10** where different concentrations of the complementary oligonucleotide were used for immobilization on the microspheres (10^{-7}–10^{-10} M) in the presence of 10^{-8} M SMART PROBE. Even concentrations of 10^{-11} M can easily be distinguished from untreated microspheres by using appropriate scaling of the fluorescence intensity at concentrations of 10^{-10} M and below (**Figure 4.10**). These results suggest that molecular harvesting on microspheres, i.e., the accumulation of unquenched SMART PROBES on the surface of microspheres, opens new avenues for the development of highly sensitive DNA-based assay formats.

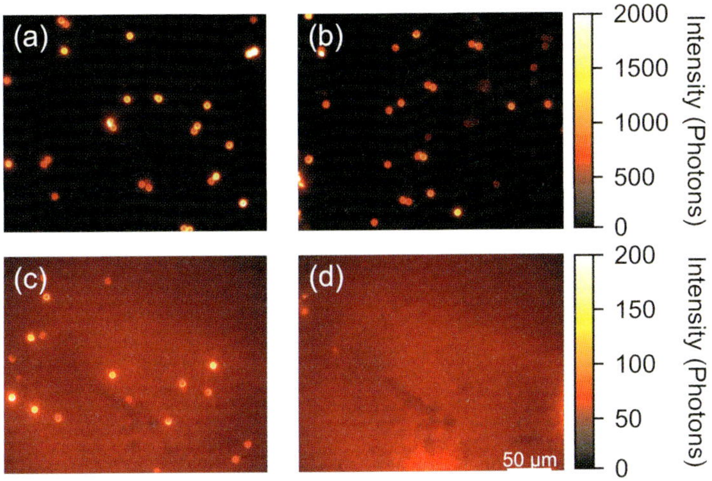

Figure 4.10 Fluorescence intensity images of microspheres modified with oligonucleotides (5′-AAAAAAAAAAAAAAA-CCCCTTGTGAGGAACTACT-3′-Biotin) in different concentrations upon addition of a 10^{-7} M solution of SMART PROBE (3′-GGGGAACACTCCTTGA TGATTCCCC-5′). Biotinylated complementary oligonucleotides were immobilized on streptavidin-coated microspheres using concentrations of 10^{-7} M (a), 10^{-8} M (b) and 10^{-10} M (c). (d): Fluorescence intensity image measured for unmodified microspheres in the presence of 10^{-7} M SMART PROBE.

4.3 Instrumentation

4.3.1 Single-molecule Fluorescence Spectroscopy

While ensemble measurements yield only information on average properties, single-molecule experiments allow the identification of individual characteristics, i.e., identification of subpopulations, in a heterogeneous mixture [39]. This capability might also be helpful for the discrimination of quenched SMART PROBES and fluorescent probe–target duplexes. Historically, the first spectroscopic investigations of single fluorescent molecules were carried out on individual pentacene molecules hosted in *p*-terphen crystals at low temperatures [61]. The chosen conditions confine the conformational degrees of freedom such that highly resolved emission spectra can be recorded indicating the individual molecular environment of each molecule. This technique has meanwhile been adapted for investigating, for example, the function of light harvesting complex 2 (LH2) as the light antenna for photosynthesis [62]. In contrast to the investigation of individual immobilized molecules at low temperatures, SMFS in solution limits the spectroscopic information gained to the time a freely diffusing molecule stays within the laser focus of approximately 1 femtoliter. (10^{-15} liter). In aqueous media at room temperature observation is restricted to $\sim 10^{-3}$ s. In order to prolong the observation time, several groups developed techniques for the immobilization of single molecules on surfaces, within gel matrices, or polymers and even within synthetic micelles [63–65]. By using total internal reflection (TIR) or wide-field microscopy in combination with back illuminated ICCDs or even video cameras, images of single chromophores can be acquired [39,66].

Fortunately, fluorescence photons carry several pieces of accessible information that can help to transmit the changes in the molecules investigated to the experimenter. Besides the fact that fluorescence intensity, lifetime [67,68], emission spectrum [69] and anisotropy [70] can be used to monitor the dynamics of an investigated system, one should recall that the number of photons detected from a single fluorescent molecule is limited. An elegant way to further improve the information obtainable from a single fluorescent molecule relies on the use of several different fluorescence characteristics, if possible in parallel [39,71,72]. Detailed spectroscopic information can be acquired from single fluorescent molecules with a confocal set-up as shown in **Figure 4.11**.

Since the set-up has already been described in detail [39], its main features will only be summarized here. For determination of the fluorescence lifetime, a time-correlated single-photon counting (TCSPC) system measures the time between excitation of the sample (laser pulse) and subsequent emission of a photon. Since the system utilizes a highly repetitive pulsed laser (80 MHz in our case using a pulsed diode laser emitting at 635 nm) the time lags between laser pulse and photon detection are collected in a histogram and the resulting data is used to fit an exponential decay and determine the fluorescence lifetime. By application of two different avalanche photodiodes (APD) for detection, the emitted light can be split into two spectrally different pathways by a dichroic mirror allowing the determination of a parameter termed the fractional intensity F_2 which can be calibrated to the emission wavelength.

The fractional intensity F_2 is given by $F_2 = \frac{I_2}{I_1+I_2}$ where I_1 and I_2 are the number of photons detected by each APD. Besides the opportunity for spectrally resolved fluorescence lifetime imaging microscopy (SFLIM), the introduction of a scanning device, such as a piezo-stage, allows the extension of the concept for the identification of individual dye molecules by their spectroscopic characteristics and hence the separation of overlapping point spread functions (PSF) of single dye molecules. Since the location of a single PSF can be determined with high precision, the positions and distances between different chromophores can be measured down to ≈ 25 nm with a precision of ≈ 7 nm which is two orders of magnitude below the optical resolution limit [73].

Using single-molecule sensitive confocal fluorescence microscopy for the detection of SMART PROBES reveals that only a few molecules exhibit sufficient fluorescence intensity for their unequivocal detection at the single-molecule level (**Figure 4.12a**). The highest fluorescence burst intensities reach

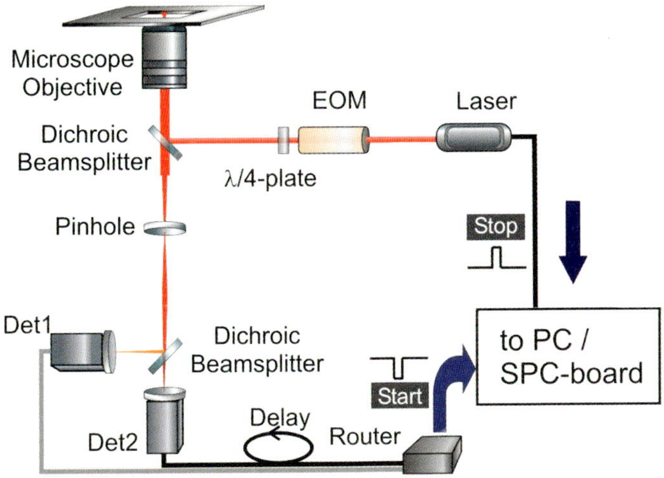

Figure 4.11 Schematic set-up for multi-parameter SMFS. The confocal detection scheme can be applied for fluorescence correlation spectroscopy (FCS) as well as for several other approaches ranging from single-molecule detection and identification in solution up to imaging of immobilized molecules, polarization modulation or coincidence analysis. The laser beam can either be circularly polarized for isotropic excitation by utilization of a quarter-wave retarder or modulated with respect to its linear polarization plane when an electro-optical modulator (EOM) is driven by a saw-tooth voltage signal. After reflection by a dichroic mirror the sample is illuminated through an oil-immersion microscope objective ($100\times$, NA 1.4) yielding a diffraction-limited excitation volume. Fluorescence emission is collected by the same objective, and transmitted through the dichroic mirror for separation from Rayleigh scattered light. Another dichroic mirror splits the fluorescence emission into two spectrally separated channels which are focused onto pinholes after passing additional emission filters prior to detection by two avalanche photodiodes (APD). For acquisition of the fluorescence lifetime, each photon detected is correlated with the subsequent laser pulse which are fed into a single-photon counting PC-card as start and stop trigger respectively for determination of the time between laser pulse and photon detection (time-correlated single-photon counting – TC-SPC). The system is also equipped with a piezo scanning stage for taking microscopic images of single molecules immobilized on surfaces or within cells. Optionally, the delay unit can be applied to shift the signal of one of the APDs in order to allow the detection of two photons at the same time, which can be used for photon antibunching experiments.

count rates of about 40 kHz. The low number of detected fluorescence bursts indicates that the fluorophore is efficiently quenched by guanosine residues in almost all SMART PROBES. Upon addition of an equimolar concentration of complementary target sequence (**Figure 4.9b**) SMART PROBES form a stable probe–target duplex that exhibits an increased fluorescence quantum yield. As in ensemble measurements, the average count rate increases. More importantly, the number of fluorescence bursts with high count rate increases significantly. This experiment clearly demonstrates that most of the unbound (quenched) SMART PROBES exhibit fluorescence quantum yields that prevent their detection at the single-molecule level.

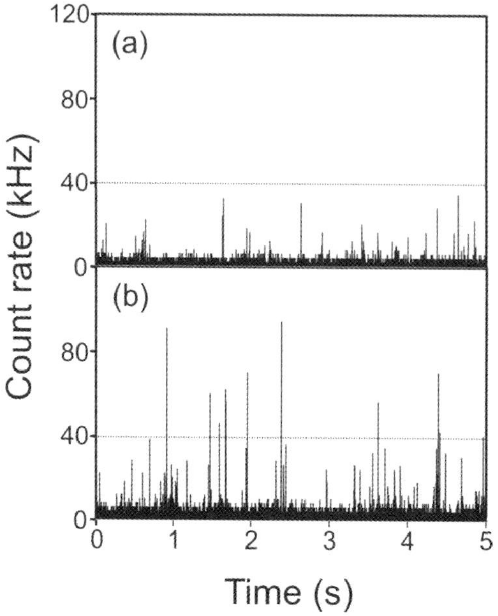

Figure 4.12 Fluorescence signals observed from a 5×10^{-10} M solution of SMART PROBES in 100 mM Tris-borate buffer (pH 7.4), 140 mM NaCl, 5 mM $MgCl_2$ in absence (a) and presence (b) of a 10^{-9} M concentration of complementary target sequence. The data were binned into 0.5 ms time intervals.

Besides the burst rate (number of fluorescence bursts per time) it is easily possible to gain additional information from the fluorescence lifetime. In the absence of a target sequence, the measured fluorescence lifetime of single-molecule events above a threshold of 40 counts/burst was measured at 1.81 ± 0.34 ns, whereas in the presence of complementary target sequence the lifetime increases to 2.89 ± 0.53 ns. The relatively long fluorescence decay time found in the absence of target DNA might be explained by at least three facts:

1. the use of a maximum likelihood estimator (MLE) algorithm without deconvolution from the laser pulse, which disregards multi-exponential fluorescence kinetics
2. efficiently quenched SMART PROBES with very short fluorescence lifetimes are not available for detection at the single-molecule level,
3. the conformational flexibility of the C_6-linker might allow the fluorophore to adopt different configurations with respect to the nucleotides in the complementary DNA strand of the hairpin system, thus changing the observed fluorescence decay time during the transition through the detection volume.

Nevertheless, the differences in fluorescence decay time can be used as an efficient additional parameter to discriminate between closed SMART PROBES and probe–target duplexes.

In order to further increase the identification sensitivity, a multidimensional analysis strategy has been developed to analyze single-molecule events [56]. The strategy takes advantage of the fact that individual SMART PROBES exhibit several different characteristics after hybridization to the target sequence. Besides higher burst sizes and longer diffusion times, probe–target duplexes exhibit significantly increased fluorescence lifetimes. This allows efficient discrimination between SMART PROBES and hybridized probe–target duplexes down to target concentrations of 10^{-12} M, i.e., identification of probe–target duplexes in the presence of a 200-fold excess of SMART PROBE molecules [56].

4.3.2
3D Fluorescence Nanoscopy

The spatial chromatin organization of the human genome in cell nuclei is not random and the architecture of chromosome territories and sub-chromosomal genome domains shows a functional correlation [74]. For instance, the radial arrangement of chromosome territories appears to be determined by gene density and/or chromosome size and is evolutionarily conserved [75,76]. On the micrometer scale, i.e., in the dimensions of whole chromosome territories and sub-chromosomal domains, differences between individuals or between homologous regions of different genetic activity are often not significant [77]. Chromosome territories and sub-chromosomal domains should not change their local position significantly when the cell mutates into a tumor cell [78]. Chromatin modifications, however, may occur during tumor genesis on the nanoscale, i.e., in functional units of the genome in the order of 100 nm in diameter. Therefore, detailed investigations of the genome architecture on the nanoscale has become more and more important since nuclear architecture has been recognized as an epigenetic factor for functional cellular mechanisms, e.g., for the formation of chromosomal mutations and translocations, and their repair.

In order to test for nanoscaled differences in chromatin organization during tumor genesis, novel concepts of high-resolution 3D-microscopy of functional genome domains specifically labelled in 3D-conserved cell nuclei are required [79]. Several advanced microscopic techniques for 3D fluorescence microscopy have been developed, as for instance Spectral Precision Distance Microscopy (SPDM) [80], 4Pi microscopy [81], Stimulated-Emission-Deletion (STED-) microscopy [82], Spatially Modulated Illumination (SMI-) microscopy [83], standing wave-field microscopy [84], etc., and applied to biological objects. Although the feasibility in principle of biological applications and the gain of resolution of these systems have been demonstrated, biological routine is still challenging. Nevertheless, very early approaches to access target sizes of the genome in the order of 100 nm in diameter by SMI microscopy have been promising [85] and have been achieved for routine structural investigations and measurements of chromatin compaction after standard FISH of tumor-correlated genes [86].

Such measurements have been performed with an accuracy of one to two nucleosomes only (typically $\pm$ 10 to 20 nm). This highly improved resolution in fluorescence microscopy, however, opens questions about the influence of the FISH labelling procedure itself on the nano-architecture of the analyzed genome domain. Heat treatment for thermal denaturation of the double-stranded target DNA, extensive use of chaotropic agents like formamide, and last but not least "huge" amounts of probe DNA bound to the chromatin target strands may modify their native structure and compaction. This may be overcome by the application of COMBO-FISH, omitting any destructive target treatment and using only a few very small oligonucleotide probes for specific targeting.

Recently, it has been demonstrated that improved measurements of the nano-architecture of the genome have become feasible using COMBO-FISH and SMI microscopy [87]. These experiments are part of COMBO-FISH investigations on blood tumor-correlated translocation breakpoint and fusion regions. For these applications COMBO-FISH probe sets for abl on chromosome 9 and bcr on chromosome 22 have been configured and applied to archival specimens from patients with haematological diseases like chronic myelogenous leukemia (CML) [88].

In the following, the principles of SMI microscopy and nano-architecture measurements after COMBO-FISH are summarized (for further details see Refs. [86] and [87]. The SMI microscope as schematically shown in **Figure 4.13** uses structured laser illumination and wide-field fluorescence detection by a sensitive CCD camera. Two counter-propagating laser beams coupled into an interferometric set-up interfere, resulting in a standing wave field of illumination light through which the specimen is piezo-electrically moved in the direction of the optical axis in steps of 20 to 40 nm. At each position a fluorescence image of the object is recorded by a CCD camera through appropriate filter settings. The detected fluorescence emission intensity of a labelled genome region through the complete image stack is modulated by the excitation wave field and enveloped by the axial point spread function of the detecting objective. This intensity distribution contains precise information about the axial position and the spatial extension of the analyzed objects in a size range of 40 to 200 nm if the modulating and nonmodulating parts of the curve are analyzed [88].

After appropriate calibration of the modulation contrast as a size measure, the diameters of the labelled sites can be determined and the chromatin com-

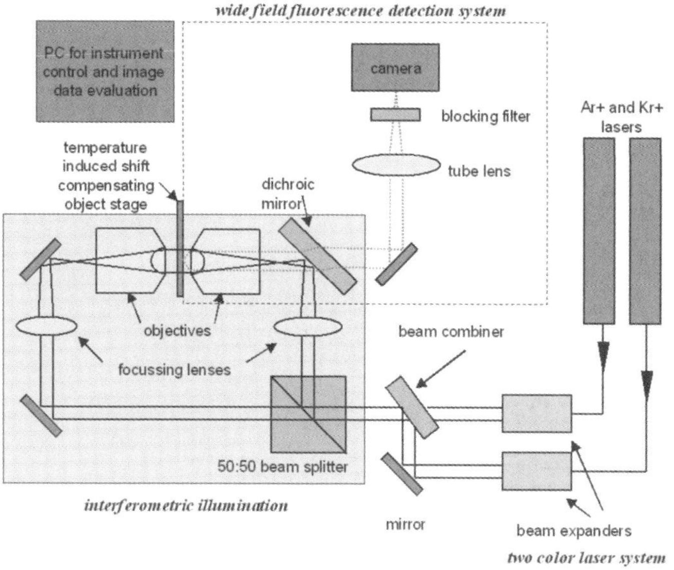

Figure 4.13 Schematic representation of the SMI microscopic set-up (for details see text).

paction can be estimated and compared to computer models. **Figure 4.14** shows a typical example in which 31 DNA oligonucleotide probes were used to label the abl breakpoint region of chromosome 9 (for the sequences of the probes see Ref. [49]). The entire length of the probes is very small compared to the length of the abl target. Only 606 nucleotides, together carrying 62 OregonGreen 488 fluorescent molecules, label 186 000 target nucleotides. This should considerably reduce any effects of modifications of the target structures by probe incorporation. Forty-two target loci in lymphocyte cell nuclei have been analyzed and have revealed an average diameter of 77 ± 22 nm using the size calibration curve described in Ref. [86]. From these data a chromatin compaction ratio of 1:821 has been estimated [87].

So far these experiments have more or less the character of a feasibility study. However, systematic investigations of abl and bcr genome regions by SMI microscopy and COMBO-FISH in blood cells of CML patients and control specimens indicate that significant differences may be found between the two types of specimen.

The application of COMBO-FISH to routine diagnostic specimens (e.g., blood smears or tissue sections) using such a probe set carrying 62 fluorochromes only, or in general using probe sets of some ten oligonucleotide probes with only 1 or 2 fluorochromes each, is a challenge in detection sensitivity, since routine specimens of cell nuclei would always show a consider-

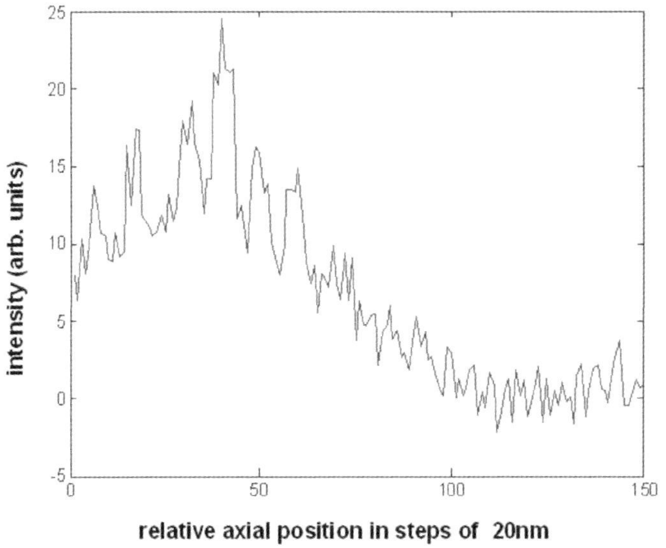

Figure 4.14 Axial fluorescence intensity distribution (AID) curve of a labelled abl genome domain. 150 images ("optical sections") have been recorded with an axial image distance of 20 nm (abscissa unit). The curve shows the AID of the labelled domain obtained from each axial "optical section" along the z-axis (abscissa). The AID is a direct measure of the spot size after calibration of the modulation contrast, i.e., the intensity ratio between the non-modulating part and the maximum modulation of the AID.

able fluorescent background caused by nonspecifically attached probe material. Therefore, SMART PROBES may be recommended to increase the signal-to-background ratio of the specifically labelled target sites.

4.4 Applications

4.4.1 Identification of Microorganisms Using Specific SMART PROBES

Although molecular probes are available for the identification of some microorganisms, routine identification in most cases uses either the traditional culture-based methods or sequencing of PCR amplified gene fragments. Both approaches are frequently used on microorganisms already grown in culture. In contrast, detection and identification of microorganisms in clinical specimens is restricted to amplification/probe detection assays, which are only available for a limited number of different bacteria.

In cases where rapidly growing microorganisms need to be identified, culture-based methods for identification are well suited to meet the requirements of medical microbiological diagnostics in a modern public health system. Bacteria such as staphylococci, streptococci, most gram negative rods or cocci can be identified by assignment of biochemical properties. Modern identification systems such as, for example, the Phoenix system (Becton Dickinson) or the VITEK2 system (BioMerieux) provide reliable identification within 12 hours or less after a pure culture of a disease-causing microorganism has been obtained from a clinical specimen. There are, however, groups of microorganisms that are too closely related to be identified by biochemical properties or that grow too slowly. In the latter case, rapid identification by molecular screening methods is highly effective and extremely important to the attending physician. This is, for instance, true for atypical mycobacteria or mycobacteria other than tuberculosis (NTM or MOTT). Members of MOTT may be saprophytic, living in different water or environmental sources, without causing disease when ingested. Other MOTTs, however, may be able to cause disease, sometimes of moderate severity (e.g., fish tank granuloma caused by *Mycobacterium marinum*), of intermediate severity (e.g., *Mycobacterium avium* lympadenitis in children) and sometimes as life threatening disease (e.g., *M. avium* infections in AIDS patients). The number of validly described MOTT species (currently 118 different species) is not stable. New species are described every year whilst others are synonymized, acknowledging new taxonomic data. The large number of different mycobacteria, many of them living in water, and different, specific and nonspecific clinical manifestations, make specific targeted therapy complicated. Accurate identification of mycobacteria is therefore crucial in the management of infectious diseases.

If identification of MOTT is required, sequencing of a signature sequence of the 16S rDNA gene is currently the method of choice. However, although DNA sequencing *per se* has evolved to a standard technique, widely available in developed countries, identification of bacteria by DNA sequencing is still restricted. For about 90% of cases in routine diagnostics, however, detection and identification of the most frequent MOTTs may be sufficient. Among the currently available assays the GenoType® identification systems (GenoType® Mycobacterium CM for common mycobacteria; GenoType® Mycobacterium AS for additional species, both Hain Lifescience) and the INNO-Lipa Mycobacteria v2 assay (Innogenetics) are two probe-based assays for the detection and identification of different mycobacteria. Species-specific probes are immobilized on membrane strips. These membranes are incubated with PCR amplicons, which may bind to their specific probe. Binding is than visualized by downstream work. The use of these assays is limited, as both tests require amplification of DNA by PCR and both are expensive. The specificity of both tests was satisfactory, although specificity was higher in the Genotype assay. If rare mycobacterial species are isolated, identification is not possible with these assays. Alternative tests or assays are consequently needed to provide state-of-the-art diagnostics, taking into account the rapid changes in MOTT taxonomy and systematics that frequently occur.

Here, we used the concept of singly labelled SMART PROBES for species-specific identification of signature DNA sequences of different atypical mycobacteria. For these purposes a homogeneous model assay was developed. Using fluorescence microscopy combined with streptavidin-coated microspheres the technique yielded hitherto unsurpassed sensitivity [89].

4.4.2
Species-specific Identification of Mycobacteria in a Homogeneous Assay

The detection and species-specific identification of different mycobacterial species in a homogeneous assay was evaluated using the microorganism *Mycobacterium xenopi* (belongs to the five most frequently isolated non-tuberculous mycobacteria [90, 91]) as a model system. For means of identification, PCR amplified fragments of the 16S rRNA gene (16S rDNA, about 1 550 bp) were used. This target is the most widely accepted gene locus in molecular identification of bacteria. The gene is an ideal choice for mycobacterial identification as it contains both conserved sequence regions for PCR amplification and highly variable regions for species identification [92, 93]. First, different reaction parameters were tested to define optimal experimental settings for subsequent hybridization experiments (specific binding of the SMART PROBE to the complementary target sequences). Therefore, a SMART PROBE ('SPxenopi4′) with a complementary sequence to a hypervariable (species-specific) region within the 16S rDNA of *M. xenopi* was designed considering the general criteria for the selection of SMART PROBES as described in Section 4.2. As a control, PCR amplicons from *Mycobacterium fortuitum* displaying eight mismatches within the loop sequence of SPxenopi4 were used (fragment amplified with the same primer pair as for *M. xenopi*). The following reaction parameters were carefully optimized:

1. hybridization temperature
2. fragment lengths of the PCR amplicons
3. hybridization efficiency
4. detection limit.

Upon completion of experimental settings, SPxenopi4 and a newly designed *M. tuberculosis* SMART PROBE (SPtuberculosis) were used successfully for the specific recognition of PCR amplicons of the respective mycobacteria and the differentiation from 15 other most frequently isolated species.

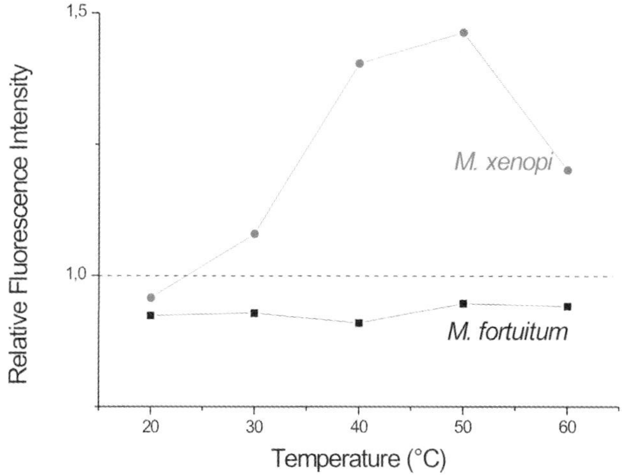

Figure 4.15 Hybridization temperature-dependent relative fluorescence intensity. PCR amplicons of *M. xenopi* and *M. fortuitum* (240 bp; 10^{-7} M) were added to SPxenopi4 in a concentration of 10^{-8} M at varying temperatures. Relative fluorescence intensities were calculated dividing the intensities after and before addition of the respective PCR amplicons. Increases <1 are due to dilution effects if no hybridization occurs.

4.4.2.1 **Hybridization temperature**

Temperatures in the range of 20 to 60 °C for binding of SPxenopi4 to the respective PCR amplicon of *M. xenopi* were tested using the mismatching *M. fortuitum* as a control. **Figure 4.15** shows that a maximum in the relative fluorescence intensity on addition of the matching PCR amplicon occurs at about 50 °C. At lower temperatures (e.g., room temperature) no fluorescence increase is recorded. At temperatures above 50 °C a decrease in fluorescence intensity occurs. No increase in fluorescence intensity is seen with the mismatching PCR amplicon of *M. fortuitum*.

In order to understand this finding the tendency of the target sequence of *M. xenopi* to form secondary structures at different temperatures which might hinder hybridization was investigated (**Figure 4.16**). Here, only the DNA strand containing the complementary sequence to SPxenopi4 was used for calculation of secondary structures, using the program MFOLD. As depicted in **Figure 4.16**, the target sequence is not necessarily accessible for hybridization with SPxenopi4 at the lower temperature. By increasing the temperature, the target sequence is increasingly unfolded (temperatures below 45 °C not shown). Finally, at a temperature of 50 °C (**Figure 4.16b**), no critical secondary structures are observed within the target region.

The fact that in hybridization experiments with SMART PROBES at temperatures higher than 50 °C the fluorescence intensity decreases although the target sequence is freely accessible can be explained by the melting curve

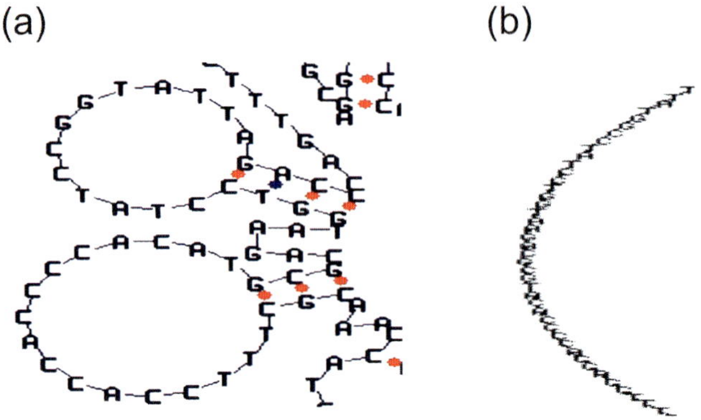

Figure 4.16 MFOLD structures (modelling parameters: 300 mM Na^+ ; 0 mM Mg^{2+}) of the target sequence of the 240-bp (cut-out) amplicon of *M. xenopi* at different temperatures. The figure shows the secondary structures of the target sequence of the PCR amplicon at temperatures of 45 °C (a) and 50 °C (b).

of SPxenopi4 (**Figure 4.17**). At higher temperatures ($\geq$ 50 °C) the relative fluorescence intensity of the DNA hairpin increases because of temperature-induced melting and accompanied unquenching of the fluorophore. Owing to this effect, the relative fluorescence increase following addition of target sequence at a temperature of about 60 °C is lower, as the hairpin already emits a relatively high fluorescence (see **Figure 4.15**). Thus, the SMART PROBE SPxenopi4 is ideally suited for hybridization experiments at 50 °C, since this temperature guarantees a maximum increase in fluorescence on addition of the complementary PCR amplicon.

4.4.2.2 **Fragment Lengths of the PCR Amplicons**

Since the number of possible secondary structures that might interfere with hybridization to SPxenopi is expected to increase with the length of the amplicon, PCR fragments ranging from 85 to 600 bp were generated for comparative hybridization experiments (**Figure 4.18**). In the following, these fragments were added to a solution of SPxenopi4 at 50 °C (**Figure 4.19**). Interestingly, hybridization occurs most efficiently at 50 °C with the 240 bp amplicon. While the decrease in fluorescence intensity for fragments with lengths >240 bp can be explained by the higher number of possible secondary structures, the reason for the decrease for shorter fragment lengths remains unclear.

4.4.2.3 **Hybridization Efficiency**

The maximum increase in fluorescence intensity of SPxenopi4 was observed after adding a 100-fold excess of the PCR amplicon of *M. xenopi* to a 10^{-8} M solution of SMART PROBE. A maximum fluorescence increase of $\approx$4.7 was

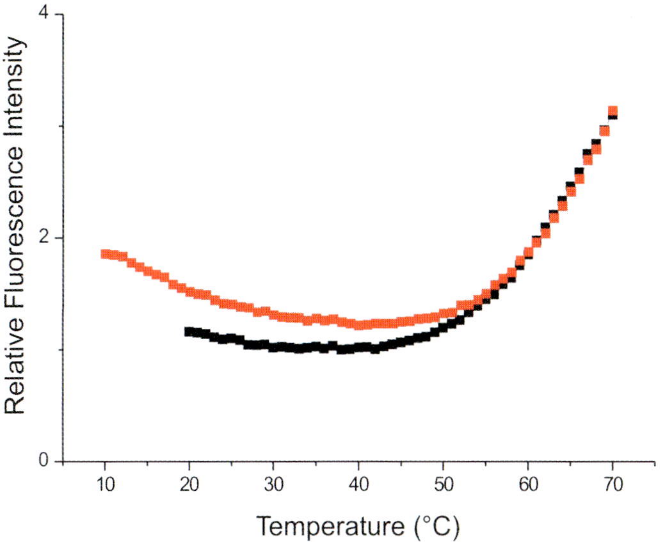

Figure 4.17 Melting curve of the SMART PROBE SPxenopi4. The figure shows the relative fluorescence intensity of a 10^{-8} M solution of SPxenopi in 10 mM Tris-HCl, pH 7.5, with increasing (black curve) and decreasing (red curve) temperature.

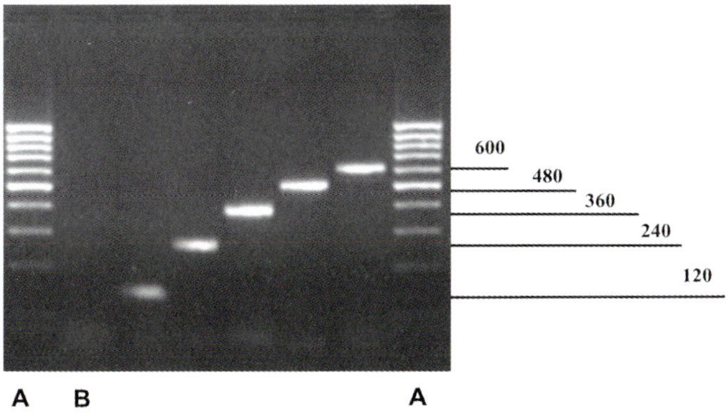

Figure 4.18 PCR amplicons of different lengths. Fragments lengths of 85 (not shown), 120, 240, 360, 480 and 600 bp of *M. xenopi* (amplicons of *M. fortuitum* with similar length are not shown) were generated in order to measure the influence of different fragment lengths upon hybridization efficiency (A = 100 bp DNA ladder; B = negative control).

recorded within the first 500 s of the experiment (**Figure 4.20**). The hybridization efficiency, calculated as the relative fluorescence quantum yield of the SMART PROBE with respect to a reference oligonucleotide without stem, was determined as 0.21. The relative fluorescence quantum yield gives information about the quenching efficiency in the closed hairpin state due to

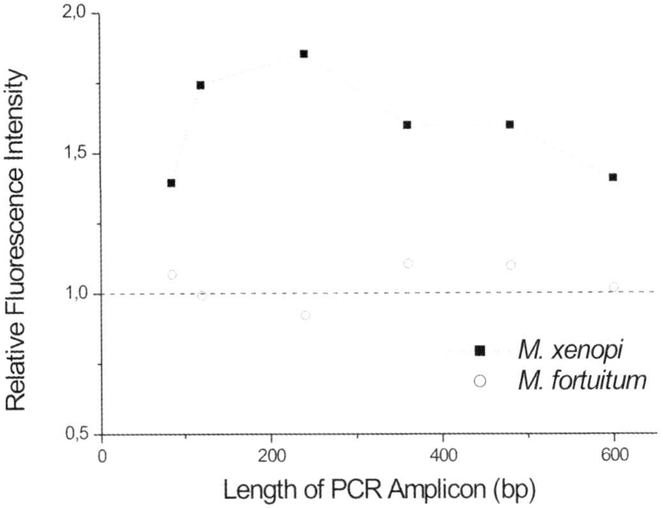

Figure 4.19 Increase of relative fluorescence intensity in dependence of different PCR amplicon lengths. PCR amplified fragments from *M. xenopi* and *M. fortuitum* (**Figure 4.18**) in a concentration of 10^{-7} M were added to a 10^{-8} M solution of SPxenopi in 10 mM Tris-HCl, pH 7.5 and the relative fluorescence increase was observed.

the guanosine residues in close proximity (0 = complete quenching, 1 = no quenching). A quantum yield of 0.21 indicates that upon complete hybridization to the target sequence a 4.8-fold increase in fluorescence intensity can be expected, which corresponds well to the experimental data (**Figure 4.20**).

4.4.2.4 **Detection limit**

In order to determine the detection limit using standard fluorescence equipment (i.e., a conventional fluorescence spectrometer) different concentrations of PCR amplicons of *M. xenopi* were added to the SMART PROBE using optimized reaction conditions (**Figure 4.21**). Using the SMART PROBE SPxenopi4 target amplicons down to concentrations of $\approx 2 \cdot 10^{-8}$ M were easily detected in a homogeneous assay. While this detection limit is not better than those of already available probes, the actual sensitivity can be further increased by 2–3 orders of magnitude (reaching a detection limit of at least 10^{-10} M by application of more sensitive detection techniques, such as single-molecule fluorescence spectroscopy or microsphere-based heterogeneous assay formats (see Sections 4.2 and 4.3.1).

To determine the specificity of SPxenopi4 hybridization, further experiments were carried out with DNA from a number of different mycobacterial species. PCR amplicons from the 16S rDNA of *M. xenopi*, *M. tuberculosis*, and 14 further NTM species containing 4 to 12 mismatches (with regard to *M. xenopi*) within the loop sequence, were used as control under the hybridiza-

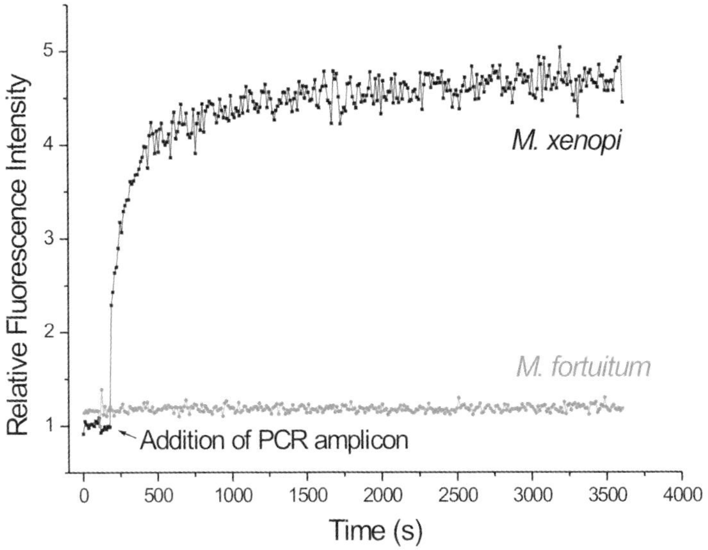

Figure 4.20 Increase in fluorescence intensity with time upon addition of a 100-fold excess of target sequence. 240-bp amplicons of *M. xenopi* and *M. fortuitum* in concentrations of 10^{-6} M were added to a solution of the hairpin SPxenopi4 (10^{-8} M) at 50 °C. Measurements were performed using a conventional fluorescence spectrometer ($\lambda_{\text{exc}} = 635$ nm, $\lambda_{\text{em}} = 680$ nm) in 10 mM Tris-HCl (pH 7.5) containing 300 mM NaCl and 1 mM EDTA.

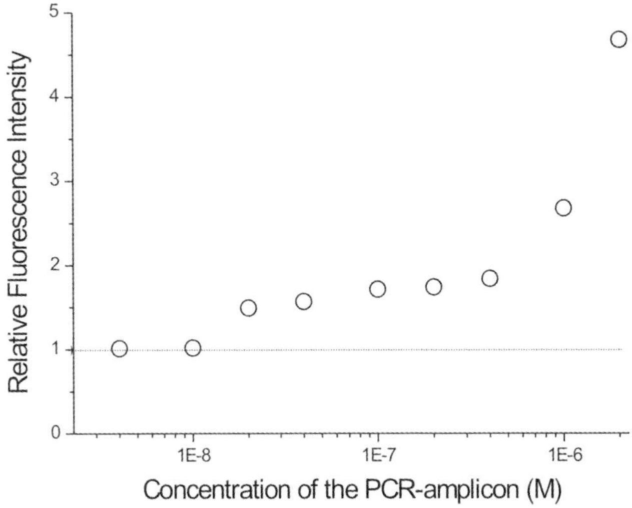

Figure 4.21 Relative fluorescence intensity in the presence of varying target sequence concentrations. PCR amplicons of *M. xenopi* (240 bp) in different concentrations (10^{-7} to 10^{-9} M) were added to a 10^{-8} M solution of the SMART PROBE at 50 °C. Measurements were performed using a conventional fluorescence spectrometer ($\lambda_{\text{exc}} = 635$ nm, $\lambda_{\text{em}} = 680$ nm) in 10 mM Tris-HCl (pH 7.5) containing 300 mM NaCl and 1 mM EDTA.

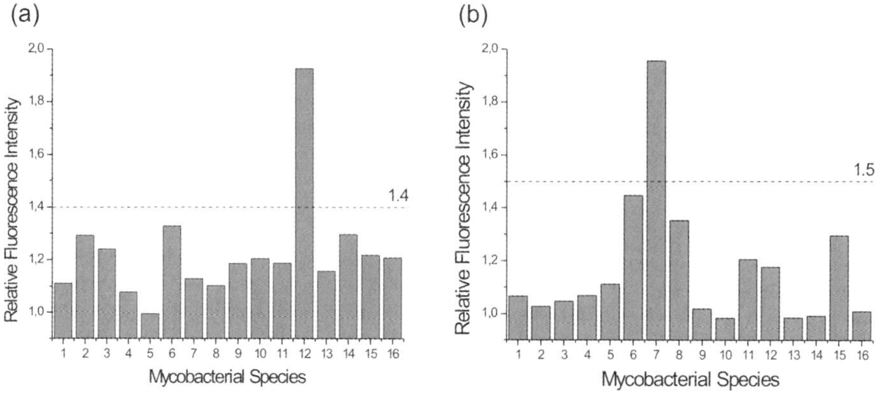

Figure 4.22 Relative fluorescence intensity of SPxenopi (a) and SPtuberculosis (b) after hybridization with a 10-fold excess of 16 different 240 bp PCR amplicons from mycobacterial strains. Amplicons of the following mycobacteria were added to 10^{-8} M solution of the respective hairpin:
1 = *M. szulgai;* 2 = *M. kansasii;*
3 = *M. abscessus;* 4 = *M. gastri;*
5 = *M. gordonae;* 6 = *M. celatum;*
7 = *M. tuberculosis;* 8 = *M. fortuitum;*
9 = *M. malmoense;* 10 = *M. peregrinum;*
11 = *M. interjectum;* 12 = *M. xenopi;*
13 = *M. chelonae;* 14 = *M. avium;*
15 = *M. marinum;* 16 = *M. intracellulare.*

tion conditions described above. Only the sense strand of the *M. xenopi* amplicon shows a perfect sequence homology to the loop sequence of SPxenopi4 within the first hypervariable region of the 16S rDNA. As demonstrated in **Figure 4.22a**, SPxenopi4 can be used advantageously for the unequivocal identification of *M. xenopi*, reflected in a significant fluorescence increase only on addition of a 10 fold excess of the perfect matching PCR amplicon. The fluorescence increase upon addition of PCR amplicons can therefore be regarded as being specific for the species *M. xenopi*. To demonstrate the general applicability of the method, we designed a second SMART PROBE, SPtuberculosis, which is exactly complementary to the antisense strand of the *M. tuberculosis* amplicon. Although the fluorescence intensity also increased slightly on addition of related PCR amplicons because of unspecific interactions, the increase never exceeded a value of ~1.5 fold, while the species-specific signal in both cases yielded an almost 2-fold increase in fluorescence intensity (**Figure 4.22b**).

These results clearly demonstrate that carefully designed and evaluated SMART PROBES are well able to distinguish among PCR amplicons of different species of mycobacteria under optimized hybridization conditions. When used in homogeneous and heterogeneous assay formats, the SMART PROBE system provides a high degree of flexibility, allowing the fast and easy design of additional probes in response to actual changes in (myco-)bacterial taxonomy.

4.4.3
Identification of Antibiotic-resistant Microorganisms on a DNA Level (Identification of Single Point Mutations)

Antibiotic resistance, the ability of a microorganism to cope with antimicrobial drugs and to escape therapy, may arise from one of several molecular causes. Additional genes may have been acquired by a bacterium (e.g., plasmids picked up by transformation events), parts of a gene may be replaced by means of horizontal transfer of genetic material (for instance genes of penicillin-binding proteins of pneumococci) or genes may be altered by mutations (rifampicin resistance). The detection of additional genetic elements like plasmids and identification of horizontally transferred subgenic fragments is comparatively easy. However, the molecular detection of antibiotic-resistance determinants is extremely difficult in cases where resistance is caused by a single point mutation in the DNA of the microorganism (**Figure 4.3**). This means that a specific probe needs to discriminate between two very similar sequence patterns, varying only in a single nucleotide. Whilst the discriminatory potential of such a probe is crucial for a diagnostic test, every false result may have terrible consequences. A false resistant result may lead to complicated therapy regimes, with increased cost and more serious side effects, while a false sensitive result may lead to therapy failure, rapid disease progress and high risks for an affected patient. These limitations, and the fact that antimicrobial susceptibility can be rapidly determined by culture-based methods for rapidly growing microorganisms, may be the reason why molecular assays for determination of antibiotic resistance are far from being in routine use.

One resistance pattern which is of particular interest is rifampicin resistance in the slow grower *M. tuberculosis*. Mycobacteria do not grow within days but within weeks. Traditional culture-based susceptibility testing is reliable but time-consuming. Resistance to rifampicin is mostly caused by a point mutation, but a plethora of different possible mutations have been reported. Most of these mutations affect a small, 81-bp region in the *rpoB* gene (coding for the beta subunit of RNA polymerase) [94, 95]. Rapid molecular assays to detect resistance as soon as possible, e.g., in a respiratory specimen, would be of extraordinary value for diagnosticians as well as for clinicians. Therefore, it is no wonder that a number of different genetic approaches for the rapid diagnosis of rifampicin resistance in mycobacteria have been published. However, only one assay is commercially available in the developed world, and this is virtually unaffordable to laboratories in the developing world.

As SMART PROBES do have the potential to open a new avenue in molecular diagnostics, their ability to discriminate between a wild type and a rifampicin resistance-conferring gene was evaluated. For detecting the worldwide most common single point mutation at the codon $\mathbf{S^{456}L}$ (TCG → TTG) in the *rpoB* gene, which causes rifampicin-resistance by an amino acid substitution from serine to leucine, ten SMART PROBES were designed with different lengths in the loop (from ten to four nucleotides) and stem (each with four and five base pairs) (see **Table 4.2**). SMART PROBES with a stem of five base pairs were constructed so that the 3′-arm of the stem was complementary to the target sequence as well. Probes with a stem of four base pairs were designed with a normal G/C stem, meaning that only the loop was able to hybridize to the target sequence. All SMART PROBES were labelled at the 5′-end with the oxazine derivate MR121. To increase quenching efficiency, the SMART PROBES were extended by three additional overhanging thymidine nucleotides at the 3′-end [58].

Table 4.2 SMART PROBES used for detection of the $S^{456}L$ single point mutation and corresponding abbreviations. Oligonucleotides were labelled with MR121 via C6 amino-modifier. Underlined nucleotides are complementary to the target sequence in the *rpoB* gene. The differences between the wild type and the mutant sequence are marked bold.

MR121-C_6-5′-CCGACAGCGCCA**A**CAGTCGGTTT-3′	[S456L_ a_ 10L5S]
MR121-C_6-5′-CCGACGCGCCA**A**CAGTCGGTTT-3′	[S456L_ b_ 9L5S]
MR121-C_6-5′-CCGACGCC**A**ACAGTCGGTTT-3′	[S456L_ d_ 7L5S]
MR121-C_6-5′-CCCCGCC**A**ACAGGGGTTT-3′	[S456L_ d_ 7L4S]
MR121-C_6-5′-CCGACCCA**A**CAGTCGGTTT-3′	[S456L_ f_ 6L5S]
MR121-C_6-5′-CCCCCCA**A**CAGGGGTTT-3′	[S456L_ f_ 6L4S]
MR121-C_6-5′-CCGACCA**A**CAGTCGGTTT-3′	[S456L_ g_ 5L5S]
MR121-C_6-5′-CCCCCA**A**CAGGGGTTT-3′	[S456L_ g_ 5L4S]
MR121-C_6-5′-CCGACA**A**CAGTCGGTTT-3′	[S456L_ h_ 4L5S]
MR121-C_6-5′-CCCCA**A**CAGGGGTTT-3′	[S456L_ h_ 4L4S]

The experimentally determined melting temperatures of all SMART PROBES investigated were between 50 and 60 °C. In a next step the application of SMART PROBES for the reproducible detection of single point mutations was investigated. Assays were performed at 50 °C with artificial, single-stranded wild type and mutant sequences (short oligonucleotides). Each SMART PROBE was adjusted to a final concentration of $5 \cdot 10^{-7}$ M or $5 \cdot 10^{-8}$ M. The artificial target sequences were subsequently added in 100-fold excess up to equimolar amounts. Of the ten SMART PROBES tested only "S456L_ d_ 7L5S" (**Table 4.2**) provided accurate and reproducible discrimination between the wild type and the mutant sequence in a test setting with short artificial oligonucleotides. Further optimization experiments (variation in temperature, probe concentration and excess of target sequence) led to an optimized protocol that allowed stable discrimination of wild type and mutant sequences, indicated by an increase in fluorescence of about 2.75-fold after addition of complementary (mutated) target sequence to the SMART PROBE (**Figure 4.23**). Preliminary results obtained with PCR amplicons confirm the obtained results and

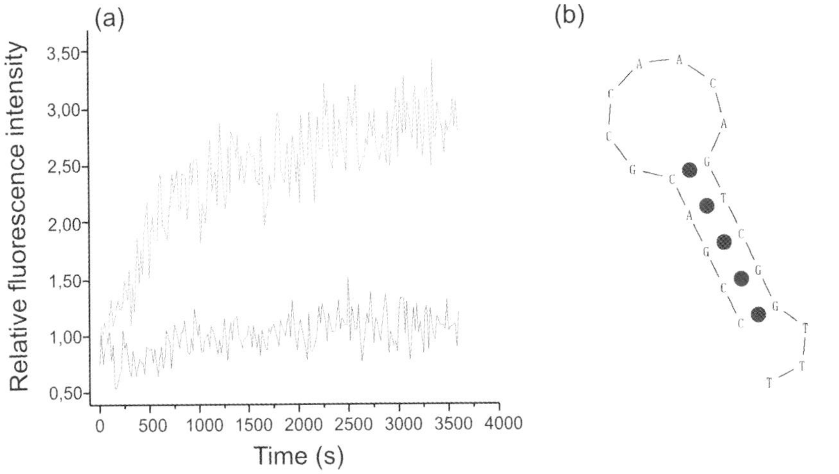

Figure 4.23 (a) Increase in fluorescence intensity with time measured for SMART PROBE S456L-d-7L5S ($4.5 \cdot 10^{-8}$ M) at 30 °C upon a 10-fold excess of artificial sequence. Upper line: addition of the mutant sequence, lower line: addition of the wild type sequence. Measurements were performed using a conventional fluorescence spectrometer ($\lambda_{exc} = 635$ nm, $\lambda_{em} = 680$ nm) in 10 mM Tris-HCl (pH 7.5) containing 300 mM NaCl and 1 mM EDTA. (b) Sketch of the folded SMART PROBE.

thus demonstrate the importance of the developed SMART PROBE technique for the highly reproducible and reliable detection and identification of single point mutations responsible for antibiotic resistance.

Antibiotic Resistance

Antimicrobial drugs fulfil the criteria of selective toxicity. This means that they act on a microorganism without causing (direct) harm to the host (commonly the human macroorganism). Selective toxicity is possible, owing to the selective action of an antimicrobial agent on a target structure that either is exclusively found in the microorganism (e.g., the bacterial cell wall) or, if the target structure is common to the micro- and the macroorganism, differs significantly between the two organisms. The toxicity of antimicrobials causes a "pressure" on the microorganism, forcing the microorganism to adapt in order to survive. This somewhat "mechanistic" view will be explained in the following. During DNA duplication prior to cell division, a few errors (commonly referred to as mutations) may occur. These mutations may be lethal for the affected, individual microorganism, may be without any measurable effect (i.e., "neutral") or may be of at least potential advantage for the individual. A coincidental mutation in a particular region of the gene encoding the beta subunit of RNA polymerase, for instance, might change the secondary structure of the mature protein in such a way that rifampicin, the drug most commonly used in the treatment of tuberculosis, is no longer able to bind to its target site. In the presence of rifampicin, the carrier of that mutation is favored over the remainder of a population of microorganisms without that mutation, rendering the carrier "resistant" to the drug. Under rifampicin treatment, a resistant population will be established. Antimicrobial resistance in microorganisms may not only be a result of a single nucleotide mutation. Small portions of a gene may be deleted, new genes may be acquired or gene cassettes may be transferred horizontally from a donor to a recipient. Among the resistance mechanisms realized by microorganisms, the following important modes may be observed (**Figure 4.24**):

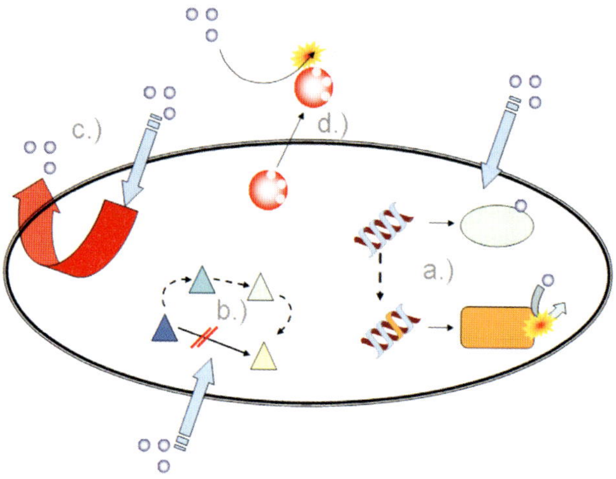

Figure 4.24 Cartoon, showing the most important principles in antibiotic resistance mechanisms in bacterial microorganisms. Antimicrobials (blue circles) may be hindered in entering the cells by altered membrane or cell wall components or by inactivating enzymes (d.), targets may be altered by mutations in the coding genes (a.), metabolic pathways blocked by antimicrobials might be detoured by a bypass in which alternative enzymes catalyze reactions leading to the metabolic product needed (b.) or efflux systems may pump antimicrobials out of the cells once they have entered (c.).

- altered target site by means of mutation (e.g., the altered subunit of RNA-polymerase [rifampicin resistance], altered penicillin binding proteins not longer bind to penicillin, [for instance methicillin resistance in *Staphylococcus aureus*] or others; the principle outlined in **Figure 4.24a**)

- alternative metabolic pathways (e.g., alternative folate pathway bypassing a metabolic pathway blocked by the antibiotic, leading to resistance against sulfonamides; **Figure 4.24b**)

- efflux mechanisms, allowing toxic or harmful substances to be removed from the cell rapidly (e.g., membrane pumps yielding resistance to hydrophobic agents in gonococci; **Figure 4.24c**)
- antibiotic inactivating enzymes (e.g., beta lactamases, commonly found in extended spectrum beta lactamase [ESBL] producing Gram negatives, inactivating penicillin and its derivatives; **Figure 4.24d**)

A given microorganism may have just one resistance mechanism against a particular drug or may be armed with more than one mechanism, leading to microbes with multiple resistance to the most important antibiotic drugs currently in use. Prominent examples are methicillin resistant *Staphylococcus aureus* (MSRA), multi drug resistant (MDR) *Mycobacterium tuberculosis*, the nearly completely resistant *Acinetobacter* species or *Pseudomonas aeruginosa*. Resistance may be already present in a microorganism infecting a host or can be acquired under treatment. The term "resistance" means that an antimicrobial given to treat an infectious disease does not reach the minimum inhibitory concentration (MIC) need to inhibit the disease-causing microbe. In the routine laboratory, resistance of microbes is measured with culture-based methods in most cases. Methods include disc diffusion techniques on agar or micro broth dilution methods in automated test systems, to mention just two important techniques. Only on rare occasions are molecular genetic methods applied to detect a resistance-conferring mutation or a resistance-conferring gene within the genome of an isolated microbe. The resistance burden varies on a geographic scale. The incidence of MRSA among clinically invasive isolates of *Staphylococcus aureus* varies widely: less than 1% for instance in The Netherlands, Denmark, or Sweden, 1–5% in Finland, 10–25% for instance in Germany or Spain, 25–50% for instance in France or Portugal to more than 50% in Greece (data for 2003, results of the EU funded European Antimicrobial Resistance Surveillance System EARSS, http://www.earss.rivm.nl/).

Resistant microbes cause economic loss in that prolonged treatment of infections may be required, mortality may increase in affected patients and additional disease management (i.e., isolation of patients or cohorts of patients, improved health care and nursing) are necessary. Although novel probes do not influence resistance development nor do they reduce resistance per se, early detection of resistant microorganisms in clinical samples enables early effective and tailored treatment, thereby contributing to a reduction in the spread of resistant microorganisms.

4.4.4
COMBO-FISH in Tumor Diagnosis (Design and Application of Highly Sensitive, Focused DNA Tumor Markers)

A major subject of research in molecular pathology is the search for new molecular markers of high specificity for the diagnosis of tumor cells. With the developments of FISH and routine FISH protocols, many DNA probes have been found that can be used to detect structural and numerical aberrations in genomes of tumor cells and thus to distinguish neoplastic (tumor) cells from non-neoplastic (normal) cells unequivocally. Nevertheless, improvements in probes and probing technologies are still required in order to be more sensitive and selective not only in tumor diagnosis but especially in tumor therapy control. One of these improved technologies is COMBO-FISH.

For specific, selective labelling of small tumor-correlated gene domains in cell nuclei, an exactly binding oligonucleotide probe set focused on the labelled target site is the first prerequisite of the COMBO-FISH technique. Therefore, combinations of appropriate oligonucleotides have to be determined. A given genome locus is screened in the human genome sequence database for all homopurine/homopyrimidine segments. On the assumption that the human genome contains 1–2% base pairs as homopurine/homopyrimidine segments of a length of about 15–35 nucleotides each, and that these segments are nearly homogeneously dispersed over coding and noncoding regions, about 150–200 of these segments can be found within a chromatin target of about 250 kb.

Then, those oligomeres can be determined that each only exist in very few copies in the whole genome each, and which all together *only co-localize* at the given genome. This means that, even if the individual sequence is not strictly unique, the COMBO-FISH approach would work as long as the co-localization of the probe set is unique in the haploid genome. With about 10–30 appropriate different oligonucleotide sequences, this results in a fluorescence signal of 20–60 fluorophores (fluorophore molecule position at the 3′- and/or at the 5′-terminus of an oligomere). With this label, the given genome locus has to be identified in a microscope by a merged diffraction image ("spot") of increased fluorescence intensity of the co-localizing oligonucleotide probes and thus to be discriminated from the background of dispersed oligonucleotide probes in a cell nucleus. This section describes an example that shows how SMART PROBES can be used advantageously to improve the signal-to-background ratio.

Computer search strategies for homopurine/homopyrimidine sequences in a given genome locus have been adapted to the whole human genome database. In a first step, a given genome region is localized using some annotation files of the chromosomes, or a contig, or some other locus information linking a DNA sequence to the cytogenetic denotation of the genome region.

The underlying filed information is processed interactively by normal text editors, because this whole search is artificially complicated by the differing structures of non-standardized file types and the mismatch of different biological, biochemical and medical names and acronyms for the same genome region. Once the location and the corresponding sequence have been identified, all homopurine and homopyrimidine stretches with a minimum length of 15 nucleotides are extracted from the respective sequence by a straightforward C-programmed algorithm. A set of appropriate stretches is selected by user involvement, excluding repetitive sequences and cutting sequences of more than 30 nucleotides. The whole genome is then screened for the occurrence of members of these candidate sequences, using a finite automaton-like searching algorithm programmed in C. The locations are stored and clusters of more than two stretches within any 250 kb of the whole genome are identified. Step by step, the number of such clusters is reduced by eliminating stretches that occur frequently in those clusters until no further clusters (except the main cluster of the given genome region) are left. Computations are performed on a Silicon Graphics workstation net (SGI O2, IRIX 6.5, IP32 Processor R10000 150 MHz Chip Rev. 2.5). Configuration of one set of stretches takes currently more than one day elapsed time, owing to user involvement as described above. CPU time for the whole genome search is several hours.

In conclusion the essential steps of COMBO-FISH are:

1. definition of a certain target region of interest (e.g., a gene)
2. systematic screening of the human genome database for all homopurine/homopyrimidine stretches between the given start and end nucleotides of the target region of interest
3. determination of all locations of each of these stretches in the whole genome
4. exclusion of all target stretches that co-localize within a genome section of 250 kb outside the given target region of interest
5. rewriting the remaining target sites that exclusively co-localize in a given genome region in homopurine or homopyrimidine probes that bind either via Watson–Crick or Hoogsteen bonding (Watson–Crick means a single-stranded oligonucleotide probe binds to a single-stranded, i.e., denatured, target sequence; Hoogsteen means a single-stranded oligonucleotide probe binds to a double-stranded target sequence; in both cases the probe orientation defines the bonding type)
6. synthesizing the probe set.

In the following, an example for the TBX-1 region on chromosome 22 (region q11) is given. The gene TBX-1 is involved in different tumors such as

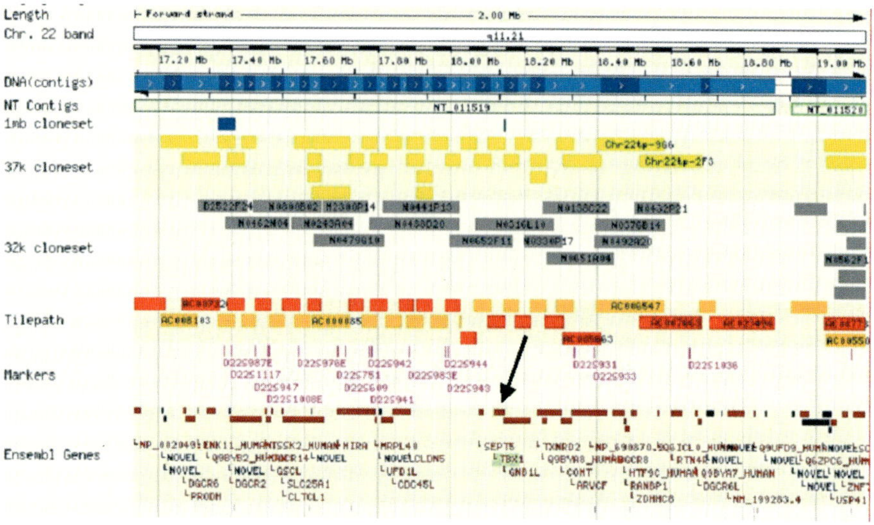

Figure 4.25 Section of the ensemble database indicating the 22q11 region of the genome. The position of TBX-1 among many other genes is highlighted by an arrow. The colored blocks represent possible DNA clones for FISH probe production.

mamma carcinoma or Barrett's esophagus carcinoma, where it shows changes in the gene copy number, especially deletions [96]. It is also involved in cardiac diseases like the DiGeorge syndrome (DGS) [97]. DGS is a hereditary disease and, with Down syndrome, one of the most frequent genetic reasons for congenital heart defects. In general, symptoms of a deletion in 22q11 include cardiac defects, backwardness or absence of thymus with T-cell defects and immunodeficiency, typical facial forms with deformities (cleft palate), etc. Congenital heart defects with an incidence of 1% are the major dysplasia of human newborns. So far it is only known that most forms of DGS are connected with a microdeletion in the genome domain of 22q11 including many different genes (see also **Figure 4.25**). This microdeletion occurs in a 3-Mb region harboring 25 relevant genes.

So far, it is unclear which of these genes is lost in case of DGS. On the molecular level the deletion of the transcription factor TBX-1 plays an important role in the cardiac defect of DGS. However, 5–10% of the patients with DGS do not show a 22q11 deletion. So, in general, for the improvement of diagnostic sensitivity for small genes like TBX-1, a specific labelling technique is necessary for detection of small chromosomal mutations, such as translocations or microdeletions and for investigations of disease-induced chromatin compaction changes. Available standard FISH probes (e.g., BAC probes) do not exactly map this region and overlap to proximately located gene domains. Therefore, the number of specifically binding clones is minimal. Particularly, in the case of microdeletions, it is a problem to find probes that for instance exclusively label TBX-1.

In **Figure 4.25**, a section of a clone database for 22q11.21 is shown (www.ensembl.org) after computer searching for appropriate clones for labelling the 26-kb long TBX-1 gene region. The figure shows a gene-dense region and although many clones are available (not only full BAC clones but also 37-kb and 32-kb sub-clones), no exactly mapping one is found. The smallest clone appropriate for labelling the TBX-1 gene region, AC000091, has a length of 44 928 bases and binds in the region between nucleotide 18 101 500 and nucleotide 18 146 427. In contrast, the TBX-1 gene is located at the positions 18 118 780 to 18 145 099 with an exact length of 26 320 bases.

The problem of exactly mapping small genes may be overcome by a computer-selected oligonucleotide probe set for COMBO-FISH. Only the relevant homopurine/homopyrimidine target stretches between the start and end positions of TBX-1 are considered for labelling. From these target sites all sequences are excluded that co-localize outsideTBX-1 in any 250-kb region in the genome. Finally a set of target sites remains that only co-localize in TBX-1. The oligonucleotide probes for these target sites are then synthesized. Since normally diagnosis is done on fixed specimens, a protocol for COMBO-FISH of an oligonucleotide probe set for Watson–Crick bonding has been developed and tested on human lymphocyte cell nuclei. Typically, a binding efficiency of ≈60% can routinely be obtained in human methanol/acetic acid fixed lymphocytes as detected by confocal laser-scanning microscopy.

In a first approach, a COMBO-FISH DNA oligonucleotide probe set for the TBX-1 region has been designed containing 30 different oligonucleotide probes with a total length of 537 bases. These probe set maps target sites dispersed over the 26 320 bases of TBX-1. Each terminus of each oligonucleotide probe is labelled with OregonGreen 488 as a fluorophore. The hybridization signal of 60 fluorophores is detectable by confocal laser-scanning microscopy but the background shows a high intensity of unspecific fluorescence inhomogeneously distributed in the cell nuclei (**Figure 4.26**).

However, the signal-to-background ratio can be drastically improved by using SMART PROBES. The probes have a stem–loop structure, which is closed at low temperature and in the absence of a complementary DNA target sequence, suppressing unspecific background from free probe material by fluorescence quenching. Owing to this considerable improvement in sensitivity, a set of only 15 SMART PROBES selected from the 30 sequences described above has been chosen for the TBX-1 gene domain. Each SMART PROBE is synthesized with one fluorophore (fluoresceine) only. After developing a SMART PROBES adapted COMBO-FISH protocol, the TBX-1 region on human lymphocytes was labelled again. **Figure 4.27** shows typical images, indicating the convincing improvement of the signal-to-background ratio.

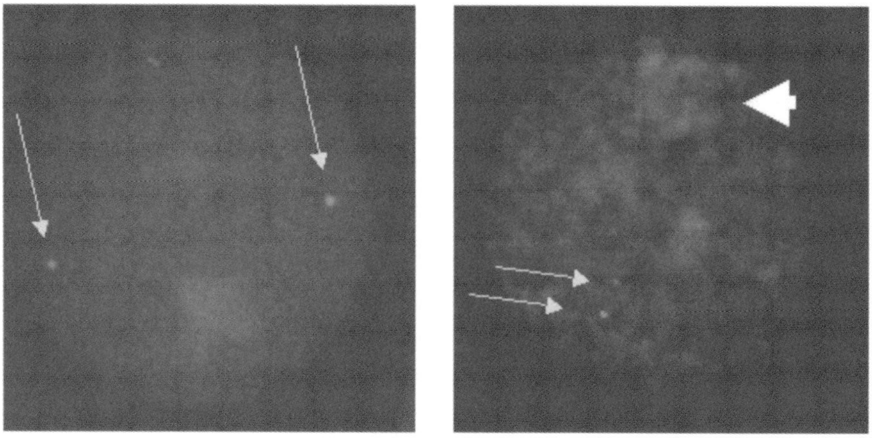

Figure 4.26 Human peripheral blood lymphocyte cell nuclei after COMBO-FISH labelling of the TBX-1 region with an oligonucleotide probe set of 30 stretches corresponding to 60 fluorescent molecules (OregonGreen 488). The arrows indicate the target sites. High unspecific background is visible which is inhomogeneously dispersed (see, for example, arrow head). Note: The cell nuclei are not counter-stained.

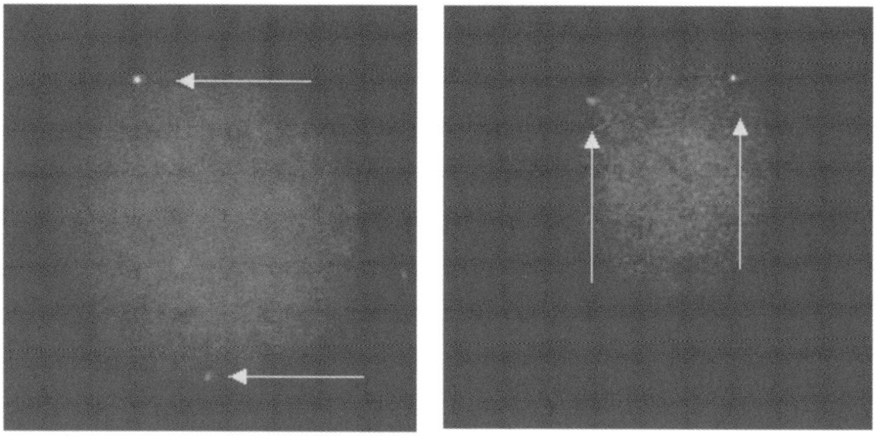

Figure 4.27 Human peripheral blood lymphocyte cell nuclei after COMBO-FISH labelling of the TBX-1 region with a SMART PROBE set of 15 stretches corresponding to 15 fluorescein molecules. The arrows indicate the target sites. The unspecific background is very low. Note: The cell nuclei are not counter-stained.

4.4.5
COMBO-FISH for Highly Specific Labelling of Nanotargets in Living Cells

Besides the applications of COMBO-FISH in molecular pathology and medical diagnosis another highly interesting field for biomedical research is the specific labelling of living cells in order to analyze dynamic processes during cellular development and differentiation, especially tumor differentiation after treatment of cells with tumor inducing, environmental noxes as for instance ionizing radiation or chemical noxes.

Second generation "SMART PROBES" have been used successfully for highly specific labelling of nanotargets in living cells. These novel DNA hairpin probes carry two fluorophores (here tetramethylrhodamine molecules) at both ends of the hairpin. Due to the formation of nonfluorescent dimeres the closed hairpin state exhibits almost no fluorescence signal. In the case of hybridized probes, dimere formation is prevented, that is, the fluorescence of the fluorophores is released. The new SMART PROBES are especially interesting for nanotarget labelling in living cells. For a first proof of principle, an oligonucleotide probe has been chosen that uniquely exists as a repetitive sequence in the centromere region of chromosome 9 [98].

Figure 4.28 shows a typical result for specific labelling of centromere 9 in methanol/acetic acid fixed human lymphocytes. The nuclei are not counterstained. No background is visible, although a high gain of the detection system has been chosen in order to detect the nuclear boundaries.

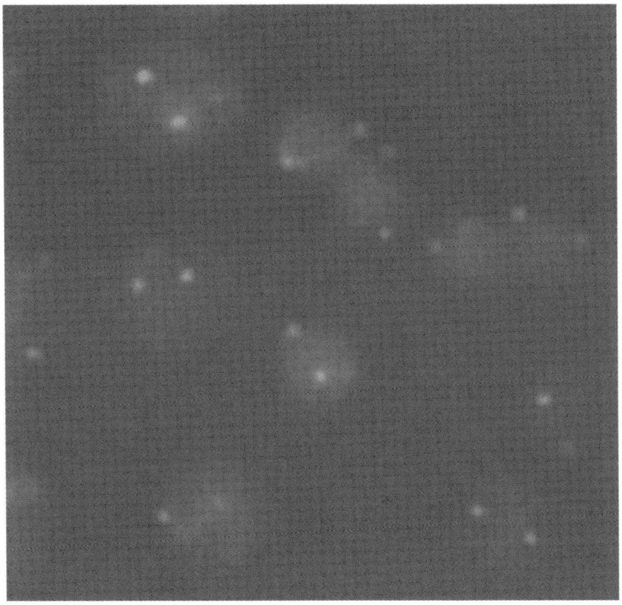

Figure 4.28 Human peripheral blood lymphocyte cell nuclei after COMBO-FISH labelling of centromere of chromosome 9 with a unique repetitive second-generation SMART PROBE. The cell nuclei have been fixed and COMBO-FISH has been performed with heat denaturation (72 °C) of the target DNA. The SMART PROBES are labelled with two tetramethylrhodamine dyes. The unspecific background is extremely low. Note: The cell nuclei are not counter-stained.

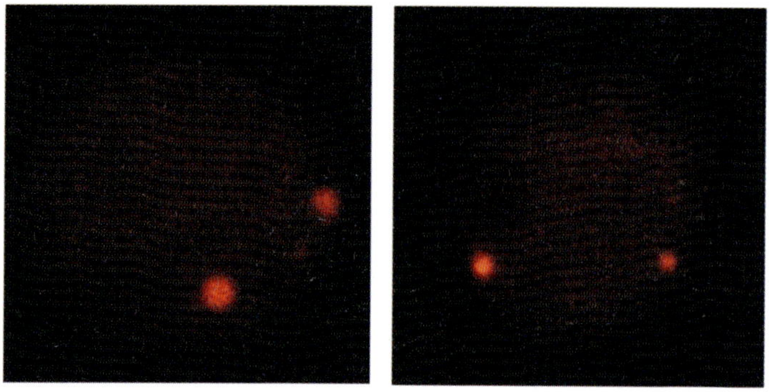

Figure 4.29 Human peripheral blood lymphocyte cell nuclei after COMBO-FISH labelling of centromere of chromosome 9 with a unique repetitive second generationSMART PROBE. The SMART PROBES were microinjected under live cell conditions. After incubation and specific probe binding the cell nuclei were fixed and imaged. SMART PROBES were labelled with two tetramethylrhodamine dyes. The unspecific background is extremely low. Note: The cell nuclei are not counter-stained.

The same probe type has been applied to demonstrate the possibilities of COMBO-FISH for live-cell imaging. For this purpose, a labelling strategy has been developed on the basis of microinjection of the SMART PROBES. Here, primary T-lymphocytes were purified from human peripheral blood by magnetic cell membrane antibodies. For accumulation of T-lymphocytes and detachment of the antibodies, the cells were cultured in a stimulating media at 37 °C with 5% CO_2 for 60–70 h. Since T-lymphocytes are nonadherent cell cultures, they were attached to a poly-l-lysine-coated cover slip and incubated for three hours. The cells were embedded in preheated media with a low concentration of HEPES buffer. Microinjection was performed with an AIS I system, a completely automated microinjection system for adherent cells. It is possible to inject approximately 200 cells in 30 min. In order to work under physiological conditions the SMART PROBES were diluted in ionic buffer at pH 7.0. After microinjection, the cell medium was changed into HEPES buffer-free medium and the T-lymphocytes were incubated in a dark cell chamber for 24 h in order to specifically bind microinjected SMART PROBES. For 3D microscopic analysis by optical sectioning with confocal laser-scanning microscopy, cells were fixed with an alcoholic buffer and washed with an ionic buffer at room temperature. Finally, microscope slides were sealed. Although the microscopic detection was done under fixed cell conditions in order to detect the low fluorescence and the cell nucleus during extended image acquisition, it should be emphasized that neither denaturation nor hard chemical treatment (e.g., exposure to chaotropic, denaturing agents) was used. The cell morphology is well maintained after microinjection, and the probe is attached to the target site *in vivo*. No background is visible in the cell nuclei (**Figure 4.29**).

4.5 Summary / Outlook

In this chapter we have described the development and future prospects of SMART PROBES – singly fluorescently labelled DNA hairpin probes – for the specific detection of target sequences of different mycobacteria and of resistance-conferring mutations as well as for advanced medical diagnosis and therapy control and live-cell imaging in biomedical research. Although several fluorescent probe-based assays are available for routine diagnosis and live-cell applications, novel probes with higher sensitivity and easier instrumentation are required to provide a versatile tool for different *in vitro* and *in vivo* diagnostic applications. Results obtained with model assays suggest that SMART PROBES can indeed fill this gap, and constitute ideal candidates on which to base the development of future molecular diagnostics. SMART PROBES are singly labelled hairpin-shaped oligonucleotides bearing a fluorescent dye at the 5′-end, which is selectively quenched by guanosine residues in the complementary stem. Upon hybridization to target sequences, a conformational change occurs reflected in an increase in fluorescence intensity. We have demonstrated that SMART PROBES can be used advantageously for the selective and sensitive detection of mycobacterial 16S rDNA signature sequences in homogeneous and heterogeneous assays. Using optimized parameters for hybridization experiments, we established a reliable method for the specific detection of *Mycobacterium tuberculosis* (*M. tuberculosis* complex) and *Mycobacterium xenopi* (a member of the atypical mycobacteria) with a detection sensitivity of $\sim 2 \cdot 10^{-8}$ M of amplified DNA in homogeneous solution. The specificity of the SMART PROBES designed was demonstrated by discrimination of *M. tuberculosis* and *M. xenopi* against 15 of the most frequently isolated mycobacterial species in a single assay. In combination with a microsphere-based heterogeneous assay format the technique is ideally suited for the detection of pathogen specific DNA sequences with hitherto unsurpassed sensitivity.

Much more complex is the discrimination of two target sequences that differ by just one nucleotide base, as is the case in certain antibiotic-resistance mechanisms. Again, we have shown in a model system, rifampicin resistance in *M. tuberculosis*, that SMART PROBES exhibit the potential to discriminate between the wild type and the very closely related mutant. Further interdisciplinary research is required to improve the performance of SMART PROBES by increasing the fluorescence intensity on hybridization, and to apply the technique in various related disciplines. In addition, robust and easy to implement instrumentation has to be developed for worldwide acceptance of the SMART PROBE principle. Nevertheless, the results obtained provide a new avenue for the development of diagnostic tools for use in molecular and medical microbiology, based on the specific recognition of genetic information.

The described applications in medical microbiology were the first for which specific SMART PROBES were developed. However, since we have gained experience with optimizing the probes themselves as well as the assay conditions, more applications are on the point of following, not only in microbiology but also in other fields of human medicine, e.g., for the detection of tumor-associated translocations in living cell nuclei using COMBO-FISH. In combination with miniaturized reaction chambers and sensitive but simple readers, SMART PROBE technology offers the realization of the "lab-on-a-chip" concept for fast and reliable identification of diseases and disease-causing conditions. With miniaturization and easy read-out instrumentation, affordable and reliable assay formats will become available to meet the global challenge of infectious diseases. Combinations of specificity with novel sensitive detection approaches permit the development of faster, cheaper and more reliable diagnostic assays that might in the near future pave the way for genetic identification without PCR.

COMBO-FISH overcomes the limitation of standard FISH techniques, both by means of probe design using computer selection of co-localizing oligonucleotides and by the possibility of synthesizing either Watson–Crick-binding or Hoogsteen-binding oligonucleotide probes. Nevertheless, a disadvantage of COMBO-FISH for fluorescence microscopy is the low number of fluorescence molecules available at the target sites. In order to improve the signal-to-background ratio under such conditions, our results promote SMART PROBES as attractive and powerful alternative probes. Moreover, with second-generation SMART PROBES it has become feasible for the first time to specifically label cell nuclei after microinjection while maintaining live-cell conditions. Together with novel high-resolution fluorescence microscopy techniques, COMBO-FISH and SMART PROBES offer the possibility for gentle, focused gene labelling of live and fixed cells. The technique therefore has the potential to become a versatile tool in biological research, medical diagnostics and therapy control. Nanoscopy by SMI microscopy of COMBO-FISH-labelled genome target sites correlated to tumors may help to detect gene conformation changes in intact cell nuclei of different individuals. This might be a chance to detect individual differences on the single-cell level, for instance in risk estimates of individual tumor sensitivity, which is often discussed for environmentally induced tumors (e.g., individual sensitivity to ionizing radiation). Future improvements in these fluorescence probing technologies will stimulate developments of new microscopic techniques and vice versa. This should introduce new approaches to the investigation of genome nano-architecture as epigenetic mechanisms into biological and medical research and applications. For example, compactness measurements on the gene level might give an indicator for the accessibility of gene domains for macromolecules and for the susceptibility to chromosomal aberrations.

Therefore the analysis of the nano-architecture of such gene domains might be used, for instance, for the investigation of (heritable) predispositions to tumor-correlated translocations.

Acknowledgement

This work was supported by a research grant from the Bundesministerium für Bildung und Forschung (BMBF), managed by the Verein Deutscher Ingenieure (VDI).

Glossary

3D-microscopy This microscopy technique generates a three-dimensional image of an object after having acquired a stack of two-dimensional image slices through the object.

APD (Avalanche photodiodes) APDs are photodetectors that can be regarded as the semiconductor analog to photomultipliers. By applying a high reverse bias voltage, APDs show an internal gain effect due to impact ionization (avalanche effect). The higher the reverse voltage the higher the gain. Avalanche photodiodes therefore are more sensitive than other semiconductor photodiodes.

BAC (Bacterial artificial chromosome) A bacterial DNA with a human insert used for DNA cloning

Chaotropic A chaotropic agent is an agent that causes molecular structure to be disrupted (salt bridges, hydrogen bonds, etc.).

CML Chronic Myelogenous Leukemia

COMBO-FISH This is a FISH technique with a probe set consisting of a combination of oligonucleotide probes that specifically co-localize at the complementary target sequences in the genome.

Confocal laser-scanning microscopy In a confocal laser-scanning microscope a laser beam passes a light source aperture and is then focused by an objective lens into a small (ideally diffraction-limited) focal volume within a fluorescent specimen. A mixture of emitted fluorescent light as well as reflected laser light from the illuminated spot is then re-collected by the objective lens. A beam splitter separates the light mixture by allowing only the laser light to pass through and reflecting the fluorescent light into the detection apparatus. After passing a pinhole, the fluorescent light is detected by a photodetection device (photomultiplier tube

(PMT) or avalanche photodiode) transforming the light signal into an electrical one which is recorded by a computer.

DNA (Deoxyribonucleic acid) DNA is a nucleic acid that contains the genetic instructions specifying the biological development of all cellular forms of life (and many viruses). DNA is often referred to as the molecule of heredity, as it is responsible for the genetic propagation of most inherited traits. During reproduction, DNA is replicated and transmitted to the offspring.

DNA libraries Collections of bacteria or yeast cells having small human DNA inserts in the genome

Donor/acceptor see FRET

FCCS (Fluorescence cross-correlation spectroscopy) In signal processing, the cross correlation or sometimes cross-covariance is a measure of the similarity of two signals, commonly used to find features in an unknown signal by comparing it to a known one.

FCS (Fluorescence correlation spectroscopy) FCS is a type of spectroscopy based on the measurement of fluorescence intensity and the analysis of its fluctuations, which can be due to the diffusion of the observed fluorophore in the excitation volume or to changes in the fluorescence quantum yield arising from chemical reactions. Measurements are usually made on only a few molecules at a time – of the order of 10 molecules – which is achieved by illuminating tiny volumes (around 1 femtoliter).

FISH (Fluorescence *in situ* hybridization) This is a technique in molecular biology in which a DNA probe (single-stranded DNA of a given sequence) carries fluorescence molecules and binds to the complementary DNA sequence of a genome *in situ*.

FRET (Fluorescence resonance energy transfer or Förster resonance energy transfer) This describes an energy transfer mechanism between two fluorescent molecules. A fluorescent donor is excited at its specific fluorescence excitation wavelength. By a long-range dipole–dipole coupling mechanism, the energy of this excited state is then nonradiatively transferred to a second molecule, the acceptor. The donor returns to the electronic ground state. The described energy-transfer mechanism is termed "Förster resonance energy transfer" (FRET), named after the German scientist Theodor Förster. When both molecules are fluorescent, the term "fluorescence resonance energy transfer" is often used, although the energy is not actually transferred by fluorescence.

Hairpin-shaped Hairpin-shaped oligonucleotides exhibit a stem–loop structure.

Homopurine DNA regions containing only purine (guanosine or adenosine) residues.

Homopyrimidine DNA regions containing only pyrimidine (cytosine or thymidine) residues.

ICCD (Intensified charge-coupled device) An ICCD is a CCD that is fiber-optically connected to a microchannel plate (MCP) to increase the sensitivity. In ICCD cameras a photocathode in front of the MCP converts photons to electrons, which are multiplied by the MCP. After the MCP a phosphor screen converts the electrons back to photons which are fiber-optically guided to the CCD. Also several cascades of MCPs are used.

MLE (Maximum likelihood estimation) MLE is a popular statistical method used to make inferences about parameters of the underlying probability distribution of a given dataset.

Molecular beacon A molecular beacon is a doubly labelled stem–loop (hairpin) shaped oligonucleotide which increases fluorescence intensity upon hybridization of the loop to its complementary target sequence.

Multi-exponential fluorescence decay A decay that can best be described using more than one exponential decay function.

Nucleosome Histone octamere with a DNA double helical strand folded around in $1\frac{3}{4}$ turns

PET (Photoinduced electron transfer) PET is the process by which an electron moves from one molecule to another molecule (oxidation or reduction, respectively) where one molecule is electronically excited to facilitate the process.

PNA (Peptide nucleic acid) PNA is a chemical similar to DNA or RNA but differing in the composition of its "backbone". PNA is not known to occur naturally in existing life on Earth but is artificially synthesized and used in some biological research and medical treatments.

Probe Probe is a generic term used to refer to a device used to gather information. Here used for fluorescently labelled molecules designed for diagnostic applications.

PSF (Point spread function) PSF defines the propagation of electromagnetic radiation from a point source. It is defined in spherical coordinates for a Lambert-type radiator. Its unit is $[m^{-2}]$. It is a useful concept in Fourier optics. In 3D microscopy (e.g., in confocal laser-scanning microscopy), a PSF is the image of a single point object. The degree of spreading (blurring) of this point object is a measure of the quality of an optical system. PSFs play an important role in the image formation theory of fluorescence microscopy.

Quencher A quencher quenches the fluorescence of excited molecules. Quenching is a general term for nonradiative de-excitation. One major reason for nonradiative de-excitation is collisions. As a consequence, the quenching is often heavily dependent on pressure and temperature. Quenching poses a problem for non-instant spectroscopic methods, such as laser-induced fluorescence.

RNA (Ribonucleic acid) RNA is a nucleic acid consisting of a string of covalently bound nucleotides. It is biochemically distinguished from DNA by the presence of an additional hydroxyl group attached to each pentose ring; also by the presence of uracil instead of thymine. One of the main functions of RNA is to copy genetic information from DNA (via transcription) and then translate it into proteins (by translation).

SFLIM (Spectrally-resolved fluorescence lifetime imaging microscopy) This technique measures simultaneously the fluorescence intensity, lifetime and spectral information of a sample.

SMFS (Single-molecule fluorescence spectroscopy) This is a technique to detect the fluorescence emitted by a single fluorophore.

SMI microscopy (Spatially modulated illumination microscopy) This system that uses a standing wave field for illumination of the specimen.

SNP (Single nucleotide polymorphism) SNP (pronounced snip) is a DNA sequence variation, occurring when a single nucleotide, adenine (A), thymine (T), cytosine (C) or guanine (G), in the genome is altered. For example, a SNP might change the nucleotide sequence AAGCCTA to AAGCTTA.

SPDM (Spectral precision distance microscopy) This microscopy technique uses point signals of different spectral signatures. These signals can be precisely localized in the microscopic image. Because of the different spectral signatures two signals can also be distinguished if their distance is below the diffraction-limited resolution. Thus an improvement of distance resolution is possible after appropriate calibration of system-dependent chromatic shifts.

STED (Stimulated emission deletion) This microscopy technique improves the resolution of fluorescence microscopy by deletion of fluorescence at the border of the focal point by means of stimulated emission.

TIR (Total internal reflection) This optical phenomenon occurs when light is refracted (bent) enough at a medium boundary to send it backwards, effectively reflecting the entire ray. An important side-effect of total internal reflection is the propagation of an evanescent wave across the boundary surface.

YAC (Yeast artificial chromosome) A yeast chromosome with a human insert used for DNA cloning

Key References

M. Hausmann, R. Winkler, G. Hildenbrand, J. Finsterle, A. Weisel, A. Rapp, E. Schmitt, S. Janz, C. Cremer, *Biotechniques*, 35 (**2003**), p. 564.

J. P. Knemeyer, N. Marmé, M. Sauer, *Anal. Chem.*, 72 (**2000**), p. 3717.

T. Heinlein, J. P. Knemeyer, O. Piestert, M. Sauer, *J. Phys. Chem. B*, 107 (**2003**), p. 7957.

K. Stöhr, B. Häfner, O. Nolte, J. Wolfrum, M. Sauer, D. P. Herten, *Anal. Chem.*, 77 (**2005**), pp. 7195–7203.

P. Tinnefeld, M. Sauer, *Angew. Chem. Int. Ed.*, 44 (**2005**), p. 2642.

N. Woodford, A. Sundsfjord, *J. Antimicrobial. Chemotherapy*, 56 (**2005**), p. 259.

V. M. Kratoch, *Indian J. Med. Res.*, 120 (**2004**), p. 418.

V. M. Kratoch, *Indian J. Med. Res.*, 120 (**2004**), p. 290.

References

1 A. E. Wright, *Brit. Med. J. London*, (**1903**), p. 1069.

2 A. Fleming, *British Journal of Experimental Pathology*, 10 (**1929**), p. 226.

3 M. Burnet, *Natural history of infectious disease*, Cambridge University Press, Cambridge, 1962.

4 B. Spellberg, J. H. Powers, E. P. Brass et al., *Clin. Infect. Dis.*, 38 (**2004**), p. 1279.

5 P. G. P. Charles, M. L. Grayson, *MJA*, 181 (**2005**), p. 549.

6 A. William, M. C. Millan, *Am. Clin. Laboratory*, (**2002**), p. 29.

7 R. M. Hagen, I. Seegmüller, J. Navai, I. Kappstein, N. Lehn, T. Miethke, *Int. J. Med. Microbiol.*, 295 (**2005**), p. 77.

8 E. M. Mascini, M. J. M. Bonten, *Clinical Microbiology & Infection*, 11 (**2005**), p. 43.

9 S. Palladino, I. D. Kay, J. P. Flexman, I. Boehm, A. M. G. Costa, E. J. Lambert, K. J. Christiansen, *J. Clin. Microbiol.*, 41 (**2003**), p. 2483.

10 A. M. Kearns, C. Graham, D. Burdess, J. Heatherington, R. Freeman, *JCM*, 40 (**2002**), p. 682.

11 R. Hakenbeck, N. Balmelle, B. Weber, C. Gardès, W. Keck, A. De Saizieu, *Infection and Immunity*, 69 (**2001**), p. 2477.

12 P. L. Winokur, R. Canton, J. M. Casellas, M. Legakis, *Clin. Infect. Dis.*, 32 (**2001**), p. S94.

13 G. F. WELDHAGEN, *Chemother.*, 48 (**2004**), p. 4059.

14 V. GRIMM, S. EZAKI, M. SUSA, C. KNABBE, R. D. SCHMID, T. T. BACHMANN, *J. Clin. Microbiol.*, 42 (**2004**), p. 3766.

15 M. A. ESPINAL, A. LASZLO, L. SIMONSEN, F. BOULAHBAL, S. J. KIM, A. RENIERO, S. HOFFNER, H. L. RIEDER, N. BINKIN, C. DYE, R. WILLIAMS, M. C. RAVIGLIONE, *N. Engl. J. Med.*, 344 (**2001**), p. 1294.

16 M. H. HAZBÓN, D. ALLAND, *J. Clin. Microbiol.*, 42 (**2004**), p. 1236.

17 D. GRYADUNOV, V. MIKHAILOVICH, S. LAPA, *Clin. Microbiol. Infect.*, 11 (**2005**), p. 531.

18 W. C. YAM, C. M. TAM, C. C. LEUNG et al., *J. Clin. Microbiol.*, 42 (**2004**), p. 4438.

19 M. LESMANA, C. I. LEBRON, D. TASLIM, P. TJANIADI, D. SUBEKTI, M. O. WASFY, J. R. CAMPBELL, B. A. OYOFO, *Antimicrob. Agents Chemother.*, 45 (**2001**), p. 359.

20 M. IEVEN, M. VAN LOOVEREN, S. Y. SUDIGDOADI, W. GOOSSENS, C. LAMMENS, A. MEHEUS, H. GOOSSENS, *Sexually Transmitted Diseases*, 30 (**2003**), pp. 25–29.

21 M. HERIDA, P. SEDNAOUI, V. GOULET, *Sexually Transmitted Diseases*, 31 (**2004**), p. 209.

22 A. MAVROIDI, L. S. TSOUVELEKIS, K. P. KYRIAKIS, H. AVGERINOU, M. DANILIDOU, E. TZELEPI, *Antimicrob. Agents Chemother.*, 45 (**2001**), p. 2651.

23 R. GONZALEZ, B. MASQUELIER, H. FLEURY et al., *Detection of human immunodeficiency virus type 1 antiretroviral resistance mutations by high-density DNA probe arrays*, 42 (**2004**) p. 2907.

24 D. HOFFMANN, I. ASSFALG-MACHLEIDT, H. NITSCHKO et al., *Biol Chem.*, 384 (**2003**), p. 1109.

25 R. F. SAIKI, S. SCHARF, F. FALOONA, K. B. MULLIS, G. T. HORN, H. A. ERLICH, N. ARNHEIM, *Science*, 230 (**1985**), p. 1350.

26 G. FISCHER, L. S. LERMAN, *Proc. Natl. Acad. Sci. USA*, 80 (**1983**), p. 1579.

27 D. H. BLOHM, A. GUISEPPE-ELIE, *Curr. Opin. Biotechnol.*, 12 (**2001**), p. 41.

28 Z. SULTAN, S. HAHAR, B. WRETLIND, E. LINDBACK, M. RAHMAN, *Genitourin. Med.*, 70 (**2004**), p. 253.

29 L. WESTIN, C. MILLER, D. VOLLMER, D. CENTER, R. RADTKEY, M. NERENBERG, P. O. O'CONNELL, *J. Clinical Microbiol.*, 39 (**2001**), p. 1097.

30 V. GRIMM, S. EZAKI, M. SUSA, C. KNABBE, R. D. SCHMID, T. T. BACHMANN, *J. Clin. Microbiol.*, 42 (**2004**), p. 3766.

31 M. H. HAZBÓN, D. ALLAND, *J. Clin. Microbiol.*, 42 (**2004**), p. 1236.

32 C. ARNOLD, L. WESTLAND, G. MOWAT, A. UNDERWOOD, J. MAGEE, S. GHARBIA, *Clin. Microbiol. Infect.*, 11 (**2005**), p. 1469.

33 T. J. AITMAN, *BMJ*, 323 (**2001**), p. 611.

34 L. HERRERA, S. JIMENEZ, A. VALVERDE, M. A. GARCA-ARANDA, J. A. SAEZ-NIETO, *Int. J. Antimicrob. Agents*, 21 (**2003**), p. 403.

35 K. A. SHAIKH, K. S. RYU, E. D. GOLUCH, J. M. NAM, J. LIU, C. S. THAXTON, T. N. CHIESL, A. E. BARRON, Y. L. CHAD, A. MIRKIN, C. LIU, *Proc. Natl. Acad. Science USA*, 102 (**2005**), p. 9745.

36 L. E. MORRISON, T. C. HALDER, L. M. STOLS, *Anal. Biochem.*, 183 (**1989**), p. 231.

37 C. T. WITTWER, M. G. HERRMANN, A. A. MOSS, R. P. RASMUSSEN, *BioTechniques*, 22 (**1997**), p. 130.

38 L. G. LEE, C. R. CONNELL, W. BLOCH, *Nucleic Acids Res.*, 21 (**1993**), p. 3761.

39 P. TINNEFELD, M. SAUER, *Angew. Chem. Int. Ed.*, 44 (**2005**), p. 2642.

40 E. MAGDE, W. W. WEBB, *Physical Review Letters*, 29 (**1972**), p. 705.

41 P. SCHWILLE, F. OEHLENSCHLAEGER, N. G. WALTER, *Biochem.*, 35 (**1996**), p. 10182.

42 M. KINJO, R. RIGLER, *Nucleic Acid Res.*, 23 (**1995**), p. 1795.

43 M. JAHNZ, P. SCHWILLE, *Nucleic Acid Res.*, 33 (**2005**), p. e60.

44 CASTRO, A., WILLIAMS, J. G. K., *Anal. Chem.*, 69 (**1997**), p. 3915.

45 I. SOLOVEI, J. WALTER, M. CREMER, F. HABERMANN, L. SCHERMELLEH, T. CREMER, FISH on three-dimensionally preserved nuclei, in: J. SQUIRE, B. BEATTY, S. MAI (Eds), *FISH: a practical approach*, Oxford University Press, Oxford, 2001.

46 J. Rauch, D. Wolf, M. Hausmann, C. Cremer, *Z. Naturforschung C*, 55 (**2000**), p. 737.

47 H. J. Tanke, R. W. Dirks, T. Raap, *Curr. Opin. Biotechnol.*, 16 (**2005**), p. 49.

48 F. Mongelard, H. C. Vour, M. Robert-Nicoud, Y. Usson, *Cytometry*, 36 (**1999**), p. 96.

49 M. Hausmann, R. Winkler, G. Hildenbrand, J. Finsterle, A. Weisel, A. Rapp, E. Schmitt, S. Janz, C. Cremer, *Biotechniques*, 35 (**2003**), p. 564.

50 J. R. Lakowicz, *Principles of Fluorescence spectroscopy*, Kluwer Academic / Plenum Publishers, New York, 1999.

51 S. Tyagi, F. R. Kramer, *Nat. Biotechnol.*, 14 (**1996**), p. 303.

52 S. Tyagi, D. Bratu, F. R. Kramer, *Nat. Biotechnol.*, 16 (**1998**), p. 49.

53 X. Fang, X. Liu, S. Schuster, W. J. Tan, *Am. Chem. Soc.*, 121 (**1999**), p. 2921.

54 M. Sauer, *Angew. Chem., Int. Ed.*, 115 (**2003**), p. 1790.

55 C. A. M. Seidel, A. Schulz, M. Sauer, *J. Phys. Chem.*, 100 (**1996**), p. 5541.

56 J. P. Knemeyer, N. Marmé, M. Sauer, *Anal. Chem.*, 72 (**2000**), p. 3717.

57 O. Piestert, H. Barsch, V. Buschmann, T. Heinlein, J. P. Knemeyer, K. D. Weston, M. Sauer, *Nano Lett.*, 3 (**2003**), p. 979.

58 T. Heinlein, J. P. Knemeyer, O. Piestert, M. Sauer, *J. Phys. Chem. B*, 107 (**2003**), p. 7957.

59 M. Zuker, D. H. Mathews, D. H. Turner, Algorithms and Thermodynamics for RNA Secondary Structure Prediction: A Practical Guide, in: J. Barciszewski, B. F. C. Clark (Eds), *RNA Biochemistry and Biotechnology*, NATO ASI Series, Kluwer Academic Publishers, Dordrecht, 1999, pp. 11–43 (http://bioweb.pasteur.fr/seqanal/ interfaces/mfold.html).

60 F. J. Steemers, J. A. Ferguson, D. R. Walt, *Nat. Biotechnol.*, 18 (**2000**), p. 91.

61 W. E. Moerner, L. Kador, *Anal. Chem.*, 61 (**1989**) p. 1217A.

62 M. A. Bopp, Y. Jai, L. Li, R. J. Cogdell, R. M. Hochstrasser, *Proc. Natl. Acad. Sci. USA* , 94 (**1997**), p. 10630.

63 J. J. Macklin, J. K. Trautman, T. D. Harris, L. E. Brus, *Science*, 272 (**1996**), p. 255.

64 E. Boukobza, A. Sonnenfeld, G. Haran, *J. Phys. Chem. B*, 105 (**2001**), p. 12165.

65 R. M. Dickson, D. J. Norris, W. E. Moerner, *Phys. Rev. Lett.*, 81 (**1998**), p. 5322.

66 H. Noji, R. Yasusda, M. Yoshida, K. Kinosita, *Nature*, 386 (**1997**), p. 299.

67 C. W. Wilkerson, P. M. Goodwin, W. P. Ambrose, J. C. Martin, R. A. Keller, *Appl. Phys. Lett.*, 62 (**1993**), p. 2030.

68 C. Zander, M. Sauer, K. H. Drexhage, D. S. Ko, A. Schulz, J. Wolfrum, L. Brand, C. Eggeling, C. A. M. Seidel, *Appl. Phys. B*, 63 (**1996**), p. 517.

69 S. A. Soper, L. M. Davis, E. B. Shera, *J. Opt. Soc. Am. B*, 9 (**1992**), p. 1761.

70 J. Schaffer, A. Volkmer, C. Eggeling, V. Subramanian, G. Striker, C. A. M. Seidel, *J. Phys. Chem. A*, 103 (**1999**), p. 331.

71 P. Tinnefeld, D. P. Herten, M. Sauer, *J. Phys. Chem.*, 105 (**2001**), p. 7989.

72 D. P. Herten, P. Tinnefeld, M. Sauer, *Appl. Phys. B*, 71 (**2000**), p. 765.

73 T. Heinlein, A. Biebricher, P. Schlüter, D. P. Herten, J. Wolfrum, M. Heilemann, C. Müller, P. Tinnefeld, M. Sauer, *ChemPhysChem*, 6 (**2005**), p. 949.

74 T. Cremer, C. Cremer, *Nature Rev. Genet.* (**2001**), p. 292.

75 M. Cremer, J. von Hase, T. Volm, A. Brero, G. Kreth, J. Walter, C. Fischer, I. Solovei, C. Cremer, T. Cremer, *Chromosome Res.*, 9 (**2001**), p. 541.

76 H. Tanabe, S. Muller, M. Neusser, J. von Hase, E. Calcagno, M. Cremer, I. Solovei, C. Cremer, T. Cremer, *Proc. Natl. Acad. Sci. USA*, 99 (**2002**), p. 4424.

77 T. Wiech, S. Timme, F. Riede, S. Stein, M. Schuricke, C. Cremer, M. Werner, M. Hausmann, A. Walch, *Histochem. Cell Biol.*, 123 (**2005**), p. 229.

78 L. A. Parada, P. G. Mc Queen, P. J. Munson, T. Misteli, *Curr. Biol.*, 12 (**2002**), p. 1692.

79 S. W. Hell, M. Dyba, S. Jakobs, *Curr. Opin. Neurobiol.*, 14 (**2004**), p. 599.

80 A. Esa, P. Edelmann, L. Trakhtenbrot, N. Amariglio, G. Rechavi, M. Hausmann, C. Cremer, *J. Microsc.*, 199 (**2000**), p. 96.

81 S. W. Hell, E. H. K. Stelzer, *Opt. Soc. Am. A* 9 (**1992**), p. 2159.

82 S. W. Hell, *Nat. Biotechnol.*, 21 (**2003**), p. 1347.

83 B. Schneider, I. Upmann, I. Kirsten, J. Bradl, M. Hausmann, C. Cremer, *Eur. Microsc. Anal.*, 57 (**1999**), p. 5.

84 B. Bailey, D. L. Farkas, D. L. Taylor, F. Lanni, *Nature*, 366 (**1993**), p. 44.

85 S. Martin, A. V. Failla, U. Spoeri, C. Cremer, A. Pombo, *Mol. Biol. Cell*, 15 (**2004**), p. 2449.

86 G. Hildenbrand, A. Rapp, U. Spöri, C. Wagner, C. Cremer, M. Hausmann, *Biophys. J.*, 88 (**2005**), p. 4312.

87 M. Hausmann, G. Hildenbrand, J. Schwarz-Finsterle, U. Birk, H. Schneider, C. Cremer, E. Schmitt, *Biophotonics Int.*, (**2005**), in press.

88 A. V. Failla, U. Spoeri, B. Albrecht, A. Kroll, C. Cremer, *Appl. Opt.*, 41 (**2002**), p. 7275.

89 K. Stöhr, B. Häfner, O. Nolte, J. Wolfrum, M. Sauer, D. P. Herten, *Anal. Chem.*, 77, (**2005**), pp. 7195–7203.

90 N. Martin-Casabona, A. R. Bahrmand, J. Bennedsen et al., *Int. J. Tuberc. Lung Dis.*, 8 (**2004**), p. 1186.

91 J. E. Clarridge, *Clin. Microbiol. Rev.*, 17 (**2004**), p. 840.

92 C. Turenne, *J. Clin. Microbiol.*, 39 (**2001**), p. 3637.

93 M. Heep, B. Brandstatter, U. Rieger, N. Lehn, E. Richter, S. Rusch-Gerdes, S. Niemann, *J. Clin. Microbiol.*, 39 (**2000**), p. 107.

94 J. D. Jin, C. A. Gross, *J. Mol. Biol.*, 202 (**1988**), p. 45.

95 B. Yang, H. Koge, H. Ohno, K. Ogawa, M. Fukuda, Y. Hirakata, S. Maesaki, K. Tomono, T. Tashiro, S. Kohno, *J. Antimicrobial Chemotherapy*, 42 (**1998**), p. 621.

96 B. Albrecht, M. Hausmann, H. Zitzelsberger, H. Stein, J. R. Siewert, U. Hopt, R. Langer, H. Höfler, M. Werner, A. Walch, *J. Pathol.*, 203 (**2004**), p. 780.

97 A. Baldini, *Curr. Opin. Cardiol.*, 19 (**2004**), p. 201.

98 C. Chen, B. I. Wu, T. Wei, M. Egholm, W. M. Strauss, *Mammalian Genome*, 11 (**2000**), p. 384.

5
Early Diagnosis of Cancer (PLOMS)

Jürgen Helfmann, Uwe Bindig, Barbara Meckelein, Katrin Wehry, Niels Röckendorf, Daniela Schädel, Marcus Alexander Schmidt, Mario Bürger, Andreas Frey[1]

> *Like with every form of cancer, early detection is what it is all about. I urge everyone to learn the facts about this condition. It can be prevented with testing, and it can be beaten if caught early!*
>
> ~ Rod Stewart, Musician ~

Cancer is a collective term for a variety of diseases whose hallmark is the uncontrolled growth of usually growth-controlled or growth-arrested cells and their eventual dissemination to other sites of the body. Once these so-called metastases are formed it is extremely difficult and often impossible to localize and eliminate the cancerous cells and the tumors they form. For that reason early diagnosis of cancer is of pivotal importance for the treatment and positive prognosis of this lethal disease.

5.1
Cancer: Epidemiological, Medical and Biological Background

5.1.1
Cancer Incidence, Prevalence and Mortality

Cancer is on the rise, despite the gigantic efforts devoted to research on the disease itself, its diagnosis and its prevention. In the year 2000, 10.1 million people were diagnosed with cancer, 22.4 million people lived with the disease and 6.2 million people died of malignant neoplasia, which represents an increase of approximately 19% in incidence and 18% in mortality since 1990 [1]. As the ratio of mortality to incidence is an indicator of prognosis, it can be concluded that on average still over 60% of people diagnosed with cancer will

1) Corresponding author, project: PLOMS – Peptidic Ligands for Optical Markers and Validation of Malign Mucosa Tumor

Biophotonics: Visions for Better Health Care. Jürgen Popp and Marion Strehle (Eds.)

ISBN: 3-527-40622-0

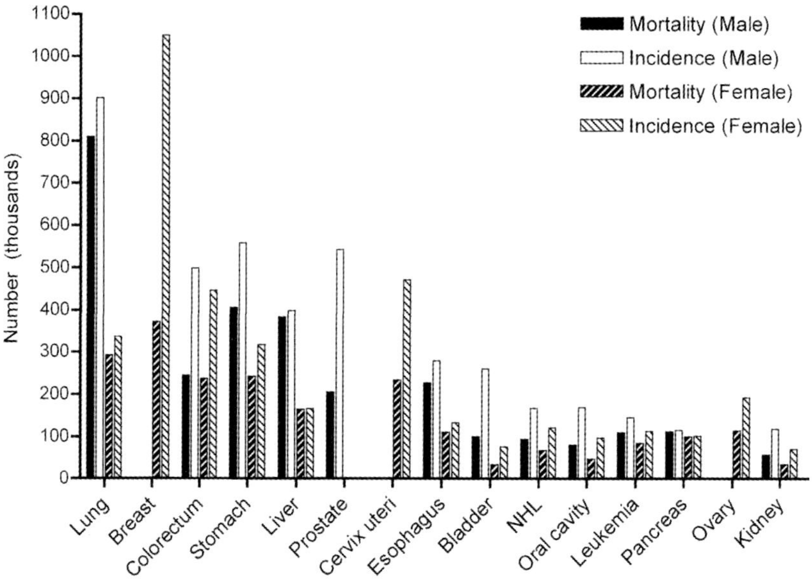

Figure 5.1 Incidence and mortality of the most common cancers worldwide in the year 2000. NHL: Non-Hodgkin's lymphoma. Source: World Cancer Report [1].

die of the disease and that the overall prognosis was barely improved within the period.

Such a collective view, however, may be misleading, because the term cancer stands for a variety of malignant neoplasias whose incidence and mortality differ considerably, not only in absolute numbers but also between sexes and societies. The most frequently diagnosed cancers worldwide are those of the lung, breast and colon/rectum. Most casualties are claimed by lung, stomach and liver cancer, mainly because they are the hardest to treat. The cancer distribution between sexes, excluding sex-specific neoplasias such as cancer of the cervix uteri or of the prostate, apparently reflects differences in lifestyle between men and women. Higher tobacco or alcohol consumption by males, for example, appears to pose an additional risk for lung, esophagus, stomach or liver cancer as indicated by the higher incidence of these disorders in men (**Figure 5.1**).

The cancer distribution also shows differences between developed and developing countries. Colorectal, prostate and bladder cancers are more abundant among members of affluent societies, which again has been attributed to differences in lifestyle. Here, dietary factors are believed to play a major role. Malignant neoplasias of stomach, liver and cervix uteri, on the other hand, are more abundant in less developed regions of the world. The higher burden of certain infectious diseases due to the poorer hygienic conditions in

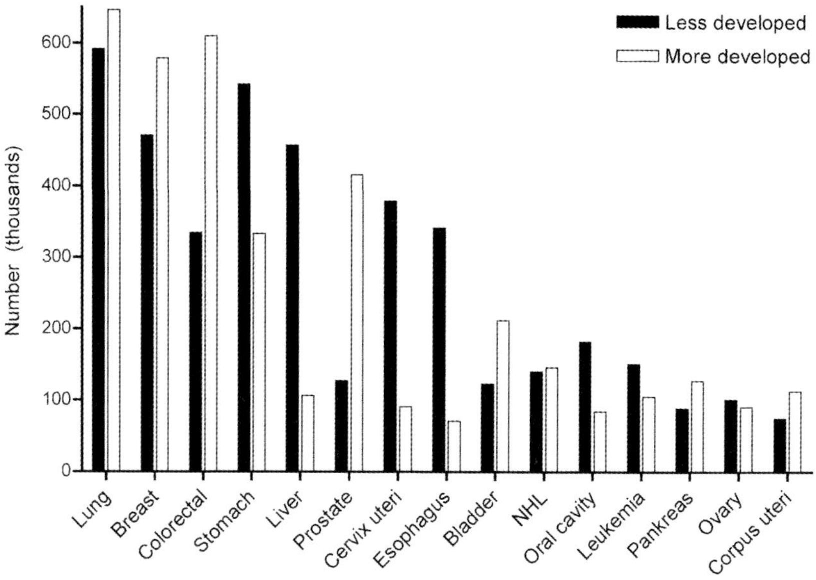

Figure 5.2 Comparison of the incidences of the most common cancers in more and less developed countries for the year 2000. NHL: Non-Hodgkin's lymphoma. Source: World Cancer Report [1].

these areas is thought to be responsible for the bias in these specific cancer incidences (**Figure 5.2**).

Nevertheless, cancer appears to be a plague of the rich. The overall mortality rate in men for all cancer types combined is 2–3 times higher in the wealthier countries of the northern hemisphere, Australia and New Zealand than in Africa, Latin America, Southern Asia and the Pacific islands. It is tempting to conclude that cancer is primarily a result of the western style of living: an unhealthier, meat-rich diet; more abundant drug abuse; more severe environmental pollution; occupational exposure to hazards and stress factors; etc. By and large this is not wrong, but it should be borne in mind that cancer mainly affects people of advanced age and the life expectancy of people in affluent countries is considerably higher than that of people in less developed regions of the world. Here, competing causes of mortality such as infectious diseases take their toll years before cancer formation usually occurs. Consequently, lower numbers of cancer cases ensue. In light of the fact that every society strives towards higher standards of living and better health care, the life expectancy is likely to increase in poorer countries, too, and so will the number of cancer cases. The World Health Organization estimates that humankind will have to cope with a cancer burden of 15 million cases by the year 2025, which would be a 50% increase on the current numbers. For that reason, it is of paramount importance to unravel the reasons for the cancer pandemic and to develop efficient countermeasures.

5.1.2 The Mechanisms of Carcinogenesis

Cancer is an ancient disease. It has existed as long as highly organized multicellular organisms have, and represents a loss of control in the society of individual cells and cell types that constitute the body. In order to prevent, diagnose and treat this aberrant behavior, it is important to understand the biological events that lead to it.

Those events are closely related to the processes which occur in the course of human development. An adult human being consists of more than a trillion cells, which are all derived from a single progenitor, the fertilized egg from which the individual developed. The reason why this huge number of cells remains together in an organism is the competitive advantage each cell gains from being a member of a society. And the key to this advantage is the allocation of tasks among the cells. To fulfil these tasks most efficiently, cells must specialize into distinct cell types and certain numbers of each cell type must be formed. The former process is called cell differentiation, the latter cell proliferation. Both processes are tightly regulated during embryonic and fetal development in order to yield a viable organism.

Although the phenomenon of cellular differentiation is of fundamental importance in biology still little is known about the mechanisms that control it. In recent years it has become increasingly clear that the first specializations in a developing embryo arise from so-called asymmetries, i.e. an uneven distribution of molecules, organelles or cells, which represents a simple binary signal and hence can be viewed as a discriminator between two cells or cell clusters [2, 3]. The type and origin of the initial imbalance that launches the entire process is still the subject of vigorous debate. An asymmetry may arise when and where the sperm enters the oocyte, it may have already been laid down by maternal influences in the unfertilized oocyte or it may occur as a result of the first cell divisions. The latter possibility is the one that is favored for mammalian embryogenesis. The concept is rather simple. As soon as a cell divides, asymmetries are inevitably created within and between its daughter cells. A between-cell asymmetry occurs, for example, if the cytoplasmic content of a progenitor cell is not evenly distributed between its daughter cells or when one of the daughter cells divides earlier than the other. A within-cell asymmetry arises, for example, if a cell is not entirely surrounded by cells or the surrounding cells are not identical. This situation is already present after the very first division of the fertilized egg. Both cells of this two-cell stage blastomer polarize each other by being on one side in contact with their sister cell, while the other part is exposed to the zona pellucida, the membrane that surrounds the early embryo. Once polarized, the plane of the next cell cleavage may be symmetrical or asymmetrical, giving rise to identical or topographically and structurally distinct cells. At the 8–16-cell stage when the blastomer compacts, some cells will be in the center of the cell cluster while others will be partially surface-exposed, and so on. All these intra- or intercell asymmetries are collectively termed topographic signals. They constitute an epigenetic factor that acts as a regulator of differential gene expression, which, in turn, is a prerequisite for cell differentiation [4]. As development progresses, other more sophisticated signal types emerge whose concert is able to fine tune the spatial and functional relationships between cells, tissues and organs. Certain signalling molecules for instance, so-called morphogens, are secreted by organizer cells in a developing tissue, spread away from their source and form a concentration gradient. Depending on the morphogen concentration, distinct target genes in receiving cells are activated by another signal cascade and the cells are thereby steered into their final shape and function depending on their specific location in the tissue [5]. Whatever the type of signal that initiates or drives the process, effective communication between the cells is indispensable for differentiation.

However, differentiation cannot occur without proliferation, and proliferation without differentiation is useless for the formation of a segmented, highly organized multicellular organism. Usually growth and development are alternating processes. Cells proliferate, then they differentiate, then the differentiated ones proliferate again, and so on. Owing to this mutual dependence proliferation must be as tightly controlled as differentiation.

In a cell division event it is particularly important that replication yields descendants without flaw. Somatic cells, which constitute the entire body except for the cells of the germline, must survive some 50 to 60 replication cycles to cover the entire lifespan of a human being. Although a differentiated somatic cell harbors a plethora of genes that its progenitor may once have used but which its descendants will never use again, it will not get rid of any of them. In each cell division, a somatic cell passes on the complete genetic blueprint that it inherited from its progenitor zygote. In addition, the imprinting pattern that governs whether a maternal or paternal copy of a gene should be read is copied, and the genes that are not yet or no longer necessary are silenced. In order to store this huge amount of information compactly the DNA strands of the genome are twisted among each other and wrapped around protein molecules like the wire in a coil. On top of that, the DNA is derivatized with methyl groups in certain areas, its packing density varies and other proteins and some RNA is embedded in this molecular tangle, which is collectively termed chromatin. As with any other information storage device, it must be handled with care and groomed in order not to lose the information stored therein. Hence, it is continuously surveyed and defects are repaired. For replication, the DNA is unpacked and unwound in parallel at numerous sites from which the copying process begins. It starts with the association of small RNA-primers to single-stranded DNA stretches and is performed by so-called DNA polymerases, which display a proofreading capacity in order to minimize mistakes [6]. Upon completion of each copy, the primers are removed and the gaps they leave are filled by the relevant upstream DNA building block during assembly. At the very 5′-end of a newly synthesized DNA strand, however, there is no upstream building block to fill the gap, which will result in a truncation at the end of each chromosome if it is not fixed by other means. A special enzyme, the so-called telomerase, fulfills this task, but in somatic cells this repair is not complete. This is one of the reasons why an individual ages on the cellular level. In this respect, more care is taken when the genome and epigenome of germline cells are copied. Here, a different type of telomerase is active, which more faithfully repairs the RNA-primer gaps at the telomeres. Why this isoenzyme is only active in germline cells is unclear, its expression in somatic cells would slow down aging and extend the lifespan of an individual.

Yet the creation of flawless copies of the genome and the epigenome is only one of numerous processes that are necessary to divide a cell into two identical and viable descendants. Cell division is a cyclic process, with no beginning and no end. Yet discrete phases are discernable. The best defined phase is the one in which the genome and epigenome are copied. It is called synthesis or S-phase and yields two identical "sister chromatids" which are still held together [7]. The S-phase is preceded by the gap-phase one or G_1-phase in which it is ensured that sufficient synthesis supplies for the following steps are at hand. After the S-phase, when DNA-copying is complete, the cell enters the gap-phase two or G_2-phase which ends when the actual cell separation process, the so-called mitosis or M-phase, begins. Then the usually loosely packed chromatin in the cell nucleus condenses and the nuclear membrane dissolves. Meanwhile, the so-called spindle-apparatus forms in the cytosol. With its help, the paired chromatids are pulled apart and transported to two opposite poles in the cell. The mitosis ends with the formation of two new nuclear membranes around the two sets of chromatids and the cleavage of the cytoplasm.

In order to go through this complicated process flawlessly, the cell initiates a fine-tuned concert of activators and inhibitors whose hallmark is the presence of so-called checkpoints [8]. At these points the cell undergoes a molecular auditing process, in which the presence of defined signals is verified in order to proceed. Of particular importance is the test whether the DNA has been replicated completely and without mistakes, which occurs before the cell can enter the M-phase [9–11]. If the control mechanisms of this self-diagnosis apparatus detect a flaw that cannot be repaired by the intrinsic damage control and repair system, a special group of genes is activated which turn on the so-called programmed cell death or apoptosis, in other words, the cell commits suicide.

As long as it remains undamaged, this tight regulatory network guarantees trouble-free control of cell proliferation. In cases of serious damage, however, a cell may no longer obey the checkpoints, may no longer be able to commit suicide and/or may no longer take orders from its surrounding cells. In most cases such havoc is lethal. If not, and the damage affects the genetic level, the cell will start proliferating in an autonomous manner, passing on the defect to all its successor cells. Although this erratic behavior is itself a loss of differentiation, loss of proliferation control is mostly accompanied by additional dedifferentiation events. Finally, both proliferation and differentiation are out of control and the cell has undergone a malignant transformation into a cancer cell. A gene whose function is activated in the course of this process and whose gene products permanently trigger proliferation is termed an oncogene. A gene whose function is deactivated in the course of transformation and whose gene product normally blocks progression of the cell cycle is called a tumor suppressor gene.

In light of the multiple safeguarding systems that are devoted to the prevention of malignant transformation, it is almost inconceivable that a malignant neoplasia may be caused by a single event within a given replication cycle. Indeed, in the vast majority of cancers this is not the case. Cancerogenesis is a multistage process in which genes responsible for the control of cell proliferation, cell death and the maintenance of genomic integrity are damaged successively over numerous cell replication cycles (**Figure 5.3**). Proof of this multistage concept was provided some years ago [12] when it was demonstrated that different, histologically distinguishable stages of colon cancer can be linked to specific successive genetic changes, including oncogene activation and loss of chromosomal regions, which were later shown to be the location of tumor suppressor genes. Subsequent studies confirmed this concept for other cancer types. It is now accepted as a cornerstone of carcinogenesis.

Because certain genes may be especially susceptible to damage, or their damage may be a prerequisite for transformation, one can observe preferential corruption of certain oncogenes and/or tumor suppressor genes. The oncogene c-myc and the gene for the tumor suppressor p53 are respectively activated and destroyed in more than 50% of all cancer cases. On a closer look, however, there is little common ground among malignant neoplasias. There are simply too many possible sites of corruption and permutations thereof not to lead to heterogeneity. If the copying machinery for the genome and the epigenome is affected, even the primordial cancer from which the disease began is not the endpoint of the transformation process. The entire genome of those cancer cells becomes instable such that new variants of the tumor are formed. This is the major reason why some tumors become refractory to cancer medicines in the course of therapy. Thus, it is a hallmark of cancer that every tumor is unique and heterogeneous in itself.

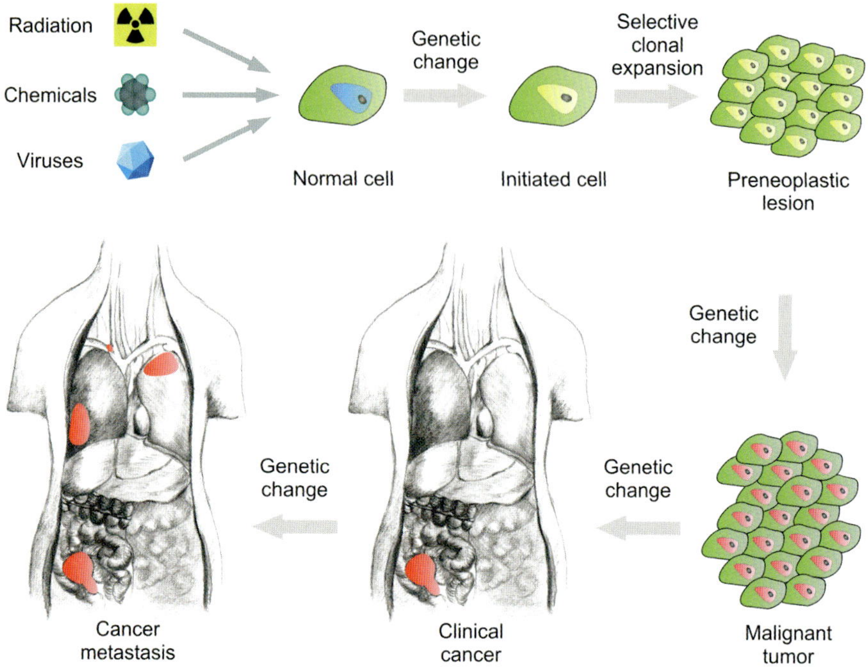

Figure 5.3 Schematic depiction of carcinogenesis as a multistage process.

This autonomous evolution towards more aggressive "subtumors" eventually leads to another atrocious feature of many tumor types: the ability to leave their place of birth and to disseminate to other sites of the body, where they give rise to daughter tumors. This event is called metastasis. It poses a serious problem to the clinician, who usually does not know whether or not this phenomenon has already occurred at the time of the initial diagnosis. If not, the so-called primary tumor can be resected in many cases and the patient is cured. If metastasis has already occurred, however, the oncologist must immediately begin a therapy in which the therapeutic agent reaches every organ, tissue and individual cell of the entire body, where it is supposed to kill the neoplastic cells. The side effects of this treatment are immense and a cure is not guaranteed, although we have seen enormous improvements in the efficacy and tolerability of this type of cancer treatment in recent decades.

5.1.3
Carcinogens and Their Mode of Action

The best cure for cancer is therefore its prevention. In light of the numerous events that are necessary for complete transformation, cancer-inducing agents must be extremely powerful, one must be continuously exposed to them and/or one must have some genetic susceptibility for the development of this disease. All of these can be the case.

For some cancers, such as certain types of breast cancer or retinoblastomas, genetic predispositions have been reported. But even then, additional environmental factors are required to initiate and complete the transformation process [13].

Such environmental factors are classified according to the impact they exert in carcinogenesis. Agents that only stimulate cell proliferation but do not harm the DNA itself are called tumor promoters. Agents that are able to damage the DNA but still are not sufficient to induce transformation to the full are termed genotoxic. Agents that do not require additional factors for induction of transformation are dubbed carcinogens. Most agents of this final category are chemicals, but physical effects such as high-energy radiation or biological agents such as certain viruses and bacteria are also able to act as carcinogens.

Almost everyone is exposed to chemical carcinogens at some time in their life (**Table 5.1**) [14]. Despite their structural diversity, most chemical carcinogens fall into two categories reflecting the two basic modes of action by which they mutate DNA. The first category comprises the ones that directly modify the purine and pyrimidine bases that represent the letters of the genetic code. Such modifications usually occur at electron-rich, i.e. nucleophilic, sites of the bases, which implies that the carcinogen is an electrophile. Substances that liberate reactive carbocations, compounds that readily undergo nucleophilic substitutions of the S_N2 or S_Ni-type, but also radical-forming and deaminating reagents, belong to this category. In this respect it is not essential that the agent is an electrophile, radical or deaminating agent *per se*. In many cases the reactive species is formed only upon metabolic conversion of a parent compound. This may happen in the course of a detoxification attempt or when the host aims to fend off microbes with the help of endogenously formed reactive oxygen or nitric oxide species. Thus, the cell intends the good but does the evil. A typical example for this mechanism are the reactions during the metabolism of benzo[*a*]pyrene [14] and aflatoxin B_1 [15] which are catalyzed by the enzyme cytochrome P450s. In both cases epoxides are formed which can bind covalently to DNA (**Figure 5.4**). Other alkylating agents will introduce less bulky methyl- or ethyl groups but the basic mode of modification remains the same. Ideally, such a modified site is quickly recognized by the DNA control and repair machinery and mended by a mechanism called excision repair [9]. However, if the damage is not fixed before the next replication cycle starts, point mutations due to incorrect base pairing will ensue. These mutations differ depending on the site of alkylation. Substitution at position N^7 of guanin, as performed for example by the aflatoxin B_1 derivative, may lead to the loss of the base from the deoxyribose in the DNA backbone, a

Table 5.1 Selected human chemical carcinogens.

Substance class	Exemplary compounds	Main sources / uses	Affected organs / cancer type
Aromatic hydrocarbons	benzene, toluene	solvent, fuel, tar, tobacco smoke	leukemia
	polychlorinated biphenyls	flame retardants, hydraulic oils	liver, skin, digestive system
Polycyclic aromatic hydrocarbons	benzopyrene, benzoanthracene, benzofluoranthene	tobacco smoke, cooked/smoked meat, coal furnace, tar	lung, skin, genitourinary tract, stomach
	nitroarenes	diesel exhausts	lung
Aromatic amines	naphthylamine	dyes, coal furnace, tobacco smoke	bladder
	benzidine	paints, dyes	bladder
N-Nitrosamines	*N*-Nitroso-dimethylamine	polymers, batteries, lubricants	liver, lung, kidney
	N-Nitroso-diethanolamine	cosmetics, antifreeze, pesticides	liver, kidney
	N′-nitroso-nornicotine	tobacco smoke	liver, lung
Metals	cadmium	fluorescent screens, paints	lung, prostate, kidney
	nickel	alloys, tobacco smoke, catalysts	lung
	arsenic	pharmaceutical compounds	skin, lung, bladder, digestive tract
Minerals and particulate materials	asbestos	roofing products, automobile brakes	lung, larynx
	diesel exhaust particulates	combustion of diesel fuel	lung
	soots	combustion of organic materials	skin, lung
Mycotoxins	aflatoxins	food contaminant	liver
Drugs	estrogens	hormone replacement therapy, contraceptive	uterus, breast
	Tamoxifen	breast cancer therapy	uterus

The substances are known or reasonably anticipated to be human carcinogens according to the Report on carcinogens, 11th edition, US Department of Health and Human Services, Public Health Service, National Toxicology Program

process called depurination. In the next replication step, an adenine is most often incorporated at the position of this gap such that, eventually, a GC base pair will be substituted by a TA base pair. Addition of the benzo[*a*]pyrenediol-epoxide at position N^2 of guanin can lead to erroneous pairing of the modified

G with T which results in the substitution of a GC by an AT base pair. Alkylation of O^4 in thymine produces an erroneous base pairing with guanine and a transition from TA to CG. Thus, transitions due to DNA alkylation do not cause a loss of letters but result in point mutations.

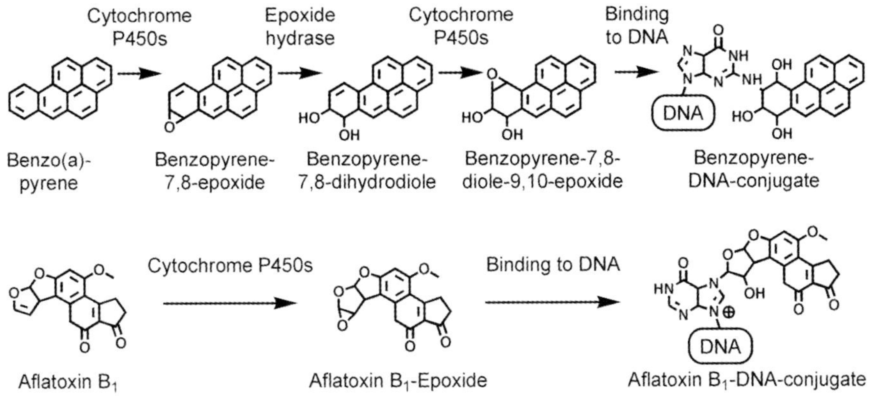

Figure 5.4 Activation of the carcinogenic substances benzo[*a*]pyrene and aflatoxin B_1 by mammalian enzymes. During the metabolism of these carcinogens, reactive epoxides are formed which bind to DNA.

A different type of mutation is the introduction of a reading frame-shift. Substances that create mutations of this type constitute the second category of carcinogens. They cause the reading frame-shift by slipping between two neighboring bases of one strand thereby pushing their neighbors apart. Certain polycyclic aromatic dyes are prototype inducers of this intercalating effect. It can cause more severe corruption than a point mutation, since entire proteins may become misassembled or prematurely truncated during biosynthesis.

Exposure to physical carcinogens is also rather frequent. An ordinary sunburn, tanning lamps and X-ray examinations are known to be potentially carcinogenic, but γ-rays or subatomic particle beams can also damage DNA. They either induce highly reactive chemical intermediates like oxygen radicals, which then act as chemical carcinogens, or they directly induce photochemical reactions in the DNA. The consequences of such photochemical events may be interstrand cross-links or single- and double-strand breakage, the formation of thymine dimers, guanosine depurination or the desamination of cytosine to uracil, which leads to a transition from CG to TA in the next replication cycle.

Certain infectious agents have also been associated with the development of malignant neoplasia. Most cancers of the liver, the cervix uteri and the stomach are caused by viruses and bacteria [16,17]. There are two principal mechanisms by which infectious agents promote the development of cancer.

They either directly corrupt the cell cycle or the genome of their target cell or they indirectly provoke transformation by causing an either unrecognized or incurable chronic inflammation, and thereby persistently induce the formation of chemical carcinogens such as reactive oxygen and nitric oxide species by the host itself.

The best-studied example of the former, direct route is that of the transformation of cervix uteri cells by certain human papilloma viruses (HPV). Here, transformation is induced by viral gene products, so-called oncoproteins, which interfere with the cell cycle and growth control machinery of the infected cell. Epstein–Barr virus, a causative agent of lymphomas, operates in a similar manner, but also causes chromosomal translocations by which the cellular oncogene c-myc is placed under the control of immunoglobulin promoters.

A prototype example of the latter, indirect mechanism is the stomach cancer-inducing bacterium *Helicobacter pylori*. It causes chronic, atrophic gastritis and evokes severe inflammatory tissue responses in the gastric mucosa. Likewise, the hepatitis viruses B and C are unable to transform human cells directly, but they induce a chronic hepatitis that is characterized by chronic liver cell necrosis and continuous regeneration efforts. In the long run, these futile attempts to clear the virus often result in the development of hepatocellular carcinoma.

All in all, it is obvious that we can minimize but hardly prevent exposure to carcinogens. Consequently, every health care system should be prepared to properly diagnose and treat the disease if a person eventually comes down with cancer.

5.1.4
Impact of Early Cancer Diagnosis

The worldwide cancer mortality-to-incidence ratio indicates that on average approximately 40% of cancer patients survive this disease. Although this is an unfavorable survival rate compared to rates for classic infectious diseases, it indicates that cancer does not equate to a death sentence. It must, however, be kept in mind that, while most infectious agents are eliminated by the immune system without the need for medical intervention, spontaneous remissions of cancer are extremely rare. Thus, cancer must be treated in order to prolong survival, which in turn requires the disease to be diagnosed properly. And as with all diseases, the earlier one is aware of a malignancy the better the odds are to keep the disease at bay (**Table 5.2**). Yet diagnosing cancer is not a simple task. It necessitates not only a state-of-the-art diagnostic infrastructure, but also enough educational, personal and financial resources to recruit the patients, to do the testing and to defray the costs. With annual per capita expenditures on health care as low as US $11, as was the case for Ethiopia in the year 2000, it is clear that cancer will be diagnosed – if at all – at very late and therefore almost incurable stages in the less developed regions of the world. Consequently, a worldwide average number of 40% survivors is artificially low in view of the success rate that may be achieved when all resources are recruited. The potential survival rates currently achievable are therefore best reflected by the cancer survival rates of US citizens, because the United States of America have the highest annual expenditure on health care at US $4540 per citizen in 2000 [18].

Table 5.2 Five-year relative survival rates and survival benefit dependent on tumor stage at time of diagnosis

Site of origin	Five-year relative survival rates[1] for tumor diagnosed at any stage	local stage[3]	regional stage[3]	distant stage[3]	Survival benefit of early diagnosis[2]
Lung and bronchus	15.2	49.4	16.1	2.1	23.5
Stomach	23.3	58.4	22.5	3.1	18.8
Urinary bladder	81.7	94.1	48.8	5.5	17.1
Kidney	63.9	91.1	59.1	9.3	9.8
Esophagus	14.3	29.3	13.3	3.1	9.5
Colon and rectum	63.4	89.9	67.3	9.6	9.4
Pancreas	4.4	15.2	6.8	1.8	8.4
Liver	8.3	18.4	6.2	2.9	6.3
Melanoma	90.5	97.6	60.3	16.2	6.0
Uterine cervix	72.7	92.2	53.3	16.8	5.5
Larynx	65.1	83.7	48.7	18.7	4.5
Breast (female)	87.7	97.5	80.4	25.5	3.8
Uterine corpus	84.4	95.8	67.0	25.6	3.7
Ovary	44.0	93.5	68.8	28.5	3.3
Prostate	99.3	100.0	–	33.5	3.0
Oral cavity	58.7	81.0	50.7	29.5	2.7
Thyroid	96.5	99.6	96.3	61.0	1.6
Testis	95.9	99.4	95.9	71.8	1.4
All sites	64.0	n.a.	n.a.	n.a.	–

1) Rates are adjusted for normal life expectancy and are based on tumors diagnosed in the United States of America from 1995 to 2000, followed through 2001.
2) Ratio of survival rate for tumor diagnosed at local stage to survival rate for tumor diagnosed at distant stage
3) Definition of tumor stages: local: invasive, malignant cancer confined entirely to the organ of origin; regional: malignant cancer extended beyond the organ of origin into surrounding organ/tissue and/or involving regional lymph nodes; distant: malignant cancer spread to remote body parts by metastases and/or to distant lymph nodes via the lymphatic system.

n.a., no data available
Source: American Cancer Society, 2005 [19].

On average, an approximately 64% five-year survival rate was recorded in the years 1995 to 2000 for US cancer patients [19]. The odds for surviving a five-year period were shown to be closely related to the stage at which the disease was diagnosed. When the condition was detected as a primary tumor, impressive survival rates of $\geq$95% were obtained for such highly prevalent cancer types as prostate or breast cancer. The likelihood to survive the next five years when diagnosed with these cancers declines over threefold when the disease is discovered at a stage where the tumor has already metastasized to distant sites. The strongest impact of an early diagnosis, however, was reported for cancer types that usually have a poor prognosis. Five-year survival of a malignant neoplasia of the lung or bronchi for instance is over 20-fold higher when the malignancy is detected as primary tumor instead of in a late metastasized stage.

In light of the fact that cancer costs an individual approximately 10 years of their life expectancy, one cannot overemphasize the importance of early cancer diagnosis. In addition to the grief and pain cancer patients, their family members and friends experience, the disease also poses an enormous economic burden on any nation. US health officials estimated the overall costs for cancer in 2000 in the US to be $180.2 billion. Of this sum, $60 billion correspond to direct medical costs (total of all health expenditures), $15 billion to indirect morbidity costs (cost of loss of productivity due to illness) and $105.2 billion to indirect mortality costs (cost of lost productivity due to premature death) [20]. This translates into $212 per capita expenditure for the direct medical costs of cancer spent by each US citizen in the year 2000. Thus, the US allocate 19-times more money for the medical treatment of cancer alone than Ethiopia spends for all medical conditions combined, again on a per capita basis. Considering this enormous bias it is conceivable that the health care systems of many societies are likely to collapse when the burden of cancer continues to increase. Consequently, every effort must be made to improve cancer prevention and early diagnosis.

5.2 Diagnosing Cancer: State-of-the-Art

5.2.1 Principles of Diagnostics

A diagnostic assay must be robust in terms of handling, design and interpretation of the results. Meeting these requirements is a difficult task. Few medical conditions disclose their symptoms so openly that they can be assayed directly. In most cases, a disorder begins and progresses in an asymptomatic manner to a stage where it may already be too late for efficient medical intervention. To overcome this problem, diagnostic medicine takes advantage of the fact that an easily detectable genetic, metabolic or morphological feature can be indicative of an underlying less-obvious medical condition. Such features are termed screening or diagnostic markers for the respective condition.

Markers are usually identified by establishing correlations between the medical condition in question and the easy-to-detect characteristic. The presence of a good correlation is a necessary, but often not a sufficient, requirement for a marker to be a good one. An ideal diagnostic marker must be in a mutual monocausal relationship to the medical condition it represents. That means a given marker must only indicate the medical condition it represents and conversely the medical condition must always give rise to the respective marker. Especially for cancer, this is quite often not the case, which is a major reason for the ongoing quest for better cancer markers.

A marker can appear or disappear on manifestation of a medical condition. In the first case it is termed a positive, in the latter a negative, marker. Both types of marker are in use and each has advantages and drawbacks. When a cancerous lesion is to be identified in a tissue sample, it is equally informative to look for the appearance or disappearance of a morphological feature, since the lesion can be directly compared to the surrounding healthy tissue. Most immunological, biochemical or molecular biological tests, however, lack such in-built controls. For this type of assay, positive markers are favored as their appearance proves that the assay set-up worked. Nonetheless, it is good diagnostic practice to include standardized positive and negative controls in an assay along with the sample to be analyzed. Yet, even when the above constraints are considered, a diagnostic assay rarely yields an absolute and unequivocal read-out. Most biological systems are simply too complex to respond with a clear-cut yes or no on investigation. Consequently, assay read-outs must be interpreted [21].

The interpretation of a diagnostic test result depends on both the setting in which the test is being used and the ability of the test to distinguish between diseased and nondiseased individuals. The former criterion is a rather soft one. It involves the rules of good medical practice, the clinical guidelines laid down by health officials of the respective state or country and other legal constraints. There are considerable between-country differences in the definition of a diseased state, in the recommendation for preemptive screening and in the section of the population encouraged to undergo testing. The discriminatory power of a given test is a rather hard, scientific criterion, albeit it will be influenced by the varying definitions of diseased and nondiseased state.

Before a test is used to estimate the posterior probability of disease it must be validated on two control groups, one containing probands known to be diseased and the second containing probands known to be nondiseased. If the test provides a dichotomous result for each subject, there will be subjects assayed positive or negative. If the subjects that appear in the test as positive belong to the diseased probands, they are classified as true positives (TP), if they belong to the nondiseased probands, they are classified as false positives (FP). Likewise, subjects with a negative test can either belong to the nondiseased probands and are therefore true negatives (TN) or to the diseased probands and are therefore false negatives (FN). The discriminatory power of a test is then expressed by four terms, the probability of correctly classifying diseased individuals, called sensitivity or true positive rate (TPR),

$$\text{TPR}\ [\%] \equiv \text{Sensitivity}\ [\%] = \frac{\text{true positives}}{\text{true positives} + \text{false negatives}} \cdot 100 \qquad (5.1)$$

the probability of correctly classifying nondiseased individuals, called specificity or true negative rate (TNR),

$$\text{TNR}\ [\%] \equiv \text{Specificity}\ [\%] = \frac{\text{true negatives}}{\text{true negatives} + \text{false positives}} \cdot 100 \qquad (5.2)$$

the probability of a type I error, termed false positive rate (FPR),

$$\text{FPR}\ [\%] = 100 - \text{TNR}\ [\%] \qquad (5.3)$$

and that of a type II error which is termed false negative rate (FNR).

$$\text{FNR}\ [\%] = 100 - \text{TPR}\ [\%] \qquad (5.4)$$

Even though any assay read-out will ultimately be dichotomized into the two categories, diseased or nondiseased, many tests provide primary read-outs on a continuous scale. In an ideal setting, diseased and nondiseased probands would yield different read-outs within a narrow bandwidth each. Unfortunately, in practice most often a remarkable within-group variability, which adopts a frequency distribution is observed for both groups, the diseased and the nondiseased, and in most cases the two distributions overlap (**Figure 5.5A**). In order to obtain the desired dichotomized classification, the medical decision-maker has to set an arbitrary threshold, termed cut-off, cut-point or decision limit (CP), from which the diseased or nondiseased status is derived. For positive markers the read-outs above decision limit are classified diseased whereas the ones below the decision limit are designated nondiseased and contrariwise for negative markers.

Setting this cut-off is the most important judgment in interpreting read-outs of continuous scale assays. The dilemma of this task resides in the fact that sensitivity and specificity are inversely related. Thus, a gain in sensitivity is inevitably accompanied by a loss in specificity and vice versa. Only in an ideal setting where two assays are available, one with high sensitivity and the other with high specificity, can one escape this dilemma by carrying out the tests consecutively and thus combining the best of both worlds [22].

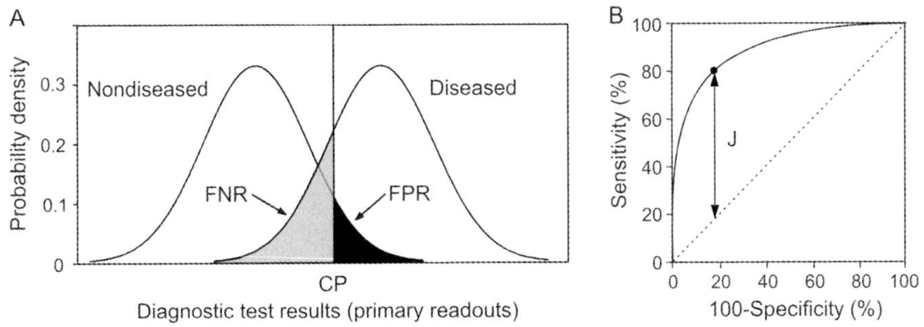

Figure 5.5 Determining the optimal cut-point for highest assay sensitivity and specificity. A: frequency distributions of diseased and nondiseased probands after read-out of a continuous scale assay. B: ROC curve and the identification of the optimal cut-point (CP) and Youden index (J). FNR, false negative rate; FPR, false positive rate.

In a mass screening campaign for a medical condition with low overall prevalence, such as female breast cancer for instance, the highly sensitive assay would be run first in order to identify every individual affected, even though some people would erroneously be diagnosed positive. The ones diagnosed positive would then be subjected to the second highly specific test in order to rescue the false positives from unnecessary treatment.

Although the above scenario is the method of choice it is not advisable to push sensitivity and specificity to the brink since the FPR and FNR will be unacceptably high. A medical decision-maker or assay-designer will therefore have to carefully evaluate which of the two parameters shall outweigh the other for the desired test environment or whether the assay should classify diseased and nondiseased individuals equally well. The cut-point that yields such a balanced assay performance is identified with the help of a so-called receiver operating characteristic curve (ROC curve).

An ROC curve is a plot of the respective TPR and FPR pairs for each possible cut-point on the primary read-out scale of the test (**Figure 5.5B**). The upper left corner of this graph represents full discrimination (TPR = 1, FPR = 0 or TNR = 1, FNR = 0) whereas on the diagonal line discrimination is no better than chance (TPR = FPR, TNR = FNR). Thus, the farther a point on the ROC curve is above the chance line the higher is the diagnostic accuracy for the corresponding cut-point. At the cut-point where the vertical distance is maximal the numbers of correctly classified diseased and nondiseased individuals will also be maximal for this particular test. Consequently, this cut-point is the one to be used when maximal, sensitivity–specificity-balanced assay performance is desired [23].

The vertical distance between the chance line and this optimal cut point is also a measure for the diagnostic accuracy of said test. It is called the Youden

Index (J) [24].

$$\text{Youden Index } (J) = \left\{ \begin{array}{l} \max\left(\frac{\text{TPR }[\%]+\text{TNR }[\%]-100}{100}\right) \\ \text{or } \max\left(\frac{\text{TPR }[\%]-\text{FPR }[\%]}{100}\right) \\ \text{or } \max\left(\frac{\text{TNR}[\%]-\text{FNR}[\%]}{100}\right) \end{array} \right\} \quad (5.5)$$

The Youden Index is a popular descriptor of accuracy for diagnostic tests, but not the only one.

The total area under the ROC curve (AUC) [25] and the proportion correct [26] are also widely used measures of diagnostic accuracy. The area under the ROC curve is calculated as the sum of rectangular areas under the plot and represents the average sensitivity over all cut-points. The proportion correct is calculated as the sum of true positives and true negatives divided by the total number of probands tested. Since it does not consider the prevalence of positives and negatives in the groups assayed, it will give little information when the proband numbers of the diseased and nondiseased groups are considerably different. For low overall prevalence a weakly sensitive and for high overall prevalence a weakly specific test will still yield high proportions correct. Unfortunately, the general term "diagnostic accuracy" is often misused for the proportion correct, such that a diagnostic test characterized this way will appear sound and powerful, albeit it may in fact be inferior to other properly validated diagnostic regimens. In view of this evident confusion, and other shortcomings in assay reporting, a guideline has been set up for reporting studies of diagnostic accuracy [27].

5.2.2 Current Techniques for Cancer Diagnosis

Considering the plethora of different malignant neoplasias that have been described, it is clear that as yet not all types of cancer can be diagnosed equally well. In the past, emphasis has been laid on the development and validation of diagnostics for the most prevalent variants of this disease. When ranked by anatomical site, more than 50% of deaths and life years lost are claimed by five cancer types, malignant neoplasias of the lung and bronchi, stomach cancer, liver cancer, cancer of the female breast and malignant neoplasia of the large intestine. Another 40% of victims succumb to 11 other cancer types. The most widely used techniques to diagnose those 90% of cancer cases are summarized in **Table 5.3** and will be appraised in the following.

Table 5.3: Sensitivity and specificity of procedures currently applied in the diagnosis of the most abundant types of cancer.

Site	Deaths[1]	TYLL[2]	Current diagnostic procedure		
	(thousands)		Method[3]	Sensitivity (%)	Specificity (%)
Lung and bronchus	1214.9	10 803	sputum analysis	14–98 [163]	90–99 [163]
			chest X-ray	54–84 [164]	90–99 [164]
			spiral CT	63–85 [28, 165]	n.a.
			LIFE bronchoscopy	94 [31]	38 [31]
			WL bronchoscopy	85 [31]	36 [31]
Stomach	834.3	7847	barium upper GI X-ray	93 [166]	77 [166]
			endoscopy	82 [167]	95 [167]
Liver	594.6	6872	ultrasonography	20–80 [34, 35]	>90 [35]
			serological (AFP)	55–69 [35, 36]	83–92 [36]
			spiral CT	52–79 [34]	n.a.
			MRI	33–78 [34]	n.a.
Breast (female)	468.6	5393	mammography	81–86 [37]	95–98 [37]
Colon and rectum	607.8	5117	FOBT	24–87 [38–40]	88–98 [38, 39]
			sigmoidoscopy	70–78 [38, 168]	84 [168]
			colonoscopy	79–100 [38, 169]	n.a.
			barium enema	62–100 [38, 169]	n.a.
Leukemia	263.0	4673	cytomorphology (CM)	n.a.	n.a.
			immunophenotyping	n.a.	n.a.
			genetic methods	n.a.	n.a.
Lymphoma and myeloma	327.3	4158	cytomorphology (CM)	n.a.	n.a.
			CM+flow cytometry	77–95 [170, 171]	85–100 [170, 171]
			CM+flow cytom.+PCR	93 [172]	n.a.
			serum κ/λ ratio [4]	90–96 [173, 174]	70–100 [173, 174]
Esophagus	429.5	4067	endoscopy	90–100 [47]	n.a.
			abr. brush cytology	74–91 [48]	81–94 [48]
Oral cavity	306.1	3335	exfoliative cytology	95 [49]	99 [49]

Table 5.3: (continued)

Site	Deaths[1] (thousands)	TYLL[2]	Current diagnostic procedure		
			Method[3]	Sensitivity (%)	Specificity (%)
			panendoscopy	95 [175]	85 [175]
Uterine cervix	231.0	2957	PAP-Smear	44–78 [51]	60–96 [51]
			HPV-testing	66–100 [51]	61–96 [51]
			cerviscoscopy	67–79 [51]	49–86 [51]
Pancreas	224.7	1895	ultrasonography	76 [176]	75 [176]
			CT	86–91 [176]	79–85 [176]
			MRI	84 [176]	82 [176]
Ovary	130.9	1453	ultrasonography	75–100 [58, 177]	40–99 [58, 177]
			CT	92 [57]	89 [57]
			MRI	92–100 [57, 58]	84–88 [57, 58]
Prostate	259.3	1315	digital rectal exam	49–89 [178, 179]	18–99 [178, 179]
			blood test (PSA)	53–100 [178, 179]	13–100 [178, 179]
			ultrasonography	60–93 [178]	27–92 [178]
Urinary bladder	172.7	1248	urine cytology	38–83 [60]	96–98 [60]
			cystoscopy	47–62 [179, 180]	98 [181]
Melanoma	64.5	639	clinical examination	70–91 [182]	77–99 [182]
			dermoscopy	86–94 [182]	74–81 [182]
Uterine corpus	70.6	638	ultrasonography	92–100 [62, 183]	80–81 [62, 183]
			endometrial biopsy	68–99 [183]	98–99 [183]
			hysteroscopy	93 [184]	99 [184]

2) Number of deaths caused by the respective cancer in 2000, worldwide. Source: *The Global Burden of Disease* [185].

3) Years of life lost (YLL): the difference between age at time of death from cancer and age corresponding to an average life expectancy. Total years of life lost (TYLL) represent the sum of all YLL for all persons who died from the respective cancer in 2000, worldwide. Source: *The Global Burden of Disease* [185].

4) For description of the different diagnostic methods, see Section 5.2.2

5) For multiple myeloma only.

n.a., not available

5.2.2.1 **Lung and Bronchial Cancer**

Malignancies of the lung and bronchi, which are predominantly caused by tobacco smoke and other airborne carcinogens, have the highest incidence and mortality among all cancer types. No reliable mass screening procedure is available to date. Diagnostic tests are usually performed on persons who either belong to high risk groups (e.g., heavy smokers, workers exposed to asbestos) or present with typical symptoms such as persistent coughs, chest pain, recurring bronchitis, etc. [28].

With the exception of a sputum analysis, where a phlegm sample is examined microscopically for the presence of cancerous cells, all other diagnostic procedures aim to detect so-called pulmonary nodules or cancerous lesions in the lung. The classical approach consists of a chest X-ray examination, which carries the intrinsic risk of the relatively high exposure to radiation. In addition, lesions often become visible only at a rather large size. Then the disease may already have progressed to an uncurable stage. A technique that reduces these sensitivity problems to some extent but requires heavy equipment is spiral (or multislice) computed tomography (CT). Spiral CT is an X-ray imaging procedure in which multiple detectors are arrayed in parallel. Because of the very fast image acquisition, interference by movements or breathing of the patient is low and a high resolution can be achieved. In addition, the radiation dose is low to moderate. The multislice process enables reconstruction of a 3D image, and pulmonary nodules of less than 10 mm size can be detected [29]. If X-ray exposure is not desired or infrastructurally impossible, the current best alternative for obtaining a definite histological analysis for lung cancer is white-light bronchoscopy. In bronchoscopic procedures the bronchi are analyzed visually with the help of a fiberoptic flexible, lighted tube ("bronchoscope", a special type of endoscope). More recently, laser-induced fluorescence endoscopy (LIFE) bronchoscopy has been developed. This procedure uses the differences in autofluorescence between normal, pre-cancerous and cancer tissue [30, 31]. Great hope is laid on this advanced imaging technique as it may be more sensitive and cost-effective and less hazardous than contemporary X-ray procedures.

5.2.2.2 **Stomach Cancer**

Malignant neoplasia of the stomach claims the second-highest number of life years among cancer patients. Although declining in affluent countries, its incidence is high in areas where *Helicobacter pylori* infection of the stomach is endemic. Screening for and eradication of *H. pylori* is a useful preventive measure, but cannot replace a still-lacking efficient screening tool for gastric cancer. In the absence of such general screening processes, diagnosis of gastric carcinoma is often delayed, because the majority of patients are asymptomatic during the early phases of this disorder. Presymptomatic screening for gastric carcinoma is mostly limited to persons with multiple risk factors, such as *H. pylori* infection, chronic gastritis or familial adenomatous polyposis. Otherwise, testing is only done once symptoms like abdominal pain, nausea and vomiting, or peptic ulcer symptoms show up [32]. As an initial diagnostic procedure, a barium-swallow X-ray may be performed. In this test, the lining of the upper gastrointestinal tract is coated by a barium sulfate suspension and subsequently analyzed by multiple X-ray radiographs. This procedure can be refined by additionally pumping air into the stomach, which will render the barium coat very thin and make abnormalities more easily visible ("double-contrast barium swallow"). The barium-swallow X-ray can identify gastric lesions and provide preliminary information about their possible malignancy. It has good predictive power to rule out cancer, and thus can help to avoid unnecessary further tests using more invasive techniques [32, 33]. However, if suspicious lesions are found endoscopy of the upper gastrointestinal tract (esophagogastroduodenoscopy) must be performed to examine the lining of the esophagus, stomach and the beginning of the small intestine. If abnormalities are noted, biopsies are taken and analyzed histologically or cytologically with high specificity. A highly sensitive laboratory assay or more comfortable imaging techniques such as autonomous cameras that would replace the X-ray test and the rather unpleasant swallowing of the endoscope would surely increase the compliance of patients for gastric cancer diagnosing.

5.2.2.3 **Liver Cancer**

As in the case of stomach cancer, most malignant liver neoplasias are caused by infectious agents. In populations with high prevalence for hepatoviridae infections patients at risk can be identified by screening for these infections, but this test cannot replace screening for the cancer itself. Nevertheless, individuals known to belong to such a liver cancer high-risk group are usually enrolled into screening programs. Apart from such campaigns, early diagnosis of this cancer is not very common. Testing is often delayed because typical symptoms usually do not show up in the early stage of this disease. The test most widely used in screening for liver cancer is ultrasonography [34, 35]. As a cheap laboratory test, the serum concentration of alpha-fetoprotein (AFP) is determined. The sensitivity of this method, however, is too low for application in general screening [36]. Spiral CT or magnetic resonance imaging (MRI) is used for diagnostic confirmation of cancerous lesions [34, 35]. Owing to the rather poor accessibility of the liver by noninvasive means, more sensitive laboratory tests appear to be the best way towards improved liver cancer diagnosis.

5.2.2.4 **Breast Cancer (Female)**

Breast cancer has a high prevalence in the female population all over the world; in fact it is the most prevalent cancer of all. Besides some inherited predispositions, the etiology relates to the reproductive life of women. Early menarche, nulliparity (i.e., women who had no births), late age at first birth, late menopause and hormonal factors are believed to play a role. This still-diffuse risk profile means that screening is in general recommended for women between 50 and 60 years of age. The standard screening procedure is X-ray mammography. A diagnostic mammogram is also performed to examine the breasts of women with specific breast conditions such as palpable lumps or generalized swelling of one breast. The procedure has a rather high sensitivity and very good specificity, but as with all X-ray techniques it also increases the cancer risk of the individual screened [1,37]. Thus, a good laboratory assay or improved non-hazardous imaging techniques would be of great benefit for early breast cancer detection.

5.2.2.5 **Colon and Rectal Cancer**

Malignant neoplasia of the large intestine is the second most common malignancy in affluent countries. Except for some inherited predispositions, the formation of colorectal cancer is primarily attributed to the unhealthy meat- and fat-rich diet of wealthier societies. Consequently, screening for colorectal cancer is recommended to everyone beyond the age of 50 in the US and other industrialized countries [1,38]. The basic screening test – apart from a digital rectal examination – is the fecal occult blood test (FOBT). It analyses a stool sample for the presence of blood, usually detecting the peroxidase-like activity of hemoglobin in stool (guaiac FOBT). Although this procedure exhibits a relatively low spot sensitivity if done only once, the result can be improved by repeating the test three to five times with different stool samples. A higher sensitivity is also obtained if an immunochemical test is used on feces (iFOBT) which looks for the presence of human globin in the excretions [39,40]. If the FOBT test is persistently positive, the cancer diagnosis is confirmed by a different technique, usually by endoscopic procedures, namely sigmoidoscopy and colonoscopy. In both cases, the colon is viewed with a flexible endoscope in order to detect cancerous lesions and adenomas. In sigmoidoscopy, a 60-cm endoscope is used, which limits the accessible portion of the colon to the distal region (approximately the last 40 cm). Colonoscopy, on the other hand, uses a 160-cm endoscope and can be used to analyze the entire colon. However, it is a more complicated procedure than sigmoidoscopy. A less-invasive yet rather specific test method is the double-contrast barium enema, which involves instillation of a barium sulfate suspension, followed by air, into the colon, and X-ray radiography [41]. In contrast to the readily accepted FOBT tests, patients are often reluctant to undergo such follow-up tests. For that reason, better laboratory assays and advanced less-embarrassing imaging techniques such as autonomous cameras would be highly desirable.

5.2.2.6 **Leukemia**

Although leukemias cause only 3 to 4% of all cancer casualties they lead on average to 17.5 years of life lost per afflicted person. This is about twice as many as most other cancer types claim and may be attributed to the fact that leukemia is the most prevalent cancer in children. Leukemias arise from a malignant transformation of white blood cells or their precursors. Depending on the cell types involved, they are classified into the subtypes lymphoid (B cells, T cells) or myeloid (granulocytic, erythroid, megakaryocytic), both of which can present as acute or chronic variants. The etiology of leukemias is poorly defined. Ionizing radiation may promote chronic myeloid leukemia, while previous chemotherapies for other malignancies or exposure to ionizing radiation or chemical carcinogens increase the risk for the acute variant. For chronic lymphoid leukemia, a genetic predisposition is suspected, while exposure to carcinogens could not be linked to this disorder. For the acute variant, all of the above, genetic predisposition, exposure to radiation or chemical carcinogens and previous cancer chemotherapy, and also infections with the human T-cell leukemia virus I (HTLV-1), endemic in Africa, Japan and the Caribbean, are suspected risk factors [42].

Leukemias rarely present with specific symptoms. Owing to the disturbance of normal hematopoiesis, myeloid leukemia may lead to anemia, general weakness, leukopenia and/or thrombocytopenia. The corruption of the body's adaptive immune system in lymphoblastic leukemias may cause increased susceptibility to infections, anemia and enlargement of liver, spleen and multiple lymph nodes [1]. In acute variants, symptoms appear rather suddenly and massively, but treatment is often successful if they are diagnosed in time, whereas chronic leukemias can remain asymptomatic for years before they turn acute all of a sudden and end fatally shortly thereafter.

Correct diagnosis of leukemic cells, which accumulate in the bone marrow and circulate in the blood, involves a cascade of diagnostic steps. A first indication is an elevated white blood cell count in peripheral blood. As such pathological cell counts may also have other causes, a cytomorphological and cytochemical examination of peripheral blood and bone marrow cells will be performed. If a neoplasia is present it will be revealed by these procedures, but their specificity is too low to correctly define the variants and numerous subvariants that are a hallmark of white blood cell cancers. As an exact typing of the cancerous cells is crucial for optimal therapy, more-specific laboratory tests must be performed with the specimens. Differentiation status and lineage commitment of the cells is obtained by immunophenotyping either via immunohisto- or cytochemistry or by flow cytometry. For identification of chromosomal aberrations, a cytogenetic analysis on the basis of (spectral) karyotyping is performed. If necessary, polymerase chain reaction (PCR) analysis and fluorescent *in situ* hybridization (FISH) may also be included [43]. Despite this battery of diagnostic tools, the correct classification of leukemias is still a daunting challenge and little is known about the sensitivity and specificity of each technique in leukemia diagnosis. The microarray analyses currently being developed offer a promising option for further improved accuracy in leukemia diagnostics.

5.2.2.7 **Lymphomas and Multiple Myelomas**

Lymphomas are a heterogeneous group of lymphoid cell neoplasias that originate in lymph nodes. The highly prevalent Epstein–Barr virus is believed to play a central role in pathogenesis but also infections with HTLV-1 and *H. pylori* have been associated with these disorders, and immunosuppressed individuals constitute a risk group. Lymphomas are categorized as either Hodgkin disease or non-Hodgkin's lymphosma. Hodgkin disease rarely metastasizes outside lymphoid organs, whereas extranodal involvement is quite common in non-Hodgkin's lymphomas. The most prevalent symptoms include painless swelling of lymph nodes in neck, armpit and groin, unexplained fever, night sweats and fatigue [44, 45]. Diagnosis usually requires a tissue biopsy, for which either an enlarged lymph node is excised completely or the suspicious lesion is punctured and a sample is acquired by fine-needle aspiration. The basic question – whether or not the sampled tissue is malignant – can be answered by morphologic examination, but verifying, typing and staging of the putative neoplasia requires additional tests analogous to leukemia. As lymphomas give rise to distinct, locally confined lesions, imaging techniques such as chest X-ray and CT of neck, chest, abdomen and pelvis are also employed in routine lymphoma diagnosis.

Multiple myelomas are caused by neoplastic antibody-producing lymphocytes, the so-called plasma cells, which infiltrate the bone marrow. As with all hematologic malignancies, exposure to ionizing radiation is the most convincing risk factor for this disorder. Hereditary factors may also contribute to some extent as may exposure to certain chemical carcinogens. The disease presents with symptoms similar to those of other hematologic cancers and most often bone pain is reported. Since the cancerous plasma cells continuously produce antibodies or fragments thereof, the disease can readily be diagnosed with high sensitivity and specificity by assaying the ratio of free immunoglobulin light chains (κ/λ-ratio) in serum. The myelomatous bone lesions allow imaging techniques such as conventional radiography, CT or MRI to be useful tools to confirm and stage this disorder [46]. While myelomas are readily detectable, detection and typing of lymphomas is rather difficult, even with the sophisticated instrumentation that is currently to hand.

5.2.2.8 **Esophageal Cancer**

Cancer of the esophagus mainly occurs in two variants. Esophageal squamous cell carcinoma is endemic to certain areas of China. Esophageal adenocarcinoma, like the cancers of the large intestine, has largely been attributed to the western style of living. Obesity and tobacco and alcohol abuse enhance the risk. Stress and dietary factors apparently facilitate the development of chronic heartburn, which is known to pose the highest risk for developing this type of cancer. While there is currently no evidence supporting population-based screening, patients with gastroesophageal reflux disease or members of certain ethnic groups that have a high incidence of esophageal squamous cell carcinoma may profit from routine screening for esophageal cancer [47, 48]. The typical diagnostic test is flexible upper endoscopy which is a very sensitive procedure. It includes the possibility to sample biopsy material for cytological analysis. Alternatively, cell sampling may also be performed by abrasive brush cytology, in which a sampler is introduced into the stomach, inflated and then withdrawn through the length of the esophagus, thereby scraping and collecting exfoliated cells. After staining, the cells are analyzed for any irregularities in staining behavior, nuclear size and shape, and chromatin distribution to identify and characterize abnormal, dysplastic cells [48]. In general, this approach is less sensitive and specific than endoscopy, and it suffers from the fact that the actual location of a malignancy cannot be identified this way. Similar to gastric cancer and colorectal cancer, compliance with the rather unpleasant endoscopy or brush sampling is low and improved laboratory and imaging tests would be of great value for keeping this disorder at bay.

5.2.2.9 **Head and Neck Cancer**

Malignancies of the oral cavity and the neck are predominantly caused by tobacco (smoking or chewing) and alcohol abuse, especially when both habits coincide. Nasopharyngeal cancer is strongly associated with Epstein–Barr virus infection, and oral infection with human papilloma virus is a risk factor for tumors of the oropharynx and squamous cell carcinomas of the neck [1]. Prognosis of squamous cell carcinomas of the oral cavity is in general poor, mostly because diagnosis of this cancer is late owing to the lack of specific symptoms. Diagnostic procedures are usually initiated by medical assessment of various pathological conditions in the head/neck region, of which a non-healing sore in the mouth is the most common but which also include dysphagia or persistent coughing. If any suspicious area is detected in the oral cavity, the first diagnostic test to be performed is exfoliative cytology, which involves preparation of several cytological smears by collecting cells with a special brush from all visible oral lesions [49]. In contrast to esophageal abrasive brush cytology, sensitivity and specificity are higher in this case, presumably because the sampling sites are better defined. Lesions in the larynx and pharynx are visualized by endoscopic examination (laryngoscopy and pharyngoscopy, "panendoscopy"), with cytological or biopsy material being sampled in most cases. While the above procedures are sufficient for the identification of superficial neoplasias, the degree of possible tumor infiltration into the cartilage or below the mucosa must be assessed with special imaging methods. Spiral CT and MRI are the methods of choice [50]. As with all inspection and sampling techniques applied via the throat, the discomfort is rather high so that further improvement of endoscopic devices is highly desirable.

5.2.2.10 **Cervical Cancer**

For the most part, cancer of the uterine cervix is a sexually transmitted disease, caused by certain papilloma viruses (HPV). This type of cancer is prevalent in all societies and very often affects younger women. For that reason, regular screening is highly recommended, with slightly differing guidelines among countries and health care organizations. For the early diagnosis of cervical precancers and cancers, several tests are available [51]. The standard procedure is the so-called Pap smear, in which cells are scraped from the cervix and smeared onto a glass slide. After a special staining procedure, the cells are analyzed microscopically to detect epithelial cell abnormalities. Owing to the high prevalence of HPV in precancerous and cancerous lesions, cervical cells may also be analyzed for the presence of HPV DNA. A pool of RNA molecules for "high-risk" HPV-types known to be associated with cervical cancer is used to specifically detect HPV DNA in cervical samples. Another diagnostic method, which yields an immediate test result, is cervicoscopy. This procedure involves inspection of the untreated cervix, examination after exposure to dilute acetic acid and inspection after application of iodine tincture (Schiller test). Application of dilute acetic acid causes a temporary dehydration of the mucosa. Because of the higher nucleus-to-cytoplasm ratio and the nuclear crowding in dysplastic and cancerous cells, this dehydration is more pronounced in diseased than in nondiseased tissue areas and thus will lead to selective opacification of the former. Punctation and mosaicism within the opacified lesions refer to vascular patterns underneath the diseased mucosa, and indicate an epithelial dysplasia (cervical intraepithelial neoplasia, CIN) or carcinoma. The lower glycogen content of such lesions means that they will lack staining after iodine treatment, whereas the glycogen-rich undiseased areas will take on a brownish color. Dysplastic or precancerous lesions whose cytoplasmic glycogen is intermediate may display an intermediate, that is ochre, color. Because ochre coloration is occasionally also observed for nondiseased cells, the Schiller test alone is not specific [52]. When the visual inspection is carried out with the help of a microscope the diagnostic regimen is called colposcopy (see also Section 5.3.1.1). Compliance for all of the above procedures is high, yet the sensitivity of those tests leaves room for improvement. Owing to the excellent accessibility of the anatomical site affected, it is likely that imaging techniques will remain the method of choice [53,54].

5.2.2.11 **Pancreatic Cancer**

Pancreatic cancer is less prevalent than prostate cancer, for instance, but nevertheless claims more life years owing to its poor prognosis. Tobacco abuse is held responsible for about 30% of pancreatic malignancies. Besides that, the western diet and occupational exposure to chemicals are considered to be risk factors for this disease. However, none of those factors is so closely related to pancreatic cancer that high-risk groups can be defined. Unfortunately, early diagnosis of pancreatic cancer is also hindered by the fact that the initial symptoms of the disease are fairly non-specific (e.g., digestive problems). In most cases, specific testing for the disease is initiated only when symptoms of the more progressed stage evolve (e.g., jaundice) [55]. Although some serological markers are being applied to the diagnosis of pancreatic malignancies, they are neither tumor-specific nor pancreatic cancer-specific. Diagnosis mainly relies on different imaging methods, namely CT, ultrasonography and MRI. Transabdominal ultrasonography is usually the initial screening investigation and can provide information about the size, site and characteristics of a potential tumor. Its diagnostic power can be enhanced by use of a sonographic contrast agent and Doppler sonography. Abdominal CT scans are usually also contrast enhanced, and the images of 3–5 mm thin cross-sections provide better tumor definition and assessment of resectability than ultrasonography [55]. All in all, however, even these staging techniques yield only mediocre sensitivity and specificity. Thus, there is an urgent need for reliable laboratory assays and/or sophisticated endoscopic tools with which at least the pancreatic duct can be inspected.

5.2.2.12 **Ovarian Cancer**

Ovarian cancer is the deadliest gynecological malignant neoplasia. A family history of ovarian malignancies is known to predispose for this disorder and the number of ovulations appears to play a role. Nulliparity, early menarche and late menopause increases whereas multiparity, late menarche and early menopause as well as the use of oral contraceptives lower the risk of an individual developing ovarian cancer. At the time of writing, no convincing screening assay for the early detection of epithelial ovarian cancer is available. Since the prevalence of the disease is low, the specificity of the currently available tests is not sufficient to justify general screening in asymptomatic women.

Consequently, examinations are usually confined to high-risk groups, for example women with a family history of ovarian cancer, and to women with abnormal results in a pelvic examination [56]. Several imaging methods are applied to detect so-called adnexal masses and to diagnose and stage ovarian cancer: ultrasonography, CT and MRI. Correct differential diagnosis is important, since the majority of adnexal masses are benign, and surgery performed for a malignant mass is radical. Transvaginally performed ultrasonography is usually the first diagnostic procedure. When suspicious masses are identified, confirmation of malignancy is either obtained histopathologically after exploratory laparatomy or by spiral CT or MRI of the pelvis [57, 58]. A relatively new laboratory assay, which displays an impressive 95% sensitivity and equal specificity, may provide the first screening tool for early diagnosis of ovarian cancer [59].

5.2.2.13 **Prostate Cancer**

Prostate cancer is a well known malignancy of men of advanced age. The higher life expectancy in affluent societies means that prostate cancer is more prevalent in those countries, but dietary factors may also play a role. Consequently, routine screening for prostate cancer is recommended for older males in these societies, although the guidelines show country-specific differences. There are two tests for prostate cancer that offer the option to screen a large population. The first is the digital rectal examination (DRE), where the prostate is palpated through the rectum by a physician. In the second, the concentration of prostate-specific antigen (PSA) in the blood is determined. If these tests reveal the possibility of cancer, transrectal ultrasonography is employed to further analyze the size, form and structure of the prostate [1]. The PSA test and the ultrasonography display the highest sensitivity and specificity, although considerable differences about the performance of these techniques have been reported in the literature.

5.2.2.14 **Bladder Cancer**

Cancer of the urinary bladder shows a bias towards the male sex, both in poor and in wealthier societies and it is more abundant in the latter. The reasons for this disparity are unclear. Tobacco consumption, certain drugs as well as environmental and occupational exposures to chemicals appear to pose a risk for bladder cancer in affluent societies whereas chronic infections with the waterborne parasite *Schistosoma heamatobium*, which is endemic to certain regions of Africa and the Middle East, are held responsible for bladder cancer in those areas. Because of this somewhat nebulous risk profile, no routine or mass screening has been recommended as yet. Tests for bladder cancer are only performed if specific symptoms, mainly hematuria, indicate its presence.

Then urine is analyzed by conventional cytology, in which sedimented cells are examined microscopically for the presence of any dysplasias and/or neoplasias. The sensitivity of this procedure can be increased by repeated testing of different samples [60]. The final diagnosis usually requires additional cystoscopy. This involves introduction of a fiberoptic tube, the cystoscope, into the bladder through the urethra and visual inspection of the bladder to detect the cancerous lesions. While the specificity of both procedures is high the sensitivity was disappointingly low in some studies. Consequently, more-sensitive assays are required. Laboratory and imaging tests will be suited equally well in this respect, as urine is easy to harvest and the bladder is readily accessible by endoscopic means.

5.2.2.15 **Skin Cancer**

The most serious variant of malignant skin neoplasia is the melanoma. The majority of melanomas are caused by exposure to UV radiation, i.e., by excessive tanning and repeated sunburns. The disease primarily affects white-skinned individuals who live in or frequently visit southern countries. There are no screening programs, only campaigns to increase awareness of this disorder. Early diagnosis of melanoma is in general initiated by a patient's self-examination of the skin. In a clinical examination, a dermatologist analyzes a cutaneous pigmented lesion with respect to asymmetries or changes in asymmetries, irregular borders, color and size ("asymmetry, border, color, diameter – ABCD-rule") [1]. A further analysis is provided by dermoscopy where the lesion is viewed with a monocular microscope and more detailed criteria, such as pigment distribution, can be judged. Upcoming, advanced optical techniques will surely add to the already high sensitivity and specificity of dermatoscopy and visual examination.

5.2.2.16 **Uterine Cancer**

Cancer of the uterine corpus or endometrium is another malignant disorder of the female reproductive tract whose etiology is poorly defined. Nulliparous women and those who undergo late menopause are at increased risk, as are breast cancer patients who are treated with the drug Tamoxifen. Except for the latter cohort, for which frequent testing is advised, routine or mass screening is not performed. Unfortunately, endometrical cancer is not detected during routine pelvic examination. Most often it is diagnosed only when women undergo testing because of specific symptoms such as postmenopausal bleeding [61]. In general, transvaginal ultrasonography is the first diagnostic means applied for detection of endometrial disorders. It is highly sensitive but only moderately specific for malignancies. Thus, if endometrial thickening is observed, cytological examination of endometrial cells must be performed either on biopsy material sampled with a flexible tube or on tissue obtained via a procedure called "dilation and curettage" in which the cervix is dilated and the endometrial tissue is scraped from inside the uterus. Here excellent specificities are achieved [61,62]. Visual inspection of the inner surface of the uterus is also possible using hysteroscopy. This technique is both highly sensitive and specific. As this anatomical site is readily accessible it is likely that imaging techniques will remain the method of choice for detecting this type of cancer.

When critically appraising all of the above techniques for cancer detection it becomes clear that in most cases our diagnostic equipment is still imperfect with regard to diagnosing cancer in its nascent stage. Many techniques suffer from a lack of diagnostic sensitivity and specificity, some are so unpleasant to the patient that they are frequently refused by asymptomatic individuals and others are so costly and complex that their application is justified only for people at risk or when neoplasia is already suspected. In addition to that, it is apparent that imaging techniques play a pivotal role in contemporary diagnosis of malignant neoplasia. Although sophisticated laboratory assays are likely to catch up to some extent, it is foreseeable that imaging techniques will remain the cornerstone of cancer diagnostics. Among those, optical techniques are particularly appealing since the low-energy radiation employed in these procedures is entirely harmless to the individual examined.

5.3 Emerging Optical Techniques in Cancer Diagnosis

Optical diagnosis of cancer is based on the variety of information carried by light after interaction with biological tissues. The properties of light that can be used for this purpose include intensity, polarization, wavelength, coherence and – using special experimental arrangement – temporal and spatial properties.

The optical methods are grouped in the following sections according to the different properties of light involved. The methods in Section 5.3.1 make diagnostic use of spatial information of light on tissue. Section 5.3.1.1 deals with the imaging of the surface of tissues with special reference to colposcopy and dermoscopy.

Comparable to normal histological findings in its scope, microscopy from inside the tissue is of particular interest. Therefore Section 5.3.1.2 discusses techniques providing optical sectioning, namely confocal laser-scanning microscopy, multiphoton microscopy and optical coherence tomography. For early diagnosis it is not feasible in most cases to take biopsies, so the alternative of using endoscopic microscopy will also be considered.

Larger tumors with altered optical parameters compared to their surroundings can also be detected with transmitted diffuse light, which is used to reconstruct a 3D distribution of the tissue's optical properties (Section 5.3.2).

The aspect of endoscopic utility is also prominent in the spectroscopic techniques following in Sections 5.3.3 (infrared spectroscopy) and 5.3.4 (fluorescence spectroscopy and bioluminescence). Targeted fluorescence, used to report the information on tumor markers, is presented separately in Section 5.4.

5.3.1 *In situ* Microscopy

5.3.1.1 Superficial Microscopy

Colposcopy with Acetic Acid and Iodine Test

Colposcopic evaluation is the first diagnostic step for women with abnormal cervical cytologic results [53], and/or a human papilloma virus infection [63].

Colposcopy, first introduced in 1925 by Hinselmann, is a simple diagnostic method, and can be defined as an examination of the uterine cervix and adjacent tissues by low-magnification microscopy [52, 64].

The colposcope itself is a stereoscopic binocular instrument, which provides magnified 3D images of the cervical portio and the distal endocervical canal. Magnification factors of commercially available colposcopes vary from 4 to 40. However, for a detailed inspection of the cervix, the magnification should be no less than 10× and preferably 16×. Similar to cervicoscopy, a comprehensive colposcopic examination involves three steps: inspection of the unprepared cervix, microscopic examination after application of dilute acetic acid and inspection after application of Lugol's iodine (see above). The clinical significance of the observed lesions can be judged by using the colposcopic classification and terminology system, which was revised recently and approved by the International Federation for Cervical Pathology and Colposcopy [65]. Typical criteria for interpretation of abnormal colposcopic findings are surface contour, color, degree of opacity, clarity of demarcation, and vascularization of the lesions. **Figures 5.6a–d** show a colposcopic finding with a low-grade cervical intraepithelial neoplasia lesion (CIN I) (see also **Figure 5.7**).

With regard to sensitivity and specificity, the figures range from 30% to 99% (sensitivity) and 26% to 97% (specificity). The large range of values is a result of different study designs. The lower values were obtained in studies where an additional differentiation between different grades was investigated. The higher values were obtained in cases of distinction between healthy and abnormal [52].

In summary, although data vary, most studies have shown that colposcopy is a reasonable evaluation tool. The method plays an important role in the diagnosis of women with abnormal Pap smears. The adequacy of the colpo-

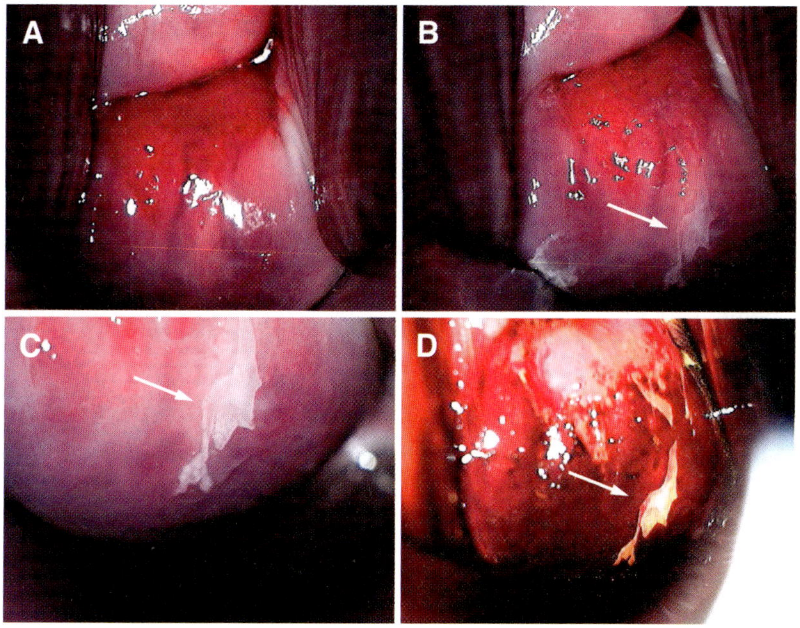

Figure 5.6 Colposcopic evaluation of a cervical intraepithelial neoplasia (CIN I). (a): Native inspection of the portio (mag. $5\times$); (b, c): Post-acetic images showing major acetowhite epithelium with major punctuation (mag. $5\times$ and $14\times$); (d): Iodine negative area after Schiller test (mag. $5\times$).

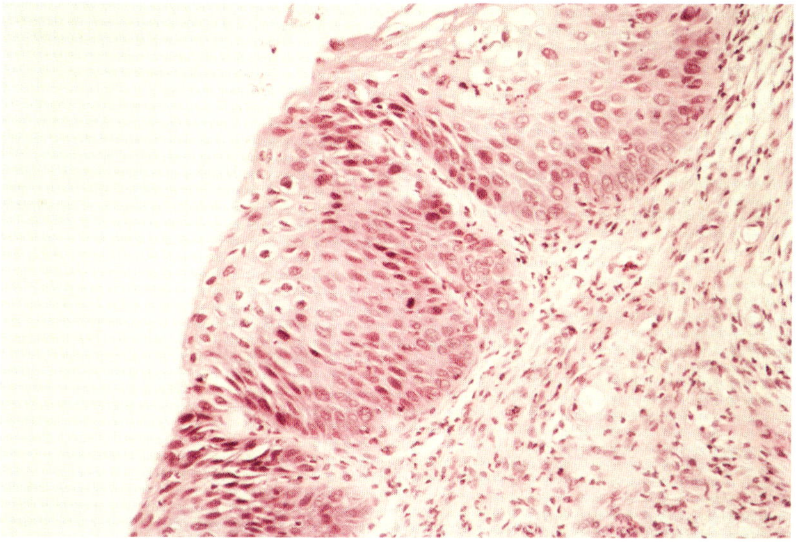

Figure 5.7 Histopathological finding of a low-grade dysplasia (cervical intraepithelial neoplasia CIN I).

scopic examination depends greatly on the ability to visualize the transformation zone (i.e., the region between the original squamous epithelium and columnar endocervical epithelium, in which the development of precancerous lesions or carcinomas usually occurs) and, if a lesion is present, the ability to visualize it entirely.

To improve the accuracy of the colposcopic diagnosis, and to decrease the inter- and intra-observer variability of the method, new technologies, e.g., digital image colposcopy and telecolposcopy, can be used.

The potential of digital colposcopy systems (a combination of a traditional binocular colposcope with digital image processing systems) is based primarily on improved image documentation, which is especially valuable for follow-ups [66, 67]. Linking a digital colposcopy system to medical data networks (telecolposcopy) will bring benefits in terms of medical access for women in rural areas, support inexperienced colposcopists by providing second opinions from specialists and increase quality control in clinical studies [68].

Skin Imaging: Dermoscopy

In order to improve the visual inspection of skin nevi and the staging (according to the ABCD rule or the 2002 AJCC staging [69]) a microscopic imaging technique, called dermoscopy is used. Based on the imaging, a grading can be done either by visual inspection of the now magnified image or supported by software algorithms.

For imaging, a homogeneous white-light illumination of the entire field of view is essential. This is achieved by an epiluminescence arrangement (epiluminescent microscopy).

The normal air–skin interface will lead to strong surface reflections of the illuminating light, which may hide the image features used for grading. For that reason two different approaches for reflection suppression are common. Most instruments use immersion with water or oil (immersion imaging) which nearly matches the refractive index of skin and thus reduces the reflection. The immersion fluid will also penetrate into the stratum corneum and fill the gaps between the corneocytes. This reduces the scattering and the imaging of deeper layers is also improved.

The second approach for the removal of the reflections takes advantage of the polarization of light which is mainly maintained in the reflections. The illumination light is linearly polarized and with a second polarizer in 90°-rotated orientation the surface reflected light in the imaging beam path, which is still linearly polarized, is removed. The advantage of the last method is the possibility of non-contact imaging without immersion. In contrast to the immersion technique not only is the reflection removed but also the light distribution from different depths is changed. As the light will lose its polarization only slowly with each scattering event, the light from superficial layers, which undergoes fewer scattering events, will be suppressed more strongly. Therefore a gain of information from deeper layers of the skin is obtained.

With the help of dermoscopy, the sensitivity and accuracy of diagnosis should be improved thus aiding in the decision whether or not to do a biopsy. Staging by a physician either on visual inspection of the skin or on the basis of the microscopic image has been compared in several studies [70–73]. Visual inspection led to sensitivities from 70% to 91% (see **Table 5.3**), whereas dermoscopy showed sensitivity values of 75% to 96%. In a meta study it was found that dermoscopy gave up to 25% higher sensitivities and specificities than visual inspection alone [74].

A further improvement of dermoscopy seems to be possible by automated diagnosis algorithms. As in application of the ABCD rule, the main features used for grading in automated diagnosis systems are geometry, color and texture. In this case sensitivities in the range from 80% to 99% and specificity values from 73% to 94% are possible [75,76].

The use of color imaging is already a spectral imaging technique with weighted integration over three overlapping wavelength ranges. The information is used for representation of the visual color impression and is strongly related to the concentration of hemoglobin, its oxygen saturation and the melanin concentration in a suspicious skin area. The analytical possibilities are further enhanced if more wavelength bands are used. This is done by the "SIAscope" which uses eight wavelength bands for spectrophotometric intracutaneous analysis [77]. Other instruments allow the collagen content to be obtained and used together with geometric features for an automated diagnosis which can yield a sensitivity of 83% and a specificity of 80%.

5.3.1.2 Microscopy of Tissue in Depth

The aim of performing microscopy deep in solid tissue is to detect pathological changes of the tissue *in vivo*. Suspicious tissue can be observed *in situ* and biopsies need only be taken in case of diagnostic uncertainty. Similar features to those used in pathology are used to obtain a diagnosis. Therefore, a spatial resolution showing the cellular structure is required. As a difficulty, cutting the tissue into thin slices, as for conventional microscopic imaging, is not possible. The optical imaging techniques used in solid tissue are often called optical sectioning.

In principle there are three possibilities for optical sectioning:

- confocal laser-scanning microscopy (CLSM)
- multiphoton microscopy (MPM)
- optical coherence tomography (OCT).

All these techniques allow a resolution in the micrometer range in all three dimensions, which is the prerequisite for obtaining useful microscopic images by optical sectioning.

For diagnostic use, good contrast in the imaging of relevant features is needed up to a certain depth. The deeper imaging is possible and the higher the relevant contrast, the better is the diagnostic value. The methods differ in what they show as the diagnostic signal. This will be discussed in detail in the next sections.

Confocal Laser-scanning Microscopy (CLSM)

The principle of CLSM is shown in **Figure 5.8**. The illuminating light from a laser is focused by a lens with high numerical aperture (NA) into the tissue. The light reflected from inhomogeneities of the tissue or the fluorescence light from endogenous or exogenous fluorophores is collected by the same lens. The detection beam path is separated from the illumination beam path by a beam splitter, and the light is focused onto the confocal pinhole. In fiber-coupled CLSM the confocal pinhole is a single mode fiber. The foci of the illumination and the detection pathways are conjugated. Therefore out-of-focus scattered light will be spread over a large spot in the plane of the detection pinhole and thus is strongly suppressed. In a table-top CLSM the 3D scanning is done by a 2D deflecting mirror arrangement and a movement of either the lens or the specimen for the third dimension. Appropriate emission blocking filters are used in the detection beam path in the case of fluorescence microscopy.

CLSM leads to some improvements in lateral resolution compared to standard wide-field microscopy and to a great improvement in axial resolution, which is needed for optical sectioning. Typical numbers for red light are a resolution of 0.4 μm in lateral and 3 μm in axial direction (NA = 0.9). Particularly for good axial resolution a high NA lens is required. For high-resolution imaging in tissue depth a lens has to be used that is corrected for the refractive index of mammalian tissue.

With this technique images can be obtained in tissue at a depth of ca. 100 μm. This is a remarkable improvement compared to conventional wide-field microscopy where only tissue slices of about 10 μm can be imaged. The limiting factors in the depth imaging capability of CLSM are the decrease of useful light and the increase of useless light with depth. Only light that has traveled to the focus without scattering, was scattered in the focal volume and travels back to the tissue surface before entering the lens aperture is useful for imaging. Most of the light will be diffused by scattering. Part of this diffuse light can enter trajectories that cross the focal volume. Although this light appears to be coming from the focal volume, it is useless for imaging. Furthermore, it will reduce the image contrast and render imaging in larger

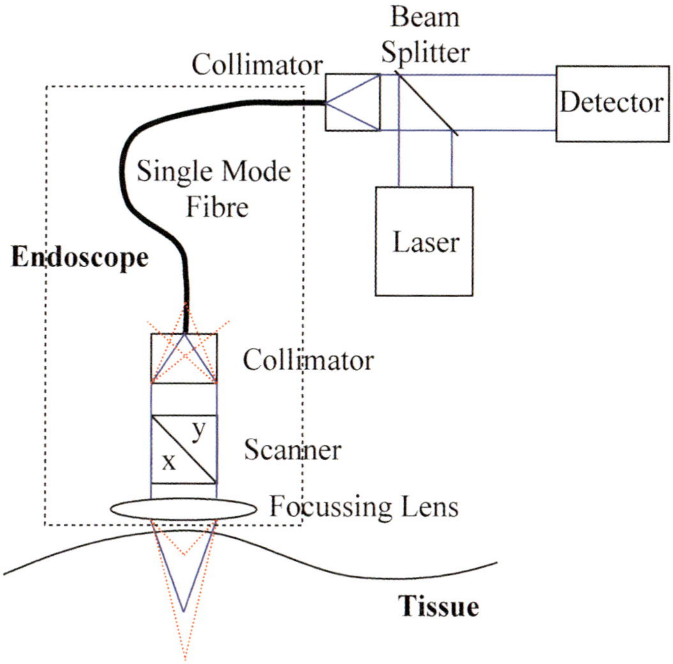

Figure 5.8 Scheme of Confocal Laser Scanning Microscopy.

depths impossible. The main aim of all techniques in microscopy in depth is the reduction of this unwanted light.

A quantification of the intensities can be done on the basis of the tissue's optical parameters, namely the absorption coefficient (μ_a) and the scattering coefficient (μ_s).

These parameters are the inverse of the mean free path for absorption and scattering, respectively. They follow Beer's Law for unscattered transmitted light. In **Figure 5.9** spectra of these coefficients are shown for a combined layer of epidermis and dermis. The penetration depth shown is the depth at which diffuse light is reduced in intensity by a factor of e (2.7).

The spectra show that apart from some prominent absorption bands of hemoglobin and water, scattering dominates over absorption up to the near-infrared (NIR) spectral range. Because scattering shows an overall decrease when using longer wavelengths, the depth that can be achieved in imaging increases with wavelength. Sensor reasons and increased absorption by water in the NIR region dictate that an optimum depth is reached between 800 and 1300 nm. Here absorption is negligible (for CLSM imaging) and the scattering coefficient reaches values of about 20 mm^{-1}. This results in an attenuation factor of the light intensity reaching the focus, being scattered and getting back to

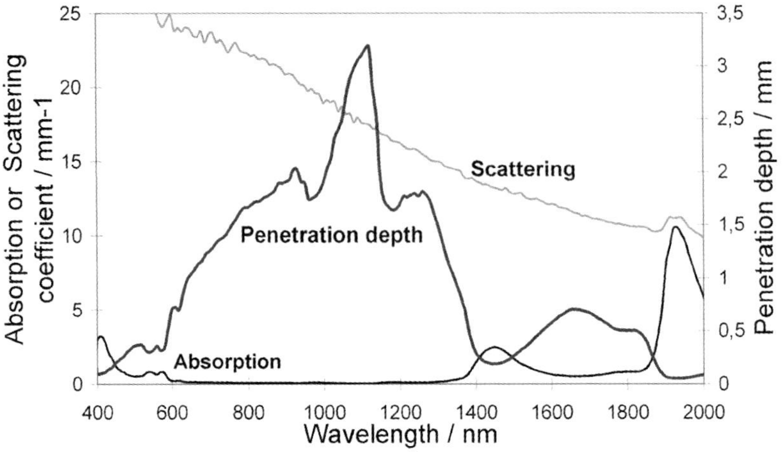

Figure 5.9 Spectra of the absorption and scattering coefficients of an epidermal and dermal layer. The penetration depth for diffuse light is calculated from these coefficients.

the tissue surface of about 1000 for an imaging depth of 100 μm. At the same time, unwanted light increasingly enters the detector, which is not shielded by the confocal pinhole. In the red spectral range, about 40% of the illuminating light is backscattered in total and a part of this reaches the detector. If the intensity of this light reaches the intensity of the light useful for imaging, no contrast will appear and imaging fails.

For the purpose of *in vivo* diagnosis, CLSM must be endoscopically available, the only exception being skin imaging where larger instruments can be used. For good imaging quality, a water immersion and high NA lens are required. Excellent images and the ability to diagnose basal cell carcinoma, squamous cell carcinoma, actinic keratosis and malignant melanoma using reflectance CLSM were shown by Gonzalez [78].

For endoscopic CLSM two different principles are used: distal scanning and proximal scanning. For proximal scanning, imaging fiber bundles with a lens, e.g., a gradient index lens, are used to transmit image information [79]. A standard CLSM scans across the proximal surface of the fiber bundle. In this case the fiber bundle limits the resolution, but the diameter of the instrument can be made quite small. A lateral resolution of 3 μm and an axial one of 17 μm was achieved with an instrument with a diameter <2 mm and fitting into the working channel of a conventional endoscope [79].

In distal scanning only a single mode fiber is needed to transmit both the illuminating light and the image signal light. The design of a scanning head is shown in **Figure 5.10** where two micro scanning mirrors do the 2D distal scanning [80]. A lateral resolution of 2 μm was achieved.

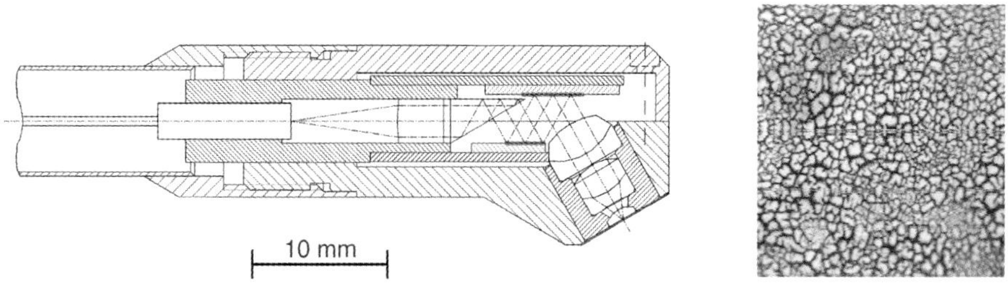

Figure 5.10 Design of a scanning head for CLSM (left). Image of a rabbit liver (right), field size 700×700 μm^2, acquisition time 2 s, wavelength 635 nm [80].

The aperture of the fiber acts both as emitting aperture, which provides diffraction limited focusing with the appropriate lens, and as receiving confocal pinhole, giving the possibility of optical sectioning. As the scanning takes place at the distal end the overall diameter of the endoscope will be larger than for proximal scanning. Scanning may be performed in different ways. In **Figure 5.10** an arrangement of two electrostatically driven mirrors is used. Other instruments use direct deflection of the single mode fiber. Commercialization has begun by the company OptiScan. Many images and references can be found at their website [81].

Only a few trials are related to the early diagnosis of tumors. Immunofluorescence staining of malignant melanoma was performed with FITC-labelled antibodies and imaged with a fiber-based CLSM up to a depth of approximately 100 μm in skin [82].

For colonoscopy a CLSM was implemented in a standard colonoscope. The microscope has a rigid part of 5 mm diameter and 43 mm length. In a prospective study healthy and inflamed sites, hyperplasia, neoplasia and cancer were imaged. The image contrast was enhanced by intravenously injected fluorescein sodium. Intraepithelial neoplasias and cancer could be predicted with a sensitivity of 97.4%, a specificity of 99.4% and an accuracy of 99.2% [83].

Multiphoton Microscopy (MPM)

The principle of multiphoton microscopy (MPM) is quite similar to that of CLSM. MPM will be used as a summarizing term for the use of processes like two-photon or three-photon absorption with subsequent fluorescence emission and processes like second and third harmonic generation. The occurrence of these processes has a probability with a quadratic or cubic dependence (according to their order) on the exciting laser power density. Therefore high NA focusing lenses and short laser pulses – typically in the femtosecond range – are required. Furthermore, this high power density is only achieved in the focal volume (**Figure 5.11**). The diffuse scattered light is distributed in time

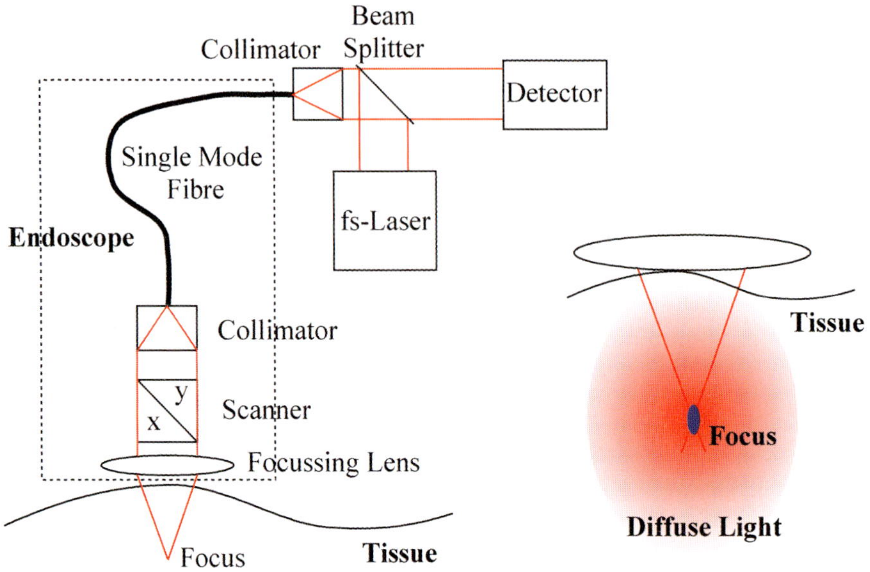

Figure 5.11 Principle of MPM (left). The intrinsic confocal effect of MPM is illustrated on the right.

and space, so that (nearly) no multiphoton process can occur in the volume surrounding the focus. This makes this kind of microscopy superior to CLSM in the suppression of scattered light and it is better suited to imaging deeper into the tissue [84]. This type of microscopy is often said to have an "intrinsic confocal effect".

In contrast to CLSM, reflective imaging is not possible with MPM. The image contrast results from fluorescence or from harmonic generation. Fluorescence lifetime imaging and time-resolved backscattering measurements are feasible with such systems, but will not be discussed here.

Although the imaging capability of MPM is better than that of CLSM as it is reaching 200 to 300 μm in tissue depth, the investigations related to tumor diagnosis are rare up to now. Indicators used for tumor diagnosis are fluorophores such as NADH [85], indicating increased metabolic activity in growing tumors or reduced activity in necrotic regions. Protoporphyrin was observed to show higher concentrations in tumors due to hampered decomposition of this compound in tumor cells. This effect can be amplified further by administration of 5-ALA (5-aminolevulinic acid) leading to a strong fluorescence from protoporphyrin [86]. The fluorescence of photosensitizers used for photodynamic therapy highlights neoplasia owing to their increased concentration in these tissues, which is caused by the strong vascularization of the neoplasm and exsudation of these drugs into the tumor. This phenomenon was microscopically observed *in vivo* for the first time by MPM [87]. With two-photon excited fluorescence and with second harmonic generation elastic fibers can be imaged [88].

Optical Coherence Tomography (OCT)

Differentiation between diffuse light and useful light for microscopic imaging is done in OCT by different path lengths. The diffuse light has a longer path length than the nonscattered light. The path lengths are discriminated by the light's ability for interference. Light from a source with short coherence length is only able to show interference if the difference in the path lengths is shorter than the coherence length.

In an interferometer (e.g., Michelson interferometer), the imaging depth is selected by the length of the reference arm (**Figure 5.12**). Only light from the signal arm that travelled the same distance as the reference arm length is able to give an interference pattern. Changing the length of the reference arm about the coherence length produces the interference pattern. A stable interferometer and endoscopically useful device can be built easily with a fiber-based beam splitter as shown in **Figure 5.12**.

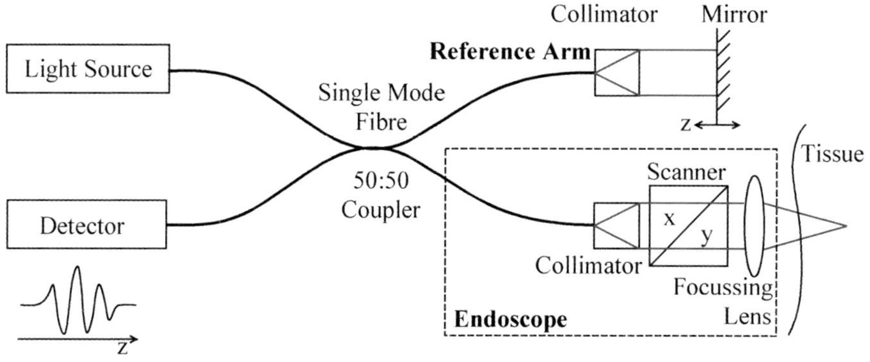

Figure 5.12 Scheme of an OCT imaging endoscope.

As tissue depth increases the useful light is dramatically reduced while the amount of light that traveled the same distance (but with scattering events on its way) increases. This restricts the depth to which imaging is possible to ca. 1–2 mm depending on the type of tissue.

Microscopic imaging with a lateral resolution given by the Rayleigh criterion should theoretically be achievable (depending on the NA). The observed lateral resolution is in the range 2–10 µm. The depth resolution is given by the coherence length and is in the range 1–10 µm [89]. Light sources for use in OCT are laser diodes with short coherence length or LEDs and fs-lasers. The high axial resolution of 1–2 μm was obtained with recently developed OCT devices with fs-lasers as light sources. With these systems built with Ti:sapphire lasers different wavelengths in the range from 400 to 1700 nm are feasible [89].

A disadvantage of OCT is the restriction to reflectance imaging. Contrast-enhancing mechanisms like fluorescence cannot be used because they destroy the coherence and make detection impossible. A way to overcome this restriction is the use of reflection instead of fluorescence labels. The use of gold nanoparticles, for example, greatly amplifies the reflection [90] and can therefore be used to improve the image contrast.

Endoscopic OCT probes are available in different designs and sizes. Typical diameters are in the range of 1–3 mm. Two-dimensional sectioning with one dimension being the depth is the main task in endoscopic applications. Three-dimensional imaging is possible but time-consuming. Often a rotating prism is used for lateral deflection because this is easily done and very suitable for certain medical tasks such as vessel inspection.

The determination of the exact size or depth of a lesion in relation to native tissue structures is an important feature of OCT to grade or stage cancer. Structured tissue gives a reflection contrast from the interfaces of different tissue structures, allowing orientation and diagnosis. OCT imaging was used early on for ophthalmic investigations. The timely detection of choroidal tumors or metastases is limited by the penetration depth to ca. 2 mm of this technique but it allows clear detection of the 3D extension of the abnormality [91]. In the upper gastrointestinal tract, submucosal invasion by cancer is an important diagnostic finding that is visualized better by OCT than by high-frequency ultrasound [92]. However, the limited penetration depth of ca. 1 mm is the main drawback of OCT. Imaging in the colon of 27 patients confirmed the findings of a penetration below 1 mm and found better resolution by OCT than from ultrasound [93]. In another 24 patients the microorganization of colonic tissue allowed the differentiation of adenomas, hyperplastic polyps and normal colon [94]. Slightly higher penetration depths of 1.5 mm were found in Barrett's adenocancer and early squamous carcinoma in a study with 40 patients [95]. For the diagnosis of cervical cancer the three-layer architecture – epithelium, stroma, basal membrane – is most important. In a pilot study with 50 woman (CIN I, II, and III, inflammation) a loss of organization in the three layers was observed in cases of invasive cervical carcinoma [96]. Early on in the disease the architecture of the tissue layers was disturbed already. Similar results regarding tissue architecture were found for invasive and early stages of cancer in the bladder wall [97]. In the main pancreatic duct, 249 OCT images were obtained and led to the conclusion that the neoplastic and non-neoplastic layer structure could readily be discriminated. Thus, OCT appeares to be a valuable, reproducible imaging technique in this respect [98]. In a rat tumor model, it was shown that an early stage induced mammary tumor could be recognized by increased backscattering and a higher nuclear-to-cytoplasmic ratio [99]. In tracheal and breast cancer in a rabbit model, healthy tissue could be distinguished from tumor tissue by the loss or change of architecture [100]. In basal cell carcinoma of the skin, the birefringence of collagen is utilized in polarization-sensitive OCT to show the disturbed skin structure [101].

5.3.2 Scattered Light Techniques – Diffuse Optical Tomography

If the region of interest lies deeper than 2 mm beneath the tissue's surface no microscopic imaging is possible so far. Nevertheless, information on the tissue's optical parameters can be gained from much deeper locations using transmitted or reflected light. If these parameters are altered in the course of neoplastic transformation, the measurement of absorption or scattering or both is of diagnostic value. Sometimes changes are secondary effects, such as vascularization or calcification in tumors. The resulting alterations of optical parameters are not very large and have to be compared to the normal spatial or inter-individual variation. Together these effects determine the possibility of detecting a tumor of given size at a certain depth. Early stages and small tumor sizes are therefore hard to detect; nevertheless it is feasible.

The main emphasis of current developments lies on the detection of mammary tumors. The female breast can be transilluminated with red and near-infrared light. The typical procedure is 2D scanning with opposite source/detector arrangements, or (for tomographic reasons) with different source/detector positions. Different wavelengths are used in order to detect blood concentration and oxygenation or to differentiate between tumors and other inhomogeneities.

For better tomographic reconstruction, additional information is needed and obtained from temporal effects. If scattered, the light will take a random walk through the tissue, which is always a longer path than the direct one. The path length's distribution of this random walk depends on the optical parameter's distribution in the tissue. As the tumor is only visible by its optical parameters, the path length distribution contains additional information about tumor size and location. A collection of optical data and basic methods is given in Ref. [102].

All of this along with 2D scanning is fed into reconstruction algorithms, which yield optical sections of the female breast. The temporal information is obtained in different ways. Either the time-dependent signal after transillumination with a pulsed laser is used or phase shifts and modulation amplitudes of intensity-modulated lasers are measured. These two strategies for image acquisition are called time domain or frequency domain techniques, respectively.

With a spatial arrangement of 32 excitation fibers and 32 detection fibers around the breast, time domain data of a breast phantom were obtained and optical parameters reconstructed. The 10-mm diameter tumor-like inserts in the phantom could clearly be reconstructed [103]. In a two-wavelength (670 nm, 785 nm) time domain transillumination set-up, 54 out of 65 tumors could be detected in a study with 90 patients [104]. As the main change of an optical parameter in tumors, an increase of absorption was measured. From the data of the two wavelengths the hemoglobin concentration in the breast could be estimated by calculation, predicting an increased hemoglobin concentration in the tumor. But the discrimination of normal breast inhomogeneities from a tumor is problematic. An attempt to overcome this is the use of additional *a priori* anatomical information. In a theoretical investigation of transillumination in the frequency domain, it was shown that spatial resolution and quantitative accuracy of the reconstruction can be significantly improved [105].

Nevertheless, the possibility of early detection of small changes seems hard to achieve with diffuse light optical tomography. Resolution is in the millimeter range and differences in optical parameters have to be high. An improvement can be expected from fluorescence markers, which give a higher contrast between tumor and normal surrounding tissue. The reconstruction of 3D data from 2D measurements is also easier in that case [106, 107]. With higher contrast, a better sensitivity even for small tumors should be possible [108]. This description here overlaps with the fluorescence techniques with markers mentioned later, but the origin of the reconstruction lies in the field of optical tomography. An extended description will be given in Section 5.4.

5.3.3 Spectroscopy

5.3.3.1 Infrared (IR) Spectroscopy

Infrared spectroscopy is a powerful noninvasive, objective optical technique, which is increasingly being employed in the study of biomedical conditions, where it has proved to be capable of detecting subtle biochemical changes within biomaterials. IR light from a broadband source is directly absorbed to excite the molecules to higher vibrational states, and spectra are recorded in the wavelength range 4000–800 cm^{-1}. The most significant characteristics of IR spectra are distinct absorption ranges that are due to specific functional groups of biomolecules (e.g., carbonyl stretching, amide-I-, amide-II-absorption, etc.) and the fingerprint region ($<1500\ cm^{-1}$) [109]. The addition of a microscope accessory to conventional Fourier-transform infrared (FTIR) spectrometers brings the potential to examine tissues at cellular resolution and to collect single-cell IR spectra with a high signal-to-noise ratio, as required for multivariate spectral analysis methods. Routine measurement modes are transmission, reflectance and attenuated total reflectance (ATR). IR transmission measurements are limited to very thin samples, IR spectra in remission (diffuse and specular reflection) mode depend on the condition of the sample surface, and ATR is typically used for nontransparent, strongly absorbing media.

IR microscopy has been used to obtain accurate, repeatable IR transmission spectra from air-dried thin unfixed tissue cryosections in short times (e.g., spot size 100 × 100 μm^2) (**Figure 5.13**) [110]. To meet that goal, IR images are recorded (mapping or imaging) and spectral information is obtained for each pixel. Standard parameters for the measurement and the procedure for tissue handling were optimized in order to evaluate distinct wavenumbers for tissue diagnostics. All spectroscopic results were compared to the pathological finding. Multivariate statistical analysis leads to classification such that diseased and healthy tissue can be distinguished (**Figure 5.14**) [111,112].

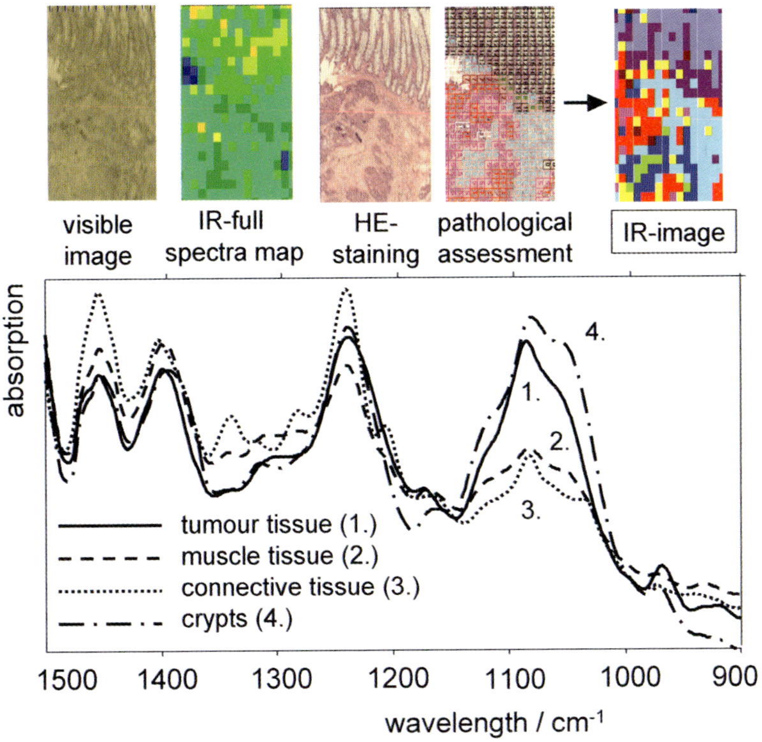

Figure 5.13 Above, sequence for IR-microscopic tissue analysis; below, superposition of mean spectra from different tissue components (IR microscope, transmission mode, 10 μm thin dried specimen of colon tissue) [110].

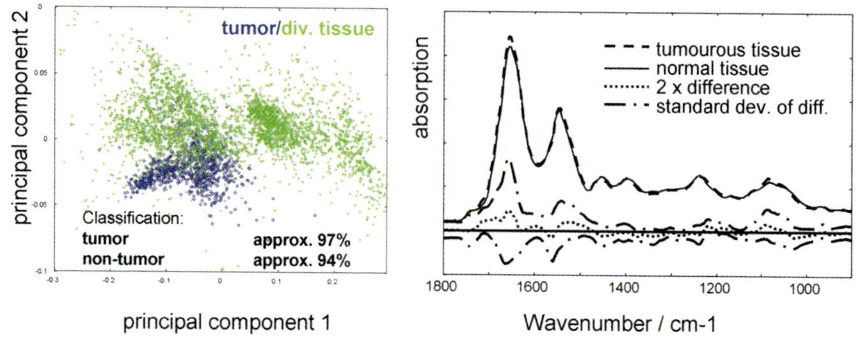

Figure 5.14 Calculated mean spectra for tumor and normal colon tissue, the difference spectrum (twice enlarged) and standard deviations for the difference, statistical distribution (right). 974 Tumor spectra and 5621 normal spectra were used for a classification leading to a sensitivity of 97% and a specificity of 94% (left) [111, 112].

IR radiation is strongly absorbed by water, so only the superficial layer of the tissue up to 10 μm in depth can be explored. In the search for a sensitive IR endoscopic application, successful IR fiber-optic investigations were carried out with silver halide waveguides in ATR mode [113, 114].

5.3.4
Fluorescence Spectroscopic Techniques – Endogenous Fluorescence

Fluorescence can occur when a molecule (fluorophore) resonantly absorbs light that promotes the molecule to an excited electronic state. Subsequent radiative relaxation of the excited states results in emission of light where some of the light is emitted at longer wavelengths than the wavelength that was absorbed (Stokes shift). Fluorescence intensity spectroscopy and fluorescence lifetime spectroscopy are well established analytical techniques. The fluorescence excitation and emission spectra can be used to identify the fluorophore and to determine its concentration, and may provide detailed information on its conformation, binding sites and interaction with its environment. In tissue a number of chromophores show fluorescence. Fluorescence can be endogenous (auto-fluorescence i.e., arising from naturally occurring materials in tissues) or exogenous (i.e., arising from a material added to allow tracking or labelling, see next section). Endogenous fluorophores may provide information on cellular energetics, as is the case for nicotinamide adenine dinucleotide (NADH), or on the amount and integrity of connective tissue, as does collagen. Low molecular weight molecules can be another important source of fluorescence. 5-Aminolevulinic acid (5-ALA), the precursor of heme, for example, is initially a nonfluorescent molecule, but it induces intracellular accumulation of protoporphyrin IX (PP IX), which is strongly fluorescent. The natural compound 5-ALA is widely used as a photosensitizer in photodynamic therapy (PDT) and diagnostics (PDD). It is better than organic fluorophores, which are usually not long-term stable and are subject to metabolic and physical effects (e.g., oxidation, bleaching) which in turn may influence their concentration *in vivo* and lead to alterations in the spectral outcome.

Fluorescence intensity measurements are affected by the attenuation of excitation and emitted fluorescent light during propagation in tissues. Richards-Kortum et al. [115] have developed a simple analytical model that relates the fluorescence of a homogeneous turbid tissue to the concentrations of chromophores that generate intrinsic fluorescence and attenuate excitation and emitted light. Pattern recognition can be used to develop and evaluate strategies to classify tissue types based on a measured spectrum [116]. The emphasis in the field has shifted towards model-based approaches. Monte Carlo techniques have been used to describe the fluorescence of inhomogeneous tissues and to examine the effects of the excitation beam profile and varying collection geometries [117]. Alternatively, analytical models have been developed wherein light propagation is modelled using the diffusion approximation to the radiative transfer equation [118–120].

Using different fluorescence spectroscopic methods, numerous investigations were done in the area of optical cancer diagnostics. Schomacker et al. [121] reported spectra obtained *in vivo* at 337-nm excitation from 170 sites in the colon. Multivariate linear regression was used to differentiate hyperplastic and adenomatous polyps. They attributed fluorescence peaks at 460 nm to NADH fluorescence and those at 390 nm to collagen fluorescence, with an intervening valley at 425 nm due to hemoglobin reabsorption [122]. Primary differences between the fluorescence of neoplastic and non-neoplastic tissues were due to collagen fluorescence and hemoglobin reabsorption. They noted dramatic differences between spectra obtained *in vitro* and *in vivo*; in particular, the contribution of NADH fluorescence was shown to decay exponentially with time following excision of the tissue, with a half-life of approximately two hours [123]. These changes call into question the relevance of *in vitro* results to clinical application [121].

In a subsequent prospective *in vivo* study by Cothren [123] at 370-nm excitation, a similar algorithm could differentiate 172 sites as adenomatous or non-adenomatous with 90% sensitivity and 95% specificity. Cothren noted significant patient-to-patient variation in the fluorescence intensity of normal and adenomatous colon. In order to achieve an acceptable accuracy, he normalized all spectra to the peak intensity of each patient's normal mucosa. D'Hallewin et al. [124] have demonstrated that autofluorescence spectroscopy at 365-nm excitation can be used to discriminate normal bladder carcinoma *in situ* and transitional cell carcinoma *in vivo* with high accuracy [121]. The discrimination is based on the decrease in fluorescence intensity as tissue progresses from carcinoma *in situ* (average decrease of 2.6 times) to transitional cell carcinoma (average decrease of 3.2 times). Endoscopic optical diagnosis of esophageal cancer using 5-ALA as fluorescence enhancer was reported to improve the detection of high-grade dysplasia in 20 patients with 77% sensitivity and 71% specificity [125].

In Ref. [126] the autofluorescence of the gastrointestinal mucosa was used to detect dysplasia. A 21% sensitivity and 91% specificity was found for the detection of cancer or dysplasia versus nondysplasia.

Initial clinical trials compared the sensitivity and specificity of optical spectroscopy to the biopsy gold standard [127]. Fluorescence-based algorithms provide sensitivity comparable with that of normal endoscopic imaging, but significantly higher specificity. In a study of 75 patients suspected to have bladder cancer, a retrospective analysis of fluorescence spectra showed that use of fluorescence could have reduced the number of unnecessary biopsies by 75% while still correctly identifying 95% of all malignant lesions found by cystoscopy [128] (see also **Table 5.4**).

Table 5.4 Comparison of six clinical studies using 5-ALA for the detection of bladder carcinomas.

Patients	Biopsies	Enhanced tumor detection (%)	Sensitivity (%)	Specificity (%)	Authors
104	433	38	95.8	63.8	[186]
34	215	76	89	57	[187]
55	130	18	87	59	[188]
123	347	30.3	96	67	[189]
52	–	25	94.6	43	[190]
49	179	69	87	63	[191]

Fluorescence diagnostics can be used to investigate the metabolic behavior of tissue, which is of particular interest in the case of tumorous diseases [129]. It is quite clear that the metabolism is different in tumorous or non-tumorous cells leading to different concentrations of certain metabolic markers. **Figure 5.15** shows the 2D plot obtained by the use of an optical biopsy system based on a laser-excited endoscopic analysis system. The time-resolved (picosecond range) fluorescence of the co-enzyme NADH is plotted versus the relative fluorescence intensity of porphyrins (in the main protoporphyrin PP IX) [130, 131].

A small trial examining the combination of diffuse reflectance spectroscopy with fluorescence spectroscopy for detection of cervical pre-cancer suggested that the two techniques provide complementary diagnostic information [132]. Similarly, Georgakoudi [133] conducted a Phase I study of diagnosing cervical pre-cancers *in vivo* using a combination of fluorescence, diffuse reflectance and

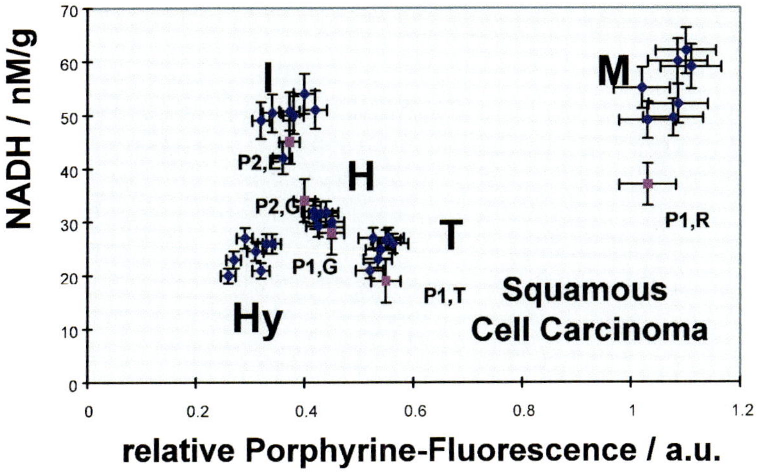

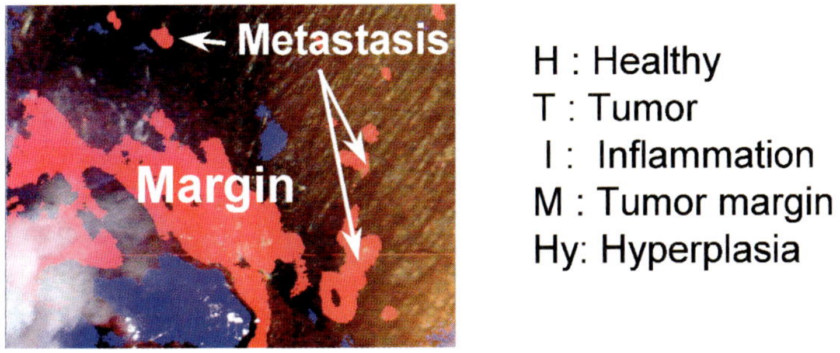

Figure 5.15 Multiple fluorescence is detected and assigned to NADH and protoporphyrin and compared with histological findings (top). An overlay of NADH fluorescence on a native image of a squamous cell carcinoma (bottom) [130, 131].

light scattering spectroscopy. The authors reported a significant improvement of sensitivity and specificity when a combination of the three techniques was used compared to any one of the single approaches. Larger trials of fluorescence spectroscopy indicate that in some organ sites (e.g., cervical tissue) fluorescence is influenced by biographic covariates, such as age and menopausal status [134]. In breast tissue, absorption is 2–3 times higher and scattering is 15–30% higher in pre-menopausal women than in post-menopausal women [135].

Multispectral fluorescence and reflectance imaging devices take advantage of the ability of optical systems to interrogate an entire organ site [136]. A device has been developed by SpectRx that incorporates the ability to measure both reflectance and fluorescence [137]. Encouraging sensitivities (97%) and specificities (70%) were reported for the diagnosis of cervical neoplasia with this device.

Bioluminescence

Mammalian tissues are opaque, so the optical properties of tissue govern the ability to detect light from the relatively weak biological light sources. Emission in the red spectral range is selectively transmitted through tissues, owing to the relatively low absorption at these wavelengths. Models of photon diffusion through tissue have indicated that as few as one hundred bioluminescent cells should be detectable at subcutaneous tissue sites, and approximately 100 cells would be required to generate signals that are detectable through two centimeters of tissue. These predictions by Rice et al. [138] were made with the assumption of a light output of 30 photons per cell per second at a wavelength of 650 nm, and have largely been supported empirically through a number of studies using different tumor cells at various tissue sites [139, 140].

In vivo bioluminescent imaging (BLI) is a versatile and sensitive tool based on the detection of light emission from cells or tissues. Optical imaging by bioluminescence allows a low-cost, noninvasive, and real time analysis of disease processes at the molecular level in living organisms. Bioluminescence has been used to track tumor cells [140], bacterial and viral infections, gene expression and treatment response [139].

Although an invasive technique, bioluminescence imaging has been used most intensively in clinical oncology, using tumor biopsies taken at the first diagnosis of the disease. It has been shown for squamous cell carcinomas of the head and neck and of the uterine cervix that accumulation of high levels of lactate in the primary lesions is associated with a high risk of metastasis formation and a reduced overall and disease-free patient survival [141]. Thus, metabolic imaging can provide additional information on the degree of malignancy and the prognosis of tumors, which may help the oncologist in adjusting specific treatment protocols to each individual malignant disorder [141].

5.4
The Future of Fluorescence Techniques in Cancer Diagnosis

The previous sections have demonstrated that the currently available methods lack either high penetration depth or high sensitivity. These drawbacks may be overcome by the use of specific optical markers. In order to fulfil their task, these markers ought to meet two criteria: They have to bind specifically to tumor cells (with a low non-specific binding to other cells) and they should carry a stable and sensitively measurable optical label.

The great advantage of optical labels resides in the fact that they do not cause any radiation damage and that the optical measurement is highly sensitive with an acceptable high penetration depth for diffuse light in the near-IR. These features are essential for methods of early cancer diagnosis that are suitable for screening campaigns.

Yet, there are further criteria for an optimal diagnosis. If specific binding to a tumor translates into binding to only one specific tumor, this in turn would imply the need for many specific tumor labels for the many different tumor types and markers. It is therefore of tremendous importance to identify labels with a "balanced" specificity, that is labels that recognize and bind to a variety of tumor cells but at the same time are capable of distinguishing tumor cells from adjacent nondiseased cells to which they should not bind. This leads directly to the problem of non-specific binding, which inevitably occurs to a certain degree. Therefore optimal conditions for the application of the label (e.g., dose, concentration, route of application, limited bioavailability due to controlled release) have to be determined. Only then is it possible to distinguish between labels that are trapped nonspecifically in healthy tissue from label that is bound specifically to early neoplasm hidden therein or underneath. This is a prerequisite for a 3D reconstruction of the fluorophore concentration in tissue.

In the next two sections tumor labels and methods for fluorescence imaging of them will be presented.

5.4.1 Markers and Labelling Strategies

A label for tumor diagnosis must implement two functions. First, it must bind specifically to a tumor, and second it must make this binding and its location visible to a diagnostic instrument or the physician's eye. These two functions may be provided by different molecular building blocks which, when combined, make up the final label. Tumor labels may be divided into different groups, according to the tumor marker which should be targeted. A tumor marker may be a molecule or any structural component that is synthesized either specifically or in altered form or concentration by the tumor cell and is recognized and reported by a label. In addition, bystander effects caused by the neoplasia may be detected and reported. The latter situation is often exploited by so-called non-specific labels.

5.4.1.1 Non-specific Labels

Non-specific labels are not divided into a labelling part and a reporting part. A good example is indocyanine green (ICG) which is used as a fluorescent label in blood. The enhanced neovascularization of many tumor types provides an improved contrast of the tumor compared to the surroundings. ICG was used, for instance, during open surgery of a glioma to improve the tumor's visibility [142]. In a canine model, Reynolds [143] demonstrated the preferential uptake of ICG in tumors. A modification of the cyanine dye (named SIDAG) with increased hydrophilicity further improved its penetration into the tumor [144]. Ebert et al. have shown that in a mammary tumor in the rat the dye accumulates up to six-fold compared to surrounding tissue [108]. All dyes used for photodynamic therapy (m-THPC, Hypericine, Photofrin, Photosan, HPD) display such tumor tropism and may serve as unspecific tumor labels [145]. However, because of their photosensitizing activity they cannot be used for screening.

5.4.1.2 **Specific Labels – Targeting of Receptors at the Cell Surface**

Specific targeting of tumors implies direct recognition of tumor markers. Receptor molecules located at the surface of a tumor cell are the preferred target for tumor labels. (see Section 5.1.1). The receptor-based targeting strategy relies on the strong binding of antibodies, peptides, hormones or other small molecules to the tumor markers. In addition to the binding specificity, pharmacokinetics, clearance time and degradation behavior of the label under *in vivo* conditions must be taken into consideration. The label has to be conjugated to a fluorophore in order to report its location via fluorescence, and this conjugation must not diminish the targeting capabilities. The fluorophore should be excited and should fluoresce in the near-IR (700–1300 nm), it should have a high fluorescence quantum yield and a large Stokes shift to allow diagnosis deep in the tissue. There are a variety of dyes available that are optimized for use in biochemical labels. In addition, fluorescent nano-particles and quantum dots are promising candidates for this purpose.

Successful site-specific targeting has already been demonstrated in animal models with different tumor markers. For example, Neri et al. have demonstrated the binding of a fluorophore-labelled antibody fragment to oncofetal fibronectin in a nude mouse [146]. Kelly et al. have shown in a murine model of a colon cancer (HT29) seven-fold higher accumulation of a fluorescence labelled cyclic peptide that was derived from a phage display library [147]. They further showed a pharmacokinetics that is suitable for endoscopic diagnosis and a preferred internalization of the label into the tumor cells. An ICG-labelled monoclonal antibody against tumor-associated antigens (anti-carcinoembryonic antigen MAb 35A7) was used by Gutowski to target peritoneal carcinomatosis (LS174T) in nude mice [148]. They could detect tumor nodules smaller than 1 mm in diameter and obtained a sensitivity of 87%–93% and specificity of 96%–99% depending on tumor size and label dosage.

In a study with 27 patients known to have colonic polypoid lesions, a fluorophore-labelled anticarcinoembryonic antigen antibody that was applied directly to the mucosal surface was detected [149]. In all cases without ulceration or bleeding, the specificity of fluorescence endoscopy was 100% and the sensitivity was 78.6%.

An essential drawback of the receptor-based tumor targeting strategy is the lack of marker molecules that are common to a variety of tumors. This requires that specific ligands for each tumor type are known and available – which is often not the case – and will impede the identification of potentially altered daughter cells and metastases. It is therefore pivotal for the success of a widely applicable targeting strategy to identify features that are characteristic for entire tumor classes, represent a cell surface property or molecule and can be exploited for this purpose. We are currently developing a possible approach to reaching this goal. In an ongoing project, we want to build on the observation that the dedifferentiated cells of intestinal tumors lack the typical microvillus structure as well as the carbohydrate coat, the so-called glycocalyx, that usually covers the surface of healthy intestinal epithelial cells (**Figure 5.16**). We have already demonstrated that cell surface molecules, which are usually shielded by the dense glycocalyx, become accessible for nano-sized particulate ligands if they are located on the surface of a very specialized, also glycocalyx-free, epithelial cell type, the so-called M cells [150]. We are now developing a targeting system that combines a ligand for a ubiquitously expressed epithelial cell surface receptor with a nanoparticle carrier and a fluorophore to specifically label intestinal tumor cells. After administration of such a fluorescent nanoparticle and washing of the intestine, only tumor cells should be stained. A subsequent endoscopic examination should allow the detection of even early tumors.

5.4.1.3 Specific Labels – Tumor Expression

Not only cell surface receptors but other molecules specifically expressed by tumor cells can be targeted by tumor labels. The metabolic changes in many tumor cells lead to increased endogenous production of the strongly fluorescent protoporphyrin IX (PP IX) after application of 5-ALA. 5-ALA is applied topically and penetration into tumors is high. But the tumor tropism is caused by the altered regulatory mechanisms in the heme pathway, which finally lead to an accumulation of PP IX in tumor cells [145]. Although it is a drug for photodynamic therapy, the use of 5-ALA in diagnosis is possible when the compound is locally administered.

Promising targets for tumor expression-specific labels are certain proteases usually involved in remodelling of the tissue structure, e.g., cathepsin B, which are overexpressed by many tumors. In a novel approach, this overexpression is detected using an activatable fluorophore [151, 152]. The label consists of an artificial peptidic substrate of cathepsin B, namely poly-L-lysine carrying a NIR fluorophore (Cy-5.5) on each lysine. If the molecule is intact, the fluorescence of the Cy-5.5 molecules is strongly quenched due to the close contact of the fluorophores. In tumor cells, the poly-L-lysine backbone is degraded by the overexpressed proteases, the fluorophores are separated from each other, and the quenching is abolished. The result is a strong increase of the fluorescence.

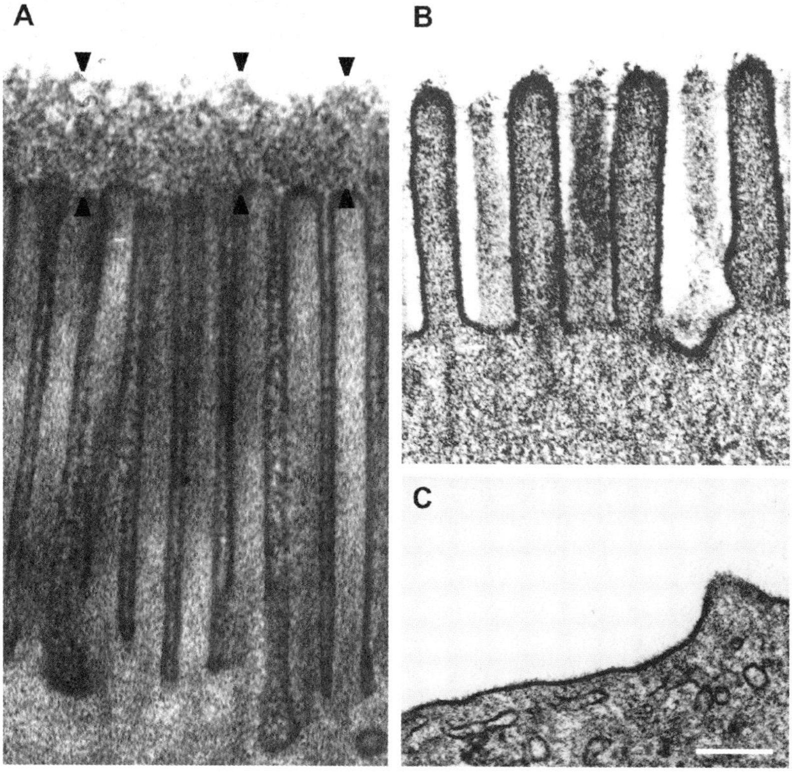

Figure 5.16 The cellular membrane of intestinal tumor cells is more readily accessible than that of healthy intestinal epithelial cells. Transmission electron microscopic views of healthy (A), slightly (B) and strongly dedifferentiated (C) human epithelial cells after staining with reduced osmium tetroxide. Healthy epithelial cells display a rich glycocalyx (arrowheads) which covers the cell membrane. With increasing dedifferentiation, glycocalyx and microvilli are reduced in size and eventually vanish completely (see also Ref. [150]). Bar, 500 nm.

5.4.2 Methods for Fluorescence Imaging – Fluorescence Tomography

Several instruments for fluorescence imaging are commercially available or at least exist as prototypes. For scientific investigations, small animal imagers can be used [153, 154]. For human patients hand-held or in-microscope [155] implemented detection devices as well as endoscopic instruments exist. However, all these devices show only the 2D raw fluorescence images. A future ideal fluorescence diagnostic technique will have to cope with two problems:

- Inhomogeneities in tissues lead to artifacts in fluorescence imaging.
- Instead of a 2D image, the 3D distribution of a tumor label must be measured.

Solving the first problem would improve the validity of 2D images and is a prerequisite for the second task. Existing instruments do not provide the necessary features for handling these problems, but there is experimental evidence that technical solutions will be available. As in the transition from simple X-ray fluoroscopy to CT, additional information is needed to do a 3D reconstruction. However, the situation is much more difficult when using diffuse light than in CT because in addition to absorption the scattering of light has to be taken into account. The necessary additional information may be obtained with spatially or temporally oriented modes of acquisition.

In a spatially oriented mode of acquisition different paths of light going through the same object will be used, as in CT. This means in the case of fluorescence imaging that several different excitation and detection sites on the surface of the tissue are used (either simultaneously with multiple detectors and excitation sources or successively by changing the relative positions of a source/detector pair). As a consequence the fluorophore distribution is acquired with different light distributions according to the different excitation positions, whereas the influence of the environment on the fluorescence light is detected by the different detection sites. Two-dimensional excitation $E(x_E, y_E)$ and detection $D(x_D, y_D)$ yields a 4D information $I(x_E, y_E, x_D, y_D)$ that allows the reconstruction of the 3D distribution of the fluorescence label's concentration.

In a temporally oriented mode of acquisition the additional information can be gained, as in ultrasound B-scan or OCT, from a knowledge of the time the light needs to travel from the excitation point on the tissue surface to the fluorophore and back. Although light travels with light velocity, the effective propagation velocity of diffuse light is slower by a factor of ten or more owing to scattering. This leads to times of flight in the picosecond range, which can be measured reliably. Acquiring the temporal distribution of the emitted fluorescence light provides the maximum information for reconstruction. Another possibility to get access to temporal information is excitation with modulated light and measurement of the phase of the emitted light together with the modulation amplitude. The phase shift between excitation and emission is a measure of the distance between tissue surface and fluorescence label.

Fluorescence emission is of course influenced by the exponential decay of fluorescence, but in the case of phase shift this leads only to a constant offset according to the fluorescence lifetime. As a result, the phase shift or the traveling time yields the additional information on the third dimension necessary for reconstruction.

Knowledge of the local optical parameters is a prerequisite for the reconstruction. These optical parameters are not measured separately but as part of the reconstruction algorithm. The result will be spatial distributions of absorption, scattering and fluorophore concentration.

Some trials have already been undertaken to use knowledge of the optical parameters to correct 2D images. With a knowledge of backscattering data, representing the optical parameters to some extent, it has been demonstrated that a fluorescence image of an F9 teratocarcinoma in nude mouse can be corrected [156, 157].

Another approach also makes use of backscattering data in a flying spot scanned imaging of a tissue phantom with fluorescence labelled vessels at different depths. The fluorescence is excited with modulated light and a backscatter signal is detected. From the modulation amplitude and the phase of the backscatter signal local optical parameters are determined, and with the phase shift of the fluorescence a reconstruction of the vessel depth is rendered possible [106, 107]. The vessel's depth could be determined down to 9 mm.

In connection with mammography, time-resolved measurements of the transmitted light of a pulsed laser were performed in an experimental arrangement with 2D scanning [108]. SIDAG [144] was used as fluorescence dye, which in a mouse tumor model showed a six-fold increase in concentration compared to surrounding healthy tissue. The same dye was located in different depths in a 50 mm thick phantom and was visible down to concentrations of 0.2 $\mu mol\,L^{-1}$. A 3D reconstruction was not performed. Normalization with the absorption improved the contrast of the images.

All experimental set-ups used for reconstruction employ multiple point sources for excitation and multiple detectors around a phantom. The detection is either achieved via pure spatial tomographic information [158] or with additional temporal information from frequency domain measurements of phase and modulation amplitude [159, 160]. These experiments demonstrated the possibility of reconstruction, even in large volumes.

In order to have a more universal arrangement for fluorescence tomography, a reflectance geometry is much better suited for medical tasks. The constraints for a well defined transillumination or arrangement of sources and detectors might be easily met for a phantom or a small animal but not for an adult patient. A set-up has been described [143] where fluorescence signals in the frequency domain could be measured on a large field simultaneously in reflection geometry. The detection of a volume of 0.1 cm^3 stained with Indocyanine green (1 $nmol\,L^{-1}$) was possible down to a depth of 7 cm in a tissue phantom [161]. However, only the modulation amplitude of the 2D fluorescence image was reported and not the phase shift, which is most important for reconstruction.

In a recent article by Ntziachristos et al. [162] the possibility of 3D reconstruction was demonstrated with resolution in the millimeter range at depths of 10 mm (tissue phantom). The resolution and sensitivity are quite impressive, with the only missing feature being a measurement in reflection geometry: The authors still use a tomographic geometry but mention the possibility of doing it by reflection as well.

5.5 Summary

Cancer imposes a continuously growing burden on humankind. Even now, the disease is responsible for twelve percent of deaths worldwide and the number is growing. The name "cancer" is a collective term for a variety of disorders whose hallmark is the uncontrolled growth of usually growth-controlled or growth-arrested cells, and their eventual dissemination to other sites of the body. Virtually all tissues and organs can be affected, symptoms are often not specific and, if untreated, cancer is lethal. As therapies for advanced cancers are elusive, it is of paramount importance to diagnose the disease in an early, asymptomatic stage when treatment is still possible. Consequently, safe, reliable, fast and cost-effective diagnostic tests that have a good patient compliance are crucial for combating this disorder. Although the death toll of all cancers combined is high, prevalence of individual cancer types can be low. Therefore, diagnostic tests must exhibit high sensitivity in order to detect every individual affected but also high specificity so as to avoid false positive diagnoses in unaffected people. None of the cancer tests available so far meets all these requirements, and screening is limited to certain cancers only. Since most cancers lead to the formation of distinct neoplastic lesions, imaging techniques represent the backbone of cancer diagnostics. Thus, progress in cancer diagnostics mainly depends on progress in imaging techniques. Among these, the optical ones are particularly appealing, since the low-energy radiation employed is harmless to the individual examined.

As light cannot penetrate the entire human body the use of optical techniques is limited to cancers that develop at or slightly underneath the body's inner and outer surfaces. However, even with this restriction about two-thirds of cancer victims could be diagnosed by optical means if proper methods were available. The inner surfaces of the body are for the most part accessible to optical examination by endoscopy, which may be substituted in the future to some extent by the use of autonomous cameras. The outer surface of the human body is readily accessible anyway, so that even larger equipment can be used. Depending on the depth down to which the tissue should be examinated, different optical techniques must be employed. The mucosal surfaces and the skin can be viewed directly by conventional optical techniques like microscopy, which reveals differences in surface texture, architecture and appearance. Infrared spectroscopy is able to gather data from a depth of several micrometers and provides a molecular fingerprint of the tissue investigated. Fluorescence spectroscopy allows an even deeper look. Depending on the fluorophore that is read out, emission sites located several millimeters underneath the surface are still detectable. The deepest tissue penetration and readout is achieved with diffuse optical procedures. An entire female breast can be scanned in this way, but the sensitivity for detection of early cancer is low. In order to improve it the cancerous lesion should be highlighted with an optical contrast agent before the patient is examined and a 3D reconstruction of the image should be the goal because it would further enhance the spatial resolution of the entire procedure. As such optical tomography would combine the best of both worlds, high penetration depth along with good resolution, it is one of the hottest subjects in the field of cancer diagnostics.

Acknowledgement

This work was supported by the German Ministry of Education and Research (BMBF), grants 13N8471, 13N8472, 13N8473.

Glossary

5′-end One end of a DNA- or RNA single strand. DNA synthesis or transcription always start at the 5′-end of a DNA segment.

5-ALA 5-Aminolevulinic acid, a precursor of heme biosynthesis

ABCD rule Examination of cutaneous pigmented lesions with respect to Asymmetries or changes in asymmetries, irregular Borders, Color and Diameter

Adenoma Benign growth of glandular origin

Apoptosis Programmed cell death, a process by which a cell commits suicide

ATR (Attenuated total internal reflection) A method for spectroscopy where the light is guided in an applicator and upon reflection probes the contacting medium of the applicator

B cells, T cells White blood cells, of specific importance for the immune defense of the body

Bioluminescence Luminescence in the course of chemical reactions in living cells

Birefringence Also named double refraction, the division of a ray of light into two rays when it passes through certain types of material, such as collagen, depending on the polarization of the light

Carcinogen Agent capable of inducing the malignant transformation of cells without the need of additional supporting factors

Cathepsin B A protease, an enzyme that catalyzes the hydrolysis of proteins

Cell cycle Cyclic process of cell division, organized in discrete phases of DNA-synthesis (S-phase), gap-phases (G1 and G2) and the actual cell separation process (mitosis, M-phase)

Cervicoscopy Microscopy of the cervix uteri

CIN (Cervical intraepithelial neoplasia) The uterine cervix is covered by a layer of epithelial cells which progress from large round cells at the basement membrane to flat (squamous) cells at the surface. A disorder of this progression of cell shape from the lower to the outermost layer is called dysplasia or cervical intraepithelial neoplasia (CIN). The closer to the surface the round cells reach the more severe the CIN is. This is reflected in the different CIN grades (I, II and III) whereby CIN III, a full thickness dysplasia, is also called carcinoma *in situ*.

CLSM (Confocal laser-scanning microscopy) Microscopy with a focused laser beam, that is sequentially scanned across the specimen by deflecting mirrors. A confocal pinhole in front of the detector is conjugated to the focus and blocks scattered light.

Coherence length The path length difference of two partial beams of a light source within which interference can be observed

Colon Large intestine

Colonoscopy Endoscopic examination of the large intestine

Colposcopy A diagnostic procedure in which a colposcope is used to examine an illuminated, magnified view of the cervix uteri, vagina and vulva

CT (Computed tomography) A special X-ray examination

Cy-5.5 (Cyanine 5.5) A fluorescence dye

Cytochemistry Examination of cells with respect to their biochemical equipment, e.g. by assaying for enzymatic activities

Cytomorphology Examination of cells with respect to their shape and architecture

Cystoscope Flexible or rigid endoscope for the investigation of the bladder (cystoscopy)

Dermoscopy Microscopic imaging of the skin with white light illumination and suppression of specular reflection

Diagnostic marker Genetic, metabolic or morphological feature of a medical condition which can be used in a screening procedure to detect the respective condition

Distal scanning Distal refers to structures further from the trunk while proximal refers to structures nearer to the trunk. In the case of using an instrument like an endoscopic scanner, the definition refers to the position in relation to the physician. Distal is further from the physician and proximal is closer.

DRE (Digital rectal exam) Manual examination of the rectum, performed by a physician

Dysphagia Difficulty or inability to swallow

Epithelium Cell layer that covers the inner or outer surface of many organs

Epstein–Barr virus A virus that integrates into the human genome and is associated with a variety of cancers, e.g. nasopharyngeal cancer, lymphomas

Fingerprint region The spectral range in the infrared (1000 to 1500 cm^{-1}) where molecules show very specific absorption

FISH (Fluorescent *in situ* hybridization) An analytical method of visualizing chromosomes or genes thereon

FITC (Fluorescein isothiocyanate) A fluorescence dye

Flow cytometry Analytical method for the characterization of different cell types. The cells are labelled with fluorescent antibodies specific for certain surface-exposed molecules and the differently fluorescing cells are detected afterwards when flowing through a special photometer.

Fluorescence lifetime spectroscopy Determination of the temporal characteristic of the fluorescence emission, which can be described by one or more time constants – the fluorescence lifetime

Fluorescence spectroscopy Determination of the wavelength-dependent intensity of the fluorescence emission after excitation

Fluorescence Emission of radiation after absorption of a photon. The emission follows the absorption with a time delay according to an exponential law with a characteristic time constant - the fluorescence lifetime. The emission is always shifted to a longer wavelength compared to the absorption wavelength.

FNR (False negative rate) The percentage of all assay-negative probands that are in truth positive for the condition analyzed

FOBT (Fecal occult blood test) Examination of stool for the presence of blood

FPR (False positive rate) The percentage of all assay-positive probands that are in truth negative for the condition analyzed

FTIR (Fourier-transform infrared)

Second (third) harmonic generation Doubling (tripling) of the light frequency by nonlinear processes in special media

Glycocalyx Carbohydrate coat covering the surface of the epithelial cell layer in the intestines

Helicobacter pylori Bacterium which is involved in the induction of stomach cancer

Hepatitis B and C virus Viruses causing chronic hepatitis and thereby often supporting the development of liver cancer

HPV (Human papilloma virus) A class of viruses which cause warts. Certain human papilloma viruses are known to cause cancer of the cervix uteri, others constitute risk factors for the development of, for example, cancer of the oropharynx.

HTLV-1 (Human T cell leukemia virus) A virus believed to be involved in the development of certain lymphomas

Hyperplasia Growth of new tissue due to erroneous cell proliferation. In contrast to neoplastic growth, a hyperplasia is in general more limited and dependent on the presence of certain triggering factors.

ICG (Indocyanine green) A fluorescence dye

Immunofluorescence Specific labelling of a tissue or cell with an antibody that is coupled to a fluorescence dye

Immunohistochemistry Examination of tissue by the use of antibodies

Incidence Number of new cases of a disease that occur in a certain time interval

Interference Superposition of waves, e.g. electromagnetic waves, which results in an intensity modulation

Interferometer An optical arrangement for the superposition of two partial beams – probe and reference beam – of a light source

IR microscopy Microscopic imaging with radiation in the infrared spectral range

LIFE (Laser-induced fluorescence endoscopy)

Monte Carlo technique Numerical method for the simulation of complex processes like the transport of light through turbid media (tissue)

MPM (Multiphoton microscopy) A method in laser-scanning microscopy, where the signal is generated by absorption of multiple photons with subsequent fluorescence photon emission from the excited electronic state. Mainly two-photon and three-photon absorptions are used.

MRI (Magnetic resonance imaging)

NA (Numerical aperture) A measure for the light cone angle, e.g. in an imaging lens. NA is the product of the refractive index and the sine of the light cone angle.

NADH (Nicotinamide adenine dinucleotide)

Neoplasia Growth of new tissue due to (uncontrolled) cell proliferation, usually describing a malignant process

Nuclear-to-cytoplasmic ratio The volume ratio of the cell nucleus to the cytoplasm helps in some cancer types to distinguish healthy from cancer cells

Nulliparous Having never given birth to a child

OCT (Optical coherence tomography) A 3D scanning microscopy, where the signal is generated from the interference signal of light backscattered from inside the specimen and reference light. Light is selected that is once-scattered at a certain depth in the specimen, and the depth is determined by the path length of the reference light.

Optical sectioning Two-dimensional imaging of a thin layer inside a 3D object

Pap smear Analytical method for the examination of cells collected from the surface of the cervix uteri. The cells are stained according to the method of Papanicolaou.

PCR (Polymerase chain reaction) An analytical method for the detection of small amounts of specific DNA (fragments)

PDD (Photo-dynamic diagnosis) This technique makes use of the fluorescence resulting from the excitation of a photosensitizer to locate a tumor

PDT (Photo-dynamic therapy) The destruction of tumors by use of a photosensitizer

Photosensitizer A molecule (dye) which absorbs light and transfers the energy to another molecule, in particular oxygen, which is in close proximity. The oxygen is transformed to its singlet state which is highly reactive.

Prevalence Number of persons in whom a disease has been diagnosed and who are alive at a certain time point

Proximal scanning See distal scanning

Rayleigh criterion An optical resolution criterion that defines spatial resolution in an imaging system. The image of a point source appears as a diffraction image with maxima and minima. According to the Rayleigh criterion the distance from the maximum in the center of the diffraction image to the first minimum is the spatial resolution.

Reflectance CLSM CLSM which uses the reflected light as signal for imaging

ROC curve (Receiver operating characteristic curve) A plot of true positive rate/false positive rate pairs which is used to determine the optimal sensitivity/specificity balance of a diagnostic assay

Sensitivity Measure of the diagnostic value of an assay, describing how well a proband positive for a certain medical condition is actually identified as such by the diagnostic assay. The sensitivity of a diagnostic assay is reflected in the true positive rate, giving the percentage of all truly positive probands that actually test positive in this assay.

Signal-to-noise ratio The ratio of the useful signal and the disturbing noise

Specificity Measure of the diagnostic value of an assay, describing how well a proband negative for a certain medical condition is actually identified as such by the diagnostic assay. The specificity of a diagnostic assay is reflected in the true negative rate, giving the percentage of all truly negative probands that actually test negative in this assay.

Spiral CT (Spiral computed tomography) A special X-ray imaging procedure

Squamous cell carcinoma A malignant tumor of squamous epithelial cells which occurs in different organs

Targeted fluorescence The combination of a specific ligand that binds to a target and a fluorescence dye. If the ligand is an antibody the description immunofluorescence is used.

Time-resolved backscattering measurement The light from a short pulsed laser which is scattered by tissue is detected in a time-resolved manner to obtain information about the travelling time of the light.

TNR (True negative rate) The percentage of all truly negative probands that actually test negative in an assay. The TNR is a measure of the specificity of an assay.

TPR (True positive rate) The percentage of all truly positive probands that actually test positive in an assay. The TPR is a measure of the sensitivity of an assay.

Tumor promoter An agent that stimulates cell proliferation without actually damaging the DNA

Vascularization Formation of blood vessels

Water immersion Filling of the gap between a microscope lens and the viewed object with water

X-ray mammography X-ray examination of the breast

Key References

B.W. Stewart, P. Kleihues (eds.), *World Cancer Report*, IARC Press, Lyon, France, (2003)

Luch, A., *Nature Rev. Cancer*, 5 (**2005**), pp. 113–125

Shapiro, D.E., *Stat. Methods Med. Res.*, 8 (**1999**), pp. 113–134

Mulshine, J.L., *Nature Rev. Cancer*, 3 (**2003**), pp. 65–73

J. Farley, J.W. McBroom, C.M. Zahn, *Clin. Obstet. Gynecol.*, 48 (**2005**), pp. 133–146

M-L. Bafounta, A. Beauchet, P. Aegerter, P. Saiag, *Arch. Dermatol*, 137 (**2001**), pp. 1343–1350

S. Gonzalez, K. Swindells, M. Rajadhyaksha, A. Torres, *Clin. Dermatol.*, 21 (**2003**), pp. 359–369

K. König, K. Schenke-Layland, I. Riemann, U.A. Stock, *Biomaterials*, 26 (**2005**), pp. 495–500

P.R. Pfau, M.V. Sivak Jr., A. Chak, *Gastrointest. Endosc.*, 58 (**2003**), pp. 196–202

D. Grosenick, H. Wabnitz, K.T. Moesta et al., *Phys. Med. Biol.*, 49 (**2004**), pp. 1165–1181

U. Bindig, H. Winter, W. Wäsche, K. Zelianeos, G. Müller, *J. Biomed. Opt.*, 7 (**2002**), pp. 100–108

M. Kriegmair, R. Baumgartner, R. Knuchel, H. Stepp, F. Hofstadter, A. Hofstetter, *J. Urol.*, 155 (**1996**), pp. 105–109

K. Licha, B. Riefke, V. Ntziachristos, A. Becker, B. Chance, W. Semmler, *Photochem. Photobiol.*, 72 (**2000**), pp. 392–398

R. Keller, G. Winde, H. J. Terpe, E.C. Foerster, W. Domschke, *Endoscopy*, 34 (**2002**), pp. 801–807

R. Weissleder, C.-H. Tung, U. Mahmood, A. Bogdanov Jr., *Nat. Biotechnol.*, 17 (**1999**), pp. 375–378

V. Ntziachristos, J. Ripoll, L.V. Wang, R. Weissleder, *Nat. Biotechnol.*, 23 (**2005**), pp. 313–320

References

1 B.W. Stewart, P. Kleihues (eds.), *World Cancer Report*, IARC Press, Lyon, France, (2003)

2 R.L. Gardner, *Development*, 128 (**2001**), pp. 839–847

3 J. Rossant, P.P.L. Tam, *Dev. Cell*, 7 (**2004**), pp. 155–164

4 D. Edgar, S. Kenny, S. Almond, P.

MURRAY, *Pediatr. Surg. Int.*, 20 (**2004**), pp. 737–740

5 E.V. ENTCHEV, M.A. GONZALEZ-GAITAN, *Traffic*, 3 (**2002**), pp. 98–109

6 A. KORNBERG, *J. Biol. Chem.*, 263 (**1988**), pp. 1–4

7 D.Y. TAKEDA, A. DUTTA, *Oncogene*, 24 (**2005**), pp. 2827–2843

8 S.M. IVANCHUK, J.T. RUTKA, *Neurosurgery*, 54 (**2004**), pp. 692–700

9 J.H.J. HOEIJMAKERS, *Nature*, 411 (**2001**), pp. 366–374

10 A. SANCAR, L.A. LINDSEY-BOLTZ, K. ÜNSAL-KACMAZ, S. LINN, *Annu. Rev. Biochem.*, 73 (**2004**), pp. 39–85

11 S. SENGUPTA, C.C. HARRIS, *Nat. Rev. Mol. Cell. Bio.*, 6 (**2005**), pp. 44–55

12 B. VOGELSTEIN, E.R. FEARON, S.R. HAMILTON, S.E. KERN, PREISINGER, A.C., M. LEPPERT, Y. NAKAMURA, R. WHITE, A.M.M. SMITS, BOS, J.L., *N. Engl. J. Med.*, 319 (**1988**), pp. 525–532

13 G.N. WOGAN, S.S. HECHT, J.S. FELTON, A.H. CONNEY, L.A. LOEB, *Semin. Cancer Biol.*, 14 (**2004**), pp. 473–486

14 A. LUCH, *Nat. Rev. Cancer*, 5 (**2005**), pp. 113–125

15 M.E. SMELA, M.L. HAMM, P.T. HENDERSON, C.M. HARRIS, T.M. HARRIS, J.M. ESSIGMANN, *Proc. Natl. Acad. Sci. USA*, 99 (**2002**), pp. 6655–6660

16 J.S. PAGANO, M. BLASER, BUENDIA, M.-A., DAMANIA, B., K. KHALILI, N. RAAB-TRAUB, B. ROIZMAN, *Semin. Cancer Biol.*, 14 (**2004**), pp. 453–471

17 L.A. HERRERA, L. BENITEZ-BRIBIESCA, A. MOHAR, P. OSTROSKY-WEGMAN, *Environ. Mol. Mutagen.*, 45 (**2005**), pp. 284–303

18 WORLD HEALTH ORG., *The World Health Report*, Geneva, (**2004**)

19 AMERICAN CANCER SOCIETY, *Cancer Facts and Figures 2005*, American Cancer Society, Atlanta, (**2005**)

20 AMERICAN CANCER SOCIETY, *Cancer Facts and Figures 2001*, American Cancer Society, Atlanta, (**2001**)

21 D.E. SHAPIRO, *Stat. Methods Med. Res.*, 8 (**1999**), pp. 113–134

22 A. LIU, E.F. SCHISTERMAN, Y. ZHU, *Statist. Med*, 24 (**2005**), pp. 37–47.

23 E.F. SCHISTERMAN, N.J. PERKINS, A. LIU, H. BONDELL, *Epidemiology*, 16 (**2005**), pp. 73–81

24 W.J. YOUDEN, *Cancer*, 3 (**1950**), pp. 32–35

25 M.H. ZWEIG, G. CAMPBELL, *Clin. Chem.*, 39 (**1993**), pp. 561–577

26 C.E. METZ, *Semin. Nucl. Med.*, 8 (**1978**), pp. 283–298

27 P.M. BOSSUYT, J.B. REITSMA, D.E. BRUNS, C.A. GATSONIS, P.P. GLASZIOU, L.M. IRWIG, D. MOHER, D. RENNIE, H.C.W. DE VET, J.G. LIJMER, *Clin. Chem.*, 49 (**2003**), pp. 7–18

28 J.L. MULSHINE, *Nat. Rev. Cancer*, 3 (**2003**), pp. 65–73

29 D. WORMANNS, K. LUDWIG, F. BEYER, W. HEINDEL, S. DIEDERICH, *Eur. Radiol.*, 15 (**2005**), pp. 14–22

30 DEPPERMANN, K.-M., *Lung Cancer*, 45(Suppl. 2)(**2004**), pp. S39–S42

31 M. SATO, A. SAKURADA, M. SAGAWA, M. MINOWA, H. TAKAHASHI, T. OYAIZU, Y. OKADA, Y. MATSUMURA, T. TANITA, T. KONDO, *Lung Cancer*, 32 (**2001**), pp. 247–253

32 J.C. LAYKE, P.P. LOPEZ, *Am. Fam. Physician*, 69 (**2004**), pp. 1133–1140

33 V.H.S. LOW, M.S. LEVINE, S.E. RUBESIN, I. LAUFER, H. HERLINGER, *Am. J. Roentgenol.*, 162 (**1994**), pp. 329–334

34 R. LENCIONI, D. CIONI, C.D. PINA, L. CROCETTI, C. BARTOLOZZI, *Semin. Liver Dis.*, 25 (**2005**), pp. 162–170

35 M. SHERMAN, *Semin. Liver Dis.*, 25 (**2005**), pp. 143–154

36 YUEN, M.-F., LAI, C.-L., *Best Pract. Res. Clin. Gastroenterol.*, 19 (**2005**), pp. 91–99

37 E. BANKS, G. REEVES, V. BERAL, D. BULL, B. CROSSLEY, M. SIMMONDS, E. HILTON, S. BAILEY, N. BARRETT, P. BRIERS, R. ENGLISH, A. JACKSON, E. KUTT, J. LAVELLE, L. ROCKALL, M.G. WALLIS, M. WILSON, J. PATNICK, *Br. Med. J.*, 329 (**2004**), pp. 477–482

38 C.S. HUANG, S.K. LAL, F.A. FARRAYE, *Cancer Causes Control*, 16 (**2005**), pp. 171–188

39 D.L. OUYANG, J.J. CHEN, R.H. GETZENBERG, R.E. SCHOEN, *Am. J. Gastroenterol.*, 100 (**2005**), pp. 1393–1403

40 B. GREENWALD, *Gastroenterol. Nurs.*, 28 (**2005**), pp. 90–96

41 J.M.E. WALSH, J.P. TERDIMAN, *J. Am. Med. Assoc.*, 289 (**2003**), pp. 1288–1296

42 M. FEURING-BUSKE, W. HIDDEMANN, C. BUSKE, *Internist*, 43 (**2002**), pp. 1179–1189

43 T. HAFERLACH, C. SCHOCH, *Internist*, 43 (**2002**), pp. 1190–1202

44 A. LOHRI, S. DELLAS, S. DIRNHOFER, R. HERRMANN, H. KNECHT, E. NITZSCHE, A. TICHELLI, *Schweiz. Med. Forum*, 34 (**2002**), pp. 773–780

45 A. LOHRI, S. DIRNHOFER, M. GREGOR, L. JOST, R. HERRMANN, M. VÖGELI, P. VON BURG, *Schweiz. Med. Forum*, 35 (**2002**), pp. 803–809

46 E. PODCZASKI, J. CAIN, *Clin. Obstet. Gynecol.*, 45 (**2002**), pp. 928–938

47 L. B. GERSON, G. TRIADAFILOPOULOS, *Am. J. Med.*, 113 (**2002**), pp. 499–505

48 L. Q. CHEN, C. Y. HU, P. GHADIRIAN, A. DURANCEAU, *Dis. Esophagus*, 12 (**1999**), pp. 161–167

49 T. W. REMMERBACH, H. WEIDENBACH, N. POMJANSKI, K. KNOPS, S. MATHES, A. HEMPRICH, A. BÖCKING, *Anal. Cell. Pathol.*, 22 (**2001**), pp. 211–221

50 M. KEBERLE, W. KENN, D. HAHN, *Eur. Radiol.*, 12 (**2002**), pp. 1672–1683

51 R. SANKARANARAYANAN, L. GAFFIKIN, M. JACOB, J. SELLORS, S. ROBLES, *Int. J. Gynecol. Obstet.*, 89 (**2005**), pp. S4–S12

52 J. FARLEY, J.W. MCBROOM, C.M. ZAHN, *Clin. Obstet. Gynecol.*, 48 (**2005**), pp. 133–146

53 T.C. WRIGHT JR., J.T. COX, L.S. MASSAD, L.B. TWIGGS, E.J. WILKINSON, *J. Am. Med. Assoc.*, 287 (**2002**), pp. 2120–2129

54 D.G. FERRIS, M.D. MILLER, *J. Fam. Pract.*, 36 (**1993**), pp. 515–520

55 A. S. TAKHAR, P. PALANIAPPAN, R. DHINGSA, D. N. LOBO, *Br. Med. J.*, 329 (**2004**), pp. 668–673

56 D. A. FISHMAN, L. S. COHEN, *Gynecol. Oncol.*, 77 (**2000**), pp. 347–349

57 A. B. KURTZ et al., *Radiology*, 212 (**1999**), pp. 19–27

58 S. A. SOHAIB, T. D. MILLS, A. SAHDEV, J. A. W. WEBB, P. O. VANTRAPPEN, I. J. JACOBS, R. H. REZNEK, *Clin. Radiol.*, 60 (**2005**), pp. 340–348

59 G. MOR, I. VISINTIN, Y. LAI, H. ZHAO, P. SCHWARTZ, T. RUTHERFORD, L. YUE, P. BRAY-WARD, D.C. WARD, *Proc. Natl. Acad. Sci. USA*, 102 (**2005**), pp. 7677–7682

60 B. PLANZ, E. JOCHIMS, T. DEIX, H. P. CASPERS, G. JAKSE, A. BOECKING, *Eur. J. Surg. Oncol.*, 31 (**2005**), pp. 304–308

61 F. AMANT, P. MOERMAN, P. NEVEN, D. TIMMERMAN, E. VAN LIMBERGEN, I. VERGOTE, *Lancet*, 366 (**2005**), pp. 491–505

62 Y. MINAGAWA, S. SATO, M. ITO, Y. ONOHARA, S. NAKAMOTO, J. KIGAWA, *Gynecol. Obstet. Invest.*, 59 (**2005**), pp. 149–154

63 W. PRENDIVILLE, J. RITTER, S. TATTI, L. TWIGGS, *Int. J. Gynecol. Cancer*, 15 (**2005**), p. 572

64 S. DEXEUS, M. CARARACH, D. DEXEUS, *Eur. J. Gynaecol. Oncol.*, 23 (**2002**), pp. 269–277

65 P. WALKER, S. DEXEUS, G. DE PALO et al., *Obstet. Gynecol.*, 101 (**2003**), pp. 175–177

66 I.J. ETHERINGTON, J. DUNN, M.I. SHAFI, T. SMITH, D.M. LUESLEY, *Br. J. Obstet. Gynaecol.*, 104 (**1997**), pp. 150–153

67 D. SCHÄDEL, A. COUMBOS, S. EY, R.-G. WILLRODT, H. ALBRECHT, W. KÜHN, *J. Telemed Telecare*, 11 (**2005**), pp. 103–107

68 D.G. FERRIS, M.S. MACFEE, J.A. MILLER, M.S. LITAKER, D. CRAWLEY, D. WATSON, *Obstet. Gynecol.*, 99 (**2002**), pp. 248–254

69 J.F. THOMPSON, R.A. SCOLYER, R. F. KEFFORD, *Lancet*, 365 (**2005**), pp. 687–701

70 A. STEINER, H. PEHAMBERGER, K. WOLFF, *J. Am. Acad. Dermatol.*, 17 (**1987**), pp. 584–591

71 F. NACHBAR, W. STOLZ, T. MERKLE et al., *J. Am. Acad. Dermatol.*, 30 (**1994**), pp. 551–559

72 M. CRISTOFOLINI, G. ZUMIANI, P. BAUER et al., *Melanoma Res.*, 4 (**1994**), pp. 391–394

73 P. CARLI, V. DE GIORGI, B. GIANNOTTI, *Arch. Dermatol.*, 137 (**2001**), pp. 1641-1644

74 M-L. Bafounta, A. Beauchet, P. Aegerter, P. Saiag, *Arch. Dermatol.*, 137 (**2001**), pp. 1343–1350

75 P. Rubegni, G. Cevenini, M. Burroni et al., *Int. J. Cancer*, 101 (**2002**), pp. 576–580

76 P. Schmid-Saugeon, J. Guillod, J-P Thiran, *Comput. Med. Imaging Graph.*, 27 (**2003**), pp. 65–78

77 M. Moncrieff, S. Cotton, E. Claridge, P. Hall, *Br. J. Dermatol.*, 146 (**2002**), pp. 448–457

78 S. Gonzalez, K. Swindells, M. Rajadhyaksha, A. Torres, *Clin. Dermatol.*, 21 (**2003**), pp. 359–369

79 J. Knittel, L. Schnieder, G. Buess, B. Messerschmidt, T. Possner, *Opt. Commun.*, 188 (**2001**), pp. 267–273

80 R. Schütz, K. Dörschel, G.J. Müller, *Proc. SPIE*, 3197 (**1997**), pp. 223–233

81 http://www.optiscan.com

82 P. Anikijenko, *J. Invest. Dermatol.*, 117 (**2001**), pp. 1442–1448

83 R. Kiesslich, J. Burg, M. Vieth et al., *Gastroenterology*, 127 (**2004**), pp. 706–713

84 W. Denk, J.H. Strickler, W.W. Webb, *Science*, 248 (**1990**), pp. 73–76

85 K. König, K. Schenke-Layland, I. Riemann, U.A. Stock, *Biomaterials*, 26 (**2005**) pp. 495–500

86 K. König, *J. Microsc.*, 200 (**2000**), pp. 83–104

87 G.M. Tozer, S.M. Ameer-Beg, J. Baker et al., *Adv. Drug. Deliv. Rev.*, 57 (**2005**), pp. 135–152

88 T. Theodossiou, G.S. Rapti, V. Hovhannisyan, E. Georgiou, K. Politopoulos, D. Yova, *Lasers. Med. Sci.*, 17 (**2002**), pp. 34–41

89 A. Unterhuber, B. Povazay, K. Bizheva et al., *Phys. Med. Biol.*, 49 (**2004**), pp. 1235–1246

90 K. Sokolov, M. Follen, J. Aaron, I. Pavlova, A. Malpica, R. Lotan, R. Richards-Kortum, *Cancer Res.*, 63 (**2003**), pp. 1999–2004

91 C.L. Shields, M.A. Materin, J.A. Shields, *Curr. Opin. Ophthalmol.*, 16 (**2005**), pp. 141–154

92 S. Gretschel, K.T. Moesta, M. Hünerbein, T. Lange, B. Gebauer, C. Stroszczinski, A. Bembenek, P. M. Schlag, *Onkologie*, 27 (**2004**), pp. 23–30

93 A. Das, M.V. Sivak Jr., A. Chak et al., *Gastrointest. Endosc.*, 54 (**2001**), pp. 219–224

94 P.R. Pfau, M.V. Sivak Jr, A. Chak, *Gastrointest. Endosc.*, 58 (**2003**), pp. 196–202

95 T. Rabenstein, O. Pech, L. Gossner, A. May, E. Guenter, M. Stolte, C. Ell, *Gastrointest. Endosc.*, 61 (**2005**), p. AB 236

96 P.F. Escobar, J.L. Belinson, A. White et al., *Int. J. Gynecol. Cancer*, 14 (**2004**), pp. 470–474

97 C.A. Jesser, S.A. Boppart, C. Pitris, D.L. Stamper, G.P. Nielsen, M.E. Brezinski, J.G. Fujimoto, *Br. J. Radiol.*, 72 (**1999**), pp. 1170–1176

98 P.A. Testoni, B. Mangiavillano, L. Albarello, P.G. Arcidiacono, E. Masci, A. Mariani, C. Dogliosi, *Gastrointest. Endosc.*, 61 (**2005**), p. AB99

99 S.A. Boppart, W. Luo, D.L. Marks, K.W. Singletary, *Breast Cancer Res Treat*, 84 (**2004**), pp. 85–97

100 N. Hanna, D. Saltzman, D. Mukai et al., *J. Thorac. Cardiovasc. Surg.*, 129 (**2005**), pp. 615–622

101 J. Strasswimmer, M.C. Pierce, B.H. Park, V. Neel, J.F. de Boer, *J. Biomed. Opt.*, 9 (**2004**), pp. 292–298

102 B.B. Das, F. Liu, R. R. Alfano, *Rep. Prog. Phys.*, 60 (**1997**), pp. 227–292

103 J.C. Hebden, H. Veenstra, H.Dehghani, E.M.C. Hillman, M. Schweiger, S.R. Arridge, D.T. Delpy, *Appl. Opt.*, 40 (**2001**), p. 3278-3287

104 D. Grosenick, H. Wabnitz, K.T. Moesta et al., *Phys. Med. Biol.*, 49 (**2004**), pp. 1165–1181

105 M. Guven, B. Yazici, X. Intes, B. Chance, *Phys. Med. Biol.*, 50 (**2005**), pp. 2837–2858

106 R. Schütz, N. Bodammer, J. Fikau, J. Helfmann, R.G. Senz, G.J. Müller, *Proc. SPIE*, 2626 (**1995**), pp. 249–257

107 J. Helfmann, R. Schütz, G.J. Müller, *Proc. SPIE*, 2927 (**1996**), pp. 450–454

108 B. Ebert, U. Sukowski, D. Grosenick et al., *J. Biomed. Opt.*, 6 (**2001**), pp. 134–140

109 U. Bindig, M. Meinke, I.H. Gersonde, O. Spector, A. Katzir, G.J. Müller, *Proc. SPIE*, 4158 (**2001**), pp. 40–48

110 U. Bindig, I. Gersonde, M. Meinke, Y. Becker, G. Müller, *Spectroscopy*, 17 (**2003**), pp. 323–344

111 U. Bindig, H. Winter, W. Wäsche, K. Zelianeos, G. Müller, *J. Biomed. Opt.*, 7 (**2002**), pp. 100–108

112 U. Bindig, F. Frank, I. Gersonde, M. Meinke, K. Zelianeos, A. Katzir, G. Müller, *Laser Phys*, 13 (**2003**), pp. 96–105

113 U. Bindig, G. Müller, *J. Phys. D: Appl. Phys.*, 38 (**2005**), pp. 2716–2731

114 U. Bindig, M. Meinke, I. Gersonde, O. Spector, I. Vasserman, A. Katzir, G. Müller, *Sens. Actuators B Chem.*, 74 (**2001**), pp. 37–46

115 R. Richards-Kortum, R.P. Rava, R. Cothren, A. Metha, M. Fitzmaurice et al., *Spectrochim. Acta*, 45 (**1989**), pp. 87–93

116 N. Ramanujam, M.F. Mitchell, A. Mahadevan-Jansen, S. L. Thomsen, G. Staerkel, A. Malpica, T. Wright, N. Atkinson, R. Richards-Kortum, *Photochem. Photobiol.*, 64 (**1996**), pp. 720–735

117 T.J. Pfefer, K.T. Schomacker, M.N. Ediger, N.S. Nishioka, *IEEE J. Sel. Top. Quant. Elect.*, 7 (**2001**), pp. 1004–1012

118 S. T. Flock, M.S. Patterson, B.C. Wilson, D.R. Wyman, *IEEE Trans. Biomed. Eng.*, 36 (**1989**), pp. 1162–1168

119 T.J. Farrell, M.S. Patterson, B. Wilson, *Med. Phys.*, 19 (**1992**), pp. 879–888

120 D.E. Hyde, T.J. Farrel, M.S. Patterson, B.C. Wilson, *Phys. Med. Biol.*, 46 (**2001**), pp. 369–383

121 K.T. Schomacker, J.K. Frisoli, C.C. Compton, T.J. Flotte, J.M. Richter et al., *Gastroenterology*, 102 (**1992**), pp. 1155–1160

122 K.T. Schomacker, J.K. Frisoli, C.C. Compton, T.J. Flotte, J.M. Richter et al., *Lasers Surg. Med.*, 12 (**1992**), pp. 63–78

123 R.M. Cothren, M.V. Sivac Jr., J. van Dam, R.E. Petras, M. Fitzmaurice, J.M. Crawford, G. Wu, F.F. Brennan, R. P. Rava, R. Manoharan, M.S. Feld, *Gastrointest. Endosc.*, 44 (**1996**), pp. 168–176

124 M.A. D'Hallewin, L. Baert, Vanherzeele, *J. Am. Paraplegia Soc.*, 17 (**1994**), pp. 161–164

125 S. Brand, T.D. Wang, K.T. Schomacker et al., *Gastrointest. Endosc.*, 56 (**2002**), pp. 479–487

126 K. Egger, M. Werner, A. Meining et al., *Gut*, 52 (**2003**), pp. 18–23

127 M. Fitzmaurice, *J. Biomed. Opt.*, 5 (**2000**), pp. 119–130

128 F. Koenig, F.J. McGovern, H. Enquist, R. Larne, T.F. Deutsch, K.T. Schomacker, *J. Urol.*, 159 (**1998**), pp. 1871–1875

129 F.F. Jobsis, J.H. Keizer, J.C. LaManna, M. Rosenthal, *J. Appl. Physiol.*, 43 (**1977**), pp. 858–72

130 J. Beuthan, O. Minet, G. Müller, *Ann. NY Acad. Sci.*, 838 (**1998**), pp. 150–170

131 J. Beuthan, T. Bocher, O. Minet, A. Roggan, I. Schmitt, A. Weber, G.J. Müller, *Proc. SPIE*, 2135 (**1994**), pp. 147–156

132 R.J. Nordstrom, L. Burke, J.M. Niloff, J.F. Myrtle, *Lasers Surg. Med.*, 29 (**2001**), pp. 118–127

133 I. Georgakoudi, E. Sheets, M.G. Muller, V. Backman, C.P. Crum, K. Badizadegan, R.R. Dasari, M.S. Feld, *Am. J. Obstet. Gynecol.*, 186 (**2002**), pp. 374–382

134 R. Drezek, K. Sokolov, U. Utzinger, I. Boiko, A. Malpica, M. Follen, R. Richards-Kortum, *J. Biomed. Opt.*, 6 (**2001**), pp. 385–396

135 N. Shah, A. Cerussi, C. Eker, J. Espinoza, J. Butler, J. Fishkin, R. Hornung, B. Tromberg, *Proc. Natl. Acad. Sci. USA*, 98 (**2001**), pp. 4420–4425

136 S. Andersson-Engels, C. Klinteberg, K. Svanberg, S. Svanberg, *Phys. Med. Biol.*, 42 (**1997**), pp. 815–824

137 D.G. Ferris, R.A. Lawhead, E.D. Dickman, N. Holtzapple, J.A. Miller, S. Grogan, S. Bambot, A. Agrawal, M.L. Faupel, *J. Low. Genit. Tract. Dis.*, 5 (**2001**), pp. 65–72

138 B.W. Rice, M.D. Cable, M.B. Nelson, *J. Biomed. Opt.*, 6 (**2001**), pp. 432–440

139 A. Rehemtulla, L.D. Stegman, S.J. Cardozo et al., *Neoplasia*, 2 (**2000**), pp. 491–495

140 M. Edinger, T.J. Sweeney, A.A. Tucker, A.B. Olomu, R.S. Negrin, C.H. Contag, *Neoplasia*, 1 (**1999**), pp. 303–310

141 S. Walenta, T. Schroeder, W. Mueller-Klieser, *Biomol. Eng.*, 18 (**2002**), pp. 249–262

142 M.M. Haglund, D.W. Hochman, A.M. Spence, M.S. Berger, *Neurosurgery*, 35 (**1994**), pp. 930–940

143 J.S. Reynolds, T. L. Troy, R. Mayer, A. B. Thompson, D. J. Waters, K.K. Cornell, P.W. Snyder, E.M. Sevick-Muraca, *Photochem. Photobiol.*, 70 (**1999**), pp. 87–94

144 K. Licha, B. Riefke, V. Ntziachristos, A. Becker, B. Chance, W. Semmler, *Photochem. Photobiol.*, 72 (**2000**), pp. 392–398.

145 K. Berg, P.K. Selbo, A. Weyergang et al., *J. Microsc.*, 218 (**2005**), pp. 133–147

146 D. Neri, B. Carnemolla, A. Nissim, A. Leprini, G. Querze, E. Balza, A. Pini, L. Tarli, C. Halin, P. Neri, L. Zardi, G. Winter, *Nat. Biotechnol.*, 15 (**1997**), pp. 1271–1275

147 K. Kelly, H. Alencar, M. Funovics, U. Mahmood, R. Weissleder, *Cancer Res.*, 64 (**2004**), pp. 6247–6251

148 M. Gutowski, M. Carcenac, D. Pourquier, C. Larroque, B. Saint-Aubert, P. Rouanet, A. Pèlegrin, *Clin Caner Res*, 7 (**2001**), pp. 1142–1148

149 R. Keller, G. Winde, H. J. Terpe, E.C. Foerster, W. Domschke, *Endoscopy*, 34 (**2002**), pp. 801–807

150 A. Frey, K.T. Giannasca, R. Weltzin, P.J. Giannasca, H. Reggio, W.I. Lencer, M.R. Neutra, *J. Exp. Med.*, 184 (**1996**), pp. 1045–1059

151 R. Weissleder, C.-H. Tung, U. Mahmood, A. Bogdanov Jr., *Nat. Biotechnol.*, 17 (**1999**), pp. 375–378

152 C. Bremer, V. Ntziachristos, R. Weissleder, *Eur. Radiol.*, 13 (**2003**), pp. 231–243

153 T. Jochum, J. Beuthan, A. Hengerer, K. Licha, T. Mertelmeier, *LaserOpto*, 33 (**2001**), p. 57

154 J.S. Lewis, S. Achilefu, J.R. Garbow, R. Laforest, M.J. Welch, *Eur. J. Cancer*, 38 (**2002**), pp. 2173–2188

155 W. Stummer, H. Stepp, G. Möller, A. Ehrhardt, M. Leonhard, H.J. Reulen, *Acta Neurochir.*, 140 (1998), pp. 995–1000

156 O. Minet, J. Beuthan, K. Licha, C. Mahnke, *J. Fluoresc.*, 12 (**2002**), pp. 201–204

157 J. Beuthan, C. Mahnke, U. Netz, O. Minet, G. Müller, *Med. Laser Appl.*, 17 (**2002**), pp. 25–30

158 V. Ntziachristos, R. Weissleder, *Opt. Lett.*, 26 (**2001**), pp. 893–895

159 D.J. Hawrysz, M.J. Eppstein, J. Lee, E.M. Sevick-Muraca, *Opt. Lett.*, 26 (**2001**), pp. 704–706

160 J. Lee, E.M. Sevick-Muraca, *J. Opt. Soc. Am.*, 19 (**2002**), pp. 759–771

161 E.M. Sevick-Muraca, J.P. Houston, M. Gurfinkel, *Curr. Opin. Chem. Biol.*, 6 (**2002**), pp. 642–650

162 V. Ntziachristos, J. Ripoll, L.V. Wang, R. Weissleder, *Nat. Biotechnol.*, 23 (**2005**), pp. 313–320

163 F. B. J. M. Thunnissen, *J. Clin. Pathol.*, 56 (**2003**), pp. 805–810

164 G. Gavelli, E. Giampalma, *Cancer*, 89 (**2000**), pp. 2453–2456

165 D. Wormanns, K. Ludwig, F. Beyer, W. Heindel, S. Diederich, *Eur. Radiol.*, 15 (**2005**), pp. 14–22

166 P. Ukrisana, M. Wangwinyuvirat, *J. Med. Assoc. Thai.*, 87 (**2004**), pp. 80–86

167 M. Bustamante, F. Devesa, A. Borghol, J. Ortuño, M. J. Ferrando, *J. Clin. Gastroenterol.*, 35 (**2002**), pp. 25–28

168 J. J. Y. Sung, F. K. L. Chan, W. K. Leung, J. C. Y. Wu, J. Y. W. Lau, J. Ching, K. F. To, Y. T. Lee, Y. W. Luk, N. N. S. Kung, S. P. Y. Kwok, M. K. W. Li, S. C. S. Chung, *Gastroenterology*, 124 (**2003**), pp. 608–614

169 I. M. De Zwart, G. Griffioen, M. P. C. Shaw, C. B. H. W. Lamers, A. De Roos, *Clin. Radiol.*, 56 (**2001**), pp. 401–409

170 B. A. Meda, D. H. Buss, R. D. Woodruff, J. O. Cappellari, R. O. Rainer, B. L. Powell, K. R. Geisinger, *Am. J. Clin. Pathol.*, 113 (**2000**), pp. 688–699

171 P. Zeppa, G. Marino, G. Troncone, F. Fulciniti, A. De Renzo, M. Picardi, G. Benincasa, B. Rotoli, A. Vetrani, L. Palombini, *Cancer Cytopathol.*, 102 (**2004**), pp. 55–65

172 B. Davidson, B. Risberg, A. Berner, E.B. Smeland, E. Torlakovic, *Diagn. Mol. Pathol.*, 8 (**1999**), p. 183–188

173 E. Bergon, E. Miravalles, E. Bergon, I. Miranda, M. Bergon, *Clin. Chem. Lab. Med.*, 43 (**2005**), pp. 32–37

174 S.-Y. Kang, J.-T. Suh, H.-J. Lee, H.-J. Yoon, W.-I. Lee, *Ann. Hematol.*, 84 (**2005**), pp. 588–593

175 E. Di Martino, B. Nowak, H. A. Hassan, R. Hausmann, G. Adam, U. Buell, M. Westhofen, *Arch. Otolaryngol. Head Neck Surg.*, 126 (**2000**), pp. 1457–1461

176 S. Bipat, S. S. K. S. Phoa, O. M. van Delden, P. M. M. Bossuyt, D. J. Gouma, J. S. Lameris, J. Stoker, *J. Comput. Assist. Tomogr.*, 29 (**2005**), pp. 438–445

177 J. R. van Nagell Jr., P. D. dePriest, M. B. Reedy, H. H. Gallion, F. R. Ueland, E. J. Pavlik, R. J. Kryscio, *Gynecol. Oncol.*, 77 (**2000**), pp. 350–356

178 J. M. Song, C.-B. Kim, H. C. Chung, R. L. Kane, *Yonsei Med. J.*, 46 (**2005**), pp. 414–424

179 K. Mistry, G. Cable, *J. Am. Board Fam. Pract.*, 16 (**2003**), pp. 95–101

180 S. Schneeweiss, M. Kriegmair, H. Stepp, *J. Urol.*, 161 (**1999**), pp. 1116–1119

181 H. G. Sim, W. K. O. Lau, M. Olivo, P. H. Tan, C. W. S. Cheng, *Br. J. Urol. Int.*, 95 (**2005**), pp. 1215–1218

182 A. Bono, C. Bartoli, N. Cascinelli, M. Lualdi, A. Maurichi, D. Moglia, G. Tragni, S. Tomatis, R. Marchesini, *Dermatology*, 205 (**2002**), pp. 362–366

183 F. P. H. L. J. Dijkhuizen, B. W. J. Mol, H. A. M. Brölmann, A. P. M. Heintz, *Cancer*, 89 (**2000**), pp. 1765–1772

184 M. Marchetti, P. Litta, P. Lanza, F. Lauri, C. Pozzan, *Eur. J. Gynaecol. Oncol.*, 23 (**2002**), pp. 151–153

185 World Health Org., *The Global Burden of Disease 2000*, 3rd ed., Geneva, (**2001**)

186 M. Kriegmair, R. Baumgartner, R. Knuchel, H. Stepp, F. Hofstadter, A. Hofstetter, *J. Urol.*, 155 (**1996**), pp. 105–109

187 P. Jichlinski, M. Forrer, J. Mizeret, T. Glanzmann, D. Braichotte, G. Wagnières, G. Zimmer, L. Guillou, F. Schmidlin, P. Graber, H. van den Bergh, H.J. Leisinger, *Lasers Surg. Med.*, 20 (**1997**), pp. 402–408

188 T. Filbeck, W. Roessler, R. Knuechel, M. Straub, H.J. Kiel, W. F. Wieland, *J. Endourol.*, 13 (**1999**), pp. 117–121

189 F. Koenig, F.J. McGovern, R. Larne, H. Enquist, K.T. Schomacker, T.F. Deutsch, *Br. J. Urol. Int.*, 83 (**1999**), pp. 129–135

190 C.-R. Riedl, E. Plas, H. Pflüger, *J. Endourol.*, 13 (**1999**), pp. 755–759

191 C. De Dominicis, M. Liberti, G. Perugia, C. De Nunzio, F. Sciobica, A. Zuccala, A. Sarkozy, F. Iori, *Urology*, 57 (**2001**), pp. 1059–1062

6
New Methods for Marker-free Live Cell and Tumor Analysis (MIKROSO)

Gert von Bally[1], *Björn Kemper, Daniel Carl, Sabine Knoche, Michael Kempe, Christian Dietrich, Michel Stutz, Ralf Wolleschensky, Karin Schütze, Monika Stich, Andrea Buchstaller, Klaus Irion, Jürgen Beuthan, Ingo Gersonde, Jürgen Schnekenburger*

6.1
Introduction

In serious diseases – especially malignant tumors – and also in life processes such as cellular growth on stem-cell matrices for tissue and organ substitution, molecular changes lead to developments of cellular features which become evident even at an early stage by micromovements and changes of elasticity at the cellular level. Therefore, these microchanges are important direct and objective parameters for modern therapy methods such as cell manipulation, cellular microsurgery, cellular and tissue engineering and for minimally invasive clinical diagnosis. Until now, tissue and cell differentiation as well as the analysis of cellular life processes have been mainly based on morphological and biochemical indicators. Features like micromovements and elasticity changes are rarely taken into account, although they are of direct and objective diagnostic and analytical value, respectively, and are used routinely on the macroscopic level, for instance in tumor diagnosis. Consequently, most of the optical methods for tissue and cell differentiation actually in use are related to morphological (e.g., microscopy) or spectroscopic (e.g., transmission, absorption, reflection, fluorescence) imaging techniques. In view of the still inadequate possibilities for diagnosing lethal tumors that need intensive and expensive treatment, there is an obvious urgent need of innovative methods for the extension of the diagnostic and therapeutic spectrum.

The importance of the microchanges described goes far beyond the medical domain and makes possible innovative techniques for the effective analysis of drugs and drug delivery in pharmacy and cosmetics. Environmental protection and agriculture are fields of application with important economic

1) Corresponding author, joint project coordinator;
project: MIKROSO – Micro interferometric optical probes for cell analysis and cell manipulation

Biophotonics: Visions for Better Health Care. Jürgen Popp and Marion Strehle (Eds.)

ISBN: 3-527-40622-0

potential, which require innovative online techniques for effective functional analysis of (living) bio and environmental indicators as well as for reducing and optimizing the use of insecticides and pesticides.

Thus, methodological and technological extensions from pure structural analysis to practical methods are required to provide functional images of spatial interactions as the well as results of micromanipulation on a cellular and sub-cellular level. Here, innovative interferometric techniques open a new field of optical probe methods for marker-free live-cell and tissue analysis.

6.2 Cellular Analysis by Interference-based Microscopy

6.2.1 Background and Motivation

By making visible things that are too small to be observable with the naked eye, light microscopy has been an indispensable tool within the life sciences for nearly four centuries. It gives detailed insight into structures and processes from the molecular to the macroscopic level and in consequence covers the spatial dimensions that are relevant for biology and medicine. Driven by advanced fabrication and refined design methods, there has been a steady improvement in the resolution and sensitivity of available optical systems.

Apart from pure magnification, different contrast methods have been introduced, which offer powerful tools for imaging cells and tissues. Many of them, e.g., fluorescence microscopy, work in combination with stains and molecular markers. These allow the specific labelling of cell organelles or molecular targets within cells. The introduction of fluorescent proteins, which are produced by cells after genetic manipulation, has made this a very powerful technique that gives insights into structure and also into the function of cellular processes [1].

Nevertheless, label- or marker-free imaging in biomedical applications is very desirable, since preparation protocols can interfere with the functionality of biological samples or can change structures within the observed objects, apart from being costly. For objects with poor natural contrast, i.e., there is only a very moderate interaction of the sample with light, contrast-enhancing microscopy techniques, such as darkfield, phase contrast or differential interference contrast (see **Figure 6.1**) are standard techniques to produce meaningful images. The advances achieved with the invention of phase contrast in biology and medicine were of such far-reaching significance that they earned Frits Zernike a Nobel Prize in physics in 1953. This technique allows the visualization of structures that are essentially translucent. By introducing a phase shift of the nondiffracted light, minor changes in the phase of the light pass-

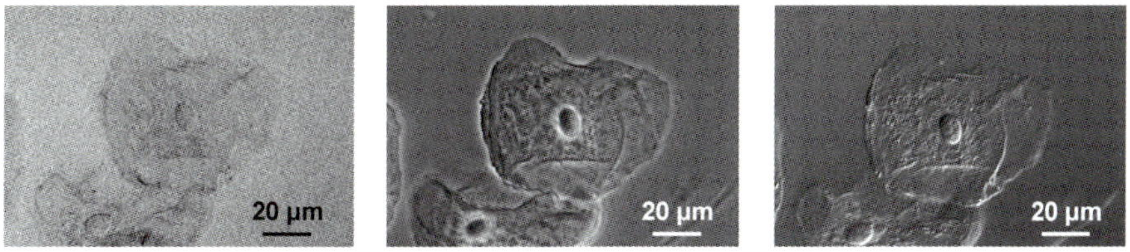

Figure 6.1 Images show cheek cells, recorded with different transmission contrasts (from left to right): brightfield, phase, differential interference contrast (DIC).

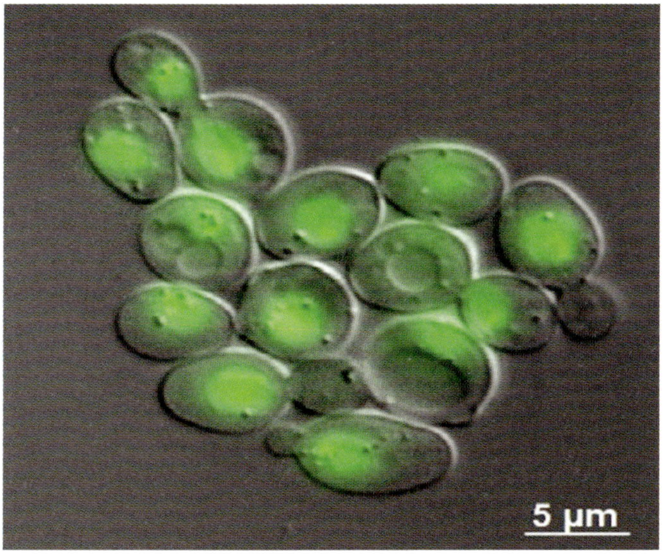

Figure 6.2 Live yeast cells imaged in DIC and fluorescence contrast (kinetochores are fluorescent by transfection with fluorescent protein). Image was acquired during the EMBO course (2003 Singapore), "Investigation of live specimens by modern optical methods", Yeast strain: Dr. Jason Swedlow.

ing through the sample are translated into intensity changes that can be seen by the eye. This contrast-enhancing method allows thin cell layers to be observed with high resolution without the need of staining. Differential interference contrast (DIC), a method introduced by Normarski in the late 1960s, complements phase contrast, especially with specimens of greater thickness.

Contrast-enhancement and marker-based techniques are often combined by subsequent or simultaneous image acquisition in two (or several) channels. As shown in **Figure 6.2**, the single channels are then overlayed to form a single image that indicates the location or distribution of specific molecules or organelles in the context of the outline of cells or tissues.

In addition the steady demand for the development of more sensitive, faster and more highly resolving imaging solutions, further conceptual tasks have to be addressed, particularly:

- quantitative imaging to allow robust and reproducible analysis of images for diagnosis and screening applications
- optimized 3D imaging, since the samples represent 3D structures
- improved functional imaging, which provides spatial mapping of biological, chemical or physical parameters in addition to simple morphological information.

Driven by the steady improvement in computer power and the progress of fast electronics and sensors, the combination of digital signal processing and adapted optics offers powerful tools for achieving significant progress. The following sections deal with efforts to develop digital holographic microscopy (DHM) and optical coherence microscopy (OCM) as two label-free "digital imaging" techniques to provide robust solutions for cellular imaging. These techniques have the potential to improve the current state of label-free microscopic imaging by promising quantitative imaging of structure- and function-related parameters as micromotions (DHM) and improved 3D imaging by increasing penetration depth and by enhancing contrast (OCM).

6.2.2 Digital Holographic Microscopy (DHM): A new Approach for Label-free Quantitative Imaging of Living Cells

Holographic interferometric metrology is a well established tool for industrial nondestructive testing and quality control [2–4]. In the fields of life sciences and biophotonics, digital holographic microscopy opens up new prospects for high-resolution analysis in both time and space, measurement and documentation in the supra-cellular, cellular and sub-cellular range by contactless and marker-free quantitative multifocus phase-contrast imaging [5–7].

6.2.2.1 Principle of DHM

Figure 6.3 depicts the principles of a digital holographic microscopy system. The transmission mode microscope arrangement in **Figure 6.3a** in combination with transmitted light illumination allows the investigation of transparent samples. **Figure 6.3b** shows an incident light set-up (reflection mode) for analysis of reflective objects or surfaces. For both arrangements the emitted light of a laser (e.g., frequency doubled Nd:YAG, $\lambda = 532$ nm) is divided into an object illumination wave and a reference wave. A condenser lens provides optimized illumination of the sample. The reference wave is guided directly by a beam splitter to a CCD sensor, which is applied for the digitization of the holograms in the hologram plane HP. Holographic off-axis geometry is

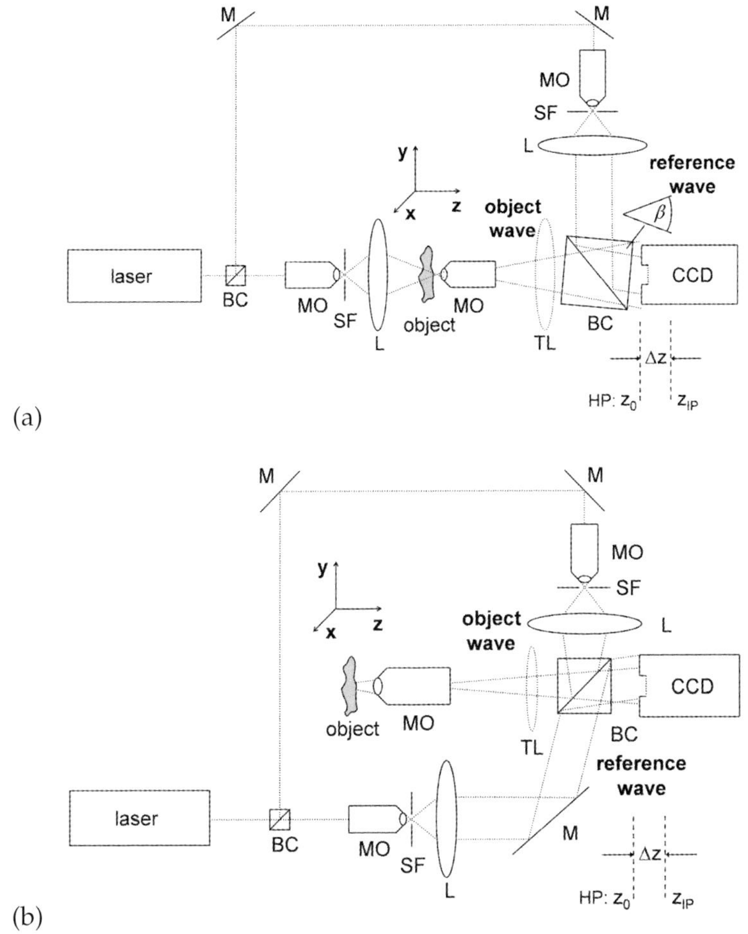

Figure 6.3 Principle of digital holographic microscopy: (a) transmission mode arrangement; (b) reflection mode arrangement. M: mirror, BC: beam splitter cube, MO: microscope lens, CCD: hologram recording device, e.g., Charged Coupled Device Sensor (CCD), SF: spatial filter, TL: tube lens, L: lens, HP: hologram plane, z_{IP}: image plane, Δz: distance between hologram HP and z_0, β: angle between object wave and reference wave at HP.

implemented by a slight tilt (angle β in **Figure 6.3a**) of the reference wave front with the beam splitter relative to the wave front of the object wave. To enhance the lateral resolution, which is restricted by the pixel pitch of the applied CCD sensor, the object wave is magnified by a microscope lens. The magnification is chosen in such a way that the recorded image of the specimen is over sampled by the image recording device. In this way, the maximum diffraction-limited resolution of the optical imaging system is not decreased by the numerical reconstruction algorithm described in Section 6.2.2.2.

6.2.2.2 Evaluation of Digital Holograms

Reconstruction of the digitally captured holograms is performed by the application of a nondiffractive reconstruction method (NDRM). In a first step the complex object wave $O(x, y, z = z_0)$ in the hologram plane HP, located at $z = z_0$, is determined by solving a set of equations that is derived from the interferogram equation [7, 8] and the intensity distribution in the hologram plane $I_{HP}(x, y, z_0)$ formed by the coherent superimposition of the object wave $O(x, y, z = z_0)$ and the reference wave $R(x, y, z = z_0)$:

$$\begin{aligned} I_{HP}(x,y,z_0) &= O(x,y,z_0)O(x,y,z_0)^* + R(x,y,z_0)R(x,y,z_0)^* \\ &\quad + O(x,y,z_0)R(x,y,z_0)^* + R(x,y,z_0)O(x,y,z_0)^* \\ &= I_O(x,y,z_0) + I_R(x,y,z_0) \\ &\quad + 2\sqrt{I_O(x,y,z_0)I_R(x,y,z_0)}\cos\Delta\phi_{HP}(x,y,z_0) \end{aligned} \tag{6.1}$$

with $I_O = OO^* = |O|^2$ and $I_R = RR^* = |R|^2$ (* denotes the conjugate complex term). The algorithm used is based on the assumption that only the phase difference $\Delta\phi_{HP}(x, y, z_0) = \phi_R(x, y, z_0) - \phi_O(x, y, z_0)$ between the phase distribution $\phi_O(x, y, z_0)$ of the object wave $O(x, y, z = z_0)$ and the phase distribution $\phi_R(x, y, z_0)$ of the reference wave $R(x, y, z_0)$ varies spatially rapidly in the hologram plane. In addition, the object wave's intensity is estimated to be constant within an area of about 5×5 pixels around a given point of interest in the hologram. With these assumptions and a mathematical model for the phase difference between object wave and reference wave at $z = z_0$ [7] the object wave amplitude is reconstructed for each pixel of a digitized hologram. Afterwards, in a second evaluation step, a further propagation of the object wave by numerical evaluation of the Fresnel–Kirchhoff diffraction integral [2] or by a convolution algorithm [9] is applied. In this way, $O(x, y, z = z_0)$ is propagated to the image plane z_{IP} that is located at $z_{IP} = z_0 + \Delta z$ at a distance Δz from HP. For a sharply focused image of the sample in the hologram plane with $\Delta z = 0$ (and thus $z_{IP} = z_0$) no further propagation is necessary. As a consequence of the applied algorithms and the parameter model for the phase difference $\Delta\phi_{HP}$ in Eq. (6.1), the resulting reconstructed holographic image $O(x, y, z_{IP})$ does not contain the disturbing terms "twin image" and "zero order". In addition to the absolute amplitude $|O(x, y, z_{IP})|$ that represents the image of the sample, the phase information $\phi_O(x, y, z_{IP})$ of the object wave is reconstructed simultaneously:

$$\phi_O(x,y,z) = \arctan\frac{\mathrm{Im}\{O(x,y,z_{IP})\}}{\mathrm{Re}\{O(x,y,z_{IP})\}} (\mathrm{mod}\, 2\pi). \tag{6.2}$$

After removal of the 2π ambiguity by a phase-unwrapping process [2], the data obtained from Eq. (6.2) can be utilized for marker-free quantitative phase contrast microscopy.

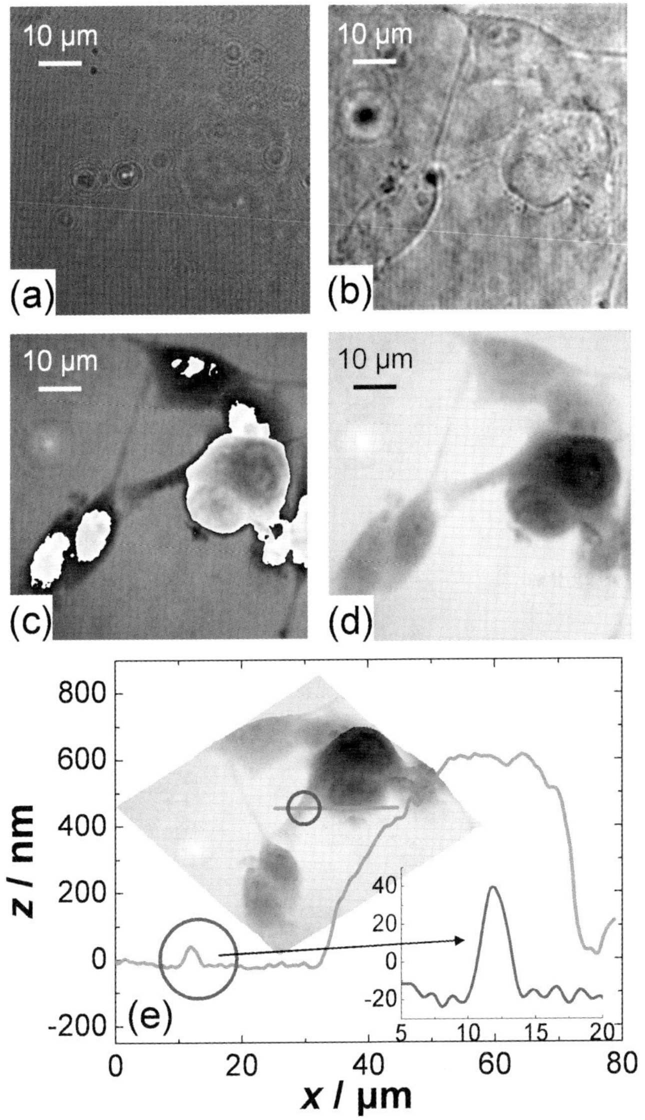

Figure 6.4 Example for evaluation of digital holograms: (a): digital hologram of human tumorous liver cells (HepG2); (b): reconstructed holographic amplitude image; (c): quantitative phase contrast image $(\mathrm{mod}\,2\pi)$, (d): unwrapped phase distribution; (e): pseudo 3D plot of the unwrapped phase image in gray-level representation including a cross-section through a single cell and one of its lamellipodia.

Figure 6.4 illustrates the evaluation process of digitally recorded holograms. **Figures 6.4a** and **6.4b** show a digital hologram obtained from living human tumorous liver cells (HepG2) with a microscope arrangement in transmission mode and the reconstructed holographic amplitude image that corresponds to a microscopic brightfield image with coherent laser light illumination. **Figure 6.4c** depicts the simultaneously reconstructed quantitative phase contrast image mod 2π obtained from Eq. (6.2). The unwrapped data without 2π ambiguity, representing the optical path length changes that are effected by the sample in comparison to the surrounding medium by the thickness and the integral refractive index, are shown in **Figure 6.4d**. **Figure 6.4e** depicts a pseudo 3D plot of the data shown in **Figure 6.4d** and its further evaluation. The cross-section through the reconstructed spatial phase distribution shows that even tiny cellular structures, e.g., lamellipodia are resolved. With information about the integral refractive index of the specimen, the reconstructed object wave phase information obtained from Eq. (6.2) can be applied to thickness measurements of semi-transparent microscopic samples, such as living single cells [10]:

$$d_{\text{cell}} = \frac{\lambda \Delta \varphi_{\text{cell}}}{2\pi} \cdot \frac{1}{n_{\text{cell}} - n_{\text{medium}}}. \tag{6.3}$$

In Eq. (6.3) d_{cell} represents the thickness of the investigated cells, $\Delta\varphi_{\text{cell}}$ is the optical path length change of the cells with integral refractive index n_{cell} to the surrounding medium with refractive index n_{medium}. For fully adherently grown cells the parameter d_{cell} is estimated as the cell shape.

6.2.3 Performance of DHM

Digital holography makes full-field, contact-free, nondestructive and marker-free multifocus microscopy possible. These features are illustrated below by results obtained at the Laboratory of Biophysics in cooperation with Carl Zeiss Jena GmbH and P.A.L.M. Microlaser Technologies, Bernried.

6.2.3.1 Lateral and Axial Resolution

Figure 6.5 illustrates the lateral and axial resolution obtained by DHM. **Figure 6.5a** depicts the holographic amplitude image of a USAF 1951 resolution chart. The image has been numerically reconstructed from a digital captured hologram (1024 × 1024 pixels, transmission mode, 40× microscope lens: NA = 0.65). The elements of group 9.6 correspond to a line width of 550 nm and demonstrate that the lateral resolution is only diffraction limited. **Figure 6.5b** shows the reconstructed quantitative phase contrast image of a reflection type phase object (chrome on chrome test chart with axial 30 nm steps, incident light arrangement, 5× microscope lens: NA = 0.1) and the corresponding pseudo 3D representation of the topography obtained from the phase data of a single 30 nm step. Due to phase noise, the axial resolution amounts to <5 nm.

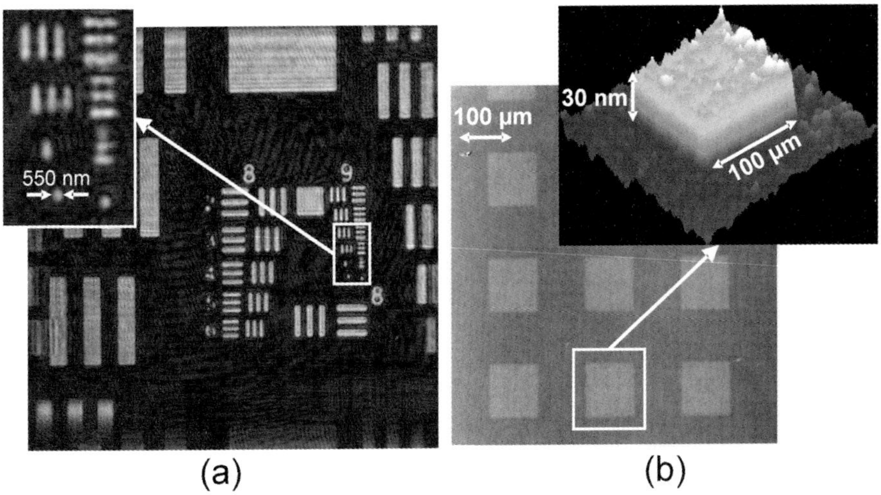

Figure 6.5 (a): Reconstructed holographic amplitude image of a USAF 1951 test chart; (b): reconstructed quantitative phase contrast image of a reflective phase object with 30-nm height steps in the axial direction.

6.2.3.2 Multi-focus Microscopy

Figure 6.6 demonstrates the digital holographic microscopy feature of subsequent numerical focus correction by variation of the reconstruction parameters. **Figure 6.6a** shows the image of a USAF 1951 test chart recorded out of focus by illumination with laser light in transmission mode. The corresponding sharply refocused holographic amplitude image reconstructed from a digitally captured hologram at the same out of focus position is depicted in **Figure 6.6b**. **Figures 6.6c** and **6.6d** show corresponding results obtained from investigations on living human tumorous hepatocytes (HepG2). The advantage of this ability is that it eliminates the need for mechanical focus adaptation, which is particularly convenient, for example, for long-term measurements on living cells and for automated measurements. Furthermore, numerical refocusing allows the reconstruction of image stacks at different focus positions ("Digital Holographic Multifocus") from a single captured hologram.

6.2.4
DHM Combined with Phase Contrast and Fluorescence Imaging

Here we focus on the construction of a DHM set-up which is specifically intended to allow direct comparison with other well established contrast methods typically applied in life sciences. Therefore, DHM contrast was made available as an add-on module that could be attached to commercial microscopes via a standardized port.

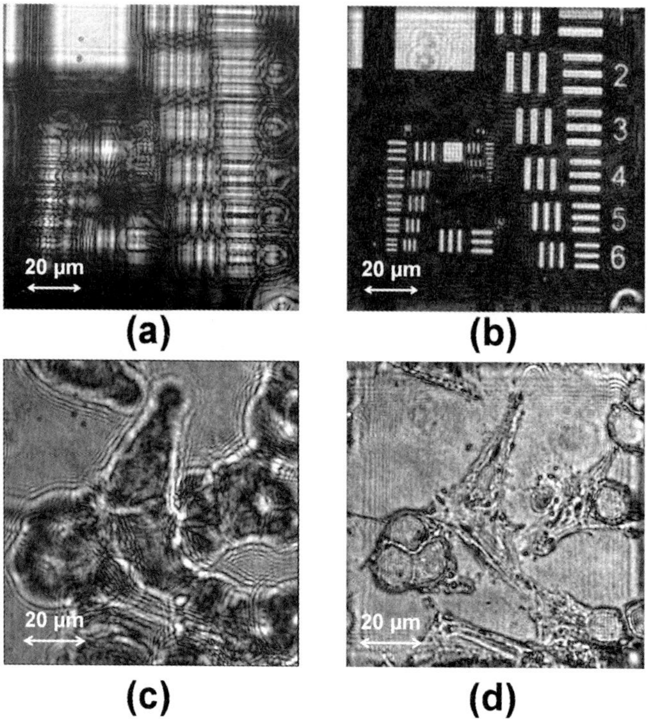

Figure 6.6 (a): USAF test chart recorded out of focus by illumination with laser light in transmission, (b): numerically refocused holographic amplitude image reconstructed from a digital hologram captured at the same focus position; (c), (d): corresponding results to (a) and (b) from investigations on living human tumorous hepatocytes (HepG2).

Care was taken that other contrast techniques, e.g., phase, differential interference, and fluorescence contrast, could be achieved simultaneously or with minimal and reproducible configuration modifications. Essentially, this allows direct comparison of the techniques since images could be acquired with different contrasts at identical locations of a sample. Furthermore, the modular design allows the installation of the DHM technique on microscopes that were already set up for certain applications and equipped with adequate optical, mechanical and electronic components.

6.2.4.1 **Modular DHM Set-up**

Figure 6.7 illustrates the concept of a modular add-on for the integration of DHM into common commercial upright microscopy systems. The modular system is based on three components:

- a coherent illumination device

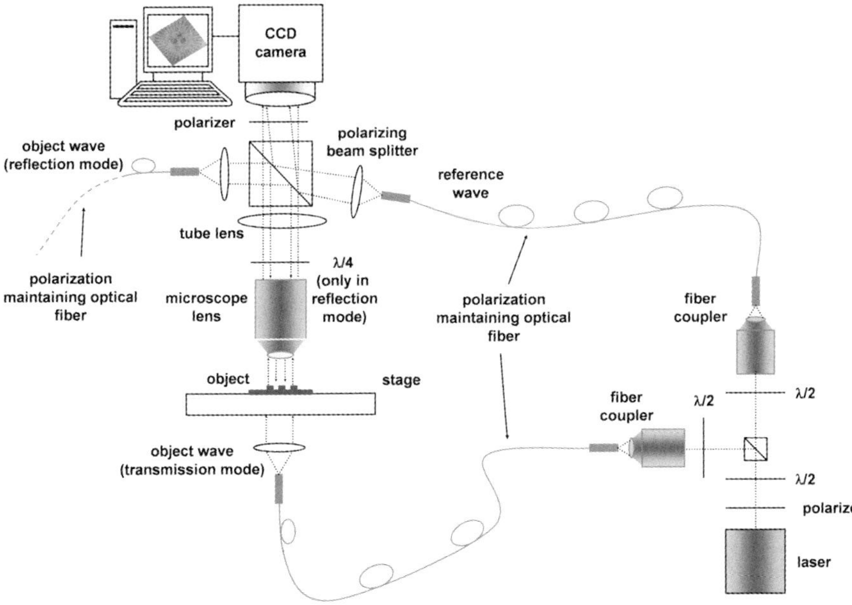

Figure 6.7 Illustration of optical components for realization of modular DHM combined with a light microscope.

- a holographic interferometric unit
- an image-processing device and evaluation software.

The coherent light source is a frequency doubled Nd:YAG laser (λ = 532 nm) or a HeNe laser (λ = 633 nm). As discussed in Section 6.2.2, holography is based on the superposition of an object wave and a reference wave. To obtain two coherent light sources, the linearly polarized laser light is divided into two parts. For variable light guidance, each beam is coupled into a polarization-maintaining single mode fiber and guided to the holographic interferometric unit. Here, a DHM set-up is chosen in which both beams are linearly polarized in orthogonal directions. Thus, optical components can be used which specifically act on one beam depending on its polarization. Depending on the imaging mode, the ends of the fibers are attached to different ports of the microscope via standardized fiber connectors. The interferogram that is formed by the interference of object wave and reference wave is recorded by a CCD camera and transferred to an image-processing system with a software module for digital reconstruction and evaluation of the digitized holograms (see Section 6.2.2).

Figure 6.8 shows the major components of a fluorescence microscope and a laser-scanning microscope with attached DHM modules. The typical hologram capture time depends on the imaging device used (here a CCD camera) and is about 1 ms. The reconstruction rate of the digital holograms is determined by the ability of the image-processing system and on the size of the digitized holograms. With current computer systems (e.g., Pentium IV 2.8 GHz) at a frame size of 512 × 512 pixels the reconstruction rate is 4 Hz. Using a region of interest with 128 × 128 pixels reconstruction rates up to 70 Hz are possible.

6.2.5
Comparison of DHM with Standard Methods of Cell Microscopy

6.2.5.1 DHM in Comparison with Brightfield and Phase Contrast Imaging

Figure 6.9 shows images of red blood cells obtained using the DHM set-up shown in the upper left panel in **Figure 6.8** in comparison with bright-field and Nomarski phase contrast images (Zeiss Achroplan water immersion 63× /0.9). The phase distribution obtained by DHM (**Figure 6.9d**) appears without "halo" and "shading-off" effects and is of higher contrast than the bright light image (**Figure 6.9a**) and the phase contrast image (**Figure 6.9b**). **Figure 6.9e** shows the 3D plot of cell thickness that is obtained from the data in **Figure 6.9d** by using Eq. (6.3) (estimated integral refractive indices $n_{\text{cell}} = 1.4$ [11], $n_{\text{medium}} = 1.33$).

6.2.5.2 DHM in Combination with Confocal Fluorescence Imaging

Confocal fluorescence microscopy has become a well established imaging technique within the life sciences. Nowadays, sophisticated and robust set-ups are commercially available. Confocal microscopy allows optical sectioning of samples, i.e., the response to a signal is restricted to a limited axial section of the sample. This is achieved by scanning samples with point illumination and using correlated descanned detection through a pinhole. In the life sciences, confocal imaging is usually conducted in an epifluorescence configuration. Typically, samples are specifically labelled with dyes, which are excited by a laser. Three-dimensional imaging is achieved by acquisition of images in different z-planes. Rendering and other advanced software tools allow the analysis and representation of the 3D structure of the sample. DHM is sensitive to changes of the optical path length when passing an object (phase information) and the degree of absorption (amplitude information) in the sample.

Figure 6.10 shows fluorescence and DHM images recorded from human erythrocytes. The acquired fluorescence image stack (see **Figure 6.10a**) consists of 20 planes (spacing 0.5 μm). It is apparent from the cross-sections that the erythrocytes adhered to the glass cover slip (mediated by a polylysine coating) and essentially form half spheres. The fluorescence images give se-

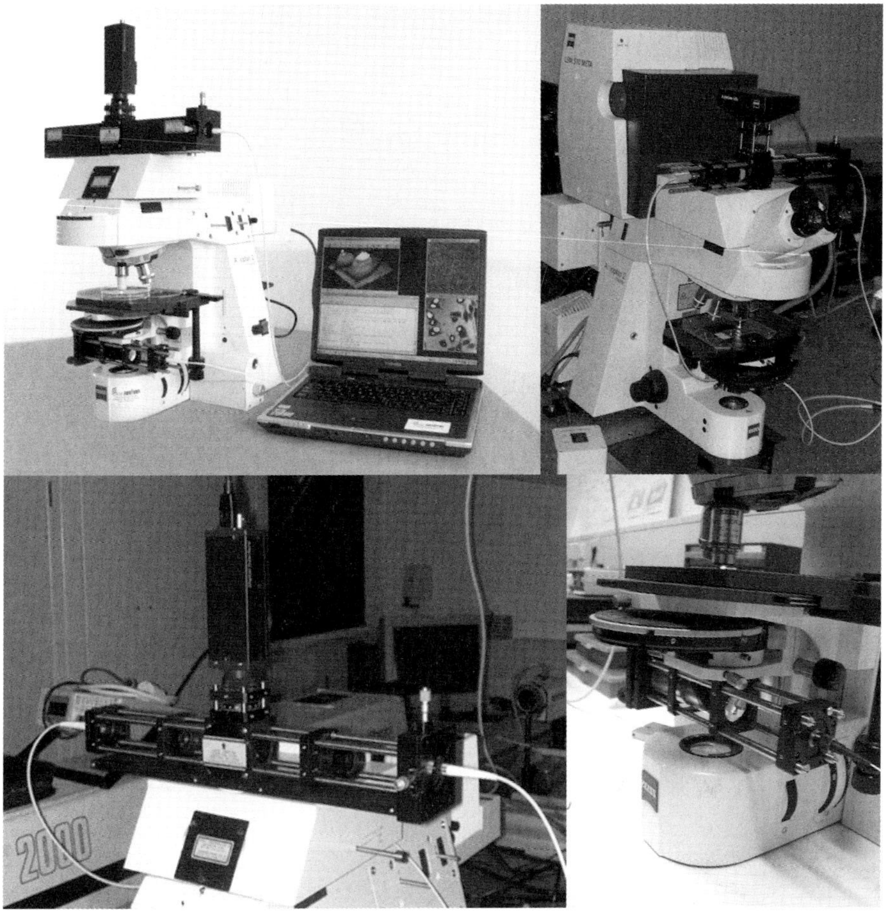

Figure 6.8 The upper left panel shows a standard light microscope (Axioplan 2, Zeiss) which is equipped with components for epifluorescence, phase and differential interference contrast microscopy. Oculars were removed to comply with laser safety standards. The DHM add-on module together with the camera (lower left panel) is on top of the microscope. The lower right panel shows a close-up of optical components which ensure proper illumination of samples in transmission mode. The beam is fed into the system via a dichroic mirror placed below the illumination condenser. The design ensures that all contrast-enhancing techniques, which are installed in the condenser wheel, remain functional. The upper right panel shows a DHM module (here in reflection mode) attached to a confocal laser-scanning system (LSM 510 META, Zeiss).

lective information on the 3D localization of the lipid membranes while the phase contrast image of the DHM (**Figures 6.10d–6.10f**) give complementary information on the optical thickness of the structures determined by their geometrical length and refractive index (estimated $n_{\text{cell}} \approx 1.4$). Here it is notable that, for complete representation of the sample, images in different planes have to be acquired for fluorescence imaging while for DHM it is sufficient

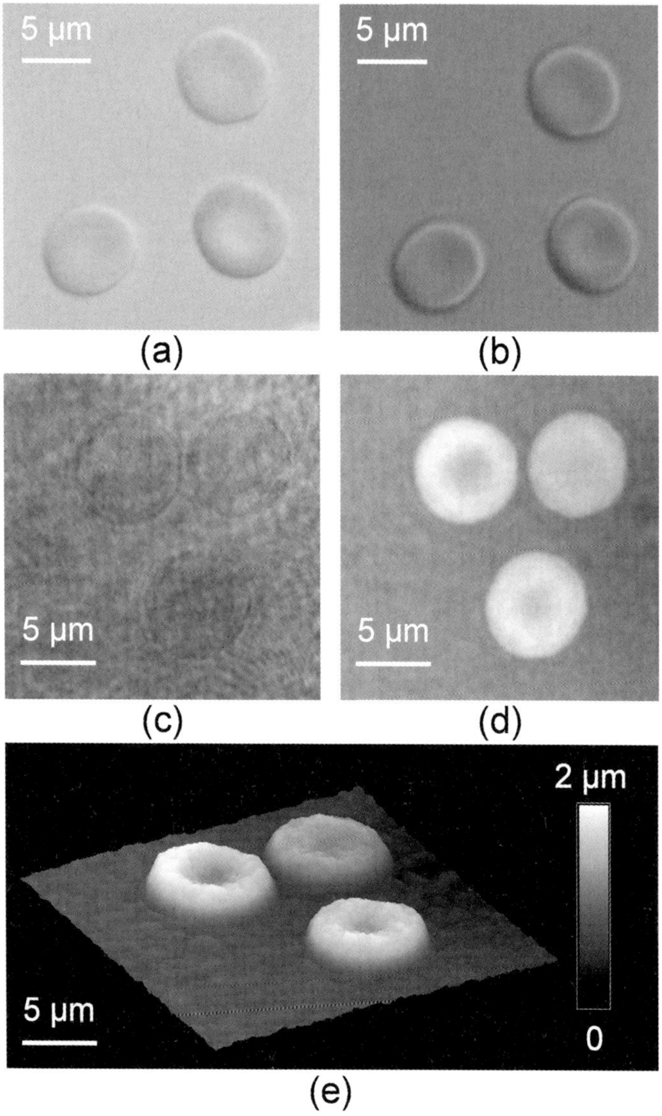

Figure 6.9 DHM phase contrast image of red blood cells compared with brightfield and phase contrast imaging obtained with a setup equipped with a Zeiss Axioplan 2 microscope and a DHM module (see **Figure 6.8** upper left panel). (a): bright light image, (b): Nomarski phase contrast image, (c): DHM amplitude image, (d): DHM phase contrast image, (e): 3D plot of the cell thickness (Zeiss Achroplan water immersion 63×/0.9).

to record one hologram. Furthermore, the adjustment of a correct objective position is less critical, since the defocus distance is a free-fit parameter and can be optimized in the reconstruction algorithm.

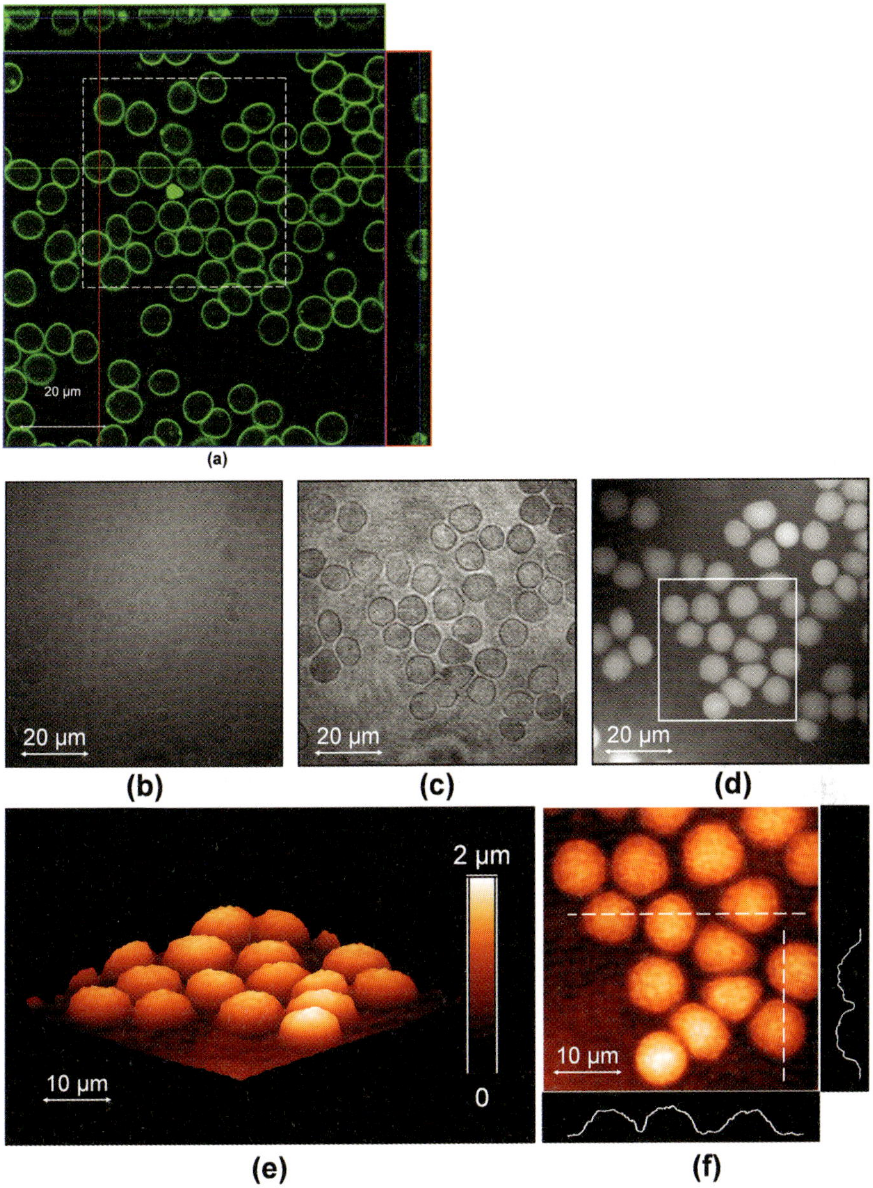

Figure 6.10 Comparison of images recorded with a set-up equipped with a confocal laser scanning unit and a DHM module (see **Figure 6.8**, upper right panel). For fluorescence imaging the cells were labelled with a fluorescent membrane marker (DiOC6). (a): Fluorescence image of human erythrocytes; the micrographs include two cross-sections in the vertical and horizontal directions; (b): corresponding digital hologram recorded in transmission mode at $\lambda = 543.5$ nm; (c), (d): amplitude and phase images reconstructed by DHM; (e): pseudo 3D plot of the cell thickness; (f): vertical and horizontal cross-section through the cell thickness.

6.2.6
Holographic Micro-interferometric Analysis in Cell Micromanipulation

6.2.6.1 Laser Microcapture Microscopy and Manipulation of Living Cells

The interaction of laser light with biological matter has a variety of applications in life sciences and medicine. Lasers focused through microscope lenses provide the unique possibility of micromanipulating biological specimens without any mechanical contact and without affecting their viability. Two different microlaser principles allow novel approaches to life science research: using *optical traps* suspended cells are caught, moved or positioned in order to segregate individuals from the bulk, and samples are stretched and held for rigidity tests or force measurements; using *laser microbeams* cell membranes can be opened to microinject drugs or genetic material, organelles or subcellular structures are ablated to study embryonic development and individual cells are fused to yield specific clones.

The method of laser microdissection and pressure catapulting (LMPC) provides a way to capture individual cells or entire tissue areas in a completely contact-free manner without the risk of contamination by unwanted biomatter. This state-of-the-art technology yields pure and homogeneous samples for various downstream analyses spanning the modern fields of functional genomic and proteomic research in animal science and plant biology, and also assisting forensic investigations to convict a suspect. With LMPC it is possible to isolate an individual single tumor cell to process it for RNA or DNA evaluation. Large tumor areas or a multitude of single rare cells can also be sampled within one collection vial. As averaging effects are avoided, this pure specimen preparation enables a more differentiated analysis and therefore improved tumor diagnosis and optimized treatment.

Up to now, LMPC has mainly been applied to fixed cells like cytospins and cell smears or to tissue slices like paraffin or cryosections. The aim was to transfer the noncontact laser microdissection and catapulting method to live cell capture. A "LifeCellHandling" system was built up as a platform to combine of laser micromanipulation with a novel 3D-imaging and measuring method for biological specimens, the so-called DHM module. Furthermore, adequate cell culture ware as well as the corresponding hardware had to be adapted, and dedicated application protocols established.

The two different laser micromanipulation procedures require different system set-ups. For laser ablation and cutting, a pulsed UV-A laser is incorporated within the PALM® MicroBeam system, whilst for trapping a continuously working red or near-infrared laser is used within the PALM® system. In both cases the lasers are coupled through the epifluorescence path into a research microscope (Zeiss Axiovert 200) and focused through the objective lens. Furthermore, the two laser systems can be used simultaneously as combined in the PALM® CombiSystem. Within the narrow laser spot (<1 μm) the focused light generates forces that in the case of laser cutting disrupt chemical bonds yielding molecular debris (photodecomposition or ablation) or in the case of laser trapping drag the samples into the spot and keep them there by radiation pressure forces. The system's accuracy, stability and precision allow the trapping of samples as small as bacteria or the preparation of subcellular components like nuclear filaments and also large tissue sections with a diameter of about 2 mm. Even entire organisms, such as the nematode *Caenorrhabditis elegans*, have been catapulted with a single laser pulse. LMPC combined with specially designed software allow thousands of cells to be harvested within a very short time as required for array and proteomic techniques.

The use of laser systems to micromanipulate living cells requires certain prerequisites regarding culture vessels. Routine plastic ware is not suitable, as the laser beams are scattered or absorbed by these materials. Thus, specialized culture ware and the necessary hardware have been developed to facilitate live-cell micromanipulation and handling. Amongst these is a modified Petriperm dish, which consists of a Teflon membrane and an upper layer of special polyethylene naphthalate (PEN), the so-called "LPC membrane". This thin membrane (ca. 2 μm) serves as a stabilizing backbone allowing the capture of difficult specimens, such as living cells, without limiting their viability or sampling large tissue areas, maintaining the morphology. As an insert to routine Petriperm dishes, a ring spanned with the LMPC membrane has turned out to be very helpful for laser capture and subsequent cultivation of living cells. For live-cell handling the use of closed compartments helps to avoid any kind of contamination. Therefore, a special magnetic collection device was developed for work with cell cultures and allows LMPC within the closed dish. For a controlled environment and a stable temperature of 37°C as recommended for long-term experiments a Zeiss/PeCon incubator was adapted to the PALM® MicroBeam system spanning the entire system platform (see **Figure 6.14** below).

Successful LMPC of living cells was first achieved by A. Mayer [12]. Working from these results we modified the method and established a protocol for the isolation of living cells with laser microbeams. The tumor cell lines Hep G2 and EJ28 were cultivated on specially developed LPC-membrane-spanned culture dishes. Single cells and whole cell areas were microdissected and catapulted into medium-filled collection tubes. The isolated cells were re-cultivated and grown to single colonies without harm to their viability and proliferation behavior [13]. The tumor cell line HCT116 was then used to demonstrate that the laser capture process does not affect the genomic content of the cell as proved by CGH (comparative genomic hybridization). Single laser-pooled cells from an adherent-growing cell culture were seeded out again in membrane dishes and after they had grown to new cell colonies the LMPC and re-cultivation were repeated five times. Then the cells of the last re-cultivation (5× catapulted) and the original cell culture were used for CGH analysis and no significant changes in genetic stability could be detected. To prove the safety of LMPC on the single-cell level M-FISH (multiplex fluorescence *in situ* hybridization) analysis was done and this approach also showed that the laser-isolated cells maintained their genetic information. All these experiments were done in cooperation with the University of Munich (LMU) [14].

This additional finding confirms the result of previous experiments such as the "comet assay" [15], proving that LMPC does not affect the viability and proliferation behavior of the captured cells, as the applied laser wavelength is not absorbed by the most abundant biomolecules such as DNA, RNA and proteins.

Thus, LMPC opens new approaches for the establishment of homogeneous cell populations from adherent-growing cell cultures, tissue cultures or fresh biopsies. The growing field of stem-cell research will, in particular, benefit from this technology.

Stem cells have the potential to differentiate into specialized cells from various lineages. Differentiation is often associated with changes in morphology and cell surface marker expression. To understand the intrinsic molecular events associated with differentiation, it is important to work with pure populations of cells and to analyze their behavior and characteristics. The selection and isolation of cells with specific features also stand at the center of stem-cell therapy and tissue engineering. The heterogeneity of cell preparations poses a serious problem to the molecular analysis and therapeutic use of stem cells. In order to develop methods for the isolation and expansion of homogeneous stem-cell clones the results obtained from live-cell LMPC were transposed to stem cells. It was first investigated whether stem cells would grow on the specific LMPC membrane and whether laser treatment would interfere with stem-cell development and differentiation. Different stem-cell lines (RM26, BalbC, P19) were plated onto culture dishes containing fibronectin-treated, UV-treated, or untreated PEN membranes. In all cases the cells adhered well and after LMPC the re-cultivated clonal cells expanded rapidly and maintained the same morphology. In a further experiment, a mixed culture of CD34 (= stem-cell marker) positive and negative cell lines (BalbC and RM26) was used for LMPC. It was successfully demonstrated that single cells could be isolated from this mixture according to morphological differences. After re-cultivation the cell clones were immunofluorescently labelled and analyzed microscopically and by fluorescence-activated cell sorting (FACS). The CD34-positive cells (RM26) maintain this stem-cell marker after LMPC and clonal expansion (**Figures 6.11–6.13**).

These experiments demonstrated that laser capture and subsequent clonal expansion of specific, individual cells yields pure and homogeneous cell

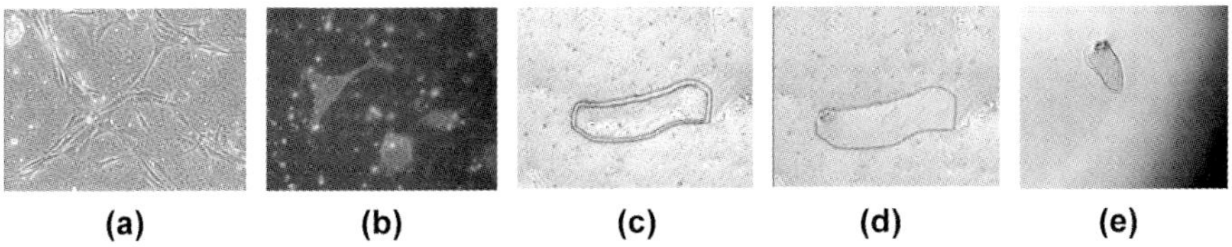

Figure 6.11 (a): BalbC cell line growing on the PEN membrane; (b): RM26 cell line fluorescently labelled for CD 34; (c)–(e): LMPC of a single cell from a mixed cell culture.

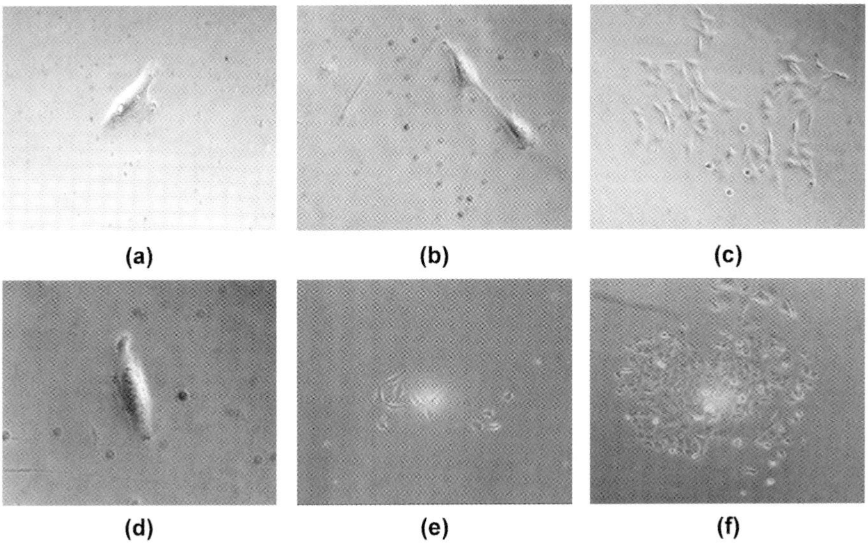

Figure 6.12 Re-cultivation of single catapulted cells. (a)–(c): BalbC cell growing to a cell clone; (d)–(f): RM26 cell proliferating to form a new cell colony.

clones. LMPC has been successfully used to purify embryonal stem-cell-derived cardiomyocytes for *in vitro* drug discovery, shortening the usual procedure from several months to a few days [16].

6.2.6.2 Set-ups for DHM in Combination with Laser Micromanipulation

For visualization, evaluation and measurement of the above described live-cell experiments a 3D imaging module, DHM, was combined with the laser micromanipulation system. This innovative approach provides a 3D delineation of a living specimen in a rapid and marker-free manner, which is especially important for live-cell observations. In principal, holographic interferometry depicts the height of a specimen via a quantitative digital holographic phase contrast. One single exposure provides the required information and

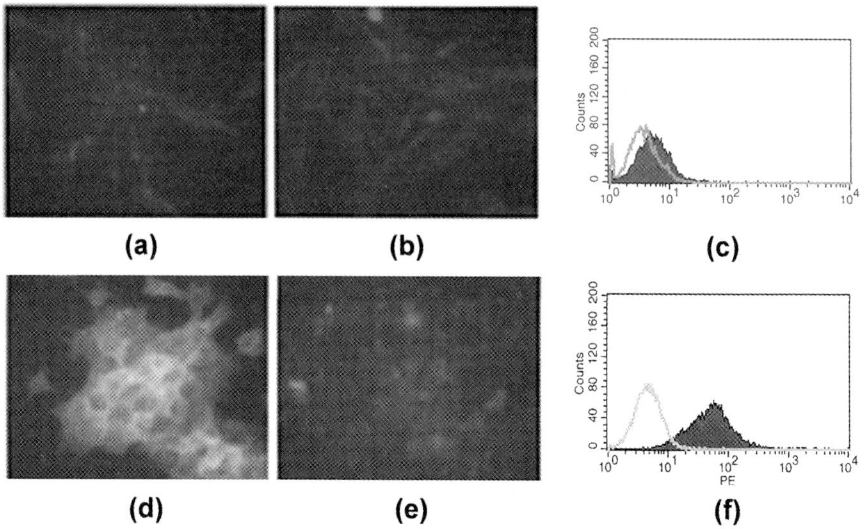

Figure 6.13 Characterization of the recultivated colonies. (a)–(c): BalbC do not express the stem-cell marker CD34; (d)–(f): RM26 cells express CD 34, fluorescently labelled; (c) and (f): results from the FACS analysis.

there is no need for constant refocusing or scanning through the specimen, as with routine imaging systems.

Figure 6.14 shows the set-up of a combined system platform including the PALM® CombiSystem with UV-MicroBeam and optical trap, the DHM module and the surrounding incubator (for details of modular add-on see Section 6.2.4).

6.2.6.3 **Monitoring of Live-cell Micromanipulation by DHM**

Live-cell processes are usually observed by phase contrast time-lapse microscopy. The resulting image sequence provides behavioral information in two spatial dimensions only. To obtain 3D information on living cells, confocal microscopy with fluorescently tagged proteins needs to be done. DHM allows monitoring of cell behavior and quantitative measurement of changes in volumes and refraction indices over time and, as a great advantage, in a marker-free way.

Apoptosis is a cellular process accompanied by marked changes in cell morphology, including cell rounding and membrane blebbing. The morphological changes triggered by UV- or chemically induced apoptosis in stem cells could be monitored by DHM (**Figure 6.15**). The initially flat cells rounded and migrated together (to each other) before they died. The degree of apoptosis over time was quantified by measurement of cell height. Using the ability to per-

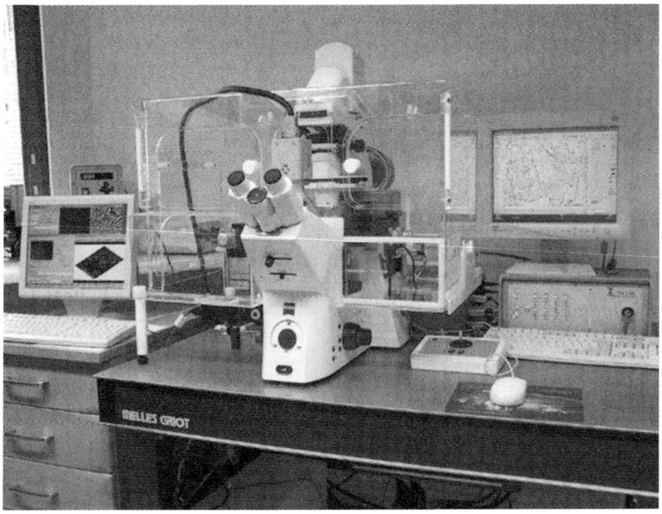

Figure 6.14 Set-up of DHM in combination with a PALM® CombiSystem.

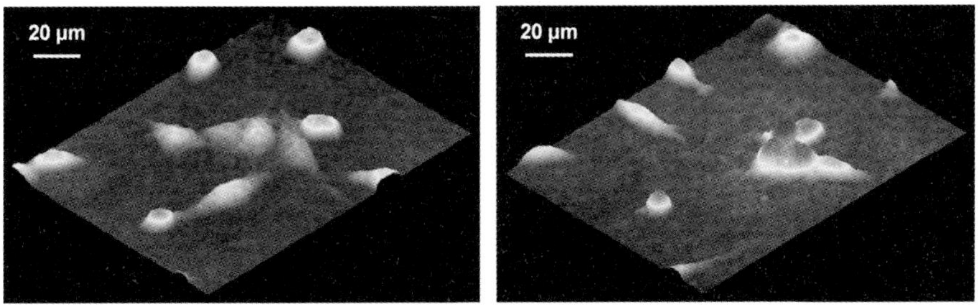

Figure 6.15 Apoptosis monitoring of RM26 stem cells. Left: before apoptosis; right: after apoptosis.

form volumetric measurements on apoptotic cells, DHM could therefore be used to screen cancer drugs in a high-throughput fashion.

With DHM the difference between spherical and flat cells (i.e., rounding-up during cell division) or piled-up cell clusters versus a plain cell sod (i.e., clone formation in a cell culture) may be detected instantaneously. Changes in cell densities occur, e.g., during differentiation of stem cells into osteoblasts depending on the incorporation of calcium phosphate into the extracellular matrix. This process is usually observed by cellular staining. The measurement of the refractive index using DHM provides a much easier and marker-free method of monitoring the differentiation process and discriminating between undifferentiated stem cell and osteoblasts.

The combination of laser micromanipulation and DHM measurements allows rapid, marker-free process control, for example the evaluation of success-

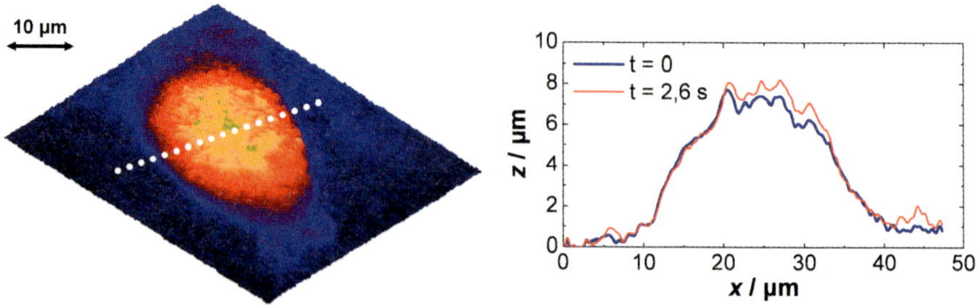

Figure 6.16 RM 26 stem cell, laser microinjection monitored by DHM; left: 3D plot of an RM 26 cell; right: cross-section though the cell before and after microinjection.

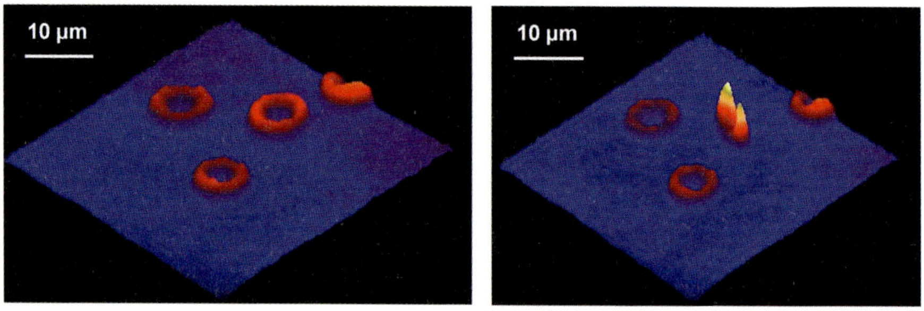

Figure 6.17 Erythrocytes manipulated by optical tweezers and monitored with DHM.

ful laser microinjection. The cell membrane of a murine stem-cell line (RM 26) was perforated with a single UV laser pulse. The uptake of medium through this transient hole led to a volume increase which was hardly detected by normal microscopy but could be monitored and quantified by digital holographic interferometry, as shown in **Figure 6.16**.

A further field of interest is the combination of optical tweezers and DHM, which allows the measurement of changes in cell morphology occurring during cell manipulation. Suspended erythrocytes were trapped and a single cell was fixed with one of the optical double beams. The visualization of this experiment was done with DHM and the changes of morphology could be observed online (**Figure 6.17**).

Anticipating new approaches in cell biology, the supplement of a newly developed software tool for force measurements in combination with optical traps and DHM will be of great interest. This method offers the possibility of comparing the elasticities of different types of cells. Tumor cells seem to be less elastic than non-tumorous cells [17].

Thus, laser microbeams are valuable tools for characterizing cell types, cell stages or states of cell differentiation, and will make noncontact LMPC an indispensable method for developmental biology studies, tumor analysis and tumor staging as well as for stem-cell research or personalized medicine. If combined with laser micromanipulation, the innovative DHM module allows 3D visualization of selected cells and marker-free quantitative measurements of volume changes, and thus can serve as a rapid process for controlling cellular mechanisms.

6.2.7 Application of DHM to Living Tumor Cells

6.2.7.1 Background

Pancreatic cancer is a devastating malignancy in western countries and the fourth most common cause of cancer-related deaths [18]. One of the reasons for the poor prognosis of pancreatic adenocarcinoma is its tendency to form micrometastases before clinical symptoms arise and before the tumor is detectable by diagnostic medical imaging techniques. The molecular mechanisms that determine the highly malignant growth and dissemination pattern of pancreatic cancer are poorly understood. Recent findings suggest that the cell–cell contact protein E-cadherin and the associated catenin complex play a key role in pancreatic cancer progression [19,20]. Since proteins of the cadherin/catenin complex are involved in the organization of the actin cytoskeleton that determines cell shape, it is a key challenge to obtain data related to the tumor state directly from analysis of the tumor cell morphology.

6.2.7.2 Shape Measurement of Living Cells

The investigated human pancreatic ductal adenocarcinoma cell lines PaTu 8988S and PaTu 8988T were obtained from the German Collection of Microorganisms and Cell Cultures (DSMZ, Braunschweig, Germany). Both cell lines were established in 1985 from a liver metastasis of a primary pancreatic adenocarcinoma from a 64-year-old woman. PaTu 8988S represents a highly differentiated carcinoma cell line, PaTu 8988T a poorly differentiated adenocarcinoma with a high metastatic potential [21]. PaTu 8988T cells were retrovirally transduced with an E-cadherin expression construct containing an E-cadherin cDNA in the expression vector pLXIN (Clontech, Palo Alto, USA). Cells were selected, cloned and analyzed for E-cadherin expression [22].

Figures 6.18 and **6.19** show results from measurements with DHM to determine the thickness of adherently grown living PaTu 8988T cells in a Petri dish with cell culture medium. For the experiments, holograms of adherent PaTu 8988T cells and PaTu 8988T pLXIN E-Cadherin cells were captured in a transmitting light arrangement (see **Figure 6.7**). From the determined phase data, which represent the optical path length change effected by the samples, the

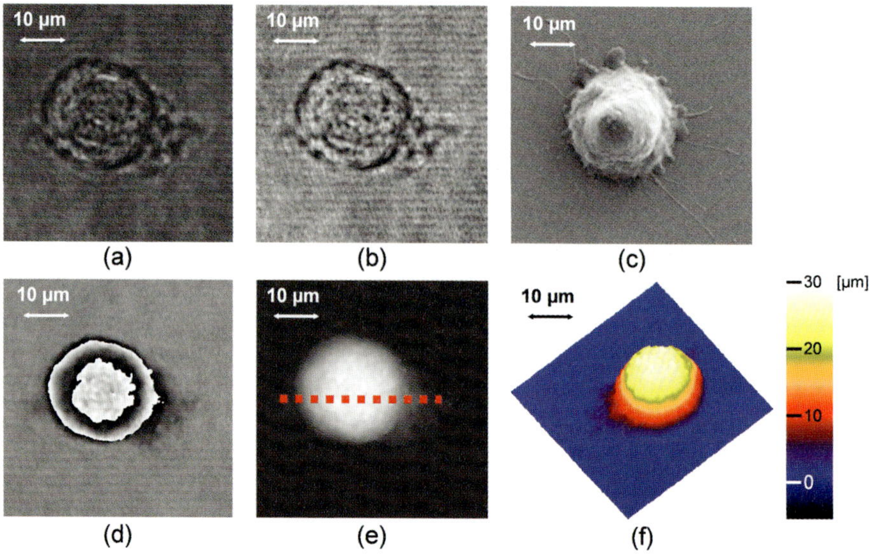

Figure 6.18 Thickness determination of living PaTu 8988T cells by DHM. (a): digital hologram; (b): reconstructed holographic amplitude image; (d): corresponding quantitative phase contrast image mod 2π; (e): unwrapped quantitative phase contrast image (for the plot of the cross-section along the dotted line see **Figure 6.20**); (f): 3D plot of (e); (c): SEM image of a PaTu 8988T cell.

thickness d_{cell} of the cells is obtained from Eq. (6.4) and an average refractive index $n_{cell} \approx 1.38$ [23].

Figure 6.18 depicts results obtained on a PaTu 8988T cell. **Figures 6.18a**, **6.18b** and **6.18d** show the captured hologram, the reconstructed holographic amplitude image and the quantitative phase contrast image mod 2π in gray-scale representation. **Figures 6.18e** and **6.18f** show the unwrapped phase data as well as the corresponding pseudo 3D representation of the calculated cell thickness in comparison to a Scanning Electron Microscope (SEM) image (**Figure 6.18c**) of a PaTu 8988T cell. **Figure 6.19** represents the corresponding results obtained for two PaTu 8988T pLXIN E-Cadherin cells. For each cell type the 3D data for the cell thickness were found in good agreement with the appearance in the SEM images. In **Figure 6.20** the cell thickness is plotted for both cell lines corresponding to the cross-sections marked by dotted lines in **Figures 6.18e** and **6.19e**. The different cell types can be clearly differentiated by the calculated thickness. Assuming that the investigated cells were grown adherently on the carrier glass, and taking into account the appearance of the pancreas cells in the SEM images, the results give a good indication of 3D shape of the cells.

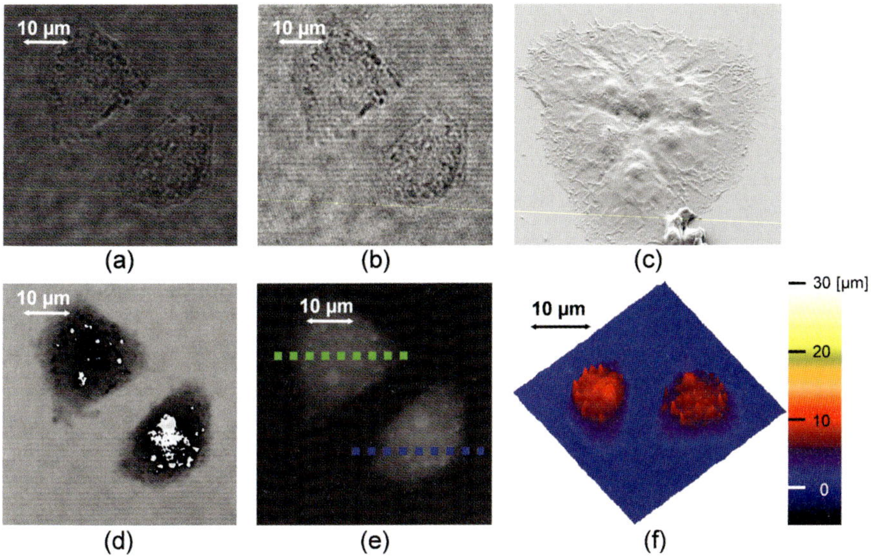

Figure 6.19 Thickness determination of living PaTu 8988T pLXIN E-Cadherin pancreas cells by DHM. (a): digital hologram; (b): reconstructed holographic amplitude image of two PaTu 8988T cells; (d): corresponding quantitative phase contrast image mod 2π; (e): unwrapped quantitative phase contrast image (for the plot of the cross-section along the dotted lines see **Figure 6.20**); (f): 3D plot of (e); (c): SEM image of a PaTu 8988T pLXIN E-Cadherin cell.

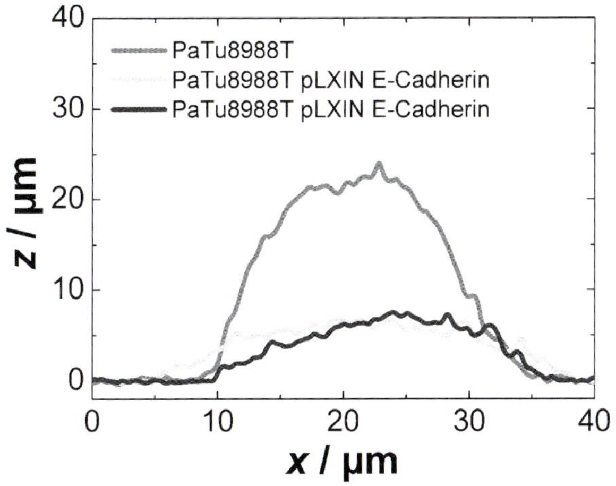

Figure 6.20 Thickness comparison of PaTu 8988T cells and PaTu 8988T pLXIN E-Cadherin cells along the cross-section lines in **Figures 6.18e** and **6.19e**.

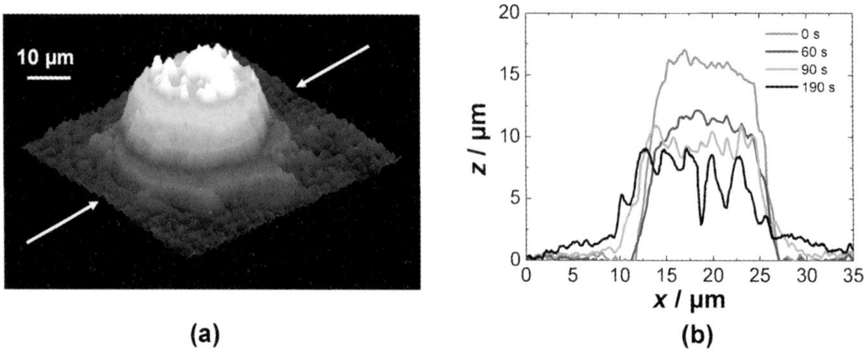

Figure 6.21 Analysis of a living PaTu 8988S cell after addition of a marine toxin (Latrunculin B) to the cell culture medium. (a): 3D plot of the cell thickness obtained from the phase contrast measurement. (b): Calculated cell thickness along the cross-sections through the middle of the cell (arrows in (a)) at $t = 0$ s, $t = 60$ s, $t = 90$ s, $t = 190$ s after the addition of Latrunculin B.

Figure 6.21 demonstrates the application of DHM for the detection of dynamic morphological changes of living cells in cell culture medium as a reaction to chemical substances. A hologram series of a single PaTu 8988S cell with a time delay between two recordings of $\Delta t = 10$ s was captured. At the beginning of the experiment at $t = 0$ s a marine cell toxin (Latrunculin B) was added to a final concentration of 10 µM to the cell culture medium to destroy the actin filaments of the cell's cytoskeleton. **Figure 6.21a** shows the 3D plot of the cell morphology calculated from the phase distribution at $t = 0$ s assuming an average refractive index of 1.37. **Figure 6.21b** represents the corresponding cell height at (a): $t = 0$ s, (b): $t = 30$ s, (c): $t = 80$ s and (d): $t = 190$ s for a cross-section through the middle of the cell (see arrows in **Figure 6.21a**). With increasing time after the addition of Latrunculin B a cell collapse is effected by the destruction of the cytoskeleton. During the collapse the cell height locally decreases up to about 50%. Furthermore, the crinkle structure of the shrunken cell is clearly shown in the plotted cross-sections.

6.2.8
Optical Coherence Tomography (OCT)

Fluorescence imaging with laser-scanning microscopy (LSM) is one of the standard techniques for analysis of the structural and functional properties of live biological samples down to the sub-cellular level [24]. A special fluorescence method that allows one to image deep in tissue is based on the use of near-infrared light using two-photon excitation [25]. One disadvantage of fluorescence imaging is the required labelling of the sample with exogenous chromophores that may affect the live sample under study as discussed above.

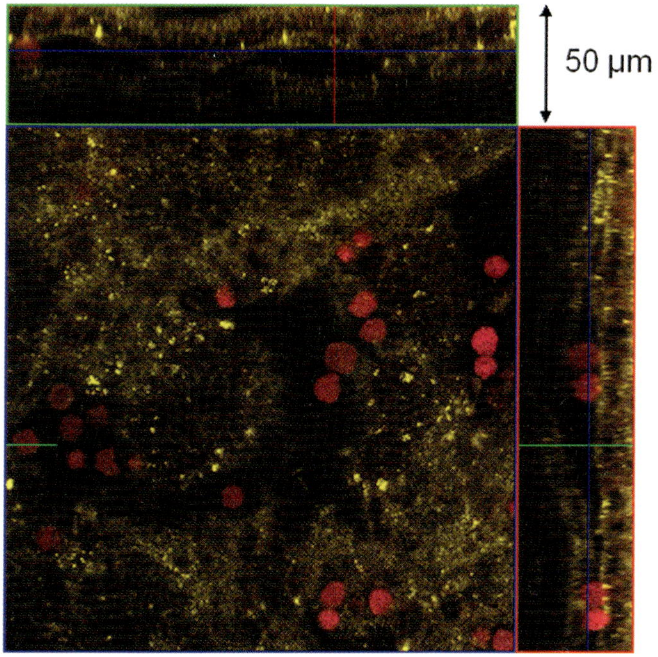

Figure 6.22 Simultaneous two-photon and confocal reflection imaging of a mouse embryo. Objective lens: Zeiss Plan-Neofluar $40\times/1.3$ Oil. Excitation/reflection: 850 nm, fluorescence: 500–550 nm (red channel) of GFP-labelled red blood cells.

Optical coherence tomography (OCT) is a label-free imaging method with 3D resolution in the micrometer range in highly scattering tissue [26]. The contrast in the images is based on refractive index changes and absorption and has been shown to yield sufficient structural information for various medical applications [27].

There have been reports on the use of OCT for microscopic imaging [28,29], referred to as optical coherence microscope (OCM). In general, the contrast of optical coherence microscopy (OCM) has been insufficient to compete with fluorescence microscopy. However, the combination of LSM, in particular with two-photon excitation of fluorescence, with OCM may allow one to obtain general structural information as a reference for fluorescence imaging without the need to label cell structures, thus minimizing the use and potential interference of exogeneous chromophores (see **Figure 6.22**).

6.2.8.1 **Principle**

The typical set-up for OCT is a Michelson interferometer with a low-coherence light source (see **Figure 6.23**). Using low-coherence light (with typical coherence lengths L_c in the range 10–30 μm) interference is obtained only within an optical path length difference L_c between the two arms. Thus the coherence length determines a zone called the coherence gate in which the backscattered light from the sample arm can produce interference with the reference light. Moving the reference mirror moves the coherence gate and thus the sample can be scanned axially.

The amplitude of the interference is directly related to the intensity of the backscattered light from the sample, and is a measure of changes of the refractive index in the sample, yielding a map of the sample structure. By scanning the beam laterally over the sample a 3D image is obtained. The interference of reference and sample arm light leads to quasiconfocal detection. The coherence gate acts as an additional discriminator against the detection of multiply scattered light in comparison to confocal reflection imaging.

This discrimination, and the optical amplification by the reference arm, which has been shown to yield shot-noise-limited detection, are the main advantages of OCM over reflection mode LSM for imaging deep in biological samples.

In the sample arm of the OCT, an LSM can be incorporated to obtain an OCM set-up. There are two main differences between OCM and OCT, one in the properties (namely the resolution) and one in the imaging method (namely scanning). These changes have technical consequences for the design of the set-up concerning in particular the modulation of the phase in the reference arm and the electronic demodulation of the signal.

Resolution and Depth Discrimination

The sample beam is focused through a high-NA microscope objective on the sample, which determines the lateral and axial resolution. The axial resolution in standard OCT is determined by the coherence gate (full-width-at-half-maximum (FWHM) = 10–30 μm) but the depth discrimination of a high-NA objective is much smaller (i.e., FWHM = 1.8 μm for NA = 0.8 at $\lambda = 830$ nm).

Nevertheless, the depth discrimination of the coherence gate is advantageous compared to confocal reflection imaging, because the rejection of out-of-focus light increases dramatically with distance from focus Δz. If we have for instance a coherence function with a Gaussian envelope the rejection is proportional to $\exp(-\Delta z)$ while it improves only moderately ($\propto \Delta z^{-2}$) in the confocal case (see **Figure 6.24**).

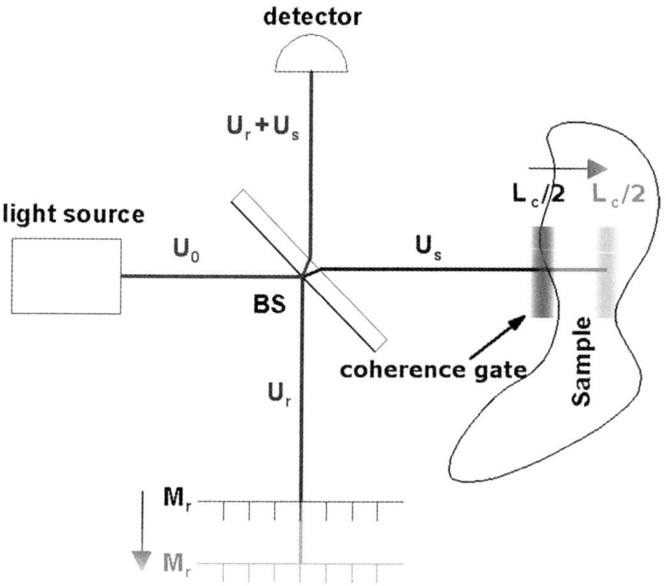

Figure 6.23 Scheme of an OCT set-up with the reference beam U_r, the sample beam U_s, and the reference mirror M_r. The light is split up by the beamsplitter BS. Moving the mirror changes the zone in the sample for which the backscattered light interferes.

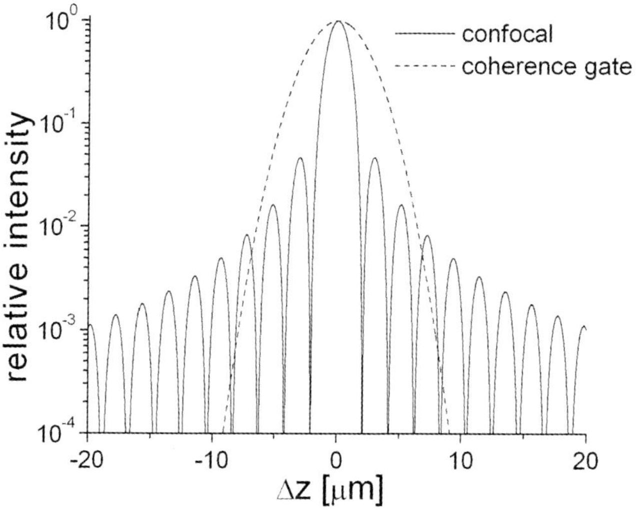

Figure 6.24 Confocal depth discrimination compared with the depth discrimination of a Gaussian coherence function in a logarithmic plot (NA = 0.8 at $\lambda = 830$ nm, $L_c = 20$ μm).

Scanning Process

In order to obtain the whole 3D information of a sample it has to be scanned in the x, y and the z directions. Unlike classic OCT, where usually an A-scan (x, z) is used in the LSM an en-face scan (x, y) is performed. Using an A-scan the interference signal is modulated and the amplitudes can be easily deduced. In contrast, using the en-face scan the position of the reference mirror for each x, y point is constant. Thus for each point a small modulation δz around the nominal depth z_0 has to be performed to detect the interference signal. The pixel acquisition time being in the order of 1 µs, the modulation has to be at least twice as fast (>2 MHz). The modulation has to be smaller than the depth discrimination of the objective lens used, which causes a complex frequency mixture in the detected signal that the electronic demodulation has to cope with [28].

6.2.8.2 Set-up for OCM

A fiber-based OCM set-up has been constructed (see **Figure 6.25**). The advantages of a fiber-based solution over a free-space set-up are a greater stability and the availability of phase modulators which fulfil the above requirements.

Light Source

There are two factors influencing the choice of the central wavelength. The higher the wavelength is, the lower the scattering in the media and the better the depth penetration. On the other hand, above about 600 nm absorption of the light, mainly by the water, increases (see **Figure 6.26**). Under this constraint near-infrared light between 800 and 1200 nm represents a good choice. Furthermore, the quantum efficiency and the noise of appropriate detectors have to be considered.

In order to have a short coherence length, the light source needs to have a broad emission spectrum. Supposing a Gaussian emission spectrum, the coherence function has a Gaussian profile and the coherence length L_c is related to the FWHM of the emission spectrum $\Delta\lambda_{FWHM}$ as:

$$L_c = \frac{(2\ln(2)/\pi)^{1/2} \cdot \lambda^2}{\Delta\lambda_{FWHM}} \approx \frac{0.66 \cdot \lambda^2}{\Delta\lambda_{FWHM}}. \tag{6.4}$$

Adequate sources are Ti:Sa femtosecond lasers, ytterbium-doped fiber lasers and super luminescence diodes (SLDs).

We used an SLD at 824 nm with a bandwidth of about 20 nm ($L_c = 22$ µm) well adapted to the avalanche photodiode (APD) with a detection efficiency maximum at 800 nm.

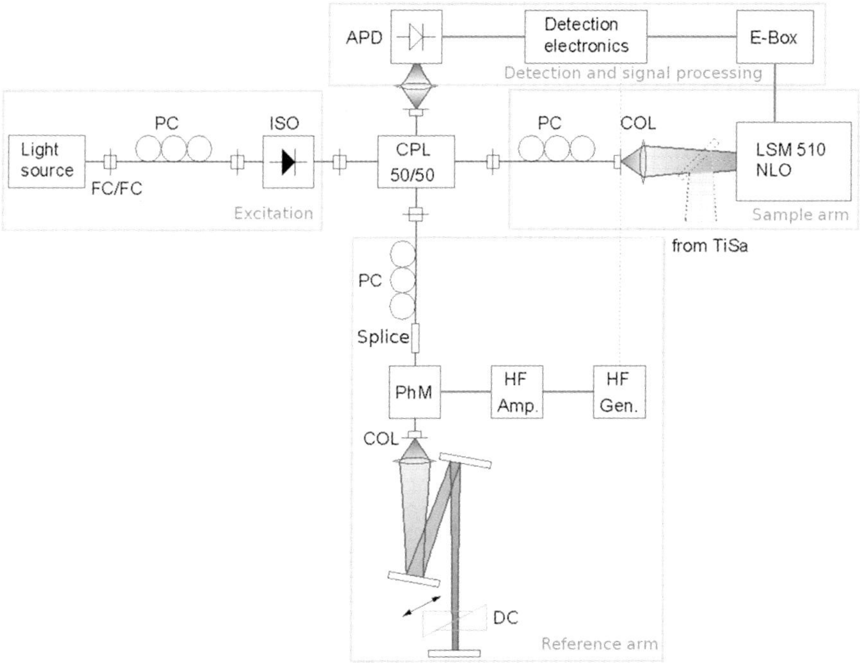

Figure 6.25 Scheme of the realized OCM set-up. All the fiber connectors are FC/APC type. The light source is a super luminescence diode (SLD). PC are polarization controllers, ISO is an isolator to prevent backscattered light from going back to the source, CPL a fused fiber coupler, COL are fiber to free space collimators, PhM is an electro-optic phase modulator, DC a dispersion controller and APD an avalanche photodiode.

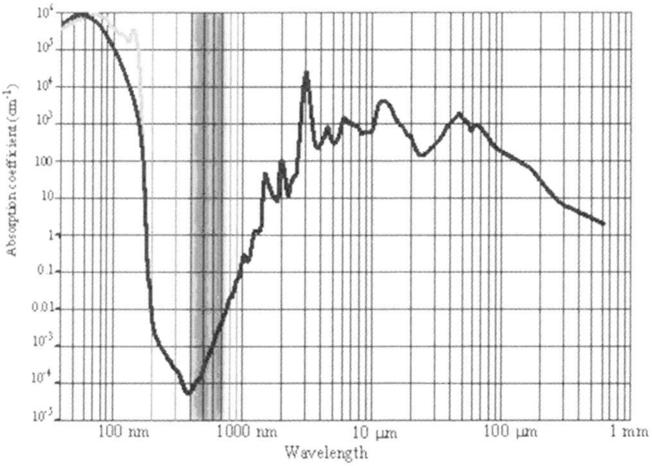

Figure 6.26 Water absorption as a function of wavelength.

Phase Modulator

As described above, the phase modulator has to be able to modulate faster than 2 MHz, so purely mechanical modulators are unsuitable. Acousto-optic or electro-optic modulators are two possible solutions. In an acousto-optic modulator it is difficult to control the amplitude of the phase shift, so a modulator based on the electro-optic Kerr or Pockels effect is an appropriate solution.

Electro-optic modulators that fulfil the requirements are usually integrated in a waveguide. Free space modulators usually require a higher voltage to generate the same electric field to produce a certain change of refractive index Δn. It is difficult to produce an alternating voltage of several hundreds volts at a frequency >2 MHz.

We used a modulation frequency in the range of 100 MHz and a phase shift optimized with regard to harmonics generation [28].

Dispersion

Dispersion is an effect occurring when polychromatic light is travelling through a medium with a wavelength-dependent refractive index $n(\lambda)$. This wavelength-dependent refractive index can be due to the material properties, so-called material dispersion, and in the case of guided light also due to so-called waveguide dispersion.

In OCM using broadband sources, it is important that the dispersion in both arms of the interferometer is the same, so that for each wavelength of the source spectrum the optical path length difference at the point of interference is exactly zero. As only light with the same wavelength can interfere, the coherence function for a system with exactly the same dispersion in both arms equals the coherence function of the light source, and its FWHM amplitude is equal to the coherence length L_C. If the dispersion in both arms is not the same, the coherence function is broadened and its amplitude is reduced (**Figure 6.27**).

Detector and Demodulation Electronics

Because of the high detection efficiency of photodiodes in the near-infrared wavelength range, we chose an avalanche photodiode (APD) as detector. Before being integrated by the LSM electronics according to the pixel dwell time the interference signal was electronically separated from the non-interfering background and demodulated. For this purpose we used a fast quadratic mixer and appropriate filters.

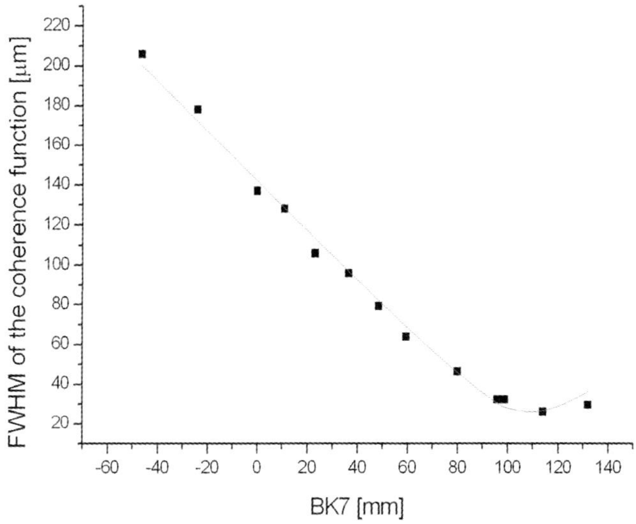

Figure 6.27 Full-width-at-half-maximum of the coherence function obtained by scanning the length of the reference arm using a fixed plane reflector as object in the microscope as a function of the dispersion compensation glass in the reference arm. Objective lens: Plan-Neofluar $20\times/0.5$. Only at the point of zero second order dispersion is the shortest possible coherence gate obtained.

6.2.8.3 **Results**

Tests: Comparison with Confocal Imaging

The optical resolution and depth discrimination are important properties of OCM. For our primary application we needed target objective lenses that provided a good compromise of NA (as high as possible for high resolution, high collection efficiency and large depth discrimination) and large working distance (for large object penetration). A good candidate, with good correction and high transmission from the visible to the near-infrared spectral range, is the water immersion objective Zeiss IR-Achroplan $40\times/0.8$W. **Figure 6.28** shows the lateral resolution and **Figure 6.29** the depth discrimination of the OCM in comparison with confocal LSM.

Images of biological samples in confocal reflection and with OCM reveal similar features but also slight differences that might be due to the different detection discrimination of scattered light (see **Figure 6.30**).

Morphological Referencing

The main application of OCM investigated in this work is the use for morphological referencing of selectively labelled fluorescing samples. Such ref-

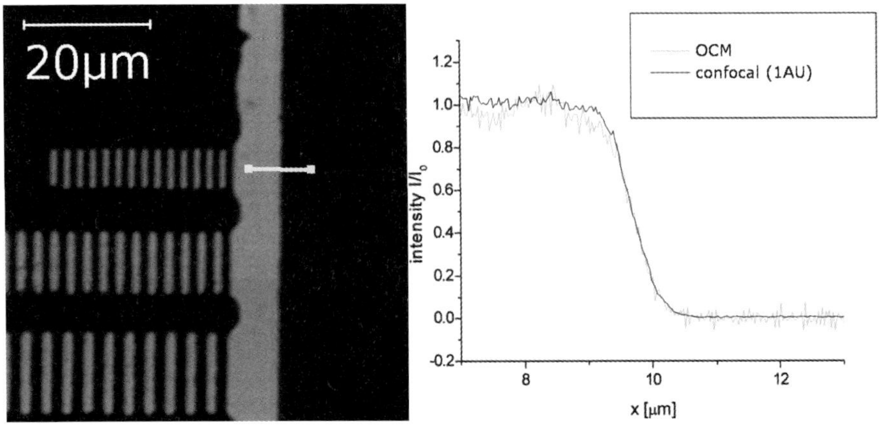

Figure 6.28 OCM image with a Zeiss IR-Achroplan $40\times/0.8$W of a grating structure (left) and intensity distribution along edge indicated in comparison to a confocal reflection image taken with a pinhole of 1 Airy unit (AU) size (right).

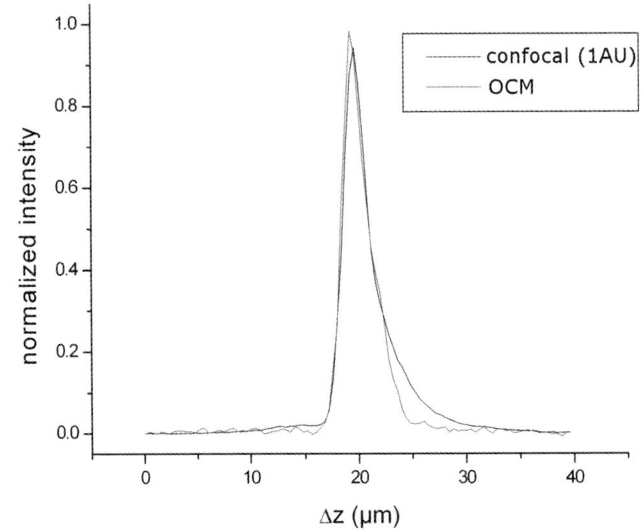

Figure 6.29 Depth discrimination (signal from a plane reflector as function of defocus) of the OCM and the confocal LSM.

erencing allows fluorescence information to be placed in the context of the morphology of the sample, which is particularly relevant for large live (developing) organisms investigated within developmental biology (see **Figure 6.31** and **6.32**). It is obvious that in the case of dynamic changes in the organism (morphology as well as expression of fluorescing proteins) referencing is required for analysis of the complex development.

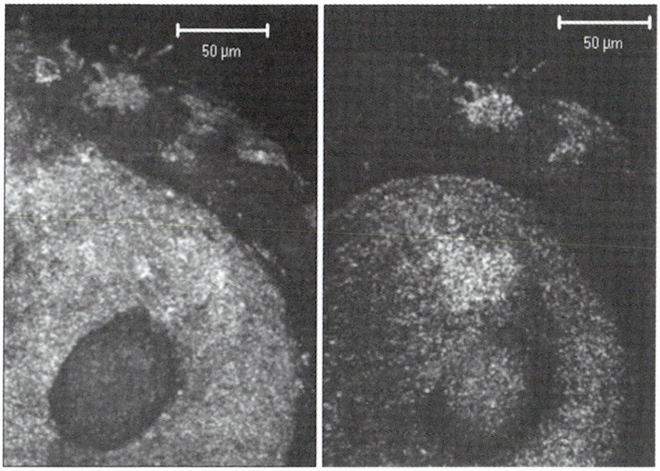

Figure 6.30 Comparison of confocal reflection (left) and OCM images of the eye of a zebra-fish embryo at 824 nm. A projection over a depth of 80 µm is shown. Objective lens: Zeiss IR-Achroplan $40\times/0.8$W.

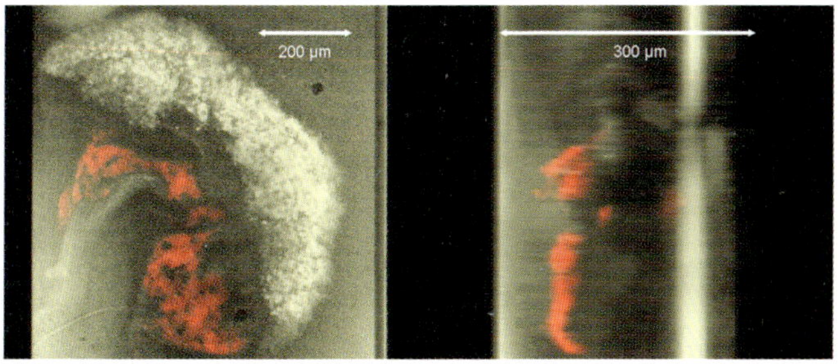

Figure 6.31 Overlayed two-photon fluorescence and reflection image of a mouse embryo. Shown is a projection of stacks from the front (left) and the side (right). Objective lens: Zeiss Plan-Neofluar $10\times/0.45$. Excitation/reflection: 870 nm, fluorescence: < 750 nm (red channel) of GFP labelled red blood cells.

One problem of morphological referencing is the fact that biologists are not familiar with the contrast associated with reflection-based imaging. It is therefore essential to develop the method for specific applications with the potential users, which is a work in progress.

6.2.9 Conclusions

Both DHM and OCM have been set up as microscopic imaging techniques in connection with standard microscopes allowing comparison of the techniques with established contrast mechanisms and to investigate the application benefit of combinations of the techniques available. This evaluation and development is a work in process.

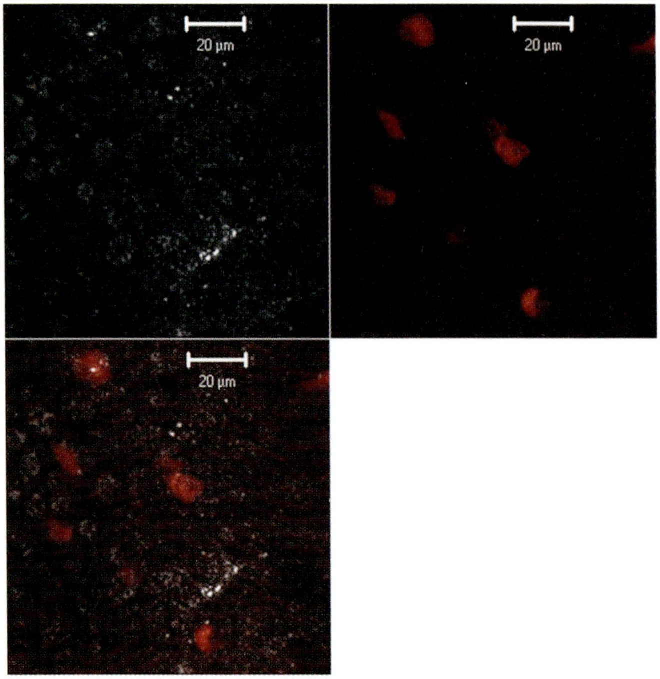

Figure 6.32 Confocal fluorescence and OCM image of a mouse embryo nine days after conception. Objective lens: Zeiss IR-Achroplan 40×/0.8W. Excitation: 488 nm, fluorescence: >505 nm (red channel) of YFP-labelled red blood cells. OCM wavelength: 824 nm.

As demonstrated in Section 6.2.2, DHM produces quantitative phase images with standard lateral and high axial resolution. The established hardware modules and software reconstruction procedures allow a convenient and robust implementation of the technique with reasonable costs and without the need of specially trained personnel.

The strengths of DHM are high resolution of the sample along the optical axis, quantitative phase measurements and the lack of need to stain samples. Nevertheless, the physical parameters obtained have to be translated into relevant biological or physiological results.

For some applications, the exact measurement of surface topology obtained in the DHM reflection mode is already sufficient to provide clear information on biological events. For example, during cell division adherent cells round up and mitotic cells can then be identified by their thickness, which allows an easy determination of cell division activity, i.e., the proliferation rate.

Interpretation of transmission measurements requires further information of the investigated specimen, e.g., the refractive index. This is based on the fact that biological samples like cells or tissues are complex composites of different compartments and soft materials. As illustrated by the comparison of confocal fluorescence and DHM images (see Section 6.3.3) the 3D topology of a biological specimen cannot be deduced directly from DHM images. For these cases it appears plausible to regard DHM as an additional imaging channel that provides complementary information. Indeed, the combination of DHM reflection and transmission modes provides already the option to separate changes due to the refractive index from changes due to the morphological structure of the specimen.

Overall, the study indicates that the highest potential for biomedical applications is in the field of live-cell imaging when variations or changes in structure have to be analyzed with high sensitivity and high temporal resolution and when minimal sample preparation is desirable.

An OCM set-up has been constructed that allows imaging in reflection mode simultaneously to fluorescence imaging including single- and two-photon excitation. It has been shown that morphological information based on label-free reflection imaging is valuable for referencing fluorescence images and allows the labelling of samples with exogeneous chromophores or fluorescing proteins to be reduced.

The technical realization of OCM on a laser-scanning microscope for simultaneous image acquisition turned out to be a challenging task. Given the proven technical advantages of OCM on one hand and the restricted capabilities of confocal imaging deep in scattering samples on the other, it is to be expected that OCM can hold its promise in such key applications as the developmental biology of model organisms.

6.3 Minimally Invasive Holographic Endoscopy

6.3.1 Background and Motivation

Endoscopy is a widespread intracavity observation technique, routinely used for industrial inspection as well as in medical minimally invasive diagnosis [30]. One of the current limitations of endoscopic imaging is the absence of tactile perception, preventing the proper assessment of elasticity during routine endoscopic examinations.

The combination of endoscopic imaging with holographic interferometric metrology has allowed the development of new tools for nondestructive quantitative detection of defects within cavities, including the analysis of shape, displacements and vibrations [31–34]. Furthermore, in combination with techniques for endoscopic reproducible stimulation of the investigated specimen new perspectives in minimally invasive diagnostics are opened up.

Thus, measurements of elasticity can give information about structural differences, even below the visible tissue surface. This information is of particular relevance to early recognition of, for example, cancer-induced tumors. Additionally, in this way the loss of subjective tactile sense impression in endoscopic surgery is compensated by objective optical (visual) information ("endoscopic taction").

Modern medical diagnostic methods require online monitoring and rapid process analysis. Here, methods of digital holographic interferometry based on digital image recording devices (e.g., CCD and CMOS cameras) and digital image processing offer, in addition to quantitative information about displacements, rapid data acquisition and data evaluation with only low demands on interferometric stability. In this way, contrary to earlier attempts at holographic endoscopy, in which single holographic interferograms were recorded with pulsed lasers [35], surface deformations can be displayed online with a repetition rate up to video frequency.

This section describes a system for minimally invasive endoscopic digital holographic interferometry combined with reproducible tissue stimulation. Silicone tissue models with areas of locally decreased elasticity are investigated using different methods of tissue stimulation to determine the detection limit of the method. Results of *in vitro* investigations on a biological specimen demonstrate the method's applicability in the field of gastroenterology.

6.3.2 Modular System for Digital Holographic Endoscopy

6.3.2.1 Principle

Figure 6.33 illustrates the set-up for a modular proximal holographic interferometric system for endoscopic displacement measurement by determination of optical path length changes due to displacements or surface deformations of the investigated object. The holographic interferometric unit is positioned outside the cavity. The advantage of this arrangement is that standard (full frame) CCD cameras (e.g., IEEE1394 standard) can be used for data acquisition. Furthermore, a connection with various flexible or rigid standard endoscopes is possible. The coherent laser light source is a cw laser (e.g., an argon-ion laser: $\lambda = 514.5$ nm or a frequency doubled Nd:YAG laser, $\lambda = 532$ nm). The laser beam is divided into an object illumination wave and a reference wave, which are both coupled into single mode optical fibers. Object illumination is performed by connecting the illumining fiber to the white light illumination system of standard endoscope optics or optionally by insertion of a separate illumination fiber with a beam expanding system into a side channel of the endoscope. Spatial phase shifting (SPS) [36, 37] is applied for quantitative evaluation of the phase differences that are effected by optical

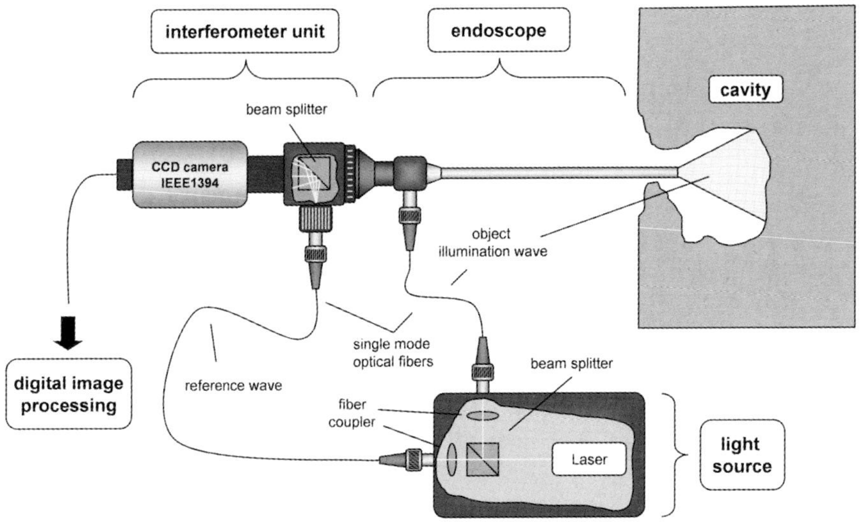

Figure 6.33 Concept of a modular proximal digital holographic endoscopy system.

path length changes due to displacements/surface deformations of the investigated object (for description in detail see Section 6.3.3). For this reason an interference pattern between object wave and reference wave is generated by positioning the end of the reference wave fiber off the optical axis of the interferometer. Thus, a nearly constant spatial phase gradient of the phase difference between the wave fronts of object wave and reference wave is generated in the image plane [37]. The resulting interference patterns are recorded by a progressive-scan (full frame) CCD camera with IEEE1394 interface that is connected to a conventional personal computer (e.g., a notebook computer) that allows the intensity patterns to be recorded. Simultaneously, the corresponding phase difference data and correlation patterns obtained by image subtraction are displayed on a monitor [31]. In this way, a stroboscopic visualization of tissue surface movements up to video repetition rate is possible.

Figure 6.34 (left) shows a photograph of a modular digital holographic endoscopy system that was developed at the Laboratory of Biophysics, Muenster, Germany in cooperation with the company Karl Storz GmbH & Co. KG, Tuttlingen, Germany. The system consists of a laser unit, a holographic interferometric unit and a notebook computer for image processing. Endoscopic imaging is performed by a laparoscope (Karl Storz, GmbH & Co. KG). The coherent light source is a frequency doubled Nd:YAG-laser (λ = 532 nm, P_{max} = 100 mW). The interferograms of different object states are recorded by a progressive scan color CCD sensor with IEEE1394 interface (Sony DFW

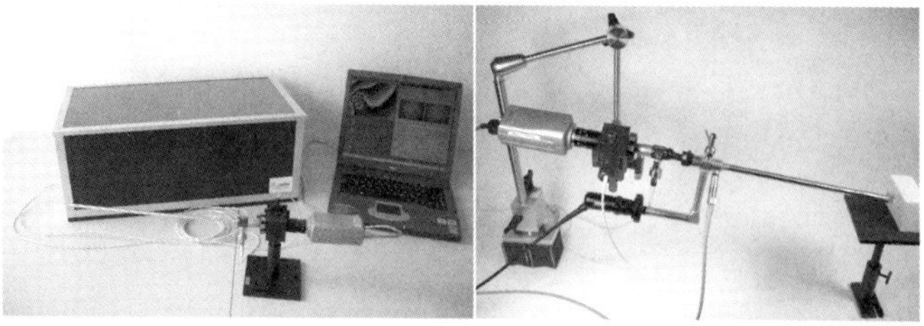

Figure 6.34 Left: Modular digital holographic endoscopy system consisting of a laser unit, a holographic interferometric module and a notebook computer for image processing. Right: Modular digital holographic endoscopy system connected to an endoscopic carrier system.

X700, resolution: 1024×768 pixels). **Figure 6.34** (right) shows the digital holographic endoscopy system connected to an endoscopic carrier system with a conventional color image acquisition system.

6.3.3 Evaluation of Spatial Phase-shifted Interferograms

Investigations on biological specimens require a method for quantitative determination of the phase differences that works even under unstable conditions. For this reason, a spatial phase-shifting (SPS) method [37] is applied for sign-correct phase difference determination that requires only a single interferogram to determine the phase distribution of one object state. Another advantage of this technique compared with other phase-shifting procedures is that neither movable parts (e.g., piezo translators) in the optical path nor additional electronic synchronization are necessary.

To achieve SPS in combination with endoscope optics, the speckle image of the investigated object surface is superimposed in the image plane with spatial carrier fringes, e.g., with vertical or horizontal orientation. This is achieved by an adequately chosen, almost constant spatial phase gradient β, which is generated by a tilt between object wave and reference wave. In digital holography, the phase distribution ϕ of speckle patterns can be determined by various methods [2]. Here, a three-step algorithm is applied that can be variably adjusted for different phase gradients β and that uses three intensities I_{k-1}, I_k, I_{k+1} recorded on three neighboring CCD pixels:

$$\phi_k + k\beta = \arctan\left(\frac{1-\cos\beta}{\sin\beta}\frac{I_{k-1}-I_{k+1}}{2I_k - I_{k-1} - I_{k+1}}\right) \pmod{2\pi}. \tag{6.5}$$

For correct phase evaluation the algorithm requires that the mean speckle size is at least the size of three pixels of the digitized interferograms. This in general is fulfilled in this case by using endoscope optics with a small aperture. Adjustment of the phase gradient β is performed by analysis of the 2D frequency spectrum of the carrier fringe pattern in the interferograms by 2D digital fast Fourier transform (FFT) [37]. The phase difference $\Delta\phi$, which is affected by different motion states of the object, is calculated by subtraction of two phase distributions ϕ, ϕ' mod 2π:

$$\Delta\phi_k = (\phi_k' + k\beta) - (\phi_k + k\beta) = \phi_k' - \phi_k. \tag{6.6}$$

Figure 6.35 illustrates the evaluation process of spatial phase-shifted interferograms for displacement detection. **Figures 6.35a** and **6.35b** show the spatial phase-shifted interferograms I, I' obtained from two different displacement states of a tilted metal plate. **Figures 6.35c** and **6.35d** show the corresponding phase distributions ϕ, ϕ' mod 2π, calculated by Eq. (6.5) from **Figures 6.35a** and **6.35b**. The resulting phase difference distribution $\Delta\phi$ mod 2π obtained by subtraction of **Figures 6.35c** and **6.35d** mod 2π and the corresponding filtered phase difference distribution are shown in **Figures 6.35e** and **6.35f**. **Figures 6.35g** and **6.35h** finally represent the unwrapped phase difference after removal of the 2π ambiguity and the related pseudo 3D representation of the data. From the data in **Figure 6.35g** the underlying displacement of the plate is calculated by taking into account the recording and imaging geometry of the experimental set-up [2].

6.3.4
Results

Investigations on silicone tissue models with areas of locally decreased elasticity have been carried out by application of digital holographic endoscopy combined with different methods of tissue stimulation in order to determine the detection limit of the method.

For the experiments, silicone tissue models were designed in such a way that their optical properties in relation to digital holography are comparable to those of biological tissue. Furthermore, the elastic properties of homogeneous muscle tissue were simulated. For a systematic investigation to determine the detection range of a digital holographic endoscopy system, impurities of different material, size and depth were implanted in the specimen. **Figure 6.36** shows for illustration three silicone specimens with different embedded objects (upper left: silicone sphere, upper right: silicone octahedron, lower right: metal sphere) in comparison to a homogeneous reference tissue model that has been investigated during the experiments.

The detection of abnormality affected differences in tissue elasticity by digital holographic endoscopy requires a reproducible endoscopic tissue stimulation. For this reason, an adapted pneumatic tube (see **Figure 6.37**) and an ultrasound tube (Karl Storz GmbH & Co. KG) were used.

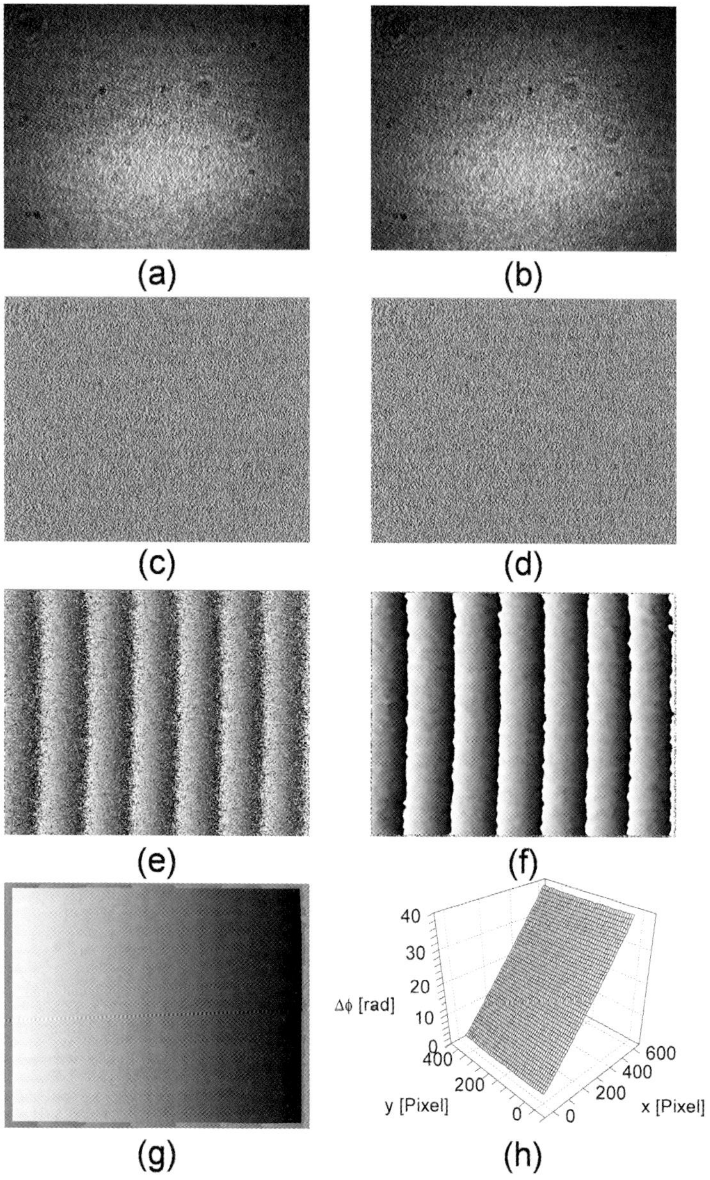

Figure 6.35 Evaluation of spatial phase-shifted interferograms. (a), (b): spatial phase-shifted interferograms I, I' of two displacement states of a tilted metal plate; (c), (d): phase distributions ϕ, ϕ' $(\mathrm{mod}\, 2\pi)$, calculated by Eq. (6.5) from (a) and (b); (e), (f): raw and filtered phase difference distribution $\Delta\phi \,\mathrm{mod}\, 2\pi$ obtained by subtraction of (c) and (d) $\mathrm{mod}\, 2\pi$; (g), (h): unwrapped phase difference $\Delta\phi$ and corresponding pseudo 3D representation of $\Delta\phi$.

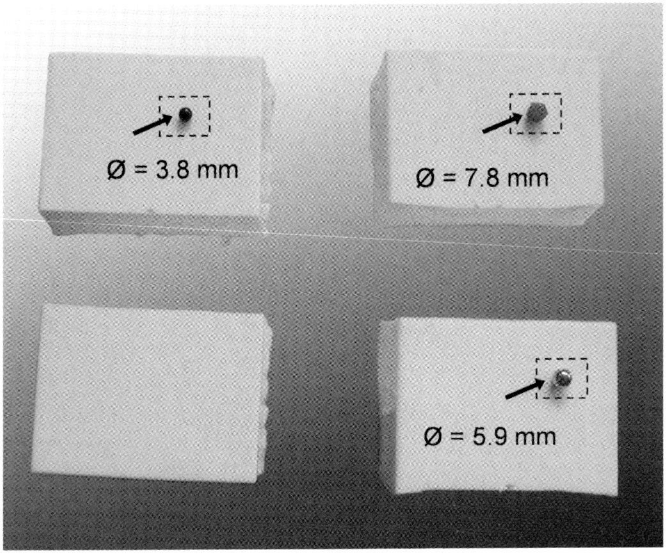

Figure 6.36 Silicone tissue models with different embedded objects (upper left: silicone sphere, upper right: silicone octahedron, lower right: metal sphere) in comparison to a homogeneous reference tissue model (lower left).

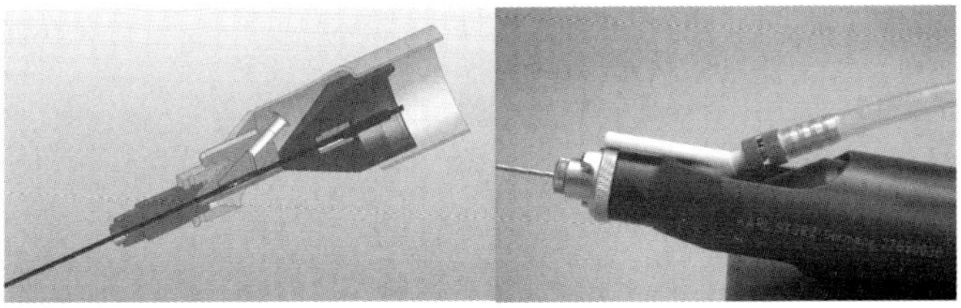

Figure 6.37 Pneumatic tube for minimally invasive endoscopic tissue aspiration. Left: sketch of the pneumatic tube. Right: pneumatic tube system built by Karl Storz GmbH & Co. KG (diameter of the tip ≈ 1.5 mm).

Figure 6.38 shows results from investigations on silicone tissue models with different elasticity distributions that are obtained with the digital holographic endoscopy system depicted in **Figure 6.34** (right). For the experiment, the tissue is stimulated by an endoscopic pneumatic tube (see **Figure 6.37**) (single aspiration, amplitude: $\approx$ 500 µm). Afterwards, during the relaxation process of the tissue, a series of stroboscopic phase difference distributions are recorded. **Figures 6.38a**, **6.38b** and **6.38c** show results obtained from investigations on a homogeneous silicone model compared with those from a model with an embedded metal sphere (circle in **Figure 6.38d**: $\varnothing = 3$ mm, depth = 1.1 mm) below the visible surface (see **Figures 6.38d**, **6.38e** and **6.38f**). For

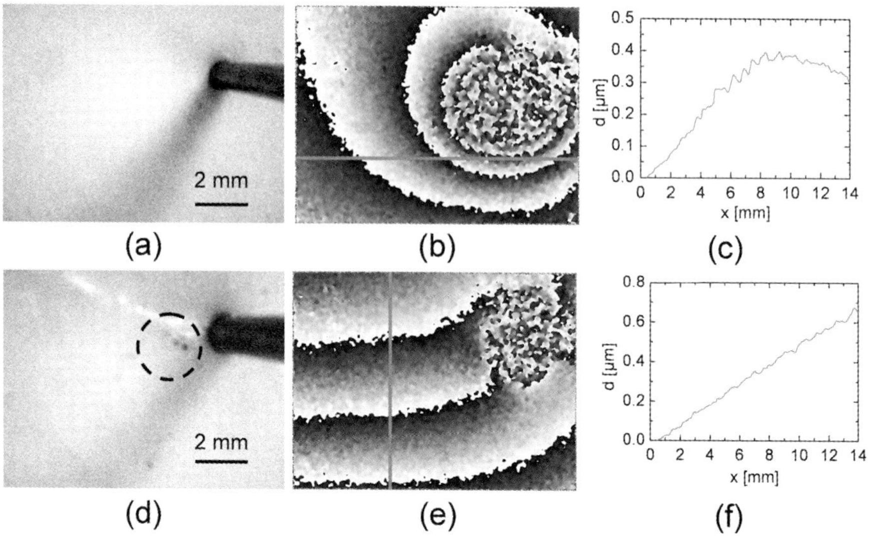

Figure 6.38 Investigations on silicone tissue models. (a), (d): White light images; (b), (e) phase difference distributions $\mathrm{mod}\, 2\pi$ (stimulation with piezo translator); (c), (f) pseudo 3D plots of the displacement along the marked cross-sections in (b) and (e). The position of the disturbance is marked by the dashed circle in (d).

the homogeneous reference model, concentric phase difference fringes around the center of elongation are observed (**Figure 6.38b**), while stimulation of the tissue model with a disturbance results in distributions of parallel fringes (**Figure 6.38e**). The corresponding calculated surface deformation along the cross-sections in **Figures 6.38b** and **6.38e** that are plotted in **Figures 6.38c** and **6.38f** visualize the response of the tissue quantitatively.

Table 6.1 shows results from systematic investigations on silicone tissue models with disturbances of different materials and size to determine the detection range of digital holographic endoscopy. The values obtained show

Table 6.1 Detection limit of digital holographic endoscopy for investigations on silicone tissue models with disturbances of different materials and size.

Disturbance	Diameter ∅ (mm)	Detection limit [depth (mm)]
Metal sphere	1.3	1
	3	2.5
	6	5.8
Silicone sphere	1.3	–
	3	1.4
	6	3.8

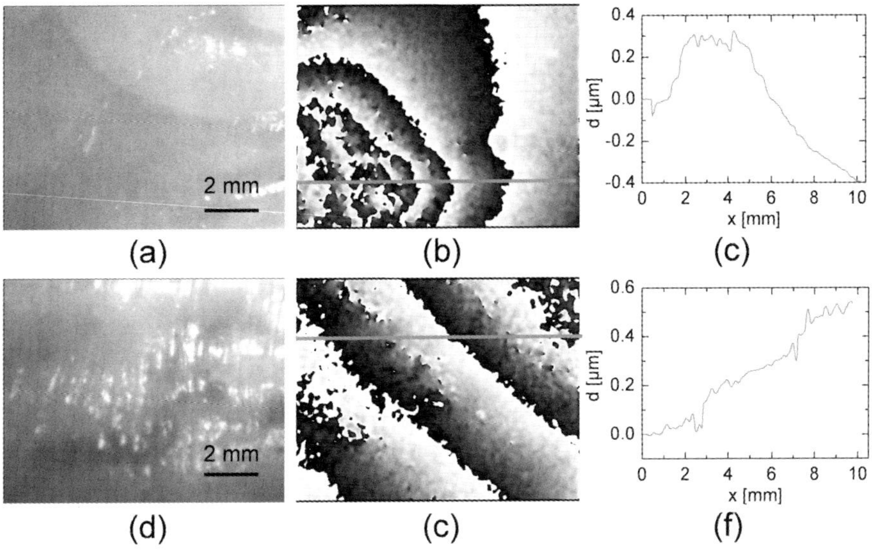

Figure 6.39 *In vitro* investigations on a human intestinal specimen by stimulation with an endoscopic ultrasound tube. (a), (d): White light images of a tissue part without pathological findings and of the tumorous tissue; (b), (e): phase difference distributions $(\bmod 2\pi)$; (c), (f) pseudo 3D plots of the displacement along the marked cross-sections in (b) and (e).

that disturbances down to a diameter of 1.3 mm and up to a depth of 5.8 mm below the visible surface are detected by elasticity differences.

In the next experimental step, *in vitro* investigations on intestinal specimens with carcinoma tissue were carried out. For the experiment, the tissue was stimulated by an endoscopic ultrasonic tube (single pulse, amplitude $\approx 200\ \mu m$). Afterwards, during the relaxation process of the tissue, a series of stroboscopic phase difference distributions were captured. **Figure 6.39** depicts characteristic results. **Figures 6.39a** and **6.39d** show the white light images of an investigated area of the specimen without pathological findings compared with a tumorous part of the tissue. **Figures 6.39b** and **6.39e** show typical results of phase difference distributions mod 2π obtained during the relaxation process of the tissue. The corresponding calculated surface deformation along the cross-sections in **Figures 6.39b** and **6.39e** is plotted in **Figures 6.39c** and **6.39f**.

Similar to the results obtained from investigations on the silicone tissue models, the phase difference shows concentric fringes for the tissue part without pathological findings (**Figure 6.39b**) and a parallel fringe distribution for the hardened tumorous tissue (**Figure 6.39e**). Thus, the tissue part without pathological findings can be distinguished from the tumorous tissue with diminished elasticity.

6.3.5
Conclusion

The results of the investigations on silicone tissue models with areas of diminished elasticity demonstrate the detection of disturbances with a diameter as low as 1.3 mm. In this way the method opens up new perspectives for the detection of tiny tumors underneath the visible surface of tissue by elasticity differences which is, for example, of particular relevance for recognition of cancer at a very early stage. Furthermore, the results of *in vitro* investigations of an intestinal specimen show that digital holographic endoscopy combined with a reproducible stimulation technique is applicable to the detection of elasticity differences in biological tissue.

6.4
Ultrasensitive Interference Spectroscopy for Marker-free Biosensor Technology (for Molecular Genetics)

6.4.1
Background and Motivation

Specific interactions of proteins play a key role in cellular information transport and in the control of cellular processes. Molecular genetics is mainly concerned with DNA/RNA–protein interactions. These interactions can be studied with affinity biosensors that use one of the interacting molecules as a receptor, which is immobilized on a transducer. As a central component of a biosensor, the transducer converts the specific activity of the receptor to a corresponding change in a physical quantity. In contrast to catalytic biosensors, most affinity biosensors contain an optical transducer. Applied optical systems and methods are surface plasmon resonance (SPR), grating couplers, gratings, ellipsometry, interferometers containing waveguides, resonant mirrors and reflectometry. For a review of optical transducers see Ref. [38].

In the case of an affinity biosensor the affinity reaction between the receptor and a target molecule is detected. Accumulation of target molecules at the transducer leads to a small change in the refractive index of the transducer or the neighboring medium. This in turn results in a change in an optical quantity of the transducer, such as plasmon resonance frequency, coupling efficiency, etc.

In interference spectroscopy, a transmission or reflection spectrum of a layer or a system of layers is analyzed to measure its optical properties. (It should be noted that methods such as Fabry–Pérot spectroscopy or FTIR spectroscopy, in which spectroscopy is done using an interferometer, are also called interference spectroscopy.) Applied to biosensors, a simple interference layer on

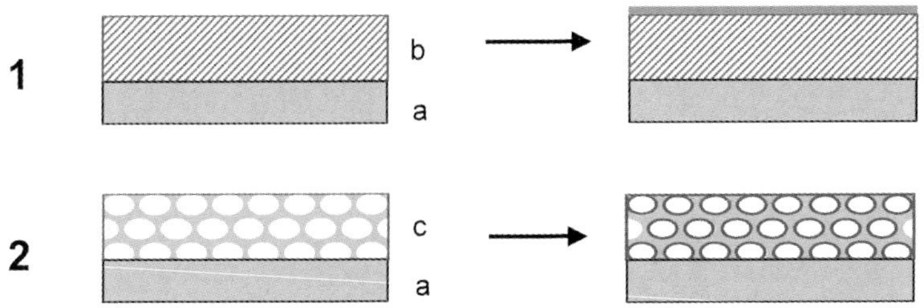

Figure 6.40 Transducers for interference spectroscopy. 1: An interference layer (b) on a substrate (a). Accumulation of target molecules on the surface of the interference layer increases the effective optical thickness ($n\,d$) of the layer. 2: A porous layer (c) on a substrate (a). Accumulation of target molecules on the pore surfaces inside the porous layer leads to a change in the effective refractive index n of the porous layer.

a substrate is chosen as a transducer. If multiple-beam interferences are neglected, the reflection spectrum of a single layer can be approximated by

$$R(\lambda) \approx I(\lambda)\left[1 + \gamma(\lambda)\cos\left(\frac{4\pi}{\lambda}n(\lambda)d\right)\right] \tag{6.7}$$

(γ describes contrast, d denotes layer thickness).

Obviously, one can determine the optical thickness $n \cdot d$ of the transducer, which is changed by accumulation of target molecules on the transducer by the binding process between receptor molecules and target molecules. The resulting method is also called reflectometric interference spectroscopy (RIfS) [39, 40]. The binding process can be monitored in real time, but the determination of rate constants is often hampered by mass transport of target molecules to the transducer. For a discussion of the evaluation of rate constants see Ref. [41]. Further applications of an affinity sensor are affinity ranking of reaction partners, the search for inhibitors of the binding process and selective detection of an analyte.

Non-specific binding of the target molecule or other substances with the transducer cannot be discriminated from the target–receptor binding under investigation. Therefore it is very important to minimize non-specific binding by chemical modification of the transducer surface, particularly if the sensor is used for specific detection of an analyte in a complex sample like a cell extract.

For reflectometric interference spectroscopy two types of transducers are used as shown in **Figure 6.40**.

1. The surface of a transparent interference layer on a substrate is functionalized with receptor molecules. Target molecules binding with the receptor molecules give rise to a small coating on the layer surface, which increases the optical thickness ($n\,d$) of the interfering layer [39, 40]. The measured optical thickness increase $\Delta(n\,d)_{\text{eff}}$ of the interference layer depends on the refractive indices of the medium (solvent) in contact with

the sensor (n_0), the coating due to accumulation of target molecules (n_1), and the interference layer (n_2). For a thin coating $n_1 d_1 \ll \lambda$ one can show that

$$\Delta(nd)_{\text{eff}} = \Delta(n_1 d_1)\eta$$

where

$$\eta = \frac{r_{01}}{r_{01} + t_{01} r_{12} t_{10}} \approx \frac{r_{01}}{r_{01} + r_{12}} = \frac{\left(\frac{n_0 - n_1}{n_0 + n_1}\right)}{\left(\frac{n_0 - n_1}{n_0 + n_1}\right) + \left(\frac{n_1 - n_2}{n_1 + n_2}\right)} \tag{6.8}$$

(r and t reflection and transmission factor, respectively).

2. Alternatively the transducer consists of a porous interference layer on a substrate. The porous layer is permeable to solvents applied to the surface of the layer so that target molecules can bind with receptor molecules that are immobilized on the pore surfaces inside the layer. If pore diameters are small compared to applied wavelengths, light is not scattered at the pores so that the porous layer is optically homogeneous, with a refractive index given by effective-medium theory. Accumulation of target molecules on the pore surfaces changes the refractive index of the porous layer corresponding to a change ($\Delta n\, d$) of the optical thickness. As the total pore surface of the layer can be orders of magnitude larger than the layer surface the transducer may contain many more receptor molecules per sensor area than the first type of transducer described above. A corresponding enhancement of sensitivity is possible as discussed below.

6.4.2 Sensitivity of Interference Spectroscopy

We consider a small change $\Delta(n\, d)$ in the optical thickness of the transducer.

If noise in the spectral data is uncorrelated one can show that a linear combination of the data

$$Y := \sum_i b_i R(\lambda_i) \tag{6.9}$$

has a maximum signal-to-noise ratio $\mathrm{S/N} = \Delta Y / \sigma(Y)$ for detection of $\Delta(n\, d)$ if the coefficients b_i are given by

$$b_i = \frac{\left(\frac{\partial R_i}{\partial (n\, d)}\right)}{\sigma^2(R_i)} \tag{6.10}$$

where $R_i := R(\lambda_i)$and $\sigma(R_i)$ is the standard deviation of R_i due to noise.

In this case the signal-to-noise ratio is given by

$$(\mathrm{S/N})^2 = \Delta(n\,d)^2 \sum_i \frac{1}{\sigma^2(R_i)} \left(\frac{\partial R_i}{\partial(n\,d)} \right)^2 \tag{6.11}$$

where the term of the sum is the squared signal-to-noise ratio if a single value R_i is used for detection.

Of course Eqs. (6.9) and (6.10) are useful for very small phase shifts only. For larger values of $\Delta(n\,d)$ a nonlinear fit with a model function like Eq. (6.7) where I, γ, and n are polynomials in λ can be used to estimate $(n\,d)$. This method is more stable concerning variations of light-source intensity. The linear method using Eqs. (6.9) and (6.10) can be used to test if a fitting algorithm introduces some additional noise in the determination of $(n\,d)$. We found that a Levenberg–Marquart algorithm for performing the fit results in the same signal-to-noise ratio as the linear method.

In accordance with other investigations experiments with a commercial spectrometer (TRIAX 320, Jobin Yvon) resulted in a sensitivity of $\Delta(n\,d)/(n\,d) \approx 10^{-6}$ which is comparable to the sensitivity of SPR-sensors (e.g., Plasmon SPR sensor, BioTul AG, Germany).

If a fixed number of target molecules per sensor area A_{sens} is applied to the sensor, an estimate shows that the two types of transducer mentioned above have a comparable limit of detection. Assuming for simplicity that the accumulation of target molecules on the transducer surface leads to a build-up of a layer with constant mass density, the thickness of this layer is $V_{\mathrm{tar}}/A_{\mathrm{sens}}$ where V_{tar} is the total volume occupied by the target molecules.

Using Eq. (6.8) the increase in optical thickness of the transducer is

$$\Delta(n\,d)_{\mathrm{eff}} = \eta\, n_1 \frac{V_{\mathrm{tar}}}{A_{\mathrm{sens}}} \tag{6.12}$$

For the porous transducer of type 2 the volume V_{tar} of target molecules is distributed among the pores. As this volume substitutes the solvent inside the pores the change in effective refractive index of the transducer is approximately

$$\Delta n \approx (n_1 - n_0) \frac{V_{\mathrm{tar}}}{v_{\mathrm{transd}}} \tag{6.13}$$

where V_{transd} is the volume of the transducer. The resulting change in optical thickness of the transducer is

$$\Delta(n\,d) = d\,\Delta n \approx (n_1 - n_0) \frac{V_{\mathrm{tar}}}{V_{\mathrm{transd}}} d = (n_1 - n_0) \frac{V_{\mathrm{tar}}}{A_{\mathrm{sens}}} \tag{6.14}$$

In most cases $\eta < 1$ so that Eqs. (6.12) and (6.14) show that both transducer types have a similar response and thereby a similar limit of detection in this case. However, if a solution with a fixed concentration of target molecules is applied which is not depleted by the sensor, the number of bound target molecules is proportional to the number of receptor molecules. In this case a porous transducer can bind substantially more target molecules because more receptor molecules can be immobilized on the large total pore surface area. This is a distinct advantage if the binding constant of the affinity reaction under study is small or target concentrations are low.

6.4.3
Porous Silicon for Affinity Sensors

Electrochemical etching of a silicon wafer results in a layer of porous silicon for a limited range of current densities. The properties of the layer – porosity, pore size, and pore morphology – vary in a wide range depending on the preparation parameters (dopant, doping density, electrolyte, current density) [42]. There are two properties that make porous silicon layers attractive for application in biosensors: The pore size can be tuned from one nanometer up to ten micrometers [42, 43], and etched under proper conditions the layers are homogeneous and transparent at visible and NIR wavelengths so they can serve as interference layers for a transducer [43–46].

The total pore surface area A_{pore} depends on the porosity p of the transducer material and the typical diameter D_{pore} of the pores:

$$\frac{A_{\text{pore}}}{V_{\text{transd}}} \sim \frac{p}{D_{\text{pore}}}. \tag{6.15}$$

The surface area of porous silicon with pore diameters in the range of 10 to 20 nanometers amounts to about 100 m^2/cm^3. Compared to the initial smooth surface of the Si wafer, the surface area is enlarged by a factor of 500 for a layer with a thickness of 5 μm. The surface area is limited because porous silicon with very high porosity (approximately $p > 0.7$) is mechanically unstable. More importantly, as the molecules under investigation have to penetrate the porous transducer, there is a lower limit for the pore diameter depending on the application. That is, for large molecules like proteins, porous silicon with a larger pore size has to be used so that sensitivity is reduced. **Figure 6.41** shows two samples of porous silicon with different pore sizes. Penetration of proteins into porous silicon also depends on protein conformation, surface charges, the isoelectric points and the pH value.

If the current density is varied during the etch process, porous silicon with stratified layers of varying refractive index is produced. With these multilayers, interference filters and Bragg gratings can be constructed [47]. To improve the sensitivity of interference spectroscopy, multilayers which represent a microcavity with two Bragg mirrors interspersed with a spacer layer have been used as transducers [44]. Using Eq. (6.5) it can be shown that the signal-to-

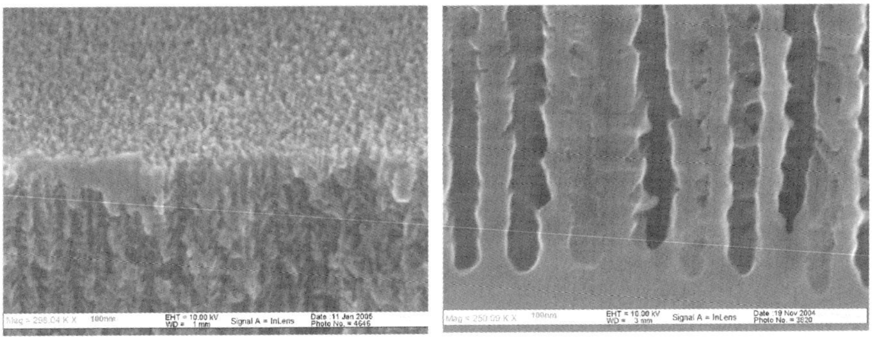

Figure 6.41 Left: Cross-section (lower half) and surface of porous silicon made of p*-Si with pore diameters in the range from $D_{\text{pore}} \approx 10$–$20$ nm. Right: Cross-section of n-type porous silicon, $D_{\text{pore}} \approx 50$–$80$ nm.

noise ratio is proportional to the square root of the finesse of the resonator (defined as the ratio of mode linewidth to mode spacing). Porous silicon microcavities in published investigations and those made in this study had a limited finesse of values about six to eight, so that the gain in sensitivity is not very high. The finesse is limited by absorption and lateral inhomogeneities of the multilayers. Moreover, at the current state-of-the-art the sensitivity of biosensors is mainly determined by drift effects and non-specific binding so that interference spectroscopy on a single porous layer is sufficiently sensitive.

In its native state the hydrogen-terminated surface of porous silicon is hydrophobic. As biological processes take place in aqueous solutions, the silicon surface must be made hydrophilic to allow filling of the pores. In addition, silicon is unstable in aqueous solutions and shows slow corrosion, leading to a drift of the sensor signal. **Figure 6.42** shows a functionalization of the porous silicon surface. At first, the surface is oxygenated by ozone whereby hydrogen is substituted by oxygen and OH groups. At the OH groups the surface is modified with aminosilane derivatives. In addition to the oxygenation process the silanization further reduces corrosion. Finally, the aminosilane derivatives are functionalized with biotin-*N*-hydroxysuccinimide ester so that biotin molecules as receptors are covalently linked to the surface by an oligomeric spacer. The spacer length has to be optimized to preserve the specific interaction of the biotin molecule with target molecules. A measurement of the specific interaction of biotin and avidin is shown in **Figure 6.43** where the refractive index of the porous transducer versus time is displayed.

As expected, the binding of avidin leads to an increase of the refractive index of the porous layer. Non-specific binding of bovine serum albumin (BSA), albeit with a smaller concentration, is also detected. Although reduced by the silanization, the corrosion of silicon appears as a distinct drift of the detec-

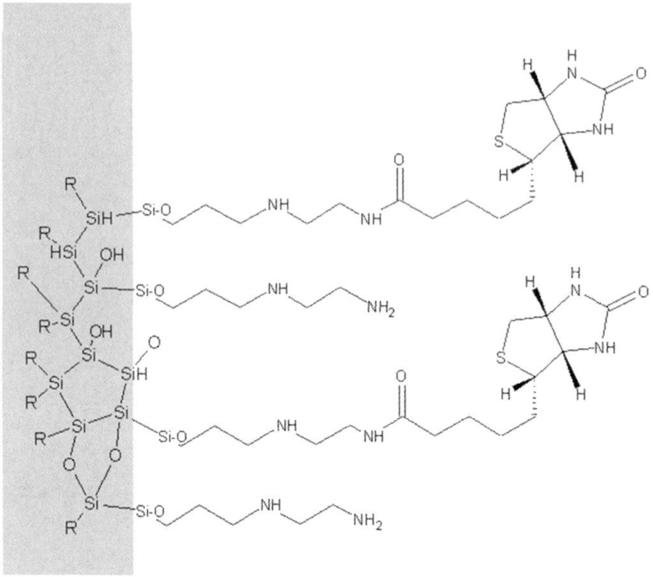

Figure 6.42 Functionalization of a Si-Surface: Immobilization of biotin with aminosilane derivatives.

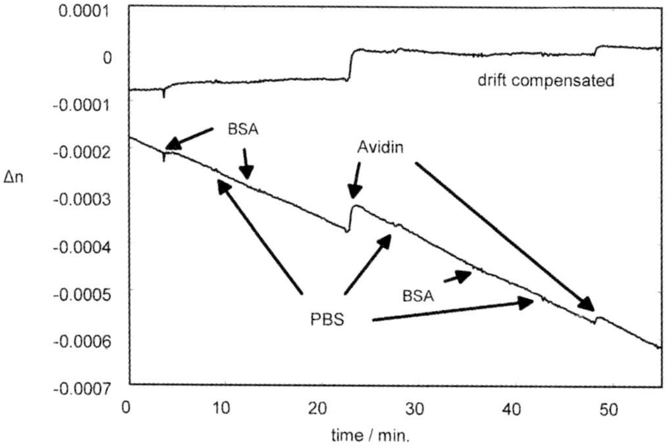

Figure 6.43 Specific binding of avidin (0.12 mg ml^{-1} in PBS buffer) to biotin immobilized on porous silicon. A small unspecific binding of BSA (1 mg ml^{-1}) is also detected. The signal drift is due to slow dissolution of the layer. Floating with pure PBS shows that the reaction is irreversible.

tor signal. Experiments show that the corrosion rate changes slightly during loading cycles of the sensor. Therefore the signal drift cannot be compensated, leading to uncertainties and even misinterpretations. To overcome these errors the chemical stability of the modified silicon surface has to be improved. A further possibility not yet examined for porous silicon biosensors is referencing with a two-beam configuration where two transducers or two positions on a single transducer are measured simultaneously. If the transducers differ only in the target molecule, which is omitted in one beam, compensation of temperature, buffer changes and non-specific binding should be possible in addition to drift compensation.

6.4.4 Conclusions

Interference spectroscopy has proved to be a reliable and sensitive method for optical biosensors. For most applications, sensor performance is limited by drift and non-specific binding, as is the case for porous silicon transducers. The large surface area of porous silicon and the possibility of tuning its properties makes it an attractive material for transducers, although it has some drawbacks in chemical stability and reproducibility.

6.5 Outlook

The techniques presented provide the basis for intelligent "label-free" 3D sensing. Furthermore, they open the way to development of a multifunctional sensor technology, combining the methods presented with other conventional and innovative techniques. Of special importance is the future development of functional imaging techniques for 3D live-cell analysis within tissue samples by combination with tomography and endoscopy ("the tumor cell in its environment"). Equally attractive is the development of multifunctional systems for high-throughput and high-content analysis. By such a "convergence of technologies" these interferometric techniques create new biophotonic tools for marker-free live-cell analysis and automated optical micromanipulation. In addition, the presented techniques and demonstrations open up basic technologies to foreseeable new application fields, such as robotic medicine, telemedicine (telehistopathology), image-supported remote control surgery, etc. This can result in an important contribution to an integration of micro-, nano- and biotechnologies.

Acknowledgements

Financial support by the German Federal Ministry for Education and Research (BMBF) within the funding program "Biophotonics" (joint research project MIKROSO: FKZ 13N8256, FKZ 13N8257, FKZ 13N8258, FKZ 13N8259, FKZ 13N8183) is gratefully acknowledged.

Glossary

APD (Avalanche photodiode) A photodetector that can be regarded as a semiconductor analog to photomultipliers. By applying a high reverse bias voltage (typically 100–200 V), APDs show an internal gain effect (around 100) due to impact ionization (avalanche effect). The higher the reverse voltage the higher the gain. Avalanche photodiodes therefore are more sensitive than other semiconductor photodiodes.

BALB/c Laboratory mouse strain

Caenorhabditis elegans Nematode (roundworm), about 1 mm in length

CCD (Charge-coupled device) A sensor for recording images, consisting of an integrated circuit containing an array of linked, or coupled, capacitors. Under the control of external circuits, each capacitor can transfer its electric charge to one or other of its neighbors. CCDs are used in digital photography and astronomy for image acquisition (particularly in photometry, optical and UV spectroscopy and high-speed techniques).

CD34+ Mesenchymal stem-cell marker

CMOS (Complementary metal-oxide-semiconductor) A major class of integrated circuits that are used as sensors in digital photography for image acquisition. A main difference to CCD sensors is a nonlinear relationship between input signal (e.g. light intensity) and the electronic output signal.

Confocal LSM (Confocal laser-scanning microscopy) This is an optical tool for processing high-resolution images and 3D reconstructions. The key feature of confocal microscopy is its ability to produce blur-free images of thick specimens at various depths by pointwise scanning of the sample through a small aperture, which suppresses scattered light.

DHM (Digital holographic microscopy) A microscopy method that is based on the holographic principle. In digital holography a digital device like a CCD camera is used for hologram recording instead of a conventional photographic film. The reconstruction process of such digital holograms is carried out by numerical computer processing. Digital holographic microscopy provides marker-free quantitative phase contrast imaging of living cells.

DIC (Differential interference contrast) A technique for imaging transparent specimens. By beam shearing interference the technique produces a monochromatic shadow cast image that displays the gradient of optical path length changes effected by the specimen. Regions of the specimen where the optical paths increase appear brighter (or darker), while regions where the path differences decrease appear in reverse contrast.

DIOC6 Fluorescent dye (dihexyloxacarbocyanine iodide)

EGFP Enhanced green fluorescent protein (fluorescence marker)

EJ28 Human bladder cancer cell line

FACS Fluorescent-activated cell sorting is a method for sorting a suspension of cells, based upon-specific light scattering and fluorescent characteristics of each cell.

FFT (Fast Fourier transform) An efficient algorithm for computing the discrete frequency spectrum of a signal/image. FFTs are applied to a wide variety of applications, from digital signal/image processing to numerical computer calculations.

FWHM (Full width at half maximum) This is an expression to quantify the extent of the maximum of a signal/function.

GFP Green fluorescent protein (fluorescence marker)

HCT116 Human colon cancer cell line

HepG2 Human liver cancer cell line

IEEE1394 Personal computer and digital video serial bus interface standard offering a high-speed data transfer rate, also denoted as Firewire.

LMPC (Laser microdissection and pressure catapult) For laser microdissection a pulsed (e.g. UV-A) laser is interfaced into a microscope and focused through the objective lens to a beam diameter in the sub-micron range for laser cutting. After microdissection, the isolated specimens are ejected out of the object plane and catapulted, for example, directly into the cap of a microfuge tube. This is performed by a single defocused laser pulse that moves the sample against gravity up to several millimeters.

NA (Numerical aperture) The numerical aperture of an object lens is a parameter to characterize the resolution of an optical imaging system; the numerical aperture depends on the aperture diameter, the focal length of the optical imaging system and the refractive index of the surrounding medium. Lenses with large numerical apertures permit high-resolution optical imaging. In addition, more light is collected from the specimen and brighter images are generated.

Nd:YAG (Neodymium-doped Yttrium Aluminum Garnet) A laser active medium used in solid state lasers for the generation of coherent laser light.

Normarski See differential interference contrast (DIC)

OCM (Optical coherence microscopy) Optical coherence topography used for microscopy applications (see also OCT).

OCT (Optical coherence tomography) OCT is an interferometric technique that uses the properties of partial coherent light for noninvasive depth selected imaging with submicrometer axial and lateral resolution. OCT is an established noninvasive imaging technique for biomedical applications, especially for ophthalmology.

Optical trap A trap formed by a focused laser beam: optical traps are used, for example, as optical tweezers for the contactless fixing, moving and manipulation of small particles such as living cells.

PaTu 8988T Human pancreas tumor cell line

RM26 Adult blood derived murine stem-cell line

RNA (Ribonucleic acid) Nucleic acid consisting of a string of covalently bound nucleotides. One of the main functions of RNA is to copy genetic information from DNA (via transcription) and translate it into proteins.

SEM (Scanning electron microscopy) A type of electron microscope for high-resolution imaging of a sample in vacuum. SEM images show a 3D appearance and are used for high-resolution characterization of surface structures.

SLD (Super luminescence diode) A special type of light emitting diode (LED) that is combined with wave guide structures to create high-intensity emission of light radiation. SLDs are applied, for example, as partial coherent light source, in optical coherence tomography (see OCT).

SPR (Surface plasmon resonance) A technique applied to measure the binding interactions of small amounts of proteins. A ligand is immobilized on a special chip and a solution of the target molecule flows across the chip. The progress of binding is then measured optically.

SPS (Spatial phase shifting) This is a variant of phase shifting that is applied in interferometry for the quantitative determination of optical path length. SPS methods provide the particular advantage that only one interferogram is necessary for phase determination and thus can be applied even under very unstable environmental conditions. For this reason, SPS methods are particularly suitable for the investigation of biological specimens.

USAF1951 test chart A standard resolution chart used for testing the resolving abilities of optical imaging systems (e.g. of microscopes and cameras).

YFP Yellow fluorescent protein (fluorescence marker)

Key References

A. BRECHT, G. GAUGLITZ, *Biosen. Bioelectronics*, 10 (**1995**), pp. 923–936.

S. THALHAMMER, G. LAHR, A. CLEMENT-SENGEWALD, W.M. HECKL, R. BURGEMEISTER AND K. SCHÜTZE: Laser microtools in cell biology and molecular medicine. *Laser Physics*, 13 (**2003**), pp. 681–691.

Y. NIYAZ, M. STICH, B. SÄGMÜLLER, R. BURGEMEISTER, G. FRIEDEMANN, U. SAUER, R. GANGNUS AND K. SCHÜTZE Noncontact Laser Microdissection and Pressure Catapulting: Sample Preparation for Genomic, Transcriptomic, and Proteomic Analysis. Microarrays in Clinical Diagnostics in *Methods in Molecular Medicine* (eds. T. Joos and P. Fortina), Humana press 2005, Vol.114, p. 1–24.

T. KREIS, *Holographic Interferometry: Principles and Methods*, Akademie-Verlag, Berlin, 1996.

P. RASTOGI, *Digital Speckle Pattern Interferometry and Related Techniques*, John Wiley & Sons, New York, 2001.

U. SCHNARS, W. JÜPTNER, *Digital Holography*, Springer-Verlag, Berlin, 2004.

U. SCHNARS, W. JÜPTNER, *Digital Recording and Numerical Reconstruction of Holograms*, Meas. Sci. Technol. 19 (**2002**), pp. R85–R101.

D. CARL, B. KEMPER, G. WERNICKE, G. V. BALLY, *Parameter Optimized Digital Holographic Microscope for High Resolution Living Cell Analysis*, Appl. Opt. 43 (**2004**), pp. 6536–6544.

References

1 R. Y. TSIEN, *Annu. Rev. Biochem.*, 67 (**1998**), pp. 509–544

2 T. KREIS, *Holographic Interferometry: Principles and Methods*, Akademie-Verlag, Berlin, (1996)

3 V. P. SHCHEPINOV, V. S. PISAREV, *Strain and Stress Analysis by Holographic and Speckle Interferometry*, John Wiley & Sons, Chichester, (1996)

4 M.-A. BEEK, W. HENTSCHEL, *Opt. Lasers Eng.*, 34 (**2000**), pp. 101–120

5 G. PEDRINI, S. SCHEDIN, H. J. TIZIANI, *J. Mod. Opt.*, 47 (**2000**), pp. 1447–1454

6 E. CUCHE, P. MARQUET, C. DEPEURSINGE, *Appl. Opt.*, 38 (**1999**), pp. 6694–7001

7 D. CARL, B. KEMPER, G. WERNICKE, G. VON BALLY, *Anal. Opt.*, 43 (**2004**), pp. 6536–6544

8 M. LIEBLING, T. BLU, M. UNSER, Complex-Wave Retrieval from a Single Off-Axis Hologram, *J. Opt. Soc. Am.*, A 21, (**2004**), pp. 367–377

9 T. KREIS, M. ADAMS, W. P. O. JÜPTNER, *Proc. SPIE*, 3098 (**1997**), pp. 224–233

10 P. MARQUET, B. RAPPAZ, J. P. MAGISTRETTI, E. CUCHE, Y. EMERY, T.

COLOMB, C. DEPEURSINGE, *Opt. Lett.*, 30 (**2005**), pp. 468–470
11 M. HAMMER, D. SCHWEITZER, B. MICHEL ET AL., *Appl. Opt.*, 37(31) (**1998**), pp. 7410–7418
12 A. MAYER, M. STICH, D. BROCKSCH, K. SCHÜTZE, G. LAHR, *Methods Enzymology, Series: Laser Capture Microscopy*, 356 (**2002**), pp. 25–33
13 M. STICH, S. THALHAMMER, R. BURGEMEISTER, G. FRIEDEMANN, S. EHNLE, C. LÜTHY, K. SCHÜTZE, *Pathol. Res. Pract.*, 199 (**2003**), pp. 405–409
14 S. LANGER, J. B. GEIGL, R. GANGNUS, M. R. SPEICHER, *Lab. Invest.*, 85 (**2005**), pp. 582–592
15 A. DE WITH, K. O. GREULICH, *J. Photochem. Photobiol. B: Biol.*, 30 (**1995**), pp. 71–76
16 K.W. CHAUDHARY, N.X. BARREZUETA, M.K. BAUCHMANN, J.A. MILICI, G.E. BECKIUS, D.A. STEDMAN, W.E. BLAKE, J.E. HAMBOR, J.D. MCNEISH, A. BAHINSKI, G.G. CEZAR, *Toxicological Sciences*, in press
17 J. GUCK, R. ANANTHAKRISHNAN, H. MAHMOOD, T. J. MOON, C. C. CUNNINGHAM, J. KAS, *Biophys. J.*, 81 (**2001**), pp. 767–784
18 A. JEMAL, R. C. TIWARI, T. MURRAY, A. GHAFOOR, A. SAMUELS, E. WARD, E.J. FEUER, M.J. THUN, *CA Cancer J. Clin.*, 54 (**2004**), pp. 8–337
19 J. MAYERLE, H. FRIESS, M. W. BÜCHLER, J. SCHNEKENBURGER, F. U. WEISS, K.-P. ZIMMER, W. DOMSCHKE, M. M. LERCH, *Gastroenterology*, 60 (**2003**), p. 949
20 J. BEHRENS, *Cancer Metastasis*, 18 (**1999**), pp. 15–23
21 H. P. ELSÄSSER, U. LEHR, B. AGRICOLA, H. F. KERN , *Virchows Arch. B. Cell Pathol. Incl. Mol. Pathol.*, 61 (**1992**), pp. 295–306
22 J. SCHNEKENBURGER, I. BREDEBUSCH, M. M. LERCH, W. DOMSCHKE, *Pancreatology*, 3 (**2003**), p. 435
23 J. BEUTHAN, O. MINET, J. HELFMANN, M. HERRIG, G. MÜLLER, *Phy. Med. Biol.*, 41 (**1996**), pp. 369–382
24 J. B. PAWLEY eds., *Handbook of Biological Confocal Microscopy*, Plenum Press, New York, (1995)
25 W. DENK, D. W. PISTON, W. W. WEBB, Two-Photon Molecular Excitation in Laser-Scanning Microscopy, in: *Handbook of Biological Confocal Microscopy*, (J.B. Pawley ed.), Plenum Press, New York, (1995), pp. 445–458
26 D. HUANG, E. A. SWANSON, C. P. LIN, J. S. SCHUMAN, W. G. STINSON, W. CHANG, M. R. HEE, T. FLOTTE, K. GREGORY, C. A. PULIAFITO, AND J. G. FUJIMOTO, *Science*, 254 (**1991**), pp. 1178–1181
27 G. J. TEARNEY, M. E. BREZINSKI, B. E. BOUMA, S. A. BOPPART, C. PITRIS, J. F. SOUTHERN, AND J. G. FUJIMOTO, *Science*, 276 (**1997**), pp. 2037–2039
28 B. M. HOELING ET AL., *Optics Express*, 6 (**2000**), pp. 136–146
29 E. BEAUREPAIRE, L. MOREAUX, F. AMBLARD, AND J. MERTZ, *Opt. Lett.*, 24 (**1999**), pp. 969- 971
30 M. A. REUTER, *History of Endoscopy*, Kommissionsverlag W. Kohlhammer, Stuttgart, (1999)
31 B. KEMPER, W. AVENHAUS, D. DIRKSEN, A. MERKER, G. VON BALLY, *Appl. Opt.*, 39 (**2000**), pp. 3899–3905
32 J. KANDULLA, B. KEMPER, S. KNOCHE, G. VON BALLY, *Appl. Opt.*, 43 (**2004**), pp. 5429–5437
33 S. SCHEDIN, G. PEDRINI, H. J. TIZIANI, *Appl. Opt.*, 39 (**2000**), pp. 2853–2857
34 S. SCHEDIN, G. PEDRINI, H. J. TIZIANI, A. K. AGGARWAL, *Appl. Opt.*, 40 (**2001**), pp. 2692–2697
35 G. VON BALLY, *Int. J. Optoelectr.*, 6 (**1991**), pp. 491–502
36 T. BOTHE, J. BURKE, H. HELMERS, *Appl. Opt.*, 36 (**1997**), pp. 5310–5316
37 B. KEMPER, J. KANDULLA, D. DIRKSEN, G. VON BALLY, *Opt. Commun.*, 217 (**2003**), pp. 151–160
38 A. BRECHT, G. GAUGLITZ, *Biosen. Bioelectronics*, 10 (**1995**), pp. 923–936
39 C. HÄNEL, G. GAUGLITZ, *Anal. Bioanal. Chem.*, 372 (**2002**), pp. 91–100
40 A. BRECHT, G. GAUGLITZ, Z. FRESENIUS, *Anal. Bioanal. Chem.*, 349 (**1994**), pp. 360–366
41 H.-M. HAAKE, A. SCHÜTZ, G. GAUGLITZ, Z. FRESENIUS, *J. Anal. Chem.*, 366 (**2000**), pp. 576–585

42 V. LEHMANN, *The Electrochemistry of Silicon*, WILEY-VCH, Weinheim, (2002)

43 A. JANSHOFF ET AL., *J. Am. Chem. Soc.*, 120 (**1998**), pp. 12108–12116

44 S. CHAN, S. R. HORNER, P. M. FAUCHET, B. L. MILLER, *J. Am. Chem. Soc.*, 123 (**2001**), pp. 11797–11798

45 M. A. ROCCHIA, E. GARRONE, F. GEOBALDO, L. BOARINO, M. J. SAILOR, *Phys. Status Solidi A*, 197 (**2003**), pp. 365–369

46 A. M. TINSLEY-BOWN ET AL., *Phys. Status Solidi A*, 182 (**2000**), pp. 547–553

47 H. G. BOHN, M. MARSO, *Phys. Stat. Solidi A*, 202 (**2005**), pp. 1437–1442

7
Regenerative Surgery (MeMo)

Volker Andresen, Heinrich Spiecker, Jörg Martini, Katja Tönsing, Dario Anselmetti, Ronald Schade, Steffi Grohmann, Gerhard Hildebrand, Klaus Liefeith[1]

7.1
Regenerative Surgery and Tissue Engineering: Medical and Biological Background

Owing to the tremendous advances in health care and the demographic changes that began in the 20th century, there is an increased awareness in the medical profession of the need to address the high demand for the replacement or repair of organs. While there has been great progress in the pharmacological treatment of patients for a variety of conditions since the 1970s, tissue engineering has initiated a new era in the field of regenerative medicine. In contrast to the repair/healing (i.e., *reparatio*) its focus is on the *restitutio ad integrum* – the restitution of the organ with the complete resumption of the organ's function. Regenerative medicine benefits from the prospering research in the fields of tissue engineering, biomaterial science, bio- and nanotechnology, stem-cell biology/cloning, proteomics and genomics.

The triad of trauma, aging and disease accounts for the either partial or complete loss of tissues and organs. Most types of tissue possess a limited regenerative potential, thus a spontaneous healing process is possible. However, when the degree of damage passes a certain level, or if the tissue does not exhibit a regenerative potential (for example nerve or cartilage cells), the damaged tissue has to be replaced.

Tissue engineering provides a new approach for regenerative medicine by using autologous tissue, while overcoming the fatal donor site morbidity. In general, autologous cells, isolated from only a small biopsy taken from the patient, are expanded *ex vitro* and are then re-implanted into the cleaned defective site. Considering the fact that tissue engineering can be regarded as the regeneration of biological tissues through the use of cells with the aid of supporting structures (scaffolds) and appropriate bioreactors, great efforts were

1) Corresponding author;
project: MeMo – Laser assisted characterization of Metabolism and Morphology of cell structures

Biophotonics: Visions for Better Health Care. Jürgen Popp and Marion Strehle (Eds.)

ISBN: 3-527-40622-0

made to extend the potential applications of tissue engineering to nearly all types of tissues:

- skin substitutes
- cardiovascular substitutes
- substitutes of the peripheral nervous system
- soft-tissue substitutes (e.g., breast implants)
- organs (like kidney, liver and lung)
- orthopedic cartilage and bone replacement

Cartilage is a highly specialized tissue with unique properties regarding stiffness, elasticity and friction. This tissue guarantees the movements of the skeletal apparatus with low friction forces at joints. Since 95% of the articular cartilage is composed of extracellular matrix (ECM) its biofunctional properties are mainly determined by the chemical composition of the ECM: water (60–80%), collagens (collagen type II: 10–20% and minor parts of collagen types V, VI, IX, X, XI), proteoglycans (aggrecan including keratan and chondroitin sulfate: 5–7%), noncollagenous proteins (for example link proteins and fibronectin) and other components like hyaluronic acid, lipids and glycoproteins [1]. The chondrocytes are embedded in a 3D network of collagen fibrils which is highly organized (**Figure 7.1**) and forms the hyaline cartilage representing healthy and biofunctional cartilage. A more detailed review is given in Ref. [2]. Thus, the biochemical structure ensures that cartilage can perform its biophysical tasks. Changes in cartilage composition due to aging or pathological processes (arthrosis) and injuries lead to loss of biofunctionality, which is one of the largest medical problems and afflictions especially of elderly people [3]. Unfortunately, the regeneration of cartilage tissue in humans is restricted by a failure of proliferation and the reduced capacity of adult chondrocytes for a turnover of matrix components. The conventional treatment of cartilage damage involves bone marrow stimulating techniques (microfracturing, Pridie drilling and abrasion plastic), the transplantation of autologous osteochondral cylinders (OCT) and eventually replacement by an artificial joint. These techniques do not enable physicians to treat arthrosis and osteoarthritis satisfyingly and in many cases additional treatment is necessary. Thus, therapies require new methods such as tissue engineering approaches.

Applying the tissue engineering approach, expanded autologous chondrocytes are either directly injected into the defective site [4] or they are cultured on 3D carrier systems (i.e., scaffolds) and/or stimulated before reimplantation into the joint (also see Section 7.3) [5–7]. The quality of engineered cartilage is determined by the cells' ability to synthesize the ECM.

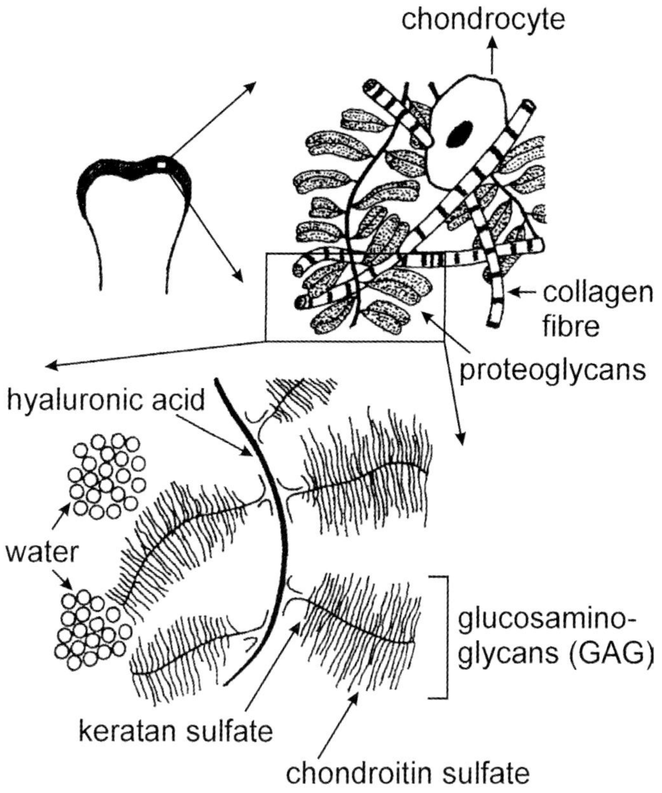

Figure 7.1 Organization of the extracellular matrix (ECM) within cartilage tissues (modified from Ref. [8]).

There has to be a well balanced production of collagens and proteoglycans in order to withstand mechanical loading in the transplantation site. From the medical point of view it is crucial to provide implantable chondrocytes that are able to rebuild new functional cartilage tissue at the defect site. Since many tissues are mechanically challenged, in principle tissue engineered constructs have to guarantee their performance. Therefore, functional requirements have to be taken into account in culture so as to engineer cartilage tissue with an optimized biological, chemical and morphological performance and, above all, with an appropriate stress–strain behavior that can tolerate the expected *in vivo* loads. To avoid the re-implantation of ineffective chondrocyte populations into the defect site it would be helpful to determine the necessary cell number and cartilage-specific differentiation of *in vitro* expanded chondrocytes before re-implantation in terms of quality control. Current techniques for quality control consist solely of dead-end procedures such as (immuno) histology, gene expression profiling by PCR and biochemical analyses. Hitherto, laser-scanning microscopy has not been considered as a promising controlling tool, but there is no doubt that this noninvasive optical technique possesses a tremendous potential in comparative validation of cartilage-specific components and cell populations. Therefore it is worthwhile to discuss the enormous impact of optical technologies and especially of the Two-photon Laser-scanning Microscopy (TPLSM) as a new alternative and promising approach for the minimally invasive and online quality control of 3D tissue-engineered constructs.

7.2 State-of-the-art and Markets

Statistical surveys have shown that approximately 20 million people have been treated with implanted medical devices. The associated costs for prostheses and organ replacement therapies exceed 300 billion US dollars per year, corresponding to nearly 8% of the total health care spending worldwide [9]. The proportion of the world's population formed by elderly people is rising dramatically. The discovery of antiseptics, penicillin, improved hygiene and vaccination on the one hand and the introduction of new immunosuppressant regimes, including improvements in post-surgical care, on the other hand, have established transplantation as the "gold standard" to successfully replace tissues, often as a life-saving procedure for patients with severe organ failure. Traditionally, organ transplantation is employed for the replacement of diseased tissue. There are four available sources of tissue for transplantation:

1. autologous (same body)
2. allogenic (intra species)
3. xenogenous (inter species)
4. artificial organs (for example implants, kidney, heart).

Autologous material is the most desirable source but it is only available in very limited amounts. Additionally, donor site morbidity is a negative side-effect, causing pain or even malfunction of the donor tissue. Though the availability of allogenic material is less limited, there are still long waiting periods for suitable donor organs. Furthermore, the restricted survival time of donor tissues reduces the supply of donor organs for patients around the world. According to the United Network of Organ Sharing (UNOS), 3216 possible donor tissues could not be transplanted in 2004 owing to the short time window for functional transport from the donor to the recipient [10]. Owing to the risk of rejection and infection, the disadvantages of an immuno-suppressive treatment have to be carefully balanced against the estimated effort of the transplantation. While the availability of xenogenous material is almost unlimited, its actual application is unfavorable for ethical reasons, difficulties with its compatibility, risk of rejection and transmission of infection and/or disease to the patient.

Although there has been good progress in stem-cell research, the use of such cells (especially embryonic stem cells) is still controversial throughout the world. Nevertheless, adult and mesenchymal stem cells may be a possible cell source for regenerative medicine in the future.

Another concept is based on the implantation of artificial materials such as high-technology polymeric, metallic and ceramic materials. Medical devices or prostheses made of these materials serve the affected patients well for extended periods by alleviating the conditions for which they were implanted. Despite the fact that the longevity and quality of life are clearly improved for patients with prostheses/implants, the long-term failure of artificial biomaterials can lead to a clinically significant event caused by adverse effects or foreign-body reactions under certain very specific conditions. According to Hench [11] the inability for self-repair and the missing potential of artificial biomaterials to respond to environmental factors such as mechanical stimuli were identified as the main drawbacks associated with a certain ratio of benefit to risk.

Taking these developments into consideration, tissue engineering offers a compelling new approach to the remaining major problems. Basically, tissue engineering can be defined as the application of engineering principles to biology, for the purpose of constructing 3D functional tissues. A more detailed definition was given by Skalak and Fox [12] as "the application of the principles and methods of engineering and the life sciences toward the fundamental understanding of structure–function relationships in normal and pathological mammalian tissues and the development of biological substitutes that restore, maintain, or improve tissue function." In contrast to the hitherto discussed approaches, tissue engineering is directed to the complete regeneration of natural tissues. According to the particular case, this process comprises *in vitro* or combined *in vitro* and *in vivo* approaches leading to the implantation of biological substitutes at the diseased site to achieve full functionality. Obviously this comprises the restoration of structure, function and metabolic and biochemical behavior as well as the restoration of biomechanical performance. Thus, there can be no doubt that the basic concept of "regenerative medicine" envisages a completely new form of therapy with the potential to change medical practice significantly.

It is of value to point out that there are two major scientific challenges to the achievement of this goal. First, the development of biomaterials that enhance the body's own reparative potential and, secondly, the availability of a technical system that allows the *ex vivo* cultivation of cell-seeded scaffold materials under conditions that mimic as closely as possible the natural process of tissue formation. Such systems, usually referred to as bioreactors, offer in principle the ability to perform a static or alternatively a dynamic cell cultivation process under controlled biochemical and biomechanical conditions. Bioreactors possess, by their very nature, a certain potential to enable a real large-scale expansion of cells because they provide excellent possibilities of guaranteeing uniform mixing and precise control over mass-transfer rates, pH values, oxygen consumption rates and the maintenance of optimal nutrient levels. However, it should be accepted that the whole process of tissue formation and regeneration remains partly unexplored [13, 14].

The remaining difficulties include an insufficiently detailed understanding of cell molecular control processes, such as the regulation of matrix formation by highly specific signalling pathways, pattern formation and tissue/organ morphogenesis.

Therefore it can be concluded that innovative noninvasive detection methods are necessary to permit a deeper understanding of all the sub-cellular (molecular), cellular and supra-cellular processes which may occur during three-dimensional tissue formation. Fortunately, optical methods such as TPLSM comply with the most important demands and provide the necessary spatial and time resolution to be able to detect metabolic pathways of newly formed tissues on a molecular level. The second scientific challenge mentioned above is the use of biomaterials as scaffolds and carriers of cells, proteins, genes and growth factors. Because the newly formed tissue compartments should gradually replace the scaffold material to allow nearly complete tissue regeneration, biodegradable or bioresorbable materials should be employed. Examples are

- polymeric scaffold materials (natural polymers [for example, collagen, hyaluronic acid], synthetic polymers [such as poly(hydroxy acid), polyphosphazenes])
- ceramic scaffold materials (CaP materials [such as hydroxyapatite, glass-ceramics, tricalcium phosphate, octacalcium phosphate])
- metallic scaffold materials (such as metallic foams based on Mg alloys)
- different composites made from the various materials.

It should be mentioned that non-biodegradable materials may also be used, especially in load-bearing situations owing to their superior mechanical properties.

However, published studies [15] have shown that an appropriate scaffold material alone is not able to provide an engineered tissue construct with a physiologically relevant architecture and composition. To obtain this, external stimuli such as electrical or mechanical stimuli are often needed, and these can be applied by means of specifically designed bioreactors. A very comprehensive and illustrative overview of the current status of bioreactor development and related biochemical and biophysical stimulation techniques is given by Müller [2].

Summarizing the state-of-the-art as described above, it can be estimated that significant progress in regenerative medicine and tissue engineering is based mainly on three major issues:

1. a detailed knowledge of molecular and cellular events during tissue formation and morphogenesis including the effect of external stimuli
2. the availability of specific scaffold materials with an appropriate architecture and composition to allow three-dimensional tissue formation
3. the availability of advanced tissue bioreactors to provide cultivation conditions appropriate to the physiological environment at the recipient site.

Bearing the above facts in mind, it is apparent that none of the remaining problems can really be solved without the use of noninvasive and image-generating measuring techniques. Hence, TPLSM, or laser microscopy in general, seems to be a powerful tool to solve the problems at least in part, owing to its unique spatial and time resolution and its ability to provide three-dimensional data describing the cellular environment with sub-cellular resolution.

Cartilage pathologies of traumatic and/or degenerative origin, among which osteoarthritis is by far most common, are a major concern in public health care. These joint ailments lead to severe articular pain for millions of individuals and, because of the lack of satisfactory repair capacity, often reach a final stage in which the affected individual is severely incapacitated, with artificial joint replacement as the only possible eventual outcome.

The world-wide scope of this problem can be clearly discerned from a few self-explanatory figures: 40 million people in the USA and Europe suffer from osteoarthritis. More than 500 000 arthroscopic procedures and total joint replacements are performed each year in the US. Every year in Europe 150 000 injured knee joints with cartilage defects requiring treatment are diagnosed (see also **Table 7.1**). Therapeutic approaches relying on bone marrow stimulation (such as drilling and microfractures) lead to a fibrocartilaginous tissue type with limited load-bearing capacity. Likewise mosaicplasty has not fulfilled expectations, because this procedure obviously requires invasive and technically demanding surgery [19].

In contrast, the autologous chondrocyte transplantation (ACT) described by Bentley [19] and Zheng [20] represents a promising method for restoring defects of hyaline cartilage in the majority of cases. However, conven-

Table 7.1 Market sizes correlated with cartilage defects/cartilage repair.

Region	Market size (EUR)	Year	Remarks	Source
Europe	2 billions	1999	Market value for joint implants (prosthesis costs only)	Biomet Merck
World	1.5 billions	1999	Market value for knee implants (prosthesis costs only)	Data-monitor
USA	5.2 billions	2001	annual spending for total knee replacement	[17]
World	6.5 billions	2001	market potential of surgical procedures for cartilage regeneration	[18]
World	25 billions	2011	market potential of surgical procedures for cartilage regeneration	[18]

Data from Ref. [16]

tional ACT suffers from some disadvantages, such as the risk of leaking out if sealing is insufficient and above all a relatively strong cell dedifferentiation, which may possibly be due to the lack of a suitable scaffold structure [21]. To overcome these problems a new therapeutic option was developed based on Matrix-induced Autologous Chondrocyte Implantation (MACI®, Verigen AG, Leverkusen, Germany) [5,21]. The MACI® technique requires the use of a 3D collagen type I/III membrane, seeded with chondrocytes to improve the structural and the biological performance of the graft. As it corresponds to the physiological environment within the natural joint cartilage tissue it is understandable that such an environment is favorable for the proliferation and differentiation of chondrocytes.

Nevertheless, the promising clinical results should not obscure the fact that any further progress in therapeutic research and clinical treatment depends strongly on a deeper understanding of the underlying biomolecular mechanisms.

Cartilage Repair Using Autologous Chondrocytes (1)

Damage to the cartilage arising from factors such as arthritis and traumatic injury cause severe pain and restrict the mobility of millions of patients worldwide. Unfortunately the ability of the cartilage tissue to regenerate the damaged area is limited. Modern therapies for cartilage reconstruction are focused on the support of the tissue to self repair by the transplantation of healthy autologous chondrocytes. Autologous Chondrocyte Transplantation (ACT) represents the basic technique for transplanting precultured chondrocytes into the defect site. The transplanted cells are retained at the site by covering them with a periosteum membrane sutured to the surrounding healthy tissue.

Cartilage Repair Using Autologous Chondrocytes (2)

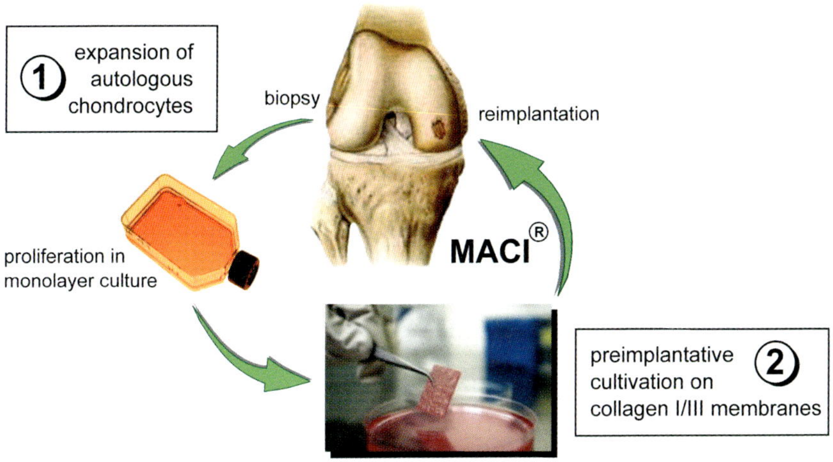

Figure 7.2 The principal stages in MACI® treatment.

A variation of this technique is the Matrix-induced Autologous Chondrocyte Implantation (MACI®) promoted by Verigen AG (Leverkusen, Germany). This technique includes two basic steps (see **Figure 7.2**): A biopsy of healthy cartilage is arthroscopically obtained from the patient. Subsequently the chondrocytes are released by enzymatic digestion of the tissue, expanded/grown *in vitro* and seeded into a collagen type I/III membrane in a clean-room facility. After debridement of the lesion the cell-seeded membrane is cut to the size and shape of the defect and glued in place with fibrin. The main difference between MACI® treatment and the original ACT is the use of a collagen type I/III membrane rather than an autologous periosteal flap. Since the MACI® membrane can be suture-free attached to the base of a prepared chondral defect with fibrin glue, this novel procedure offers the following surgical advantages:

- access to the lesion can be gained through a mini-arthrotomy
- no requirement for periosteal harvesting and therefore a reduction of the number of grafts and graft sites
- no risk of leakage of chondrocytes and uneven distribution because there is no injection of a cell suspension below a membrane
- substantially reduced operating theater time and, because it is performed by minimally invasive surgery, shortened rehabilitation period.

Current research findings have revealed that the success of a therapy employing tissue engineering products can be enhanced when specific stimuli are applied in the preimplantative cultivation phase. A chondrocytic differentiation status can be achieved through 3D cultivation techniques, as well as biochemical and mechanical stimulation. The MACI® technique has promising prerequisites for this purpose. The necessity to monitor the cell performance in response to the applied stimuli is not yet solved satisfactorily. Noninvasive optical techniques such as multiphoton microscopy are a promising approach for an online quality control tool.

7.3 Cell and Tissue Culture Technologies

The cultivation techniques for cells and tissue-engineering constructs are diverse and a wide range of culture approaches have been developed for many clinical applications. From a methodical point of view they can be distinguished generally into two different methods: static and dynamic culture techniques. The classical technique for static cultivation of cells is the monolayer technique established in plastic dishes. In 1994 Brittberg et al. [4] published the first approach for tissue engineering of cartilage – autologous chondrocyte transplantation (ACT). Autologous chondrocytes of a healthy cartilage biopsy are expanded *in vitro* as a monolayer and are then re-injected into the defect site. In order to minimize donor site morbidity, only a small autologous specimen can be sacrificed and a cell expansion is inevitable. To ensure an effective cell expansion, cultivation chambers were developed to minimize the amount of medium needed (and so the costs of expensive medium supplements) and to enlarge the available growth surface. However, the extensive proliferation of autologous cells *in vitro* is correlated with a progressive dedifferentiation (**Figure 7.3**) and may eventually result in the formation of fibro-cartilage with inferior mechanical properties [22–24]. According to the requirements of functional tissue engineering, the scientific focus is now directed to the physiologically relevant stimulation of these cells to redifferentiate into a phenotype typical for native chondrocytes.

As proven by many scientists and our own results (**Figure 7.4**) 3D culture technologies can enhance the cell response with respect to the differentiation status compared to 2D culture techniques [25]. Different approaches to static 3D techniques utilize agarose, alginate and hydrogels as scaffolds [23,26,27]. Current biomaterials, based on either synthetic or natural polymers, are designed to be able to mimic the native extracellular matrix. In addition to the chemical composition, the microstructure (for example pore size, porosity, interconnecting pores, elasticity, stability) of a scaffold material may also stim-

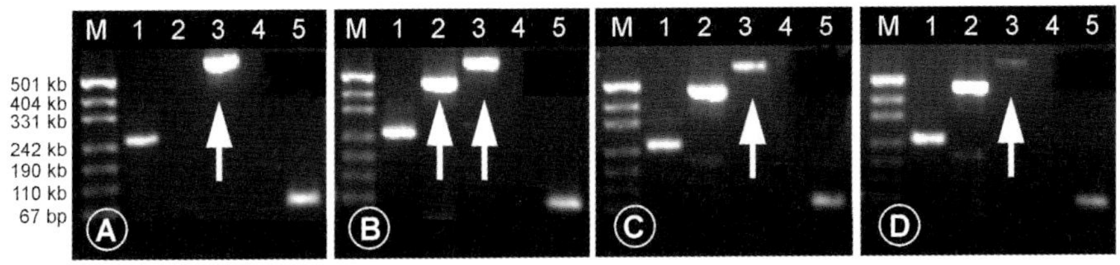

Figure 7.3 Dedifferentiation during the monolayer cultivation indicated by starting of nonspecific collagen type I synthesis coupled with decreasing expression of collagen type II (RT-PCR; GAPDH (1), collagen type I (2), collagen type II (3), collagen type X (4), aggrecan (5); fresh isolated chondrocytes (A), chondrocytes without subcultivation (B) and after one (C) and two (D) subcultivations).

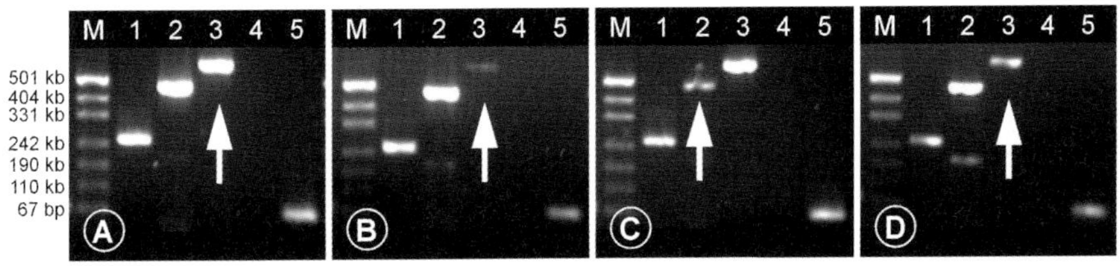

Figure 7.4 Delayed dedifferentiation of chondrocytes depending on the substrate. On 3D scaffolds (C, D) the gene expression pattern indicates a prolonged synthesis of cartilage-specific ECM components (RT-PCR; GAPDH (1), collagen type I (2), collagen type II (3), collagen type X (4), aggrecan (5); chondrocytes in a monolayer without (A) and after two subcultivations (B), chondrocytes on a 3D scaffold without (C) and after two (D) subcultivations).

ulate the inherent cells. Thus, the choice of a suitable substrate is a crucial issue [28].

MACI® is a promising 3D cultivation technique derived from the classical ACT technique. A porcine collagen type I/III bilayer seeded with cultured chondrocytes is subsequently glued (suture free) into the debrided defect site. The regenerated cartilage appears hyaline to hyaline-like and shows satisfactory biomechanical properties [5–7].

A common disadvantage imposed by the static culture conditions are the diffusion gradients of oxygen, nutrients and metabolites that occur, especially within the deeper layers of a 3D scaffold construct. Some culture systems, such as the wave bioreactor [29], implement a mixing of the media through the wave motion of the whole system to avoid any such gradients. Systems providing dynamic culture conditions enable the researcher to monitor critical components in the influx and efflux media and are thus preferable. For instance, it was found that a supply with a low-oxygen gas mixture can enhance the synthesis of ECM-specific cartilage components to a certain extent [30,31]. Another important aspect is the simulation of the natural conditions of cells in tissues to induce cellular interactions using co-cultures of different cell types and their cross-talk by emitting signalling molecules. This cross-talk can be supported by adding appropriate molecules, such as vitamins, growth factors, hormones and beta-glycerophosphate [32,33], originating from a better understanding of the metabolic processes, genomics and gene regulation. However, the progress in tissue engineering since the mid-1990s has not only arisen from increased knowledge of the biological aspects of functional tissues but also from the embedding of biotechnological procedures and devices into the *in vitro* cultivation process. A broad spectrum of different dynamic cultivation systems has been developed and adapted to the requirements of cartilage tissue engineering, ranging from simple flow chambers up to more elaborated culture systems including the hollow fiber reactor [34], the flat membrane reactor [35], gradient containers [36–38] and bioreactors with rotating components and air–liquid phases [39–41]. The rotating wall vessel bioreactor is one example allowing the harboring of cells or cell-seeded scaffolds and assuring dynamic controllable culture conditions [42]. A comprehensive description of the different approaches and their applications is given in Ref. [2].

Especially for load-bearing tissues like cartilage, the aforementioned techniques are not sufficient for the *in vitro* construction of completely redifferentiated and functional tissues. Since 1998 a new discipline called functional tissue engineering (FTE) seeks to combine biomechanical considerations with the tissue engineering techniques [43]. Biomechanical stresses, strains and strain rates have to be imposed on the cells during cultivation in order to stimulate redifferentiation. The aim is the stimulation of redifferentiation indicated by an increased synthesis of typical tissue-specific ECM components like proteoglycans and collagen type II. Bioreactors equipped with diverse (bio)mechanical loading tools are currently being designed and successfully applied [44,45]. The generation of stress/strain can generally be achieved by hydrostatic and hydrodynamic forces, mechanical loading and magnetically imposed stress with different frequencies and amplitudes as well as with and without rest intervals. The combination of reduced oxygen tension and intermittent hydrostatic pressure can enhance the response of chondrocytes [30]. However, the frequency and amplitude necessary to stimulate an *in vitro* tissue engineering construct in an optimal way has not yet been established. Interestingly, the results reveal that static compression over a long period is counterproductive compared to intermittent mechanical loading [44,46,47]. These facts illustrate the complex spectrum of scientific challenges that has to be taken into consideration in cartilage tissue engineering.

All of the approaches described were designed to generate implantable tissue constructs. But the structure and organization of these constructs range from expanded but dedifferentiated cells to redifferentiated cell populations and structural reconstructed cartilage tissues with various qualities, depending particularly on the patient's own cells. So it would be useful to detect (a) the quality of biopsies (including the number of cells) before the expanding phase and (b) the effects of culture conditions on the cell response. However, it has to be admitted that clinically applicable approaches as a special kind of quality control are currently not available. In this context, the embedding of noninvasive optical online techniques to detect corresponding cell parameters is a crucial issue with increasing importance. The most important points are:

- cell cultivation under reproducible and controllable culture conditions
- options to stimulate the cell populations during the cultivation process
- coupling of optical laser-scanning techniques with the tissue engineering process
- applicability of the optical method for the detection of autofluorescent signals in thick and strongly scattering samples
- distinct analysis of fluorescent signals of tissue-typical marker components.

The coupling of biotechnological devices including bioreactors, flow chambers and mechanical manipulation ports with microscopic laser scanning techniques is a very promising approach to meet the requirements of online analytical systems in regenerative medicine. The system described here uses an adapted tissue culture chamber with an optical window to enable direct access to the TPLSM as shown in **Figure 7.5**.

7.4 Controlled Tissue Cultivation Through Laser Optical Online Monitoring

To control the cultivation of tissue in order to generate better and patient-specific tissue-engineering products, parameters that are relevant for their quality must first be identified. In case of cartilage repair tissues according to the MACI® technique, two aspects are of special interest: the number and morphology of chondrocytes on scaffolds in the preimplantative cultivation stage and the capability of chondrocytes to synthesize ECM components indicating the differentiation status, which depends strongly on extrinsic stimulations corresponding to the local functional environment. A set of suitable noninvasive measurement techniques and adequate detection methods has to be defined to evaluate those parameters during the growth process. At present the only measurement technique that allows the investigation of optically dense tissues with sub-cellular resolution and a low damaging potential is two-photon microscopy. Since very weak autofluorescence signals must be recorded, arising from endogenous fluorophores as well as from the rapid responses of cartilage to mechanical stimulation, the scanning process must

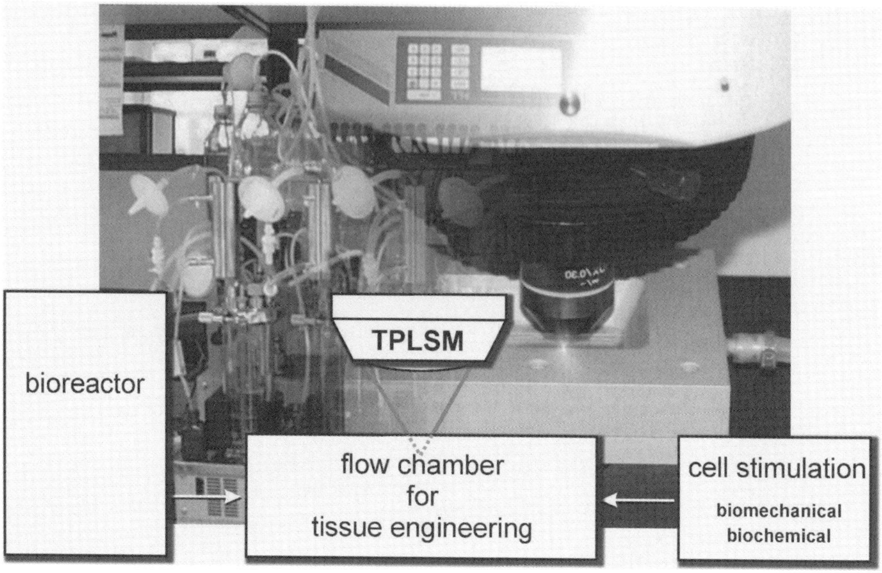

Figure 7.5 Scheme of a tissue-engineering flow chamber coupled with online-TPLSM.

be parallelized to enhance the resulting fluorescence signal. Over the same time period, the time needed to acquire a 3D image stack of the sample can be reduced. Owing to the complex tissue architecture and composition the detection method must be able to distinguish between the different endogenous fluorophores, such as NAD(P)H, flavine, elastin and ECM-specific collagens as well as chondrocytes. Using spectra-resolved detection, the fluorescence from endogenous fluorophores, ECM and chondrocytes can be separated. However, the fluorescence of several ECM components and the collagen membrane frequently show a large spectral overlap, so that further contrast modes have to be applied. For this purpose Fluorescence Lifetime Imaging Microscopy (FLIM) can be used, which provides more-specific results. Additionally, Second Harmonic Generation Imaging Microscopy (SHIM) is used in single-beam scan mode to highlight non-centrosymmetric structures.

To sum up it can be concluded that the combination of optical laser scanning measurements with highly specialized tissue bioreactors and perfusion chambers adapted for tissue engineering provides excellent conditions for continous monitoring of the cultivation process. In addition, flow or perfusion chambers allow the application of mechanical stress or strain to the cell-seeded scaffolds. Thus, the influence of the most important cultivation parameters on the resulting tissue composition and morphology can be directly observed during the cultivation process under real time conditions. Using this feedback, specific cultivation protocols can be developed that enhance the quality of the final tissue-engineered product and potentially adapt it to the needs of the patient.

7.5 Characterization and Evaluation of Tissues by Innovative Biophotonic Technologies

In biophotonics [48–50] organic and biological materials consisting of molecules, cells and tissue are imaged, analyzed and manipulated utilizing photons.

There are several biomedical (*in vivo*) imaging methods for thick tissue sections, which differ significantly in terms of spatial resolution and maximum achievable imaging depth. Techniques that feature a resolution in the order of several millimeters include X-ray imaging, magnetic resonance tomography, positron emission and ultrasound imaging. Other techniques such as scanning electron and atomic force microscopy allow a higher resolution (down to 0.1 nm). However, imaging is restricted to the surface and therefore mechanical slicing of the sample is mandatory. Light imaging methods are very promising alternatives because of their high spatial resolution, large penetration depth and noninvasive nature. Consequently, these techniques are more frequently used to image, analyze and manipulate the structure and function of molecules, cells and tissues in biomedical applications.

Several endogenous fluorophores contained in tissues such as NAD(P)H, flavine, elastin and collagen show autofluorescence, allowing the direct visualization of morphology, cell metabolism and disease states (e.g., Alzheimer's disease, cancer). Therefore no fixation or staining procedures are required. As these components are best excited with light in the 260–400 nm range up to now most investigations have used UV light sources. Two considerable drawbacks are the very low penetration depth, due to strong scattering and absorption, and the sample destruction and heating caused by this type of light.

A new approach to the study of tissues and inherent endogenous fluorescent species is the use of nonlinear microscopic methods such as TPLSM [51] and SHIM [52]. TPLSM has already been successfully used to perform optical sectioning of various biological tissues, such as brain, lung or intestine slices [51, 53–55]. Besides the intrinsic high spatial resolution, scattering and absorption are clearly reduced by the longer NIR excitation wavelengths in the spectral range 700–1100 nm, enhancing the penetration depth significantly. As a result, autofluorescence of biomedical tissue sections (collagen, cells) can be excited and detected down to a depth of 1000 μm. SHIM is based on the homonymous nonlinear optical effect [56, 57] in reference to a frequency doubling of the incident light. Similar to TPLSM, the amplitude of second-harmonic generation (SHG) is proportional to the square of the incident light intensity. SHIM therefore also offers intrinsic 3D sectioning. The application of SHIM is restricted to the imaging of highly non-centrosymmetric molecular assemblies such as cellular membranes or collagen fibrils.

In conclusion, the two nonlinear imaging methods TPLSM and SHIM provide enhanced in-depth information at a high spatial resolution in contrast to common imaging techniques including confocal laser microscopy. TPLSM has clearly demonstrated the potential of this technique for both scientific investigation and clinical diagnosis and is becoming an indispensable tool for noninvasive observation of tissue features *in situ*.

In addition to tissue morphology imaging, further parameters like emission wavelength, fluorescence lifetime and the emitted light's polarization provide complementary and essential information for tissue characterization. Therefore they represent a powerful method of identifying endogenous fluorescent species. The relative occurrence of these species is related to tissue physiological and pathological states. Fluorescence spectroscopy has also been used to characterize different tissue types such as cartilage and skin [53,57].

7.5.1 Microscopy Basics and Techniques

7.5.1.1 Conventional Microscopy

In conventional brightfield or epifluorescence microscopy the sample is illuminated with a homogeneous light source. A combination of stray and fluorescence light from the complete sample is collected by the microscope's objective lens, resulting in a microscopic image that consists of the focal plane and blurred off-focus optical planes. This means that, for example, all depths of a cell contribute to its microscopic image, creating a blurred projection of the complete cell. Therefore, it is impossible to distinguish between different objects that lie in series on the optical axis within the sample without applying deconvolution image processing.

Excitation, Fluorescence and Second Harmonic Generation (1)

Fluorescence techniques are important tools for the study of a large variety of applications in biology and medicine. In particular, this is due to recent advances in the development of more selective, specific, stable, efficient and user-friendly fluorescent probes (e.g., cyanine dyes, GFP, RFP, quantum dots). The principal physical mechanism of excitation and fluorescence is illustrated in the Jablonski diagram (**Figure 7.6**).

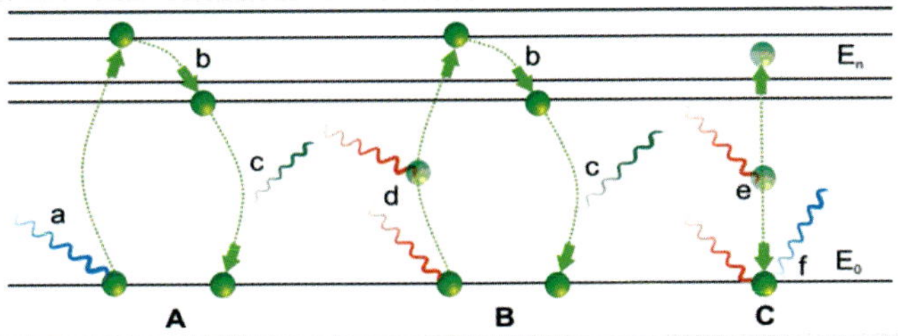

Figure 7.6 (A): fluorescence emission after one-photon excitation; (B): fluorescence emission after two-photon excitation; (C): second harmonic generation.

A fluorophore (i.e., atom, molecule or fluorescent probe) in its energy ground state E_0 is excited by a photon to a higher energy state E_n (see **Figure 7.6A**). This photon holds the energy difference $E_d = E_n - E_0$, which is connected to its frequency or wavelength. The fluorophore first relaxes by nonradiative transitions to a lower energy state via inter- or intramolecular collisions. From this energy state the molecule returns to its ground state, emitting a photon. As there are many unoccupied energy states in molecular fluorophores, both the absorption spectrum (range of wavelengths suitable for excitation) and the emission spectrum of these molecules are rather broad ($\sim$100 nm).

The average time a molecule takes to relax from the excited state E_n to the ground state E_0 is called the fluorescence lifetime (typically 1–5 ns). In addition to its emission spectrum, the fluorescence lifetime of a molecule is an important parameter, since it carries not only information about the molecule itself but also about its local chemical environment and its bonding conditions.

In case of two-photon excitation, the energy transfer is performed by two photons, each carrying half of the required energy E_d and therefore twice the required wavelength (**Figure 7.6B**). According to Heisenberg's uncertainty principle this absorption takes place within approximately 10^{-16} s. Therefore, two-photon excitation is an extremely improbable process. From the excited state E_n the fluorophore then thermally relaxes and emits fluorescence light in the visible spectrum just as in the one-photon excited case. It has to be mentioned that owing to thermal relaxation the emission wavelength λ_{em} is always longer than the excitation wavelength λ_{ex} (Stokes shift) in the one-photon excitation case.

Excitation, Fluorescence and Second Harmonic Generation (2)

The conversion of two photons with wavelength λ_{ex} into a single one with wavelength $\lambda_{ex}/2$ is called second harmonic generation (**Figure 7.6C**). This effect happens in the vicinity of highly organized, crystal-like specimens that exhibit a local polarization. The strong electrical fields of intense light waves (i.e., laser light) induce an oscillation of the electrons in the sample. As these electrons are influenced by the non-harmonic potential of their nuclei, their oscillation generates electromagnetic waves not only with the incoming (light) wavelength but also with one-half (quarter, eighth,...) of this wavelength (non-vanishing Fourier-Terms of higher order harmonics). From the physical point of view the process of SHG is more comparable to the effect of Raman scattering than to the effect of fluorescence, as it has neither a lifetime that underlies Heisenberg's time uncertainty nor does it require free energy states in a molecule.

7.5.1.2 **Three-dimensional Laser Scanning Microscopy**

The first light microscope that offered true 3D resolution was introduced by M. Minsky [58] and the technique is called confocal laser-scanning microscopy (CLSM). The basic idea of the confocal principle is to place an aperture with a very small diameter (10–50 μm) in front of the detector and focus the fluorescence light on it. Only light from the focal volume of the objective lens is able to pass, whereas scattered light and light arising from out-of-focus planes is almost completely blocked. With this simple but ingenious set-up it is possible to get an image of a point-like object within the sample.

In order to generate a 3D image of the sample it is necessary to scan the focus of the objective lens point by point across the focal plane. Thus, the size of the focus defines the resolution of the measurement. The fluorescence intensity of a sample is detected point by point by scanning the focused laser beam across a Cartesian coordinate system – point-by-point in the x-direction, line-by-line in the y-direction, and section-by-section in the z-direction – through the region of interest. Thus a series of x-y-planes at different z-positions is acquired to represent a 3D fluorescence map of the sample. There are two commonly used scanning methods. In a stage-scanning microscope the complete object under investigation is moved in all three spatial directions. This method allows for very large scan fields in the order of several square centimeters. The most important drawback is the slow scan speed and the disturbance of the sample by the inherent acceleration of the stage. In a beam-scanning microscope the exciting laser beam is moved by two galvanometric scanning mirrors or acousto-optic modulators across the sample. To generate a 3D image the detected fluorescence intensity at each time point has to be correlated with the position of the exciting laser beam in the sample (i.e., the position of the scan mirrors). The maximum size of the field of view of this type of scanning is limited to the field of view of the objective lens. To overcome the disadvantage of a relatively small scanning area, beam-scanning systems are often combined with mechanical-scanning stages. The great advantage of a beam-scanning set-up is the high scanning frequency (up to ca. 7 kHz for resonant scanners), which allows for the detection of fast (e.g. intracellular) processes and does not disturb the sample.

The variation of the imaging depth inside the sample is usually done by moving the objective lens along its optical axis. This can be performed either through the microscope's mechanical focus drive or a piezoelectric focus drive. The 3D scanning concept is identical for both confocal and two-photon laser-scanning microscopy.

7.5.1.3 Two-photon Laser-scanning and Second Harmonic Generation Imaging Microscopy

The simultaneous absorption of two photons was first predicted in 1931 by Maria Göppert-Mayer in her doctoral thesis. As this effect requires extremely high photon densities, it is no surprise that the experimental proof in 1961 [59] involved a laser, which was developed only one year earlier by Theodore H. Maiman at Hughes Research Laboratories. Also in 1961 the first second harmonic generation experiment was reported by P. A. Franken et al. [60], again using a ruby laser to generate the necessary excitation power.

Denk et al. [61] presented an impressive example of the improbability of a two-photon excitation process for an excellent one- and two-photon absorber molecule of rhodamine B in bright sunlight. Such a molecule is excited about once every second by a one-photon process but only once every 10 million years by a two-photon process. This calculation illustrates the fact that two-photon excitation requires power densities in the range of GW cm^{-2} to achieve sufficient fluorescence. This photon density can be achieved by focusing a pulsed Ti:Sa laser with a high NA objective lens. Two-photon excited fluorescence is then generated only in the attoliter focal volume of the objective lens. Using this principle, filtering the exciting laser light, detecting the fluorescence and scanning the focus through the sample, Denk et al. [51] introduced the first two-photon laser-scanning microscope in 1990, allowing the generation of fluorescence images from deep inside the specimen with a high 3D resolution.

Most two-photon laser-scanning microscopes use a pulsed femtosecond titanium:sapphire (Ti:Sa) laser with a (tunable) wavelength range between 710 and 1050 nm, in the NIR spectrum. This prevalent choice is justified by several advantages of this type of laser source. First of all, mode-locked Ti:Sa lasers are both commercially available and, nowadays, reliable turn key systems. They deliver output pulses with a pulse duration of typically 120 fs at a repetition rate of 80 MHz and the time averaged laser power ranges between 1 and 2 W. Furthermore, IR laser light is well capable of two-photon excitation of a great variety of fluorescent probes and native biological fluorophores, such as NAD(P)H, flavine, elastin and collagen [56]. As biological tissue has an absorption window [62] in the NIR spectral range, such light has, indeed, a higher penetration depth and a lower (out of focus) photodamage potential than the visible excitation light used in confocal laser-scanning microscopy.

The probability n for a fluorophore to absorb two photons simultaneously during one laser pulse is given by Denk [51] and Diaspro [63]:

$$n = \frac{s_2 P_{\text{ave}}^2}{\tau f_{\text{p}}^2} \left(\frac{\text{NA}^2}{2\hbar c \lambda} \right)^2 \qquad (7.1)$$

where s_2 is the two-photon cross-section of the fluorophore, P_{ave} the average laser power, τ the laser pulse duration, f_{p} the laser repetition frequency, NA the numerical aperture of the objective lens, c the speed of light, λ the wavelength, $h = 2\hbar\pi$ Planck's constant.

Equation (7.1) shows that it is not only the laser characteristics (wavelength, repetition frequency and pulse duration) that play an important role in the excitation probability but also the characteristics of the objective lens used to focus the laser beam to a very tiny spot. This is because the two-photon excitation probability is proportional to the laser intensity squared, and is therefore dependent on the focal volume that the incident laser light is confined to. The focal volume is also connected to the resolution of a TPLSM, in that it essentially states the capability of separating two point objects next to each other. Many aspects of the experimental set-up [64] contribute to a detailed description of a fluorescence microscope's resolution but a good estimation is [63]:

$$r_{\text{lat}} = 0.7 \cdot \frac{\lambda_{\text{em}}}{\text{NA}} \qquad r_{\text{ax}} = 2.3 \cdot \frac{\lambda_{\text{em}} \cdot n}{\text{NA}^2}$$

where r_{lat} is the lateral resolution, r_{ax} the axial resolution, λ_{em} the emission wavelength, NA the numerical aperture of the objective lens, n the refractive index.

The resolution of high NA objective lenses can be calculated more precisely according to Born and Wolf [65].

For a GFP-expressing cell (emission peak at 508 nm, $n_{\text{water}} = 1.33$) the resolution is calculated to be $r_{\text{lat}} = 0.25$ µm and $r_{\text{ax}} = 0.79$ µm when using a 1.4 NA objective lens and an excitation wavelength of 800 nm. It has to be mentioned that two-photon cross-sections usually have broader absorption spectra than one-photon cross-sections so that in many cases fluorescence of different fluorophores is induced simultaneously by two-photon excitation. The main difference between these two microscopy methods is that in the confocal case only light from the focal volume is *detected*, whereas in the two-photon case light is only *generated* in the focal volume (**Figure 7.7**). Hence, photobleaching and photodamage are restricted to this small volume.

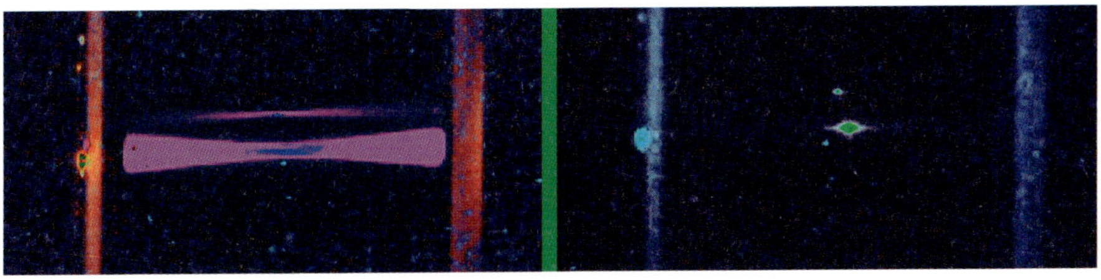

Figure 7.7 Excitation of samples: Left: one-photon excitation of Pyridin 2 at 560 nm, Right: two-photon excitation of Coumarin at 770 nm.

SHIM is based on the homonymous nonlinear optical effect [57,66] in reference to a frequency doubling of the incident light. Similar to TPLSM, the amplitude of SHG is proportional to the square of the incident light intensity. SHIM therefore also offers intrinsic 3D optical sectioning without the need for a confocal aperture. Since SHG is based on photon scattering and does not involve excitation of molecules, out-of-plane photobleaching and phototoxicity, which often limits the usefulness of fluorescence microscopy for imaging of living specimens, are significantly reduced. However, it differs from other nonlinear microscopy modes, since SHG is restricted to specimens that are highly non-centrosymmetric molecular assemblies, such as cellular membranes or collagen fibrils. In practice, SHIM can be performed with TPLSM instruments by using detection filters that transmit half of the exciting laser wavelength and therefore use the advantage of low NIR scattering losses and high penetration depths in thick tissue samples. Recent studies of the 3D *in vivo* morphology of native and unstained, well-ordered protein assemblies, such as collagen, microtubules and muscle myosin, have proven the superior applicability of SHIM as a nondestructive and label-free imaging method [52,67].

7.5.2 Multifocal Multiphoton Microscopy

The most important drawback of single-beam laser-scanning microscopy is the small yield of fluorescence per time unit generated by a single focus. This is particularly true for two-photon microscopy as all dyes feature very low two-photon absorption cross-sections. Since endogenous fluorophores exhibit even weaker signals than most fluorescent dyes used for staining, up to tens of seconds are required to record a single plane of a sample. As a result the observation of fast dynamics as well as the acquisition of 3D data stacks of unstained living samples is nearly impossible with a single-beam laser-scanning microscope. In addition, sample movement or morphological changes within the sample during the acquisition can degrade the spatial resolution considerably.

To overcome these problems a common approach is to increase the excitation laser power in order to generate more fluorescence photons per time unit. But above a critical power level ($> \sim 5$ mW) the induced photodamage rises more rapidly with increasing laser power than the number of fluorescent molecules – thus it does not solve the problem. The only way to increase the amount of fluorescence per time unit without raising photodamage is to use more excitation beams in parallel. The simultaneous application of N excitation foci results in an N-fold increase of the excited molecules, such that the sample emits more fluorescence light in the same time interval (even with a lower laser power in each focus).

In the late 1990s, two different parallelized two-photon microscopes were invented [68, 69]. The first one uses a microlens disk to split the laser beam, typically into 25–36 beams. As the microlenses are arranged in spirals on the disk the scanning process is accomplished simply by rotating the disc. In the second approach a single laser beam is split into 64 beams by multiple transitions through a 50% beamsplitting substrate (see **Figure 7.10**). A two-axis galvanometric scanner rasters all foci simultaneously across the object plane. This method has a very high optical throughput and offers uncompromised optical sectioning quality, which is due to the exclusive use of flat optics to divide the incoming laser beam and also to the fact that all beams arrive at slightly different points in time at the sample (no cross-talk). The latter of these two parallelized two-photon microscopy techniques is the basis for the tissue imaging microscope that was developed in the context of this study. Subsequent publications on multifocal multiphoton microscopy have shown the large potential for biomedical applications and the superior sectioning and image quality compared to confocal Nipkow Disk laser-scanning microscopes [70].

7.5.3 Detection Methods

7.5.3.1 Descanned and Non-descanned Detection

In TPLSM and SHIM it is possible to use two fundamentally different detection arrangements to record the sample fluorescence. In the non-descanned mode, excitation and fluorescence are separated by a dichroic mirror located directly behind the objective lens (see **Figure 7.8**). The dichroic mirror is tuned to transmit a signal from the sample to the detector or the eyepiece of the microscope, while the excitation NIR light is reflected. The position of the fluorescent focal volume in the sample is directly imaged on the detector, resulting in a direct image of the sample if the exciting beam is scanned. Therefore, it is possible to use a camera (2D detector), the eyepiece or a large-field point detector such as a PMT as detector. In the latter case the scan field must be scaled down by intermediate optics to fit the PMT surface. The advantage of non-descanned detection using a PMT is its great sensitivity due to the fact that even strongly scattered fluorescence photons contribute to the signal.

In the second detection mode (descanned mode) the fluorescence from the sample is directed back via the scanning mirrors. As the average fluorescence lifetime of a molecule is much shorter than the time needed to move the scanner from one position to the next the fluorescence signal is directed back onto the excitation beam axis. A dichroic mirror (see **Figure 7.8**) located in front of the scanning mirrors separates the excitation laser light and the fluorescence signal. The main difference to the non-descanned mode is the fact that the fluorescence is always directed onto the same spot. In order to construct an image of the sample under investigation, it is necessary to correlate the time-dependent fluorescence with the time-dependent position of the scanning mirrors. Therefore the use of point detectors is much easier in this measurement mode and in addition it is possible to introduce a confocal pinhole to improve the resolution.

7.5.3.2 Spectral-resolved Imaging

As fluorophores have different emission properties, it is reasonable to provide contrast mechanisms for fluorescence detection that go beyond mere intensity registration. For measurements that involve fluorescent probes with known and distinguishable emission spectra, it is effective to use different filters adapted to these spectra, for instance in epifluorescence microscopy. This allows fast measurements with a very effective use of the generated fluorescence signal. For weak native fluorescence signals, it is advantageous to use broad filters to improve the signal-to-noise ratio or to distinguish between fluorophores and SHG signals.

The use of spectrographs or spectrometers allows for the measurement of complete emission spectra with a much higher spectral resolution. When using a spectrometer in the non-descanned arrangement, line scans in the sample are projected onto its entrance slit and an emission spectrum for each point along the line scan is generated perpendicular to the slit axis. Therefore a 2D camera image is generated that consists of the emission spectra in the x-axis along different positions in the line scan, which are represented in the y-axis. As the line scan is imaged onto the spectrographs entrance slit it must remain at the same position inside the sample. Therefore, it is only possible to perform beam scanning along this axis. To acquire a spectral-resolved image of a sample plane, stage scanning along the direction perpendicular to the entrance slit is required. To overcome the disadvantage of sample scanning it is also possible to perform spectral-resolved measurements in the descanned mode. Through the use of a camera up to 64 spectra (corresponding to the number of foci in the sample) can then be measured simultaneously. Nevertheless, this measurement mode is comparatively slow as it requires a slow scanning process because the generation of complete camera pictures takes milliseconds.

Generally, spectral-resolved measurements are a trade-off of the factors resolution (local and spectral), time, photodamage, fluorescent yield and data volume. Different samples and the scientific questions connected to them require different techniques of spectral measurements.

7.5.3.3 **Fluorescence Lifetime Measurements**

The average time a molecule remains in the excited state is determined by the number of deactivation pathways and their competing rates. Using fluorescence lifetime measurements, information about complex photophysical processes can be obtained to determine the rates of deactivation. Lifetime measurements are extremely sensitive to the molecular environment of fluorescent molecules.

Fluorescence lifetime measurement techniques can principally be divided into time and frequency domain approaches. In frequency domain measurements [71] the fluorophores are excited with sinusoidally modulated light at high frequencies (20–80 MHz). The emitted fluorescence signal has the same frequency, but undergoes a phase shift and a decrease in amplitude (demodulation) with respect to the excitation radiation. These dynamic parameters can be related directly to the lifetime of the emission. The maximum temporal resolution is determined by the modulation frequency and is roughly 1 ns.

Far more common are time domain measurement methods that use a short pulsed light source and detect the time-dependent fluorescence with respect to the excitation pulse. One of them uses an intensified CCD camera as a time-gated detector [72]. The lifetime image is generated by recording the intensity of the fluorescence at a series of different time points after the excitation pulse. The maximum temporal resolution is determined by the minimum gate width of the intensified CCD camera ($\sim$ 50 ps). Although many photons are lost by the use of gates, this method delivers FLIM images much faster than all other methods, as lifetimes are recorded simultaneously in each point of a whole sample section. A different technique is time-correlated single-photon counting (TCSPC) which is probably the most widespread method [73,74]. In this case a point detector (PMT) delivers an electrical pulse for each registered photon and a fast counting electronic device measures the time difference between the excitation radiation and the detected photon. Thus the fluorescence decay curve can directly be measured with a maximum temporal resolution of $\sim$ 10 ps. The only drawback of this method is the long time required to generate a FLIM image ($\sim$ 30 s), which is due to the point-by-point acquisition process. An interesting new time domain approach is a streak camera FLIM system [75], which uses a sweep electrode to deflect photoelectrons to different positions on a phosphor screen depending on their arrival time at the detector. This method has an acceptable temporal resolution ($\sim$ 20 ps) and the potential to acquire fast FLIM images. But at present it still needs several tens of seconds to generate a FLIM image.

7.6 Results and Application

7.6.1 Optics

7.6.1.1 Development of a Parallelized Two-photon Measurement System for Rapid and High-resolution Tissue Imaging

The desire to monitor tissue-engineering products for cartilage regeneration during the *in vitro* cultivation process required the invention of a very special measurement system. The demand for sub-cellular resolution deep inside dense optical material could only be fulfilled using TPLSM. In addition, weakly fluorescent endogenous fluorophores contained in the ECM of cartilage should be observed, as well as the response of unstained cartilage tissue to mechanical or biochemical stimulation. Therefore the two-photon microscope had to be parallelized to acquire as much light as possible in a given time interval. The underlying principle of splitting a single beam into 64 beams that are scanned simultaneously across the object plane was developed in the framework of the BMBF-project "Non-Linear Laser Raster Microscopy". Using this technique, the acquisition of images or 3D volumes can be accelerated by a factor of 64.

The scheme in **Figure 7.8** illustrates the principle of operation of the parallelized two-photon laser-scanning microscopes that are adapted to the requirements of strongly scattering samples.

The incoming laser beam first passes an attenuator, a telescope and the prechirp arrangement. It is then split into up to 64 beams by multiple transitions through a 50% splitting substrate (**Figure 7.9**).

The split beams are coupled into the microscope through intermediate optics and focused onto the sample by the objective lens. A single line of foci is generated that is rastered by a galvanometric x, y-scanner in the object plane. Together with the microscope z-drive, this enables the 3D imaging process. The fluorescence can be observed through the eyepieces, or imaged onto a CCD-camera or a PMT in a non-descanned arrangement. Furthermore, it can be detected with a parallelized descanned PMT detector.

All optical elements were optimized with regard to the highest transmission achievable, as this determines the maximum degree of parallelization, which in turn limits the image acquisition speed. A throughput of over 75% was

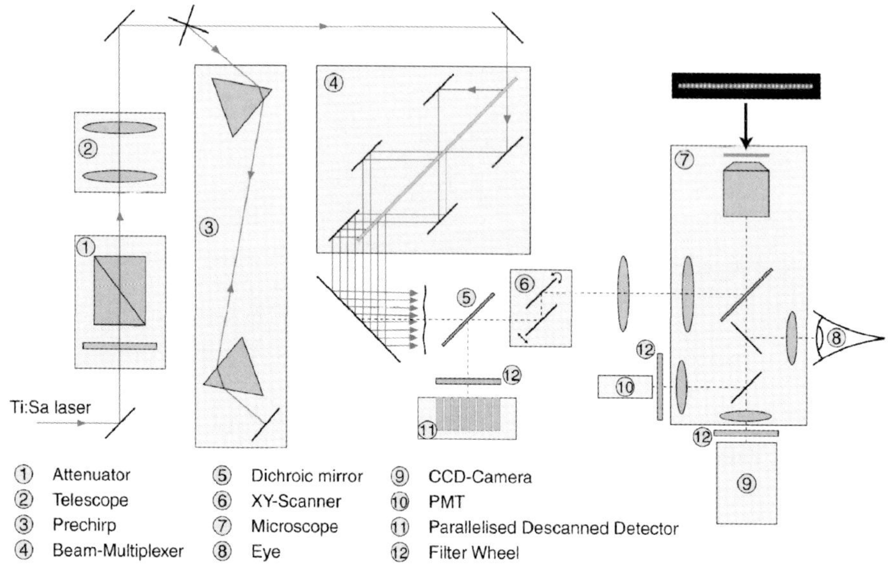

Figure 7.8 Schematic set-up of the parallelized two-photon laser-scanning microscope.

achieved for the whole Ti:Sa wavelength area of 710–980 nm. Since the two-photon generated fluorescence decreases linearly with increasing excitation pulse length the dispersion compensation was set to completely equalize the pulse broadening introduced by the optical elements. As a result, the temporal length of the pulse is 120 fs in the sample. Special scan optics were developed that, in combination with an innovative objective lens (20×, 0.95 NA), allow the observation of very large sample sections (up to 500 × 500 μm) at high resolution. A key feature is the possibility of reducing the degree of parallelization from 64 to 32, 16, 8, 4 or even to a single beam (**Figure 7.10**).

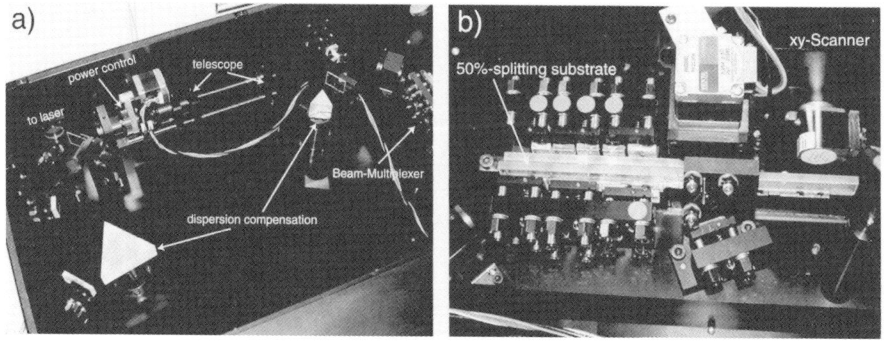

Figure 7.9 Photographs of the inside of the scan head: (a): the beam shaping and steering elements and (b): the beam-multiplexer.

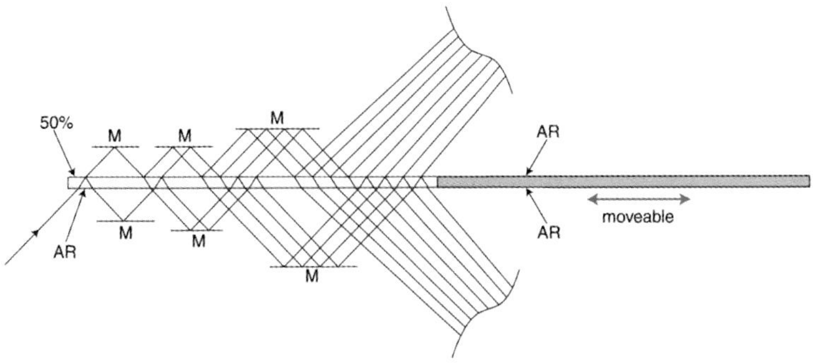

Figure 7.10 Principle of changing the degree of parallelization.

This is done through a stepwise exchange of the 50%/AR coated substrate with the AR/AR coated one. Both substrates are arranged in line and mounted on a motorized holder, allowing a simple and software-controlled conversion. Each reduction of the number of beams by a factor of two increases the power in each remaining beam by the same factor, yielding an overall gain in fluorescence of two. The increase of laser power per beam is crucial if deep sample planes are to be imaged. It permits roughly 20% additional imaging depth in dense tissue. By replacing the 50%/AR substrate completely with the AR/AR substrate, only one beam passes the set-up. It seems to be clear that this is extremely useful, since very deep sample planes cannot be observed with multiple beams and a field detector such as a CCD camera owing to the strong scattering. Thus a single-beam scan mode was developed featuring a PMT in the non-descanned arrangement as detector. In addition, SHIM can be performed, which provides an important tool to discriminate between collagen matrix structures, chondrocytes and autofluorescence signals, but cannot be performed in parallelized mode.

The maximum penetration depth in parallelized mode was measured to be roughly 80% of the single-beam mode while the resolution is nearly equal during the first 60%. At the surface the resolution of both modes is 310 nm in the lateral and 900 nm in the axial direction when using a 1.4 NA objective lens and 800-nm light for excitation. The maximum achievable frame rate of the instrument is roughly 1000 Hz but is nearly always limited by the small number of photons emitted by the sample.

7.6.1.2 **Control and Automatization of the System**

All components of the system, such as the power control, x, y-scanner, shutter, parallelization selector, z-stepper, filterwheel, x, y-sample stage, CCD-camera and the A/D-converter to read out the PMT are software controlled. To observe *in vitro* tissue cultivation automatically over several hours or days, up to six-dimensional datasets (x, y, z, P, t,) can be recorded, in which all measurement settings are saved together with the appendant dataset.

A synchronization module has been developed that enables the exact timing of all system components that take part in a complex measurement. It assigns a time point to each acquired image, thus allowing the exact determination of the time elapsed between interesting molecular or cellular events.

To observe dynamics such as the response of cartilage to mechanical stimulation with high temporal resolution, a fast time-lapse mode was arranged that reads out the CCD camera during the time needed to acquire a new image. For many typical applications, this speeds up the imaging process by a factor of two.

7.6.1.3 **Development of New Measurement Methods to Image Strongly Scattering Tissues**

As already mentioned, 3D tissues often possess a very complex architecture and composition. To be able to acquire informative data from strongly scattering tissues, additional optical methods have to be employed as described below.

Sequential Imaging of Different Sample Positions

Through the use of a motorized x, y-sample stage it is possible to scan different regions within the sample in a single measurement each with its own settings (e.g., scan mode, emission filter). The number of recurrences and the time between two scans can be freely chosen. In addition this concept allows high-resolution imaging of very large sample sections or volumes in the order of several square or cubic millimeters. In this way, images of neighboring sample sections are recorded sequentially and patched together by the software into the final image.

Spectral Unmixing

If emission filters are used to spectrally separate different fluorophores and endogenous species, a common problem is cross-talk between these filters caused by the broadness of the emission spectra. To overcome this problem and to visualize weak contrasts, a spectral unmixing mode was developed: data points in an acquired image that contain characteristic spectra are chosen and used as reference points; afterwards a software algorithm searches for these spectra in all other pixels and assigns only one spectrum to each (**Figure 7.11**).

z-Drop

A fundamental problem in 3D microscopy is the decrease of fluorescence signal as a function of penetration depth (z-drop) caused by the focus degrada-

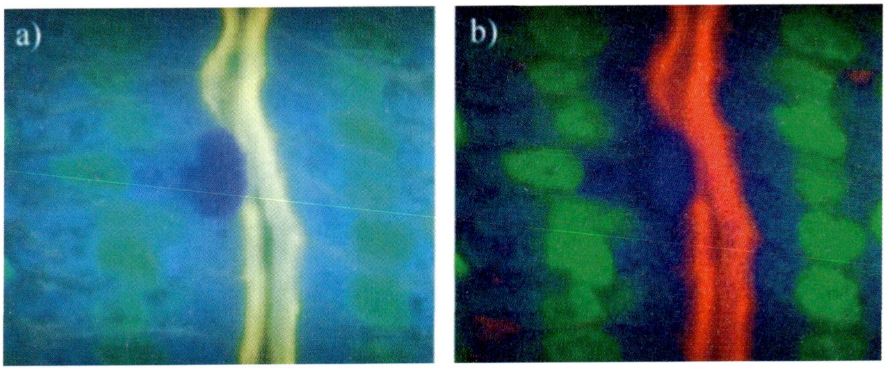

Figure 7.11 Spectral-resolved measurement of a mouse intestine section using a filterwheel and three different detection filters; (a): Overlay of the three color channels and (b): spectral-unmixed image.

tion and increased scattering of the photons on their way back through the sample. To compensate for these effects it is possible to automatically reduce the number of beams or/and to increase the excitation power with increasing depth. The aim is to keep the camera integration time and the recorded fluorescence intensity constant. Otherwise it is hardly possible to measure a large depth range and to perform a subsequent 3D reconstruction of the imaged volume. Using this method it was possible to measure to a depth of 1 mm into a collagen gel in a single measurement (**Figure 7.12**).

High Throughput Spectral Measurements

The spectroscopical non-descanned mode of operation to determine native fluorescence spectra (see **Figure 7.17**) is compromised by the fact that these spectra are rather weak and broad. The total fluorescence intensity arising from one point inside the sample is not simply collected on one camera pixel, as in imaging measurements, but is dispersed to a complete line on the CCD chip, generating the emission spectrum of the fluorescent point. This, however, requires long acquisition times for a complete analysis. To optimize the optical throughput of this type of spectral measurement, a new spectral measurement mode was developed which uses a straight vision prism instead of a spectrograph. As a line scan in the sample works as an entrance slit for the prism, no fluorescence light is blocked by a mechanical slit and all of the fluorescence from the line scan can contribute to the spectrum. Furthermore, losses due to absorption in the detection pathway are minimal, as only the prism is additionally introduced. The advantages of this high throughput, easy and affordable set-up and wide detection spectrum (180 nm for 8 mm $\times$ 8 mm CCD chip) are only compromised by a relatively high sensitivity to stray light.

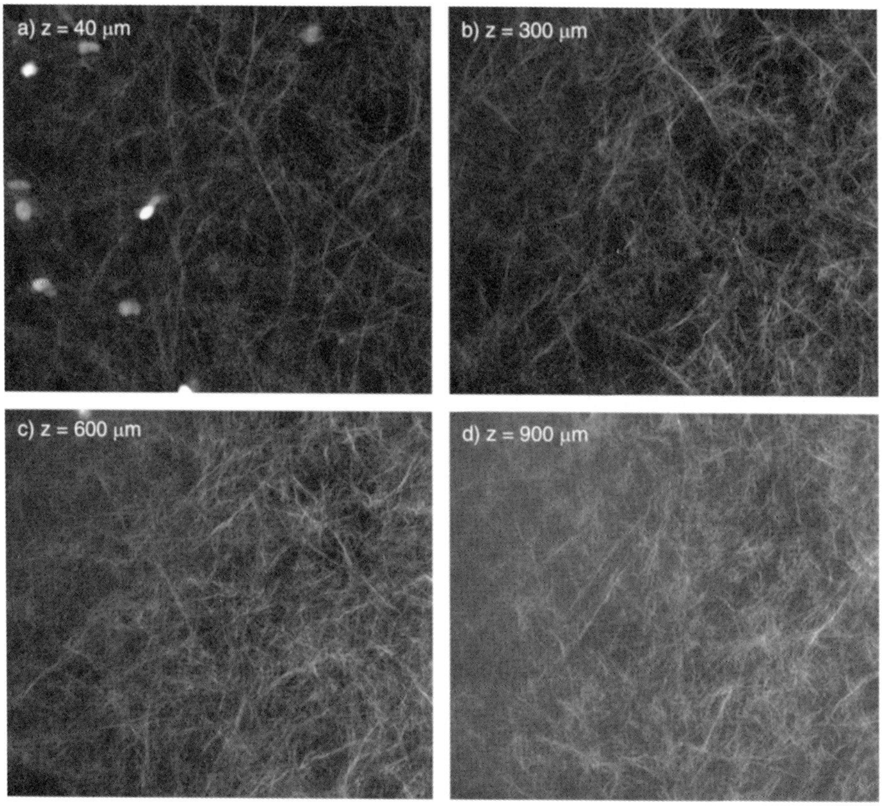

Figure 7.12 Adaptation of the number of beams and the excitation power as a function of depth inside the sample to compensate for the z-drop.

7.6.2 Cartilage and Chondrocytes

7.6.2.1 Human Cartilage Tissue

Owing to the intrinsically dense structure of its extracellular matrix (ECM), cartilage is a tissue that strongly scatters light but does not absorb much. Hence, articular cartilage has an opaque appearance and is called hyaline. Penetration depth of visible light into this tissue is rather low, making characterization with conventional brightfield microscopy difficult, as it requires staining and/or microtome cuts.

Figure 7.13 shows the 3D autofluorescence reconstruction of unstained healthy human cartilage measured with TPLSM [76]. Measurements with a tissue penetration depth of up to 200 μm (up to 460 μm are possible for arthritic bovine cartilage) were performed at an excitation wavelength of

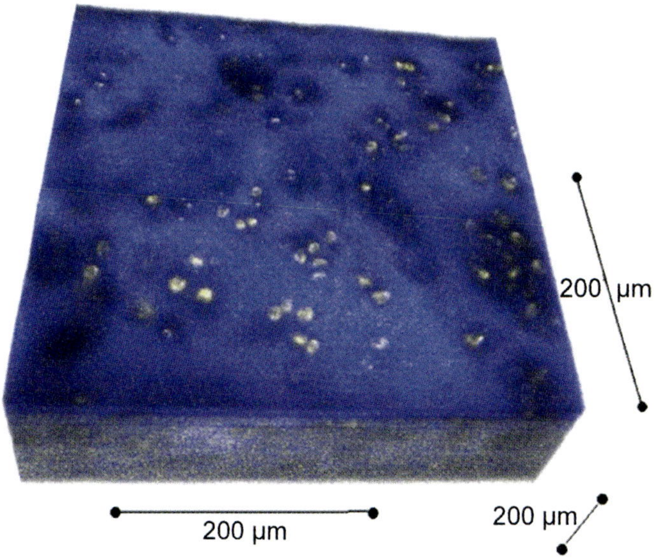

Figure 7.13 3D autofluorescence reconstruction of unstained healthy human cartilage tissue. The ECM (HQ 525/50) and the chondrocytic cells (HQ 575/50) are represented in blue and yellow. The presented plane lies 50 µm below the tissue surface of the sample (cartilage sample courtesy of Dr. M. Dickob, Bielefeld).

800 nm with 64 parallel foci at a total laser power of 260 mW, keeping laser power at 4 mW per focus to prevent photodamage [77]. The spectral discrimination between ECM and chondrocytic cells was achieved by recording two separate image stacks with fluorescence emission filtering for ECM (HQ 525/50) and for the chondrocytes (HQ 575/50). As **Figure 7.13** demonstrates, spectral discrimination between ECM and chondrocytes is possible. Furthermore, complete mapping of the tissue allows a direct estimate of the corresponding chondrocyte density in the tissue region (in this case approximately 20×10^6 cells cm^{-3}). It has to be mentioned, however, that the chondrocyte density varies vastly for different samples, even within the same sample, depending on the relative position of the region of interest within the cartilage. We found chondrocyte densities that range from approximately 2×10^6 cells cm^{-3} to 20×10^6 cells cm^{-3} in the same cartilage sample, which is indeed in accordance with data derived from healthy cartilage by other research groups [78]. Furthermore, in order to compare healthy and arthritic tissue, samples from the same patient were investigated (**Figure 7.14**).

Both tissue samples were characterized with TPLSM at 800 nm using a laser power of 240 mW in 64-foci parallel operation mode. For mapping the surface morphology, only the green fluorescent emission filtering for the ECM (HQ 525/50) was recorded.

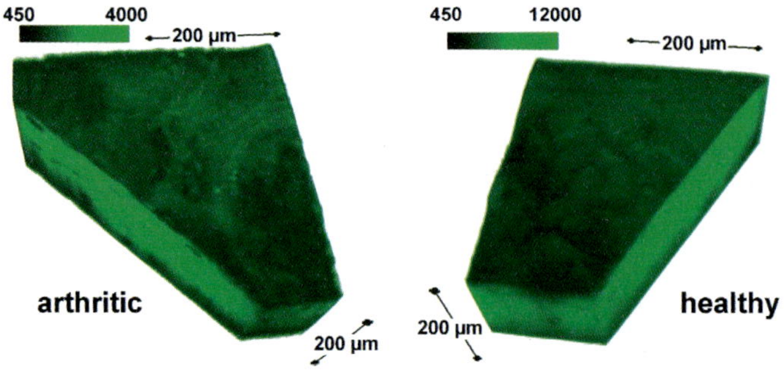

Figure 7.14 3D autofluorescence reconstruction of the surface of unstained healthy and arthritic human cartilage tissue from the same patient (cartilage sample courtesy of Dr. M. Dickob, Bielefeld).

Considering the autofluorescence images two aspects that are related to the macroscopic diagnosis are quite evident (**Figure 7.14**). First, healthy cartilage tissue displays a much higher autofluorescence emission from ECM than does arthritic tissue, which can be interpreted as an indication of reduced tissue density in the arthritic case. Second, the two outer surface structures differ significantly in respect of smoothness and morphology. Whereas the surface of healthy tissue is smooth and isotropic, the arthritic surface is fibrous and rather rough. This change in arthritic tissue morphology to a rough, fibrous surface is consistent with increased frictional resistance and consequently increased wear damage between articulating cartilage surfaces.

These experiments reveal that native hyaline cartilage from a human knee joint can directly be investigated with TPLSM without using additional staining or labelling protocols. It is important to note that this technique can potentially be used in future diagnostic applications, for example for a better quantitative definition of different stages of arthritis or osteoarthritis of articular cartilage.

7.6.2.2 **Chondrocytes on Collagen Scaffolds**

The influence of scaffold materials and structures on cell performance has been described above. To demonstrate the ability of appropriate scaffold materials to enhance the chondrocyte response in the tissue-engineering process on the one hand and to show the potential of laser-scanning microscopy associated with the ability to monitor relevant process parameters on the other, two different collagen scaffold structures were used (**Figure 7.15**).

Based on native fluorescence and SHG properties, impressive images could be taken from the different collagen matrices via TPLSM. **Figure 7.16** shows a collagen I/III fleece. The autofluorescent signals provide a high-resolution image of the scaffold structure similar to that seen in non-native SEM investi-

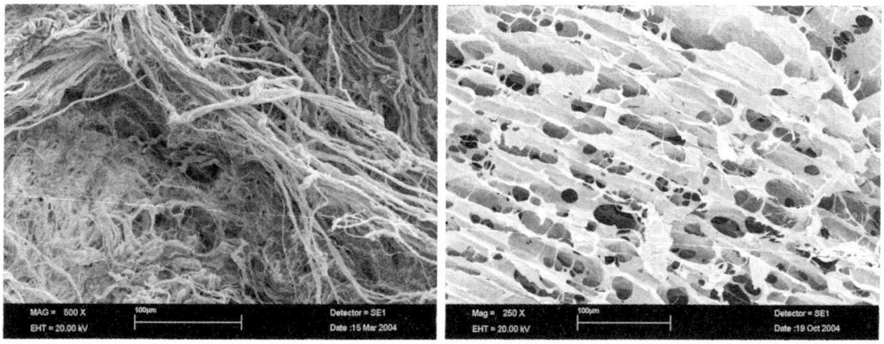

Figure 7.15 Collagen I/III scaffolds for cultivation of chondrocytes based on porcine collagen (SEM). Left: fleece (ACI-Maix™, Matricel GmbH), Right: sponge-like structure (Matricel GmbH, Germany).

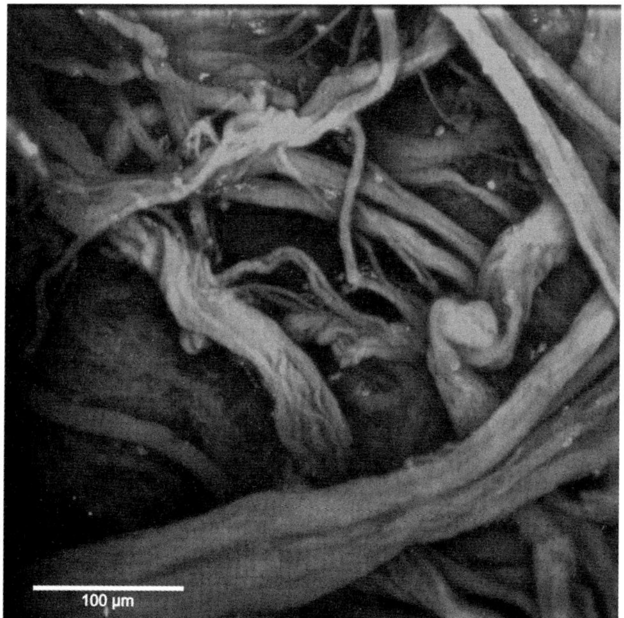

Figure 7.16 Projected image of a 170 µm thick and unstained collagen membrane, two-photon-excitation at 800 nm taken in parallel beam mode (64 beams; bar: 100 µm).

gations. A distinct fibrous structure, providing a 3D scaffold for chondrocytes is apparent.

In **Figure 7.17** a sponge-like collagen scaffold is presented. According to preliminary test results in this project, the sponge-like scaffold provides higher viability for chondrocytes than fibrous membranes. This could be shown by comparative investigations on the influence of scaffold structures on the chondrocyte response.

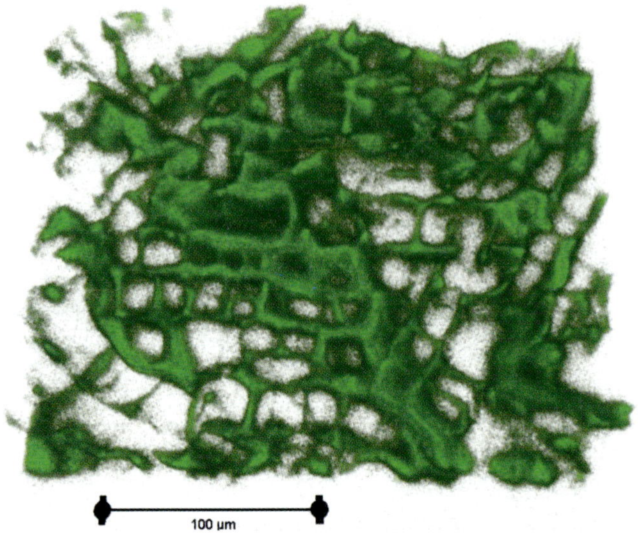

Figure 7.17 Reconstruction of sponge-like collagen membrane, two-photon-excitation at 800 nm, 64 beams, <5 mW per focus, native fluorescence with filter HQ 525/50 (bar: 100 µm).

The crucial point of an online analysis of cells and the dependence of their differentiation on the incubation conditions is the visualization of cells and their metabolic products within an engineered tissue construct. Investigations by laser-scanning microscopy require signals that are specific for the distinct components. In this context, implantable functional chondrocytes are a suitable example to demonstrate the potential of laser-scanning microscopy as a tool for quality control. Primary bovine chondrocytes (healthy femoral knee joint) cultured on collagen I/III scaffolds were used as a model system to detect the fundamental components of cartilage tissue-engineering constructs: scaffold material, cells and synthesized ECM. **Figure 7.18** shows chondrocytes on a collagen I/III fleece cultivated for seven days detected with a common CLSM which is applicable for thin and low-scattering samples. Autofluorescent signals of the collagen fleece can be separated from the cells (labelled with Syto 83) and the synthesized proteoglycans (keratan sulfate, labelled with a monoclonal antibody for keratan sulfate conjugated with FITC). The cells are attached to the collagen fibers. The proteoglycans are synthesized by the chondrocytes and appear as released compounds around the cells. The quantity of synthesized ECM components around the cells can be an essential marker for cell differentiation and cell stimulation. Interestingly, it could be shown that the cell response can be enhanced by cultivation on collagen I/III scaffolds with a sponge-like structure. **Figure 7.19** shows the appearance of differentiation markers depending on the scaffold structure. In contrast to the fleece

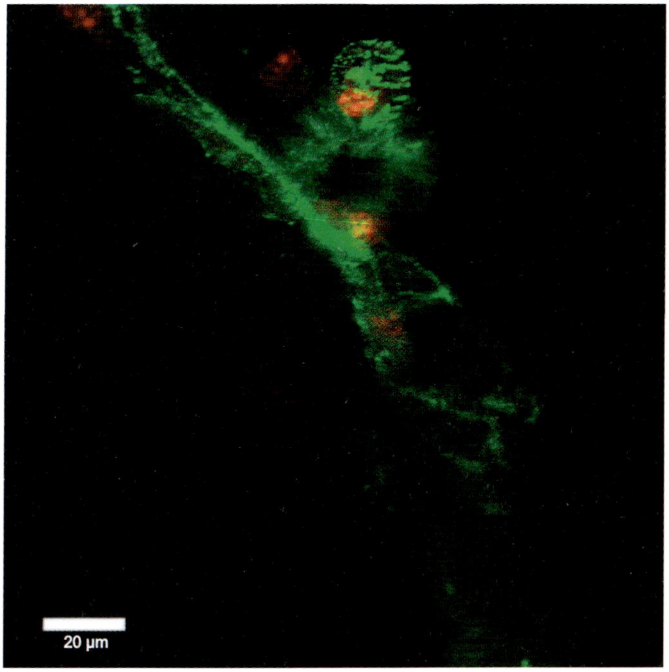

Figure 7.18 Chondrocytes on collagen I/III fleece synthesizing keratan sulfate (red: cells, green: keratan sulfate around the cells, the strung-out green structure represents an autofluorescent collagen fiber; CLSM, bar: 20 μm).

scaffold the cell morphology on the sponge-like scaffold is similar to native chondrocytes embedded in cartilage tissue. The synthesis of ECM components such as chondroitin sulfate and collagen type VI appears more homogeneous and is enhanced on sponge-like scaffolds, indicating an improvement of cell redifferentiation.

The cell-promoting effects of structured collagen scaffolds detected by laser-scanning microscopy correlate with quantitative biochemical data concerning the amount of released proteoglycans (**Figure 7.20**). Despite the fact that a progressive dedifferentiation during the expansion phase cannot totally be prevented, the sponge-like scaffolds reveal a clear enhancement of the cartilage-specific ECM synthesis.

The results show the tremendous potential of laser-scanning microscopy to detect cells and ECM components directly within deep tissue regions engineered under *in vitro* conditions. However, a real online analysis of tissue-engineering constructs during the incubation process requires the visualization of the most important components (scaffold, cells, ECM marker) based on autofluorescence signals. In fact, a wide spectrum of biological samples show autofluorescent signals generated from a wide range of molecules present.

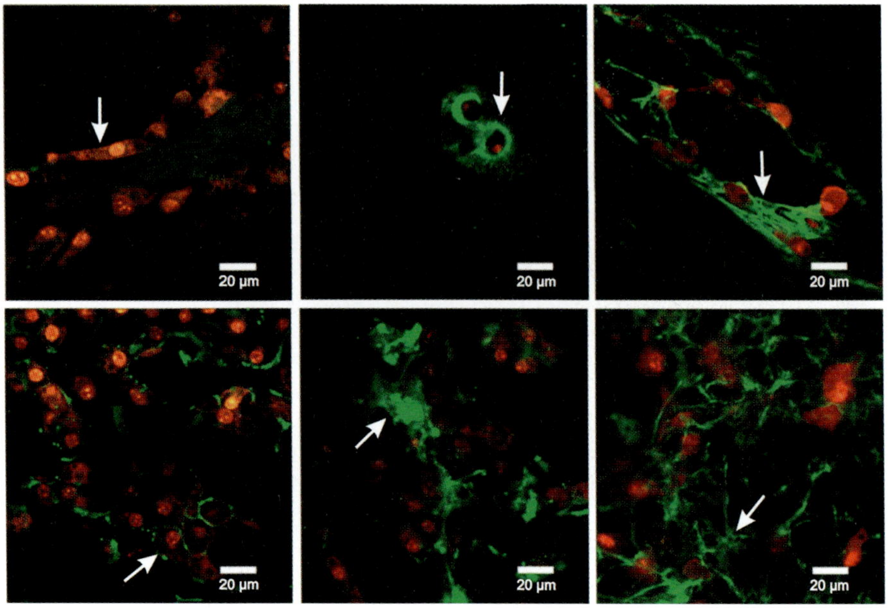

Figure 7.19 Performance of chondrocytes (after one subcultivation) on collagen fleeces (top) and sponge-like collagen scaffolds (bottom). Left: morphology, Middle: synthesis of chondroitin sulfate, Right: synthesis of collagen type VI (Syto 83-staining of cells, FITC-labelled Anti-CS and Anti-Coll VI; CLSM; bars: 20 µm); arrows indicate the main results described in the text.

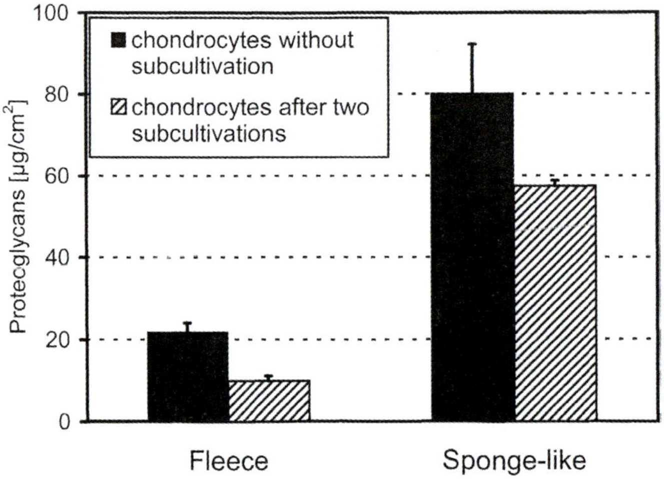

Figure 7.20 Synthesis of proteoglycans depending on the scaffold structure. Sponge-like collagen scaffolds promote ECM synthesis by chondrocytes seeded on scaffolds without and after two subcultivations in the cell expansion phase ($n = 6$).

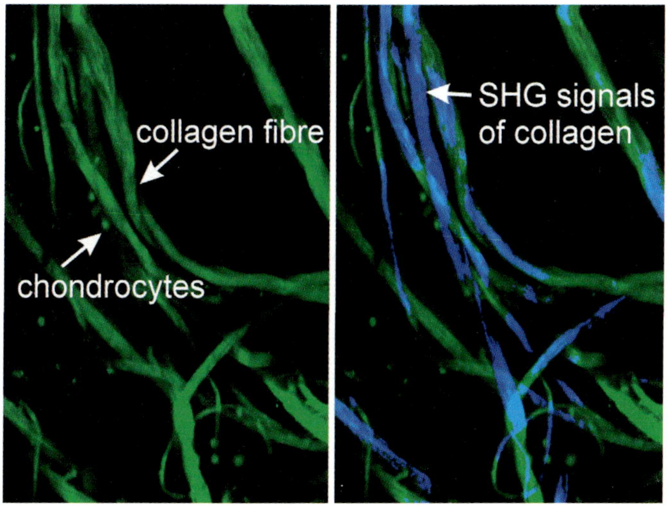

Figure 7.21 Chondrocytes on collagen fleece: TPLSM, @ 800 nm; Left: autofluorescence, Right: autofluorescence and SHG of collagen fibers (blue, HQ 10/20; merged image).

The excitation of autofluorescent molecules in biological samples is very effective using TPLSM, because the employed NIR excitation wavelengths allow observation in the VIS region arising from the simultaneous absorption of two photons. Parallelized TPLSM provides a wide range of flexibility to control the excitation energy by splitting the beam into up to 64 single beams. Thus, the excitation of autofluorescence can be maximized without damaging the sample. The basic task in online analysis of tissue constructs represents the visualization of the three main components described above within a sample. However, besides the laser optical conditions necessary for a sensitive autofluorescence monitoring, the selective detection of distinct components requires further analytical methods. The excitation of autofluorescence within a tissue construct is realizable using the advantages of TPLSM. Chondrocytes as well as collagen fibers can be detected (**Figure 7.21**, left).

SHG signals of collagen fibers (**Figure 7.21**, right) allow the separation of scaffold structures from cells by spectral unmixing and enable a quantitative estimation of cell numbers. Furthermore, the actual number of vital chondrocytes in the tissue matrix often represents a key parameter for the optimization of tissue-engineering technologies. In order to obtain primary information, a spectral discrimination of the chondrocytes from the collagen matrix and the synovial fluid and the culture medium is mandatory. Accordingly, we analyzed the spectral response of isolated chondrocytes, collagen matrix and culture medium by two-photon excitation in the accessible wavelength range from 450 to 700 nm. All the investigated systems exhibited spectra that

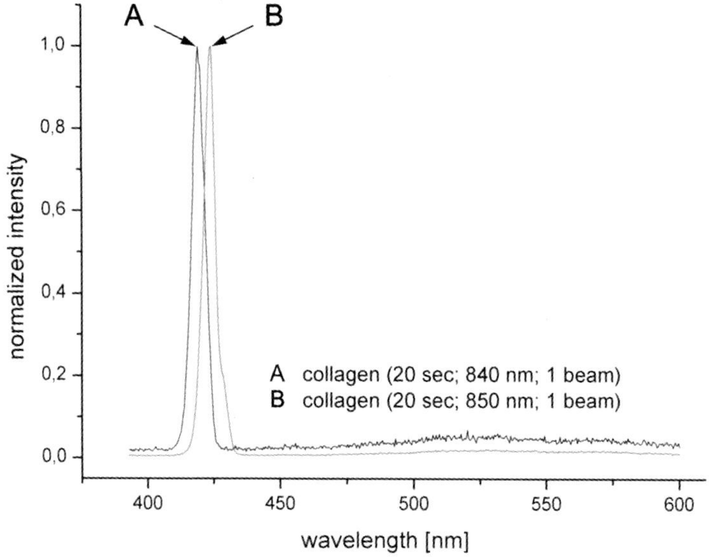

Figure 7.22 SHG of collagen I/III fleece.

were very similar and therefore not helpful for proper spectral discrimination. However, collagen fleece scaffolds generate distinct SHG signals, as displayed in the spectrum shown in **Figure 7.22**. This property possibly provides the background to the development of an effective contrast mechanism to detect the collagen scaffold and to discriminate the scaffold material from the chondrocytes.

By using two-photon excitation at 820 nm with a single laser beam (power <15 mW) and a set of filters, a three-color image was generated in which a single chondrocytic cell (yellow) is clearly visible on a collagen matrix (blue) (**Figure 7.23a**). In **Figure 7.23b**, the same dataset has been digitally reconstructed in a pseudo-3D representation displaying the chondrocyte in red.

The advanced separation of ECM-specific autofluorescent signals provides the basis for a qualitative evaluation of the status of the cartilage-specific differentiation of cells in dependence on the cultivation conditions, including the observation of the effect of biochemical and mechanical stimulation. Furthermore, fluorescence decay times are currently being investigated to provide detailed data referring to the appearance and possible kinetic changes of synthesized ECM components within cultured tissue constructs. In this context FLIM is expected to be a powerful tool to detect and analyze biological tissues and molecular interactions quantitatively [79,80].

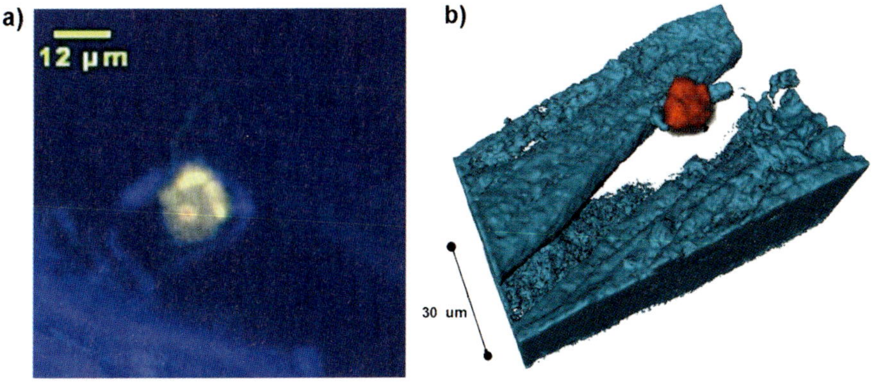

Figure 7.23 (a): Spectral discrimination of chondrocyte and collagen matrix (filter set: blue: SHG with HQ 410/20, red: native fluorescence with HQ 525/50, green: native fluorescence with HQ 575/50). (b): Same dataset in digital pseudo-3D representation.

7.7 Summary and Outlook

Regenerative medicine and tissue engineering are exciting new fields taking advantage of both engineering and biology. The process of creating living, physiological, 3D tissues utilizes specific biomaterials as scaffolds to guide tissue growth *in vivo* and *in vitro*. The most appropriate scaffolds are the ones that provide the intricate hierarchical structure (e.g., 3D architecture, chemical composition) that characterize the native tissue to be replaced. In the framework of the present study, two chemically equivalent but structurally different collagen scaffolds were investigated and it was found that the analyzed sponge-like collagen membrane inhibits cellular dedifferentiation of the chondrocytes seeded on the membrane. Obviously the sponge-like membrane offers favorable conditions for tissue formation and tissue regeneration. It is of value to point out that the whole process of a 3D tissue formation is a highly orchestrated set of sub-cellular (molecular), cellular, and supra-cellular events that are far from being well understood. That is why noninvasive measuring methods with appropriate spatial and temporal resolution are necessary.

Besides the necessity to provide suitable scaffolds consisting of advanced bioactive materials, a second major challenge in tissue engineering was identified. The cultivation process itself needs a technology platform to guarantee reproducible and controllable conditions for tissue growth and cell differentiation according to the nature of the tissue-engineered product. It is widely accepted that bioreactors and flow chamber systems offer a tremendous potential to ensure that all relevant aspects are fully considered. In this context, growth conditions (e.g., pH-value, pO_2, temperature, nutrient supply), scale-up and sterility issues are important factors if safe, clinically effective and, last but not least, competitive tissue-engineered products are to be launched onto the market. Keeping in mind the concept of functional tissue engineering, it is essential, especially in the field of cartilage repair, to establish a biomechanical stimulation during the cultivation process so as to mimic the native mechanical environment within the bioreactor.

The whole process of bioreactor design and bioreactor-based tissue and cell cultivation will be like a "mission impossible" without using proper measuring methods, with appropriate spatial and temporal resolution, to obtain reliable feedback from the cultivation process.

The present study has shown that noninvasive imaging methods such as laser-scanning microscopy provide striking advantages over conventional fluorescence microscopy and appear to be a novel detection tool for 3D resolved fluorescence imaging. Of special importance is the possibility of monitoring and controlling cell cultivation processes in the field of regenerative medicine to ensure a high quality of tissue-engineered constructs that can be used successfully to treat affected patients. The application of these methods benefits from the fact that being noninvasive they do not disturb the cultivation process itself. A spatially resolved analysis of tissue engineering constructs will be extended by kinetic data (4D analysis) to obtain the necessary temporal resolution. In this way NIR multiphoton excitation laser-scanning microscopy will become a powerful tool in the area of quality control in biomedical applications. The present study is focused on tissue-engineering approaches and the basic feature of selective imaging of thick biological samples allows in principle an equivalent application in therapeutic and diagnostic fields as well as in biofilm monitoring.

Acknowledgements

The authors would like to thank the BMBF (Federal Ministry of Education and Research) for granting this work in the frame of the "Biophotonics" network (grant numbers 13N8432, 13N8434, 13N8435).

Glossary

MACI®/ACT Matrix-induced Autologous Chondrocyte Implantation / Autologous Chondrocyte Transplantation. Tissue engineering techniques for the regeneration of cartilage. Autologous chondrocytes are isolated, expanded *in vitro* and subsequently re-implanted into the defect site with and without supporting collagen scaffold materials, respectively.

TPLSM Two-photon Laser-scanning Microscopy. In TPLSM a focused pulsed laser beam is used to excite (native) fluorophores through the simultaneous absorption of two NIR photons. The two-photon absorption process requires a high-power density and therefore only takes place in the focal volume of the microscope's objective lens. This intrinsic sectioning property allows the generation of 3D fluorescence images deep inside living samples.

SHG/SHIM Second Harmonic Generation/Second Harmonic Generation Imaging Microscopy. Second harmonic generation is a nonlinear optical effect that generates one photon out of two, carrying the total energy of both incident photons. This conversion requires high optical power densities and the vicinity of polarized structures. Therefore, SHG can be used to image certain materials such as collagen fibers, resulting in a new microscopy technique called SHIM.

FLIM Fluorescence Lifetime Imaging Microscopy. The average time a molecule remains in the excited state is called its fluorescence lifetime. Fluorescence lifetime measurements are extremely sensitive to the molecular environment and provide information about complex photophysical processes. In FLIM typically a short pulsed light source and a detector that registers the time-dependent fluorescence with respect to the excitation pulse is used to generate the fluorescence decay curves.

Key References

K. KÖNIG, *Journal of Microscopy* 200 (**2000**), p. 83.

W. R. ZIPFEL et al., *Nature Biotech.* 21 (**2003**), p. 1369.

A. DIASPRO, *Confocal and Two-Photon Microscopy: Foundations, Applications and Advances*, Wiley-Liss Inc., New York, 2002.

W. DENK, K. SVOBODA, *Neuron* 18 (**1997**), p. 351.

R. SKALAK, C. F. FOX, in: *Frontiers in Tissue Engineering* (Eds.: C. W. Patrick, A. G. Mikos, L. V. McIntire), Elsevier Science, New York, 1998.

J. M. POLAK, L. L. HENCH, P. KEMP, *Future Strategies for Tissue Engineering and Organ Replacement*, Imperial College Press, London, 2002.

D. SHI (Ed.), *Biomaterials and Tissue Engineering*, Springer-Verlag, Berlin, Heidelberg, 2004.

References

1 V. MARTINEK, *Deutsche Zeitschrift für Sportmedizin*, **2003**, p. 166.

2 A. MÜLLER, *Diplomarbeit, Westsächsische Hochschule Zwickau*, 2004.

3 P. SARZI-PUTTINI et al., *Semin. Arthritis. Rheum.* 35 (**2005**), No. 1, p. 1.

4 M. BRITTBERG et al., *N. Engl. J. Med.* 331 (**1994**), No. 14, p. 889.

5 M. RONGA et al., *Foot Ankle. Surg.* 11 (**2005**), p. 29.

6 E. BASAD et al., *Orthopädische Praxis* 40 (**2004**), p. 6.

7 M. RONGA et al., *Arthroscopy: J. Arthro. Rel. Surg.* 20 (**2004**), No. 1, p. 79.

8 D. LUPPA, *KCS* 1 (**2000**), No. 12, p. 29.

9 M. J. LYSAGHT, J. A. O'LOUGHLIN, *ASAIO J.* 46 (**2000**), p. 515.

10 M. B. ROTH, T. G. NYSTUL, *Spektrum der Wissenschaft* **2005**, Sept., p. 42.

11 L. L. HENCH, *Biomaterials* 19 (**1998**), p. 1419.

12 R. SKALAK, C. F. FOX, in: *Frontiers in Tissue Engineering* (Eds.: C.W.Patrick, A.G. Mikos, L.V. McIntire), Elsevier Science, New York, 1998.

13 J. M. POLAK, L. L. HENCH, P. KEMP, *Future Strategies for Tissue Engineering and Organ Replacement*, Imperial College Press, London, 2002.

14 D. SHI (Ed.), *Biomaterials and Tissue Engineering* Springer-Verlag, Berlin, Heidelberg, 2004.

15 S. NAGEL-HEYER, *Ingenieurtechnische Aspekte bei der Herstellung von dreidimensionalen Knorpel-Träger-Konstrukten*, Books on Demand GmbH, Hamburg, 2004.

16 B. HÜSING, B. BÜHRLEN, S. GAISSER, *Human Tissue Engineered Products-Today's Markets and Future Prospects*, Fraunhofer Institute for Systems and Innovation Research, Karlsruhe, 2003.

17 J. RUSSELL, S. CROSS, *Commercial Prospects for Tissue Engineering*, Business Intelligence Program-SRI Consulting, 2001, pp. 1–15.

18 *Landesbank Baden-Württemberg Equity Research: Tissue Engineering.*, Stuttgart: Landesbank Baden-Württemberg Equity Research, 2001, p. 62.

19 G. BENTLEY et al., *J. Bone Joint Surg. [Br]* 85B (**2003**), p. 223.

20 M. H. ZHENG et al., *Int. J. Molecular Med.* 13 (**2004**), p. 623.

21 E. GENOVESE et al., *J. Ortho. Surg.* 11 (**2003**), No. 1, p. 10.

22 R. STREHL et al., *Tissue Eng.* 8 (**2002**), No. 1, p. 37.

23 P. D. BENYA, J. D. SHAFFER, *Cell* 30 (**1982**), p. 215.

24 M. BRITTBERG et al., *J. Bone Joint Surg.* 85A (**2003**), No. 3, p. 109.

25 N. C. ZANETTI, M. SOLURSH, *J. Cell Biol.* 99 (**1984**), p. 115.

26 K. MASUDA et al., *J. Orthop. Res.* 21 (**2003**), No. 1, p. 139.

27 I. MARTIN et al., *J. Cell. Biochem.* 83 (**2001**), p. 121

28 P. M. VAN DER KRAAN et al., *Osteoarthritis and Cartilage* 10 (**2002**), No. 8, p. 631.

29 S. VIJAY, *Cytotechnology* 30 (**1999**), p. 149.

30 U. HANSEN et al., *J. Biomech.* 34 (**2001**), No. 7, p. 941.

31 J. MALDA et al., *Crit. Rev. Biotechnol.* 23 (**2003**), No. 3, p. 175.

32 T. M. HERING, *Frontiers in Bioscience* 4 (**1999**), p. 743.

33 M. JAKOB et al., *J. Cell. Biochem.* 81 (**2001**), No. 2, p. 368.

34 R. KNAZEK, P. GULLINO, *Patent US3821087, 1974-06-28* (**1974**).

35 L. DE BARTOLO, A. BADER, *Annals of Transplantation* 6 (**2001**), p. 40.

36 W. MINUTH, *Patent DE19530556, 1996-09-05* (**1996**).

37 W. MINUTH, *Patent DE4443902, 1996-04-18* (**1996**).

38 W. MINUTH, *Patent DE19952847, 2001-04-19* (**2001**).

39 L. E. FREED, G. VUNJAK-NOVAKOVIC, *Cell Biology and Biotechnology in Space* **2002**, p. 177.

40 C. DODD, C. D. ANDERSON, *Patent WO125396, 2001-04-12* (**2001**).

41 M. HULS et al., *Patent US5155034, 1992-10-13* (**1992**).

42 J. S. Temenoff, A. G. Mikos, *Biomaterials* 21 (**2000**), p. 431.

43 D. L. Butler et al., *J. Biomech. Eng.* 122 (**2000**), p. 570.

44 O. Demarteau et al., *Biochem. Biophys. Res. Commun.* 310 (**2003**), No. 2, p. 580.

45 J. B. Fitzgerald et al., *J. Biol. Chem.* 279 (**2004**), No. 19, p. 19502.

46 O. Demarteau et al., *Biorheology* 40 (**2003**), p. 331.

47 R. L. Sah et al., *J. Ortho. Res.* 7 (**1989**), p. 619.

48 G. Marriott, I. Parker (Eds.), *Methods in Enzymology*, Volume 360, Biophotonics, Part A, ISBN 0-12-182263-X, Academic Press, San Diego, California, 2003.

49 G. Marriott, I. Parker (Eds.), *Methods in Enzymology*, Volume 361, Biophotonics, Part B, ISBN 0-12-182264-8, Academic Press, San Diego, California, 2003.

50 N. Prasad, *Introduction to Biophotonics*, John Wiley & Sons, Hoboken, New Jersey, 2003.

51 W. Denk et al., *Science* 248 (**1990**), p. 73.

52 S. Fine, W. P. Hansen, *Applied Optics* 10 (**1971**), p. 2350.

53 C. Buehler et al., *IEEE Eng. Med. Biol.* 18 (**1999**), p. 23.

54 W. R. Zipfel et al., *Proc. Natl. Acad. Sci. USA* 100 (**2003**), p. 7075.

55 W. R. Zipfel et al., *Nature Biotech.* 21 (**2003**), p. 1369.

56 I. Freund et al., *Biophys. J.* 50 (**1986**), p. 693.

57 P. J. Campagnola, L. M. Loew, *Nature Biotech.* 21 (**2003**), p. 1356.

58 M. Minsky, *Patent US 3,013,467* (**1961**).

59 W. Kaiser, C. G. B. Garrett, *Phys. Rev. Lett.* 7 (**1961**), p. 229.

60 P. A. Franken et al., *Phys. Rev. Lett.* 7 (**1961**), p. 118.

61 W. Denk, K. Svoboda, *Neuron* 18 (**1997**), p. 351.

62 W. F. Cheong et al., *IEEE J. Quantum Electron.* 26 (**1990**), p. 2166.

63 A. Diaspro, *Confocal and Two-Photon Microscopy: Foundations, Applications and Advances*, Wiley-Liss Inc., New York, 2002.

64 E. H. K. Stelzer, *J. Microsc.* 189 (**1998**), p. 15.

65 M. Born, E. Wolf, *Principles of Optics*, 6th edition, Pergamon Press, 1993.

66 T. M. Ragan et al., in: *Methods in Enzymology* (Eds.: G. Marriott, I. Parker), Academic Press, San Diego, California, 2003, p. 481.

67 R. M. Williams et al., *Biophys. J.* 88 (**2005**), p. 1377.

68 T. Nielsen et al., *J. Microsc.* 201 (**2002**), p. 368.

69 J. Bewersdorf et al., *Opt. Lett.* **1998**, p. 23.

70 A. Egner et al., *Journal of Microscopy* 206 (**2002**), p. 24.

71 J. R. Lakowicz et al., *Fluorescence lifetime imaging of free and protein bound NADH*, PNAS 89, **1992**.

72 M. Straub, S. W. Hell, *Appl. Phys. Lett.* 73 (**1998**).

73 D. V. O'Connor, D. Phillips, *Time Correlated Single Photon Counting*, Academic Press, 1984.

74 W. Becker et al., *Proceedings of SPIE* 4431 (**2001**), pp. 249–254.

75 C. Biskup et al., *Nat. Biotechnol.* 22 (**2004**).

76 J. Martini et al., *Proc. SPIE* 5860 (Confocal, Multiphoton, and Nonlinear Microscopic Imaging II, Ed.: T. Wilson), **2005**, pp. 16–21.

77 K. König, *J. Microsc.* 200 (**2000**), p. 83.

78 E. B. Hunziker et al., *Osteoarthritis and Cartilage* 10 (**2002**), p. 564.

79 T. C. Voss et al., *Biotechniques* 38 (**2005**), No. 3, p. 413.

80 V. Ulrich et al., *Scanning* 26 (**2004**), No. 5, p. 217.

8
Microarray Biochips – Thousands of Reactions on a Small Chip (MOBA)

Wolfgang Mönch[1], *Johannes Donauer, Bernd M. Fischer, Rüdiger Frank, Günter Gauglitz, Carsten Glasenapp, Hanspeter Helm, Paul Hing, Matthias Hoffmann, Peter Uhd Jepsen, Thomas Kleine-Ostmann, Martin Koch, Holger Krause, Nicolae Leopold, Tina Mutschler, Frank Rutz, Titus Sparna, Hans Zappe*

8.1
Introduction

The cell is the fundamental unit of life and common to all organisms. The nucleus of each cell contains chromosomes, which serve as carriers of genetic information. Chromosomes are macromolecules made of deoxyribonucleic acid (DNA). DNA encodes the genetic material needed for the formation of thousands of different genes according to an alphabet that consists of four bases. A gene, in turn, is a DNA fragment that contains the instructions on how to produce a specific protein. Each chromosome is made of hundreds to thousands of genes. For the cell to be able to use the genetic information contained in a gene, the information needs to be read and the inherent protein construction plan translated. For this purpose, DNA fragments are first transcribed into messenger ribonucleic acid (mRNA), a process which, to put it simply, corresponds to the copying of the construction plan for the relevant protein. Thereafter, each of these construction plans is transported from the cell nucleus to the sites of protein synthesis in the cell, the so-called ribosomes, where the specific protein is built.

Based on the idea that all intracellular processes are mediated through the action of proteins, it seems plausible that, compared to healthy cells, the pathological state of a cell may be characterized by changes in composition, quantity or structure of cellular proteins. However, from a methodological point of view, it is disproportionately more complex to detect changes directly on the protein level than to measure them indirectly via the produced RNA.

With some restrictions, the quantity of transcribed mRNA of a gene permits us to infer the quantity of protein synthesized therefrom. Laboratory methods to detect the expression of a single gene, or small numbers of genes, have been

1) Corresponding author;
project: MOBA – Micro-Optical Biochip Analysis

Biophotonics: Visions for Better Health Care. Jürgen Popp and Marion Strehle (Eds.)

ISBN: 3-527-40622-0

available for a long time. Yet, it was only the late 1990s that simultaneous detection of the expression of thousands of genes, up to the whole transcriptome of a cell or cell structure, in a single experiment became possible [1,2]. This was due to the development of so-called microarrays, or biochips. Since it was first described in 1995 [3] by Schena et al., this technology has revolutionized gene expression analysis (gene profiling). Hence, today "gene expression analysis" in the narrower sense of the term refers to the "microarray experiment."

8.2 Microarrays: Biological Background

8.2.1 Principle of the Microarray Experiment

The concept of comprehensive gene expression analysis by means of the microarray experiment is rather simple (see **Figure 8.1** and for a review, Ref. [4]. A very good animated website illustrating the conduct of a microarray experiment can be found at http://www.bio.davidson.edu/courses/genomics/chip/chip.html): cells generated from tissues, blood, bone marrow or even cell cultures can be used as starting material. In a first step, RNA is isolated from cells or cell structure. RNA contains the whole transcribed genetic information. Secondly, RNA pieces are transformed into stable cDNA or, depending on the procedure employed, undergo amplification. During this step, labelled nucleotides are incorporated into the growing RNA or DNA strands, a procedure that allows for later detection of the sample. For labelling, either a fluorescent dye coupled to a nucleotide (direct staining) or a molecule to which a dye can be coupled (indirect staining) is used. Thereafter, so-called hybridization occurs, which means that the labelled sample specifically binds to the sequences on the chip. The chip itself consists of a solid matrix, e.g., glass or plastic, on which thousands of different DNA sequences are fixed following a known pattern. Each of these DNA sequences is a PCR copy of a transcribed gene fragment and represents a gene to be assessed. During hybridization, the labelled DNA or RNA sections of the sample attach to the complementary or matching DNA sequences on the chip. This procedure is followed by a washing step in order to remove any residual, unbound material. The intensity of the sample's color-labelling indicates the concentration of bound DNA. As the samples are labelled with fluorescent dyes, the arrays are read using fluorescence scanners. The measured intensity of staining indicates the RNA concentration of the sample. In this respect, it is assumed that the amount of gene expression in the target material is associated with the amount of sample staining on the microarray, and, consequently, with the measured intensity of the staining reaction. Accordingly, microarray analysis facilitates parallel assessment of the expression of thousands of genes in a single experiment.

Two different techniques of microarray analysis have become established [5]. Both will be outlined in the following.

8.2.2
Oligonucleotide Microarrays

In this kind of array, short, single-stranded oligonucleotides (30 to 70 bases in length) are brought onto a chip. Oligonucleotides are synthesized in accordance with sequence data retrievable from databases, and represent short fragments of known gene sequences. They are either fixed on a matrix in a fully synthesized form, or they are actively synthesized on a chip. The market-leading manufacturer of commercially available microarrays is Affymetrix®. In these products, specific oligonucleotides (25 mers) are synthesized on the chip in a photolithographic process [6]. For each gene, there are up to 20 different oligonucleotides on the chip. In addition, for each oligonucleotide, a further oligonucleotide is included, which is identical except for one single base (a so-called "mismatch" oligonucleotide). This design facilitates subtraction of non-specific bonds and of local background staining after hybridization. In this manner, a single chip is loaded with several hundreds of thousands of different oligonucleotides. Approximately 5 μg of total RNA or 0.2 μg of mRNA are used as sample. RNA is first transcribed into cDNA and then amplified via *in vitro* transcription (with the help of the enzyme T7 RNA polymerase). During the process, biotin-labelled ribonucleotides are incorporated into the growing RNA strands. Of these biotinylated RNA fragments (35 to 200 bases in length), 15 μg are then hybridized onto the microarray. Eventually, after a washing step, the remaining DNA–cRNA hybrids are stained with a fluorescent dye (streptavidin-phycoerythrin) and the intensity of the staining reaction is assessed with the help of an argon ion laser scanner.

The use of a large number of different short oligonucleotides for detection of a single gene even distinguishes splice variants and closely similar sequences. The inclusion of mismatch oligonucleotides allows the elimination of the effects of non-specific bonds and local non-specific background staining. There are some drawbacks to the system, and these include the costs of the system, which are often unaffordable for scientific institutions, the fact that direct comparison of two samples on a single chip is impossible and the lack of flexibility of the premanufactured GeneChips®.

Alternatively, presynthesized oligonucleotides (50 to 70 mers) may be spotted on coated slides. This is when the cDNA microarray technique is adopted (see below) [7].

8.2.3 cDNA Arrays

cDNA arrays consist of PCR products (cDNAs) which are transferred onto nylon membranes, or plastic or glass surfaces. In this case, the spotted cDNAs are longer than oligonucleotides present on other arrays (e.g., those provided by Affymetrix®), approximately 100 to 2 000 base pairs in length. For hybridization of nylon arrays, radioactively labelled cDNAs are employed. As a result, nylon arrays are very sensitive and economical, since they can be used repeatedly. However, they are not suitable for parallel evaluation of the regulation of more than a few hundred genes (which is why they are referred to as macroarrays). For this reason, they are commonly arranged to address selective questions (e.g., apoptosis arrays, signal transduction arrays etc.).

For a more comprehensive gene expression analysis, up to several tens of thousands of oligonucleotides (50 to 70 mers) [7] or PCR fragments (approximately 100 to 2 000 base pairs in length) [8] are brought onto specially coated glass or plastic slides (so-called microarrays). As DNA has to be available in excess, compared to the sample, it will be spotted as a high concentration solution.

8.2.4 Production of cDNA Microarrays

In most cases, cDNA on microarrays is derived from so-called cDNA databases. They consist of thousands of bacterial cultures, each of them containing a plasmid into which the specific DNA sequence is cloned. Every sequence corresponds to a fragment of the gene of interest. At present, we are using two such cDNA libraries in our laboratory: one human library representing about 7 000 genes as well as one murine library covering 22 500 genes. A human library that is supposed to comprise 36 000 genes is currently in preparation.

The production of cDNA microarrays starts with the preparation of plasmids from bacteria. For this purpose, every single bacteria culture is grown overnight, followed by lysis of the bacteria. Thereafter, plasmids are purified from the lysate (see **Figure 8.1**).

The purified plasmids are then used as starting material for PCR amplification of the cloned DNA fragments. Finally, the yielded PCR products are purified again and transferred onto coated glass slides (see **Figure 8.2**) [1,9].

For spotting of cDNA, both ink jetting and contact pin printing techniques are applied, the latter being used in our institution. During contact pin printing (please refer to http://cmgm.stanford.edu/pbrown or Ref. [10]), each pin draws up an aliquot of a few microliters of cDNA by capillary action. Several of these pins (24 to 96) are arranged within one printer head. Afterwards,

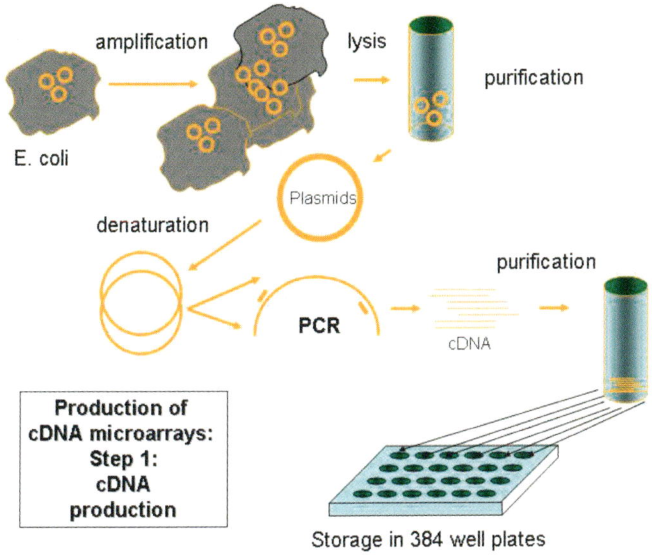

Figure 8.1 Production of cDNA microarrays: Step 1: cDNA production.

the printer head moves from slide to slide, and upon contact with the glass surface the pins release a cDNA aliquot. After each run, the pins are rinsed and subsequently reloaded. In general, the concentration of the spotted cDNA is in the range 100 to 500 $\mu g\,\mu l^{-1}$. For each single spot, only a few nanoliters

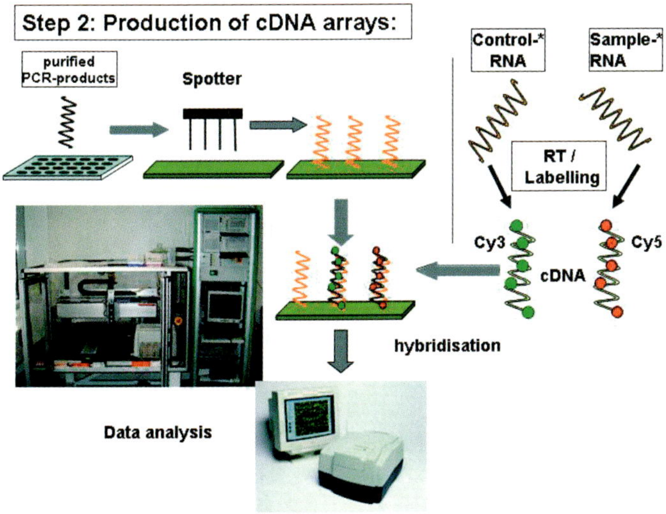

Figure 8.2 Step 2: Production of cDNA arrays.

of the probe are applied, which is why the size of the spots is as small as approximately 50 to 100 µm. Therefore, up to 20 000 different DNA sequences per square centimeter can be spotted on a single chip.

8.2.5
Hybridization to cDNA Microarrays

We have now reached the point where the actual microarray experiment begins (see **Figure 8.2** or http://cmgm.stanford.edu/pbrown). RNA is isolated from two samples and purified. Both RNA samples are then reversely transcribed into single-stranded cDNA. Reverse transcription occurs through an oncoretroviral enzyme called reverse transcriptase. During this phase, fluorescence-labelled nuclides are incorporated into the growing cDNA strands. Typically, two fluorochromes with different fluorescence spectra are employed (e.g., Cy-3-dUTP and Cy5-d-UTP), one for staining the test or target sample and the other for staining the reference RNA. At this point, an advantage of this technique over the Affymetrix® oligonucleotide arrays becomes obvious: prior to hybridization onto the array, target and reference samples are mixed and then co-hybridized onto the chip. Thus, direct comparison of two samples in a single experiment is possible. The hybridization reaction then takes place over several hours in a controlled chamber protected from light. Finally, the hybridized chips are washed to remove any unbound DNA. After a drying step, detection of hybridization signals is performed. Again, a fluorescence scanner is used. Staining intensities are measured separately for the two fluorescent dyes and expressed as numerical values. Furthermore, an image consisting of two color channels is obtained (see **Figures 8.3** and **8.4**). If the sample contains the transcript of a target gene, the stained cDNA molecules bind to the respective PCR products on the array and, accordingly, the fluorescence image indicates a staining (e.g., green) at the binding site. If the gene is not expressed, the site where the matching DNA sequence is applied remains dark. If the gene is expressed in the target as well as the reference sample, the ratio of the measured staining intensities indicates the respective ratio of RNA concentrations (so-called gene expression ratio, see **Figure 8.5**). The color image exhibits a mixture of red and green, which indicates the concentration ratio of the bound samples.

Unlike "on-chip" synthesized oligonucleotide arrays, cDNA microarray technology is quite economical. Experiments can readily be conducted, and no major technical equipment is required. Obviously, given the fact that individual arrays can be tailored to the scientific problem addressed, which is not the case with commercially available premanufactured solutions, this system is particularly interesting for research laboratories. The length of the capture probe (i.e., the probe on the array) ensures a high sensitivity of detection. Low

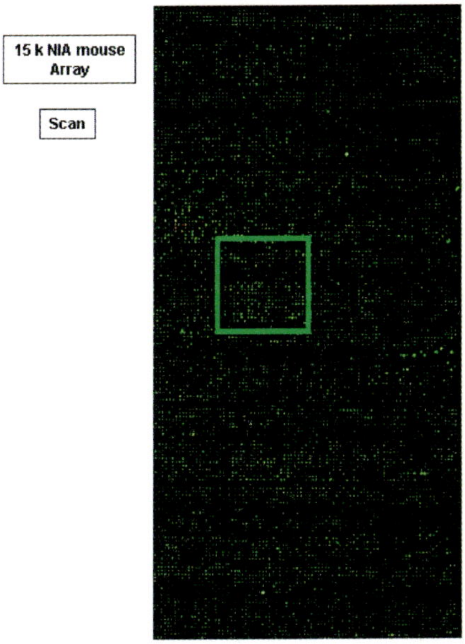

Figure 8.3 NIA 15 k mouse microarray, complete view.

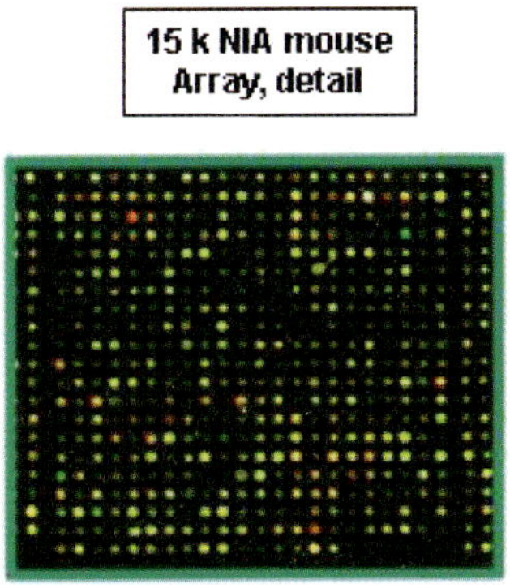

Figure 8.4 NIA 15 k mouse microarray, detailed view.

specificity, for example as a result of the binding of several closely similar, related genes to a capture probe, may be reduced by using PCR fragments located within the untranslated 3′ end of the mRNA. These regions usually exhibit gene-specific sequence diversity. Unfortunately, cDNA microarray technology requires a large quantity of RNA to generate adequate signals. This, in turn, limits the technology's applicability to those experiments for which large quantities of RNA are available. Likewise, differences in expression patterns are not detectable for low-abundance transcripts. This issue may be tackled with linear mRNA amplification using T7 RNA polymerase [11]. With this procedure, an up to 10^6-fold amplification of RNA quantity can be reached and, in our experience, the procedure has proved to be reliable.

8.2.6 Analysis of Data Generated from Microarray Experiments

8.2.6.1 Low-level Data Analysis

Gene expression analysis by means of cDNA microarrays allows tens of thousands of hybridization experiments to be conducted simultaneously. Thus, a single microarray experiment yields a dataset containing several tens of thousands of single pieces of information. For sound interpretation and comparison of the datasets with other microarray experiments, preliminary normal-

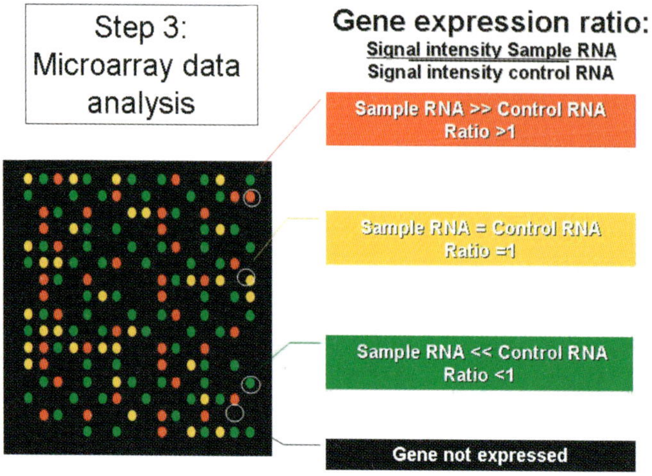

Figure 8.5 Step 3. Microarray data analysis.

ization of data is required. In this procedure, measurement artifacts as well as any unspecific background staining and quality fluctuations of single arrays are mathematically corrected [12]. The appropriate automated normalization tools are part of the image analysis software of the fluorescence scanner. Furthermore, to enhance the specificity of the results, the investigator may establish predefined "minimum criteria" for signal interpretation (e.g., minimum distance from local background). Even for fully identical samples, significant differences in terms of the assessed spot intensities in both color channels cannot be ruled out. Such differences are induced by variations in the incorporation rate for the respective fluorescent dyes and/or by a potential difference in the performance of the two lasers used for scanning. Assuming that the sum of the measured staining intensities should be identical for a given sample and the respective reference, which means that both hybridized samples exhibit the same number of up- or downregulated genes, the actual ratios of all measurements have to demonstrate a geometric mean of 1, with an arithmetic mean of their logarithms of 0. If the mean value of the log ratios of the measured staining intensities is set to 0 by adding a constant, the data are normalized in terms of the labelling efficiency of the dyes (see **Figure 8.6**) [13]. With "lowest-subgrid" analysis, even regional variations within one array, e.g., due to inaccuracy of the pins, may be normalized.

To keep all microarray experiments comparable, all samples may be hybridized against a reference RNA [8,14]. The reference RNA used in our laboratory is generated from a pool of up to ten different cell lines. Its composition is always identical. All samples are hybridized on the arrays against the refer-

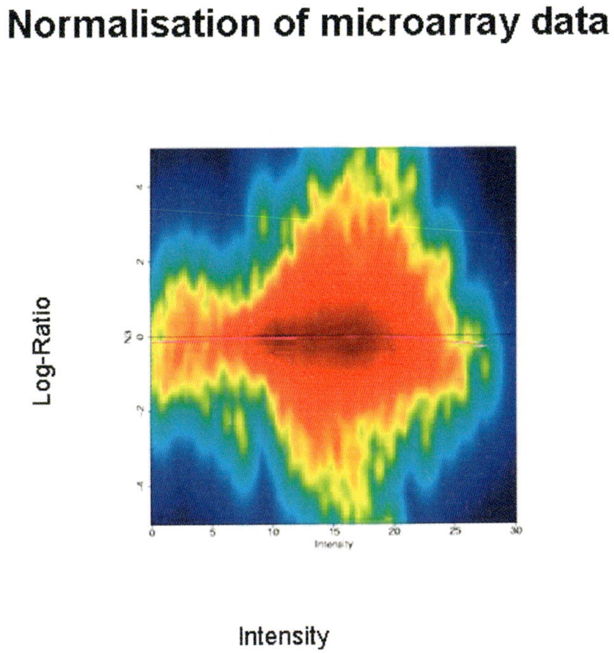

Figure 8.6 Data normalization: the figure shows data of a representative microarray from our own studies of chronic allograft nephropathy [14]. The logarithm of the ratios for the measured staining intensities (y-axis) for each gene is plotted against signal intensity (x-axis). To compensate for different detection of the total staining intensities for each fluorescence, the 0-line (black) has been appropriately adjusted (red).

ence RNA. As a matter of principle, the transcripts from the samples of interest are compared to the same concentration of the reference transcript, which is why correlations can be calculated across different arrays.

In order to support worldwide comparability of microarray experiments, the Microarray Gene Expression Database Consortium has developed guidelines for the documentation of microarray experiments. To this end, the so-called MIAME (Minimum Information about a Microarray Experiment) standard defines the minimum information necessary for interpretation and verification of microarray experiments [15]. Thus, the data generated in our institution are archived in a central database in accordance with MIAME criteria.

8.2.6.2 **High-level Analysis**

High-level data analysis or data mining refers to the process of identifying relevant gene expression patterns from normalized microarray data. The easiest way of interpretation is provided by the ratio of the two samples co-hybridized onto the array, i.e., the gene expression ratio, with any ratio defined as "significant" (e.g., two-fold regulation of the gene with respect to gene y). Indeed, data interpreted in this way may possess low sensitivity and specificity, if any. We therefore combine this criterion with a univariate statistical analysis, for example, a t-test. In addition, all our microarray experiments are routinely performed with at least three parallel setups. As thousands of single experiments are performed on one array, a so-called multiple testing correction (Buonferroni correction) has to be applied to the results of the t-tests. We use the method of Benjamini and Hochberg [16] to estimate the false discovery rate indicating the significance level at which a given gene is correctly detected as regulated.

Beyond the determination of single, significantly regulated genes, interpretation of microarray experiments may also involve grouping of genes with identical expression patterns. In this regard, unsupervised and supervised approaches are distinguished. Unsupervised analysis aims to group regulated genes solely on the basis of their expression behavior observed in the experiment, whereas supervised analysis uses biological information (e.g., diseased vs. normal) to test the established classification of genes into regulation groups. A commonly used method for unsupervised classification is cluster analysis [17]. For the majority of clinical questions, hierarchical cluster analysis is applied. Here, genes are grouped according to the similarity of their regulation, comparable to a genealogy tree (see **Figure 8.7**). Non-hierarchical clustering techniques include K-means clustering [18] or self-organizing map [19]. A frequently used method of supervised clustering is linear discriminant analysis, for which initially a classification based on a training set of samples is established. Thereafter, this classification is validated with the help of independent samples [20].

Apart from statical analysis of gene expression, microarray technology provides the ability to conduct time-series gene expression studies [21].

8.2.7
Application of Microarrays in Medical Research

In contrast to the majority of experiments in medical research, gene profiling does not follow a hypothesis-driven approach. However, a disinterested analysis of a given sample's transcriptome may lead to the development of hypotheses, which can then serve as starting points for further experiments. In principle, there are several possible fields of application for comprehensive microarray-based gene expression analysis (see Ref. [4]).

8.2.7.1 Deciphering of New Gene Regulations

Microarray experiments allow for identification of specific genes whose regulation is involved in physiological or pathological processes.

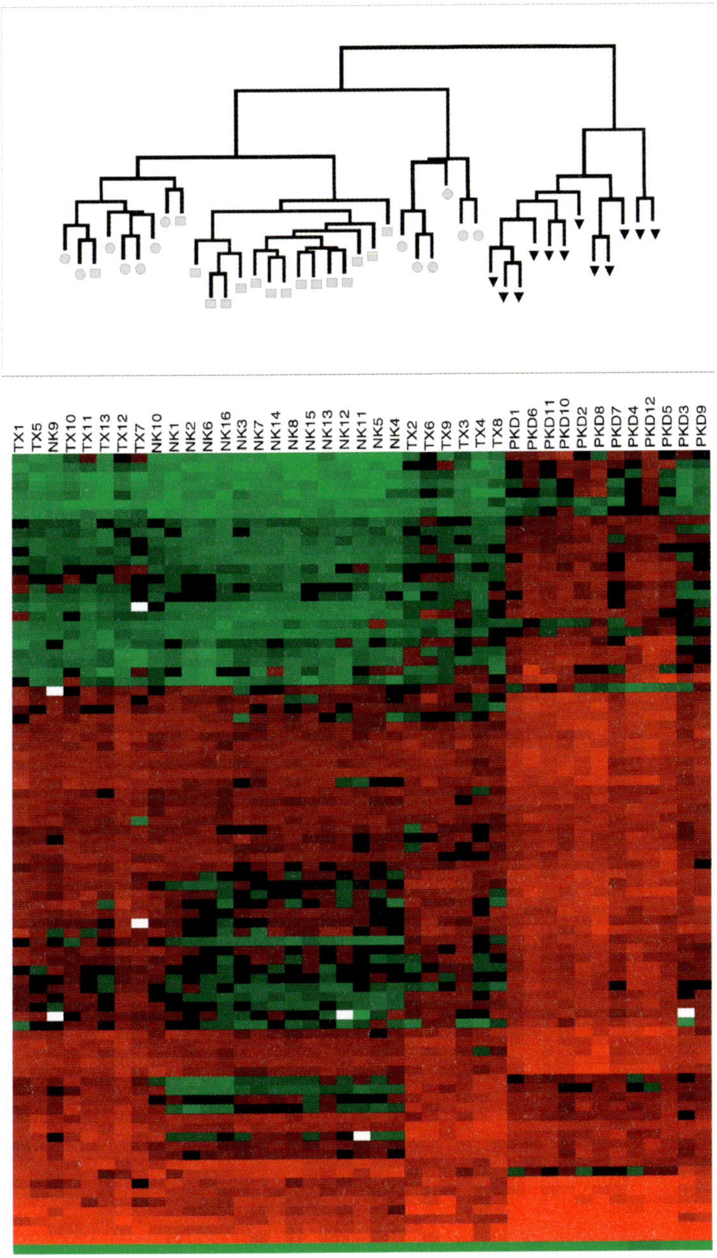

Figure 8.7 Hierarchical cluster analysis performed on 16 normal kidneys, 13 renal grafts and 12 kidneys affected by polycystic kidney disease [14]. Cluster analysis and the respective dendrogram demonstrate that gene expression in ADPKD (autosomal dominant polycystic kidney disease) differs significantly from that found in either of the other renal tissues.

8.2.7.2 **Time-series Gene Regulation Data**

Based on the assumption that genes displaying similar expression over time participate in the same molecular processes, new hypotheses on regulatory networks may be generated from time-series experiments [21].

8.2.7.3 **Diagnostics**

Samples collected from diseased patients may reveal typical gene expression patterns when compared with those from healthy individuals. In a fundamental work, Golub [22] was the first to provide the methodological and statistical background for the characterization of tumors on the basis of specifically regulated genes (so-called class prediction), and, moreover, for the identification of previously unknown subtypes (so-called class discovery). In this context, with the help of microarray analysis, Alizadeh et al. [23] were able to observe two subtypes, associated with significantly different survival rates, in a groundbreaking study in patients diagnosed with diffuse large B-cell lymphoma (DLBCL; the most frequently encountered form of non-Hodgkin's lymphoma). Their work was followed by countless others – mostly related to oncological diseases – which, based on gene expression studies, have led to new classifications or prognostic assessments. Typical examples include Refs. [24] and [25].

Microarrays are also used to investigate mutations in genomic DNA. For example, DNA arrays may reveal point mutations, deletions or insertions. Thus, the occurrence of specific medical conditions may be correlated with the presence of specific alleles. Here, the so-called SNiP (single nucleotide polymorphism) analyses performed with the help of oligonucleotide microarrays [26] are of particular interest, as they allow easy detection. For analysis, oligonucleotides are used, which, compared to the original sequences, carry a modification (A, G, T or C) in one position. Upon hybridization with genomic DNA, the matching DNA strand from the sample will attach to the four different oligonucleotides with varying stringency. Signal intensity will be higher for the matching oligonucleotide than for mismatch constellations, which permits detection of the specific modification in the genomic DNA. In order to test for all possible SNiPs of a given single-stranded DNA with a length of 5 000 bp, an array consisting of $4 \times 5\,000 = 20\,000$ spots is required. Given today's technologies, this can easily be attained.

8.2.7.4 **Pharmacogenomics [27, 28]**

Gene expression analysis may be used to characterize desirable and undesirable effects of medicinal products. For example, the cytochrome p-450 enzyme system located in the liver was detected in that way. Approximately 90% of all drugs undergo hepatic metabolism by enzymes pertaining to the cytochrome p-450 system. Now, with the help of microarrays, it is possible to elucidate which of the numerous enzymes in this system is responsible for metabolism of a specific drug, since transcription of the respective enzyme is increased [29]. This question is of significant clinical relevance, because, for example, inactive alleles of the respective gene might induce accumulation and life-threatening side effects in a subpopulation of patients. This is, for instance, a well known problem associated with the anticoagulant agent warfarin. Screening of gene expression for known target proteins in various tissues under the influence of pharmacological substances constitutes another example of the application of microarrays in drug research. In this context, a highly selective expression exclusively in the target tissue (e.g., tumor tissue) is considered beneficial, as the relevant medicinal product exerts its action at the target site only and undesirable effects in other tissues are avoided [30].

8.3 Fluorescence Techniques

8.3.1 CCD Sensors in Fluorescence Analysis: State-of-the-Art [31–33]

Ever since the sequencing of the human genome, CCD technology has proven its capability and sensitivity for low-level optical measurement. The great advantage of providing millions of individual sensors allows the level of parallelization and multiplexing necessary to enable very high-throughput analysis of samples. As a result, CCDs are playing an increasingly important role in fluorescence- and label-free analysis techniques in the areas of biotechnology and medical diagnostics. Not only do state-of-the-art CCDs now achieve the speed and sensitivity to cover the majority, if not all, of the various measurement techniques, but the image sensing industry is now flexible and mature enough to accommodate applications ranging from demanding research needs to low-cost "patient-near" diagnostic testers.

CCD image sensors differ vastly in their capabilities, depending on the application for which they are designed. Here we concentrate on "scientific grade" CCDs, which have the sensitivity and performance suitable for fluorescence measurements. In the following, some of the important features of scientific CCDs will be described in the context of fluorescence applications. The ability to visualize select biological samples and processes is an extremely important capability, but it is important to note that the use of image sensors for such analytical measurements goes far beyond visualization. CCDs are demanded to measure precisely the photonic and spectral composition of fluorescence emissions that make up the essential information content of an image.

On the one hand, to image the extremely low light levels emitted through chemiluminescence, where almost no extraneous background emission exists, a CCD must be able to collect incoming photons over exposure periods ranging from minutes to hours, and slowly read the values with extremely low noise. In this case, modern scientific grade, back-illuminated CCDs, which are cooled to 100 °C below ambient, can achieve this goal. Back-illumination is a technique for achieving very high quantum efficiency, over 90%, over a wide spectral range from the UV to NIR. This technology involves the thinning (etching) of the back side of the silicon CCD device such that photons of all energies that are incident on this back surface are efficiently collected. Such CCDs must also intrinsically exhibit extremely low dark response (i.e., thermal response) in order to be exposed for such long times. A common technology used to achieve this is called Multi-Pinned-Phase (MPP). "Pinned", alternatively called "inverted", operation of the MPP-CCD lowers the dark current by forcing the collection of electrons deeper into the silicon crystal lattice, where lattice defects are minimal. As a result, fewer electron–hole pairs are generated thermally.

Cooling reduces the low dark response of MPP-CCDs further. In general, cooling is necessary for fluorescence measurements, owing to the increased exposure times necessary to measure the weak emission. The most common and cost-effective technique for cooling CCDs is the use of thermo-electric Peltier elements. These can be either forced air cooled or liquid cooled. For deep cooling ($\Delta T = 100$ °C), the cooled CCD is typically hermetically sealed in a high vacuum chamber. It is important that cameras for fluorescence applications incorporate robust, maintenance-free hermetically sealed cooling since condensation can occur as soon as the CCD is cooled below dew point.

There are other methods of improving sensitivity (quantum efficiency) which are not as process-sensitive and expensive as back-thinning. One of these is the use of more transparent gates, typically fabricated with indium tin oxide (ITO). "Gates" are the electrical contacts to the pixels, through which the incoming photons must normally pass. Typically, these gates are fabricated using polysilicon, which is transparent, although not as transparent as ITO. A second method of increasing sensitivity is to leave a portion of the pixel completely open to incoming light. This is called "open phase" or "virtual phase", and allows more photons to be collected, at least in this area of the pixel. A third method of further increasing sensitivity is to combine the previous two methods with the use of microlenses on each pixel. Microlenses increase the "fill-factor" of the pixel by focusing incoming light into the more sensitive portion of the pixel.

We are all most familiar with CCD cameras as devices used to record images, for example in fluorescence microscopy. CCDs are now capable of allowing visualization at high speeds when monitoring kinetic events, such as intracellular processes and molecular binding. However, in many techniques the CCD is used in quite different modes to perform quantitative measurements. Its two-dimensional array of sensors enables acquisition of spectral profiles with high spectral resolution – hyperspectral images can be generated. Similarly, CCDs can extend the capability of confocal and multiphoton scanning systems by providing more spectral information – digitizing the emission spectrum of fluorescence dyes, auto-fluorescence or of Raman scatter.

CCDs are now applied to automated analysis, such as high-throughput screening for drug development and automated microscopy in clinical testing. The demands of such applications go beyond imaging, requiring an approach that focuses on the samples, and the sifting through vast amounts of data for rare events and for specific information. We note that as fluorescence analysis techniques mature from the research laboratory to commercial use, CCDs must be the enabling technology to provide the necessary sensitivity at increased speed.

With this in mind, the challenge for CCDs today is to enable fast fluorescence imaging with very high resolution and high sensitivity. Detection systems must accommodate the variety of sample carriers: gel plates, microplates, microscope slides, miniaturized biochips, microfluidic lab-on-a-chip and very large proteomics analysis plates. Moreover, they must be capable of performing data processing and informatics in real time, at speeds that can match the desired throughput. In such cases the images are no longer the result of interest, but are simply a means of obtaining the desired information content.

Particularly to address this challenge, the Electron Multiplied CCD (EMCCD), also known in the industry as "Charge Multiplied CCD", "Single Photon Dectection CCD", as well as other trade names, offer the latest performance in sensitivity and speed. When these sensors are deeply cooled (to approximately $-50\ ^{\circ}\mathrm{C}$) they are capable of detecting single photons while imaging at video speeds. To emphasize: each and every one of the sensor's million pixels is capable of detecting one photon. This is accomplished by integrating a large amplification of the detected signal directly on the silicon chip, similar in function to the gain of a photomultiplier tube (PMT) or avalanche photodiode (APD).

One of the main limitations of the sensitivity of a CCD is its read noise, which has traditionally posed a limit to the usefulness of CCDs. Even for slow a scientific CCD this read noise is equivalent to tens of photons, thereby establishing a fundamental limit of detection. Another complication is that read noise increases with speed (at which pixels are read). Yet another limitation is that the frame rate decreases as the size (pixel resolution) of the CCD increases. The concept of the EMCCD is to amplify the signal, before it gets to the device's output, to such a level that it is now significantly larger than the read noise of the output. In this way a signal of one electron can be read with a signal-to-noise ratio much greater than one. Hence a higher read-out noise can be tolerated, thereby enabling higher read-out speeds. The EMCCD uses the architecture of a conventional scientific grade CCD, but adds an "electron multiplication register" just before the output. Incident photons generate electrons in the active (exposed) area of the CCD, which form charge packets corresponding to pixels. During the read-out of the pixels, these charge packets are transferred ("shifted") serially toward an output, where they are converted to a measurable voltage signal. It is at this output node that the charge packet first encounters noise – electronic $1/f$ noise, capacitive "KTC" noise – which all collectively contribute to "read-out noise".

In the EMCCD, the charge packet is amplified (multiplied) before it reaches the output, by clocking the charge through regions of high electric field. The multiplication occurs via an "impact ionization" process, similar to an avalanche effect. Gains in excess of 1000 are possible, thereby enabling the detection of extremely small charge packets, down to one single electron.

Now one approach in using EMCCDs for fluorescence imaging is to try to produce the largest, highest resolution EMCCD possible, and to then achieve multifluorophore imaging by the use of interchangeable optical filters. This is the traditional way in which epifluorescence microscopes operate. Although this may be adequate for research or applications with lower throughput requirements, the following bottlenecks are recognized: high-resolution image sensors take longer to read-out, implying lower frame rate; multiple images are necessary to generate a multicolor image; and very large EMCCDs are not yet available and are prohibitively expensive. There is a fundamental limit to the size of image sensor that can be practically fabricated at reasonable cost. In the end, this limits the size and resolution of sample carrier that can be imaged.

A new approach is to use a multitude of small, spectroscopic EMCCDs in a spectrally resolving scanning mode. This would function in a similar way to a flatbed document scanner. In this case, one axis of the EMCCD is used as a spectral axis in order to resolve the spectral characteristics of the fluorophores. The EMCCD is therefore combined with a spectrograph, which spreads the incoming fluorescence emission into its spectrum. The other axis of the EMCCD is spatially resolving the width of the sample carrier to be scanned. In this way, a line across the width of the sample carrier (or a row of samples) can be spectrally acquired. Now, to maximize speed and cost, the CCD size is kept small, and a multitude of them are butted together to modularly enable the scanning of any arbitrarily wide sample carrier. Scanning is performed either by moving the sample carrier (e.g., on a conveyer belt) past the detector, or by scanning the detector across the sample carrier. With the high speeds at which the small EMCCDs are read, minimal cooling of the devices is necessary – also an important cost factor for commercial use. A "hyperspectral" image results, which has a digitized spectral content. Higher spectral resolution is often useful in better discriminating overlapping spectral profiles of dyes, or to allow increased multiplexing of tests through the use of increased number of dyes. The latter is in increased demand owing to the latest improvements in fluorescence labels, for example semiconductor particle labels and precious metal labels, which afford much narrower emission bands and are more efficient.

Since the scanning system described above eliminates the need for large imaging optics, the optical system can be miniaturized using refractive and diffractive micro-optics technology. The result is an extremely compact and robust solid-state detector, which can handle the widest variety of sample carrier formats, applications and assays, and measurement techniques.

8.4
Optical Systems for Fluorescence Analysis

Biochip fluorescence analysis relies on the precise measurement of the fluorescence light intensity emitted from biochemical species immobilized at the surface of a biochip. The laboratory device for this measuring task is usually called a "biochip reader". In this subchapter, we will first give a short overview of the current state-of-the-art in commercial biochip readers and then proceed to new developments for a highly versatile biochip reader unit developed within the biophotonics project MOBA (micro-optical biochip analysis).

8.4.1
Biochip Readers: State-of-the-Art

A biochip reader has to measure the intensity of light emitted from fluorescence-labelled biochemical species in the spots of a biochip, which have been printed on a chemically surface-treated substrate [34, 35]. A number of techniques have been developed for this purpose. The experimental techniques differ with respect to excitation mechanism (direct irradiation or via evanescent waves; single-spot, multispot or full-area excitation) and detection technique (type of sensor; temporally, spatially and/or spectrally resolved). In the following, we will give a brief overview of current technologies in this field and discuss their specific advantages and disadvantages. **Table 8.1** summarizes this discussion.

In order to obtain a high throughput from biochips, a full-area flood illumination of the fluorescence samples is advantageous. Equipped with a CCD-based detection unit, these diagnostic systems are rapid and cost-effective [36, 37]. Parallel read-out of the biochip allows the screening of many fluorescence samples simultaneously. Also, these systems are versatile with respect to the printing format of the biochips, i.e., they do not require a well defined pitch of the spots of the microarray. On the other hand, a comparatively high level of background fluorescence significantly reduces their sensitivity.

Highly sensitive fluorescence detection is accomplished by excitation with a focused laser beam [38]. The fluorescence light from the spot is then detected by a photomultiplier tube (PMT), where the background fluorescence is reduced by a confocally mounted pinhole in front of the PMT. Here, an increased signal-to-noise ratio allows an improved detection limit for the fluorescence light. Combined with fluorescence excitation by means of evanescent waves (also called the total internal reflection fluorescence method, TIRF) [39, 40], this method is even capable of detecting single fluorescent molecules.

However, in order to read out a complete microarray, a mechanical scanning system is required, because only one spot can be read at time. Owing to the sequential read-out of the biochip, the throughput of these systems is very low. Furthermore, the high intensity level of the laser beam at the focal point leads to photobleaching of the fluorescence dyes. Photobleaching reduces the fluorescence intensity significantly, such that a second read-out procedure of the biochip may become impossible.

In order to combine highly sensitive fluorescence detection with high throughput, multispot excitation by using a diffractive optical element ("holographic excitation") has been successfully demonstrated by Mehta et al. [41] and Blom et al. [42]. Holographic excitation provides two essential advantages: highly parallel excitation and excitation only at the position of the spots of the biochip. Highly parallel multispot excitation increases the throughput of samples. Excitation specifically at the position of the spots reduces background fluorescence from unbound marker molecules in the analyte surrounding the microarray, which means an increased excitation efficiency and improved signal-to-noise ratio. However, when using static diffractive optical elements for holographic excitation, matching the diffraction pattern to the spot pattern requires additional imaging optics and mechanical translation stages. Therefore, the adaptation of the excitation unit to new spot patterns requires modification in the hardware of the set-up. Today, the availability of dynamic diffractive optical elements (DDOEs) allows this limitation to be overcome, i.e., arbitrarily pitched spot arrays can be excited without need for any further optical components.

In the following section, a biochip reader is presented that provides a dynamic holographic excitation unit in combination with a CCD-based spectrally resolving detection unit for simultaneous fluorescence analysis at several wavelengths.

Table 8.1 Comparison of established techniques for fluorescence analysis of biochips.

	Scanning techniques	Parallel techniques	
Excitation	single spot	Multispot excitation at fixed pitch	Full-area flood excitation
Detection	photomultiplier tube	CCD sensor	
Advantages	independent of spot pitch	high contrast	high throughput
	dynamic range	high throughput	independent of spot pitch
Disadvantages	low contrast	spot pitch fixed	low sensitivity
	low throughput	low dynamic range of the sensor	low dynamic range of the sensor

8.4.2 New Developments

8.4.2.1 Dynamic Holographic Excitation of Microarray Biochips

Within the framework of the "MOBA" biophotonics project, dynamic holographic excitation has been developed as a highly versatile tool for fluorescence excitation of microarray biochips of arbitrary format and spot pitch. This novel technique is based on the use of dynamic diffractive optical elements (DDOEs). The central idea is to diffract excitation light from the DDOE in such a way that the diffraction pattern is exactly matched to the spot array to be excited. In order to achieve this, the DDOE has to be addressed with a computer-generated image file (computer-generated hologram, CGH) encoding the required 2D phase modulation at each pixel of the DDOE. Any desired intensity profile in the plane of the biochip may be generated by an appropriate phase modulation on the DDOE [43].

In the work presented here, reflective liquid crystal displays (LCDs) were used as DDOEs. LCDs provide the possibility of two-dimensional phase modulation of the incident light and can be addressed with a time constant of a few milliseconds. Moving intensity profiles may even be obtained. Furthermore, the technique can easily be extended to excitation at several wavelengths by using two or more DDOEs and light sources.

Figure 8.8 shows the optical set-up of a microarray biochip reader with holographic excitation unit at two wavelengths (here 404 and 532 nm). Two DDOEs are illuminated by two expanded laser beams. Using a dichroic mirror, two separate light paths are combined in order to provide the same intensity pattern on the biochip at both wavelengths simultaneously. **Figure 8.9** shows a photograph of the light intensity profile generated for a one-dimensional array of spots with a pitch of 500 μm.

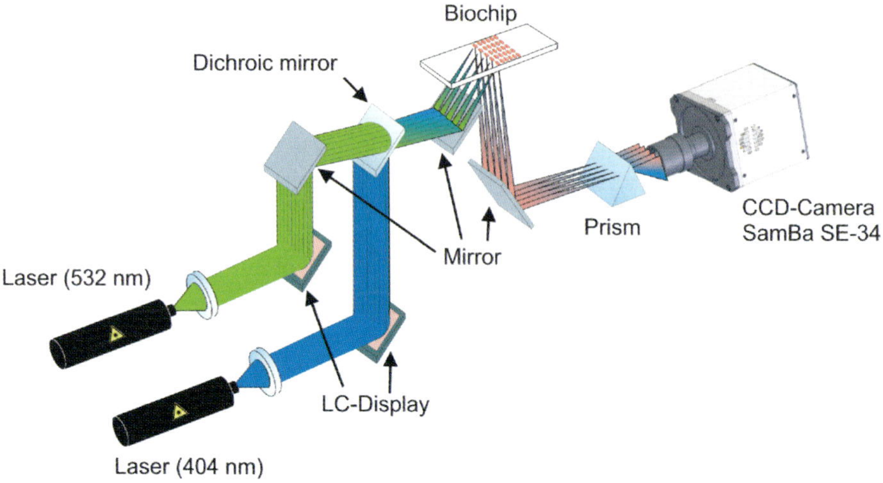

Figure 8.8 Experimental set-up for holographic excitation using two lasers and two LCDs.

Figure 8.9 Photograph of the biochip stage illuminated by a one-dimensional array of 24 focused beams generated with a DDOE. The pitch of the beams is 500 μm.

In addition to the above-mentioned advantages of holographic excitation, the use of DDOEs for dynamic holographic excitation allows purely software-based adaptation of the biochip reader to new biochip formats. Therefore, this development represents an essential step towards a new generation of efficient and highly versatile universal biochip readers. In the following sections we will discuss briefly the procedure for calculation of the CGHs.

8.4.2.2 Diffractive Optical Elements: Fundamental Considerations

The calculation of the required phase modulation for the DDOE is based on the reversibility of the light paths, i.e., light passes through an optical system backward in the same way as it does forward. For example, let us assume a point source emitting spherical waves in the positive direction on the optical axis [44]. When tracing back the wave fronts in the negative direction, we find the spherical wave fronts contracting to a focal point.

The electric field at a distance r from a point source located at the origin of the coordinate system is given by the following expression:

$$E(x, y) \propto \frac{\exp(jkr)}{r} \tag{8.1}$$

where $r = \sqrt{x^2 + y^2 + z^2}$.

Here, x and y are the coordinates in a plane at a distance z from the source, and k is the wave number. For large z, the factor r^{-1} may be approximated by z^{-1}. The amplitude of the electric field then depends only on z. Appropriate modulation of the phase (jkz) at any point is therefore sufficient for reconstruction of the electric field at any other arbitrary point.

The illumination holograms are based on spatially distributed phase shifts imposed on an incident plane wave in order to obtain the desired intensity distribution at some distance from the phase-modulating element. Such a spatially distributed phase shift is generally introduced by a transparent object exhibiting a locally varying optical thickness $(n\,d)$ (the product of refractive index n and the physical thickness d). The optical thickness may therefore be modulated either by the topography of a diffractive element, as is the case with Fresnel lenses, or, alternatively, by a phase shift, obtained by a spatially modulated refractive index, as in the case of an LCD.

For multispot biochip excitation, a phase modulation pattern of the LCD has to be calculated from which an intensity pattern is generated that is matched to the extension and pitch of the biochip spots at the position of the biochip. The most straightforward consideration for this application is stitching together subholograms, each of which generates a single focus point. In this case, the subholograms essentially have the structure of Fresnel zone plates. The centers of these subholograms are spaced a distance equal to the desired pitch of the spots; see **Figure 8.10**. We will refer to this phase modulation pattern in the following as a "stitched hologram".

The other extreme would be calculation of the full wave front at the position of the LCD, such that this wave front reconstructs all spots simultaneously. This is done by tracing the waves emitted from the spots backwards, i.e., the spots are assumed to be point sources and their electric fields are superposed at a distance equal to the distance from the biochip to the DDOE. The full

Figure 8.10 "Stitched hologram" for generation of an intensity pattern matched to eight spots on a biochip. The gray scale of the image encodes phase modulation of the DDOE (black = 2π, white = 0).

wave front of a discrete number N of spots is then given by:

$$E(x,y) \propto \sum_{i=1}^{N} \frac{\exp(jkr_i)}{r_i} \tag{8.2}$$

where $r_i = \sqrt{(x - x_i)^2 + y^2 + z^2}$.

Here, x_i is the position of each spot (or point source, respectively) and z the distance between the spots and the DDOE. **Figure 8.11** shows a calculated phase modulation pattern for eight spots with a pitch of 500 µm ($x_i = i \times 500$ µm). We will refer to the phase modulation pattern from this calculation method as a "full hologram".

In this case, because of the superposition procedure, the amplitude of the full wave front is no longer constant over the area of the phase modulation pattern. Since an LCD can only modulate the phase of an incident wave, this leads to a decreased efficiency of the wave front reconstruction. Furthermore, when compared to the stitched hologram, more Fresnel zones of increasingly narrower spacing are included in the full hologram. Therefore, the full hologram requires a higher resolution of the DDOE, but the efficiency is increased with the number of Fresnel zones considered in the hologram.

8.4.2.3 **Optimization of Holographic Excitation Using DDOEs**

LCDs used as diffractive optical elements allow dynamic variation of spatial light modulation in real time. The pitch of the reconstructed intensity focus spots and their distance from the LCD can be adjusted by changing the phase modulation pattern on the LCD. This means that a modified spot pitch or dis-

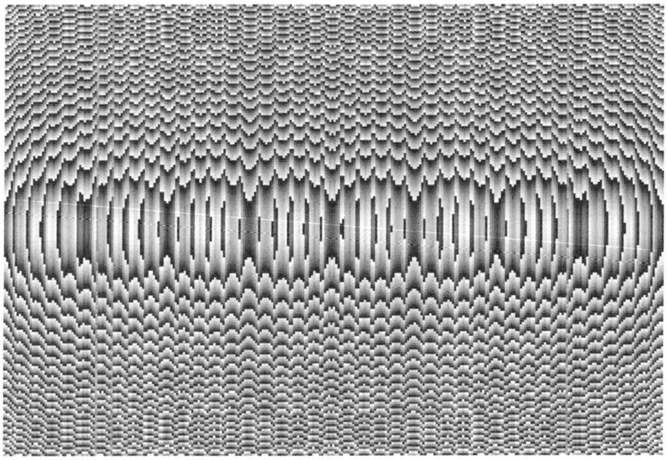

Figure 8.11 "Full hologram" for generation of an intensity pattern matched to eight spots on a biochip.

tance to the biochip may be compensated for without any mechanical movement by using a DDOE.

An important fact to consider when using any type of DDOE is the limited resolution due to the finite number of illuminated pixels on the display. For best reconstruction of the desired intensity pattern, the DDOE should be illuminated completely. The aspect ratio of typical LCDs is 5:4 or 4:3, but never 1:1; on the other hand, the beam profile of the excitation lasers used here is circularly symmetric. Consequently, in order to avoid an intensity loss due to truncation of the expanded laser beam, oblique illumination of the LCD at an angle of incidence equal to the arc cosine of the aspect ratio is advantageous. This oblique illumination yields a symmetric intensity profile in the reflected light.

However, oblique illumination of a reflective DDOE results in different path lengths between the DDOE and the plane where the intensity is to be reconstructed, i.e., the position of the biochip. This path length difference has to be taken into account when calculating the hologram. Otherwise the intensity pattern at the position of the spots would be blurred.

Usually, the pixels of an LCD are square. To a planar wave front obliquely incident onto the LCD, however, the pixels appear rectangular. This effect also has to be included in the calculation of the hologram. When compensating for the oblique incidence on the DDOE, the electric field E is given by Eq. (8.3). Here, AR is the aspect ratio of the display, p is the distance between two pixels on the display, and n_x and n_y are the number of pixels in the x- and

y-directions of the DDOE.

$$E(n_x, n_y) \propto \sum_{i=1}^{N} \frac{\exp(jkr_i)}{r_i} \tag{8.3}$$

where

$$r_i = \sqrt{\left(z + x \cdot \sqrt{1 - \mathrm{AR}^2}\right)^2 + \left(n_x \cdot \frac{p}{\mathrm{AR}} - x_i\right)^2 + \left(n_y \cdot p - y_i\right)^2}.$$

Because of the Gaussian intensity profile of the laser light source, the intensity of the excitation light has a maximum on the optical axis and decays to zero perpendicular to the optical axis. When illuminating a stitched hologram (**Figure 8.10**) with this Gaussian intensity distribution, the central spots of the reconstructed intensity pattern have the highest intensity within the one-dimensional array. However, a uniform intensity of the spots is desired. A more uniform intensity profile is obtained by reduction of the phase modulation depth in the central part of the hologram structures. A reduced phase modulation depth causes a reduction of the diffraction efficiency, i.e., a dimmed intensity of the reconstructed central spots. On the other hand, the background intensity increases and the total efficiency of the illumination is reduced.

Better results are usually achieved by the full hologram (**Figure 8.11**). To understand this, the effect of the finite pixel resolution of the DDOE has to be considered in more detail. Noting that the radius R_m of the mth Fresnel zone of a point source as a function of the wavelength λ and the focal distance z is given by

$$R_m = \sqrt{m \frac{\lambda}{2} z}, \tag{8.4}$$

we see that the radius difference between two adjacent zones, $(R_m - R_{m+1})$, gets increasingly smaller as the zone index m increases. Consequently, because of the finite pixel resolution of the DDOE, Fresnel zones can only be represented by the DDOE up to a certain finite index m. Zones of even higher index need not be considered in the calculation since they cannot be resolved. This truncation, however, leads to a reduced diffraction efficiency of the hologram.

Best performance is obtained through a combination of both phase image calculation methods and compensation for the laser intensity profile. In a first step, several subholograms are calculated up to a reasonable number of Fresnel zones and are stitched together using standard graphics software. For the overlap regions of adjacent subholograms, the superposition of the fields is used. In a last step, the phase modulation depth for each spot is adjusted individually in order to achieve uniform spot intensity. A typical result is shown in **Figure 8.12**.

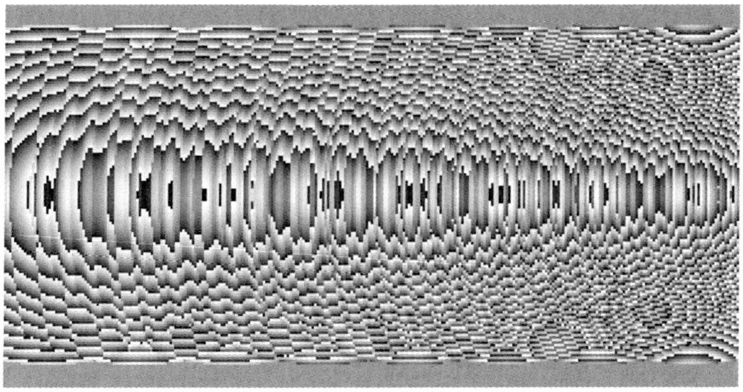

Figure 8.12 Optimized hologram for eight spots, consisting of a combination of "stitched" and "full" hologram, together with compensation for the excitation laser intensity profile.

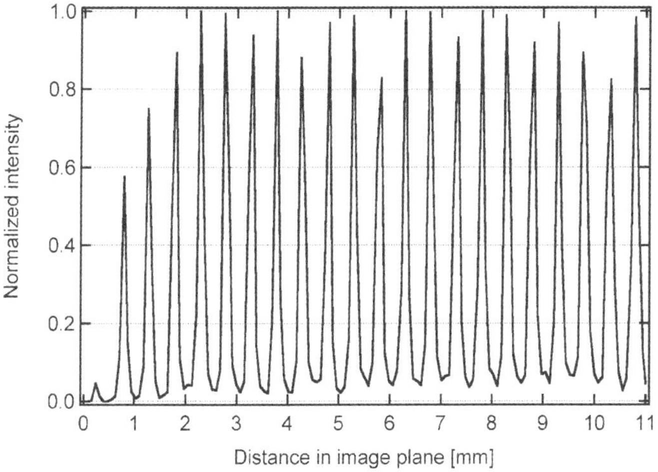

Figure 8.13 Intensity profile of a one-dimensional array of 21 focused spots, generated with a DDOE. The intensity per spot is 10 μW, the beam diameter in the focal plane is 160 μm. Pitch is 500 μm. Pixel pitch of the DDOE 13.7 μm.

Figure 8.13 shows the intensity profile of 21 foci generated with an LCD as the diffractive optical element. The contrast between bright and dark zones is 0.93, and the standard deviation of the intensity maximum over the 21 foci is 10%.

8.4.3 Spectral Fluorescence Detection

8.4.3.1 Instrumentation

In the biochip reader developed within the MOBA project, the holographic excitation unit was combined with a detection unit allowing hyperspectral read-out of the biochip. Hyperspectral imaging means that the positions, fluorescence intensities and fluorescence spectra of each single spot are measured simultaneously. To achieve this, the fluorescence light of a one-dimensional array of spots is imaged onto a CCD-sensor (a 2D object), with one coordinate on the CCD for the spatial information and the other coordinate for the spectral information from the fluorescence light [45].

The essential elements of the hyperspectral detection unit, as seen in **Figure 8.14**, is an objective lens with high numerical aperture, a dispersion prism for spectral separation, and a CCD-sensor. The sensor in the detection unit used here was a commercial CCD-camera (type SamBa SE-34, provided by Sensovation AG, Stockach, Germany).

The typical spectral width of a fluorochrome emission spectrum is about 50 nm [46]. Therefore, in order to distinguish five different fluorochromes in the range from 400 to 800 nm, a spectral resolution of 10 nm is sufficient. The optical design of the detection unit was optimized with respect to maximum sensitivity to fluorescence signals. Though diffraction gratings are frequently used in spectrometers, their overall performance turned out to be inferior compared with the dispersive prism used here, owing to the strongly wavelength-dependent diffraction efficiency of gratings. Therefore, an anti-reflection-coated 50-mm prism (N-BK 7 glass) was used in this demonstrator. The detection unit shown in **Figure 8.14** achieved a spectral resolution of 2 nm.

The fluorescence emission peak is shifted towards higher wavelengths ("Stokes shift") with respect to the excitation wavelength [46]; this shift is typically about 20 nm. In order to detect weak fluorescence signals, the excitation wavelength must thus not reach detection unit. Therefore, efficient filters for blocking the excitation light are required. In the set-up of **Figure 8.14**, a special filter combination is used to measure the fluorescence of a set of fluorochromes. A long-pass filter blocks the shortest excitation laser wavelength (404 nm), and a notch filter blocks the second excitation laser line at 532 nm.

8.4.3.2 Quantitative Detection of Concentrations

The goal of fluorescence detection is to determine the fluorochrome concentrations in the biospots, since fluorochrome concentrations are associated with the biochemical reactions to be investigated. By measuring the intensity of the fluorescence light, dye concentrations can be determined.

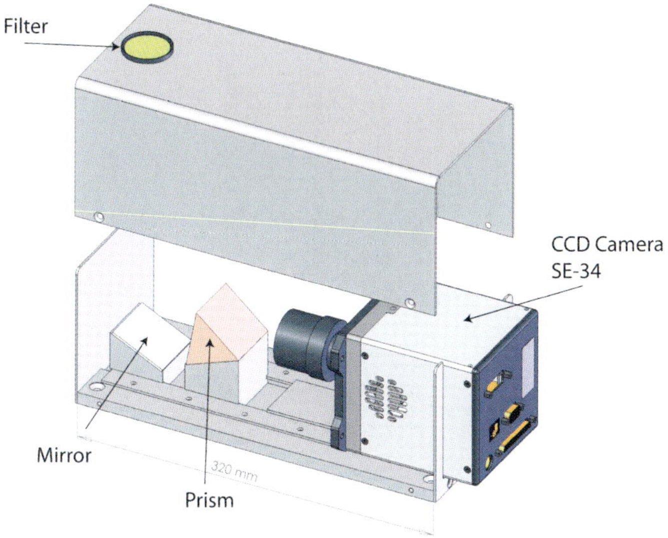

Figure 8.14 Fluorescence detection unit for hyperspectral detection, consisting of filters for blocking the excitation light, a dispersive prism and a highly-sensitive CCD camera, equipped with a high numerical aperture objective lens.

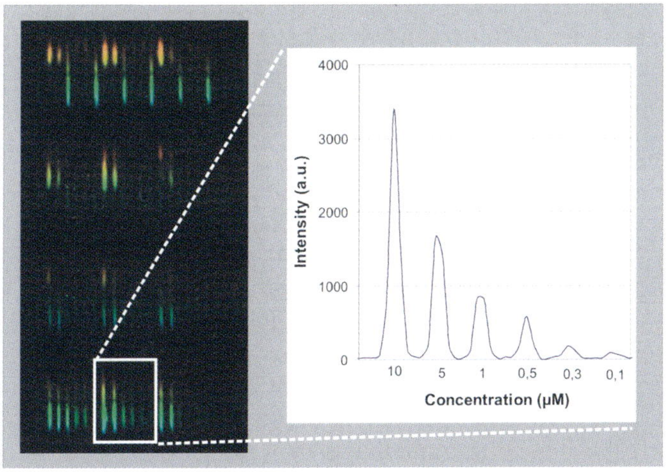

Figure 8.15 Detection of fluorescence dye molecules at several concentrations. On the left is the image on the CCD. The spots are aligned on the lateral axis, and the vertical axis corresponds to the fluorescence spectrum of the spots.

Figure 8.15 shows the intensity signal from test spots with different concentrations from 10 to 0.1 μM. In this experiment, fluorochromes Dy 631, Dy 550 and Dy 415 (Dyomics GmbH) were spotted onto a glass substrate with a

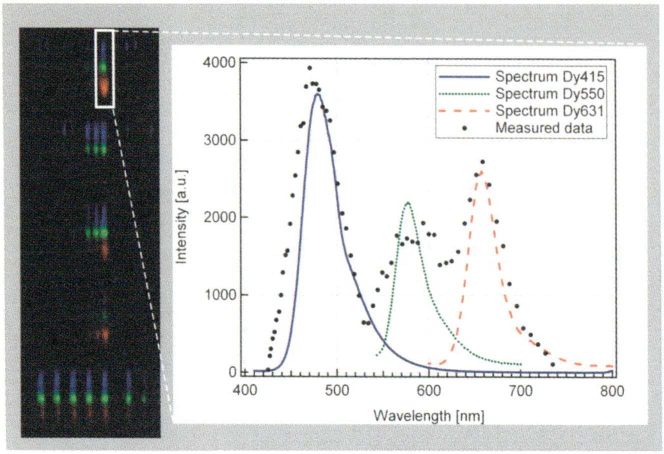

Figure 8.16 Spectrally resolved detection of three different dye molecules. The spots are aligned on the lateral axis, and the vertical axis corresponds to the fluorescence spectrum of the spots.

commercial needle printer. After drying, the resulting surface concentrations ranged from 2×10^3 to 2×10^5 molecules per μm^2. The left side of **Figure 8.15** shows the spectra of four one-dimensional arrays of fluorescence spots. In the bottom line, six different concentrations of the dye Dy 415, ranging from 10 to 0.1 μM, were spotted. The plot on the right shows fluorescence intensity as a function of the concentration. With the dynamic range of 12 bits of the CCD, three orders of magnitude in dye concentrations can be measured simultaneously.

8.4.3.3 **Detection of Emission Spectra**

Modern biochips are able to assay complex biochemical reactions. To obtain more information from a single spot, several fluorochromes are used simultaneously. This requires a biochip reader capable of spectral detection of the fluorescence signal in order to discriminate different fluorochromes.

As can be seen from **Figure 8.16**, it is possible to detect the fluorescence intensity of three different dyes in each individual spot of the microarray biochip. The figure shows five spot lines on a test chip with different concentrations of the fluorochromes Dy 415 (blue), Dy 550 (green) and Dy 631 (red). The spectral intensity of the first line is plotted on the right side of **Figure 8.16**. The three maxima at the wavelengths 470, 580 and 660 nm have intensity distributions closely matching those of the dyes measured independently (colored lines in plot). The resolution of 2 nm is sufficient to separate up to five spectra in the visible range.

8.4.4
Advantages of Dynamic Holographic Excitation of Microarray Biochips

In this section, a new approach for hyperspectral fluorescence spectroscopy was presented. The key feature of the new instrument is a flexible, point-like excitation of fluorescence biochips by using spatial light modulators. This new approach has four notable advantages.

1. The holographic multipoint illumination focuses the light only on the spots under investigation; background fluorescence is thereby reduced and the measurement shows a higher contrast than flood illumination.
2. An adaptation to arbitrary spot lattices can be achieved by changing the holograms, which is accomplished by software adaptation only; modifications of the hardware are not required.
3. Parallel illumination and read-out of one-dimensional spot arrays increases the throughput compared to single-point scanning techniques.
4. Continuous spectral fluorescence detection allows discrimination of various dyes without modification of the optical set-up, providing additional information on the chemical reactions. The system presented represents a fast, universal and reliable prototype biochip read-out unit.

8.5
Label-free Techniques

The two main features of a sensing device are sensitivity and specificity (selectivity). The sensitivity is prescribed by the sensor modality and design, whereas the degree of specificity is dependent on the detailed nature of the chemical surface of the receptor part. Thus, the receptor specificity is gained by minimizing non-specific binding at the transducer surface, in particular when using complex sample matrices. At the same time functional sites must be provided that allow controlled immobilization of chemical or biochemical compounds. The native properties of immobilized compounds should be maintained at the surface, and the surface accessibility must not be affected by the interfacial layer. Furthermore, stability of the interfacial layer is required for long-term operation. However, providing these properties by surface modification must not affect the sensitivity of the signal transducer. Thus, optical transducers typically require thin and homogeneous films to allow efficient detection [47].

Ellipsometry can be used for physical characterization of attached layers. This technique allows for an absolute and nondestructive assessment of the thickness and the dielectric function of thin organic and biochemical layers in atmospheric and aqueous environments [48]. These parameters give information about the surface loading and morphology of the layers.

The properties of electromagnetic radiation can be characterized by amplitude, frequency (wavelength), phase, polarization state and time dependence. Molecular modifications at the biosensor sensitive layer lead to changes in the properties of the interacting electromagnetic radiation, opening the way to different optical sensing principles. The dependence on the thickness of the layer and/or the refractive index influences the phase and/or amplitude of the electromagnetic radiation penetrating this layer or being reflected. Hence, reflectometric interference spectroscopy (RIfS) and surface plasmon resonance (SPR) sensing techniques will be discussed. Also, Mach–Zehnder interferometer-type waveguides determining the phase difference between two waves will be introduced.

8.5.1
Surface Chemistry

The chemistry of surface immobilization generally involves three main steps: cleaning and activation of the support surface, surface modification and surface functionalization. Performing these steps sequentially leads to a layer-by-layer development of the surface.

Silica (SiO_2) surfaces are technologically important since these are found on oxidized silicon, quartz and many types of glass. The structure is mainly composed of Si-O-Si and Si-OH bonds.

Usually, cleaning of the support surface occurs with a strong acid or plasma treatment. Piranha is a trade name for a chemical mixture consisting of sulfuric acid (H_2SO_4) and hydrogen peroxide (H_2O_2). It is used to remove organic contaminants from glass or Si surfaces. Also, the charge of a silica surface in an aqueous medium can be positively charged at very low pH < 1 and negatively charged at higher pH owing to deprotonation:

$$\mathrm{SiOH} \rightleftarrows \mathrm{SiO^-} + \mathrm{H^+}$$

An alternative cleaning and surface activation method, often used, is oxygen plasma treatment to remove organic residues and native oxides.

After the cleaning and surface activation step, modification of the surface by silanization is necessary. Glass, quartz or Si support surfaces can be chemically modified by silanization with reactive silanes containing alkyl chains terminated with functional groups [49]. Polysiloxanes are known for their useful properties, such as flexibility, low glass transition temperature, T_g, (about 146 K for poly(dimethylsiloxane)) and low surface energy. Polysiloxanes are characterized by linear Si-O-Si chains and can be bonded through functional groups to the Si atoms. At this working stage the layer-by-layer surface chemistry can be generated by using polysiloxanes terminating in different functional groups, for example:

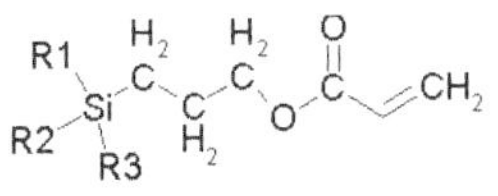

acrylic-oxy-propyl-silane

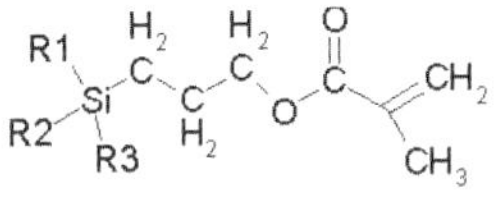

methacrylic-oxy-propyl-silane

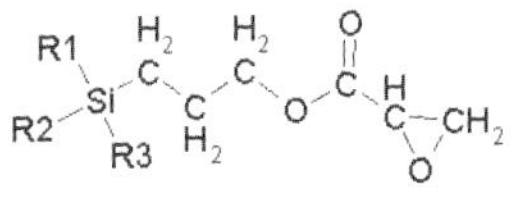

glycid-oxy-propyl-silane

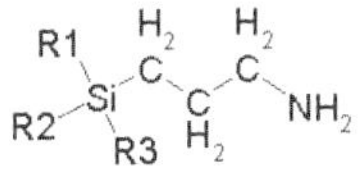

3-amino-silane

Physical factors, such as duration of silanization, solvent or temperature, influence the layer thickness and homogeneity. A schematic example of a silica surface after silanization is presented in **Figure 8.17**.

The functionalization of silica surfaces occurs by covalent binding of various biopolymers, supplying reduced non-specific binding properties. Dextrans are hydrophilic and non-charged natural polymeric carbohydrates, which are soluble in water in any proportion and form highly hydrated hydrogels. Dextrans show very low non-specific interactions with proteins or other macromolecules. Owning to the high concentration of hydroxyl groups in the dextran molecule, chemical modification of these polymers is possible without significantly affecting their hydrophilicity. The essentially unbranched polymer chains are highly flexible and ligands immobilized in dextran matrices are readily accessible to the receptor molecules. Dextran hydrogels supply a large number of functional sites distributed over the entire volume of the dextran layer. This brings the advantage of a large number of binding sites, but also the disadvantage that not all binding sites are equally accessible. Thus, hydrogel-based biosensitive layers are more suitable for thermodynamic measurements.

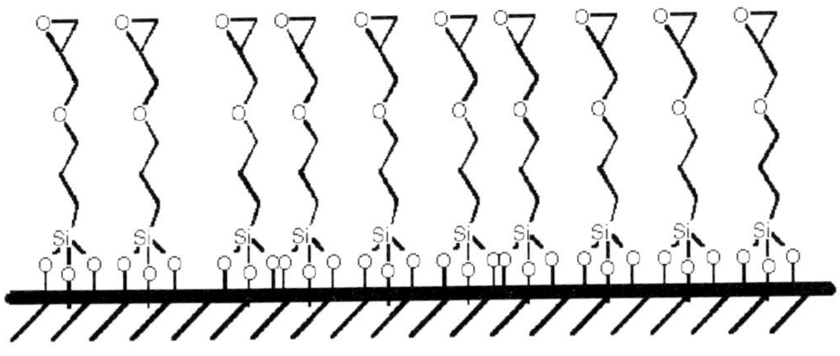

Figure 8.17 Model of a silanized silica surface.

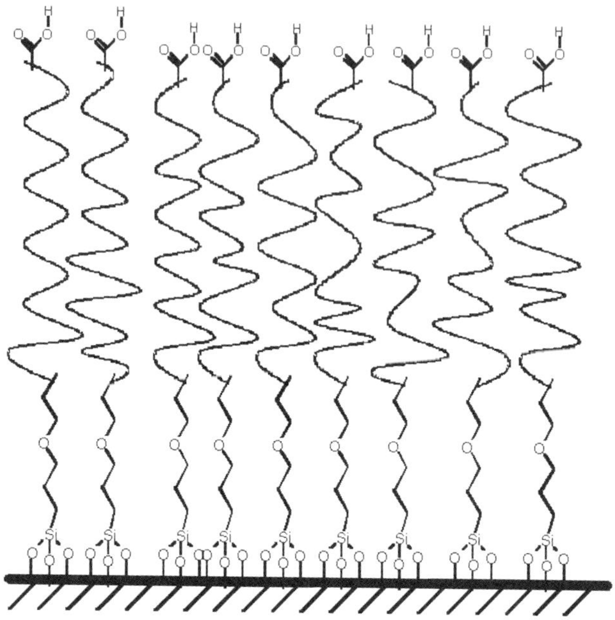

Figure 8.18 Model of a silanized silica surface functionalized with PEG.

Often, however, especially when observing protein interactions, non-specificity is not reduced sufficiently by the dextran layer. Another approach, therefore, is to bind poly(ethylene glycol) (PEG) of different chain length to the silanized surface to produce a kind of a polymer brush. PEGs are unbranched polymers, which have high exclusion volumes due to high conformational entropy and therefore repel (bio-)polymers. Thus, surface-attached PEG layers substantially decrease non-specific binding but these layers have a reduced number of interaction sites, because they are restricted to the surface and not in the volume. PEG- and dextran-functionalized silica surfaces are schematically shown in **Figures 8.18** and **8.19**.

At this stage of surface development, different ligands needed to bind biosensitive receptors can be covalently bonded through amino or carboxyl functional groups to the dextran or PEG biopolymers [47].

8.5.2 Surface Characterization

Each step of the layer-by-layer surface chemistry development can be characterized by different physical methods, such as contact angle measurements, atomic force microscopy, or ellipsometry [50]. In general, characterization of the sensitive layers by these different approaches allows us to understand the

Figure 8.19 Model of a silanized silica surface functionalized with PEG and dextran.

chemical or biochemical interaction process and thus to tailor the sensitive layers for optimal properties and find parameters for an optimal layer performance for various applications. This section is restricted to ellipsometric investigations of dextran- and PEG-functionalized silica surfaces [51,52].

Ellipsometry is a nondestructive optical method to determine the refractive index and the thickness of thin layers separately. Light is radiated at a defined angle of incidence with a well defined wavelength and polarization state onto the sample. Thus in ellipsometric measurements we determine the two ellipsometric angles $\tan \Psi$ and $\cos \Delta$. These angles describe the change of the state of polarization of the incident beam due to reflection at the probe interfaces. The product ρ of the two ellipsometric angles is defined as the ratio of the complex coefficients of reflection of parallel and perpendicular polarized light, R_P and R_S, relative to the plane of incidence:

$$\rho = \tan \Psi \cdot e^{i\Delta} = \frac{R_P}{R_S}. \tag{8.5}$$

Hence, small changes in thickness (<2 nm) and refractive index (<0.005) can be easily detected.

If the ellipsometric measurements are carried out in a wavelength range where the studied surfaces are nonabsorbing, the imaginary part of the refractive index can be neglected and the dispersion of the refractive index of the surface is satisfactorily described by a two-parameter Cauchy model [53]:

$$n(\lambda) = A + B/\lambda^2. \tag{8.6}$$

The index of refraction and film thickness are determined from the measured ellipsometric data by varying the model parameters until the best experimental data fit is obtained.

Dextrans of various molecular mass were covalently attached onto silanized planar silica surfaces and investigated by ellipsometry. The layer thicknesses found in air were (2.5 ± 0.3) nm, (4.1 ± 0.3) nm and (6.2 ± 0.2) nm, corresponding to dextrans with the molecular masses 10 kDa, 40 kDa, and 300 kDa, respectively. For evaluation of the surface density, the refractive indices of compact dextran layers of several 100 nm thickness prepared by spin-coating were determined. The dispersion of the refractive indices was numerically simulated by using a two-parameter Cauchy model. The refractive indices of all covalently attached dextran layers were in the range of the refractive index of the densely packed, spin-coated layers, indicating that a high coverage and homogeneous distribution is achieved by the covalent immobilization technique [53].

For closer understanding of ligand interaction with dextran layers, investigations were carried out by ellipsometry under phosphate-buffered saline aqueous medium (PBS). A strong swelling of dextran in PBS solution was observed [53]. The thicknesses of the dextran layers were found to be (19 ± 0.1) nm, (28 ± 8) nm, and (36 ± 6) nm corresponding to the molecular masses 10 kDa, 40 kDa, and 300 kDa, respectively.

Comparable results were obtained for layer thicknesses of aminodextran (AMD) and carboxymethyldextran (CMD) (200 kDa) functionalized surfaces [54]. The ellipsometric measurements were evaluated by setting a constant refractive index at 1.52 of the biolayer in air, and at 1.38 in PBS, on the assumption that in PBS solution a rise in volume of the hydrogel occurs. For both CMD and AMD layers, thicknesses in the 4–6 nm range were determined in air. Under PBS, however, the two layers clearly behave differently. CMD shows a swelling up to 150 nm while AMD swells up to 65 nm. Since CMD has a lower number of binding sites than AMD, CMD possesses fewer binding sites to the surface than AMD. Thus, CMD can build to a higher layer thickness than AMD.

Likewise, PEG polymers with chain lengths corresponding to molecular masses of 2000 Da, 3000 Da, 6000 Da, and 10000 Da were covalently bonded to silanized silica surfaces [54]. These PEG polymers correspond to theoretically calculated lengths of 20 nm, 30 nm, 60 nm, and 100 nm, respectively. Similarly, the refractive index was set as constant at 1.52 for measurements in air, whereas in PBS a constant refractive index of 1.37 was used.

Since during the ellipsometric measurement over the entire measured spot is averaged, any inhomogeneity or domain building causes a lower result for the layer thickness, while the layer thickness of a homogeneous polymer layer would give rise to a thickness close to the calculated chain length, because the chains tend to straighten up to their full length.

In air, all the investigated PEG layers collapsed on themselves and formed layers of thickness 3–7 nm. The layer thicknesses reached by the PEG layers in PBS are more informative, as already mentioned above. PEG-2000 and PEG-3000 reached (19 ± 1) nm and (28 ± 2) nm, respectively, the measured layer thicknesses closely matching their calculated chain lengths. The layers thus form a very homogeneous smooth surface, with no building of domains or islands. Equally, one can assume that very few of these PEG chains bind with both ends to the surface.

The PEG-6000 and PEG-10000 layers do, however, not achieve their calculated chain length, with layer thicknesses of no more than (52 ± 4) nm and (76 ± 5) nm, respectively. With increasing chain length, the PEG chains bond with both ends and thus cause island formation or domain building, so that the molecules in these layers can no longer straighten up to their full chain length.

In conclusion, ellipsometric investigations are suitable for the detection of high functionalization of the surface. They supply information that supplements data available from contact angle measurements and atomic force microscopy.

8.5.3
Reflectometric Interference Spectroscopy (RIfS) Sensing

This sensing, in principle, is based on monitoring biomolecular interactions using the modified surfaces described above and various types of transduction element. These can be categorized according to their detection principles, including thermal, electronic, mass-specific, electrochemical and optical principles. Optical principles can be classified into label-free methods and methods using labels such as fluorophores. Label-free methods require the use of either reflectometric or refractometric methods. The product of the reflective index n and the physical thickness d of a thin layer represents the optical thickness. Direct optical detection techniques measure this optical thickness during biomolecular interaction processes.

Changes in optical thickness can be measured using interferometric methods in microreflectometry. A typical example is ellipsometry (Section 8.5.1), which is preferably used to characterize the chemical modification of the sur-

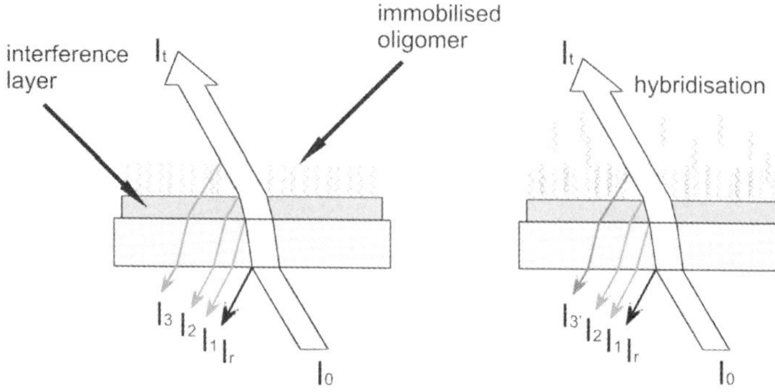

Figure 8.20 White light interference in RIfS.

faces. However, this method requires polarized radiation and is rather complex with respect to measurement and evaluation of the data. Reflectometric interference spectroscopy represents a simplified principle which does not use polarized light, but is based simply on white-light Fabry–Pérot interferometry. At the interface of such a thin layer, the incident radiation is usually partially reflected. The other part of the radiation penetrates the layer and is reflected at the other interface. These two partial reflected beams superimpose and form an interference pattern, which results in either constructive or destructive interference, depending on the angle of incidence, the wavelength, and the optical thickness of the layer.

These changes in optical thickness are demonstrated in **Figure 8.20** in the case of hybridization studies. Complementary sequences of DNA interact with the immobilized sequence at the transduction surface; this causes the physical thickness to change, which in turn shifts the modulation pattern as can be seen in **Figure 8.21**. The advantage of this microreflectometric method is that the main influence on the modulation is caused by the change in physical thickness rather than in refractive index, which is in contrast to microrefractive methods based on evanescent field techniques. Since the refractive index is very sensitive to temperature, this dependence can turn out to be a problem.

Performing a series of measurements over time, as is done in **Figure 8.22**, it is possible to monitor the association and dissociation rates during these hybridization studies. In this specific case, the loading of the surface should not be too high, otherwise a mass-transport-controlled dependence of ligands from solution to the immobilized receptors controls the signal. It will be linear with time. A high concentration of receptor molecules at the surface shifts the signal to monitor the time dependence of the diffusion-controlled motion of ligands to the surface. A low concentration of receptors at the surface allows measurement of the time dependence to reach equilibrium at the surface.

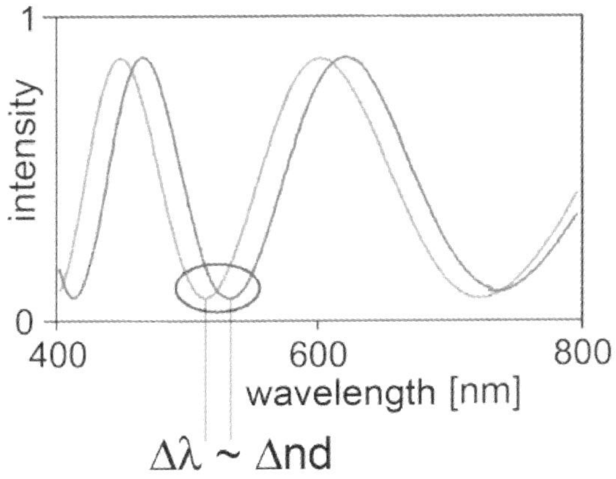

Figure 8.21 Modulation pattern for RIfS, shifted by changes in the physical thickness. The wavelength shifts in this pattern at the extremes (minima or maxima) or even at the point of inflection correlate to the amount of binding.

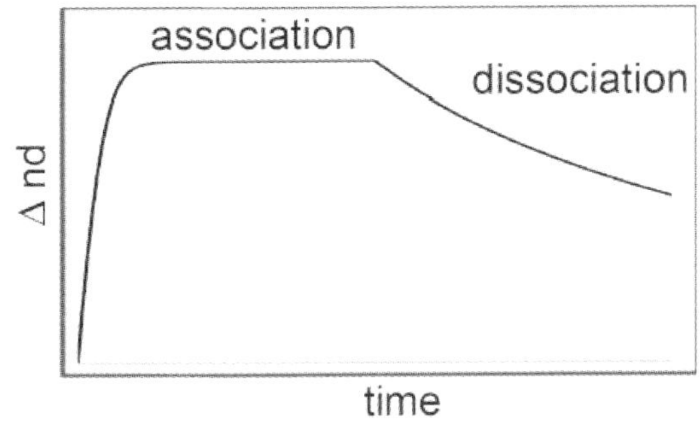

Figure 8.22 Measurement of RIfS in dependence on time, giving association and dissociation rates [55, 56].

Reflectometric interference spectroscopy has found wide application in monitoring biomolecular interaction processes such as DNA/DNA hybridization, protein/protein interaction, antibody/antigen measurements and other binding studies [57]. Typical examples are the characterization of antibodies against benzo[*a*]-pyrene [58] or the monitoring of glycopeptide antibiotic fermentation using white light interference as an optical biosensor principle [59].

These methods can be parallelized either for assays measured in microtiter plates or on planar layers. Since CCD-cameras can be used to monitor "intensity" changes on areas, RIfS supplies an easy approach to high-throughput screening. An example is the label-free parallel screening of combinatorial triazine libraries using RIfS [60] or the epitope-mapping of transglutaminase with parallel label-free optical detection [61]. This specific reflectometric method has been compared with other instruments of label-free optical detection using evanescent field techniques [62].

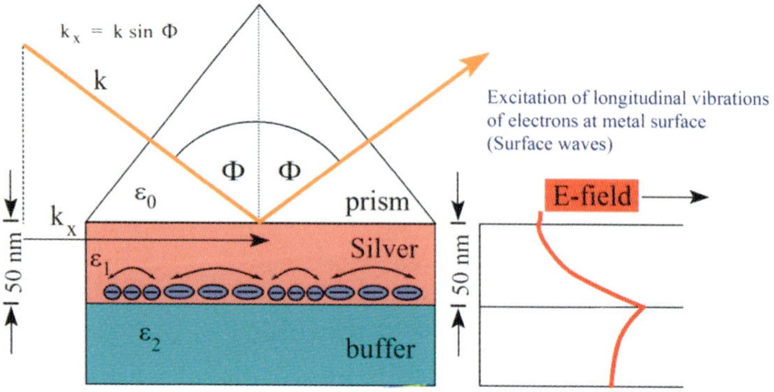

Figure 8.23 Schematic drawing of surface plasmon resonance set-up.

8.5.4
Surface Plasmon Resonance (SPR) and Mach–Zehnder Techniques

As mentioned above, in label-free detection methods the measurement of microrefractometry is another approach. This encompasses methods based on gold layers and methods based on glass transducers. These methods rely on evanescent field effects by changes in the refractive index of an analyte solution or interaction layer close to a waveguide surface (the transducer element).

Many years ago, SPR [63] was commercialized by Biacore [64]. The principle is based on total internal reflectance of an incident radiation on a prism to achieve the optimal ratio of refractive indices; this prism is coated at its base by a thin metal film of ca. 50 nm. Depending on the angle of incidence and the wavelength, the incident radiation excites plasmons within this metal film, in resonance on the surface opposite to the waveguide interface adjacent to the analyte medium. This medium determines the resonance condition of these plasmons, depending on its refractive index. This means that changes in the refractive index will influence the resonance of the plasmons, reducing the reflected intensity of the polarized light and resulting in a reduced reflectance. Several reviews have been published regarding this SPR technique [65]. The principle is shown schematically in **Figure 8.23**.

Whereas this method uses a gold-coated transducer, other transducers based on microrefractometry such as a resonant mirror [66], a grating coupler [67] and integrated Mach–Zehnder chips use glass-type transducers. Mach–Zehnder chips use interferometric-type waveguides. In these chips, the phase difference between two waves traveling in the two arms of the waveguide is measured [68]. One arm is inert to its surroundings, the other arm is in contact with the analyte. These chips allow very sensitive detection, since the phase difference depends on the interaction length of the waveguide arms and the quality of the waveguide structure. In principle, a theoretical quality of up 10^{-9} differences in refractive indices can be calculated. This means it should be possible to measure very low concentrations, down to less than picomolar. However, theoretical calculations and measured limits of detection differ, especially in bioanalysis.

8.6 Terahertz Spectroscopy

The terahertz (THz), or far-infrared region is among the least explored parts of the electromagnetic spectrum. Until recently it was rather difficult to generate and detect THz waves efficiently. Most THz sources were either low brightness emitters such as thermal sources, or bulky, single-frequency gas lasers. Detection usually relied on bolometric methods, which required cryogenic cooling and generally provided low sensitivity. More recently, however, there has been a revolution in THz technology, as a number of newly discovered or rediscovered generation and detection schemes have revitalized the field. With the advent of femtosecond laser technology in the late 1980s, opto-electronic generation and detection of broadband pulses for THz time domain spectroscopy became possible [69]. Many sensing applications using THz waves have been discussed since then, ranging from medical diagnostics, through packaging inspection, quality control and chemical composition analysis, to security applications and gene tests [70]. However, femtosecond laser-based spectrometers are still expensive and require a laboratory environment. More practicable spectrometers require fiber-coupled THz antennas and advanced optical delay lines for fast measurements. A possible compact and cost-effective alternative is a spectrometer based on the generation of continuous-wave (cw) THz radiation by photomixing [71]. After a description of the principles of pulsed THz time domain and cw THz spectroscopy in Section 8.6.1, we describe experimental advances for pulsed and cw THz spectrometers achieved within the MOBA project in Section 8.6.2.

Independent work from the groups of Markelz [72] and Kurz [73] shows that the complex dielectric constant of DNA depends on its hybridization state. This raised the hope that THz waves could be used as a marker-free method to probe the results of hybridization experiments. Nagel et al. first demonstrated a scheme for on-chip gene tests with THz pulses propagating along a stripline with an incorporated filter structure loaded with DNA [74]. In Section 8.6.3 we describe the design and fabrication of efficient filter structures for future on-chip gene tests.

For small biomolecules in the solid state, such as saccharides and amino acids, distinct absorption bands due to intermolecular vibrations have been observed. These absorption spectra represent unique fingerprints of the substances that can be used for identification. They also provide information about conformations and binding strengths. One of the goals in this project was to examine a wide range of biomolecules with THz time domain spectroscopy. The results are shown in Section 8.6.5

THz time domain spectroscopy measurements on artificial RNA show that standard free space THz spectroscopy can also be used to distinguish between different molecules spotted onto a substrate. We show spatially resolved measurements in Section 8.6.6 that allow for the identification of spots of different samples. The high future potential of THz spectroscopy is concluded in Section 8.6.7.

8.6.1 Principles of THz Spectroscopy

8.6.1.1 THz Time Domain Spectroscopy

The best performance in broadband THz spectroscopy in terms of bandwidth, dynamic range, and signal-to-noise ratio is obtained with time domain methods. In the following we will outline the principles behind THz time domain spectroscopy, or THz TDS.

THz TDS is based on coherent generation and field-resolved detection of an ultrashort burst of electromagnetic radiation. The radiation is generated by femtosecond optical excitation above the band gap in an ultrafast semiconductor [75] or below the band gap in a nonlinear crystal [76]. Similarly, the generated THz radiation is detected by either a photoconductive receiver [75] gated by a second, delayed replica of the femtosecond laser pulse, or an electro-optic crystal [77–79].

In the following we will focus the attention on photoconductive generation and detection of ultrashort THz pulses. **Figure 8.24a** is a schematic diagram of a THz TDS system showing the beam path of the femtosecond laser pulse, the THz emitter and receiver and the THz beam path, including collimating and focusing optics [80, 81]. The inset illustrates a typical arrangement of the electrodes on the THz emitter and detector.

Femtosecond excitation of the biased photoconductive gap on the emitter antenna results in acceleration of the photogenerated charges. This acceleration leads to a burst of electromagnetic radiation, with frequency components determined by the speed of the photoconductive material. It is possible to grow semiconductor materials with a carrier lifetime in the range of 100 fs [82], resulting in an intrinsic bandwidth of the generated electromagnetic transient in the range of the inverse of this lifetime, i.e., approximately 10 THz.

The THz radiation can be propagated with reflective or transmissive optics. Typically we use elliptically or paraboloidally shaped reflective optics in order to minimize reflection losses and absorption in lens materials. Most traditional

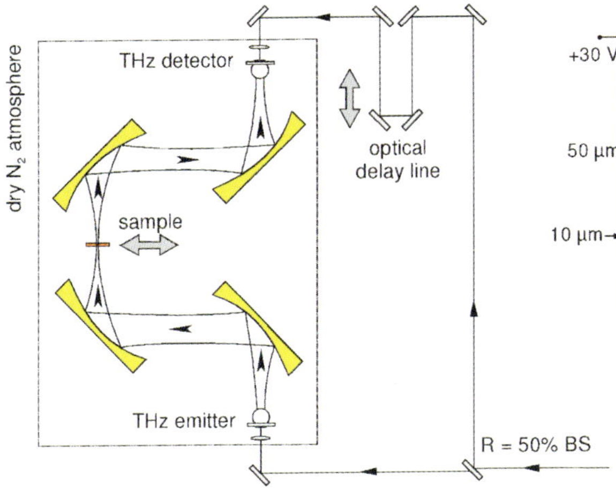

Figure 8.24 Schematic diagram of a THz time domain spectrometer and detailed view of the arrangement of the electrodes on a pulsed THz emitter and detector.

optical materials, such as optical-grade glass and transparent polymers, are not optimized for good performance in the THz region.

Photoconductive detection of ultrashort THz pulses is performed in a semiconductor structure similar to the THz emitter antenna. The main difference is that no bias is applied across the photoconductive gap defined on the semiconductor surface. Instead the electric field of the incoming THz signal is used to drive a photocurrent in the external circuit. By gating the detector with a second, delayed portion of the femtosecond laser beam there is an intrinsic synchronization between the THz signal and the read-out of the photocurrent. With an ultrashort lifetime of the photogenerated carriers in the detector it is possible to measure the instantaneous electric field of the THz signal. Under these conditions, time-resolved measurement of the electric field of the THz signal is then possible by a simple mechanical scanning of the arrival time of the gate pulse with respect to the THz signal.

A THz TDS measurement is carried out by the recording of two time domain signals, the reference signal $E_{\mathrm{ref}}(t)$ and the sample $E_{\mathrm{sam}}(t)$ signal. A subsequent transformation to the frequency domain and comparison of the reference and sample spectra yields the absorption coefficient and index of refraction of the sample material,

$$\begin{aligned}
&\frac{E_{\mathrm{sam}}(\omega)}{E_{\mathrm{ref}}(\omega)} \equiv A\,\mathrm{e}^{i\phi} = \frac{4n}{(n+1)^2}\,\mathrm{e}^{-\alpha d/2}\,\mathrm{e}^{i(n-1)\omega d/c} \\
&\Downarrow \\
&n(\omega) = 1 + \frac{\phi \cdot c}{\omega d}, \\
&\alpha(\omega) = -\frac{2}{d}\ln\left(\frac{(n+1)^2}{4n}A\right).
\end{aligned} \tag{8.7}$$

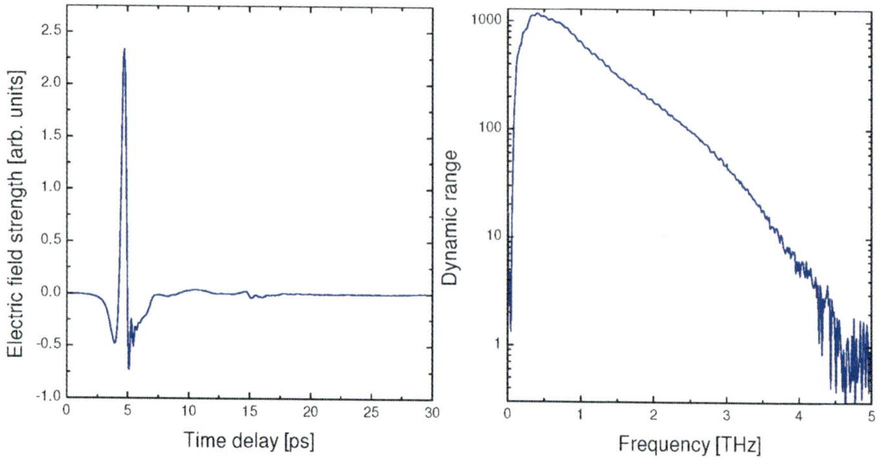

Figure 8.25 The left panel shows the recording of a typical THz pulse in the time domain. The right panel shows the corresponding frequency spectrum of the pulse, covering the range 0.1–4 THz.

Figure 8.25 shows a typical THz pulse and its associated frequency spectrum. In this case the bandwidth spans the range of 0.1–4 THz, with a dynamic range exceeding 60 dB at the peak of the frequency spectrum.

8.6.1.2 **Photomixing**

The generation of cw THz radiation by photomixing requires the spatial overlap of two laser modes of wavelengths $\lambda_1 = \frac{2\pi c_0}{\omega_1}$ and $\lambda_2 = \frac{2\pi c_0}{\omega_2}$ with $\lambda_1 < \lambda_2$ in a nonlinear device with a frequency difference

$$\Delta f = \frac{\omega_{\mathrm{THz}}}{2\pi} = \frac{\omega_1 - \omega_2}{2\pi} = c_0 \cdot \left(\frac{1}{\lambda_1} - \frac{1}{\lambda_2}\right) \tag{8.8}$$

in the terahertz range. The superposition of the two laser modes $E_1(t) = E_0 \cos \omega_1 t$ and $E_2(t) = E_0 \cos \omega_2 t$ as shown in **Figure 8.26** leads to a frequency beat of

$$\begin{aligned} E(t) &= E_1(t) + E_2(t) = E_0 \cos(\omega_1 t) + E_0 \cos(\omega_2 t) \\ &= 2E_0 \cdot \cos\left(\frac{\omega_1 + \omega_2}{2} t\right) \cdot \cos\left(\frac{\omega_1 - \omega_2}{2} t\right) \end{aligned} \tag{8.9}$$

that contains half of the sum frequency $\frac{\omega_1+\omega_2}{2}$ and half of the difference frequency $\frac{\omega_1-\omega_2}{2}$ of the two laser modes. At the photomixer it generates a photocurrent $I(t)$ that is proportional to the optical power and hence to the square

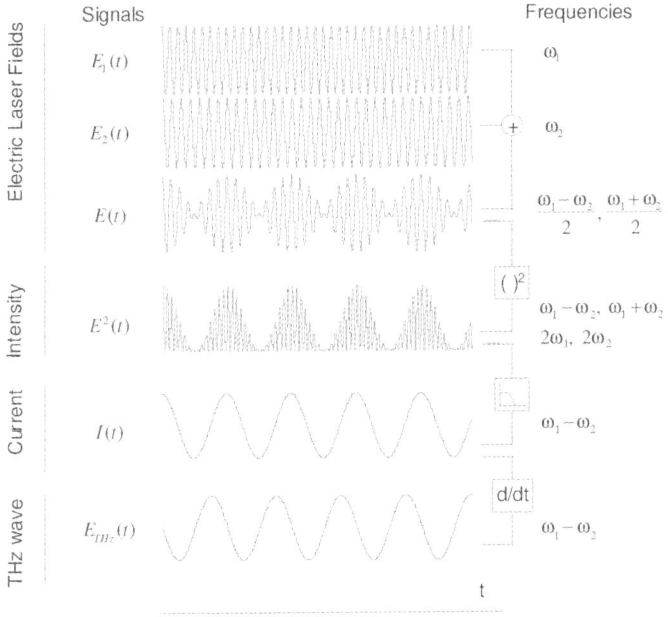

Figure 8.26 Principle of photomixing. In the photomixer a photocurrent is generated at the difference frequency of the two incident laser modes that radiates the THz wave.

of the electric field. This photocurrent

$$I(t) \propto E^2(t) = E_0^2 \Big[1 + \cos(\omega_1 + \omega_2)t + \cos(\omega_1 - \omega_2)t + \frac{1}{2}\cos(2\omega_1 t) + \frac{1}{2}\cos(2\omega_2 t) \Big] \quad (8.10)$$

contains the desired THz frequency $\omega_{\text{THz}} = \omega_1 - \omega_2$. According to Maxwell's equation the photocurrent acts as a source for the radiated THz wave that is proportional to the first time derivative of the photocurrent.

A photomixer consists of a region where the photocurrent is generated and an antenna that efficiently radiates the generated THz waves. In the simplest case, the free carriers are generated in the biased gap of a photoconductive dipole antenna [83]. More advanced photomixers consist of an interdigital structure that feeds a logarithmic spiral antenna or a double dipole element [84].

If the photon energy $\hbar\omega_1 \approx \hbar\omega_2 \approx \hbar\frac{\omega_1+\omega_2}{2}$ of the laser radiation is above the energy of the direct transition, each "impulse" of the light beat generates free electrons and holes in the semiconductor substrate. The free carriers are accelerated in an external electrical field, which gives rise to a photocurrent. The decay time of the current impulses depends firstly on the screening of

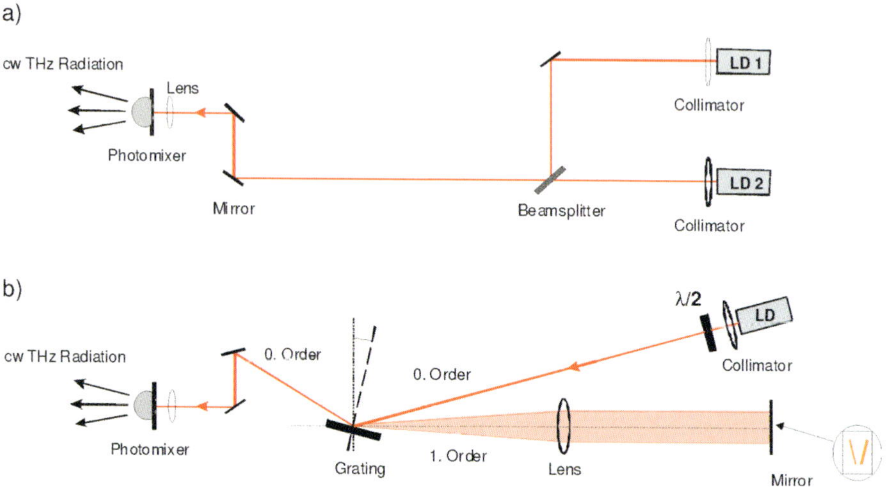

Figure 8.27 Laser sources for photomixing. a) The laser beams of two separately tunable laser diodes (LD 1 and 2) are spatially overlapped in a beamsplitter and focussed onto a photomixer. b) An antireflection coated laser diode in an external cavity with frequency selective feedback is used to generate two laser modes simultaneously.

the external electric field caused by the electron–hole pairs that are already separated and secondly on the free carrier lifetime τ. In order to enable the photocurrent to follow the THz beat, $\omega_{\text{THz}} < \frac{2\pi}{\tau}$ has to hold. According to Ohm's law the resulting photocurrent density $j(t) = -e\mu n(t)E(t)$ depends on the effective electric field $E(t)$, the mobility μ and the concentration $n(t)$ of the carriers with charge e. For this reason photomixers are produced on substrates of low-temperature-grown GaAs (LT GaAs) [82]. Apart from a carrier lifetime as short as 0.1 ps, it has a high mobility in the order of 200 cm^2 (V s)$^{-1}$. In addition, it has a breakdown field strength of 100 kV cm^{-1} that allows the application of an appropriately high bias without causing an electrical breakdown or significant dark currents, which would increase the noise background of the THz signal.

In the simplest case, the two laser modes required for photomixing are generated by two separate laser diodes (LD 1 and LD 2) and overlapped with a beamsplitter as shown in **Figure 8.27a**. Of the two laser sources, at least one has to be tunable in order to generate any frequency difference. Typically, external cavity lasers in Littman configuration [85] or distributed feedback (DFB) laser diodes that are temperature tunable are used.

Using two separate lasers requires a very precise overlap of the two beams. Easier to construct and also more compact is the two-color laser shown in **Figure 8.27b**. It consists of an anti-reflection coated laser diode that is operated in a special kind of external resonator. The incident light from the laser diode is diffracted from a grating. The first diffraction order beam is collimated. After the lens each parallel beam corresponds to a certain wavelength in the gain spectrum of the laser diode. With a V-shaped mirror two single wavelengths are reflected selectively. This causes the laser diode to oscillate on two external resonator modes simultaneously. By changing the height of the V-mirror, the wavelength difference can be adjusted to the desired value.

This concept of a two-color laser with an external resonator based on a spatial Fourier transform (Fourier-transform external cavity laser – FTECAL) was invented by M. Hofmann and coworkers [86]. A similar concept was realized independently by Gu et al. in Japan [87]. A full mirror with a transparency controllable liquid crystal matrix in front can be used, instead of the V-shaped mirror, as the frequency selective element (electronically tunable external cavity laser diode – ETECAL). This concept allows for purely electronic tuning of the laser without any moving mechanical parts. However, using a liquid crystal matrix as the dissipative element in the external resonator limits the output power significantly. Hence, the V-mirror version is better suited for cw THz generation.

In a cw THz spectrometer the detection of the radiation can either be realized with a second photomixer that is coherently pumped with part of the two-color laser light or by a bolometer. Using a second photomixer requires a significant amount of two-color laser power and is very difficult to adjust, but also quite compact. A bolometer on the other hand, is easy to align but requires liquid helium cooling.

8.6.2 Experimental Advances

Applying THz spectroscopy as a tool for the marker-free examination of biomolecules requires significant enhancements of the spectrometers beyond today's state-of-the-art. We present three significant advances in this subsection, including the transmission of femtosecond laser pulses via optical fibers as required for a portable spectrometer, a new optical delay line for faster data acquisition and the stabilization of a two-color laser, which is necessary for reliable long-term cw measurements.

8.6.2.1 Femtosecond Optical Pulses Guided by Optical Fibers

THz spectrometers are, with few exceptions, optically fed by free space optics. In terms of stability and reliability, this solution is only useful in the laboratory. For this reason the development of fiber-coupled spectrometers is desirable. The THz spectrometer itself can then be constructed in a highly compact manner, and driven basically by a set of optical fibers from a control unit containing the laser source and a compact optical set-up with delay lines and fiber coupling units.

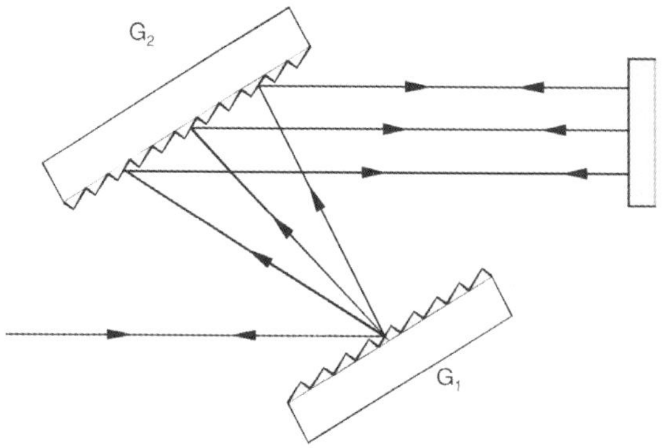

Figure 8.28 Principal components of a dispersion compensator for ultrashort laser pulses. G_1 and G_2 are reflective diffraction gratings.

For a THz TDS system the main technological challenge to overcome is to propagate ultrashort pulses over a significant distance without temporal dispersion through the fiber. A standard optical fiber is notoriously dispersive, leading to severe pulse lengthening even after propagation through just a few meters of fiber. We have developed a pre-compensator unit for a THz TDS system that overcomes the dispersion problems by proper pre-shaping of the optical pulses before they are launched into the optical fiber. The principle behind this is that the dispersion experienced by the pulse upon propagation through the fiber can be compensated before the pulse enters the fiber. This is possible because the pulse dispersion is caused by different propagation speeds of the different frequency components of the laser pulse. The bandwidth required to support a given pulse length is inversely proportional to the pulse length. Hence, a femtosecond laser pulse has a spectrum spanning from 10 to more than 100 nm around its center wavelength. In a normal fiber the long-wavelength ("red") components of the pulse propagate slightly faster than its short-wavelength ("blue") components, an effect referred to as Group Velocity Dispersion (GVD). Hence the fiber dispersion can be compensated by a device that exactly reverses the GVD. Such a pre-compensator based on a pair of gratings is well known in the field of ultrashort optical pulse generation. The schematic layout of the device is shown in **Figure 8.28**.

The different frequency components are diffracted into different angles by the first diffraction grating G_1 onto the second grating G_2 and further onto a mirror which reflects the beam back onto itself. The arrangement of the optical beam path assures that the blue components travel a shorter distance than the red components, thereby compensating the GVD experienced by the laser pulse in the optical fiber. The exact amount of dispersion introduced can be controlled by adjusting the distance between the two gratings.

With this device it has been possible to launch optical pulses at a wave-

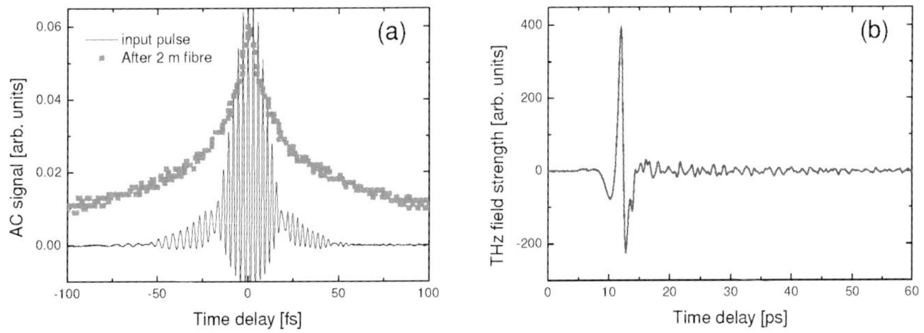

Figure 8.29 (a): Autocorrelation traces of the input optical pulse (central wavelength 800 nm) and output pulse after dispersion pre-compensation and propagation through 2 m of single-mode optical fiber. (b): THz pulse generated and detected with a fiber-coupled THz emitter and detector.

length of 800 nm into several meters of optical fiber while retaining short pulse duration at the output of the fiber. In **Figure 8.29** the result of such an experiment is shown. The optical pulse duration before and after propagation through the fiber was measured with a standard autocorrelator for ultrashort optical pulses. The pulse duration after the fiber is comparable to the input pulse duration. The short, compressed pulse (ca. 40 fs) is overlaid with a slower (ca. 300 fs) signal. This signal is caused by higher-order dispersion in the fiber, which cannot be compensated by the pre-compensator unit. However, for the purpose of THz pulse generation this slow background signal is inconsequential. Subsequently the optical fibres were coupled to the photoconductive THz antennas of a THz TDS system. The resulting THz signal is also shown in **Figure 8.29**. As can be seen, the THz signal resembles that generated and detected by a free space coupled spectrometer.

8.6.2.2 **Spiral Optical Delay Line**

Optical delay lines are widely used to control the temporal positions of ultrafast laser pulses in all kinds of pump–probe experiments, optical tomography, interferometry measurements and THz time domain spectroscopy [88]. The most common optical delay lines are the stepper motor and the shaker, both of which are used to move a retroreflector. While the stepper motor is very accurate it is inherently slow and cannot support fast data acquisition. For imaging purposes, for example terahertz imaging where speed is a crucial issue the oscillating shaker is more favorable. Yet, it moves back and forth in a sinusoidal motion and therefore suffers from a nonlinear time axis, resulting in a rather low precision and more important a "dead time" since only the nearly linear fraction of the shaker oscillation period is of practical use. In addition, the scanning rate is limited to a few hertz owing to the translated mass

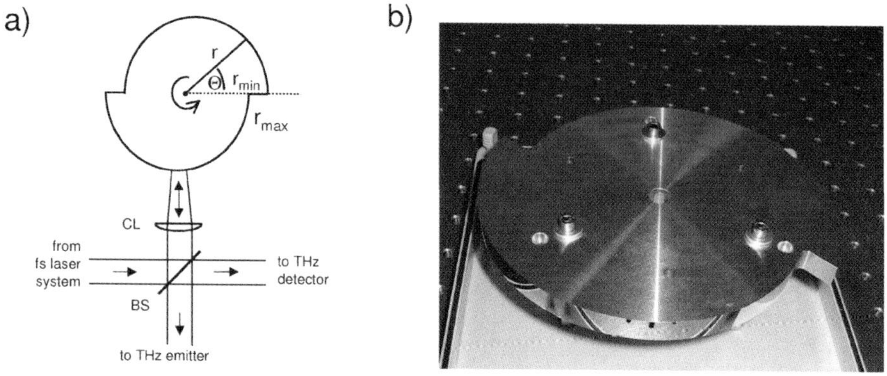

Figure 8.30 (a): Spiral optical delay line as part of the THz time domain spectroscopy set-up. (b): Photograph of the spiral optical delay line.

of the attached retroreflector, which needs to be accelerated periodically. This limits the scanning frequency as well as the stability of the beam. A rotary movement at a constant speed does not cause these complications and is thus superior to a linear one.

Here, we present a spiral optical delay line that supports high data acquisition rates, has no "dead time" and is simple to produce [89,90]. We note that, simultaneously but completely independently of our work, a similar design has been developed by Zhang and coworkers [91]. As the spiral reflector rotates, the path length of the reflected beam alters in such a way that the time delay depends linearly on the angular position of the spiral. Hence, our device provides a sawtooth-like delay function when rotating with a constant speed. To demonstrate the working principle, the delay line was incorporated into a THz time domain spectroscopy system. Our data shows that it provides a better signal-to-noise ratio than the shaker commonly applied for THz imaging.

Figure 8.30a shows how the spiral delay line is implemented in an optical set-up. A photograph of the device is shown in **Figure 8.30b**. The incoming laser beam is retroreflected from the curved surface in the direction of incidence. The length of the optical path depends on the azimuth position of the spiral. For an ideally sawtooth-like delay function, the radius of the spiral reads: $r(\theta) = a + b\theta$. In order to avoid being out-of-balance, the device should be centro-symmetric to the rotational axis. In Cartesian coordinates the spiral is described by

$$x = \left(a + \frac{\Delta r}{\pi}\theta\right) \cdot \cos(\theta) \qquad \text{and} \qquad y = \left(a + \frac{\Delta r}{\pi}\theta\right) \cdot \sin(\theta) \tag{8.11}$$

where $a = r_{\min}$ is the minimum radius of the spiral, θ is the azimuth angle between 0 and π, and Δr is the difference between the maximum and min-

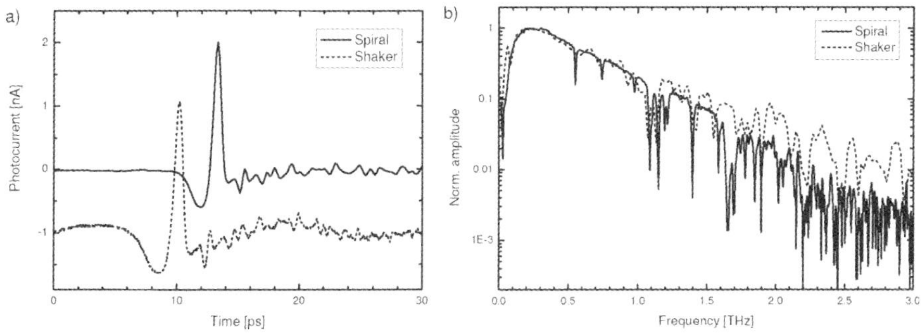

Figure 8.31 (a): THz waveforms obtained with spiral delay line (solid line) and shaker (dashed line). (b): The corresponding THz spectra.

imum radius. For a prototype we have chosen $a = 64$ mm, $\Delta r = 16$ mm, which corresponds to a delay range of 106 ps, and a spiral height $h = 10$ mm. The alternative approach by Zhang and coworkers [84] uses a more complicated involute curve design. Yet, the discrepancy in delay time to our model is below 10 fs which is beyond the manufactural surface accuracy. Hence, for practical devices with a diameter of more than a few centimeters it does not matter which of the two mathematical descriptions is chosen. For both approaches, the surface curvature will of course result in a divergently reflected beam. To compensate for this a thin cylindrical lens is inserted into the beam immediately in front of the reflector at a distance of its focal length f (here $f = 100$ mm) from the rotational axis. Owing to its focusing effect in one plane, every part of the initially parallel laser beam profile hits the curved surface of the reflector at a right angle. Hence, no beam expansion occurs at any angular position of the delay spiral.

To test the functionality of the delay line it was incorporated into the emitter arm of a standard THz time domain spectrometer using a 20 fs Ti:S laser. The emitter arm also contained the linear shaker which was disabled when using the spiral delay line and vice versa. Thus, a direct and fair comparison between the two delay schemes was possible. While the shaker operated at a frequency of 4 Hz, sampling with the spiral delay line was performed at 30 Hz, merely limited by the motor available. **Figure 8.31a** shows two THz waveforms obtained with the spiral delay line and with the shaker. Each waveform is acquired by 128 averages. The corresponding spectra are plotted in **Figure 8.31b**. Although the spiral measurement takes only 4.26 sec whereas the shaker measurement takes 32 sec, the signal-to-noise ratio obtained with the spiral is significantly higher than that obtained with the shaker. Besides, in the case of the spiral no correction for a nonlinear delay (as in the case of the shaker) is required. In principle, the spiral can be rotated with much higher frequency, which makes it ideal for applications such as THz imaging.

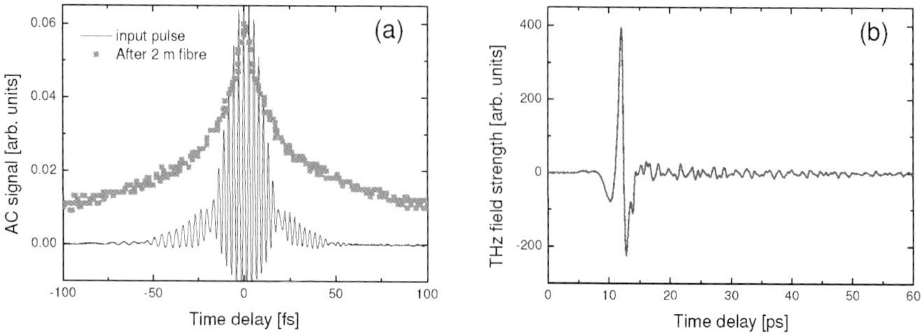

Figure 8.32 Balance of the two laser modes after initial adjustment. (a): without control, (b): with control unit activated.

8.6.2.3 **Stabilization of the Two-color Lasers**

In order to make reliable long-term measurements, a two-color laser source that is stable for several hours is needed. As can be seen in **Figure 8.32a**, an unstabilized FTECAL configuration that has been initially adjusted in such a way that both laser modes at 834.944 nm and 834.221 nm (frequency spacing $\Delta f = 311.4\,\text{GHz}$) have equal intensity can lose its balance after a while. In this case, chaotic behavior occurs after 400 s.

In order to stabilize the two-color laser we first implemented a temperature control of the laser diode support. Using a Peltier element and a temperature sensor, the temperature of the laser diode mount was made adjustable with an accuracy of less than 0.01 °C. The program controllable laser driver was used to remotely adjust the laser diode temperature between 5 and 30 °C. After optimizing the thermal design of the mount, fast tuning of the temperature within a few seconds and excellent long-term stability were achieved.

Even after temperature stabilization, dynamic control of the mode balance remained necessary. First, adjustment of the V-mirror position and steering were considered as control parameters. With the piezo-driven translation and rotation stage available, the reproducibility turned out to be insufficient. Instead, the laser diode injection current proved to be suitable for changing the mode balance in a defined way. The current value for which both modes are balanced changes with time. After incorporating a control algorithm based on measurements of the two-color laser spectrum into the control program, stabilized operation of the FTECAL for several hours was achieved. **Figure 8.32b** shows the intensity of the two laser modes, with the injection current as control parameter, over time with the control algorithm activated. The two laser modes remain balanced even if a region of instability occurs, as indicated in **Figure 8.32b**.

8.6.3 On-chip Techniques

Free space detection of biological molecules is restricted by the amount of substance available under practical test conditions. In order to enhance the sensitivity of a test it is necessary to increase the interaction between THz wave and substance. Kurz and coworkers have demonstrated an approach for marker-free DNA analysis at the femtomol level using electrical THz pulses that propagate along thin-film microstrip lines [74]. These structures incorporate a resonant structure, which is detuned when loaded with genetic material of different dielectric properties. Using the low-loss polymeric material benzocyclobutene (BCB) as substrate material with a low dielectric constant guarantees that a considerable part of the electromagnetic wave is guided outside the substrate. The DNA is deposited simply by pipetting it from an aqueous solution onto the resonant structure. After water evaporation the DNA forms a thin film with a thickness of only a few tens of nanometers. A train of electrical THz pulses is generated using a Ti:S laser with a repetition rate of 78 MHz. The laser pulses gate a biased photoconductive switch, which causes an electric signal to propagate along the microstrip line and through the DNA-loaded parallel-coupled quarter-wavelength microstrip line resonator which, is a standard microwave filter. In more recent work, Kurz and coworkers used more efficient, high-quality factor resonators as well as schemes for hybridization of the DNA material directly on the chip in order to further increase the sensitivity of the detection scheme [92]. The transmitted pulses are detected via electro-optic sampling [93]. As an alternative to electro-optic sampling, the pulses can also be detected by sampling the photocurrent in a detector gap that is also illuminated with part of the laser radiation.

We performed a numerical evaluation of the sensitivity of standard as well as newly developed filters, using commercially available simulation packages. We found that the shift of the resonance frequency is higher for hybridized DNA (HDNA) than for denatured DNA (DDNA). In a second step we also designed enhanced filter structures that have a higher sensitivity than standard filters. We used Momentum™, which is part of the design environment Advanced Design System (ADS)™ from Agilent Technologies, to calculate and design the characteristics of our microstrip line filter structures. It uses the momentum method which is based on a numerical solution of the integral equation that connects the radiated electric field with the current distribution of a given structure via Green's function. Discretization of the currents yields a linear system of equations. Since this method is completely frequency independent, we see no reason why it should fail at THz frequencies.

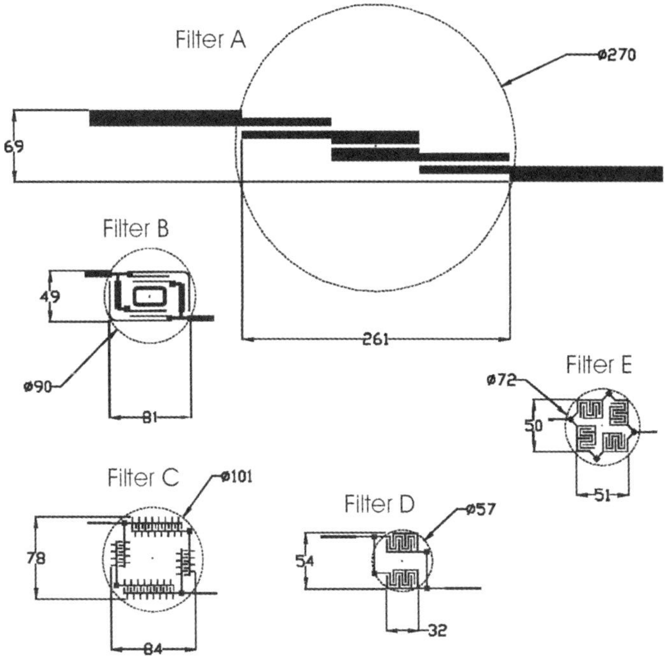

Figure 8.33 Optimized filter geometries (B–E) in comparison to the geometry of a standard parallel-coupled quarter-wavelength microstrip line resonator (A). All diameters given in micrometers.

8.6.4
Design and Fabrication of Efficient Filters

In the following, we present a new class of filters (see **Figure 8.33**, Filters B–E) which require much less DNA material. They are more quadratically shaped and hence mimic the round form more closely. In addition, they are of much smaller dimensions than standard filters. All exhibit sharp resonances in the range of several hundred gigahertz, which shift when loaded with DNA material. In addition, they also have higher transmission than Filter A. However, the most important parameter is the achievable difference between the filter curves (in dB) when the filter is loaded separately with hybridized and denatured DNA. This parameter, which we call differential transmittivity, should be as large as possible in order to obtain a reliable distinction between the two types of DNA. Also in this regard, all the proposed structures are superior to the parallel-coupled filter.

The new filters B–E are compared to Filter A, a standard parallel-coupled quarter-wavelength microstrip line resonator. In **Figure 8.33** the area of the resonant structure that needs to be covered by DNA is marked with a dashed circle. It can be seen that all new filters need to be covered with much less DNA than the parallel-coupled filter, e.g., Filter D with the smallest spot di-

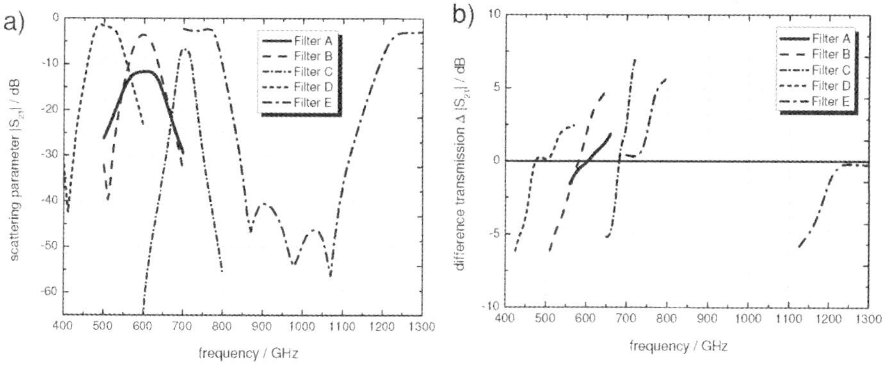

Figure 8.34 (a): scattering parameter S_{21} of the different filters; (b): difference of the scattering parameter S_{21} between the unloaded filter and the filter loaded with a 3 μm thick layer of hybridized DNA.

ameter of 57 μm requires only 4.5% of the amount of DNA deposited on Filter A with a diameter of 270 μm. **Figure 8.34a** shows the transmission characteristics of all filter designs. Filters B–D are bandpass filters whereas Filter E is a stopband filter. It can be seen that the maximum transmission of all filters exceeds the maximum transmission of -11 dB of Filter A. This is due to stronger coupling between input and output and in the case of Filter E due to a direct conductive connection. Furthermore, all new filters have steeper flanks than Filter A, which promise increased signal changes for measurements restricted to a single frequency. However, the most important parameter regarding the filter performance for DNA on-chip analysis is the achievable change of the filter curves once loaded with different types of DNA. **Figure 8.34b** depicts the differential transmittivity curves for the filter first loaded with a 3 μm dielectric layer of $\varepsilon_r = 1.44 + i0.095$, corresponding to hybridized DNA (HDNA) followed by a dielectric load of $\varepsilon_r = 1.21 + i0.095$, corresponding to denatured DNA (DDNA). It can be seen that all new filters show a shift superior to that of the parallel-coupled quarter-wavelength filter. Hence, all new filters are more sensitive. Measurements can be performed at the frequency that corresponds to the maximum of the difference curve. Usually, this is located close to the steepest point of the unloaded filter curve.

We fabricate the new filter designs in a clean-room environment. A silicon wafer metallized with a thin gold layer is used as substrate. The BCB dielectric layer is deposited with a spin coater. Afterwards the filter metallization is structured using photolithography techniques. As a last step, spots of a low-temperature-grown GaAs epitaxial layer are van der Waals bonded on top of the excitation and detection gaps, using a lift-off technique [94]. **Figure 8.35** shows photographs of the chips from a light microscope at different stages of the production process.

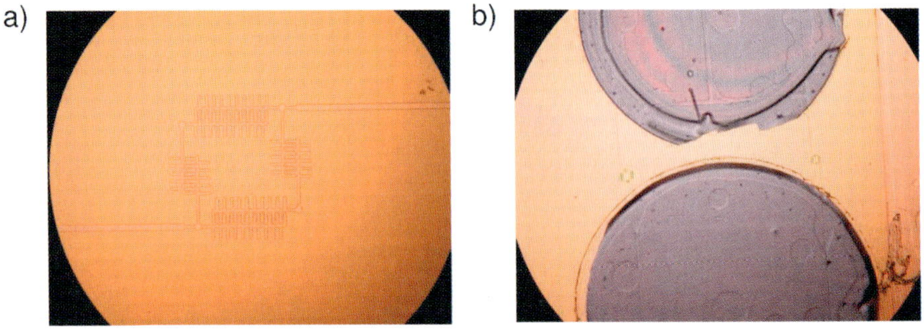

Figure 8.35 Light microscope photograph of (a): filter structure C metallized on BCB and (b): a complete chip with two filter structures and epitaxial layers of LT-GaAs placed over the excitation and detection gaps.

8.6.5
THz Spectroscopy on Biomolecules

The absorption spectra of rather large and complex macromolecules exhibit distinct resonance features in the far-infrared (FIR) region of the electromagnetic spectrum [95]. These absorption features arise predominantly from intramolecular vibrations and represent a characteristic fingerprint of a substance. While strongly localized oscillations of single atoms in a molecule have resonance frequencies in the mid-infrared (MIR), delocalized vibrations involving a larger number of atoms are usually resonant in the FIR or THz range. The same holds for intermolecular motions of more than one molecule coupled via hydrogen bonds or van der Waals interactions. Since large moving masses and weak potential forces result in low vibrational frequencies, biomolecular spectra are expected to contain features between 100 GHz and a few terahertz [96]. This information will be useful for directly determination of conformations and binding strengths [97] or for simple identification of different substances in a mixture [98].

We performed independent studies on selected biomolecules in two different laboratories in Braunschweig and in Freiburg. In Braunschweig we measured the absorption spectra in the region between 100 GHz and 2.5 THz of four different molecules at room temperature and, in one case, at 77 K in a nitrogen cryostat. We used a conventional THz time domain spectrometer purged with dry nitrogen as described earlier. The results are shown in **Figure 8.36**. As an example of a rather small biomolecule we chose dimethyluracil (1,3-dimethyluracil, Fluka, product number 41730), which is a derivative of the nucleobase uracil with two methyl groups attached to the nitrogen atoms. As larger molecules we examined the tripeptides gly-gly-gly (Sigma-Aldrich, product number G1377), leu-gly-gly (Sigma-Aldrich, product number L-9750) and the ionic complex enalapril maleate (Sigma-Aldrich, product

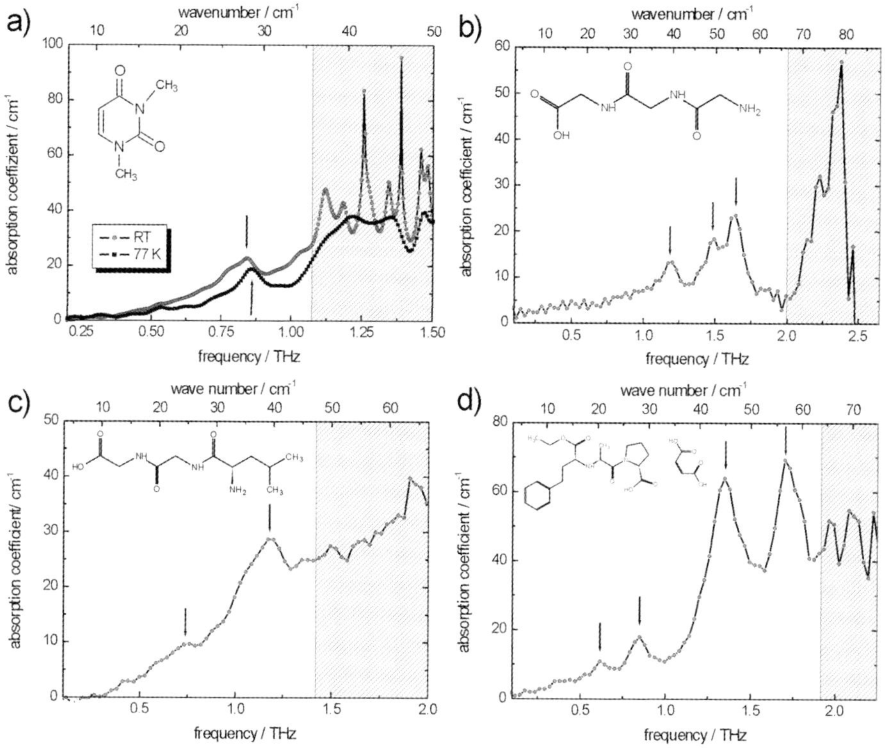

Figure 8.36 Far infrared absorption spectra of biomolecules. (a): Dimethyluracil, (b): Gly-Gly-Gly, (c): Leu-Gly-Gly, (d): Enalapril Maleate. The structural formulas of the molecules are shown in the insets. The part of the measured spectrum that exhibits an uncertainty due to a low signal-to-noise ratio is hatched.

number E6888), which is also of pharmaceutical relevance as an inhibitor of the angiotensin converting enzyme.

Dimethyluracil, which was measured at room temperature as well as at 77 K, exhibits a distinct resonance that shifts from 860 GHz at 77 K to 840 GHz at room temperature. This is the conventional red shift that is expected for an anharmonic molecular potential and that has been observed earlier [99]. For gly-gly-gly we observed distinct resonance peaks at 1.3, 1.5 and 1.64 THz. Leu-gly-gly has at least two absorption peaks at 0.75 and 1.2 THz. The larger enalapril maleate ion complex has more degrees of freedom and as many as four resonance peaks below 2 THz at frequencies of 0.6, 0.85, 1.35 and 1.7 THz. The results on enalapril maleate are consistent with measurements by Strachan et al. [100]. The measurements clearly indicate that biomolecules exhibit defined resonance frequencies that can be used for identification, even at room temperature.

Further studies that clarify the origin of the resonance peaks and their dependence on the biomolecular conformation require measurements at low temperatures over a larger bandwidth. Such measurements have been performed in Freiburg.

Investigation of the temperature dependence of the THz absorption spectra of biomolecular systems offers insight into the fundamental microscopic interactions responsible for the observed spectral features. It is generally accepted that the hydrogen bond and other low-energy interactions, such as dipole–dipole and dispersive interactions between molecules are of utmost importance in biochemistry and biology. To some extent sugars represent prototype systems for the investigation of hydrogen-bonded networks both in the crystalline, amorphous and liquid states. In the solid state, saccharides are linked by a rigid network of hydrogen bonds which in crystals are of long-range order [101,102], resulting in highly regular lattice vibrations, or phonon modes, of the whole crystal structure. In contrast, amorphous systems are characterized by the lack of long-range order, and hence the regularity of the intermolecular vibrational modes is destroyed [103].

In **Figure 8.37** the absorption spectrum of polycrystalline sucrose is shown for temperatures between 10 K and 300 K. From this figure it is possible to observe the shift of the spectral position of the peaks as a function of temperature. Under normal circumstances, the spectral position of a vibrational transition will shift to slightly lower frequencies when the temperature is increased. This shift is caused by the nonharmonic shape of the vibrational potential.

However, in the case of polycrystalline sucrose the transition frequencies of the lowest modes are observed to shift to higher frequencies as the crystal temperature increases. Only at the highest investigated temperatures did we observe the expected red shift of the transitional frequencies. The absorption lines at higher frequencies all shift to lower frequencies with increasing temperature. We tentatively interpret this unusual blue shift of the lowest transitions in sucrose as a direct manifestation of intermolecular forces much weaker than the hydrogen-bond network between the molecules. At low temperatures these weak forces (such as van der Waals forces) dominate over thermal motion of the lattice, and hence they can be directional. They can lead to a softening of the intermolecular hydrogen bonds, in analogy to the well known softening of covalent bonds by the presence of hydrogen bonds, often observed in mid-IR spectroscopy [104], and used as an indicator of the presence of hydrogen bonds. At higher temperatures, these weak interactions are effectively switched off by thermal motion of the crystal lattice, leading to a tightening of the hydrogen bond and a resulting blue shift of the vibrational frequency of the bond. At still higher temperatures the nonharmonic potential becomes important, and the regular red shift of the vibrational frequency is observed.

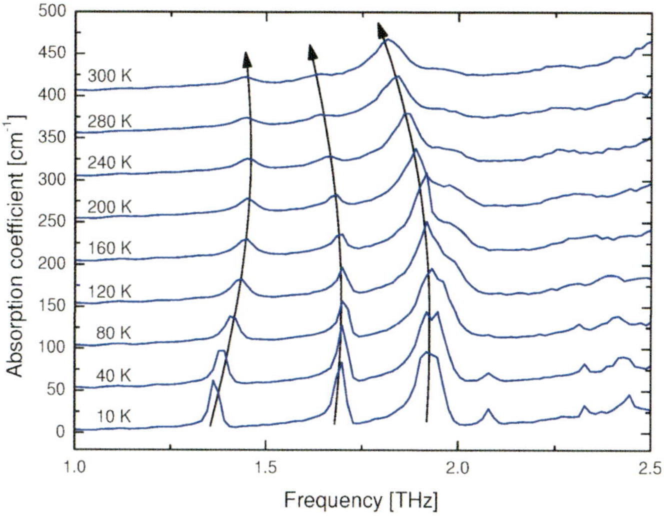

Figure 8.37 Temperature dependence of the absorption spectrum of polycrystalline sucrose, in the temperature range 10–300 K. The arrows indicate the shift of the position of the absorption peaks with temperature. For the lowest-frequency modes we observe an unusual blue shift of the peak position with increasing temperature.

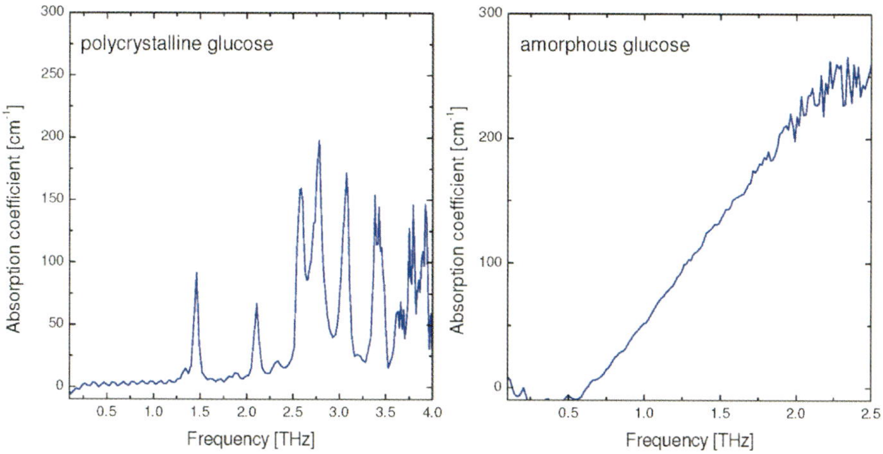

Figure 8.38 Comparison of the absorption spectrum of polycrystalline glucose (left) and amorphous glucose (right).

The importance of the crystalline structure for the appearance of the THz absorption spectrum of a molecular system is demonstrated in **Figure 8.38**, where the absorption coefficient of polycrystalline glucose and amorphous glucose in the frequency range 0.2–4 THz is shown. It is clear that while the polycrystalline material displays a rich and highly specific dielectric spectrum the amorphous sample displays no specific absorption features. This observation leads us to an important conclusion regarding the feasibility of THz spectroscopy for the identification of biological substances. If the substance is crystalline there is a good chance that THz spectroscopy can be used for identification, but if the substance is amorphous the specificity of the THz absorption spectrum to the individual substance is lost.

8.6.6
THz Spectroscopy of Spotted RNA

In contrast to the crystalline biomolecules that typically exhibit strong and specific absorption features in their FIR spectra [105], biological systems with little or no long-range ordering of the constituent molecules exhibit little or no structure in their THz dielectric properties [106]. This may arise firstly from the lack of defined phonon modes and secondly from an inhomogeneous broadening of the intramolecular vibrations.

However, in spite of the lack of characteristic absorption features, the absorption coefficients of different samples typically still differ considerably from each other, offering a tremendous potential for identification purposes. Yet, if we want to compare absolute absorption values instead of characteristic fingerprint features, precise sample preparation becomes very important. To avoid any artifacts arising from inappropriate sample preparation, the experiments were performed independently in two laboratories, using a variety of sample preparation methods.

In this study we investigated the possibility of using the THz absorption spectrum of two different RNA polymer strands, polyadenylic acid (poly-A) and polycytidylic acid (poly-C), for identification purposes. We were particularly interested in applications where the material is spotted in arrays on a disposable substrate.

We used commercially available poly-A and poly-C potassium salts with polymer chains with lengths of approximately 600–2000 A or C units. The sample material was delivered in a low-density powder form.

In order to establish the spectral characteristics of the samples we measured the absorption coefficient and the index of refraction of poly-A and poly-C pressed in a die into plane pellets of thicknesses in the range from 150 to 300 μm. In **Figures 8.39a** and **8.39b** the absorption and index of refraction spectra recorded in Braunschweig are shown, and in **Figures 8.39c** and **8.39d** the spectra recorded in Freiburg are shown. The results obtained in the two laboratories on poly-A and poly-C pellets are similar. The absorption coefficient of both materials increases in an almost linear fashion in the region between 0.1 and 2.5 THz. In particular, the data show that poly-C absorbs more strongly than poly-A at all frequencies. The index of refraction of poly-C is markedly larger than that of poly-A. Apart from etalon artifacts not related to the sample material (see below) no distinct spectral features are observed.

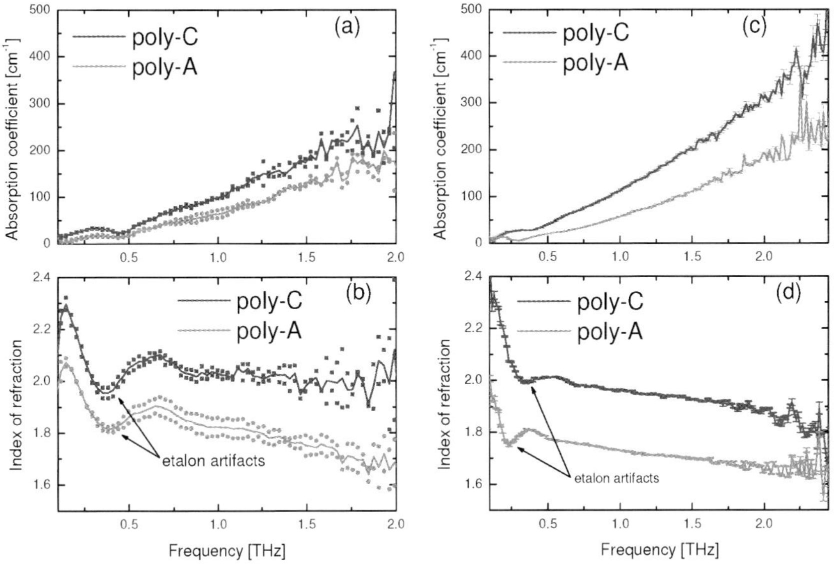

Figure 8.39 Absorption coefficient and index of refraction of poly-C and poly-A, measured in Braunschweig (panels (a) and (b)) and in Freiburg (panels (c) and (d)).

A constant scaling factor can overlap the absorption spectra of poly-A and poly-C completely. This shows that although we observe a difference in the absorption of poly-A and poly-C the density of the sample material must be known before any identification of the material based on its THz absorption spectrum can be made.

The index of refraction of poly-C is approximately 10% larger than that of poly-A. The strong oscillatory feature at low frequencies observed in all measurements is due to multiple reflections of the sample beam inside the relatively thin samples. We see this etalon effect only at low frequencies where the absorption is low. At higher frequencies, where the absorption is substantial, the THz field is absorbed before multiple reflections can occur. In Freiburg we performed additional spectroscopic measurements on poly-A and poly-C as freestanding films. These were obtained by dissolving the powder material in deionized water, distributing this solution over a 1-cm^2 area on a nonpolar polymer substrate and lifting off the film after drying the solution by evaporation. The thicknesses of the films were in the range of 50 to 100 μm. Although we obtained slightly lower absorption coefficients the results are in good agreement with the data shown in **Figure 8.39**; in particular, the overall stronger absorption of poly-C was clearly reproduced [107]. Finally, in Braunschweig we recorded the absorption coefficient of thick spots of dried poly-A

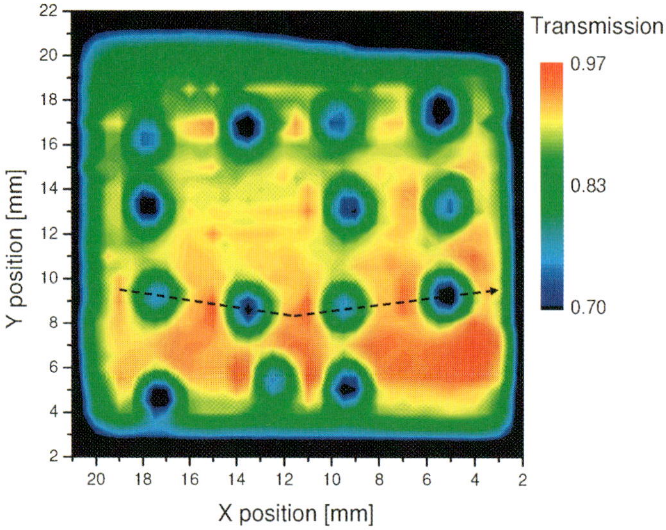

Figure 8.40 THz transmission image of a chip with spots of alternating poly-A and poly-C. The spot in the top, left corner of the image is poly-A.

and poly-C, which, apart from some strong oscillations due to the small sample geometry, reproduced the same results.

Having established that we could use THz time domain spectroscopy to distinguish between different biopolymers we were interested to determine if this differentiation can be used in THz imaging as well. For this purpose we prepared targets for imaging that challenges the spatial resolution obtainable with a THz-TDS imaging system based on free space propagation and apertureless focusing of the THz beam. In **Figure 8.40** we show an image of a sample that was prepared for THz imaging by spotting small liquid volumes in a 4×4 array of alternating poly-A and poly-C on a TOPAS substrate. Each spot was deposited with 2 μL of deionized water containing 0.2 mg of material.

The diameter of each spot was 1 mm. Nominally the sample consisted of a 4×4 array of spots, two of the spots were, however, removed from the substrate in order to identify the orientation of the substrate in the image. The spot in the top left corner of the image contains poly-A. The THz image was formed by frequency-integrated transmission over the full bandwidth of the probe pulse. The poly-C spots lead to a substantial transmission drop of 30%, whereas the poly-A spots reduce the transmission by 20%. In order to verify that the differences observed in the image between the different spot types is related to differences in the sample material and not caused by differences in spot size or thickness we performed surface height scans with a mechanical

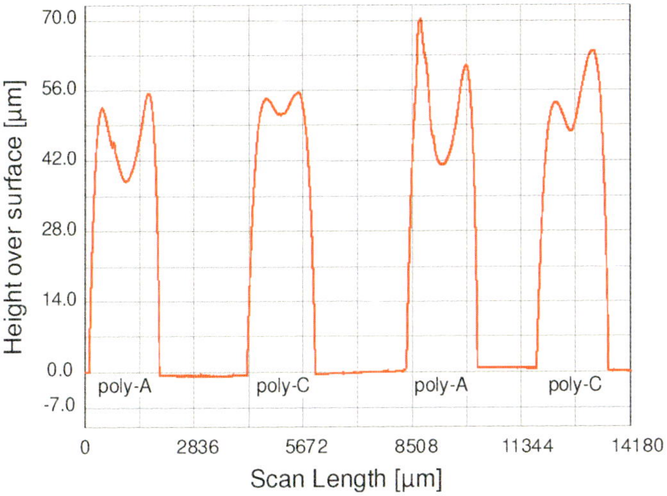

Figure 8.41 Surface profile scan of the chip with RNA spots, along the path indicated in **Figure 8.40**. Notice the different vertical and horizontal scales. The average height of the RNA spots is ca. 50 µm.

surface profiler of the sample after the measurement.

In **Figure 8.41** we show the surface profile of the sample along the path indicated in **Figure 8.40**. Notice that the horizontal scale of the plot is much smaller than the vertical scale. The spots all have approximately the same height, diameter and topology. The average height is 40–50 µm, and the diameter is slightly larger than 1 mm. There is a pronounced height enhancement at the edges of the spots, which is normal for this type of spotting. We expect that if the topology of the spots is not similar then diffraction effects may influence the appearance of the image strongly.

Finally, we recorded a series of diluted spots as shown in **Figure 8.42**. The sample was prepared in analogy to the sample shown in **Figure 8.40**, however, the amount of sample material in each spot is indicated in the figure, and ranges from 20 to 200 µg. As indicated by the strong signal attenuation observed in the image of the 4×4 spot array in **Figure 8.40** we can reduce the amount of sample material by approximately an order of magnitude before the signal disappears in the present form of the experiment. The contrast between poly-A and poly-C degrades rapidly with decreasing amount of sample material.

8.6.7
THz Spectroscopy: A Technology with High Future Potential

The field of THz science and technology is developing rapidly. Owing to promising initial reports concerning the application of THz radiation for the

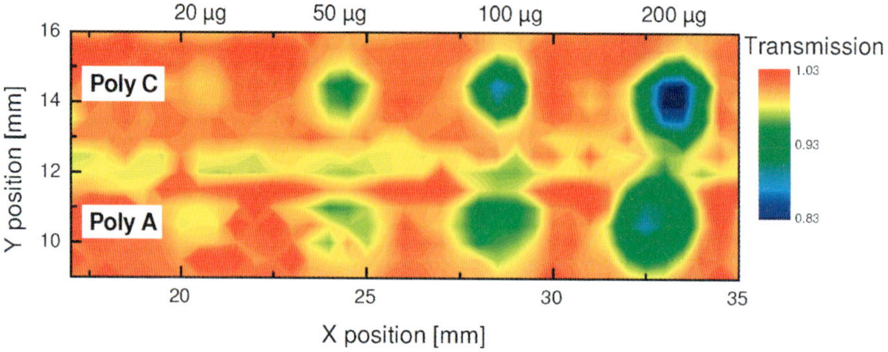

Figure 8.42 THz image of a chip with spots of poly-A and poly-C. The amount of material in each spot varies from 20 μg to 200 μg as indicated in the figure.

detection of chemical substances and biological material, there is currently considerable interest from industry, funding agencies and research groups in further investigation of this exciting new spectral region. In this work we have demonstrated the effectiveness of THz radiation for fundamental investigations of intermolecular interactions in well-ordered solid-state materials of biological relevance. We demonstrated that long-range order, or more specifically the crystallinity of the material, is a crucial parameter in the determination of the spectral properties in the THz range. This observation is of profound importance for biological applications of THz radiation. Since most biological material is in a state with little or no long-range order on the molecular scale it cannot be expected that the THz region can be used as a generally applicable "fingerprint" region for the identification of biological material. However, we have demonstrated that in spite of the lack of reproducible spectral features in the THz range of complicated biological material, THz spectroscopy and THz imaging can be used for certain biological applications. The sensitivity to the presence of hydrogen bonds in a material is a crucial contrast mechanism that will become important in biological sensing applications. This sensitivity allows the distinction between, for example, hybridized and denatured DNA [73], and, as demonstrated here, the distinction between different types of RNA strands [107]. For larger-scale biological systems the strong absorption of water forms a stark contrast to the low absorption of fatty tissue, opening up the avenue of, for example, noninvasive control of water/fat content in food products.

The development of practical instrumentation for such potential applications requires simplification of the equipment used for THz spectroscopy. We have developed new components to aid this development, including novel, fast, delay line systems, fiber-coupled THz antennas, stabilized two-color laser sources for continuous-wave (cw) THz generation and new designs for on-chip detection of minute amounts of biological material. The first commercial THz TDS spectrometers are already on the market. Within the next few years we will witness the development of instruments relying on spectral information in the THz range for biological and industrial sensing.

8.7 Outlook

The use of microarrays for genome-wide expression profiling is a standard procedure throughout academic and pharmaceutical research. The technology is intended to add a dynamic understanding of how and under what circumstances gene expression is regulated, thereby extending the "static" view of the plain sequence data derived by genome-wide sequencing projects. However, the full benefit derived by comparing experimental results with expression data stored in various databases is still hampered by concerns about the reliability and reproducibility of results between different microarray technologies [108–111]. This is mainly due to noise added at a multitude of steps necessary for completing microarray experiments, and may sum up to a level equal to the signal of low-abundance gene expression. Several noise sources are intrinsic to the technology, but may be further elevated by different handling procedures, labelling protocols and methods of RNA preparation – all adding variability to the signal. Attempts to standardize the experimental annotation (MIAME [112]) have been successfully implemented, but they can mainly provide documentation for the different procedures rather than providing solutions for more accurate comparison and analysis of changes observed in RNA abundance.

One of the main differences between platforms used for gene expression profiling is the DNA template immobilized on the solid substrate. Systems use DNA as short oligonucleotides (15–25 bp), long oligonucleotides (50–120 bp) or cDNAs (300–2000 bp). Moreover, immobilization of the capture probe on the solid substrate may be directed by an anchor molecule or just left to chance. Surface chemistry may enable direct, close-to-surface binding or high-density loading by means of branching macromolecules. It is obvious that length and secondary structure of the capture probe, will strongly influence the availability of a certain capture sequence for hybridization of the target probe and therefore determine hybridization kinetics. However, this will directly influence signal strength and thereby yield incorrect expression ratios. Since this is a feature of any single capture probe sequence, it cannot be controlled by experimental conditions that allow only an average setting for temperature and ionic strength, as is common for today's hybridization experiments.

In addition, the binding kinetics at the fluid-to-solid phase border may reduce hybridization rate constants as much as three orders of magnitude in comparison to solution phase experiments [113]. This is supported by the observation that surface-bound DNA-duplexes are generally less stable [114]. The use of longer cDNA sequences may circumvent this problem to a certain extent. However, an important drawback is the introduction of possible cross-hybridization between unrelated sequences, thus making it more difficult to identify errors in expression data systematically.

Efficient hybridization requires the close proximity of the capture probe and the target probe. However, most target probe molecules will have no possibility of reaching their capture probes by plain diffusion using the conventional coverslip method – and this is especially true for large microarrays [115, 116].

Labelling of the target probes for fluorescence-based systems is another rich source of variability added to the analysis. Besides well known differences in incorporation rates of different dyes for two-color microarray experiments [117–121], most labelling protocols use a stoichiometric ratio between labelled and nonlabelled nucleotides for incorporation into the newly synthesized cDNA strand. Thus, measured intensities are a function of RNA length and RNA secondary structure in addition to the RNA abundance.

It was the initial aim of the research project presented here to minimize variation in gene expression measurement by technical means and enhance the reproducibility of microarray experiments significantly. Two main innovations have been implemented towards this goal: instead of the commonly used fluorescence read-out of the washed and dried slide after hybridization, a time-resolved measurement was implemented that allows time-resolved recording of the spectrally resolved fluorescence signal. This not only avoids further processing steps but allows the utilization of the hybridization kinetics for analysis. To enhance the quality of measurement, the hybridization chamber was designed in a way that permits dynamic temperature control and mixing, enabling recordings of melting curves to be made for critical applications such as single-nucleotide mismatches.

The second innovation is the combination of alternating measurement of labelled fluorescence and label-free RIfS measurement, which adds another quality control possibility beyond the label-free detection. Combining the two independent methods is expected to allow quantitative measurement of RNA abundance.

Pioneering the feasibility of terahertz spectroscopy for characterization of biological macromolecules was another invention originally not expected to be miniaturized for use on microarray-sized samples.

In parallel to the efforts described, two commercial initiatives towards time-resolved gene expression profiling became available; the more advanced system of PamGene International B.V. (Netherlands), and the system of MetriGenix (USA). Both systems take advantage of a porous substrate containing microchannels, in which capture probes are immobilized on the wall of the pores, allowing the use of 100–500 times greater surface area, thus allowing rapid diffusion and short hybridization times of less than 1 h. The porous substrate is reminiscent of that of a filter, and active mixing by applying air pressure differences force the hybridization solution through the substrate [122, 123]. For time-resolved optical read-out, a fluorescence-based instrument was developed that permits hybridization, mixing and scanning of microarrays in a single process.

After a decade of exponential growth of microarray technology, and a multitude of applications, it seems the technology has become more mature and may enter a commodity phase in the near future. Most likely, this will be accompanied with more robust instrumentation and a smaller diversity of platforms in the long run. The primary detection technology will continue to be fluorescence based. There might also be a substitution of organic fluorophores by nanocrystals, owing to their inherent robustness and custom-made excitation wavelength. For high-density microarrays covering whole genome transcription profiling, highly automated solutions will be developed that may cover the whole production process, from light-directed synthesis to time-resolved optical read-out [124]. In the low-density (i.e., diagnostic) segment, there is still the need for a more sensitive system. It seems possible that homogeneous, bead-based assays will replace the solid-substrate format because of their increased sensitivity and ease of use [125].

Acknowledgements

We gratefully acknowledge funding of this work within the project MOBA – Micro-Optical Biochip Analysis – by the German Ministry of Education and Research (BMBF) within grant references (FKZ) 13N8309, 13N8310, 13N8311, 13N8313, 13N8314, 13N8330, 13N8375. We thank our project partner Sensovation AG (Stockach, Germany) and our colleague Thorsten Neumann (IMTEK, Laboratory for Chemistry and Physics of Interfaces) for support and fruitful discussions. We would also like to thank Rafal Wilk, Lars Beckmann and Felix Stewing for their contributions to this work.

Glossary

Biochip Efficient tool for highly parallel investigations of biochemical binding reactions, e.g., hybridization of DNA. A biochip consists of a glass or plastic substrate (microscope object slides), coated with a functional layer in order to allow immobilization, and printed with an array of probe molecule spots (up to several tens of thousends).

Biochip reader Apparatus for excitation and read-out of fluorescence signals generated from marker molecules, which have successfully reacted with immobilized probe molecules at the surface of a biochip. From the fluorescence intensity of each spot, conclusions on the kinetics and the success of binding reactions can be drawn.

CCD (Charge-coupled Device) A device for detection of photons based on metal oxide semiconductor capacitors usually arranged in a two-dimensional array. CCD chips are the photosensitive element in many digital cameras.

cDNA (copy-DNA or complementary DNA) DNA obtained from an mRNA molecule by an enzymatic reaction (enzyme reverse transcriptase).

DOE (Diffractive Optical Element) Optical element based on diffraction of light for the generation of a desired intensity pattern. The structure of the DOE has to be calculated from the desired intensity pattern.

Ellipsometry Ellipsometry is an nondestructive optical technique using polarized light for probing the physical film thickness and optical properties of thin layers.

Fluorescence Emission of light from a molecule upon a radiating transition from a vibrational level (usually the lowest) of an excited electronic state (usually the first) into a vibrational level of the ground state. Fluorescent dye molecules are frequently used in biochip analysis.

Hybridization Reaction of two (at least partially) complementary single-stranded DNA molecules.

Immobilization Fixation of molecules on a surface by covalent bonds

Mach–Zehnder interferometry A coherent beam is split by a beam splitter in two arms. The Mach–Zehnder interferometer is used to determine the phase shift caused by a small sample which is to be placed into one of the two beams. The difference in refractive indices is caused by the analyte, which binds on a sensitive surface that covers only one arm of the Mach–Zehnder interferometer.

Marker molecule A fluorescent molecule that is linked to an unknown sample molecule of interest in order to measure the binding reactions of the latter by the fluorescence light intensity.

Microarray Pattern of spots on a biochip. Each spot contains a certain species of probe molecules (e.g., oligonucleotides), which are immobilized at the surface of the biochip. The spots have typical diameters and distances in the range of several hundreds of micrometers. For printing a microarray, needle printers or contact-free printing systems (comparable to inkjet printers) are used.

PCR (Polymerase chain reaction) An experimental technique for the rapid production of millions of copies of a particular sequence of a DNA molecule by an enzymatic reaction.

RIfS (Reflectometric Interference Spectroscopy) RIfS is a label-free detection method based on the multiple reflection of white light at thin interfaces (white light interferometry according to Fabry–Pérot). Time resolved interactions at solid surfaces can be observed, allowing thermodynamic and kinetic constants of these processes to be determined.

Surface modification The chemical modification of surfaces controls the immobilization of biopolymers. On the one hand, the surface should not bind irrelevant biomolecules non-specifically, while on the other hand the surface should be prepared for the immobilization of ligands and receptors.

SPR (Surface Plasmon Resonance) SPR is an optical method for measuring the refractive index when light is reflected under certain conditions from a very thin layer of material adsorbed on a metal. Changes of the refractive index are monitored and thereby molecular interactions at the layer are detected.

Terahertz time domain spectroscopy Phase-sensitive spectroscopy using terahertz waves. Both real and imaginary parts of the dielectric function are measured simultaneously with this technique. This has only be possible since femtosecond pulse lasers became available.

Terahertz waves Electromagnetic waves with a wavelength in the range of several hundreds of micrometers, corresponding to a frequency in the range of 10^{12} Hz (terahertz). Typically generated by firing femtosecond laser pulses onto a biased semiconductor surface, or by photomixing two continuous-wave laser wavelengths in a biased semiconductor structure.

Key References

M.B. EISEN, P.O. BROWN, *Methods Enzymol.*, 303 (**1999**), pp. 179–205

H.C. KING, A.A. SINHA, *Jama*, 286 (**2001**), pp. 2280–2288

M. SCHENA, D. SHALON, R.W. DAVIS, P.O. BROWN, *Science*, 270 (**1995**), pp. 467–470

D. MURPHY, *Adv. Physiol. Educ.*, 26 (**2002**), pp. 256–270

D. GERSHON, *Nature*, 416 (**2002**), pp. 885–891

M. SCHENA, *Microarray Biochip Technology*, Eaton Publications, Natick, MA, 2000

P. BALDI, G.W. HATFIELD, *DNA Microarrays and Gene Expression*, Cambridge University Press, Cambridge, 2002

J. PIEHLER, A. BRECHT, R. VALIOKAS, B. LIEDBERG, G. GAUGLITZ, *Biosens. Bioelectron.*, 15 (**2000**), p. 473

H. ARWIN, *Sensor. Actuat. A-Phys.*, 92 (**2001**), p. 43

T. MUTSCHLER, B. KIESER, R. FRANK, G. GAUGLITZ, *Anal. Bioanal. Chem.*, 374 (**2002**), p. 658

G. GAUGLITZ, *Anal. Bioanal. Chem.*, 381(1) (**2005**), p. 141

G. GAUGLITZ, *Rev. Sci. Instrum.*, 76(6) (**2005**), p. 062224/1 (online publication)

VAN M. EXTER, C. FATTINGER, D. GRISCHKOWSKY, *Appl. Phys. Lett.*, 55 (**1989**), pp. 337–339

M. KOCH, L. BECKMANN, F. RUTZ, P. KNOBLOCH, T. KLEINE-OSTMANN, K. PIERZ, G. HEIN, H. NIEMANN, B. GÜTTLER, *VDI Berichte*, 1844 (**2004**), pp. 11–22

E.R. BROWN, K.A. MCINTOSH, K.B. NICHOLS, C.L. DENNIS, *Appl. Phys. Lett.*, 66 (**1995**), pp. 285–287

B.M. FISCHER, M. HOFFMANN, H. HELM, R. WILK, F. RUTZ, T. KLEINE-OSTMANN, M. KOCH, P. U. JEPSEN, , *Opt. Express*, 13 (**2005**), pp. 5205–5215

References

1 M. B. EISEN, P. O. BROWN, *Methods Enzymol.*, 303 (**1999**), pp. 179–205.

2 H. C. KING, A. A. SINHA, *Jama*, 286 (**2001**), pp. 2280–2288.

3 M. SCHENA, D. SHALON, R. W. DAVIS, P. O. BROWN, *Science*, 270 (**1995**), pp. 467–470.

4 D. MURPHY, *Adv. Physiol. Educ.*, 26 (**2002**), pp. 256–270.

5 D. GERSHON, *Nature*, 416 (**2002**), pp. 885–891.

6 S. P. FODOR, J. L. READ, M. C. PIRRUNG, L. STRYER, A. T. LU, D. SOLAS, *Science*, 251 (**1991**), pp. 767–773.

7 M. D. KANE, T. A. JATKOE, C. R. STUMPF, J. LU, J. D. THOMAS, S. J. MADORE, *Nucleic Acids Res.*, 22 (**2000**), pp. 4552–4557.

8 J. C. BOLDRICK, A. A. ALIZADEH, M. DIEHN et al., *Proc. Natl. Acad. Sci. USA*, 99 (**2002**), pp. 972–977.

9 E. SOUTHERN, K. MIR, M. SHCHEPINOV, *Nat. Genet.*, 21 (**1999**), pp. 5–9.

10 A. ALIZADEH, M. EISEN, R. E. DAVIS et al., *Cold Spring Harb. Symp. Quant. Biol.*, 64 (**1999**), pp. 71–78.

11 R. N. VAN GELDER, M. E. VON ZASTROW, A. YOOL, W. C. DEMENT, J. D. BARCHAS, J. H. EBERWINE, *Proc. Natl. Acad. Sci. USA*, 87 (**1990**), pp. 1663–1667.

12 T. C. KROLL, S. WÖLFL, *Nucleic Acids Res.*, 30 (**2002**), p. e50.

13 S. DUDOIT, Y. H. YOUNG, T. SPEED, M. CALLOW, *Statistica Sinica*, 12 (**2002**), p. 111.

14 J. DONAUER, B. RUMBERGER, M. KLEIN et al., *Transplantation*, 76 (**2003**), pp. 539–547.

15 A. BRAZMA, P. HINGAMP, J. QUACKENBUSH et al., *Nat. Genet.*, 29 (**2001**), pp. 365–371.

16 Y. BENJAMINI, Y. HOCHBERG, *J. Royal Stat. Soc. B*, 57 (**1995**), p. 289.

17 M. B. EISEN, P. T. SPELLMAN, P. O. BROWN, D. BOTSTEIN, *Proc. Natl. Acad. Sci. USA*, 25 (**1998**), pp. 14863–14868.

18 J. C. VARELA, M. H. GOLDSTEIN, H. V. BAKER, G. S. SCHULTZ, *Invest. Ophthalmol. Vis. Sci.*, 43 (**2002**), pp. 1772–1782.

19 P. TAMAYO, D. SLONIM, J. MESIROV et al., *Natl. Acad. Sci. USA*, 96 (**1999**), pp. 2907–2912.

20 Y. HAKAK, J. R. WALKER, C. LI et al., *Proc. Natl. Acad. Sci. USA*, 98 (**2001**), pp. 4746–4751.

21 V. R. IYER, M. B. EISEN, D. T. ROSS et al., *Science*, 283 (**1999**), pp. 83–87.

22 T. R. GOLUB, D. K. SLONIM, P. TAMAYO et al., *Science*, 286 (**1999**), pp. 531–537.

23 A. A. ALIZADEH, M. B. EISEN, R. E. DAVIS et al., *Nature*, 403 (**2000**), pp. 503–511.

24 L. J. VAN'T VEER, H. DAI, M. J. VAN DE VIJVER et al., *Nature*, 415 (**2002**), pp. 530–536.

25 W. K. HOFMANN, S. DE VOS, D. ELASHOFF et al., *Lancet*, 359 (**2002**), pp. 481–486.

26 J. G. HACIA, F. S. COLLINS, *J. Med. Genet.*, 36 (**1999**), pp. 730–736.

27 P. A. CLARKE, R. TE POELE, R. WOOSTER, P. WORKMAN, *Biochem. Pharmacol.*, 62 (**2001**), pp. 1311–1336.

28 D. E. HECK, A. ROY, J. D. LASKIN, *Adv. Exp. Med. Biol.*, 500 (**2001**), pp. 709–714.

29 D. GERHOLD, M. LU, J. XU, C. AUSTIN, C. T. CASKEY, T. RUSHMORE, *Physiol. Genomics*, 4 (**2001**), pp. 161–170.

30 C. DEBOUCK, P. N. GOODFELLOW, *Nat. Genet.*, 21 (1 Suppl) (**1999**), pp. 48–50.

31 P. HING et al., *Biophotonics International*, 10 (**2003**), p. 52.

32 W. DES JARDIN, *Photonics Spectra*, August (**2002**).

33 K. WETZEL, *Advanced Imaging*, October (**2004**).

34 M. SCHENA, *Microarray Biochip Technology*, Eaton Publications, Natick, MA, 2000.

35 P. BALDI, G. W. HATFIELD, *DNA Microarrays and Gene Expression*, Cambridge University Press, Cambridge, 2002.

36 F. J. STEEMERS et al., *Nature Biotechnology*, 18 (**2000**), p. 91.

37 G. VALENTINI, C. D'ANDREA, *Optics Letters*, 25 (**2000**), p. 1648.

38 W. DENKL, D. W. PISTON, W. W. WEBB, in: *Handbook of Biological confocal microscopy* (Ed.: J.B. Pawley), Plenum Press, New York, 1990.

39 W. M. JAMES, C. GU, M. GU, *Applied Optics*, 43 (**2004**), p. 1063.

40 H.-P. LEHR, A. BRANDENBURG, G. SULZ, *Sensors and Actuators B: Chemical*, 92 (**2003**), p. 303.

41 D. S. MEHTA, C. Y. LEE, A. CHIOU, *Optics Communications*, 190 (**2001**), p. 59.

42 H. BLOM, M. JOHANSSON, A. S. HEDMANN, *Applied Optics*, 41 (**2002**), p. 3336.

43 L. B. LESEM, P. M. HIRSCH, J. A. JORDAN JR., *IBM Journal of Research and Development*, 13 (**1969**), p. 150.

44 J. W. GOODMAN, *Introduction to Fourier Optics*, 2nd edition, McGraw-Hill, New York, 1996.

45 R. A. SCHULTZ, T. NIELSEN, J. R. ZAVALETA et al., *Cytometry*, 43 (**2001**), p. 239.

46 S. SVANBERG, *Atomic and Molecular Spectroscopy*, 2nd edition, Springer Verlag, Berlin, 1992.

47 J. PIEHLER, A. BRECHT, R. VALIOKAS, B. LIEDBERG, G. GAUGLITZ, *Biosens. Bioelectron.*, 15 (**2000**), p. 473.

48 H. ARWIN, *Sensor. Actuat. A-Phys.*, 92 (**2001**), p. 43.

49 S. H. BEHRENS, D. G. GRIER, *J. Chem. Phys.*, 115 (**2001**), p. 6716.

50 M. RAITZA, M. HEROLD, A. ELLWANGER, G. GAUGLITZ, K. ALBERT, *Macromol. Chem. Phys.*, 201 (**2000**), p. 825.

51 T. MUTSCHLER, B. KIESER, R. FRANK, G. GAUGLITZ, *Anal. Bioanal. Chem.*, 374 (**2002**), p. 658.

52 K. SPAETH, G. KRAUS, G. GAUGLITZ, *Fresenius J. Anal. Chem.*, 357 (**1997**), p. 292.

53 J. PIEHLER, A. BRECHT, K. HEHL, G. GAUGLITZ, *Colloid. Surfaces B*, 13 (**1999**), p. 325.

54 T. MUTSCHLER, Dissertation, University of Tübingen, Germany, 2004.

55 A. BRECHT, W. NAHM, G. GAUGLITZ, *Analusis*, 20 (**1992**), p. 135.

56 G. GAUGLITZ, *Anal. Bioanal. Chem.*, 381(1) (**2005**), p. 141.

57 G. GAUGLITZ, *Rev. Sci. Instrum.*, 76(6) (**2005**), p. 062224/1 (online publication).

58 K. LÄNGE, G. GRIFFIN, T. VO-DINH, G. GAUGLITZ, *Talanta*, 56 (**2002**), pp. 1153.

59 R. TÜNNEMANN, M. MEHLMANN, R. D. SÜSSMUTH, B. BÜHLER, S. PELZER, W. WOHLLEBEN, H.-P. FIEDLER, H. WIESMÜLLER, G. GAUGLITZ, G. JUNG, *Anal. Chem.*, 73 (**2001**), p. 4313.

60 O. BIRKERT, R. TÜNNEMANN, G. JUNG, G. GAUGLITZ, *Anal. Chem.*, 74 (**2002**), p. 834.

61 K. KRÖGER, J. BAUER, J. FLECKENSTEIN, G. RADEMANN, G. JUNG, G. GAUGLITZ, *Biosens. Bioelectron.*, 17 (**2002**), p. 937.

62 C. HÄNEL, G. GAUGLITZ, *Anal. Bioanal. Chem.*, 372 (**2002**), p. 91.

63 B. LIEDBERG, C. NYLANDER, L. LUNDSTRÖM, *Sens. Actuators B*, 4 (**1983**), p. 299.

64 http://www.biacore.com/home.lasso

65 J. HOMOLA, S. S. YEE, G. GAUGLITZ, *Sens. Actuators B*, 54 (**1999**), p. 3.

66 T. KINNING, P. EDWARDS, in: *Optical Biosensors: Present and Future* (Eds.: F. S. Ligler, C. A. Rowe Taitt), Elsevier, Amsterdam, 2002, p. 253.

67 J. VOROS, J. J. RAMSDEN, G. SCUCS, I. SZENDRO, S. M. DE PAUL, M. TEXTOR, N. D. SPENCER, *Biomaterials*, 23(17) (**2002**), p. 3699.

68 R. G. HEIDEMAN, R. P. H. KOOYMAN, J. GREVE, *Sens. Actuators B*, 10 (**1993**), p. 209.

69 M. VAN EXTER, C. FATTINGER, D. GRISCHKOWSKY, *Appl. Phys. Lett.*, 55 (**1989**), pp. 337–339.

70 M. KOCH, L. BECKMANN, F. RUTZ, P. KNOBLOCH, T. KLEINE-OSTMANN, K. PIERZ, G. HEIN, H. NIEMANN, B. GÜTTLER, *VDI Berichte*, 1844 (**2004**), pp. 11–22.

71 E. R. BROWN, K. A. MCINTOSH, K. B. NICHOLS, C. L. DENNIS, *Appl. Phys. Lett.*, 66 (**1995**), pp. 285–287.

72 A. G. MARKELZ, A. ROITBERG, E. J. HEILWEIL, *Chem. Phys. Lett.*, 320 (**2000**), pp. 42–48.

73 M. BRUCHERSEIFER, M. NAGEL, P. H. BOLIVAR, H. KURZ, A. BOSSERHOFF, R. BÜTTNER, *Appl. Phys. Lett.*, 77 (**2000**), pp. 4049–4051.

74 M. NAGEL, P. H. BOLIVAR, M. BRUCHERSEIFER, H. KURZ, *Appl. Phys. Lett.*, 80 (**2002**), pp. 154–156.

75 M. VAN EXTER, C. FATTINGER, D. GRISCHKOWSKY, *Appl. Phys. Lett.*, 55 (**1989**), pp. 337–339.

76 L. XU, X.-C. ZHANG, D. H. AUSTON, *Appl. Phys. Lett.*, 61 (**1992**), pp. 1784–1786.

77 Q. WU, X.-C. ZHANG, *Appl. Phys. Lett.*, 67 (**1995**), pp. 3523–3525.

78 A. NAHATA, D. H. AUSTON, T. F. HEINZ, C. WU, *Appl. Phys. Lett.*, 68 (**1996**), pp. 150–152.

79 P. U. JEPSEN, S. R. KEIDING, *Opt. Lett.*, 20 (**1995**), pp. 807–809.

80 P. U. JEPSEN, R. H. JACOBSEN, S. R. KEIDING, *J. Opt. Soc. Am. B*, 13 (**1996**), pp. 2424–2436.

81 P. U. JEPSEN, R. H. JACOBSEN, S. R. KEIDING, *J. Opt. Soc. Am. B*, 13 (**1996**), pp. 2424–2436.

82 I. S. GREGORY, C. BAKER, W. R. TRIBE, M. J. EVANS, H. E. BEERE, E. H. LINFIELD, A. G. DAVIES, M. MISSOUS, *Appl. Phys. Lett.*, 83 (**2003**), pp. 4199–4201.

83 M. BREEDE, S. HOFFMANN, J. ZIMMERMANN, J. STRUCKMEIER, M. HOFMANN, T. KLEINE-OSTMANN, P., KNOBLOCH, M. KOCH, J. P. MEYN, M. MATUS, MS. W. KOCH, J. V. MOLONE, *Opt. Comm.*, 207 (**2002**), pp. 193–203.

84 S. M. DUFFY, S. VERGHESE, K. A. MCINTOSH, Photomixers for Continous-Wave Terahertz Radiation, in: *Sensing with Terahertz Radiation* (Ed.: D. Mittleman), Springer, Berlin, 2002.

85 M. G. LITTMAN, H. METCALF, *Appl. Opt.*, 17 (**1978**), pp. 2224–2227.

86 J. STRUCKMEIER, A. EUTENEUER, B. SMARSLY, M. BORN, M. HOFMANN, L. HILDEBRAND, J. SACHER, *Opt. Lett.*, 24 (**1999**), pp. 1573–1574.

87 P. GU, F. CHANG, M. TANI, K. SAKAI, C.-L. PAN, *Jpn. J. Appl. Phys.*, 38 (**1999**), pp. 1246–1248.

88 X. LIU, M. J. COBB, X. LI, *Opt. Lett.*, 29 (**2004**), pp. 80–82.

89 M. SALHI, F. RUTZ, T. KLEINE-OSTMANN, V. PETUKHOV, C. METZ, M. KOCH, *Spiral Optical Delay Line*, OSA

Workshop Optical Terahertz Science and Technology, Orlando, USA, 2005.

90 German patent application # 10 2005 011 045.2

91 J. Xu, Z. Lu, X.-C. Zhang, *Electron. Lett.*, 40 (**2004**), pp. 1218–1219.

92 M. Nagel, P. Haring Bolivar, H. Kurz, *Semicond. Sci. Technol.*, 20 (**2005**), pp. S281–S285.

93 J. A. Valdmanis, G. A. Mourou, *IEEE J. Quantum Electron.*, 22 (**1986**), pp. 69–78.

94 E. Yablonovitch, D. M. Hwang, T. J. Gmitter, L. T. Florez, J. P. Harbison, *Appl. Phys. Lett.*, 56 (**1990**), pp. 2419–2421.

95 J. W. Powell, G. S. Edwards, L. Genzel, F. Kremer, A. Wittlin, *Phys. Rev. A*, 35 (**1987**), p. 3929.

96 M. Walther, P. Plochocka, B. Fischer, H. Helm, P. U. Jepsen, *Biopolymers (Biospectroscopy)*, 67 (**2002**), p. 310.

97 M. B. Johnston, L. M. Herz, A. L. T. Khan, A. Köhler, A. G. Davies, E. H. Linfield, *Chem. Phys. Lett.*, 377 (**2003**), p. 256.

98 K. Kawase, Y. Ogawa, Y. Watanabe, H. Inoue, *Opt. Express*, 11 (**2003**), p. 2549.

99 M. Walther, B. M. Fischer, P. U. Jepsen, *Chem. Phys.*, 288 (**2003**), pp. 261–268.

100 C. J. Strachan, T. Rades, D. A. Newnham, K. C. Gordon, M. Pepper, P. F. Taday, *Chem. Phys. Lett.*, 390 (**2004**), pp. 20–24.

101 G. M. Brown, H. A. Levy, *Acta Chryst. B*, 35 (**1976**), pp. 656–659.

102 J. A. Kanters, G. Roelofson, B. P. Albas, I. Meinders, *Acta Cryst. B*, 33 (**1977**), pp. 665–672.

103 S. Soderholm, Y. H. Roos, N. Meinander, N. Meinander, M. Hotokka, *J. Raman Spect.*, 30 (**1999**), pp. 1009–1018.

104 H. O. Desseyn, K. Clou, R. Keuleers, R. Miao, V. E. Van Doren, N. Blaton, *Spectrochim. Acta A*, 57 (**2001**), pp. 231–246.

105 B. M. Fischer, M. Walther, P. U. Jepsen, *Phys. Med. Biol.*, 47 (**2002**), pp. 3807–3814.

106 B. Fischer, M. Hofmann, H. Helm, G. Modjesch, P. U. Jepsen, *Semicond. Sci. Technol.*, 20 (**2005**), pp. 246–253.

107 B. M. Fischer, M. Hoffmann, H. Helm, R. Wilk, F. Rutz, T. Kleine-Ostmann, M. Koch, P. U. Jepsen, *Opt. Express*, 13 (**2005**), pp. 5205–5215.

108 T. Bammler, R. P. Beyer, S. Bhattacharya, G. A. Boorman, A. Boyles, B. U. Bradford, R. E. Bumgarner, P. R. Bushel, K. Chaturvedi, D. Choi et al., *Nat. Methods*, 2 (**2005**), pp. 351–356.

109 J. E. Larkin, B. C. Frank, H. Gavras, R. Sultana, J. Quackenbush, *Nat. Methods*, 2 (**2005**), pp. 337–344.

110 E. Marshall, *Science*, 306 (**2004**), pp. 630–631.

111 P. J. Park, Y. A. Cao, S. Y. Lee, J. W. Kim, M. S. Chang, R. Hart, S. Choi, *J. Biotechnol.*, 112 (**2004**), pp. 225–245.

112 A. Brazma, P. Hingamp, J. Quackenbush, G. Sherlock, P. Spellman, C. Stoeckert, J. Aach, W. Ansorge, C. A. Ball, H. C. Causton et al., *Nat. Genet.*, 29 (**2001**), pp. 365–371.

113 M. M. Sekar, W. Bloch, P. M. St John, *Nucleic Acids Res.*, 33 (**2005**), pp. 366–375.

114 A. W. Peterson, R. J. Heaton, R. M. Georgiadis, *Nucleic Acids Res.*, 29 (**2001**), pp. 5163–5168.

115 S. A. Allison, S. H. Northrup, J. A. McCammon, *Biophys. J.*, 49 (**1986**), pp. 167–175.

116 V. Chan, D. J. Graves, S. E. McKenzie, *Biophys. J.*, 69 (**1995**), pp. 2243–2255.

117 A. B. Goryachev, P. F. Macgregor, A. M. Edwards, *J. Comput. Biol.*, 8 (**2001**), pp. 443–461.

118 T. Ideker, V. Thorsson, A. F. Siegel, L. E. Hood, *J. Comput. Biol.*, 7 (**2000**), pp. 805–817.

119 M. K. Kerr, M. Martin, G. A. Churchill, *J. Comput. Biol.*, 7 (**2000**), pp. 819–837.

120 G. C. Tseng, M. K. Oh, L. Rohlin, J. C. Liao, W. H. Wong, *Nucleic Acids Res.*, 29 (**2001**), pp. 2549–2557.

121 X. Wang, S. Ghosh, S. W. Guo, *Nucleic Acids Res.*, 29 (**2001**), p. e75.

122 N. Kessler, O. Ferraris, K. Palmer, W. Marsh, A. Steel, *J. Clin. Microbiol.*, 42 (**2004**), pp. 2173–2185.

123 Y. Wu, P. de Kievit, L. Vahlkamp, D. Pijnenburg, M. Smit, M. Dankers, D. Melchers, M. Stax, P. J. Boender, C. Ingham et al., *Nucleic Acids Res.*, 32 (**2004**), p. e3637.

124 Y. A. Dolginow, *Drug Discovery World*, 6 (**2005**), pp. 56–62.

125 J. Lu, G. Getz, E. A. Miska, E. Alvarez-Saavedra, J. Lamb, D. Peck, A. Sweet-Cordero, B. L. Ebert, R. H. Mak, A. A. Ferrando et al., *Nature*, 435 (**2005**), pp. 834–838.

9
Hybrid Optodes (HYBOP)

Dirk Gansert[1], Mathias Arnold, Sergey Borisov, Christian Krause, Andrea Müller, Achim Stangelmayer, Otto Wolfbeis

9.1
Introduction

Scientific progress in elucidation and understanding of life processes from the molecular structure to the organism level strongly depends on methods and techniques that provide minimally invasive access to bioprocesses on a high-resolution scale in space and time. Hence, in life sciences, medicine or biotechnology there is a great need for analytical tools that adequately satisfy the requirements for miniaturization, high sensitivity, good measurement performance, longevity and versatility. Moreover, adaptive low-cost measuring systems are essential requirements for future analytical tools, particularly for online measuring routines such as automated feed-back control of steady-state life conditions in biotechnology or tissue engineering. At present, however, progress in bioprocess analysis *in vivo* towards the sub-cellular level is hampered by a severe lack of such analytical tools, even for such basic quantities as oxygen, carbon dioxide, pH or temperature, used as measures of metabolism and physiological conditions of heterotrophous and autotrophous organisms and their respective compartments.

Conventional gas analysis systems used for investigations of bioprocesses such as oxidative respiration, photosynthesis, fermentation, denitrification, methanogenesis and many others are technically restricted to measurement in a single state of aggregation, either in the gaseous or in the liquid phase. Generally, oxygen and carbon dioxide dissolved in the aqueous phase are electrochemically measured using Clark-type or Severinghaus electrodes, respectively. The Clark-type electrode consists of a platinum cathode and a silver anode both immersed in an electrolyte solution, and is coated by a gas permeable PTFE membrane [9]. Molecular oxygen that permeates through the membrane will be reduced to hydroxy anions at the cathode, while the electrons being consumed by the reduction reaction stem from the oxidation of

1) Corresponding author

Biophotonics: Visions for Better Health Care. Jürgen Popp and Marion Strehle (Eds.)

ISBN: 3-527-40622-0

silver at the anode, where silver chloride is produced. The resulting current intensity is proportional to the number of O_2 molecules consumed by this redox process. The electrochemical measuring principle of all CO_2 electrodes is based on measuring pH changes as a function of CO_2 concentration via the formation of carbonic acid in the aqueous phase of the electrolyte solution of the electrode [8]. In order to accelerate the relatively slow formation of carbonic acid, and thereby to reduce the response time of the electrode to changes of the CO_2 concentration in the sample, the system was modified by addition of the enzyme carbonic anhydrase in the electrolyte solution [2]. Such an electrochemical CO_2 biosensor can be minimized to 2 μm at the sensor tip, and is capable of a CO_2 resolution down to 50 ppm. However, several limitations, including extreme sensitivity to mechanical stress, short life, expending calibration or insufficient standardization, particularly with respect to fluctuating enzyme activity, prevent its wide-ranging application for bioprocess analysis. These examples illustrate that analyte consumption by the electrochemical measurement of gases affects the accuracy of bioprocess analysis because an artificial gas sink is formed during the measuring process.

In the gas phase, CO_2 measurement is excellently performed by high-resolution infrared gas analyzers (IRGA). Using the optical infrared absorption characteristics of the CO_2 molecule (main absorption band at $\lambda = 4.26$ μm) the measuring principle is based on the decrease of quantum flow density of infrared light as a quantitative measure of CO_2. An accurate infrared CO_2 analysis requires a constant molar gas flow passing through the measuring cell of the detector. This is obtained by a precision logic and control unit equipped with mass-flow meters. Consequently, IRGA-based CO_2 analytical devices are expensive, demanding much training, time and servicing, and, because of their open gas pathways, they are unsuitable for measurements in sterile conditions.

Oxygen in the gas phase can be most accurately measured by analyzers using the strong paramagnetic character of oxygen as the measuring principle. Oxygen molecules that enter an inhomogeneous magnetic field cause differential partial pressures inside an optical bench. The pressure difference causes a body of revolution to rotate and, accordingly, causes the deflection of a light beam between two photodiodes, which is quantified by the resulting potential difference. The potential difference induces a current flow, which is used to compensate for the rotation of the body, hence the designation of the principle as magneto-mechanical compensation. The compensation current flow is proportional to the number of oxygen molecules passing through the magnetic field.

In contrast to the electrochemically based methods of gas analysis applied in the liquid phase, those used for CO_2 and O_2 analysis in the gas phase are free from analyte-consumption flaws. However, as already mentioned, these open-flow systems require precision logic and control units to generate constant gas flows through the cuvettes that house the samples, tubes and analyzers. Therefore, like the electrodes, the open-flow systems are also inappropriate for *in vivo* bioprocess analysis at the microscale, particularly when sterile conditions need to be maintained.

Optical sensors, particularly optodes that are based on fluorescence lifetime as the quantitative measure of a physicochemical parameter such, as temperature, gases or pH value, are, among other special qualities, distinguished by their outstanding measuring performance in both gaseous and liquid phases. The use of light as the carrier of information allows decoupling between the analyte-specific sensor and the detector, providing noninvasive investigations of bioprocesses. This represents a methodical breakthrough, because real time measurement and experimental manipulation of bioprocesses can be performed without contamination of sterile samples and media. Miniaturization of analytical tools is the third major requirement for minimally invasive process analysis that can also be performed by optodes. According to the optical measuring principle, optodes can be designed as fibrous microsensors merely a few micrometers in diameter. In comparison to microelectrodes of similar size, the sensing performance of optodes in respect of resolution per time unit, analyte specificity, sensor drift and durability is much more favorable. Optodes thus represent a versatile analytical tool not only for laboratory use but also for dynamic bioprocess analysis under natural environmental conditions.

In this chapter, we introduce a new generation of optical sensors, the hybrid optodes. These sensors are able to measure simultaneously two parameters – e.g., oxygen and temperature – at the same sensor spot. Hence, for the first time, physiological processes can be noninvasively measured in the corresponding effective thermal condition of the microenvironment where the bioprocess takes place. Owing to their hybrid character the new optodes represent a highly efficient analytical and cost-saving tool.

9.2 Optical Sensors – State-of-the-Art

9.2.1 Optical Sensing Technologies

In general, a sensor can be defined as a system capable of monitoring continuously a physicochemical parameter such as gas partial pressure, pH or temperature [10]. In an optical sensor the optical properties of a given parameter are analyzed and monitored spectroscopically over a certain distance. Typically, optical properties such as absorbance, reflectance and luminescence

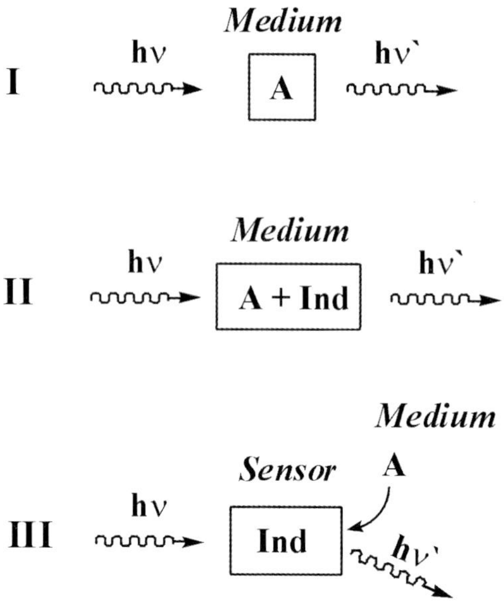

Figure 9.1 Scheme of three different principles of optical sensing (see text).

are monitored. Compared to other sensing techniques, optical sensors are superior because they use light for information transfer over considerable distances. Optical sensing provides a unique possibility of noninvasive measurement because no physical contact is needed between the sensor material, which is placed in the medium to be analyzed, and the detection system.

Monitoring of analytes by optical sensing is usually performed in three different ways, which are illustrated in **Figure 9.1**.

In the first type an analyte 'A' which is contained in the analyzed medium is measured directly via its optical properties. For example, carbon dioxide can be determined by infrared spectroscopy, chlorophyll by its specific fluorescence emission and many other biomolecules by circular dichroism.

In the second type of optical sensing an indicator 'Ind' is added to the analyzed medium, where it reacts with the analyte. The change of the optical properties of the indicator will then be measured. This type of optical sensing is widely used (e.g., pH determination via absorption spectroscopy) but is certainly invasive. If the reaction between the indicator and the analyte is not reversible, this kind of sensing is often referred to as 'probing'.

Finally, the third type of sensing is performed by an indicator which is embedded in a matrix. The analyte diffuses from the medium into the matrix, and chemically or photophysically affects the optical properties of the indicator. The interaction should be fully reversible and the indicator should return to its initial state when the analyte is removed.

Absorption spectroscopy is still one of the most commonly used methods in optical sensing. Some analytes can be monitored directly via absorption spectroscopy, such as CO_2 and CO by means of infrared spectroscopy. A chemical reaction of an indicator with the analyte often results in a specific change of the absorption spectrum of the latter, which can be monitored – typically by UV-Vis spectroscopy. The most common examples are absorption-based pH indicators, distinguished by different absorption spectra of the protonated and deprotonated form. However, if the analyzed medium contains substances with absorption properties similar to the indicator, the measurement will be compromised. Even in the simplest case, the precision of absorption spectroscopy is limited by the fact that the amount of light which passes through the sample is always compared with the amount of light when no sample is present.

Reflectance spectroscopy uses light being scattered by the analyte or the indicator as a quantitative measure. Again, measurements will be erroneous if several substances with similar optical properties are present. Raman spectroscopy has become very popular for determination of biomolecules and organic particles such as pollen grains. The method relies on different energy levels between scattered photons and photons of incident light. This difference, called the Raman shift, is dependent on the vibrational-rotational levels of a molecular species.

Circular dichroism (CD) spectroscopy is an important method for determination and sensing of biomolecules (e.g., proteins, carbohydrates, fatty acids). Nearly all biomolecules are optically active and differently absorb left- and right-handed components of the polarized light. Thus, the resulting elliptically polarized light can provide information about the molecular species to be analyzed. However, the results are affected by perturbations and aggregation of the substance. Therefore, the method is unsuitable for more detailed structural analysis. Optically inactive inorganic and organic analytes, of course, cannot be determined at all by circular dichroism.

Fluorescence spectroscopy is one of the most powerful methods in analytical chemistry, and is thus used in a wide range of applications. Fluorometry is also preferred in optical sensing because of the extremely high sensitivity of the method, where even single photons can be detected. Numerous parameters can be measured, including luminescence intensity, decay time, polarization, quenching efficiency, radiative and nonradiative energy transfer and combinations of these parameters. The processes which occur upon light absorption by a molecule can be illustrated schematically by the Jablonski diagram (**Figure 9.2**). For reasons of simplicity, upper electron excitation levels and vibrational levels are not shown.

Typically, absorption of light of a certain energy ($h\nu_1$) promotes one electron from a ground state level (S_0) to an excited vibronic level of an excited

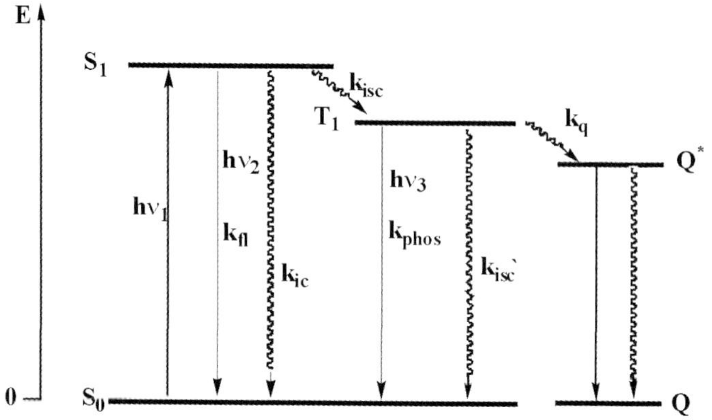

Figure 9.2 Jablonski diagram of energy levels of a molecule which are involved in light absorption (see text). The ordinate indicates increment of energy (E) from a given ground level (0).

singlet state (S_1). After vibration relaxation (overall time for absorption and relaxation is about 10^{-11} s) the thermally equilibrated S_1 state of the molecule is populated. The typical lifetime of this fluorescent state is 10^{-9}–10^{-8} s. The excitation energy then dissipates in radiationless thermal deactivation to the ground state (with rate constant k_{ic} or via emission of a photon ($h\nu_2$) which is called fluorescence. Chemical interaction between indicator and analyte (e.g., pH indicator and protons) results in changes of the energy levels, and therefore can be monitored via absorption and fluorescence spectroscopes. An intersystem crossing, which is the alternative to thermal deactivation and fluorescence, populates the lowest triplet excited state (T_1) but is spin-forbidden. The probability of the process (k_{isc}) increases dramatically in the presence of heavy atoms (e.g., iodine) or metal ions. In some metal complexes (platinum group metals) this probability is close to unity [3] so that after excitation all the molecules will finally populate the lowest triplet excited state. In the absence of a quencher either phosphorescence ($h\nu_3$) or thermal deactivation can occur. For phosphorescent temperature indicators, the rate constant of thermal deactivation (k'_{isc}) is very temperature dependent, and thermal quenching is observed at higher temperatures. In the presence of a quencher (Q) energy or electron-transfer processes can occur which, in the simplest case, is described by the Stern–Volmer equation:

$$\frac{I_0}{I} = \frac{\tau_0}{\tau} = 1 + k_q \cdot \tau_0 \cdot [Q], \qquad (9.1)$$

where I_0 and I are luminescence intensities, τ_0 and τ are decay times in the absence and presence of the quencher, respectively. The quenching constant is k_q, and [Q] is the concentration of the quencher.

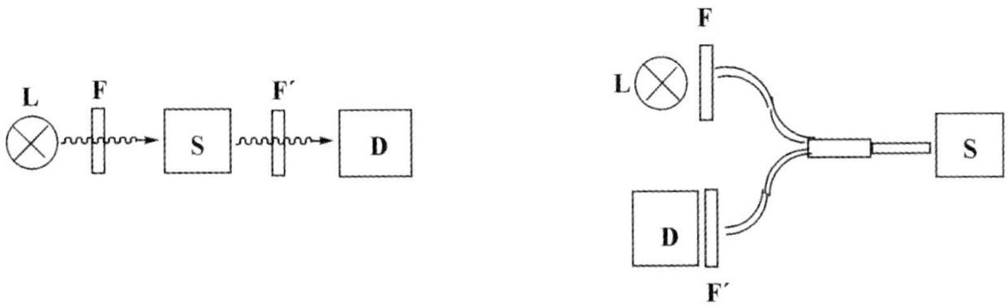

Figure 9.3 Scheme of two principal set-ups for optical sensing. A linear optical set-up (left) is used for optical sensors based on light absorption, while an angular set-up (right) is preferred for luminescent sensors (see text).

Phosphorescence quenching is much more efficient than quenching of the fluorescence because of the relatively long decay time ($>10^{-7}$ s) of the triplet state. Quenching of the phosphorescence plays a very important role in sensor chemistry. For example, all optical sensors for oxygen detection are based on energy transfer from the triplet excited state of the oxygen indicator. Since this kind of quenching is a photophysical process, the effects of the quencher on intensity and decay time are fully reversible. The sensitivity of the sensor is determined by the decay time of the triplet state of the indicator (the longer the decay time the more efficient the quenching) but also by the concentration of the quencher.

Decay time can be measured either by the time domain or by the frequency domain method. In the time domain method, luminescence emission is measured as a function of time after a short excitation light pulse. In the frequency domain method a sample is excited by a sinusoidally modulated light. Luminescent light has the same waveform but is modulated and phase-shifted from the excitation light. Thus, the decay time can be determined by measuring the phase shift or demodulation. The phase modulation technique allows precise decay time determination in a micro- and millisecond range with small equipment. However, the equipment becomes more complicated, bulky and expensive for determination of short decay times (in the order of nanoseconds) which is the case for fluorescent indicators (e.g., pH indicators). Dual lifetime referencing (DLR) [4,5] is an elegant technique to overcome the problem. Here, a long-lived phosphorescent luminophore is used for referencing of the short-lived fluorescent indicator. A detailed description of the frequency domain method and the new DLR technique is presented in Section 9.2.2.

The main components of a typical optical sensor are schematically shown in **Figure 9.3**. Light from the excitation light source (L) passes through the excitation filter (F) and is absorbed by the sensor spot (S), containing an indicator. The light from the indicator (transmitted, reflected or emitted) is registered by the detector (D). In case of a luminescent sensor an emission filter (F′) is placed between the sensor spot and the detector in order to avoid interception of excitation light by the detector. The angle between the excitation light source and the detector may vary. For absorption-based sensors a linear scheme of the optical components is usual (**Figure 9.3** left). For luminescent optical sensors, however, either 90°or 180°is preferred. In the latter case, excitation and emission light can be guided by means of fiber bundles (**Figure 9.3** right).

The most important component of an optical sensor system is the sensor spot, which contains an analyte-specific indicator. The configuration of the light source, excitation and emission filters as well as the modulation frequencies are optimally fitted for the best sensing performance of an indicator. The indicator is embedded in a polymer matrix, which not only provides the shape of the sensor spot but also tunes sensor properties such as sensitivity, response times and, particularly, cross-sensitivity to other analytes. The requirements of the properties of the polymer matrix can vary considerably for different sensor types. For example, the polymer matrix of an optical pH sensor must be permeable to protons, while for an oxygen sensors ion permeability can be a serious drawback since ions can cause undesirable quenching of the oxygen indicator.

9.2.2
Optodes – Their Principle of Measurement

Optodes based on fluorescence lifetime proved to be superior to many other optical sensors in terms of reliability and robustness. The advantages to measure the luminescence decay time, an intrinsically referenced parameter, in comparison to conventional intensity measurement, can be seen in the textbox on page 489.

For some analytes like molecular oxygen, a direct relationship exists between the analyte concentration and the decay time of a sensor ($\tau = f([O_2])$).

The principle of measurement is based on the effect of dynamic luminescence quenching, with molecular oxygen as a quenching molecule. The following scheme explains the principle of dynamic luminescence quenching by oxygen (**Figure 9.4**).

The collision between the O_2-sensitive luminescent indicator molecule (O_2 luminophore) in its excited state, and the quencher (molecular oxygen) results in radiationless deactivation, and is called collisional or dynamic quenching. After collision, energy transfer takes place from the excited indicator molecule to oxygen which consequently is transferred from its ground state (triplet state) to its excited singlet state. As a result, the indicator molecule does not emit luminescence light and the measurable luminescence signal decreases. A relation exists between the oxygen concentration in the sample and the luminescence intensity as well as the luminescence lifetime which is described by Eq. (9.1), the Stern–Volmer equation. The graph shown in **Fig-**

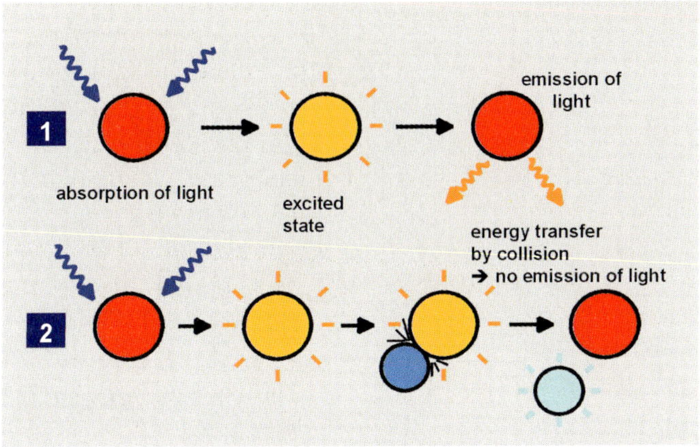

Figure 9.4 Principle of dynamic luminescence quenching by molecular oxygen. (1): Luminescence process in absence of oxygen, (2): Deactivation of the luminescent indicator molecule by molecular oxygen.

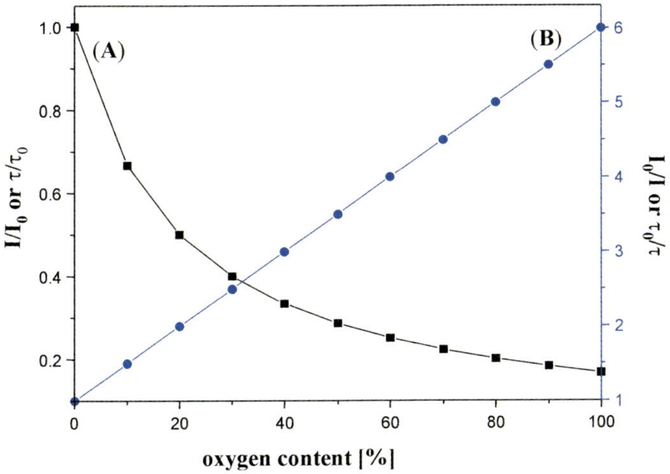

Figure 9.5 Decrease of luminescence in presence of molecular oxygen (A), and the corresponding Stern–Volmer plot (B).

ure 9.5 represents an example of the luminescence decrease of an O_2-sensitive luminophore as a function of oxygen concentration.

The phase-modulation technique is most advantageous to evaluate the luminescence decay time. If the luminophore of an optode is excited by a sinusoidally intensity-modulated light, its decay time causes a time delay in the emitted light signal. In technical terms, this time delay is the phase angle between the excitation and the emitted light signal. The phase angle is shifted as a function of the analyte concentration. It therefore represents a quantitative measure of the substance (molecules or ions) to be analyzed. The relation be-

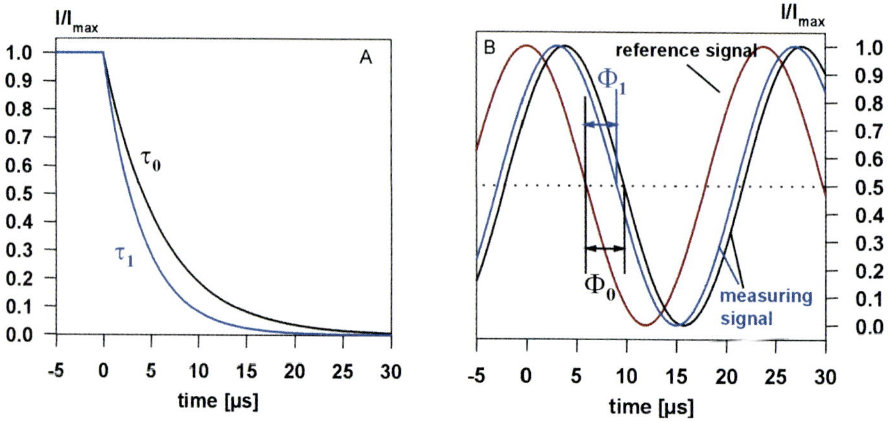

Figure 9.6 Scheme of the luminescence decay time τ in the absence (τ_0) and presence (τ_1) of oxygen (left). The luminophore is excited by sinusoidally modulated light (reference signal). Relative to the excitation signal, emission light is delayed in phase by the decay time of the excited luminophore (right). Phase delay is expressed by the phase angle Φ in the absence (Φ_0) and presence (Φ_1) of oxygen.

tween luminescence decay time τ and the phase angle Φ is given by Eq. (9.2):

$$\tau = \frac{\tan \Phi}{2\pi \cdot f_{\text{mod}}} \tag{9.2}$$

with f_{mod} as the modulation frequency.

In the presence of an analyte, the luminescence decay time of a luminophore is smaller than in the absence of the analyte ($\tau_1 < \tau_0$). According to Eq. (9.2), the phase angle therefore decreases with increasing analyte concentration (**Figure 9.6**).

For some analytes, such as pH and CO_2, no indicator exists which changes its decay time in the range of microseconds (µsec). The measurement of luminescence intensity would be simple in terms of instrumentation but its accuracy is often compromised by adverse effects such as drifts of the optoelectronic system and variations in the optical properties of the sample, including fluorophore concentration, turbidity, coloration and refractive index. Therefore, efficient referencing methods are required for quantification of intensity signals.

The measurement of the fluorescence decay time, an intrinsically referenced parameter, is hardly affected by fluctuations of the overall fluorescence intensity. However, the decay time of most pH-sensitive indicator dyes is in the nanosecond (nsec) time-scale, requiring sophisticated and expensive instrumentation which limits their use in sensor application. Therefore, a new measuring principle, the Dual Lifetime Referencing (DLR) method [5], based on decay time measurement, has been introduced. This method uses a couple of luminophores with different decay times and similar excitation spec-

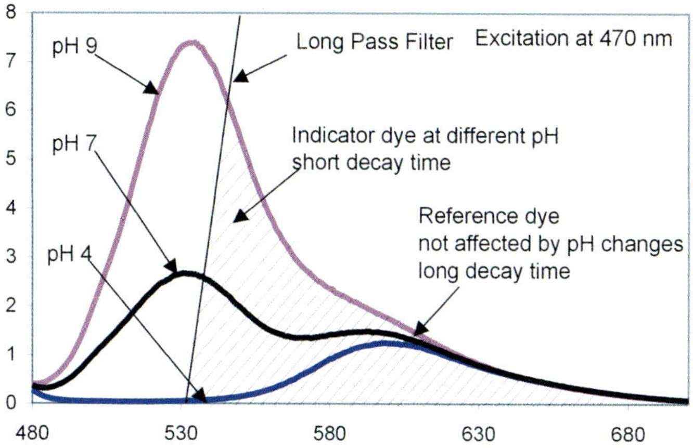

Figure 9.7 Fluorescence spectra of a DLR pH optode. The intensity of the pH indicator increases with pH, while the intensity of the reference dye is not affected by pH changes. The hatched area is detected.

tra. An analyte-insensitive, μsec-lifetime luminophore is combined with an analyte-sensitive, nsec-lifetime fluorophore, and fluorescence intensity is converted into a phase shift. Preferably, the reference dyes display decay times in the microsecond or millisecond time domain to simplify the opto-electronic system [7]. According to Eq. (9.2), the phase angle Φ measured at a single modulation frequency f_{mod} reflects the luminescence decay time τ provided that the decay is single-exponential. In the DLR method the analyte-sensitive fluorophore, referred to as the indicator, has a short decay time (τ_{ind}), while the analyte-insensitive luminophore serves as the reference standard with a decay time in the µs range (τ_{ref}). Ideally, the two luminophores have overlapping excitation and emission spectra so that they can be excited at the same wavelength and their fluorescence can be detected using the same emission window and photodetector (**Figure 9.7**).

The phase shift Φ_{m} of the overall luminescence obtained at a single frequency depends on the ratio of intensities of the reference luminophore and the indicator dye. Φ_{m} can be represented as the superposition of the single sine wave signals of the indicator and the reference luminophore (**Figure 9.8**). The reference luminophore gives a constant background signal (ref) while the fluorescence signal of the indicator (ind) depends on the analyte concentration. The average phase shift Φ_{m} directly reflects the intensity of the indicator dye and, consequently, the analyte concentration. The modulation frequency is adjusted to the decay time of the reference dye.

Equations (9.3) and (9.4) give the superposition of the phase signals of the reference dye, which has a constant decay time and luminescence intensity,

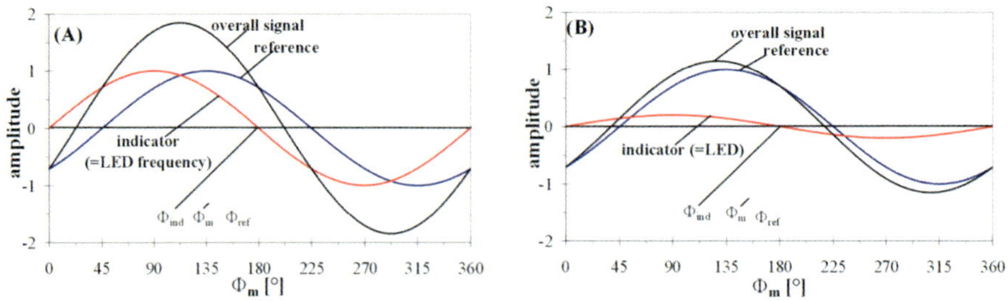

Figure 9.8 Phase shift of the overall (Φ_m), the reference (Φ_{ref}) and the indicator (Φ_{ind}) luminescence. Fluorescence of the indicator occurs in the absence (A) and presence (B) of the analyte.

and the indicator:

$$A_m \cdot \cos \Phi_m = A_{ref} \cdot \cos \Phi_{ref} + A_{ind} \cdot \cos \Phi_{ind} \tag{9.3}$$

$$A_m \cdot \sin \Phi_m = A_{ref} \cdot \sin \Phi_{ref} + A_{ind} \cdot \sin \Phi_{ind} \tag{9.4}$$

where A is the amplitude of either overall signal (m), luminophore (ref), or indicator (ind), and Φ is the phase angle of either the overall signal (m), the luminophore (ref), or the indicator (ind), respectively. If the modulation frequency (f_{mod}) is optimal, $\tan \Phi_{ref}$ is described by Eq. (9.5):

$$\tan \Phi_{ref} = 2\pi \cdot f_{mod} \cdot \tau_{ref} = 1 \tag{9.5}$$

and Φ_{ind} can be given as

$$\tan \Phi_{ind} = 2\pi \cdot f_{mod} \cdot \tau_{ind} = \frac{2\pi \cdot \tau_{ind}}{2\pi \cdot \tau_{ref}} = \frac{\tau_{ind}}{\tau_{ref}}. \tag{9.6}$$

The reference luminophore has a decay time that is orders of magnitude longer than that of the indicator. Consequently, Φ_{ind} can be set equal to zero in Eq. (9.7), because there is no phase shift at low modulation frequencies in the kilohertz range

$$\tan \Phi_{ind} = \frac{\tau_{ind}}{\tau_{ref}} \quad \tau_{ind} \overrightarrow{\ll} \tau_{ref} \quad 0 \Rightarrow \Phi_{ind} \longrightarrow 0\,. \tag{9.7}$$

The decay time of the reference luminophore is not affected by the analyte, hence Φ_{ref} remains constant and $\tan \Phi_{ref}$ is also constant. Therefore, Eqs. (9.3) and (9.4) can be simplified as

$$A_m \cdot \cos \Phi_m = A_{ref} \cdot \cos \Phi_{ref} + A_{ind} \tag{9.8}$$

and

$$A_m \cdot \sin \Phi_m = A_{ref} \cdot \sin \Phi_{ref} \tag{9.9}$$

respectively. Dividing Eq. (9.8) by Eq. (9.9) results in a correlation of the phase angle (Φ_m) and the intensity ratio of the indicator dye (A_{ind}) and reference luminophore (A_{ref}):

$$\begin{aligned}\frac{A_m \cdot \cos\Phi_m}{A_m \cdot \sin\Phi_m} = \cot\Phi_m &= \frac{A_{ref} \cdot \cos\Phi_{ref} + A_{ind}}{A_{ref} \cdot \sin\Phi_{ref}} \\ &= \cot\Phi_{ref} + \frac{1}{\sin\Phi_{ref}} \cdot \frac{A_{ind}}{A_{ref}} .\end{aligned} \tag{9.10}$$

In this equation, $\cot\Phi_m$ reflects the referenced intensity of the fluorescence indicator. A linear relation is obtained between $\cot\Phi_m$ and the ratio of A_{ind}/A_{ref} because the phase angle of Φ_{ref} of the reference luminophore is assumed to be constant. Therefore, this method is referred to as Dual Lifetime Referencing.

Application of Decay Time Measurements in Analytical Approaches

The measurement of the luminescence decay time, an intrinsically referenced parameter, has the following advantages compared to conventional intensity measurements.

- The decay time does not depend on fluctuations of the intensity of the light source and the sensitivity of the detector.
- The decay time is not influenced by signal loss caused by fiber bending or by intensity changes caused by changes in the geometry of the sensor.
- The decay time is, to a great extent, independent of the concentration of the indicator dye in the sensitive layer. Therefore, photobleaching and leaching of the indicator dye have no influence on the measured signal.
- The decay time is not influenced by variations in the optical properties of the sample, including turbidity, refractive index and coloration.

In respect to hybrid optodes, measuring decay time displays a further advantage: it allows signals from two dyes with different decay times to be distinguished, and therefore to be separated for measurement. This principle is also used to minimize background fluorescence. Usually any fluorescent background or insufficiently filtered excitation light has a very short decay time. Thus, the measurement of a long decay time is not affected by these interferences if detection is started shortly after an excitation flash. This principle is widely used in assays that use europium complexes as phosphorescent probes, for instance. The oxygen–temperature hybrid optode also makes use of different decay times. The signals are separated by using different optical filters and different modulation frequencies.

The prerequisite for the application of decay time measurement in analytical approaches is a significant change of the decay time corresponding to a change of analyte content. There are several ways to construct an optical sensor using decay time measurements. The four most common principles are briefly described here.

1. Thermal Deactivation

   As for nearly every parameter, the decay time of a phosphorescent dye is also temperature dependent, which often results in an undesirable cross-sensitivity to temperature. However, if the temperature dependence of the decay time of a dye is distinct then it can be used as a temperature sensor. This principle is described in detail for the oxygen–temperature hybrid optode.

2. Dynamic Quenching

   This means there is an immediate change of the decay time in response to a change of analyte concentration. It is the preferred opto-chemical interaction between dye and analyte. The principle is described for the oxygen–temperature hybrid optode. Unfortunately, for several analytes of biological relevance, such as pH or CO_2, dyes displaying a significant change in decay time are not available. To overcome this limitation a second analyte-sensitive dye must be used, which affects the measured decay time.

3. Energy Transfer (ET)

   The term fluorescence energy transfer refers to a nonradiative transfer of excited-state energy from a donor (D) to an acceptor (A), and results from a dipole–dipole interaction between donor and acceptor. Nonradiative energy transfer does not involve the emission and reabsorption of photons called the radiative transfer or inner filter effect. The rate of nonradiative energy transfer (k_T) depends on (i) the fluorescence quantum yield of the donor, (ii) the overlap of the emission spectrum of the donor and the absorption spectrum of the acceptor and (iii) their relative orientation and distance. The theory was derived by Förster, who gave a quantitative expression of k_T between a donor and acceptor pair at a fixed separation distance r (Eq. (9.11)).

$$k_{\mathrm{T}} = \frac{8.71 \cdot 10^{23} \cdot \kappa^2 \cdot \Phi_{\mathrm{d}}}{r^6 \cdot n^4 \cdot \tau_{\mathrm{d}}} \cdot J = \frac{1}{\tau_{\mathrm{d}}} \left(\frac{R_0}{r} \right)^6 \tag{9.11}$$

   with

$$J = \int_0^{\infty} F_{\mathrm{D}}(\lambda) \cdot \varepsilon_{\mathrm{A}}(\lambda) \cdot \lambda^4 \cdot \mathrm{d}\lambda \tag{9.12}$$

where Φ_d and τ_d are the quantum yield and lifetime of the donor in absence of the acceptor, respectively, n is the refractive index of the medium, r is the distance between donor and acceptor, and κ_2 is a factor that gives the relative orientation in space of the transition dipoles of the donor and acceptor. The overlap integral J, given in Eq. (9.12), expresses the degree of spectral overlap between the emission of the donor and the absorption by the acceptor. $F_d(\lambda)$ is the corrected fluorescence intensity of the donor in the wavelength range λ to $\lambda + d\lambda$ with the total intensity normalized to unity, and $\varepsilon_A(\lambda)$ is the extinction coefficient of the acceptor at λ. The Förster distance R_0 is the critical donor–acceptor transfer distance at which radiative decay and nonradiative energy transfer are equally possible. Remarkably, the efficiency depends on the sixth power of r.

The principle of energy transfer was used in many immunoassays [11, 12] because a donor-marked protein can react with an acceptor-marked antigen which results in ET. Many optical sensors exploit the principle of energy transfer. The potential of ET-based sensors relies on the fact that wellinvestigated absorbance-based indicators can be applied by adding an analyte-insensitive fluorescent donor with a sufficiently large spectral overlap. Sensors for pH, CO_2 and NH_4^+ were developed utilizing the energy transfer from a pH-insensitive donor to a pH-sensitive acceptor. Additionally, luminescence energy transfer is a convenient way to overcome the lack of suitable lifetime based indicators, since the color change of an absorber can be converted into decay time information.

Energy transfer measurements require a constant separation distance between the D–A pair which does not vary during the excited state lifetime of the donor. This can be achieved by covalently binding the donor and acceptor via spacer groups or alternatively, by the formation of donor–acceptor ion pairs.

4. Dual Lifetime Referencing (DLR)

This method has been applied in the O_2–pH, and O_2–CO_2 hybrid optode. Again, an analyte-insensitive phosphorescent dye (reference dye) is combined with a fluorescent analyte-sensitive dye (indicator dye). If the luminescence decay time is measured by the phase-modulation method, the measured overall decay time will be related to the ratio of the reference dye and indicator dye. An important prerequisite for such a measurement is a sufficient spectral overlap of reference and the indicator dyes. Because this method represents a measurement of ratio rather than lifetime it is less insensitive to photobleaching and leaching than the other methods unless indicator and reference dyes are influenced in the same way. The DLR is advantageous because it remains unaffected by fluctuations of the light source intensity.

9.3
Planar and Fibrous Optodes – New Optical Tools for Non- and Minimally Invasive Process Analysis

Optodes can be designed as fibrous microsensors of only a few micrometers in diameter but also as planar sensor spots several millimeters wide or even as large sensor foils. Their free shaping offers the advantage of a broad range of application for *in situ* process analysis in tissues (e.g., tissue engineering), organs or in cell culture. Free shaping is possible owing to the elaborate process of manufacture where analyte-specific indicator microparticles are homogeneously embedded in a polymer matrix composed of several organic compounds that represents a specific sensor "cocktail". This cocktail can be individually designed to be appropriate to the purpose of bioprocess analysis (**Figure 9.9**).

Optodes offer two substantial advantages over other sensors: firstly, the optical measurement is independent of the physical aggregation state, i.e., mea-

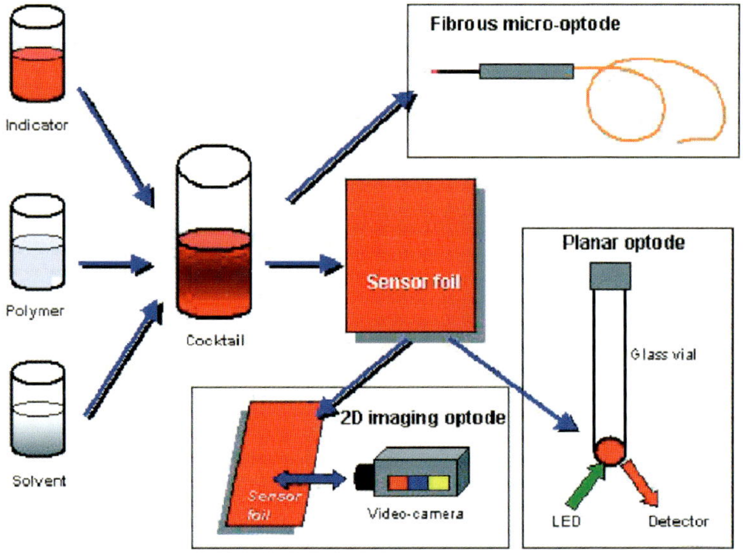

Figure 9.9 Scheme of the manufacturing method of fibrous micro-optodes, planar optodes or analyte-specific 2D imaging optodes.

surements can be performed in the gaseous and liquid phases without any change in sensor configuration or analyzer and, secondly, they can be used for noninvasive measurements. The latter allows the maintenance of sterility during the measuring process. Whenever an optically transparent window is present in a system, a sensor spot can be applied inside and optically measured from outside through this window (**Figure 9.10**). This measuring principle opens an extremely broad variety of applications, including:

1. recording of oxygen ingress in PET bottles
2. quality control in packaging processes
3. monitoring of cell growth in disposables for biotechnology
4. in-line measurement in bioreactors and breweries.

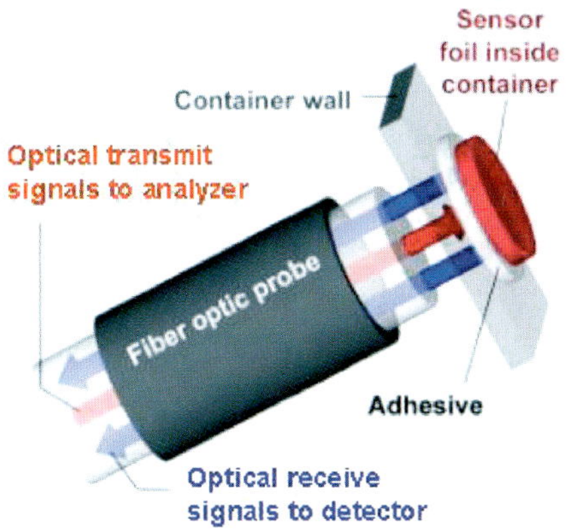

Figure 9.10 Principle of noninvasive optical measurement through a transparent container wall. An analyte-sensitive foil (planar optode), which is fixed inside the container is optically scanned from outside by a fiber-optic probe. Both excitation and emission light to and from the optode must pass the transparent window without spectrum shift.

A schematic diagram for optical measurement of oxygen partial pressure in a sterilized cell culture medium is shown in **Figure 9.11**. The oxygen-sensitive luminophore (sensor spot) is fixed inside a transparent shaking flask that contains a cell culture. The optical fiber is positioned near the outside surface of the flask opposite to the sensor spot. If glass vessels are used, the system is fully autoclavable. The fluid dynamics of the cell culture will not be affected by a planar optode sensor spot as they would be by electrodes. Owing to the small dimensions of an optode, measurements in Petri dishes with fill heights of about one millimeter or less are also possible.

Optodes offer a third advantage over conventional sensor systems: they can be easily extended to highly parallelized systems. As an example, this can

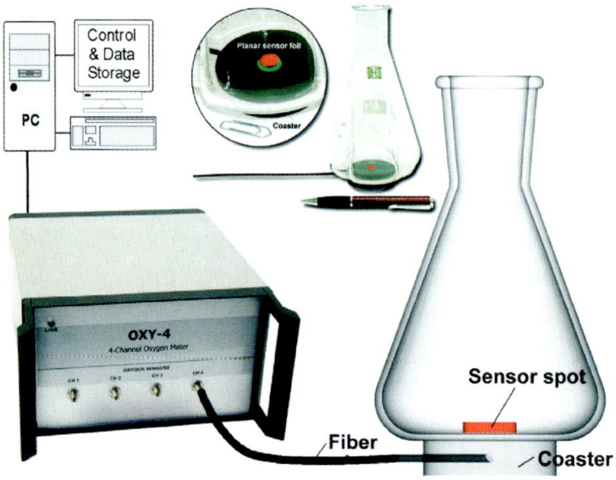

Figure 9.11 Application scheme of noninvasive optical measurement of oxygen partial pressure and/or pH value in cell culture.

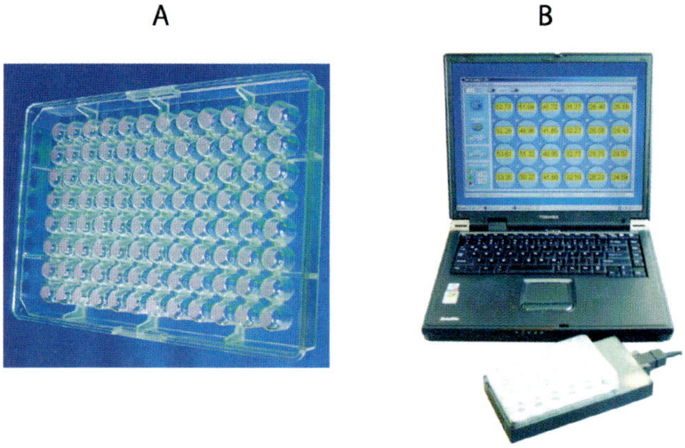

Figure 9.12 A 96-well microtiter plate (A) equipped with individually integrated analyte-specific sensor spots. The microplates are scanned from outside at the bottom by a Sensor Dish Reader system (SDR, B).

be applied in microtiter plates. Each well is equipped with a single sensor spot, which can be scanned separately and recurrently in very short time intervals. This method can even be used for High-throughput Screening (HTS). **Figure 9.12** illustrates a 96-well microtiter plate (A) with individually integrated calibration-free sensor spots for quantification of dissolved oxygen or pH in the physiological range. No additional reagents for measurement are required. These microplates are scanned from outside at the bottom by a Sensor Dish Reader system (SDR, B). Internal calibration allows quantification of

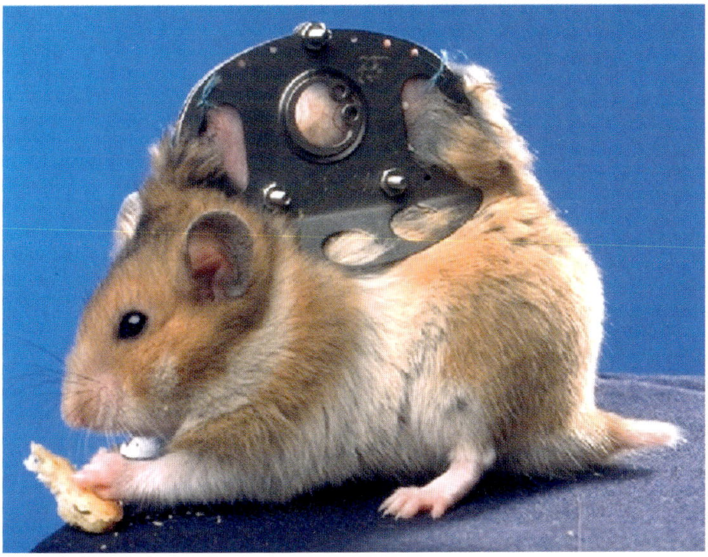

Figure 9.13 Hamster carrying a dorsal skinfold chamber for noninvasive optical measurement of oxygen partial pressure in the skin.

pH or partial pressure of oxygen even in scattering or colored media. Potential fields of SDR application are enzyme screening, controlled cell cultivation (Section 9.5), toxicological tests or HTS.

The methodical principle of a series of sensor spots condensed in a small area leads to another field of sensor application which is also nearly exclusively occupied by optodes, the two-dimensional imaging (2D imaging) of physicochemical parameters. Using cameras or microscopes for read-out of sensors leads to manifold applications such as diagnostic tools in medicine and also in environmental research.

The oxygen partial pressure in the skin of a mammal can be measured by a sensor foil pressed onto the skin (**Figure 9.13**). 2D optical imaging of oxygen partial pressure may facilitate the detection of skin cancer to the abnormally enhanced oxygen consumption of cancer cells. Spatio-temporally high resolution 2D imaging of several physicochemical parameters such as pH, O_2 and CO_2 partial pressure as well as temperature will also provide new insights into metabolite transport processes and signal transfer between microorganisms, mycorrhizal fungi and fine roots in the rhizosphere of higher plants.

Optodes manufactured as fibrous microsensors based on glass fibers of various diameter represent a minimally invasive and highly efficient microanalytical tool. Sensors with a tip size of less than 50 μm can be implanted even in small insects (**Figure 9.14**) or the tissues and organs of animals such as an isolated heart of a rat (**Figure 9.15**).

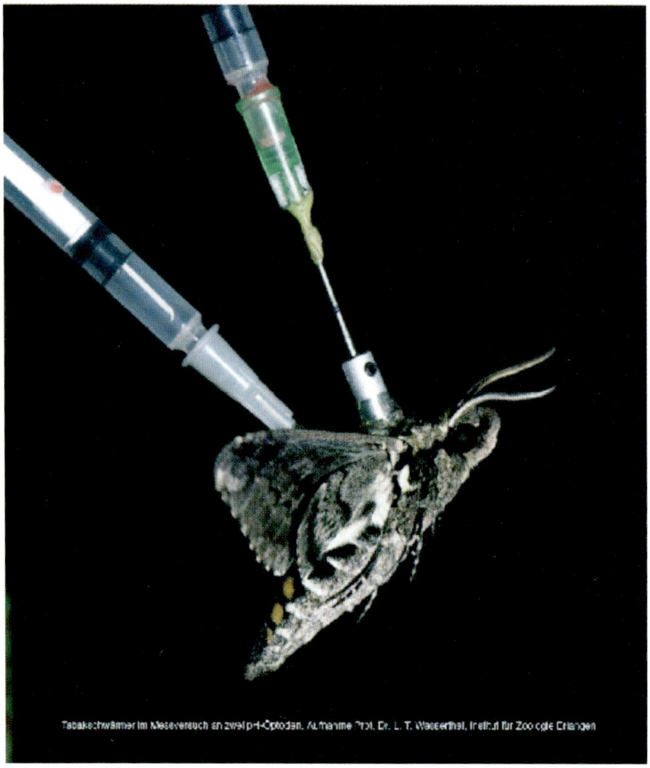

Figure 9.14 Moth with two implanted pH-sensitive micro-optodes

Figure 9.15 Isolated heart of a rat with an implanted oxygen optode

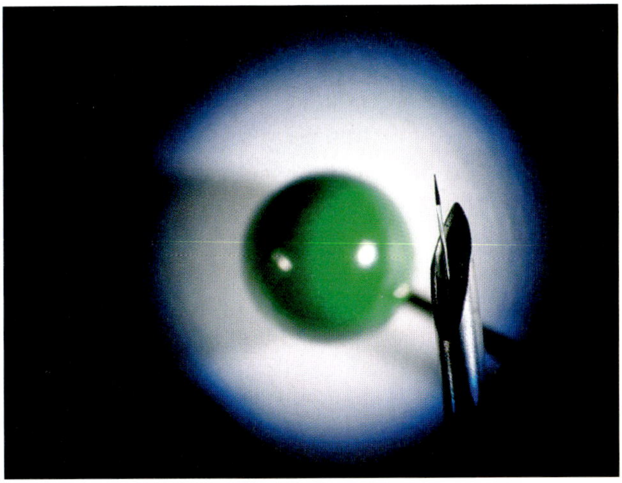

Figure 9.16 Sensor tip of a fibrous oxygen micro-optode housed in a stainless cannula.

If an appropriate sensor housing is chosen, a penetration probe can be designed (**Figure 9.16**). The glass fiber with its oxygen-sensitive tip is housed in a stainless steel cannula, and can be extended for measurement by pushing the piston forward. With the sensor tip sheltered inside the needle, it can easily penetrate through a septum rubber or any other harsh material.

The use of optodes as fibrous microsensors offers new perspectives for quantitative metabolic bioprocess analysis because of its spatially ($< 50\ \mu m$) and temporally ($t_{90} < 1$ s) high-resolution measurement, high sensitivity (parts per billion), longevity, independence of aggregate state and analyte consumption-free measurement. Moreover, the measuring signal is independent of changes in flow velocity, which is of major importance for *in situ* measurements in live compartments. Moreover, optodes are insensitive towards electrical interferences and magnetic fields, which makes them appropriate for use during computer tomographic investigations.

9.4 Fluorescence Optical Hybrid Optodes – the Technology of Tomorrow

9.4.1 Principles of Measurement

Hybrid optodes are intended to provide independent information on two analytes or combinations of analytes and variables like temperature at the same location and at the same time. The composition of a hybrid optode is therefore more complicated than that of an optical sensor for a single analyte. One of the major problems to avoid, or at least to minimize, is cross-sensitivity of the indicators. For example, oxygen indicators are not only quenched by oxygen but also exhibit sensitivity to temperature because of the temperature-dependent photophysical properties of the oxygen-sensitive dye and quenching kinetics. In the oxygen–temperature hybrid optode, independent information about the temperature is obtained from temperature-sensitive particles, and the measured oxygen concentration is temperature-corrected. However, temperature correction becomes much more complicated if the temperature indicator shows cross-sensitivity to oxygen. In hybrid optodes, cross-sensitivity of an indicator "I1" to the analyte "A2" can also arise from poor separation of the emission spectra of the indicators. Two different approaches – spectral and multifrequency – can be used to separate the information about the concentrations of two analytes using the frequency domain method.

In the spectral approach, the absorption and emission spectra of indicator "I1" should be shifted significantly from the spectra of the indicator "I2". In this case, separate excitation light sources are used for the two indicators, and different filter set-ups for isolating the emission. This approach has been successfully applied in the oxygen–temperature hybrid optode. As an oxygen indicator it contains a porphyrin complex, and a ruthenium polypyridyl complex is used as the temperature indicating dye. The spectral properties and optical set-up are shown in **Figures 9.17** and **9.18**, respectively. The temperature indicator is excited by a blue LED (470 nm-peak wavelength), and a FITCA filter is used to filter the excitation light. A Chroma 580 filter is used for isolating the emission. The oxygen indicator is excited by a green LED (525 nm peak wavelength), a FITCE filter is used for filtering the excitation light, and a Chroma 680 filter for isolating the emission.

In the multifrequency approach, the two indicators should have overlapping absorption and emission spectra, so that, simultaneous excitation of both indicator dyes is possible with one light source. One emission filter is chosen to pass the emission of both indicators. In accordance with this principle, an extended DLR sensor was designed. The only difference is that the reference dye – the analyte-insensitive luminophore (p. 487) – is now sensitive to one of both analytes. Thus, for separation of information phase shift measurements are performed at two or more different modulation frequencies, and mathematical models are used for data separation. Signal discrimination is possible because the dyes, having different decay times, are modulated differently. Compared to the spectral approach, the advantage of multifrequency measurement lies in the simplified optical set-up (**Figure 9.18**). The calibration procedure is, however, more complicated. It is necessary that the decay times of the two indicators differ significantly (e.g., the combination of fluorescent and phosphorescent indicators used in the DLR method).

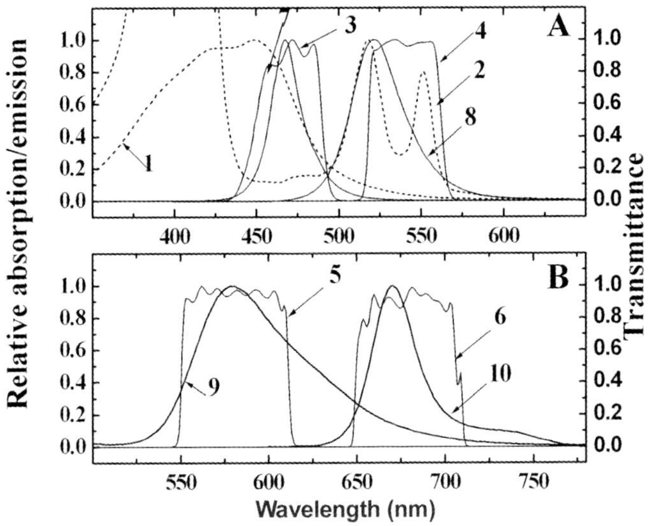

Figure 9.17 Spectral properties of an oxygen–temperature hybrid optode. Excitation (A) and emission (B) optical set-ups are shown. 1 and 2 give the absorption spectra of the temperature and oxygen indicators, respectively. 3, 4, 5 and 6 indicate the respective transmittance spectra of the interference filters FITCA, FITCE, Chroma 580 and Chroma 680. 7, 8, 9 and 10 are the emission spectra of the LED 470, LED 525, temperature and oxygen indicators, respectively.

9.4.2 Examples of Hybrid Optodes

9.4.2.1 Oxygen–Temperature Hybrid Optode

A knowledge of the temperature is a prerequisite for the precise determination of oxygen concentration because all oxygen indicators exhibit temperature-dependent luminescence properties (**Figure 9.19**). As mentioned in Section 9.2.1, the sensing properties of hybrid optodes are not only determined by the chemical and photophysical properties of the indicators but also by the gas and ion permeability of the polymers in which the indicator microparticles are incorporated. The necessary polymer properties are easier to obtain when a so-called "one layer–two particle system" is used. The composition of such a system, which was adapted for the oxygen–temperature hybrid optode, is illustrated in **Figure 9.20**. The oxygen indicator is immobilized in oxygen-permeable polymer nanoparticles. The temperature indicator is embedded in oxygen impermeable nanoparticles, thus precluding cross-sensitivity. Both types of nanoparticles are homogeneously dispersed in the polymer matrix which has excellent mechanical and tolerance properties. Absorption and emission spectra of the indicators, and the optical set-up for this hybrid optode were discussed earlier (Section 9.4.1). Calibration plots of the oxygen–temperature hybrid optode are presented in **Figure 9.21**.

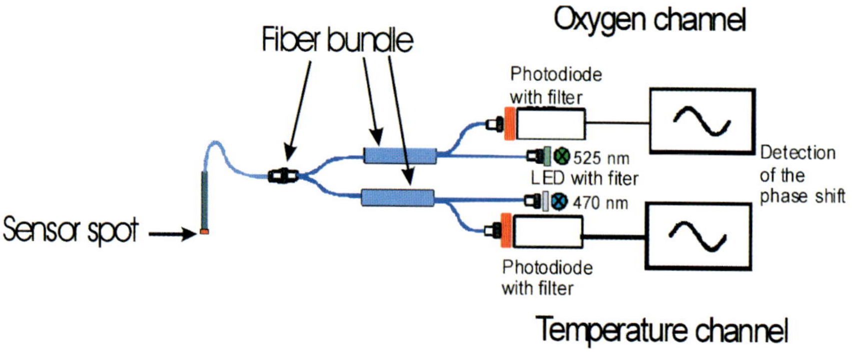

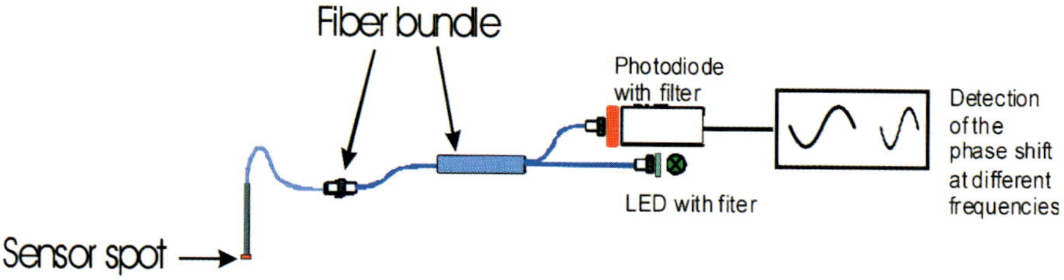

Figure 9.18 Optical system components for spectral (upper), and multifrequency (lower) measurement principles used in hybrid optode technology.

Figure 9.21A shows a linear decrease of the luminescence decay time τ of the temperature indicator with increasing temperature. The graph also proves that cross-sensitivity of the temperature indicator to oxygen is negligible even at 100% oxygen saturation. The decay time drops by only 1.5% compared to the oxygen-free condition. The oxygen indicator is excellently quenched by molecular oxygen, indicated by the exponential decrease of τ with rising O_2 concentration (**Figure 9.21B**). However, the quenching is more efficient at higher temperature, and, therefore, the actual temperature is required if correct oxygen concentration values are to be obtained. With respect to temperature sensing, the calibration plots are ideally described by an Arrhenius-type equation [6], while a modified Stern–Volmer equation has proven to give the best results for oxygen sensing [1]. The images shown in **Figure 9.22** illustrate the luminescence performance of the hybrid optode. The luminescence of the temperature indicator (left picture) drops with rising temperature, indicated by the color shift from light-red at 0 °C to nearly dark at 60 °C. After correction for temperature effects the dark-red luminescence of the oxygen indicator (right picture) reflects areas of low oxygen where nitrogen is blown in, resulting in high luminescence (light-red colored spot), and areas of high oxygen concentration where luminescence is low.

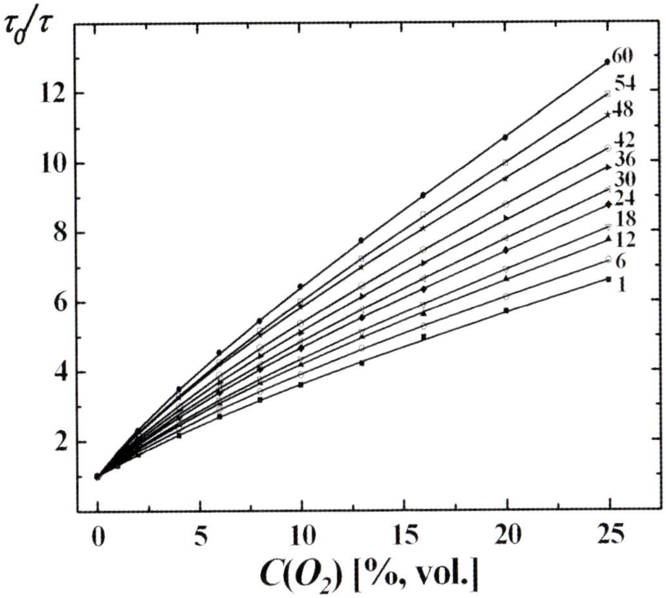

Figure 9.19 Temperature-dependent quenching of the oxygen indicator of a hybrid optode.

9.4.2.2 **Oxygen–pH Hybrid Optode**

In life science, simultaneous monitoring of oxygen and pH is of major importance for any investigation of metabolic processes, because these parameters not only determine the oxygenation state and enzyme activity in living compartments but also affect the redox potential, and, thus, the overall energy state in these compartments. Hypoxia and anaerobiosis concurrently result in a low-energy charge of cells, and a decrease of the cytoplasmic pH. Acidification of the cytoplasm not only has detrimental effects on enzyme activity but also induces a rise in redox potential, which in turn affects electron transport

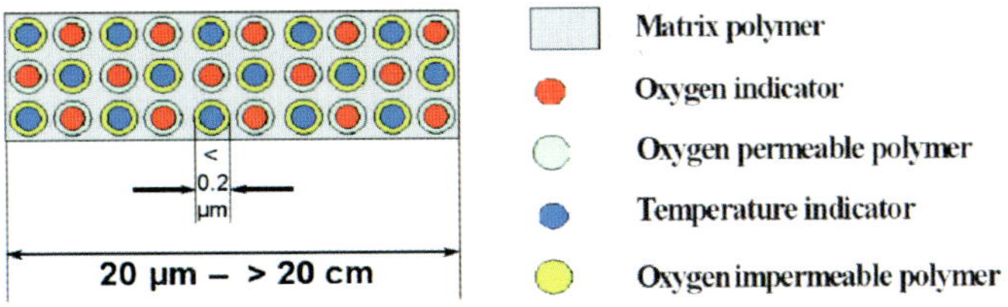

Figure 9.20 Scheme of the chemical components of the "one layer–two particle" system of the oxygen–temperature hybrid optode. An O_2 sensitive particle consists of the O_2 indicator coated by an O_2 permeable polymer. The temperature-sensitive particle consists of the T indicator coated by an O_2 impermeable polymer. These particles of less than 0.2-μm diameter are incorporated in a matrix polymer of variable dimensions (from μm to cm).

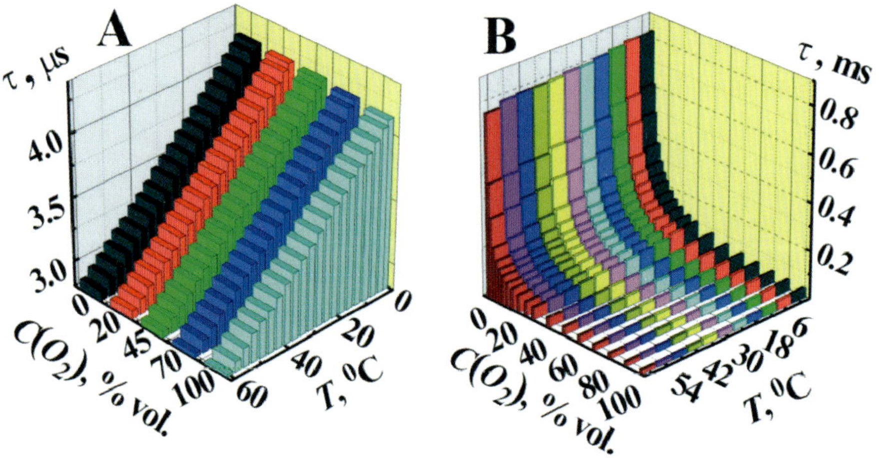

Figure 9.21 Calibration plots of the oxygen–temperature hybrid optode. Temperature sensing from 0–60 °C is not affected by an O_2 concentration up to 100% Vol. (A). The oxygen indicator shows excellent sensitivity to molecular oxygen, indicated by the exponential decrease of luminescence decay time τ with increasing O_2 concentration (B). $C(O_2)$ is the oxygen concentration in the gas phase at normal barometric pressure (101.3 kPa).

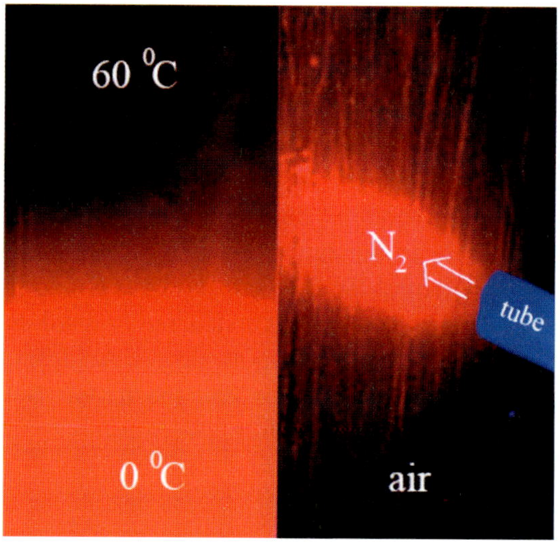

Figure 9.22 The luminescence performance of the oxygen–temperature hybrid optode. Temperature sensitivity of the temperature indicator is visualized by a color shift from light-red at $T = 0$ °C to nearly dark at $T = 60$ °C (left). Temperature insensitivity of the oxygen indicator is visualized by the dark-red color being homogeneously distributed over the entire temperature range (0–60 °C, left). N_2 injection creates an oxygen-free spot indicated by the light-red colored area.

processes across plasma membranes. Therefore, in biotechnological cell cultivation and in tissue engineering, online but sterile control of oxygen partial pressure and pH is most important for successful cell multiplication, growth

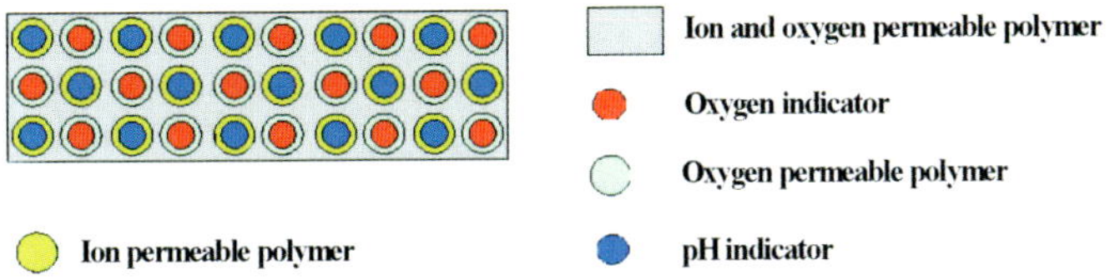

Figure 9.23 Scheme of the chemical components of the "one layer–two particle" system of the oxygen–pH hybrid optode. An O_2-sensitive particle consists of the O_2 indicator coated by an O_2 permeable polymer. The pH-sensitive particle consists of the pH indicator coated by an ion permeable polymer. These particles are incorporated in an ion and oxygen-permeable matrix polymer.

and differentiation of tissues, or biogenic production of metabolic substances such as podophyllotoxines used in chemotherapy.

A number of different oxygen–pH hybrid optodes have been developed. All of them are based on a "one layer–two particle system" (**Figure 9.23**) but they have different dynamic pH ranges. The hybrid optode for investigations in near-neutral media consists of a pH indicator that is covalently attached to ion-permeable polymer particles. The oxygen indicator is electrostatically bound to the oxygen-permeable polymer nanoparticles. These pH- and oxygen-sensitive particles are dispersed in the ion- and oxygen-permeable polymer layer. The O_2 indicator particles are not only used for determination of oxygen but also serve as a reference for short-lived fluorescence of the pH indicator. Simultaneous excitation of both indicators is performed by a blue 470 nm LED, and emission light of the indicators is filtered through an OG 530 long-pass filter.

For the oxygen–pH hybrid optode the principle of multifrequency measurement is applied. The calibration plots for oxygen and pH sensitivity are referred to different modulation frequencies. **Figure 9.24** presents an example of a 3D oxygen–pH calibration plot. Mathematical fitting of several 3D plots results in empirical calibration equations for oxygen and pH.

9.4.2.3 Oxygen–Carbon Dioxide Hybrid Optode

Among other applications, simultaneous monitoring of oxygen and carbon dioxide is required for quantitative measurement of the two basic metabolic processes on which the existence of higher life forms on earth depends: photosynthesis and oxidative respiration. The biochemical processes involved in respiration mean that oxygen consumption provides the measure of choice for respiratory activity rather than carbon dioxide production. Hitherto, measurement of respiratory oxygen consumption is primarily confined to laboratory *in vitro* investigations, particularly those in the aqueous phase, while for *in vivo* measurements under natural environmental conditions carbon dioxide is used as the measured parameter. The latter is for technical reasons, because

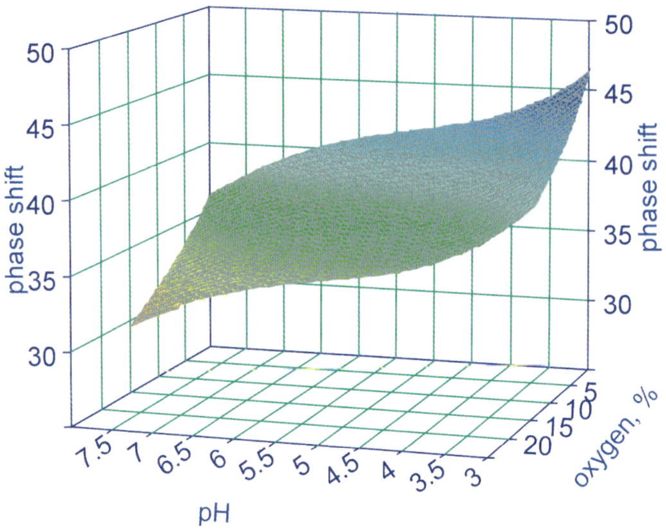

Figure 9.24 3D calibration plot of the oxygen–pH hybrid optode at a given modulation frequency. The numerical phase shift is given by the phase angle Φ.

low rates of CO_2 release (nmol g^{-1} dry weight s^{-1} or μmol $m^{-2}s^{-1}$) from live tissues can reliably be measured at low atmospheric partial pressure currently 37 Pa (370 ppm), whereas the measurement of reciprocal oxygen consumption is more complicated, involving a high atmospheric reference value of 21 kPa (210,000 ppm). Optodes have overcome this metrological limitation and provide a high-performance technique, able to detect even small changes of a given parameter independently of the background concentration level. The use of hybrid optodes, allowing simultaneous measurement of respiratory O_2 consumption and CO_2 production, and vice versa for photosynthesis, is a milestone in analytical life science. Henceforth, it will be possible to discriminate between different CO_2-releasing biochemical processes *in vivo* such as oxidative respiration, fermentation or photorespiration, which under natural conditions can take place in tissues and organs at the same time.

Optical sensing of carbon dioxide is based on the same principle as was applied successfully for optical pH measurement. Changes in pH result in protonation or deprotonation of the indicator, and subsequently changes in the absorption and fluorescence characteristics. For CO_2 optodes, however, one additional component is essential: a specific phase-transfer agent, i.e., an organic or an inorganic base that deprotonates the pH indicator. The base reacts with CO_2 and the pH indicator is protonated. This chemical challenge was overcome by invention of a "two layer–three particle system" (**Figure 9.25**) in combination with the spectral measurement technique.

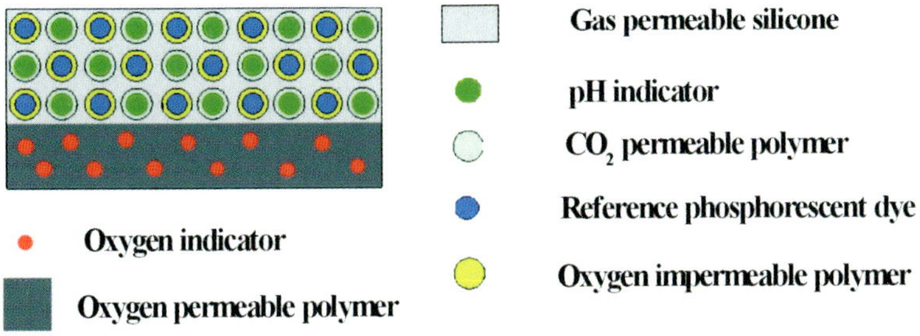

Figure 9.25 Scheme of the chemical components of the "two layer–three particle" system of the oxygen–carbon dioxide hybrid optode. The O_2 sensing layer consists of O_2 indicator particles incorporated in an O_2 permeable polymer. The CO_2 sensing layer consists of CO_2 indicator particles, and O_2-insensitive reference particles. These particles are embedded in a highly gas permeable silicone matrix. The CO_2 indicator is a pH indicator modified by addition of a phase-transfer agent, and is coated by a gas permeable polymer.

The oxygen sensing layer consists of oxygen-sensitive indicator particles incorporated in an oxygen-permeable polymer. The carbon dioxide sensing layer contains CO_2-sensitive nanoparticles (pH indicator + phase transfer agent in a gas permeable polymer) and oxygen-insensitive reference particles. The matrix polymer for the CO_2 sensing layer is a highly gas permeable silicone. Excited by blue LED light, both CO_2 and reference indicator show overlapping emission and absorption spectra (**Figure 9.26A**). The spectral properties of the oxygen indicator (**Figure 9.26B**) allow excitation with a different light source (green LED). Light emissions from the CO_2 sensing components and the oxygen indicator are isolated by means of filters, and measured separately. By this elaborate combination of indicator particles, polymer matrices and optical devices a complete spectral separation of the CO_2 and O_2 sensing systems was achieved.

Calibration plots for CO_2 and O_2 derived from the hybrid optode are shown in **Figure 9.27**. It clearly demonstrates that CO_2 sensing over a range of six orders of magnitude is independent of oxygen concentration. The CO_2 phase angle changes sigmoidally with increasing CO_2 concentration. Similarly, the O_2 sensing characteristic is not affected by CO_2, and the phase angle decreases exponentially with increasing O_2 concentration.

In summary, **Table 9.1** presents a survey of the principal qualities of the hybrid optodes. According to their hybrid character of measuring simultaneously two physicochemical parameters spatially in a fraction of a micrometer, preclusion of reciprocal cross-sensitivity is the prime task to be technically solved. In this regard, the different types of hybrid optodes introduced in this chapter give evidence of the success of the new optical-chemical measuring principle applied. A knowledge of indicator particles, and the optophysical

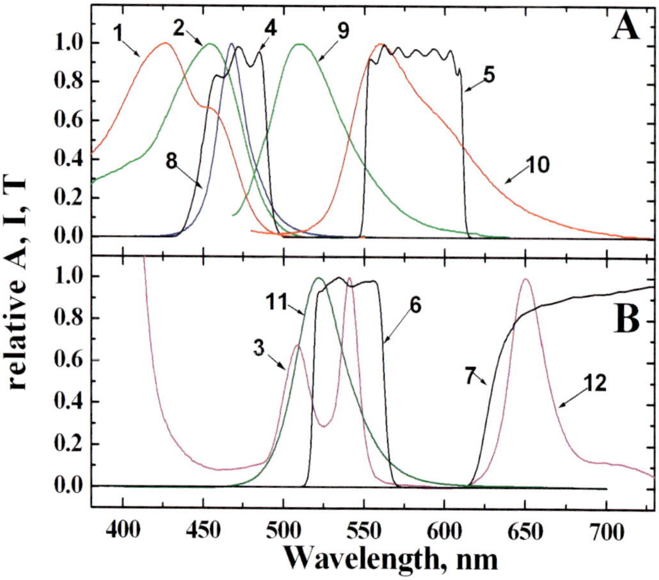

Figure 9.26 Spectral properties and optical set-ups of the CO_2 (A) and O_2 (B) sensing components of the oxygen–carbon dioxide hybrid optode. 1, 2, and 3 represent the absorption spectra of the reference complex, the CO_2 and the O_2 indicator, respectively. 4, 5, 6 and 7 show the corresponding transmittance spectra of the filters FITCA, Chroma 580, FITCE and Chroma 680. 8, 9, 10, 11 and 12 are the emission spectra of LED 470, the CO_2 indicator, the reference complex, LED 525 and the O_2 indicator, respectively.

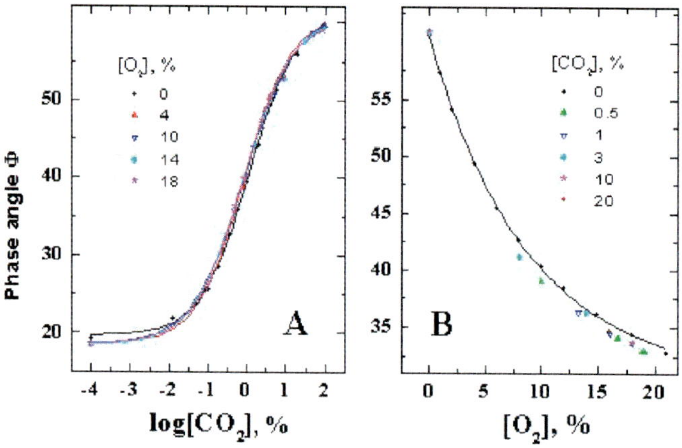

Figure 9.27 Calibration plots of the oxygen–carbon dioxide hybrid optode. CO_2 sensing over a range of six orders of magnitude of CO_2 concentration is not affected by oxygen up to at least 18% Vol. (A). Similarly, the exponential O_2 sensing characteristic of the O_2 indicator is not influenced by a CO_2 concentration up to 20% Vol. (B). Note that the phase angle Φ of the CO_2 indicator is positively correlated with CO_2 concentration, while Φ of the O_2 indicator shows an inverse correlation.

Table 9.1 Criteria used for qualification of three different types of hybrid optodes.

Properties	Type of hybrid optode		
	Oxygen–Temperature	Oxygen–pH	Oxygen–Carbon dioxide
Sensitivity to oxygen	+++	+	+++
Sensitivity to the second parameter	++	+++	+++
Response times	+	+++	++
Homogeneity	++	+++	+
Photostability	+++	+	++
Long-term stability	+++	+++	–

+++ excellent, ++ good, + average, – low

know-how of optical-chemical sensing have given rise to the development of a new generation of multipurpose optical sensors for future non-, and minimally invasive quantitative process analysis.

9.5 Hybrid Optodes: Applications and Perspectives in Biotechnology

9.5.1 Introduction

9.5.1.1 Biotechnology

Biotechnology, in particular the pharmaceutical and food industries as well as fine and speciality chemistry, is always looking for new highly productive microorganisms and cells. Comprehensive screening of cell lines and studies into the optimization of biotechnological processes are increasingly important. This is true for research as well as for biomanufacturing.

In white or industrial biotechnology, strains like bacteria, yeast or fungi are used as production factories for products on a large scale. Typical products include speciality and fine chemicals, antibiotics, amino acids, vitamins, enzymes and biopolymers. New applications and production processes desperately require optimization.

Red biotechnology concentrates on pharmaceutical issues and, therefore, deals with human medical applications from diagnoses up to therapy.

Biotechnological processes aid in detecting diseases and genetic defects rapidly and reliably. Additionally, they open new perspectives: the battle against cancer, the renewal of injured body parts via tissue engineering and the cultivation of patient-own tissue are only a few examples of huge fields of application. Stem cell-based therapies and other patient-customized treatment represent innovative approaches to treatment and cure.

There is one common feature in all the applications mentioned above: for their development as well as their implementation microorganisms such as bacteria, yeasts, fungi and particularly higher cells (mammalian and plant cells) need to be cultivated under individually controlled conditions.

Different media need to be investigated, the features of a vast array of genetically modified clones require characterization and their growth and expression under different environmental conditions such as temperature, pH value, and oxygen concentration have to be analyzed.

9.5.1.2 State of Cultivation Techniques

Usually, standard systems like Petri dishes, T-flasks, spinners or shake flasks are used for a broad range of procedures such as screening experiments or optimization steps. In such simple cultivation systems, a great number of parallel experiments can be executed easily and within a short time. But only a small number of process conditions such as temperature and CO_2 concentration can be controlled globally. However, the different cultures are not monitored and controlled individually at this small scale.

Cells are cultivated under conditions different from those found *in vivo* or on a production scale. On the other hand, fully equipped bioreactors can monitor and control ambient conditions, such as temperature, pH, dissolved oxygen, and agitation. However, it is hard to reach the required throughput within economically justifiable expenditure with such bioreactors. Furthermore, for some experiments only a limited number of cells are available, such as those derived from a biopsy. Standard bioreactors are too large to allow cultivation of such cells.

To close the gap, a tool is required which combines the benefits of small parallel cultivation systems with those of a controlled bioreactor, while overcoming the disadvantages mentioned above (**Figure 9.28**). The lack of miniaturized sensors for monitoring physiologically relevant parameters, e.g., pH and dissolved oxygen, has, so far, prevented the development of such a miniature parallel and controlled cultivation system.

Optical hybrid sensors offer the potential to accomplish such a cultivation system. In the following the concept and some fundamental achievements for its development are outlined.

9.5.1.3 System Concept

As shown in **Figure 9.29**, the entire system consists of cultivation and control modules. Standard disposable 24-well plates are used as the cultivation system. The standard cultivation plates are characterized by their ease of use.

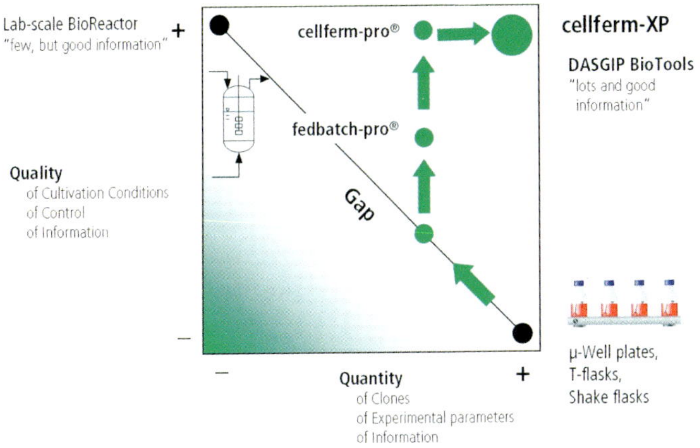

Figure 9.28 Optimization of cultivation technique and classification of the new system.

They are additionally equipped with optical hybrid sensors (planar hybrid optodes), sterile protected and gamma sterilized. The sensor spots are attached to the bottom inside of the wells. Arranged beneath the cultivation system there is an analyzing unit consisting of optical and electronic devices. To mea-

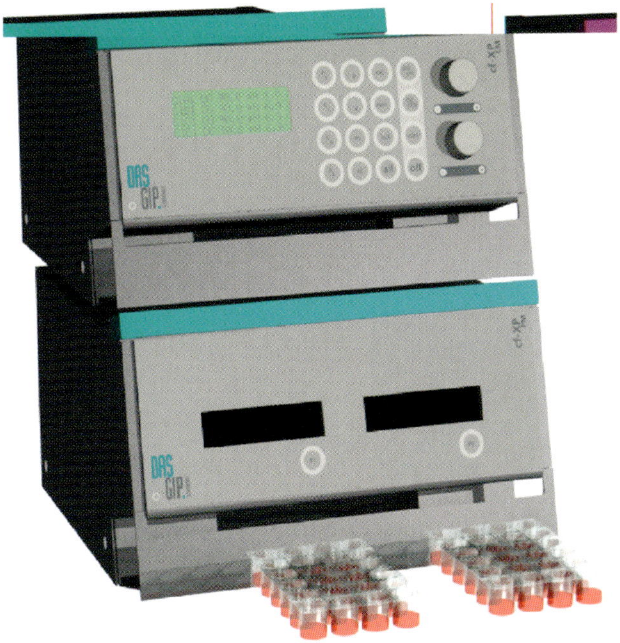

Figure 9.29 Integrated system for cell cultivation, monitoring and online control of physicochemical parameters.

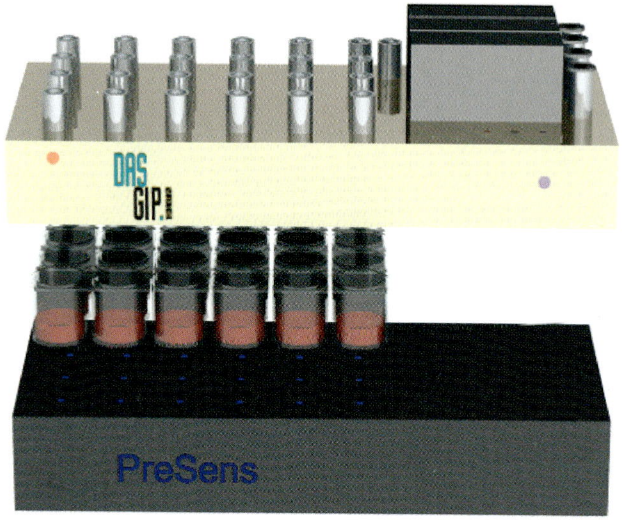

Figure 9.30 Scheme of the Process Module (PM, bottom), and the Controller Module (CM, top).

sure the process parameters, the hybrid optodes are activated with blue LED light for a short period of time, and simultaneously the phase shift of the fluorescence response is measured at high resolution. From this measurement the actual process figures, e.g., pH and dissolved oxygen levels, are determined. Online control of oxygen concentration and pH value of the cultures is carried out by variation of the inlet gas composition. Depending on the requirements of the different cultures the O_2 and CO_2 concentrations are adjusted individually. Gassing is pursued via a sequential gassing module with several inputs (air, N_2, O_2, CO_2) and 24 gas outputs. The module is attached on top of the cultivation plate. Thus, the gas mixture for each well can be adjusted individually. Sterile membranes, with good diffusion characteristics serve as sterile barriers between culture and gassing system. The membrane should show a low evaporation rate, because the osmolarity of the culture media can be a critical factor in cultivation of higher cells. Temperature control is achieved by an integrated heating system.

Complete functionality for cell cultivation, monitoring and control is achieved in two specific devices (**Figure 9.30**): the Process Module "PM" contains sensors, gassing systems and the temperature control for the cultivation plates. The Controller Module "CM" can address one or two Process Modules. Optionally, enhanced features like graphical display or data logging can be performed on an attached process computer. Online monitoring of cell growth can be accomplished via the calculation of actual respiration rates (OTR, CTR, RQ).

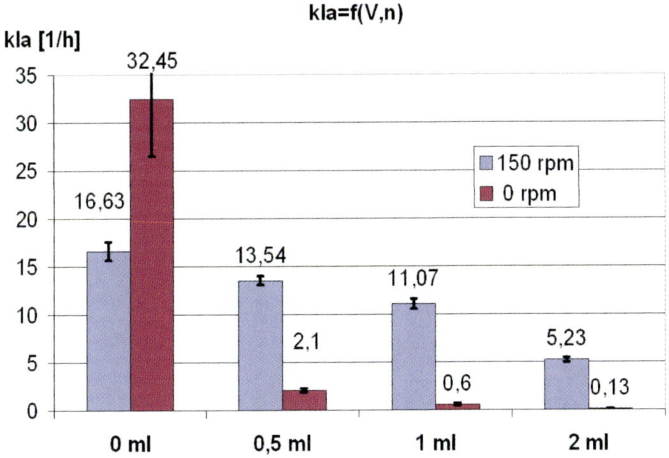

Figure 9.31 kLa-characterization of a 24-well plate.

9.5.2 Status of Development

The new miniature parallel cultivation system is based on the integration of two modules

1. the cultivation and monitoring system
2. the control system.

After establishing a suitable measurement method, a process engineering characterization of the cultivation system was carried out. **Figure 9.31** shows the relationship between oxygen transfer kLa [$1\,h^{-1}$], working volume of the wells and agitation speed. Five potentially suitable sterile membranes were identified for sterile covering of the cultivation plates (**Figure 9.32**). They were evaluated for up to 10 d, in respect of handling and adhesion characteristics as well as sterility and evaporation. Further experiments on oxygen transfer were applied to judge diffusion characteristics. After evaluating several different criteria, one membrane was selected as suitable. As shown in **Table 9.2**, handling, sterility and existing material characteristics were considered as important factors for the selection.

Subsequently, the overall system consisting of cultivation plate, sterile membrane and sensors was evaluated for its toxicity and the long-term stability of the sensors using different cell cultures (CHO, ceratinocytes (HaCat), fibroblasts), and primary cells (ceratinocytes NHK). No negative reactions regarding cell growth, viable cell count and morphology of cells were observed (**Figures 9.33** and **9.34**). Sterile conditions were maintained during the experimental period of up to six weeks.

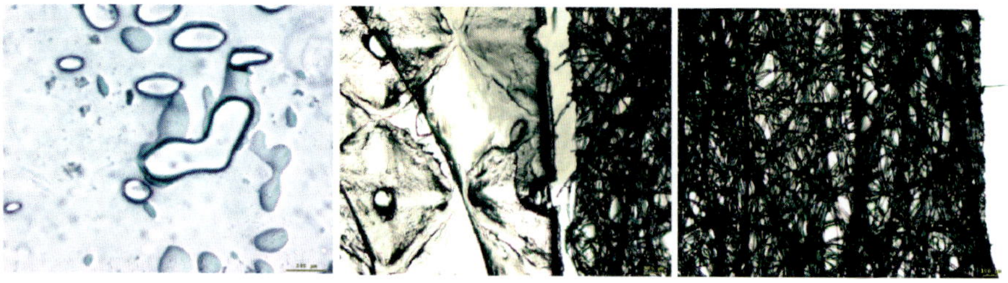

Figure 9.32 Suitable membranes for sterile covering of cell cultivation plates. Foil a (left), perforated foil b (middle), fabric c (right).

Table 9.2 Qualification of five potentially suitable sterile membranes for sterile covering of the cultivation plates. (– low, ◯ indifferent)

	Foil a	Foil b	Foil c	Foil d	Foil e
Handling	✓	◯	✓	◯	◯
Adhesion	–	◯	✓	✓	✓
Sterility	✓	–	✓	✓	✓
Diffusion	–	✓	✓	✓	✓
Toxicity	✓	✓	✓	✓	✓
Selection			✓		

After identifying suitable components, such as miniaturized valves and flow sensors, a first version of the control electronics was developed. Using 3D CAD methods, a manifold that integrates the different components was sketched and manufactured. The overall system, which covers functions like

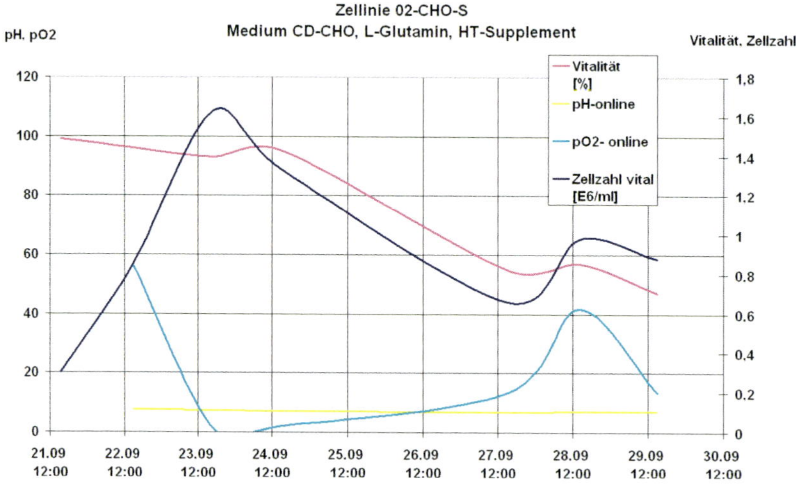

Figure 9.33 Batch experiment with CHO cells.

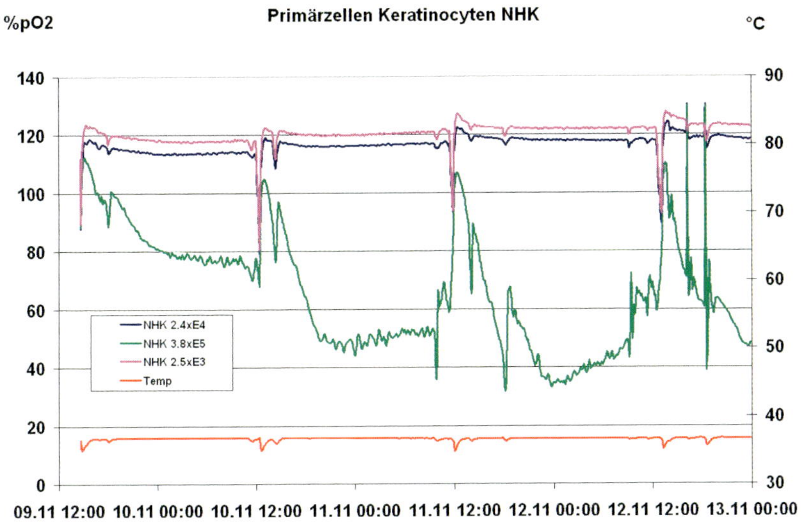

Figure 9.34 Repeated batch with ceratinocytes; primary cells.

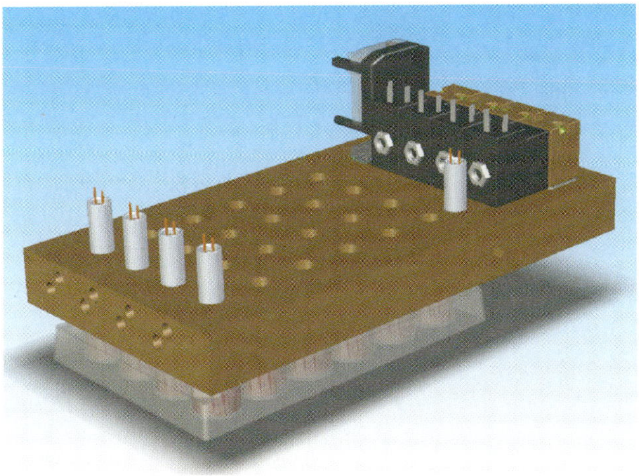

Figure 9.35 Manifold

gas mixture, gas distribution, temperature control, communication with the sensors and super ordinate control, was assembled (**Figure 9.35**). During the course of various experiments, algorithms for gas mixing relating to the control of pH and dissolved oxygen were implemented, evaluated and optimized.

Current and future investigations cover aspects such as cell expansion and product yield under controlled conditions. First results underline the huge potential of miniaturized, parallel cell cultivation in comparison to standard methods.

9.5.3
Applications and Perspectives

Hybrid optodes are the core enabling technology for miniaturized, parallel, controlled-cultivation systems as described above. With the unique features of such a system, e.g., individual monitoring and control of pH and dissolved oxygen, it can serve as a high-throughput tool in different cell culture disciplines.

While overcoming the restrictions of established uncontrolled methods, miniaturized, parallel controlled-cultivation systems open the bottleneck in cell line and clone screening as well as media screening under predefined conditions. Using such systems, the quality of insights and results in the early steps of a bioprocess development chain can be improved, thus saving costs and saving critical time to market.

Owing to the small number of cells available for inoculation, cultivation of primary and stem cells is typically performed in uncontrolled small-volume devices like T-flasks. In future, with the help of optical sensing technologies in combination with advanced control systems, the process conditions can be matched much more closely to *in vivo* conditions. Controlling the course of certain process parameters, such as pH, dissolved oxygen, and temperature during cultivation might be an aid to influencing cell differentiation and other biological aspects.

In the emerging fields of autologous tissue engineering as well as autologous cell therapy, an increasing demand for controlled cell expansion devices can be foreseen. Such devices can be made up from application-specific disposable bioreactors with integrated hybrid optodes and advanced process control systems similar to the one described above.

The need for easy process validation leads to an increasing number of manufacturing processes in biotechnology being performed in disposable cultivation systems. Optical sensors allow online monitoring of major process parameters in disposable bioreactors, such as spinner flasks or cultivation bags. Beyond the applications mentioned above, hybrid optodes have a huge potential in diagnostics, for example in the wide field of point-of-care devices.

9.6
Outlook

Hybrid optodes represent a new generation of optical sensors for non-, and minimally invasive quantitative process analysis in a broad field of application that comprises all disciplines of natural, life, and environmental sciences, and their industrial spin-offs. In particular, the multipurpose design of hybrid optodes over orders of magnitude of dimension, and their spatio-temporally high-resolution sensing performance in gases and liquids, distinguish these sensors as an enabling technology for the near future. Two major strands of progress should be sketched for future optode sensing technology.

Firstly, hybrid optodes with novel analyte-specific indicators will widen the spectrum of analytes towards the establishment of different analyte families relevant to life and environmental science, such as gases (CO, NO, NO_x, N_2O, H_2O_2, etc.), inorganic molecule ions (NO_x^-, SO_4^{2-}, PO_4^{3-}, HCO_3^-, etc.), organic molecules (hormones, energy equivalents, redox compounds etc.) or metabolites (sugars, amino acids, fatty acids, etc.). Concurrently with the developement of a broad analyte spectrum, further miniaturization of hybrid optodes will be an ongoing challenge for optical sensor research. Both miniaturization and highly specific analyte detection are essential for progress in bioprocess analysis. For example, information on metabolite transport and chemical signal transmission within cells, and between cells in tissues and organs, is a prerequisite for understanding the communication of cells in live organisms. The suitability of optodes for noninvasive measurement of several parameters at the same time opens up the perspective of a multifactorial *in vivo* communication process analysis. With respect to preventive medicine, this opens the way for improved early diagnoses of tissue and organ malfunction as well as to novel therapies by controlled interference in cell-to-cell communication.

Secondly, planar and spatial hybrid imaging (2D-, 3D-HY-Imaging) will provide an innovative and powerful tool for the elucidation of coordinate process dynamics such as species-specific growth, differentiation and senescence of tissues and organs, organismic interrelationship (e.g., symbiosis or parasitism) or regeneration of body parts after injury. At present, 2D or 3D imaging performance has concentrated on the spatio-temporally high-resolution visualization of coordinate process dynamics, while the underlying anabolism and transport processes of metabolites and signal transmitters have yet to be quantified in terms of pools and fluxes. Therefore, 2D and 3D visualization of quantitative process dynamics, concomitantly with qualitative imaging performance, will provide a platform for new insights into the various strategies of cell formation towards complex tissues, organs and organisms. For example, in tissue engineering the online control of cultivation conditions during cell differentiation, particularly during spatial tissue formation, has still to cope with uncontrollable detrimental or even lethal effects. Noninvasive, multifactorial 2D and 3D high-resolution quantitative optical sensing may thus provide a high-performance analytical tool for future tissue and organ cultivation.

Acknowledgement

The authors wish to express their gratitude to the BMBF Ministry for funding the research on hybrid optodes within the HYBOP cooperation: Univ. Düsseldorf (13N8482), Univ. Regensburg (13N8489), DASGIP AG (13N8483), PreSens GmbH (13N8484).

Glossary

Arrhenius equation This equation describes the temperature dependence of a chemical reaction rate. In order for reactants to be transformed into products, they first need to acquire enough energy to form an "activated complex". This minimum energy is called the *activation energy* E_a for the reaction. In thermal equilibrium at an absolute temperature T, the fraction of molecules that have a kinetic energy greater than E_a is proportional to $e^{\frac{-E_a}{RT}}$, where E_a is measured in molar units (joules per mole) and R is the gas constant. This leads to the Arrhenius formula for the reaction rate constant k:

$$k = Ae^{\frac{-E_a}{RT}}$$

A is a constant specific to a particular reaction.

Denitrification The biological process of reducing nitrate (NO_3^-) into gaseous ammonia (NH_3) and nitrogen (N_2). The process is performed by heterotrophic bacteria such as *Pseudomonas fluorescence* from all main proteolitic groups. Denitrification takes place under special conditions in both terrestrial and marine ecosystems. It occurs under oxygen-deficient conditions, and bacteria turn to nitrate in order to respire organic matter.

DLR (Dual Lifetime Referencing) DLR uses two luminophores with different decay times but similar excitation spectra. An analyte-insensitive, microsecond-lifetime luminophore (reference standard) is combined with an analyte-sensitive nanosecond-lifetime fluorophore (indicator), and fluorescence is converted into a phase shift.

Fluorescence A luminescence which is mostly found as an optical phenomenon in cold bodies, in which a molecule attains an excitated singlet state (S_1) by absorption of a high-energy photon and re-emits it as a lower energy photon with a longer wavelength (return to the ground state level S_0). The typical lifetime of the fluorescent state is 10^{-9}–10^{-8} s. The energy difference between the absorbed and emitted photons ends up as molecular vibrations or heat. Usually the absorbed photon is in the ultraviolet, and the emitted light is in the visible range.

Fluorophore A functional group in a molecule which will absorb energy of a specific wavelength and re-emit energy at a different (but equally specific) wavelength. The amount and wavelength of the emitted energy depend on both the fluorophore and the chemical environment of the fluorophore. Fluorophores are of importance in biochemistry and protein studies, e.g. in immunofluorescence and immunohistochemistry.

HTS (High-throughput screening) High-throughput screening is a method for scientific experimentation especially used in drug discovery and relevant to the fields of biology, chemistry, pharmaceutics and medical research.

Hybrid optode Based on the same measuring principle used for optodes, hybrid optodes provide independent information on the concentration of two variables, e.g., pH and O_2, at the same location and at the same time.

Luminescence Luminescence is light not generated by high temperatures alone. It usually occurs at low (ambient) temperatures. Luminescence can be caused by chemical or biochemical changes, electrical energy, subatomic motions, reactions in crystals or stimulation of an atomic system. Examples are photoluminescence, including fluorescence and phosphorescence, chemoluminescence including bioluminescence, crystalloluminescence, electroluminescence, radioluminescence, sonoluminescence, thermoluminescence and triboluminescence.

Optode A sensor that uses light for detection and as the carrier on information of physical and chemical parameters, such as temperature, gases, inorganic and organic molecules or ions. Optodes provide noninvasive measurements of analytes in the gaseous and liquid states of aggregation on the microscale and at the same time.

Phosphorescence A specific type of photoluminescence, related to fluorescence, but distinguished by slower time-scales (10^{-7} s) of the transition from the excited triplet state (T_1) to the ground state (S_0) of a phosphorescent molecule.

Stern–Volmer equation In the presence of a quencher (Q), energy or electron-transfer processes from an excited molecule can occur, which, in the simplest case, is described by the Stern–Volmer equation:

$$\frac{I_0}{I} = \frac{\tau_0}{\tau} = 1 + k_q \cdot \tau_0 \cdot [Q]$$

where I_0 and I are luminescence intensities and τ_0 and τ are decay times in the absence and presence of the quencher, respectively. The quenching constant is k_q and [Q] is the concentration of the quencher.

Key References

Ch. Huber, I. Klimant, Ch. Krause, O. S. Wolfbeis, Dual Lifetime Referencing as applied to an optical chloride sensor, *Anal. Chem.* 73, (**2001**), 2097–2103.

I. Klimant, Ch. Huber, G. Liebsch, G. Neurauter, A. Stangelmayer, O. S. Wolfbeis, Dual Lifetime Referencing (DLR) – a new scheme for converting fluorescence intensity into a frequency-domain or time-domain information, in: B. Valeur, J. C. Brochon (Eds.), *New Trends in Fluorescence Spectroscopy: Application to Chemical and Life Sciences*, Springer, Berlin, 2001, pp. 257–275.

References

1 E. R. Carraway, J. N. Demas, B. A. DeGraff, J. R. Bacon, *Anal. Chem.*, 63 (**1991**), p. 337.

2 S. Hanstein, D. de Beer, H. H. Felle, *Sensors and Actuators B*, 81 (**2001**), pp. 107–114.

3 K. Kalyanasundaram, *Photochemistry of Polypyridine and Porphyrin Complexes*, Academic Press, London, 1992.

4 Ch. Huber, I. Klimant, Ch. Krause, O. S. Wolfbeis, *Anal. Chem.*, 73 (**2001**), p. 2097.

5 I. Klimant, Ch. Huber, G. Liebsch, G. Neurauter, A. Stangelmayer, O. S. Wolfbeis, in: *Springer Series in Fluorescence Spectroscopy*, Vol. 1 (Eds.: B. Valeur, J. C. Brochon), Springer, Berlin, 2001, Chap. 13, p. 257.

6 G. Liebsch, I. Klimant, O. S. Wolfbeis, *Adv. Mater.*, 11 (**1999**), p. 1296.

7 G. Liebsch, I. Klimant, Ch. Krause, O. S. Wolfbeis, *Anal. Chem.*, (**2001**), p. 4354.

8 M. A. McGuire, R. O. Teskey, *Tree Physiol.*, 22 (**2002**), p. 807.

9 D. J. von Willert, R. Matyssek, W. Herppich, *Experimentelle Pflanzenökologie*, Thieme, Stuttgart, 1995, p. 344.

10 O. S. Wolfbeis, *J. Mater. Chem.*, 15 (**2005**), p. 2657.

11 J. M. Kürner, I. Klimant, C. Krause, E. Pringsheim, O. Wolfbeis, *Analytical Biochemistry*, 297 (**2001**), pp. 32–41.

12 J. Szöllösi, S. Damhjanovich, L. Matyus, *Cytometry*, 34 (**1998**), pp. 159–179.

10
Digital Microscopy (ODMS)

Andreas Nolte[1], *Christian Dietrich, Lutz Höring, Nicholas Salmon, Ernst H. K. Stelzer*[1], *Alfons Riedinger, Julien Colombelli, Philip Denner, Gernot Langer, Karsten Parczyk, Claude-Dietrich Voigt, Stefan Prechtl, Udo Löhrs, Joachim Diebold, Volker Mordstein*

10.1
Introduction

Digital microscopy has developed continuously since the mid-1980s [1]. Computing power and consequently the ability to capture and process digital images and handle huge amounts of data have continued to improve. Many new tools and features have become available. These development have had a significant impact on the daily routine of scientists and research professionals, and also that of every consumer. In particular, they are important since they influence the expectations of a younger generation who look for modern user interfaces and an instrument-based modern research environment. This chapter focuses on digital microscopy for applications in clinical and industrial routine. It describes the concept of an ocular-free digital microscope system (ODMS) and its applications in cell biology, modern drug research and telepathology. The common goals of diagnosis and treatment are improved with the prospects of a more efficient health care system (see **Figure 10.1**).

The authors are confident that major advances in health care applications will become possible by realizing even a fraction of their technical concept. Many of their ideas will become incorporated in the basis of future generations of fully digital microscope systems. As an introduction, and in order to explain and illustrate the benefits and the many different but converging goals of a technology that relies on digital microscopy, it is perhaps helpful to address our novel approach from the application point of view.

In the following, routine applications and their requirements in cell biology, drug research and pathology are explained. A "routine" application is defined by a workflow, where certain steps or actions are repeated over and over again. Therefore, it is possible to define a standard operation procedure (SOP), which describes every step of a routine's application. In the following

1) Corresponding authors

Biophotonics: Visions for Better Health Care. Jürgen Popp and Marion Strehle (Eds.)

ISBN: 3-527-40622-0

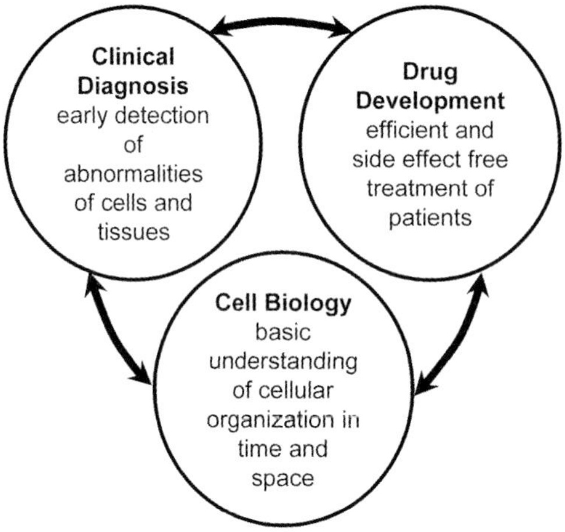

Figure 10.1 Circle of health care applications.

section, a brief overview over some routine applications and their requirements are provided. For a detailed discussion refer to Sections 10.2 to 10.3.1.

The basis within the application circle depicted in **Figure 10.1** is **Cell Biology**, which attempts to understand cell organization and cell function in the spatio-temporal domain (see Section 10.2). Understanding the structure, the functionality and the interaction of all cell components is a prerequisite for any clinical effort, and for the production of reliable results in the later processes of clinical diagnosis and drug development. In different experiments the cellular components will be manipulated by light, e.g., by laser cutting, by changing the environmental parameters, such as the temperature or the atmosphere, or by adding active agents, e.g., growth inhibitors and siRNA, to see how these manipulations affect the behavior and the morphology of a living cell or a cluster of cells. Therefore, living cells of interest are cultivated in a sterile environment, provided by micro-well plates or Petri dishes with the proper culture medium, and kept suitably for later observation with an optical system [4].

To conduct meaningful experiments with live cells, various requirements have to be fulfilled. The different phenomena, e.g., the movement of the cell along the bottom of the culture dish (essential for wound healing) or transport processes within the cell (cargo transport along microtubules, for example) have to be resolved in time and space. The extent of the observed field of interest ranges from a fraction of a micron to a few millimeters while the time-scale of these experiments ranges from a few 100 milliseconds to several days. For long-term experiments, e.g. observing the growth of a cell culture involving cell division (mitosis), an observation system has to be extremely stable. Vibrations, thermal drift and the incubation system, which maintains environmental parameters, have to be tightly controlled. On the other hand, the system has to be responsive at each level of its components, for example when positioning and picture acquisition are concerned. This is necessary to resolve fast processes, such as the change of ion concentrations in a cell or the transport of molecules in a cell membrane. Meanwhile, elaborate protocols and reagents are available that allow the specific labelling of molecules, organelles or cell compartments with markers that have different spectral properties, i.e., they are excited and emit at different wavelengths. Since it is desirable to simultaneously stain a specimen with different markers, it is necessary to distinguish markers by different colors. This allows us to identify and to co-localize different cell types and cell components [2].

Understanding cells and tissues enables and provides methods for the **Clinical Diagnosis** of diseases, e.g., cancers (see Section 6.4). The earlier a cancer is detected, the higher are the patient's chances of survival. The faster a diagnosis is obtained, e.g., during surgery or prior to therapy, the smaller is the patient's exposure to stress. Sensitive and reliable medical and optical methods for the analysis and assessment of cells and tissue are required to achieve high success rates for disease treatment. Under ideal circumstances malignant cells are correctly localized, characterized and removed.

For intra-operative frozen section assessment the three goals "earlier", "faster" and "more secure" lead to well defined pathology processes. Suspicious tissue will be removed during surgery or biopsy, sent to a pathology department, cut into slices a few microns thick, stained with different dyes, mounted on microscope slides for optical analysis and finally evaluated by an expert pathologist. However, the pathology department is usually located in a separate building or even institute. In fact, it is often located at a significant distance from the surgical theater and, therefore, it is quite common that a patient has to be kept under narcosis for long periods until results return to the surgeon and allow a continuation of the surgery.

Optimization of the three goals requires the consideration of different factors. The whole tissue is necessary for a secure detection of cancer, since both a sub-micron (high spatial resolution) and a lower resolution overview are required. Sometimes several dyes have to be applied in parallel, since different types of tissues, cells and cell components have to be distinguished. The number of slides a pathologist has to evaluate during a day varies between a few dozen and hundreds. Therefore, the speed of the evaluation process is necessarily high and has to be designed ergonomically. In case of uncertainties, it is desirable that a pathologist requests a colleague's second opinion. Since experts for different pathology disease classes are only found internationally, i.e., they are usually not located in the office next door but in other countries or even continents, efficient communication is necessary. Finally, for legal reasons, all results have to be stored for at least ten years.

The final member of our circle (**Figure 10.1**) uses cells for the **Development of Drugs** for the treatment of diseases (see Section 6.5). For example, upon diagnosis of cancer and, if possible, before as well as after the surgical removal of the affected tissue, an efficient treatment with drugs is necessary to suppress further growth of those cancer cells that remain in the patient. The more specific a reagent is, e.g., for the inhibition of cell growth (mitosis) and motility to avoid formation of metastases, or for inducing programmed cell death (apoptosis), the lower is the exposure of the patient. It is even more important to achieve a high selectivity to address only the malfunctioning cells while leaving all other cells essentially unaffected. This minimizes the adverse effects (toxicity) of a drug. Efficient and specific active agents allow the patient to vanquish and survive a cancer. The goal is to find natural or synthetic substances and to optimize them for intravenous application to achieve an effective cancer drug. Many experiments have to be performed with a huge number of substances to develop drugs with such attributes.

Drug development involves long series of experiments, which start with the identification of plausible molecular targets and end with final clinical trials. Cell-based assays are an established and currently indispensable part of this effort. Culture cells are exposed to chemical substances of a "substance library" while cultivated in micro-wellplates. A substance of interest affects the cell growth or kills the defective cells reliably, without affecting the healthy ones. Since the response of cells depends very sensitively on environmental parameters, such as composition, temperature and pH of the culture medium many requirements have to be met to obtain meaningful results on the effectiveness of a reagent. The needs are well known from academic studies in cell biology. The environmental parameters in the culture medium have to be kept constant over time, for all observed cells and up to a few days, because every deviation causes a different behavior and adds to the complication in interpreting the results. Regarding the temperature, this is quite obvious when one keeps in mind how closely temperature is controlled in mammals.

Images of cells have to be resolved on a sub-micron range to allow for subcellular imaging. It has to be kept in mind that the visualization of internal structures opens a wide range of options to assess the state of a cell e.g., to resolve changes in the functionality of cellular components caused by an added agent. Nevertheless, regarding the experiment as whole, areas of a few millimeters have to be covered to observe parallel effects in many cells and to obtain sufficiently well defined statistical parameters for reliable results.

This overview of health care applications and their needs from the application point of view purposely ignores the technical details of an optical observation system. Prior to the dawning age of digital microscopy, all the steps of an application process (and not only the observations) were managed manually. This means, the entire control, the execution and the analysis were accomplished by humans with their subjective slant. Therefore, the procedure and also the results of an experiment or an examination depended on the person conducting the experiments. Furthermore, throughput was limited by labour cost, which were and remain a relevant factor.

The processes of drug development are complex and financially demanding. High-throughput screening, which has only evolved since the mid-1990s, was triggered by the human genome project. It provides thousands of potential drug targets and generates huge reagent libraries. Currently, significant efforts are directed towards an increase of the quality of the screening process, mainly to reduce costs. Important topics are better validated targets and reagents obtained from meaningful and sensitive experimental results, as well as automation to increase throughput and reproducibility while maintaining a flexibility that offers options, which address the correct questions. This contributes to the attempts to stop the cost explosion within European health care systems while improving their quality.

For further progress of routine health care applications a paradigm change has to be induced. Qualified staff must be relieved of routine work. This reduces errors caused by human fatigue, helps to standardize results, brings down cost and creates space for patient-oriented efforts. As we will see, digital technologies enable the scientist to automate application processes to a considerable degree. SOPs allow a machine to perform certain steps of an application unattended.

Developments in digital microscopy over recent years have led to tremendous progress in all applications using optical microscope systems for analysis. Today's microscope systems use digital signal processing. This affects the control of the actual microscope hardware and the microscopy process, image capture and processing, data handling such as documentation and report generation and the embedding of the microscope into a higher level application process. In this manner, digital technologies are integrated step by step, thus optimizing the application process and the quality of the results. **Table 10.1** shows some examples of the changing requirements in an optical system on its way from a manual microscope to an automated digital microscope system. Since the digital microscope system does not need to be designed ergonomically, there is an important degree of freedom to meet the application needs (process, specimen, costs, etc.).

As addressed above, a digital microscope system has to fulfil many needs to become a part of a system solution for a complete application process. In **Table 10.2** the requirements of a digital microscope system and the related goals are listed. In general these features are all valid for all applications, but the prioritization depends on the application and varies between different experiments. In general, a digital microscope is more specialized than current devices.

Table 10.1 Changing the requirements – from a manual microscope to an ocular-free digital microscope system.

Manual microscope	Digital microscope system
Ergonomic design to meet the needs of the user; for manual usage only	Designed to meet the application needs; usage via application software only
Open device; system status undocumented	Closed device; system status documented
Automation is add on	Automation only
Documentation and reporting separate	Documentation and reporting are included

Table 10.2 Key features of a digital microscope system and their goals.

Requirement	Goal
Application-oriented hardware	Enable a more efficient work flow
Stable system, not influenced by varying environmental parameters	Reproducible results, particularly for long-term and independently performed processes
High optical resolution	Resolution of sub-cellular components
Fast image acquisition and data handling	High throughput of specimen, time resolution for short-term processes
Closed automated system solution	Documentation, objective results, optimized work flow
Flexible interfaces in hard- and software	Easy embedding into existing application processes, data export to image processing software
High fidelity of color and structure of cells and tissues	Evaluation and image processing of digital images of cells and tissues
Application-oriented graphical user interface	Easy handling, work flow oriented
Stable incubation system	Long-term experiments with living cells
Interfaces for web-based applications	Remote access, networking via internet
Extendable for new technologies	Suitable for new methods and applications

The main focus for an application in pathology lies on an increased throughput in image capture and data handling, on a fast unattended processing of slides, on a high fidelity of colors and structures in cells and tissues in the digital images, on an application-oriented graphical user interface for fast and reliable evaluation of slides and on the ability to become part of web-based applications such as telepathology.

For live-cell applications, the main focus lies on a high-performance stable incubation system for cell cultivation to perform long-term experiments, on a high optical resolution to resolve sub-cellular components and processes, on flexible interfaces in hard- and software that adapt the system to the ever-changing requirements of different experiments and on a fast export of the image data to an image-processing program.

10.2 State-of-the-Art: Digital Microscopy

10.2.1 A Real Multipurpose Digital Microscopy Platform: The Ocular-free Digital Microscopy System (ODMS)

The main goal of the ODMS cooperation was to build a digital microscope system that meets at least the requirements listed in **Table 10.2**. Many current limitations, some of them mentioned above, had to be overcome.

Each member of the cooperation is one of the innovation leaders in their field:

1. the European Molecular Biological Laboratory (EMBL) in Heidelberg, for basic understanding of the cell organization in the spatio-temporal domain
2. the Pathology Institute of the Ludwig-Maximilian University of Munich, in the clinical diagnosis of cancer
3. Schering AG in Berlin, in the drug development for cancer treatment
4. Carl Zeiss AG in Göttingen, for the development of the actual microscope system solutions.

It is obvious that some of the requirements are incompatible. A chimera with capabilities far beyond the possible would be required to meet all the requirements to the same degree.

Therefore, in a complex development the application processes of the three application partners (1–3 in the above list) were analyzed to determine their microscopy requirements. These were collected, prioritized and grouped to interlink the different applications. For each requirement a feasible technological concept was found. The ODMS thus became a demonstrator for a multipurpose digital microscopy platform. It is not a product, but owing to its forward-looking system concept and technical solutions it will influence future generations of microscope systems. For a better understanding of the following section, we must point out what the ODMS is not and why it lacks certain capabilities.

The ODMS is not a monolithic system with a high degree of specialization and optimization for a single application. Such systems could not be easily interlinked with other applications or adapted to further developing new applications. The costs for a monolithic system are higher than the costs for a comparable system with a platform concept, which can be adapted to a few different applications.

Also, it is not a system solution for applications in research, which has to be very flexible to adapt the components of the system to many different experiments/applications. Scientists tend to develop their own procedures to run their experiments. Such a system would have to be ergonomically designed to allow manual interaction at many components.

In building an ODMS most of the attributes in **Table 10.2** were not obtained in a single system. Rather, the ODMS was built applying a platform concept, which in turn yields a multipurpose solution with the features listed in **Table 10.2**. A layer model with platform- and application-specific modules is shown in **Figure 10.2**.

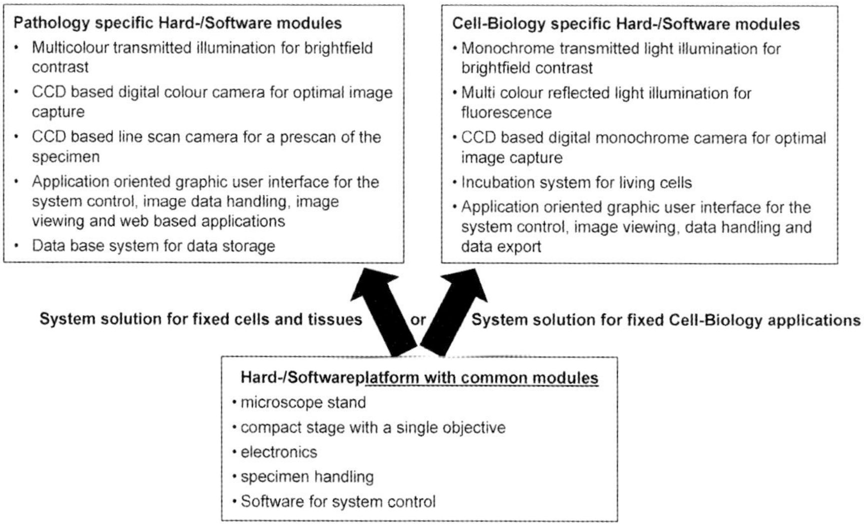
Pathology specific Hard-/Software modules
• Multicolour transmitted illumination for brightfield contrast
• CCD based digital colour camera for optimal image capture
• CCD based line scan camera for a prescan of the specimen
• Application oriented graphic user interface for the system control, image data handling, image viewing and web based applications
• Data base system for data storage
Cell-Biology specific Hard-/Software modules
• Monochrome transmitted light illumination for brightfield contrast
• Multi colour reflected light illumination for fluorescence
• CCD based digital monochrome camera for optimal image capture
• Incubation system for living cells
• Application oriented graphic user interface for the system control, image viewing, data handling and data export
System solution for fixed cells and tissues
or
System solution for fixed Cell-Biology applications
Hard-/Softwareplatform with common modules
• microscope stand
• compact stage with a single objective
• electronics
• specimen handling
• Software for system control

Figure 10.2 Layer model with platform- and application-specific modules in hard- and software.

10.2.2
The Platform Concept of ODMS for Different Applications

In this context platform means that some of the requirements are common amongst all addressed applications. Those needs are addressed by the same technical solution in the ODMS demonstrator. These common technical solutions are built into the platform. In addition, the whole system was set up to be as modular as possible with various module-based hard- and software interfaces. As a result, the application-oriented system solution takes advantage of the common platform and adds application-specific modules as required. As indicated in **Figure 10.3**, two different ODMS versions were constructed, a pathology ODMS for the analysis and assessment of fixed cells and tissues and a cell biology ODMS for live-cell applications. The two versions of the ODMS can only be distinguished when looking from behind. The connections for the incubation control units are only available with the cell biology ODMS.

Figure 10.3 The ODMS with imaging unit, client PC and monitor. The separate parts are described in Section 10.2.3

10.2.2.1 **Features of the ODMS Platform Modules in General and for Specific Applications**

The platform consists of the following modules:

- a microscope stand with a high degree of symmetry in the optical path to achieve the necessary stability to thermal and vibrational disturbances. The stand incorporates all the other modules and ensures a proper thermal decoupling of the specimen from the environmental parameters in the laboratory.
- a compact stage for the translation of a specimen in two dimensions perpendicular to the optical axis of the system, with one single high-aperture objective. The objective lens is embedded into the coupling plate of the stage for focusing and magnifying the region of interest in the specimen. Owing to the special coupling of the objective to the stage, and therefore to the specimen, this module is not sensitive to thermal

and vibrational disturbances and is, therefore, suitable for long-term experiments.

- the electronics for the power supply, for controlling the components and the image data handling
- the specimen transfer between a flexible interface to a superior laboratory process for specimen handling and the position for optical observation. This module is capable of handling all specimen holders based on the footprint of a micro-well plate.

The pathology-specific requirements have been met by the following ODMS modules:

- a multicolor transmitted light illumination for brightfield contrast, with a high numerical aperture for illumination of the specimen and a high fidelity with respect to the specimen's color and structure
- a module consisting of a CCD-based digital color camera and further optics, fitted to the resolution of the optical imaging path, for optimal imaging of the specimen
- a CCD-based line-scan camera for pre-scanning the specimen to save imaging time by not imaging empty space on the slide
- an application-oriented graphical user interface for controlling the system, viewing the digital images, handling patient-related data, reporting the result of the evaluation and enabling web-based applications such as telepathology
- a database system for the efficient storage of very large amounts of data.

The live-cell specific requirements have been met by the following ODMS modules:

- a monochrome low-aperture transmitted illumination for brightfield contrast, which enables the user to take images without exposing the live cells to intense light, while the specimen is incubated in the system
- a module that contains a fast switching multicolor reflected light illumination for fluorescence microscopy, with the appropriate filters for excitation and emission color selection (also applicable in the pathology version of the ODMS)
- a module consisting of a CCD-based digital monochrome camera with further optics, adapted to the resolution of the optical imaging path, for efficient observation of the specimen

- a module for incubation of the specimen, which controls the atmosphere's temperature, humidity and CO_2 level
- an application-oriented graphical user interface for controlling the system, viewing the digital images and handling the data export to an image processing program.

10.2.3 Innovative Technologies for Digital Microscopy

The performance of a digital microscope system in relation to a certain application depends strongly on the interaction of all components and on the ability to embed it into a higher level application process. In this sense ODMS is an application-driven system solution. This is among the most innovative aspects of ODMS. Some of the technical solutions are emphasized to illustrate the statement.

- Compared to existing solutions, the ODMS is a compact stable desktop device shaped like a cube intended for partially unattended use in a laboratory (**Figure 10.3**). Owing to its compact structure it is relatively insensitive to external environmental disturbances. All components are inside a solid frame. Only the interfaces for the specimen transfer (1), the power supply and the data handling (2) have contact with the laboratory's environment. All controls occur through the client's PC (3).
- All illumination elements are based on light emitting diode (LED) technology. This technology has some tremendous advantages over conventional halogen or high-pressure lamps: (1) no thermal problem, i.e., the light source can be incorporated into the system; (2) ability for fast switching/triggering, i.e., a shutter is no longer necessary, the specimen's exposure to light is smaller and the light source can be synchronized exactly to the image capture device; (3) automation, i.e., the light source is fully automated and controlled by the PC; (4) the lifetime of LEDs is about a factor of 10–50 higher than alternatives, i.e., a very small need for service and small maintenance costs [3]. **Figure 10.4** shows a multicolor LED array with matrices of different colors for transmitted light illumination in the pathology ODMS. Additive mixing and the separate intensity control of each diode generate essentially every color in the visible spectrum.
- The optical path is optimized along both directions. Towards the image capturing device (usually the camera) the resolution is designed to fulfil Nyquist's theorem [1]. Every resolved point in the specimen is imaged by four picture elements. Towards the specimen the resolution is designed to resolve the sub-cellular components the application needs. A

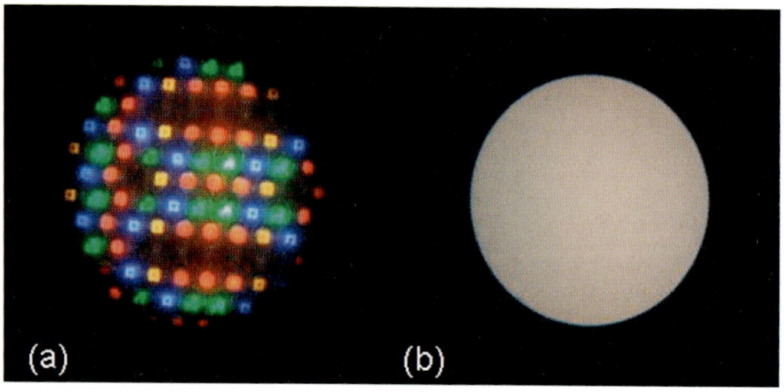

Figure 10.4 (a): Multicolor LED array for transmitted light illumination; (b): with additional glass in front to destroy source structure and to improve color mixing.

single objective lens with an aperture of 0.8 is used with two different tube lenses. The combinations provide magnifications of 16× and 32×. The single objective has the benefit of very high thermal and mechanical stability. Images (1) and (2) in **Figure 10.5** were taken with the pathology ODMS using different contrast methods: (1): brightfield contrast in transmission, slice of tissue with H/E staining, (2): polarization contrast in transmission, polarization active plastic on a slide. Images (3) and (4) were taken with the cell-biology ODMS using different constrast methods: (3): brightfield contrast in transmission of incubated living cells, (4): three-fold fluorescence contrast in reflection of fixed cells.

- The incubation system allows long-term observations of live cells. The application of a double chamber principle ensures that the space around the specimen is decoupled from the outer atmosphere in the laboratory. In this double chamber principle the inner incubator chamber contains the specimen, and the proper atmosphere for the live cells. The outer chamber can be flooded with dry air of the proper temperature to decouple the inner chamber from the environment. The incubation system is thus not sensitive to thermal disturbances, e.g., open windows or doors, or air condition. The temperature deviation of arbitrary points on a micro-well plate is smaller than 1° (see **Figure 10.6**), which is necessary for a uniform growth of live cells. The results of a 72 h test with cells illustrate the quality of the incubation: growth and division only stopped when the bottom of the well was fully covered with cells. Healthy cells do not grow over each other.

- The software enables the user to control application-oriented processes for the digital image capture on the ODMS, to handle the huge amount of image data and to run net-based applications for communication with partners.

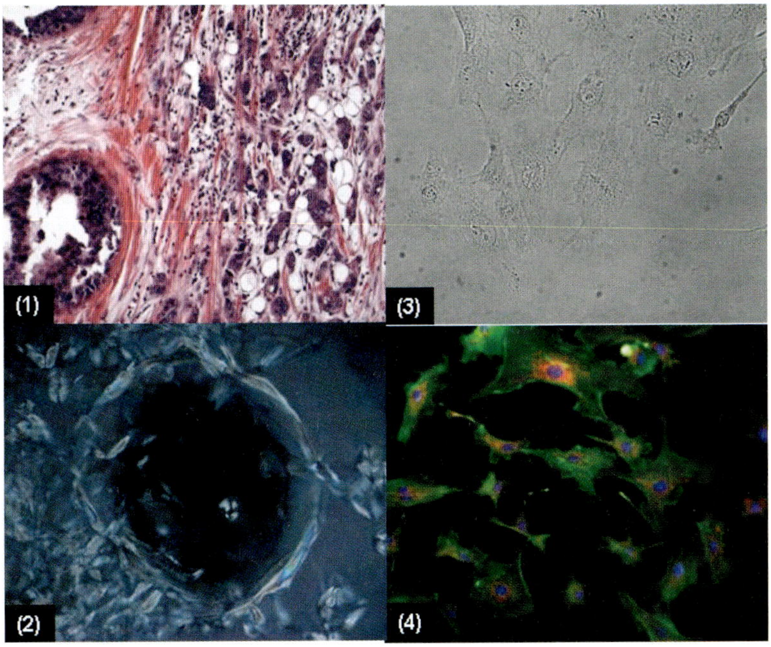

Figure 10.5 (1): slice of tissue with H/E staining on a slide, brightfield contrast in transmission; (2): polarization active plastic on a slide, polarization contrast in transmission; (3): incubated living cells in a microwell plate, brightfield contrast in transmission; (4): fixed cells on a slide (for used dyes and marked cellular components see specification of Molecular Probes: slide Molecular Probes # 1), three-fold fluorescence contrast in reflection.

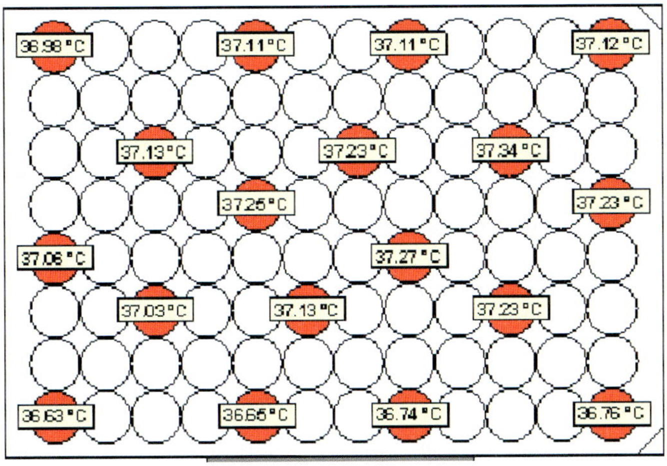

Figure 10.6 Temperature distribution on a 96-micro-well plate incubated in the ODMS.

10.2.3.1 Structure of the Software

The software is built around different layers responsible for controlling the ODMS. One layer manages the functions of the ODMS components, one layer is responsible for scheduling and executing application-oriented processes on the ODMS and a third layer is responsible for handling the system by the user (see **Figure 10.7a**).

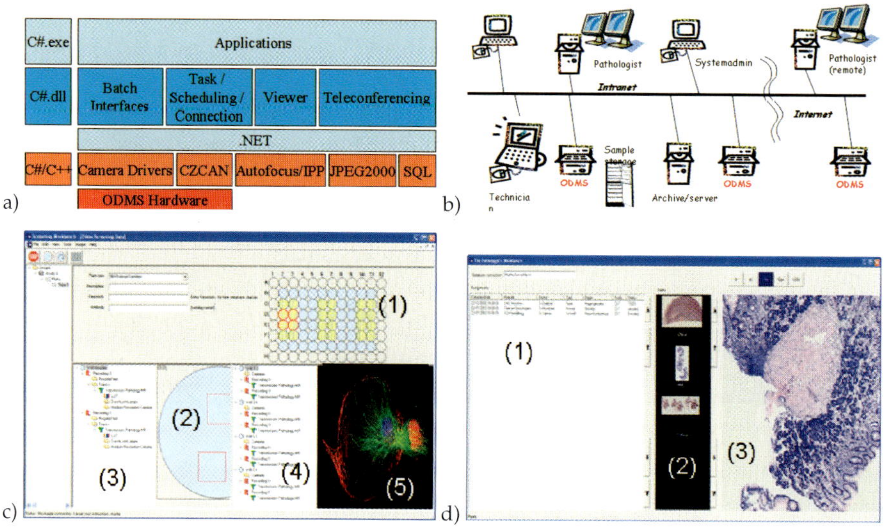

Figure 10.7 (a): The layer structure of the ODMS software; (b): The system philosophy of the ODMS: a distributed system; (c): The ODMS user interface for scheduling a batch to run and control an application; (d): The ODMS user interface for viewing and evaluating the image data.

The structure of the software allows the ODMS to be used in a net-based client–server mode via inter- or intranet (see **Figure 10.7b**). Multiuser operation on one ODMS is possible, as is one user running several ODMSs.

10.2.3.2 Handling of the System

The graphical user interfaces for the both versions of the ODMS were built application-oriented, with an intuitive handling of the system. Two different goals were achieved. For the cell biology version the goal was to enable the user to schedule and to execute different processes on the ODMS and to intervene with the image capturing process when having to decide on the further strategy of the experiment based on late data. The handling of the user interface, shown in **Figure 10.7c**, enables the user to schedule and run application-oriented batches for an experiment. At any time the user is able to stop the batch, change it and then continue the batch. The batch is made by choosing a location where digital images are captured somewhere in the specimen holder (here a micro-well plate) in windows (1) and (2). The user defines the well (1) and the area to be scanned in each well (2). All sub-processes available on the ODMS are shown in window (3) of the interface. For each location a chain of sub-processes is defined in window (4) by drag and drop from window (3). Every sub-process can be parameterized separately for an optimal application process (e.g., contrast method, exposure time, number and specification of fluorescence channels and so on). This process enables the user to choose the batch very flexibly and to fit it very well to his application process. After the process is defined, the ODMS is able to run the batch unattended. While the batch is running on the ODMS a progress monitor shows the captured pictures (5), which are usually saved to hard disk.

For the pathology application the goal was to relieve the user from defining every sub-process of the application. The handling of the user interface to schedule and control the ODMS image capture process is analogous to the one for the cell biology ODMS. In contrast to cell biology applications the pathology process is well defined. Therefore, many defaults can be set. The user chooses only a few settings, e.g., scan a slide with a certain resolution and a well defined contrast method, and determines if a z-stack is required. Thereafter, the system is able to run unattended. At the end of the process all digitized slides are saved to a database for further processing. In a next step, an image viewer layer (see **Figure 10.7d**) allows the pathologist to access the digital slides and inspect all data concerning an examination: (1) patient data, (2) thumbnails of the digital slides of an examination and (3) the high-resolution image data for evaluation. The pathologist accesses the data with a few mouse clicks by drag and drop along all three dimensions. A digital zoom is available by using the mouse wheel. With a remote connection to a partner using the same software this image viewer surface is used for telepathology.

10.3 Life Science Applications of the ODMS

10.3.1 Combining Laser Manipulation with Wide-field Automated Microscopy

In this section we discuss the application of manipulation tools at the sub-cellular level. In particular, we concentrate on the description of a laser nano-dissection system for *in vivo* and *in situ* manipulation of biological tissues. A pulsed laser beam operating at a wavelength of 355 nm generates diffraction-limited damage and is particularly valuable for intra-cellular nano-surgery. Coupled into a digital or an inverted microscope and scanned across a field of up to $100 \times 100\ \mu m^2$, a nano-scalpel takes advantage of sub-nanosecond pulsed UV light and performs *in vivo* plasma-induced ablation inside organisms, which can range from intracellular organelles up to embryos. In combination with a microscope, the system allows the use of conventional microscopy contrasts and methods, fast dissection with up to 1000 shots per second and simultaneous dissection and imaging. We outline an efficient set-up with a small number of components. The instrument provides an improvement over previously described instruments with a ratio of plasma volume to beam focal volume of around five. Information can be directly written at the sample location by plasma glass nano-patterning. Essentially the same technology can be used to operate systems that provide fluorescence recovery after photobleaching (FRAP).

10.3.1.1 **Introduction**

The main application of lasers in the biological sciences is certainly confocal fluorescence microscopy [5]. The lasers provide the light source and replace xenon lamps or, more recently, LEDs. However, even before lasers were used in laser scanning microscopy they were applied in the manipulation of biological material. IR lasers are applied with optical tweezers (e.g., Ref. [6]), while pulsed UV lasers can be used for severing biological material [7]. Continuous (CW) lasers operating in the visible range are mainly used for photobleaching, e.g., with FRAP [8,9].

Interestingly, the latter are usually used in confocal fluorescence microscopes although this makes essentially no sense, since in fluorescence a confocal microscope is relatively inefficient. All these techniques are currently best employed either in a regular wide-field system or combined with a single plane illumination microscope (SPIM) [10]. In the context of ODMS we evaluated a general approach to combining laser manipulation with wide-field, i.e., camera-based, imaging. For the sake of brevity we concentrate on the application of the laser nanoscalpel and on how it is fitted into an ODMS. However, the technology is easily adapted to other applications such as optical tweezers and FRAP.

Laser microsurgery of biological tissues [11–13] has been studied for over 30 years but is still a field of thorough research. The mechanisms of laser–tissue interactions have been extensively investigated [13–20]. The fundamentals of the underlying physical principles are quite well described as "Plasma-induced Ablation". Plasma formation, i.e., the ionization of matter inside the laser focal volume, occurs above a certain energy or power density threshold [13,15,21,22] when the ionization of the medium starts by thermal means or multiphoton absorption. The energy density threshold for ionization has been found to depend on the pulse duration [21,22]. Below the nanosecond range, the threshold decreases and microplasmas can be induced with significantly less energy than with the pulses lasting tens of nanoseconds commonly utilized in commercially available microdissection systems [23]. These longer pulses result in mechanical side effects such as shockwave or cavitations, which damages the tissue structure even at some distance from the ablating spot area.

With this in mind, a nano-dissection system was initially constructed [7] on a conventional inverted microscope. Using a pulsed UV laser with a 500-ps pulse width and a high NA lens, low-energy ablation [24] (due to the short pulse) in a highly confined volume (due to the short wavelength of 355 nm) is possible. Peak power densities up to 10 TW cm^{-2} are achieved with a low-energy laser source (8.8 μJ per pulse). Potential mechanical or thermal damage to the living samples outside the focal volume are avoided during this ablation process. Our set-up combines features from available commercial systems and overcomes many of their limitations. In particular, since we achieve diffraction-limited focusing we avoid severe thermal and mechanical damage to living samples. Fast beam scanning and optimal coupling allow simultaneous *in vivo* dissection and image acquisition. All microscopy modes including fluorescence and a port for a confocal module remain available. We present a detailed description of this device and discuss its technical accuracy and speed and its potential biological applications and present the further option to store information directly at the sample site by means of plasma-induced glass patterning inside the sample holder (e.g., cover slip).

10.3.1.2 **Instrument Overview**

The main parts of the instrument are listed in **Table 10.3** together with their specifications.

Table 10.3 Main components used in the nanodissection set-up with their corresponding manufacturer and main specifications.

Part	Company- Part number	Specifications
Pulsed Nd:YAG laser	JDS Uniphase Power Chip PNV-001025-050	$\lambda = 355$ nm, energy per pulse 10 μJ, pulse width 500 ps, repetition rate 1 kHz
UV acousto-optical tunable filter (AOTF)	AA optoelectronique AOTF.4C-UV	Transmission at 355 nm: 92%
UV beam expander	Sill Optics S6ASS3103/075	Magnification x3
Galvanometer mirrors	GSI Lumonics XY10	Optical scanning angle ± 175 mrad, position accuracy 50 μrad[1], repeatability 10 μrad[1]
Piezoelectric objective positioner	Physik Instrumente PIFOC	Positioning range 100 μm, travelling time <1 ms

1) Values specified by the company.

In the set-up, a pulsed UV laser Nd:YAG at 355 nm ($\lambda = 1064$ nm/3) is expanded by means of telecentric optics to meet the diffraction limit. The theoretical beam diameter in the focal plane depends on the objective but can be as low as 361 nm. A galvanometer pair scan unit guides the laser beam across a field of around 100×100 μm^2. Coupling into the microscope is achieved via an external illumination/detection path. The laser power is controlled by an acoustic-optical tunable filter (AOTF). Because most objective lenses are not designed for UV-A illumination, chromatic aberrations are expected and will cause the UV laser focus to be in a different z-plane compared to the visible light focus. These focal shifts can be corrected by using, for example, a piezoelectric positioner. Both the AOTF and the piezoelectric positioner are driven via a computer's interface communicating with a Scanning Controller DSP board triggering the laser pulses and synchronizing them to the galvo-mirror steps. A user interface is provided through a graphical software application [25]. To control the irradiation sequence, the user may adapt the following parameters: pulse energy (up to 8.8 μJ), number of pulses, repetition rate (up to 1 kHz), focus z-position correction (up to 100 μm). A custom target shape is defined on a live window across the full irradiation field, and the scan controller board converts the graphical coordinates to angular coordinates that position the laser beam pulses inside the sample. The set-up offers all commonly used microscopy contrast modes (fluorescence, Differential Interference Contrast (DIC) and phase contrast as well as confocal microscopy) *in vivo* and also permits the use of different objective lenses.

10.3.1.3 Coupling and Scanning the UV Beam

The coupling of UV-A light into a conventional microscope does not interfere with the available optical contrasts. The beam is coupled into the fluorescence path. In order to allow simultaneous dissection and imaging, the field aperture slider of the microscope contains a dichroic mirror with a high reflectance at 355 nm and high transmittance in the visible spectrum. For standard epi-fluorescence the classical excitation lines in the visible spectrum must be efficiently transmitted. We apply 70% transmission at 400 nm (Optosigma, USA) and more than 95% from 450 to 700 nm. Excitation filters are removed from the filter block and inserted into a filter wheel between the arc lamp and the microscope chassis to allow the propagation of UV laser light from the field aperture to the objective lens. Using the C-Apo 1.2 W lens the diffraction limit is achieved with a three-times beam expander and scanning lenses (magnification $\times 2$), leading to a scan field of 80×80 μm^2 (with an optical angle of $7.5° \approx 131$ mrad at the galvanometric mirrors). The accuracy and repeatability of the scanning mirrors depend mainly on the electronics. In the ODMS we achieve a positioning accuracy of 10 μrad (corresponding to 15 nm in the sample plane) with a digital control of 16-bits and stable analog drivers. The thermal noise of the electronics results in a steering accuracy of 50 μrad (25 nm). Thus, the positioning accuracy is significantly lower than the beam diameter in the focal plane (with an expected ratio of 1/60).

In a previous version of this set-up [26,27], the stage was controlled to move the sample relative to the stationary beam. This former solution had the advantage that the beam remains steady at the center position of the field and, therefore, is less likely to suffer from spherical aberrations. Moreover, irradiation can also be performed across a larger field of view. Combined with a regular Zeiss Axiovert 200, we achieve a diffraction limit using the Zeiss C-Apo 63x/1.2 W water immersion lens. A simple method for measuring the beam dimension in the object plane is illustrated in **Figure 10.8**.

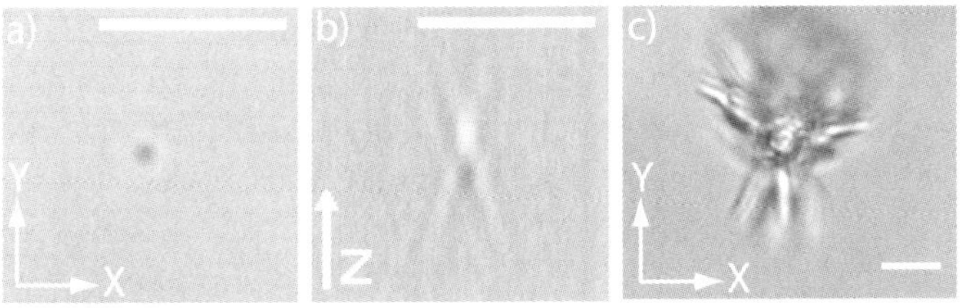

Figure 10.8 Visible effect of pulsed UV laser–glass interaction. Above a certain peak power density threshold (a,b) of ca. 65 GW cm^{-2}, glass properties change inside the focal volume of the focused beam and the irradiation volume becomes visible. At higher powers in (c), mechanical side effects induced by photo disruption provoke glass fracture. The three pictures are made in brightfield transmission; therefore, the dark and light zones do not provide direct information about the material property but only represent the reflection of light through the modified glass material. In (b), the image was extracted from a stack of x, y images recorded along the optical axis. Scale bar 5 μm.

The UV beam was focused inside the volume of a conventional glass cover slip and the visible effect of one emitted pulse is observed in brightfield transmission mode for two different optical powers. The minimum effect occurred with a pulse energy of 0.4 ± 0.1 μJ deposited in the medium, or 65 GW cm^{-2} of peak power density threshold. **Figure 10.8a** and **10.8b** show a spot with a diameter d_{xy} = 0.45 ± 0.05 μm in the x, y plane and a length of $l_z = 2.50 \pm 0.25$ μm along z. The maximum effect in **Figure 10.8c** at 8.8 ± 0.1μJ shows an extended glass fracture across a few microns in three dimensions. To characterize the beam focus quality, the latter measurements can be compared to theoretical values. To estimate the focal volume dimensions of the beam, we consider it to be an ellipsoid with lateral extent D_{xy} and elongation L_z according to the formulae applicable to high NA lenses [28]:

$$L_z = \frac{2\lambda}{n(1 - \cos\alpha)} \qquad (10.1)$$

and

$$D_{xy} = \frac{2\lambda}{(3 - 2\cos\alpha - \cos 2\alpha)^{\frac{1}{2}}} \tag{10.2}$$

where α is the angle used in the definition of the numerical aperture NA $=$ $n \sin\alpha$ and n the refractive index of the sample medium. The resulting volume $V_f = \frac{\pi D_{xy}^2 L_z}{6}$ of the focal ellipsoid is calculated, i.e., $V_f = 0.050\ \mu m^3$, and compared to the affected measured volume in glass $V_g = 0.265\ \mu m^3$. We therefore characterize the beam quality with thc factor $\frac{V_g}{V_f} = 5.2$.

Table 10.4 lists the accuracy and working conditions of the system with different Zeiss objective lenses, chosen according to a minimum of 50% transmission at 355 nm in order to minimize the UV losses. The difference in focal extent is directly due to differences in numerical aperture and back aperture diameter, conserving the beam expansion. Using a NA $=$ 0.6 air immersion lens instead of a NA $=$ 1.2 water immersion lens, one loses in spot accuracy with a twice more extended beam but gains a factor of seven in working distance.

Table 10.4 Different objective lenses used for 355-nm dissection system. The numerical aperture is a key point for achieving a high-accuracy focus spot. However, many lenses have a reasonable behavior and are sufficiently efficient for nano-surgical purposes, especially in thick samples owing to their long working distance.

Objective lens (Zeiss)	Theoretical Airy Disc (nm)	Measured x, y spot accuracy (nm)	Free working distance (mm)	Focus aberration at 355 nm (mm)	Transmission %	Scanning field (μm^2)
C-Apo 63x/1.2W	361	< 450	0.24	4.0 ± 0.5	> 50	$> 80 \times 80$
Achroplan 63x/0.75 Air	577	< 750	157	42.0 ± 0.5	> 50	$> 70 \times 70$
C-Apo 40x/1.2W	361	< 450	0.23	21.5 ± 0.5	> 70	$> 110 \times 110$
Achroplan 40x/0.6 Air	722	< 800	1.8	44 ± 0.5	> 60	$> 100 \times 100$

Finally, in order to improve the coupling of the 355 nm laser line into the microscope and to optimize system transmission, the beam polarization must be considered. Conventional filter sets are not designed to fit a special requirement down to 355 nm and their polarization behavior can result in unspecified reflectivity values at this wavelength. In general, the reflectivity is higher with an incoming s-polarized beam on the beam splitter. In **Figure 10.9**, the arrows represent the orientation of the UV beam's polarization and the coupling and scanning optical elements have been organized in order to hit the beam split-

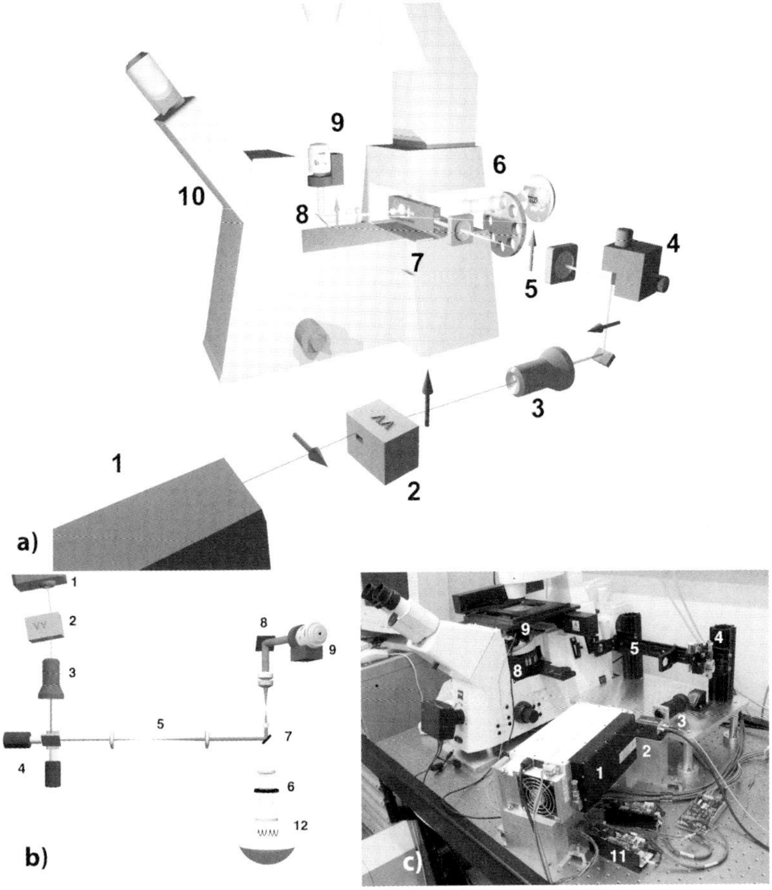

Figure 10.9 Two schematic representations (a,b) and a photograph (c) of the UV irradiation set-up. The second scheme (b) shows the optical path from above. The power of the frequency tripled Nd:YAG laser (1) is controlled by an AOTF (2). The beam is expanded by a telecentric UV lens (3, magnification x3), hits a galvo-mirror pair (4) and is coupled into an Axiovert 200M microscope (10) with two telecentric scanning lenses (5, magnification x2) and a dichroic mirror mounted on a slider (7) at the field aperture location. Arrows indicate the orientation of the laser light's polarization. The system transmission is optimized using an s-polarized beam hitting the fluorescence beam splitter (8). Simultaneous dissection and fluorescence image acquisition is possible by placing the excitation filters at the back of the fluorescence port in a filter wheel (6). Other elements: (9) objective + piezoelectric positioner, (11) DSP board and the galvanometer's analog drivers, (12) fluorescence lamp + shutter.

ter with an s-polarized beam. Nevertheless, commercial filters will not always reflect sufficient power to perform surgery.

Two solutions achieve high reflectance. First, the beam splitter is coated with a special layer that reflects the 355 nm line. Second, double-band filter sets are used, specified for two chromophores, one with absorption spectrum in the close UV such as 4′,6-diamidino-2-phenylindole (or DAPI), to ensure high reflectance of the beam splitter, and the other for the detected chromophore. In our set-up, the maximum laser power in the focus can reach up to 40% of the total output laser power, where we use an optimized beam splitter for the 355 nm line and an objective lens with high transmission. This means that up to 8.8 μJ can be deposited in the sample per pulse, corresponding to an estimated peak power density of about 10 TW cm^{-2} with a diffraction-limited beam.

10.3.1.4 Synchronization and Software

The software interface of the system has been embedded in the software for the Compact Confocal Camera [7]. Users outline a target on a live picture and control laser output power, number of laser pulses, objective defocus and mirror pair scanning speeds. The dynamic behavior of the system is limited by the optimal laser repetition rate of 1 kHz. Any point in the scan field can thus be reached and irradiated once every millisecond. The digital controller for the mirrors, the AOTF and the piezoelectric positioner are controlled via the PC. The microscope is independent of the irradiation protocol, i.e., its hardware does not have to be controlled. In **Figure 10.10**, a block diagram shows the logical sequence of a typical irradiation procedure together with the hardware connections.

Figure 10.11 presents the timing structure of this sequence. After the user has set the parameters for dissection, target and the acquisition, e.g., how many pictures before and after the irradiation are captured, the nano-surgery procedure is started.

The irradiated positions are downloaded on the Scan Controller board SC2000 at a speed of 115.2 kB s^{-1} to convert the graphical coordinates to angle values for the mirrors. The duration of this download was measured to last $\Delta t1 = (6.1n + 48)$ ms, with n being the number of shots included in the target. The piezoelectric objective positioner is then requested to place the beam focus on the initial image plane. The surgical sequence is executed by the SC2000, which triggers and synchronizes the laser pulses to the mirror steps at the required frequency f of the laser up to 1 kHz, therefore lasting a time period of $\Delta t1 = \frac{n}{f}$. The positioner is then moved back to the image plane position. The total time interval between the irradiation command and the end of the dissection procedure is calculated to be $\Delta t = (6.1n + n/f + 15)$ ms. For example, the whole protocol for irradiating a hundred pulses along a line at 1 kHz would last 668 ms, with only 10ms accounting for the optical dissection sequence. A single pulse irradiation is achieved in about 70 ms. The acquisition runs in parallel with an IEEE 1394 CCD camera, triggering a shutter for fluorescence mode as shown in **Figure 10.10**. One alternative to this system is to directly drive the stage of the microscope in order to move the sample around

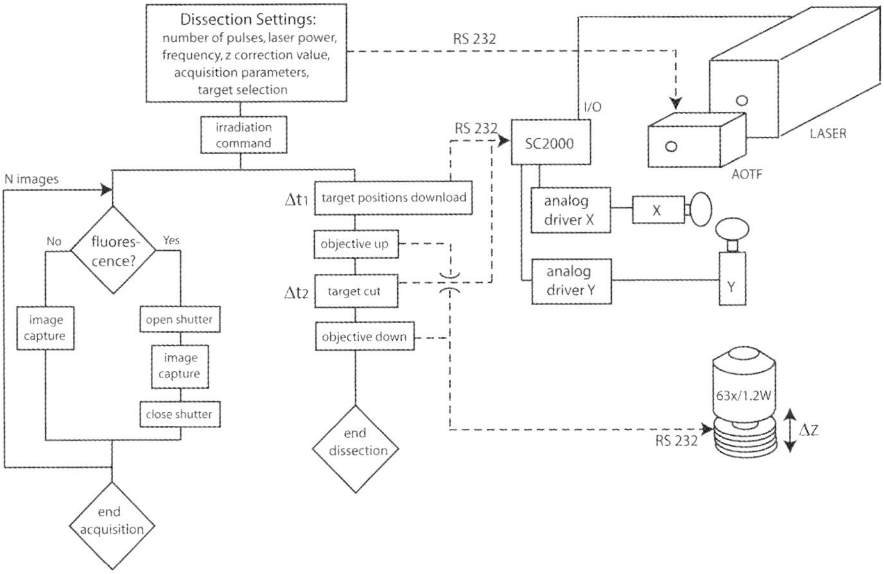

Figure 10.10 Block diagram and schematic representation of the hardware control and dissection sequence. The acquisition and irradiation protocols run in parallel. Image grabbing occurs through an IEEE1394 interface whereas the other components are driven via the PCs RS232 port. The irradiation positions are downloaded onto the Scan Controller board and executed after the objective positioner has placed the beam on the sample plane.

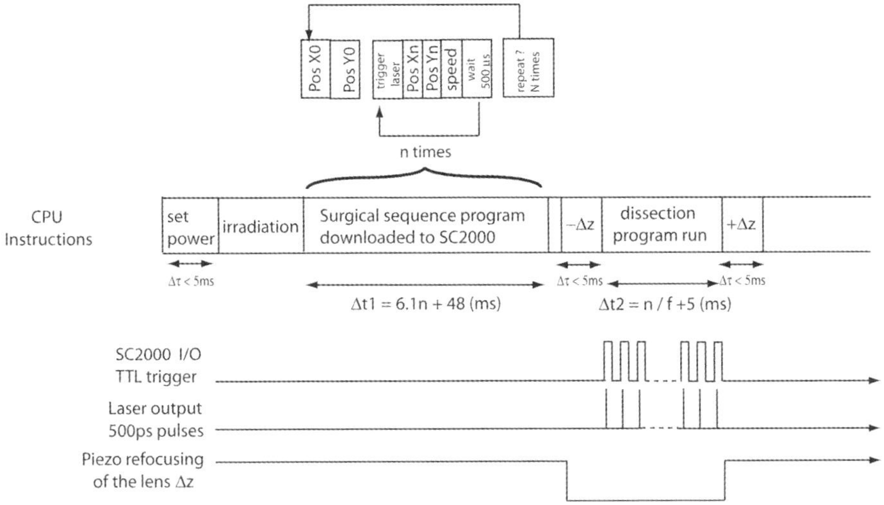

Figure 10.11 Time diagram for an irradiation sequence. $\Delta t1$ refers to the period for downloading the mirror positions for n laser pulses (Pos X0, Pos Y0 . . . Pos Xn, Pos Yn), $\Delta t2$ represents the irradiation period. The minimum period for the execution of a single shot irradiation is about 70 ms.

a steady beam. This solution offers a perfect beam quality because spherical aberrations are avoided, but is considerably slower for multiple-pulse dissection since the stage has to be moved and stabilized before irradiation. The duration for a single pulse shot with this first solution was measured to be larger than 400 ms, where the exact duration depends on the distance traveled. Furthermore, imaging of the sample at the time of irradiation is not possible, because the stage is moved around. Another drawback is the stage control, which, since it is mechanical and suffers from unavoidable backlash issues, tends to involve a non-trivial software modification.

10.3.1.5 **Performance**

Laser–glass interaction was used to measure the diffraction-limited UV spot dimensions. According to Snell's law, the interface between two media, 1 and 2, conserves the numerical aperture:

$$n_1 \cdot \sin(\alpha_1) = n_2 \cdot \sin(\alpha_2) = \text{NA}, \tag{10.3}$$

Therefore, using a different medium to measure the system accuracy makes sense because the spot diameter as a function of the NA is not affected by a refractive index change. This assumes that the experiment is not performed too deep inside a glass volume, which would result in geometric aberrations. Each objective lens is corrected only for specific working conditions, which affect the beam spot diameter and the irradiation efficiency. This glass patterning process is a simple method to visualize the effect of pulsed laser-induced plasma formation and allows us to find reasonable conditions for processing biological tissues.

As shown previously in **Figure 10.8**, glass properties are modified at the threshold of plasma formation with low-energy pulses. At higher energy ranges, plasma shielding is likely to occur and to provoke mechanical and thermal stress, as shown in **Figure 10.8c** where a wide glass fracture can be observed. To characterize the strength of a plasma-induced ablation process, the plasma volume at ionization threshold V_p, equivalent to the affected volume V_g previously measured, was compared to the beam focal volume V_f. The obtained ratio $\frac{V_\text{p}}{V_\text{f}} = 5.2$ can be compared to data from Venugopalan et al. [13] who reported plasma formation in water media with 63x/0.9 objective and 6-ns pulses at 532 nm and 1064 nm. Their experiments resulted in a ratio $\frac{V_\text{p}}{V_\text{f}}$ of 117 and 16.3 at 1064 and 532 nm, respectively, at plasma threshold formation and with a focal volume calculated with a Gaussian approximation using the Rayleigh range as the half height of a cylinder representing the focal volume. However, with high NA lenses, typically with $\sin\alpha > 0.5$, the Gaussian approximation is no longer valid. Moreover, considering the focal volume to be cylindrical results in a volume over-estimation by 50% compared to an ellipsoid. Using Eqs. (10.1) and (10.2) and an ellipsoid geometry, the same measurements lead to even higher $\frac{V_p}{V_f}$ ratios of 207 and 28.3 at 1064 and 532 nm, respectively. Our experimentally determined ratio $\frac{V_p}{V_f}$ of 5.2 shows that using the frequency tripled pulsed Nd:YAG laser and a high NA immersion lens improves the irradiation accuracy dramatically.

When performing intracellular surgery, one wants to avoid mechanical side effects such as cavitation or shockwave formation. Those physical effects arise with high-energy plasmas and can result in an extended ablation volume causing biological tissue damage far outside the focal volume. Using water or biological media, cavitation and shockwaves are generated by the formation of bubbles or the explosion of cell membranes due to high internal pressure caused by shockwave propagation. Both those physical effects were induced above a threshold in irradiance ranging from about 100 GW cm^{-2} to 120 GW cm^{-2}. Those values are similar to data from Venugopalan et al. [13] who reported thresholds of 77 GW cm^{-2} and 187 GW cm^{-2} at 532 and 1064 nm, respectively. However, the pulse energy involved in cavitation and shockwave formation is much lower in our case, i.e., about 0.25 μJ, to be compared with 1.89 μJ and 18.3 μJ at 532 and 1064 nm, respectively, and with a 6-ns-pulse duration. This difference of a factor of 7 to 70 in pulse energy allows us to assume that the mechanical strength of shockwave or cavitation is reduced because much less energy is transferred to the microplasma.

In a biological application, the older of the two versions of the set-up presented here was used to ablate centrosomes in a living one-cell stage *Caenorhabditis elegans* embryo [27]. This version does not use galvanometric mirrors but moves the stage around to irradiate different sample locations. Furthermore, it does not contain a piezoelectric positioner; instead, the laser beam is defocused to compensate for the chromatic aberrations of the lens. This defocusing causes the optical set-up to be non-telecentric. While these simplifications are concomitant with a decrease in versatility of the system and in beam quality at the focus, we were still capable to perform reproducible centrosome disintegration experiments with this set-up [27–30].

Finally, short pulsed lasers induce microplasmas and, in particular applying femtosecond pulsed lasers, applications have developed quickly in optical communications [31] and information storage [32]. By focusing short laser pulses into glass volumes, patterns can be written and read and material properties can be modified. Working with 500-ps pulses, our system offers a direct application of this process in biology by hard coding *in situ* information about single cells or organisms inside the glass volume of their support (cover slip). **Figure 10.12** illustrates this application.

A sample of fission yeast cells lies at the surface of a cover slip and information is generated in the glass volume below the sample. The laser beam

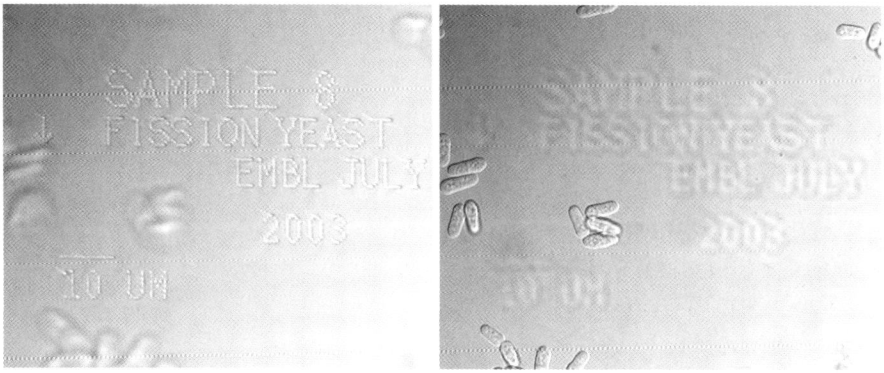

Figure 10.12 Pulsed laser-induced microplasmas are used to encode information at the sample location, inside the glass cover slip (6 μm below the water/glass interface). Observed in DIC with a condenser aperture of 0.15, the sample and the information planes are visualized simultaneously. In the middle left of each picture, an arrow points towards the sample of interest. Below the sample, a 10 μm scale bar has been created.

was focused 6 μm below the yeast cell plane and text coordinates were coded and sent to the scan controller. During a cell dissection experiment, being able to track back the processed samples is possible with this technique, as well as storing all kinds of information at the sample site (experiment parameters, date, scale bar, bar codes of various type). Glass patterning is also used in **Figure 10.13** to demonstrate the three-dimensional flexibility of the system.

Figure 10.13 Demonstration of the 3D flexibility of the system. A pyramid is patterned inside a glass slide by designing concentric square shapes at an axial distance of 3 μm from each other. In (a) and (b), two x, y images of two different squares about 18 μm apart along the optical axis. (c) x, z view extracted from a center position of the stack.

This 3D feature makes the instrument an optimal tool for developmental biology applications. Thick samples can be penetrated by the UV beam as long as the absorption and scattering of the medium are sufficiently low.

By combining UV-A irradiation at a pulse width of 500 ps with fast scanning, high numerical aperture and large working distance lenses, this versatile nano-dissection set-up is an optimal tool to perform laser ablation with improved plasma confinement and very low-energy levels. Delicate ablation is possible within a broad range of *in vivo* applications from intracellular nanosurgery in cell biology to embryo surgery in developmental biology. In general, the high level of molecular characterization in a number of biological model organisms has opened the door to a further study of the mechanical details involved in processes such as spindle positioning, chromosome segregation, cleavage furrow ingression, etc. However, mechanical statements require mechanical perturbation experiments, and the severing of a biological structure to test for mechanical tension within that structure by the use of a pulsed UV laser has proven to be, and will remain, a powerful tool to illuminate some of the micromechanical details involved in these complex processes.

10.3.2 ODMS Technology is Opening New Options for Modern Drug Research

10.3.2.1 Cellular Assays in Modern Drug Research

Following a period of decline, during which molecular and biochemical technologies were regarded as offering a comprehensive solution to the mysteries encountered in modern drug research, cellular assays are now back again at the top of the list of highly recommended scientific tools. They are widely requested during the drug-finding process for specifying and elucidating discrepancies uncovered during modern pharmaceutical research. They are used during primary high-throughput screening (HTS) and also during *in vitro* toxicology studies. The sticking point is that cellular model systems need to be selected very carefully and should be established within an appropriately short time period in order to provide the pharmacist with statistically relevant data.

In an attempt to cope with ever-increasing downstream costs from persistently high clinical failure rates, almost all pharmaceutical companies now devote a significant amount of their resources to the substantial profiling of potential drug candidates at the earliest stages possible. The most common tactic is to acquire rapidly an (almost) complete picture of the pharmacological profile of at least two lead series in order to select the most suitable clinical drug candidate. This is primarily achieved by

1. applying a disease-linked targeted strategy and
2. taking advantage of quite a large number of state-of-the-art-methods and technologies to provide all decision-relevant data efficiently.

Early drug discovery is initiated once a disease state-associated protein target has been identified and validated with appropriate bioinformatics and genomics tools (e.g., siRNA knockdown studies) and a therapeutically driven working hypothesis as to how to affect its actual function is devised. It starts with a "hit discovery", which is aimed at the identification of novel, low molecular weight molecules that modulate the activity of the target protein. The whole process depends heavily on using multiple state-of-the-art technology platforms as well as biochemical or cell-based functional approaches that allow for both accurate and fast HTS of large compound libraries. At the end of this "exercise", several lead structures are nominated, based on well founded, knowledge-driven decisions concerning their compliance with a given pharmacological profile (i.e., potency and selectivity as well as apoptotic effect, solubility, permeation, metabolic stability, etc.).

10.3.2.2 **Meeting the Needs**

Given the enormous competitive pressures for this scenario, the demand for innovative, high-throughput, precise and top quality (functional) technology platforms is quite obvious. The ODMS system satisfies several of these criteria and has been used to implement functional cellular screening platforms previously not available for functional screening at a higher throughput.

Even though they are disease centered, target-based biochemical approaches clearly help to improve drug discovery as well as the potency, drug-likeness and selectivity of leads; target validation, compound screening and pharmacological characterization requirements all increasingly call for more complex cellular functional approaches. ODMS specifications allow for microtiter plate-based screening modalities, multiple wavelength operation and high-resolution imaging and have been used to quantify complex drug actions at the level of the individual cell using sub-cellular resolution and the aid of appropriate image analysis algorithms.

The use of the ODMS system for high-content analysis (HCA) approaches will thus permit us to gain more detailed insights into sub-cellular processes through the use of cell culture systems. This is becoming increasingly important, not only during attempts to meet current technology requirements, but also for rapidly coping with and incorporating scientific progress into future drug discoveries.

10.3.2.3 **High-content Analysis: an Innovative Tool for Effective Lead Compound Identification and Drug Profiling**

It is only in cell culture systems that pharmacologically relevant mechanisms can be studied efficiently. For that reason, a detailed insight into cells that allows the quantification of structural and metabolic changes occurring there following compound interference (**Figure 10.14**) has become indispensable in modern drug research. HCA provides the industry with a novel drug-discovery tool for the rapid screening of drug effects in these cell culture systems. This technique is based on automated high-resolution and high-throughput microscopy. In combination with automated image analysis, HCA permits the simultaneous acquisition and quantification of multiple cellular signals. Complex correlations of biochemical and morphological parameters can be analyzed and quantified within mixed cell populations at the single-cell level. Thus higher informational content HCA contributes significantly to improved drug-discovery efforts. In the twenty-first century, interest in HCA has been increasing exponentially. This is not only mirrored by the growing HCA community currently emerging in pharmaceutics and academia, but also by the self-confidence that HCA technology has gained. This is due to the stability and reliability that HCA assays have demonstrated in the everyday routine of drug-discovery processes in the pharmaceutical industry. HCA is generally accepted in the scientific community as a powerful tool capable of supporting experienced and open-minded cell biologists in challenging the current limits of cell biology. This corroborates the versatility of the HCA tool for providing statistically secured data from cellular and sub-cellular events.

10.3.2.4 **High-content Analysis Contributes to Effective Drug Discovery Processes**

When an HCA system is used for screening, drug researchers find themselves in a beneficial situation for obtaining information concerning unpredictable compound interactions within the cellular context. This provides a very important and cost-saving potential for the identification of a drug candidate's possible side effects within the cell at a very early stage of the lead discovery process. Another advantage of HCA assays is that they are also applicable throughout almost all of the stages of the drug-discovery sequence. They can be applied to target validation, lead discovery and lead optimization, as well as to detailed functional studies during preclinical development. HCA assays have the potential of significantly accelerating the drug-finding process by opening up the currently existing bottlenecks that can be found in all stages of the drug-discovery pipeline. HCA relieves these bottlenecks by enabling higher-throughput cellular assays that provide multidimensional, multiparametric, high-quality data. Because they are designed in this kind of a multilinear approach, HCA assays offer not only scientific advances but also great potentials for saving time and costs. In addition, HCA helps to identify genes and proteins that are involved in diseases and HCA is capable of qualifying side effects and the therapeutic benefits to be derived from selected compounds.

10.3.2.5 **High-content Analysis Approach at Schering AG**

A sub-cellular imaging project was initiated in 2001 at the Assay Development and High-throughput Screening Department at Schering AG for the purpose of evaluating and implementing HCA. Where applicable, HCA is to be integrated into the lead discovery process as a novel application offering highly sophisticated HCA assays as standard applications capable of analyzing drug

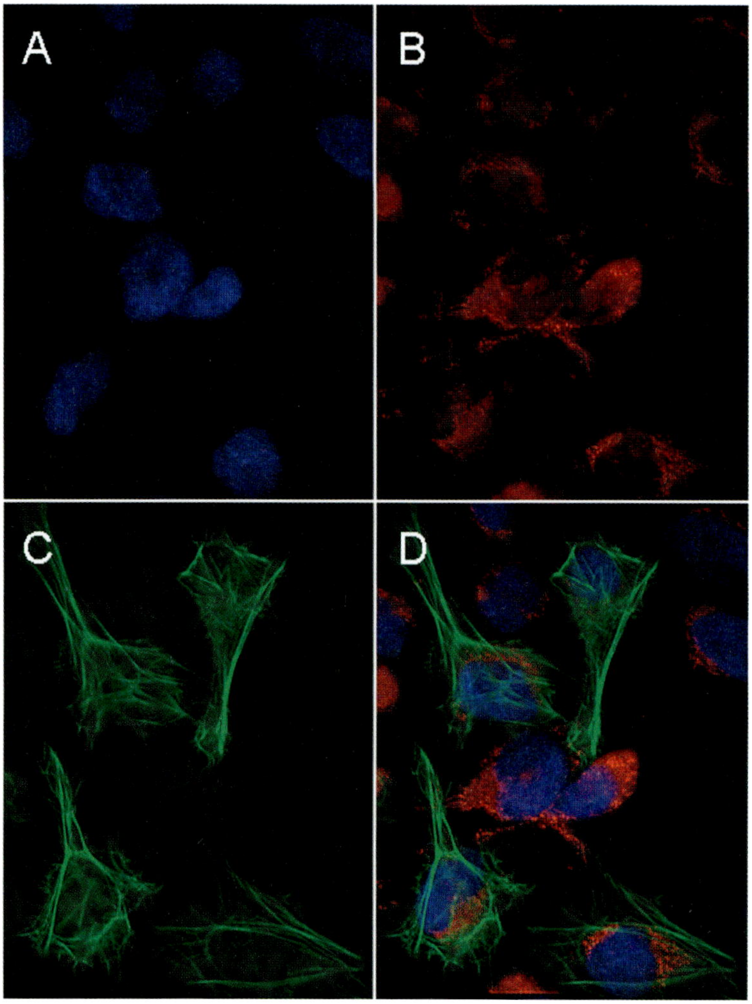

Figure 10.14 Multiparametric HCA assays allows the simultaneous analysis of such sub-cellular structures as chromatin (blue - A/D), mitochondria (red - B/D) and tubuline (green - C/D) at the single cell level. After compound treatment with inhibitors targeting such things as cell cycle kinases, pathological phenotypes and the complex relationships of the individual structures can be quantified using sophisticated image analysis routines. HCA assays may thus contribute to the uncovering of unpredictable side effects in cell culture samples that occur after compound treatment.

impact at the level of individual cells. Meanwhile, HCA has become deeply integrated into the drug-finding process at Schering AG, and our HCA group offers a broad panel of HCA applications that can be used during Target Validation and Lead Optimization (**Figure 10.15**). In the context of the ODMS project, the task has been to establish cellular *in vitro* screening procedures for automated image acquisition and automated image analysis. Cellular model systems with high informational content were established for this purpose. The design of these cellular model systems permits the simultaneous acquisition of multiple, sub-cellular parameters as well as phenotypic identification and characterization of distinct sub-populations within mixed cell cultures. An array of highly sophisticated HCA applications was established, resulting in an important quality improvement in the drug-discovery and development process. The Discovery 1 system, an HCA system provided by Molecular Devices, is in place at Schering AG and is used for the development of microtiter-based high-resolution, high-throughput microscopy approaches that can be deployed with the Zeiss ODMS prototype. During the course of the project, the well established assays are transferred from the Discovery 1 system to the ODMS prototype. Discovery 1 serves as a reference HCA system for comparing the novel ODMS prototype with systems already available on the market and currently being used in the pharmaceutical industry. Complex image datasets of sub-cellular structures are analyzed with the use of high-resolution image analysis tools. High-resolution image analysis routines were developed with the aid of an existing graphics user interface and individually adapted to specific experimental set-ups. For cell biologists, this requires intensive initial skill adaptation training, but the investment in time and specialized personnel pays off when one is confronted with novel and challenging HCA applications. In addition, this solution offers a great degree of freedom with respect to image analysis approaches, which is indispensable for meeting widely varying prerequisites. These kinds of routines are capable of distinguishing between different sub-populations in mixed cell culture samples and provide the researcher with statistically secured data leading to reliable scientific conclusions. Once the decision is made to follow this HCA approach, one is able to establish a wide range of HCA applications and a broad panel of cell culture systems. During the course of the ODMS project we were able to establish 16 different HCA applications which are now in place and which can be applied to 14 different cell systems (**Figure 10.15**). These HCA applications have been presented to researchers at Schering AG and are in great demand during the lead discovery process. It is particularly during critical and difficult stages of lead discovery projects that HCA assays are to be highly recommended for their reliability and improved biological relevance over standard cell culture assays. This is reflected by the enormous increase in requests for the use of HCA technology and by the exponential increases in images recorded, among other things (**Figure 10.16**). In addition, the mushrooming of requests relating to complex functional HCA assays requiring numerous images at the highest possible resolution that also permits a statistically secured analysis of protein–protein interactions at the sub-cellular level is also contributing to the data explosion previously described.

HCA applications

Proliferation	nucleus	⇒	total cell number
	cytoplasm	⇒	living cell number
	BrdU incorporation	⇒	DNA replication
Apoptosis	nucleus	⇒	nuclear area
	mitochondria	⇒	membrane potential
	phosphatidylserine	⇒	membrane structure
	TUNEL	⇒	DNA fragmentation
	caspase 3	⇒	"apoptotic" protein
Cell cycle	DNA	⇒	DNA content
	γ-tubuline	⇒	centrosome duplication
	histone H3	⇒	phosphorylation
Cell differentiation	protein	⇒	translocation
	cytoplasm	⇒	tube formation
GFP technology	substrate	⇒	phosphorylation
	cell cycle proteins	⇒	tumor progression
Live-Cell Imaging(cell cycle)	DicII / Ph / fluoresc.	⇒	phenotype characterisation

Figure 10.15 During the course of the ODMS project, 16 different HCA applications were established and are now in place for application to 14 different cell systems (HeLa, PC3, CHO, LNCap, MCF7, T47D, BT472, SKOV, U-2 OS, HUVEC, MVEC, PAEC, S49, Jurkat). These applications can be multiplexed to address up to five different parameters in parallel fashion.

10.3.2.6 **Conclusions**

HCA has revolutionized the way fluorescence microscopy is used in modern drug discovery and in basic research approaches. This has been made possible by prodigious developments in both the miniaturization of technical components and computer technology. HCA offers scientists a tool that enables them to analyze complex pathway relationships through an optical approach that is effectively suited to human physiology. In addition, the ODMS approach not only holds its own among current HCA approaches, it also outperforms all the contemporary HCA equipment currently available. The assembly of modern, newly developed technical components not yet available on the market has given rise to the ODMS, producing an integrated complex of novel innovative tools. This will result in a quantum leap in HCA technology, if the OMDS approach meets technical expectations with respect to speed, reliability and stability in the context of everyday drug-discovery routines. The ODMS was developed in a way that was not optimized with human physiology in mind but instead for the interaction of all the integrated technical components. The ODMS thus incorporates the best available recently developed technical equipment, making it an inimitable and innovative milestone in today's HCA world. If it is able to fulfil approximately 50% of the requisitions demanded, then it will have more than met expectations.

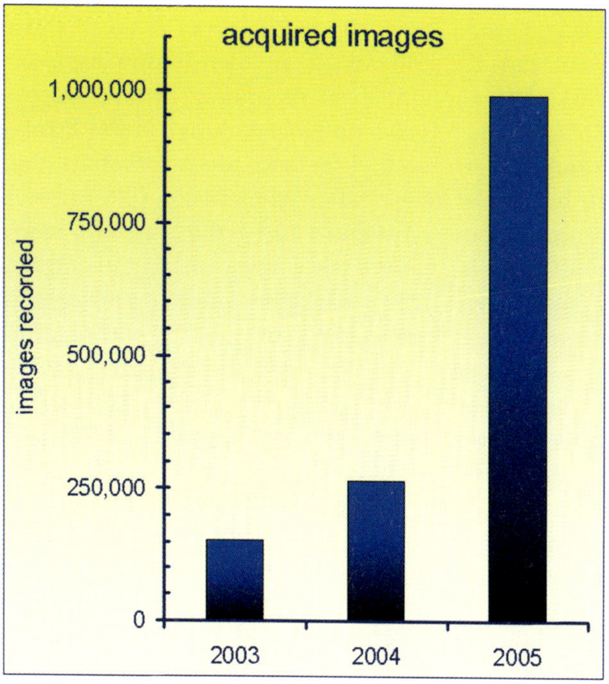

Figure 10.16 The amount of stored HCA data undergoes an exponential progression. The figure shows the number of images archived per year in our laboratory. This development goes hand in hand with increasing requests for complex functional assays addressing complex protein–protein interaction analysis.

10.3.2.7 Live-cell Imaging for Target Validation and Compound Qualification

Live-cell imaging technologies are used for the phenotypic characterization of living cells in the drug-discovery process. It is not only the morphological changes of cells resulting from drug treatment but also the function of specific proteins within the cell that are of high interest, and which can be analyzed with this tool. The living cell provides an environment that allows the analysis of protein function under native cellular conditions. As a result, a large number of predictable and unpredictable biological parameters can be investigated, which are all involved in the functionality of the target protein. Such assays differ substantially for that reason from common biochemical assays, which are limited to a small set of biological parameters dependent on the given assay conditions.

A prerequisite for live-cell studies is to use visualization techniques other than immunofluorescence procedures in order not to harm the cells. For that reason, proteins need to be tagged with alternative signals that can be recorded by microscopic systems such as auto-fluorescent proteins (AFPs).

Fusion proteins need to be generated in which the protein of interest is then coupled with an AFP. This permits the study of protein distribution or protein translocation and even the quantitative analysis of protein–protein interactions within living cells. In addition, it is also possible to derive unpredictable aspects of native protein function from cell cycle-dependent processes, thus opening up new perspectives for the investigation of more complex intracellular pathways. In the example shown in **Figure 10.17**, the gene of a target protein has been tagged with an AFP gene and the whole construct then transfected into recipient cells. The target protein can be detected following protein expression without affecting the cell because of the fluorescent nature of the AFP. This offers essential insights into the cellular function of the protein during cell cycle progression (**Figure 10.18**).

In the early 1990s AFPs were established from native bioluminescence proteins found in a variety of organisms such as reef corals, jellyfish, fish and insects. Mutated forms of such proteins are commercially available and because of their molecular structure feature the ability to absorb and emit light of specific wavelengths. The most popular and best-investigated auto-fluorescent proteins are the members of the GFP family (Green Fluorescent Protein) isolated from *Aequorea victoria*. Mutations in the amino acids of GFP lead to the blue- and yellow-shifted variants CFP (Cyan Fluorescent Protein) and YFP (Yellow Fluorescent Protein). Nowadays the entire spectrum from 400 to 750 nm can be used for fluorescence applications since red- and far-red-shifted AFPs have also been established. In theory, up to four different cellular proteins, each tagged with an AFP of a different color, can all be visualized and analyzed at the same time in a single living cell, but the overlapping absorption and emission spectra of these proteins lead to fluorescence cross-talk. As a result, such applications are limited to a smaller number of AFPs that can be used simultaneously. The combination of optimal absorption and emission filter sets, together with the selection of optimal light sources, can minimize cross-talk problems and need to be recalculated for each application.

Cross-talk between different AFPs is utilized to analyze the interaction of two proteins in some experiments. When a donor AFP and an acceptor AFP (e.g., CFP and YFP) come together in close spatial contact, a fluorescence resonance energy transfer (FRET) occurs. The energy emerging from the emission of the excited donor fluorophore (CFP) is shifted to the absorber fluorophore (YFP) and initiates a photo energy transfer, resulting in the loss of fluorescence intensity in the donor AFP and the emission of the acceptor fluorophore. It is only if both AFPs are located within a narrow range of a few nanometers that this energy transfer takes place and can be used as a signal for direct protein–protein interaction. In addition, the expression and distribution pattern of two or more AFP-labelled proteins which fluoresce at different wavelengths and which are applied simultaneously can be utilized for the analysis of the dy-

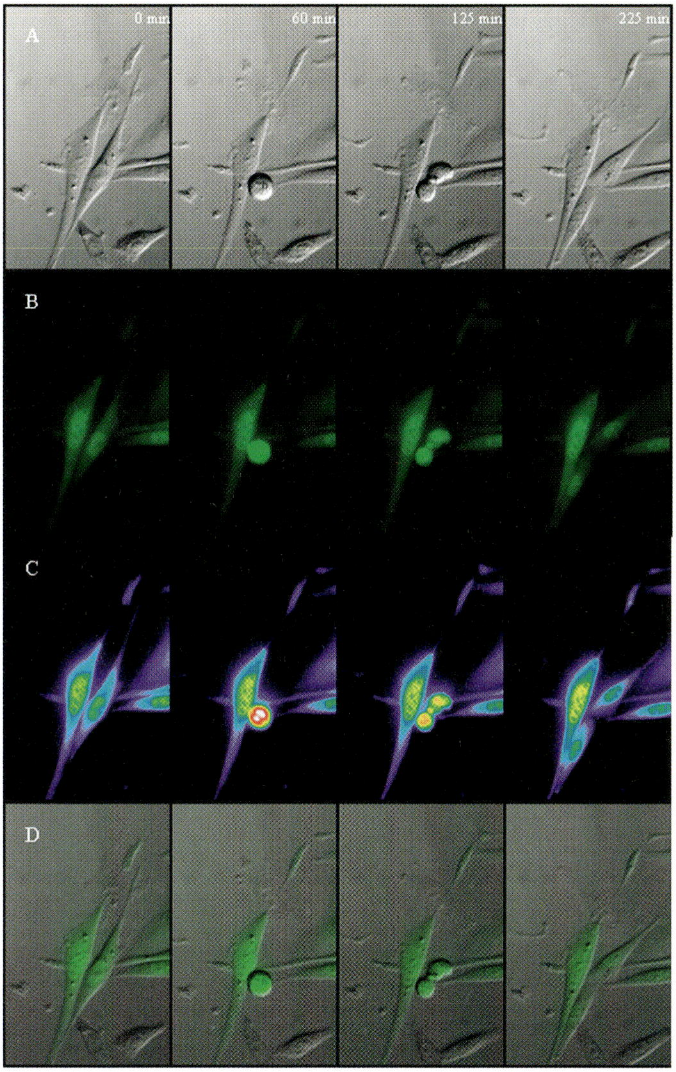

Figure 10.17 Cell division of a stably transfected CHO cell. The images show an abridgement of a live-cell experiment revealing the intracellular localization of a GFP-labelled kinase in stably transfected CHO cells. Series A shows the differential interference contrast (DIC) images of the time series. Series B shows the distribution of the kinase-GFP fusion protein in the GFP channel. Series C illustrates GFP fluorescence as intensity images. Series D superimposes series A and B. The cells show a predominantly nuclear localization of GFP-labelled kinase during the interphase of the cell cycle. During mitosis, the fusion protein can be found in specific regions of the cell. In the metaphase (60 min), the protein is located at two opposite poles in the rounded cell. During cytokinesis (125 min), the fusion protein can be found in the area of the reassembling nuclei and in the region of the contractile ring.

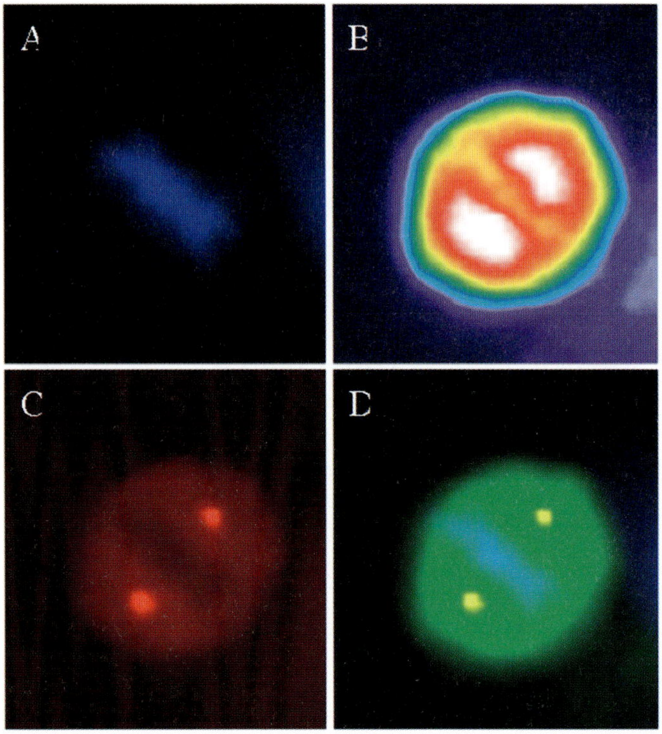

Figure 10.18 Immunofluorescence staining of a stably transfected CHO cell in the metaphase of mitosis. The images illustrate the distribution of a GFP-labelled kinase in stably transfected CHO cells during the metaphase of mitosis. Image A: Chromosomes (stained with Hoechst dye) are located on the metaphase plate. Image B: GFP fluorescence of the fusion protein, shown in different intensities. Image C illustrates the immunofluorescence staining of gamma tubulin, revealing the localization of the spindle poles. Image D superimposes Images A, B and C. These images show a predominant localization of the kinase-GFP fusion protein around the spindle poles in the metaphase of mitosis.

namic processes within living cells that occur following such cellular perturbations as drug treatment. This can be done via comparison studies for which analysis is made of the intracellular localization, interaction and dynamics of these proteins.

In the drug-discovery process, live-cell imaging technologies are deployed for target validation, phenotypic characterization of RNAi knockdown studies and lead optimization. During target validation and RNAi knockdown studies, live-cell experiments help to establish new cellular assays to evaluate the (dys)function of potential drug targets. Such applications can reveal specific time points during cell cycle progression at which the target protein is located in defined intracellular compartments. From such observations it is possible to derive aspects of native protein function and to identify potential substrates. During lead optimization, cellular models are utilized to study the effects of new compounds on target proteins in living cells on a functional basis. Here additional side effects of the compounds, such as cell toxicity or strong morphological changes, can be readily measured. The application of live-cell assays in this part of the drug-discovery process contributes to a reduction of time and costs for drug development.

10.3.3
From Telepathology to Virtual Slide Technology

10.3.3.1 Why Do We Need Telepathology?

One of the main fields of work for the pathologist is the review of histological and cytological specimens including immunohistochemical and molecularpathologic or zytogenetic preparations. Most of today's diagnostic routine work is done using a conventional light microscope. The "dynamic" assessment of a large number of images of one specimen in different magnifications during one session and the integration and interpretation of the image content leads to the diagnosis. Time plays a crucial role because further medical treatment of a patient will often depend on the pathologist's diagnosis. This is especially true with intra-operative frozen section assessment, when the course of the operation depends on a reliable histological result. If no local pathologist is available within a reasonable time, telepathology can be helpful. In this setting a small laboratory will be near the operating theater where tissue can be prepared by trained medical personnel and frozen sections of relevant areas can be produced. Together with the macroscopic images, these will be tele-transmitted by an appropriate microscope to the pathologist on duty, and evaluated on a monitor. At least for the time being, for quality assurance and for legal reasons, it is necessary for the pathologist to review the original specimens afterwards (usually one day later) under a conventional light microscope and to prepare further paraffin-embedded specimens from the original excised tissue.

Moreover, it has to be kept in mind that a correct pathological diagnosis principally requires the possibility of macroscopical sight and palpatory findings as well as histological assessment. It has to be explored, whether and to what extent, these facts produce limitations with regard to pragmatic and legal aspects and how they could be overcome.

In this context, telepathology can mean exchange of information between different continents as is demonstrated, for example, in a project between different South-east Asian countries, Germany and Switzerland [36]. Particularly in developing countries the situation of having a nearby pathologist is very improbable and telepathology can be a valuable tool to overcome the shortage of experts.

Telepathology will produce some advantages that are not strictly time dependent. It has become a suitable technique for getting a second opinion from a distant expert without the need for shipping paraffin-embedded tissue or glass slides, which is costly in both time and money. This practice yields direct benefits in terms of quality assurance and health care.

Telepathology is a locally, nationally and even internationally applicable tool for continuing medical education (CME) and can be used for teaching histology and cytology to the local medical staff as, for example, presented at the 7th European Congress on Telepathology, 2004, by T. Kuakpaetoon from Bangkok [36].

But above all, the relatively low technical requirements on infrastructure for data transmission and on the equipment itself helps to contain the costs of medical services.

10.3.3.2 The Limitations of Telepathology – Results from an Evaluation Study

To examine the efficacy, reliability, speed and acceptance of telepathology we performed a study questioning 13 pathologists from the Institute for Pathology of the University of Munich. It is important to note that not only experienced pathologists were included in this study but also young trainee pathologists, to get a variety of different evaluations on this technique. We had a ZEISS Axiopath system at our disposal (**Figure 10.19**) comprising a fully motorized ZEISS Axioplan microscope (objectives: 2.5×/10×/20×/40×/63× Plan-Neofluar) with a Hitachi CCD-camera attached to it (HV-C20AP 3-CCD Color Camera, Hitachi Kokusai, Tokyo, Japan), the appropriate software for navigation and viewing the images (AutoCyte LINK 2.0.0.77, TriPath Imaging, USA) and a computer running Windows 2000 at each of the remote sites as well as at the core institute (Intel Pentium 4, 1.700 MHz, 1 GB RAM, Fujitsu Siemens, Augsburg, Germany; monitor: 21" T3, Fujitsu Siemens, Augsburg, Germany).

Each participant had to assess eight different cases from a total of sixteen. The whole set of cases comprised 29 slides taken from different fields of pathology, which means frozen sections, bioptical, surgical and autoptical diagnostics and transplantation pathology. We paid attention to cutting particularly thin tissue sections for more convenient focussing [38]. There were nine cases comprising one slide, four cases with two slides, two cases with three slides and one case comprising six slides. We paid particular attention to having a variety of different stainings to evaluate the display of color shading. For red tones we used hematoxylin-eosin and periodic acid staining, for bluish spectrum Giemsa and hematoxylin-immunohistochemistry and for yellowish Elastica van Gieson staining. For each case, diagnostic questions had to be answered and a questionnaire had to be filled in to check out the assessor's opinion of hardware, software, handling and especially image quality. The

Figure 10.19 Telepathology with the ZEISS Axiopath system as it is carried out at the Institute for Pathology of the University of Munich. The two parts of the institution are located about 10 km apart. The finding of the diagnosis at the campus "Grosshadern" (right) can be supported by a senior pathologist at the remote site in the "city center" (left). The two parts of our institution are linked by glass-fiber cable.

different topics had to be marked on a five step scale from 1 = "very good" to 5 = "very bad". Further annotations were welcomed and integrated in our final judgments.

The series of 104 sessions for assessment of the specimens took 19 h and 20 min. The mean time spent for one slide ranged from 3 to 10 min depending on the case. The minimum time for diagnosis was 2 min for a single slide case, a *pseudomembranous colitis*, and the maximum time for diagnosis was 40 min for the six slides case, which was a follicular B-cell lymphoma. The mean time spent on finding a diagnosis ranged between 5 and 18 min. The correct diagnosis was obtained in 59 assessments, 56.7%. A further 35 estimations yielded the right diagnosis in a set of two or more differential diagnoses. Taken together, this means a maximum of 90.4% correct diagnoses. Wrong diagnoses occurred 10 times (9.6%). In 36.5% the particular pathologist referred the final diagnosis to conventional light microscopic assessment.

In **Figure 10.20** it can be seen that average skills in use of computers led to rather good marks within the judgment of hard- and software. In contrast, the image quality, comprising sharpness, contrast and truth of colors, yielded less encouraging results (**Figure 10.21**). Really bad marks were given for system speed. During the evaluation phase of about 20 h there were six system crashes that required shutting down and restarting the system.

Searching the PubMed library (http://www.pubmed.com), the first article on "telepathology" dates to 1986 [42]. By 2005, 505 further publications on this topic followed. Many of them are trials that compare telepathology-derived diagnoses to conventional light microscopic ones [40, 41, 43]. Most studies

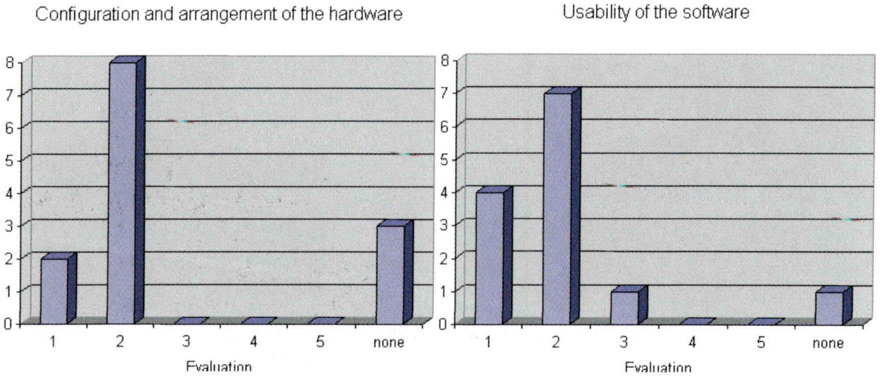

Figure 10.20 Marks given for hard- and software by the 13 questioned pathologists from the Institute for Pathology of the University of Munich on a five-step scale from 1 = "very good" to 5 = "very bad". Results are mainly in the range of good marks with average knowledge in use of computers with a graphical user interface. Some of the assessors did not even realize any need to address these topics. This is reflected by the missing judgment in three cases for the hardware components and in one case for the handling of the software.

find high concordances between 95 and 100% [38, 41, 43] but others report a much lower agreement with the gold standard diagnosis, dropping as low as 80% [34] or finding significantly higher accuracy in diagnosis with light microscopy [40]. The lack of accuracy in our series is surely due to the inclusion of inexperienced colleagues with as little as half a year's further education in pathology. However, the main focus of our study was the assessment of hard- and software. Therefore, we accepted the lower diagnostic correctness.

Leong et al. stated in 2002 that the time spent for evaluation of bronchoscopic biopsies was 14 times longer than for conventional light microscopy. The average time spent per slide was 7 min 21 sec, compared with 32 sec per slide with conventional light microscopy. For open lung biopsies and resections telepathology was five times slower at 6 min 13 sec compared with 1 min 10 sec [38]. This is in agreement with our findings that demonstrate an average of 6 min 53 sec for evaluation of one slide. Taking into account that each participant only diagnosed eight cases, one can expect that there will be an increase in speed with more experience in this system. Leong at al found a decrease in time after approximately 60 sessions or cases [37].

Dunn et al. have shown that over the course of 2200 telepathology cases the time spent per case was reduced from 10 min 26 sec to 3 min 58 sec. Viewing times per slide were also reduced by a factor of three from 4 min 44 sec to 1 min 13 sec [35].

Our study provides valuable results concerning hardware, software, image quality and handling of the system. The fact that in 36.5% of the cases a need for conventional light microscopic assessment was claimed, already points to

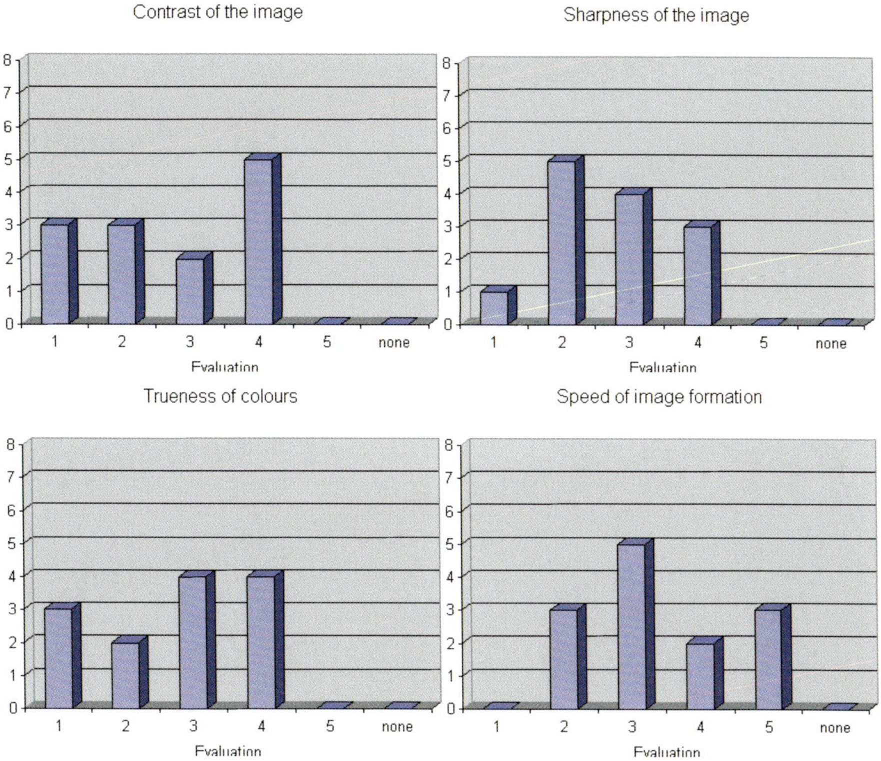

Figure 10.21 Image quality, measured by sharpness, contrast and trueness of colors were subject to less positive opinions and the given marks are located mainly in the middle of the scale, ranging from "very good" to "bad". The last topic – speed – yielded even "very bad" results and showed to be, together with the dissatisfactory image quality, the most annoying subject during remote assessment of the specimens.

problems with this mode of assessment. Narrowing down the reasons, loss of image quality compared to light microscopy plays a central role as can be seen in **Figure 10.21**. Six system crashes in 104 sessions might have led to dissatisfaction with system stability. Finally, the overall tardiness of the remote microscope and slowness of transmission of images was regarded as one of the most annoying problems. It is most important to notice that slowness does not only mean boredom in front of the screen but has an influence on the assessment of the specimen. The pathologist is used to a quick overview, appreciation of problematic areas in low magnification and rapid changing to higher magnifications. Coarse focusing is done in an instant and during examination fine focus is used to perceive fine structures (e.g., chromatin) in the 3D tissue sample. All these requirements are constrained by interposition of camera, transmission cable, robotic stage, motorized objective revolver and z-positioning for focusing.

In summary, it can be stated that slowness and poor image quality in particular led to loss of the plot and thereby to loss of confidence in the pathologist's own diagnosis.

10.3.3.3 Overcoming the Limitations of Telepathology

Meanwhile, the history of telepathology is long. The first proof-of-concept demonstration was in 1968, initially called "television microscopy", following the idea of teleradiology, which had its proof-of-concept nine years earlier [34,44]. At that time only static images were transmitted. It was not possible to operate the microscope over a distance. A remarkable example is cited by Cross [34]:

> In 1973 static black and white images of peripheral blood and bone marrow smears were transmitted by satellite from a ship docked in Brazil to Washington. The patient was in acute respiratory distress and the remote diagnosis of mediastinal lymphosarcoma (histopathological classification has also changed in the past 30 years!) enabled immediate treatment to be given.

During the following decades, microscopes became robotic instruments that included the facility of remote control. This technology evolved from a direct cable-bound connection to a comfortable and almost place-independent connection via the World Wide Web. The common principle of all systems was the generation of a live view of a particular slide, so called "dynamic telepathology". Sophisticated autofocus algorithms enabled these devices to become more and more integrated into routine work, but lack of speed and deficiencies in image quality prevented the technology from being widely accepted.

With the expansion of the World Wide Web and the possibility of fast and easy sending and receiving of image data, specialized consultation centers arose, such as the UICC-TPCC (Union international contre le cancer Telepathology Consultation Center; http://pathoweb.charite.de/UICC-TPCC/default.asp) or the AFIP Department of Telemedicine (Armed Forces Institute of Pathology; http://www.afip.org/Departments/telepathology/ or https://www3.afip.org/). Unfortunately, in contrast to dynamic telepathology, as described above, there is only limited slide access. The images for teleconsultation are preselected and sampling errors are likely to occur when deploying so-called "static telepathology" [39]. However, a much higher standard of image quality can be achieved by the use of a high-resolution digital camera for still images instead of a live CCD-video-cam.

To overcome the deficiencies of conventional telepathology with a remote controlled robotic online microscope and those of static image takes, we decided to combine the advantages of both technologies. So called "virtual slides" promise to be the method of choice. This new technology has attracted

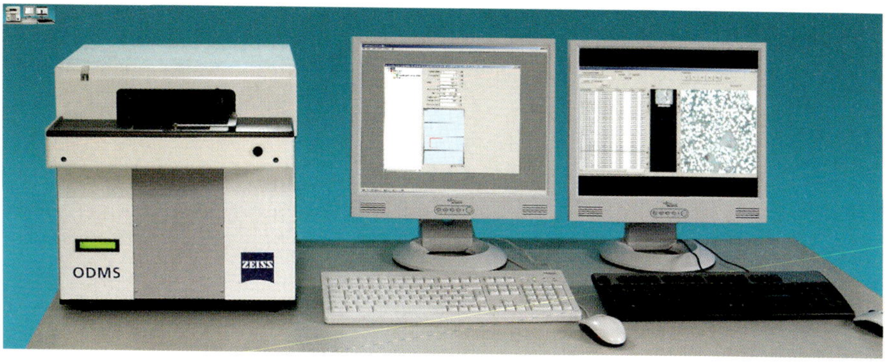

Figure 10.22 The ODMS as configured at the LMU, Munich, comprising two work stations. The left screen shows the acquisition software for control of the ODMS and the right screen displays the so-called "Pathologist's Workbench", a newly developed GUI viewer for convenient assessment of the images. This user interface is divided into three principal units. There is the database of cases to the left, a column for overviews of the appropriate slides in the middle (here only one) and the assessment window to the right.

widespread interest, e.g., evidenced at the 1st International Congress on Virtual Microscopy which was held together with the 7th European Telepathology Congress in Poland, 2004. Basically a "virtual slide" means a large data file composed of thousands of single images stitched together as a mosaic. To introduce this kind of technology makes high demands on the underlying infrastructure. Fast data throughput and the possibility of handling an enormous amount of data has to be provided. There has been such an increase in storage capacities and in the speed of personal computers at affordable costs that "virtual slide technology" has become possible. Paralleling the achievements in teleradiology and storage of digital radiological images this process will soon enter an exponential growth phase in the field of pathology. However, the demands on technical equipment for the complex field of pathology are much higher: In contrast to radiological needs, pathology is done with stained slides. Color images are more complex than black-and-white images. Different magnifications have to be available. A histological specimen is not a 2D image but a 3D object, with its depth of a maximum of 4–5 μm playing an important part in arriving at a reliable diagnosis. Furthermore, the whole system and its environment must not be inferior to that of a conventional microscope with regard to ease of handling and comfort. The viewing device, i.e., the monitor of the system, must satisfy demands that far exceed those of radiology.

With these prerequisites in mind, ZEISS, Germany, has taken part in the development of a new image capturing device for the generation of virtual slides, ODMS. Since it was co-developed with Schering (Berlin) and the EMBL at Heidelberg, the ODMS features criteria that meet the needs of cell biology such as high-throughput live-cell imaging. For the pathology application a special viewer was developed to archive the image folders for each patient, to get an overview of the folder content and to provide an extra window for assessment (**Figure 10.22**). This software is called the "Pathologist's Workbench" and can be integrated in nearly all PACS environments or data-storage systems. In low magnification a rapid overview of a specimen is ensured. Changing the magnification can be done either in defined steps, imitating changing objectives of a conventional microscope, or fluently in terms of a zoom function. When using the zoom function the observer is always informed about the level of magnification so that the feeling of size will never get lost when a slide is viewed.

What is also innovative about the ODMS and the "Pathologist's Workbench" is the possibility of having 3D images. There is nothing more annoying than having only one plane for assessment, since it is essential for the pathologist to assess specimens in depth. Furthermore, slight deficiencies in sharpness can be modulated by hand, and regional defocus issues can easily be adjusted.

10.3.3.4 **Virtual Slides – A Tool not only for Teleconsultation but for Multipurpose Use**

Taking the first steps into the field of "virtual microscopy" is challenging but the advantages of pathology based on digital image assessment are obvious.

In contrast to glass slides, virtual slides cannot break; they cannot deteriorate in color quality and can be stored for an infinite period of time. The slides can be stored and redistributed easily. In respect of electronic medical files, virtual slides add to improvements in health care and to cost saving. Thinking about rare diagnoses, virtual slides are a powerful tool in continuing medical education. In medical training, "virtual microscopy" will provide each student with the same optimal specimen, and serial cuts of real tissue specimens are no longer needed.

With regard to legal aspects and quality management, virtual slides can be integrated into the pathologist's report for documentation. Furthermore, electronic tracking of the assessed fields of view ensures that all regions on the slide that contain tissue are viewed by the pathologist. Annotations can be made easily and stored with the slide.

The future pathologist probably will no longer sit uncomfortably at a microscope but lie back in an armchair in front of a wide-view screen changing magnifications at will or simply zooming into or out of the specimen while a variable scale tells him or her the exact degree of magnification.

One day telepathology will no longer be a technique that requires a special work station. The possibility of teleconsultation is built into the assessment software and pathologists can call up their colleagues – near or far – for a second opinion without leaving their seats. Since virtual slides are always provided by a local server and everyone who has access to it can view the appropriate image data, the borders between telepathology and locally based assessment will fade.

10.3.3.5 Telepathology – What the Future will Bring

Since work flow in the health care system is becoming more and more integrated into a digital environment, today the field of pathology is confronted with the challenges and opportunities of technical evolution. The virtual slide technique offers totally new starting points for standardization of laboratory workflow, distribution and assessment of specimens. There will be a direct influence on quality assurance, as the whole path of a specimen can be tracked from the time it enters an institute of pathology to the time the final report leaves the assessment center. Even the way the pathologist looked at the virtual slide is traceable and will be documented, together with the annotations for each slide. It can be traced back whether there are areas in a slide that were ignored and which parts were viewed at higher or at lower magnification levels.

Recent innovations have shown that object-based image analysis systems are becoming more and more capable of handling complex structures in biological research as, for example, demonstrated by Biberthaler et al. with transmission electron micrographs [33]. There are ongoing studies at our own Institute that deal with image analysis in histological slides to demonstrate that even more complex image content with 32-bit color depth, in contrast with 256-step gray scale images from transmission electron micrographs, can be interpreted automatically with valid outcome.

In summary virtual slides will open the way towards routine use of image analysis in diagnostic pathology. This may lead to a change of timing, along with the infrastructural and institutional conditions of pathologists' work.

10.4 Summary and Future Trends

As an intelligent part of an application process, the ODMS enables scientists to save and protect life through an earlier, faster and more secure diagnosis, and by enabling the development of more specific agents with smaller adverse effects for the treatment of diseases such as cancer and Alzheimer's disease.

There are some trends influencing further developments in the fields of application and technology. Health care applicable experiments with live cells with quantitative results will become routine. Therefore, and because of growing cost pressures, standardization of the processes has to be further developed. The further impact of digital technology on our world (e.g., World

Wide Web, automation, signal detection and processing) catalyzes the paradigm shift in health care applications. Scientists will be relieved of time-consuming work, which can be delegated to an intelligent system solution for an application process. The profile of users of personal computer applications has changed from a technician with experience in programming to a user of standardized applications with a few options. Analogously, the user profile in the health care applications will also change. Digital technology supports the scientist by providing more space for creative work and by saving costs in a rapidly growing health care system.

Acknowledgments

Carl Zeiss Göttingen wishes to thank the whole ODMS project team for the fruitful cooperation. Due to many constructive discussions with our application partners, CZG were able to build a forward-looking digital microscope system. The ODMS project was triggered by suggestions made by Dr. Ernst H.K. Stelzer in the early 1990s. All teams wish to thank Carl Zeiss Jena and Göttingen for providing the microscope apparatuses and many fruitful discussions. At EMBL we wish to thank Georg Ritter and Wolfgang Dilling for their contributions to the electronics and mechanics, respectively, and Stefan Grill for some early decisions. The Schering team wishes to greatly acknowledge the work of Alexander Dimerski and Sebastian Schaefer, whose practical contributions were instrumental in achieving the progress made in this project. We would also like to thank our colleagues and mentors for the fruitful discussions that furthered the advancement of HCA at our facilities.

The collaboration with SLS Software-Technologies GmbH, i.e., with Nicholas J. Salmon, David Richter and Fabian Haerle, is particularly appreciated by all members of all teams. Finally, we also thank the BMBF/VDI-TZ for financial support. To run a risky project like this, concerned with application and technology, without it would have been unrealistic.

Glossary

Apoptosis The process of suicide of a cell. This occurs naturally during development but can also be induced by dramatic defects detected by a cell's internal controls.

Cell The cell is regarded as the basic structure of all animals and plants. The cell is essentially a complex machine that performs specific functions in collaboration with other cells, or imports certain material while exporting other materials. Cells are subdivided into cell types.

Clinical study Among the last steps in drug development. The drug is used to treat patients in a carefully monitored manner.

Drug A pharmaceutical with a specific effect. Usually employed to improve the pathological status of an individual.

Ergonomics The adaptation of an instrument or a device to the requirements of its human operator. The adoption of digital technology tends to reduce the requirement for a human to directly touch and interact with an instrument, and reduces the ergonomic requirements.

Fluorescence Some molecules are able to store the energy of a photon for a short period of time. Due to internal effects a lower amount of energy remains available to emit the photon again. This means that, for example, a blue photon is absorbed while a green photon is emitted. The distinction between excitation wavelength and emission wavelength is used to enhance the contrast in microscopy.

Fluorophore A molecule that is able to generate the effect of fluorescence. Fluorophores come in all sizes. Some such as FITC are quite small, others such as GFP are very large.

HCS (High-content screening) See Screening

HTS (High-throughput screening) See Screening

Laser cutter The laser nano-scalpel uses pulsed light to sever, i.e. damage, cellular structure. This is used to break structures in a cell of to generate an effect to which the structures in a cell have to react.

Laser line Lasers operate at various wavelengths. In contrast to most light sources the laser has a very narrow spectrum. A laser line will be less than a nanometer wide while halogen lamps have spectral widths in the range of hundreds of nanometers and LEDs of tens of nanometers. Laser lines cover the whole range from the ultraviolet to the infrared.

Microscope A microscope is an optical instrument that is used to collect a magnified view of a microscopic object. Typical magnifications for cells are in the range of 300 to 1200 times. The typical object size is in the range of 200 μm down to 50 μm. In modern microscopes cameras are used to generate a digital image. Of particular importance is the use of fluorophores and fluorescence effects. Modern microscopes come with many different light sources and detectors.

NA (Numerical aperture) Optical instruments are limited in their resolution capability by the collection angle of the optical system. The numerical aperture very much like the f-number of a camera lens describes this property. The resolution is a physical and not a technical limit.

NIR (Near-infra red) Usually the range above 780 nm up to about 1200 nm. This range is particularly interesting for optical tweezers.

Objective lens Usually the lens closest to an object in a microscope. Modern microscopes require an objective lens, a tube lens and a camera. In fluorescence microscopy one would add a filter between the two lenses. While the microscope objective lens is usually quite complicated and expensive the tube lens is relatively simple.

Organelle Sub-cellular features are often describes as organelles, i.e. little organs. Understanding of the existence or the importance of organelles has changed during the past few years. Typical organelles are the nucleus, the Golgi apparatus, the endoplasmic reticulum or the nucleolus.

Pathology The process of characterizing cells that are involved in a disease or in the process of recovering from an accident, surgery or a disease. Classical pathology relies on morphological distinctions and is supported by labelling techniques that affect the ratio of DNA to protein. More modern approaches take advantage of antibody labelling and fluorescence.

Screening The process of searching for a specific property. A typical search would be for drugs that kill cells. In general, pharmaceutical companies search for a drug that compensates deficiencies in a cell or a tissue. A current distinction is applied between high-throughput screening, i.e. checking thousands or hundreds of thousands of drugs looking for dramatic effects, and high-content screening, in which various subtle effects are evaluated.

UV (Ultra violet) Usually the invisible range below 380 nm. In microscopy only the range between 400 nm and 350 nm can be used easily. Deeper UV requires special optical systems.

Key References

J. HUISKEN, J. SWOGER, F. DEL BENE, J. WITTBRODT, E.H.K. STELZER, *Science*, 305 (**2004**), pp. 1007–1009

J. WHITE, E.H.K. STELZER, *Trends Cell. Biol.*, 9 (**1999**), pp. 61–65

S.W. GRILL, P. GÖNCZY, E.H.K. STELZER, A.A. HYMAN, *Nature*, 409 (**2001**), pp. 630–633

J. COLOMBELLI, E.G. REYNAUD, J. RIETDORF, R. PEPPERKOK, E.H.K. STELZER, *Traffic*, 6(12) (**2005**), pp. 1093–102

T KUAKPAETOON, G. STRAUCH, K.D. KUNZE, M. OBERHOLZER, K. ATISOOK, CH. VATHANA, B. SAMOUNTRY, *7th European congress on telepathology and 1st international congress on virtual microscopy*, Lecture, Poznan, Poland, 2004

R.S. WEINSTEIN, *Hum. Pathol.*, 36 (**2005**), pp. 317–319

References

1 S. INOUÉ, K.R. SPRING, *Video Microscopy, The fundamentals*, Plenum Press, New York, 1986

2 B. HERMAN, H. J. TANKE, B. HERMAN, *Fluorescence Microscopy, 2nd Edition*, Springer, (1998)

3 F. SCHUBERT, *Light-Emitting Diodes*, Cambridge University Press, New York, (2003)

4 J. INGLIS, K. JANSSEN, *Live Cell Imaging – A laboratory manual*, Cold Spring Harbour Laboratory Press, Cold Spring Habour, New York, (2005),

5 J.E.N. JONKMAN, J. SWOGER, H. KRESS, A. ROHRBACH, E.H.K. STELZER, , (**2003**), 360 pp. 416–446

6 A. ROHRBACH, E.H.K. STELZER, *J. Opt. Soc. Am. A*, 18(4) (**2001**), pp. 839–853

7 J. COLOMBELLI, S.W. GRILL, E.H.K. STELZER, *Rev. Sci. Instr.*, 75(2) (**2004**), pp. 472–478

8 J. COLOMBELLI, E.G. REYNAUD, J. RIETDORF, R. PEPPERKOK, E.H.K. STELZER, *Traffic*, 6(12) (**2005**), pp. 1093–102

9 J. WHITE, E.H.K. STELZER, *Trends Cell. Biol.*, 9 (**1999**), pp. 61–65

10 J. HUISKEN, J. SWOGER, F. DEL BENE, J. WITTBRODT, E.H.K. STELZER, *Science*, 305 (**2004**), pp. 1007–1009

11 M. W. BERNS ET AL., *Science*, 213(4507) (**1981**), pp. 505–513

12 K.O. GREULICH, G. PILARCZYK, A. HOFFMANN, G.M.Z. HÖRSTE, B. SCHÄFER, V. UHL, S. MONAJEMBASHI, *J. Microscopy*, 198 (**2000**), p. 182

13 V. VENUGOPALAN, A. GUERRA, K. NAHEN, A. VOGEL, *Physical Review Letters*, 88(7) (**2002**), art. no. 078103

14 R. SRINIVASAN, *Science*, 234 (**1986**), pp. 559–565

15 M. NIEMZ, *Laser–Tissue Interactions. Fundamentals and Applications*, 2nd Edition, Springer – Biological & Medical Physics Series, Springer, Berlin, 2002

16 V. VENUGOPALAN, N.S. NISHIOKA, B.B. MIKIC, *Biophysical Journal*, 69 (**1995**), pp. 1259–1271

17 V. VENUGOPALAN, *SPIE*, 2391 (**1995**), pp. 184–189

18 B.M. KIM, M.D. FEIT, A. M. RUBENCHIK, E.J. JOSLIN, P.M. CELLIERS, J. EICHLER, L.B. DA SILVA, *J. Biomedical Optics*, 6(3) (**2001**), pp. 332–338

19 J.A. IZATT, D. ALBAGLI, M. BRITTON, J.M. JUBAS, I. ITZKAN, M.S. FELD, *Lasers in Surgery and Medicine*, 11 (**1991**), pp. 238–249

20 D. ALBAGLI, M. DARK, L. PERELMAN, C. VON ROSENBERG, I. ITZKAN, M.S. FELD, *Optics Letters*, 19(21) (**1994**), pp. 1684–1686

21 J. NOACK, A. VOGEL, *IEEE Journal of Quantum Electronics*, 35(8) (**1999**)

22 A. VOGEL, J. NOACK, K. NAHEN, D. THEISEN, S. BUSCH, U. PARLITZ, D.X.

HAMMER, G.D. NOOJIN, B.A. ROCKWELL, R. BIRNGRUBER, *Applied Physics B*, 68 (**1999**), pp. 271–280

23 E. WILLINGHAM, *The Scientist*, 16(10) (**2002**), p. 42

24 J. COLOMBELLI, E.G. REYNAUD, E.H.K. STELZER, *Medical Laser Application*, 20 (**2005**), pp. 217–222

25 N.J. SALMON, J.E.N. JONKMAN, E.H.K. STELZER, *Proceedings of the 12th IEEE International Congress on Real Time for Nuclear and Plasma Sciences*, Valencia, Spain, 2001

26 S.W. GRILL, P. GÖNCZY, E.H.K. STELZER, A.A. HYMAN, *Nature*, 409 (**2001**), pp. 630–633

27 S.W. GRILL, J. HOWARD, E. SCHÄFFER, E.H.K. STELZER, A.A. HYMAN, *Science*, 301 (**2003**), pp. 518–521

28 S.W. GRILL, E.H.K. STELZER, *JOSA A*, 16(11) (**1999**), pp. 2658–2665

29 D.J. SHARP, G.C. ROGERS, J. M. SCHOLEY, *Nature*, 407 (**2000**), p. 41

30 K. OEGEMA, A. DESAI, S. RYBINA, M. KIRKHAM, A.A. HYMAN, *Journal of Cell Biology*, 153 (**2001**), p. 1209

31 MING LI, KIYOTAKA MORI, MAKOTO ISHIZUKA, XINBING LIU, YOSHIMASA SUGIMOTO, NAOKI IKEDA, KIYOSHI ASAKAWA, *Applied Physics Letter*, 83(2) (**2002**), pp. 216–218

32 GUANGHUA CHENG, YISHAN WANG, J. D. WHITE, QING LIU, WEI ZHAO, GUOFU CHEN, *J. Appl. Phys.*, 94(3) (**2003**), pp. 1304–1307

33 P. BIBERTHALER, M. ATHELOGOU, S. LANGER, B. LUCHTING, R. LEIDERER, K. MESSMER, *Eur. J. Med. Res.*, 8 (**2003**), pp. 275–282

34 S.S. CROSS, T. DENNIS, R.D. START, *Histopathology*, 41 (**2002**), pp. 91–109

35 B.E. DUNN, H.Y. CHOI, U.A. ALMAGRO, D.L. RECLA, E.A. KRUPINSKI, R.S. WEINSTEIN, *Telemed. J.*, 5 (**1999**), pp. 323–337

36 T. KUAKPAETOON, G. STRAUCH, K. D. KUNZE, M. OBERHOLZER, K. ATISOOK, CH. VATHANA, B. SAMOUNTRY, 7th *European congress on telepathology and 1st international congress on virtual microscopy*, Lecture, Poznan, Poland, 2004

37 F.J.W.-M. LEONG, A.K. GRAHAM, P. SCHWARZMANN, J.O.'D MCGEE, *Telemed. J. E. Health*, 6 (**2000**), pp. 373–377

38 F.J.W.-M. LEONG, A.G. NICHOLSON, J.O.'D. MCGEE, *J. Pathol.*, 197 (**2002**), pp. 211–217

39 B. MOLNAR, L. BERCZI, C. DICZHAZY, A. TAGSCHERER, S.V. VARGA, B. SZENDE, Z. TULASSAY, *J. Clin. Patho.l*, 56 (**2003**), pp. 433–438

40 M.B. MORGAN, M. TANNENBAUM, B.R. SMOLLER, *Arch. Dermatol.*, 139 (**2003**), pp. 637–40

41 J. SZYMAS, G. WOLF, W. PAPIERZ, B. JAROSZ, R.S. WEINSTEIN, *Hum. Pathol.*, 32 (**2001**), pp. 1304–1308

42 R.S. WEINSTEIN, *Hum. Pathol.*, 17 (**1986**), pp. 433–434

43 L.J. WEINSTEIN, J.I. EPSTEIN, D. EDLOW, W.H. WESTRA, *Hum. Pathol.*, 28 (**1997**), pp. 30–35

44 R.S. WEINSTEIN, *Hum. Pathol.*, 36 (**2005**), pp. 317–319

11 Outlook: Further Perspectives of Biophotonics

Susanne Liedtke, Michael Schmitt, Marion Strehle, Jürgen Popp[1]

Since the 1960s, photonics has evolved as one of the most intriguing technologies worldwide. Thus, the twenty-first century is often called the century of light. Photonics has become an enabler and catalyst, and, therefore, the invention of photonic technologies has had deep consequences and triggered a revolution in many fields of science and technology. Prominent examples are microscopy and imaging, spectroscopy, astronomy, nonlinear optics and information processing. Modern photonics has a strong impact on all fields of natural science, e.g., chemistry, material science, biology, environmental science, pharmacy and medicine. However, most of the achievements of photonics are based on the rapid technical development of micro- and nano-technology.

Biophotonics, as part of photonics, has proved to be an important enabling technology for accelerated progress in medicine and biotechnology. Light can be used as a universal tool for the investigation and manipulation of biological samples. This book provides some insights into the work and the outcome of nine network projects funded and organized within the Biophotonics research framework supported by the German Federal Ministry of Education and Research (BMBF). This volume certainly does not summarize all Biophotonics activities either in Germany or in Europe, not to mention America, Asia or Africa, but it attempts to present an overview of the huge possibilities of Biophotonics.

Now the questions arise: where is the Biophotonics train going? Where are the further perspectives of Biophotonics? In order to answer these quite complicated questions, we first have to differentiate between the various fields of application, such as medicine, biotechnology, the pharmaceutical and food industries and environmental protection. Currently, the most important topic of Biophotonics is its medical applications. In this Outlook we outline the scientific progress of light-based technologies in the life sciences. Its not an exaggeration to say, as indicated before, that Biophotonics is on the way to solving the most important problems in biology and medicine. It can do so because it arose at the interface of the three most innovative academic disciplines of

1) Corresponding author

Biophotonics: Visions for Better Health Care. Jürgen Popp and Marion Strehle (Eds.)

ISBN: 3-527-40622-0

the twentieth century, i.e., photonics, biotechnology and nanotechnology. The combination of the three does not simply triple the scientific and technological output, it increases them exponentially. But, and this is at least as important as technological improvement if not of even of more significance, Biophotonics also provides economic progress. However, the new discipline therefore has to bridge another gap, that between academia and industry. As Lothar Lilge, director of the Biophotonics Facility at Photonics Research Ontario (Canada), suggested in 2001, the breakthrough of Biophotonics applications is limited not only by the shortage of scientists and technicians with multidisciplinary education but also, and more fundamentally, by the lack of awareness among the CEOs of start-up companies. This appraisal is shared by the German Federal Ministry for Education and Research, and this is the reason for its effort to bring engineers and academic scientists together. This effort is motivated by the desire to encourage the former to listen keenly to biologist or physicians formulating their daily needs in the laboratory and the latter to look into the "lowlands" of technical details. Looming even larger is the need to challenge investors to spend money even on unusual, and maybe eccentric, ideas. To overcome the shortage of well educated scientific personnel in Germany as well as in other countries, new training programs and study courses must be set up.

11.1
Future Research Topics

The first topic for future research activities concerns fundamental research in the field of light–matter interaction. To unravel the secrets of life, a comprehensive understanding of molecular structure and of the inherent dynamic and kinetic processes taking place within biological matter is required. Therefore a successful symbiosis of photonics and life sciences is one of the main research topics of Biophotonics; although not the only one. Biophotonics will have a major impact on health care since it offers the great potential for the early detection of diseases, which is especially important these days, because an increasingly aging world population leads to unique health care problems. Thus, fundamental insights in the light–biological matter interaction are essential for Biophotonics. One distant goal is a deeper understanding of the molecular processes occurring inside living cells. Furthermore, chemical reactions and functions in cells driven by metabolites or bioactive compounds, as well as cell–cell communication processes, need to be studied. Building on such knowledge, the origin of diseases can be resolved, therapies can be optimized and the occurrence of diseases might be prevented or at least minimized. Biophotonics will open pathways to momentous progress in the life sciences such as biotechnology, medicine, environmental science, etc. The derivation of structure–dynamics relationships is one of the most challenging topics within life science since a detailed knowledge of light-driven processes in nature is indispensable for the rational design of new and innovative photoactive matter. Utilizing advanced nanostructuring technologies, artificial bioinspired materials with new promising properties can be developed.

To achieve all these ambitious goals, innovative optical components and sophisticated frequency-, time- and spatially resolved innovative laser spectroscopic methods need to be developed, along with systems ranging from the XUV to the THz with unparalleled functionalities. Therefore, future research activities have to focus on the development of innovative optical spectroscopic techniques such as, for example:

- frequency-resolved spectroscopic techniques exploiting novel spectral ranges to access innovative matter structures
- time-resolved methods ranging from nanosecond to sub-femtosecond time resolution to study the entire range of time-dependent structural changes
- frequency- and time-resolved reaction control schemes utilizing specially tailored light pulses
- highly specific and long-lasting labeling techniques that allow *in vivo* activity measurements while leaving the biological processes within the living sample widely uneffected
- optical detection systems representing substantial improvements of state-of-the-art techniques providing parallel and quantitative analyses of dynamic processes in living cells. Therefore, the emphasis must be on the perfecting of optical components such as light sources and beam lines
- innovative optical methods for the preparation, control and manipulation of intracellular activities as well as for cell sorting and cell positioning.

With an armamentarium like this, we shall be able to close the gaps in our understanding of dynamic molecular processes in cells and tissues and to provide a basis for the early detection of and tailor-made therapies for cancer and other diseases.

The second research topic is mainly concerned with the technical development of innovative photonic instrumentation. Much research to date has been performed on single cells between two glass plates and under a microscope. Thus, biological objects have more or less been treated as 2D objects.

When dealing with tissue the situation is quite similar. Before investigation, the sample has very often been cut in micrometer-sized slices and afterward these slices have been stained. This means we are looking at a more or less 2D sample, which is already dead. Time-dependent and 3D spatially resolved information cannot be obtained from those samples. Now one might respond that with the new laser-scanning microscopes it is possible to investigate living cells with video repetition rate. Although this is true, one still has to stain the cells. Without staining it is quite hard to get decent contrast or even molecular information. If we use a microscope with high magnification our field of view is quite limited, but if we study a large area as a whole, without scanning, we lose the view of the details. No matter which way we turn, we very often face the same limits. One major limitation in microscopy, for example, is the diffraction limit postulated by Abbe. We could now start a discussion about the advantages and disadvantages of various optical methods, but the thrilling goal of "a fundamental understanding of life" can hardly be achieved by the optical techniques commercially available today. Since an understanding of life processes is seen as essential in order to protect people from diseases and to come up with new more effective therapies, we need innovative optical techniques. New spectroscopic methods and correspondingly innovative photonic instrumentation need to be optimized and applied, based on an increased understanding of biological processes as well as light–matter interactions (see also the first research area, above). Examples of such developments include

- functional and molecular imaging – CARS microscopy
- high-efficiency spectral imaging for diagnostics
- applied plasmonics for ultrasensitive diagnostics
- lensless microscopes
- THz spectroscopy and technique.

The potential of optical fibers for innovative diagnostic methods must also be utilized. Optical fibers can be used for guiding light efficiently to and from a place of measurement even under difficult conditions, owing to their small size and their flexibility. The concepts of microstructured and photonic crystal fibers lead to the development of fibers with extended and new specific properties relating to wavelength range, numerical aperture, bending properties, dispersion properties, etc. Therefore, it is necessary to develop and apply specifically designed fibers based on the concept of a microstructured photonic crystal for microscopic and diagnostic measurement methods. These would include fibers for delivering specific illumination conditions (e.g., pulsed light) as well as efficient light collection for measurement purposes (especially in the case of weak signals).

A third research topic will combine the achievements of Biophotonics with those from micro- and nanotechnology. The main goal of this topic is the development of new industrial products. In particular, existing achievements in nanobiotechnology will be used to develop a new generation of optical accessible microchips. One major topic in this area will be the development of a powerful "Point-of-Care" diagnostic based on biochips. Compared to conventional laboratory diagnostics, biochips have many advantages:

- considerable reduction of sample preparation
- miniaturization of the sample volume and the time required for the diagnostic
- extreme parallelization of the diagnostics (high-throughput diagnostics)
- Point-of-Care diagnostic
- standardized fabrication in mass production
- high potential for automation.

All three topics are of great importance for future-oriented Biophotonics research. Biophotonics, in combination with the modern micro- and nanotechnology, will be a key to the resolution of many scientific problems and will therefore be of particular importance for society.

11.2 Promising Innovative Microscopy Techniques

Let's start with a scientific "daydream". What are the features a genuinely innovative microscope should exhibit? We are quite sure that you will agree with the following attributes: The microscope should provide 3D images with high axial and lateral resolution (if required with nanometer resolution); label-free *in vivo* imaging should be contactless (without distortion), "full-field" (no scanning), and "online" (video repetition rate); it should be capable of handling samples ranging in size from micrometers to centimeters; and last but not least it should provide molecular information with high contrast. Every scientist dreams of possessing such a microscope. However, we believe that in this respect the future has already started.

One example has already been addressed in this book. Have another look at the MIKROSO chapter (Chapter 6). Here, a fairly new microscopic technique, Digital Holographic Microscopy, has been put into practice. But other promising techniques are also under development.

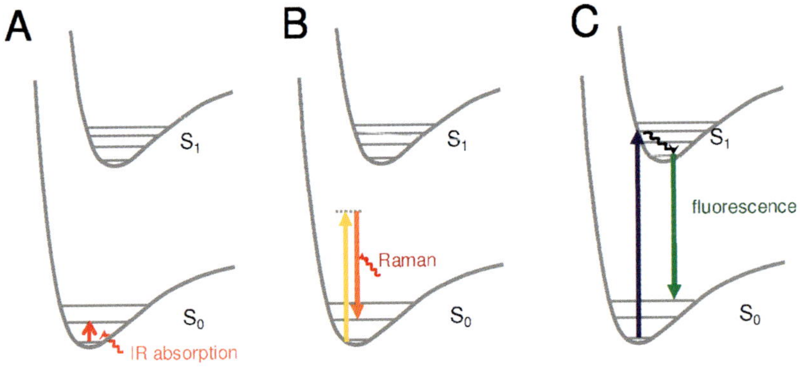

Figure 11.1 Physical processes commonly applied for contrast creation in optical microscopy.

Conventional light microscopes principally use differences in the absorption, transmission and reflection of white light in biological tissues to create the necessary contrast. However, in addition to these conventional microscopy techniques, several other light–matter interaction phenomena (see, for example, Chapter 1) are gaining more and more importance for generating microscope images. Fluorescence, IR absorption or Raman spectroscopy provide detailed molecular information about biological systems. Basic information on these techniques was outlined in Chapter 1. **Figure 11.1** compares the above mentioned three light–matter interaction phenomena.

The emission of light following electronic absorption is called fluorescence. Fluorescence light is red-shifted compared to the electronic absorption excitation wavelength. The detection of fluorescence instead of the weak absorption changes of white light in optical microscopy offers several advantages. Fluorescence detection is background free and extremely sensitive, so that even single molecules can be investigated. Since many biological samples exhibit no auto-fluorescence when excited with visible light and direct UV excitation very often leads to sample degradation, the technique of fluorescence labelling has been established. Samples are labelled with specially developed dyes and afterwards the position of the flourescent dyes within the sample is detected. The introduction of the confocal measurement procedure into fluorescence microscopy allows us to record 3D microscopic fluorescence images. In this confocal technique the sample is successively scanned and by applying a confocal pinhole located within the fluorescence detection optical path the signal of only a small volume segment is recorded. By continuously optimizing these optical systems, processes taking place within biological cells can be monitored nowadays, together with spatial information. Recent innovative concepts within confocal fluorescence microscopy allow real time investigation of living samples. These innovative technologies permit scientists to get a deeper insight into cellular processes by a unique combination of scanning speed, image quality and sensitivity. Fluorescence microscopy has developed as a method capable of resolving a multitude of questions in life sciences and medicine. Confocal microscopes can be found in almost every biology laboratory. Scientists from the European Molecular Biology Laboratory recently developed a particularly smart fluorescence microscopy technology to perform 3D live-cell analysis. These scientists, under the leadership of Ernst Stelzer, developed a Selective Plane Illumination Microscopy (SPIM) capable of displaying living systems up to a size of a few millimeters in a 3D fashion with an extremely high spatial resolution [1]. In doing so, the living sample is embedded into a fixation material which does not influence the viability of the cells, and illuminated by a so-called light plane (see **Figure 11.2A** and **B**). Therefore, fluorescence light is only excited within this small illumination plane. Simple fluorescence light-collecting optics image the illuminated area onto a camera. To suppress stray light caused by the illumination laser, a fluorescence emission filter is introduced within the detection optical path. By rotating the sample, information from totally different perspectives can be monitored and later processed to create a 3D image. **Figure 11.2C–G** shows various illustrative SPIM images (for further details see figure caption). This innovative SPIM technology is the only method capable of recording such images.

Recent progress in the development of high-intensity ultrashort laser sources has revolutionized microscopy by using nonlinear optical phenomena to create higher microscopic contrast. These ultrashort lasers generate a moderate amount of energy within an extremely short time-scale (pico or femtoseconds), and the ultrashort pulses can be focused down to generate extremely high light intensities. Such high intensities lead to certain nonlinear light–matter interactions at which the observed response, for example, a fluorescence signal, no longer scales linearly with the initially applied light intensity. The nonlinear matter response allows us to observe special processes of light multiphoton absorption. Multiphoton absorption leads to a simultaneous absorption of $m \geq 2$ photons by an atom or molecule (see **Figure 11.3A**). Since m photons need to be present simultaneously, this process depends on the mth power of the incident radiation. Such a nonlinear dependence can be utilized for applications in which a reaction should be initialized out of only a small spatial area. Thus multiphoton absorption is especially suited for microscopic applications. By the application of a femtosecond titanium sapphire laser providing excitation wavelengths in the area of 800–960 nm fluorophores absorbing at 400–480 nm (absorption of two photons) or 270–320 nm (absorption of three photons) can be excited and their red-shifted fluorescence can then be detected. The simultaneous absorption of two or three photons results in a very good localization of the fluorescence light, since such a nonlinear

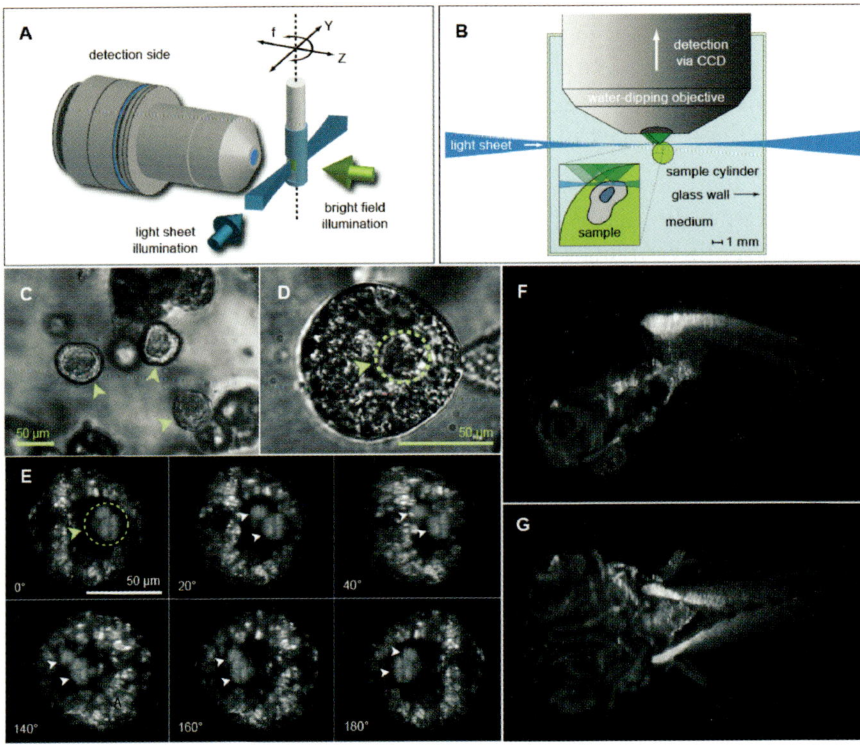

Figure 11.2 Single Plane Illumination Microscopy (SPIM) for modern 3D life sciences. The panels at the top illustrate the basic principles of SPIM while specific imaging examples are presented at the bottom.
(A), (B): Three-dimensional sample preparation and the optical arrangement for the study of dynamic processes in modern life sciences with SPIM. (A): The illuminating light sheet enters from the side, while fluorescence light is detected perpendicular to the illumination plane. (B): Drawings of the SPIM imaging chamber. The detection objective lens is immersed in the medium-filled chamber. The sample cylinder is placed in front of the objective. All drawings are to scale.
(C)–(E): Madin–Darby Canine Kidney (MDCK) cells form hollow clusters when grown in Matrigel. (C): Cysts were obtained by culturing MDCK cells in Matrigel for 7–10 days. The average diameter of mature cysts is 50–60 µm. The image was recorded in brightfield illumination using a Carl Zeiss W 10x/0.3. (D): Single MDCK cyst imaged with SPIM in brightfield at a higher magnification (Carl Zeiss W 40x/0.8). A small cluster of cells (dotted circle) is visible inside the hollow cavity of the cyst. (E): Maximum projection of 29 slices with 1 µm spacing rotated to different angles. The 3D spatial arrangement of the small cell clusters inside the lumen can be readily visualized. Labelling: Syto61 (absorbance 628 nm, emission 645 nm).
(F), (G): A Medaka embryo imaged with SPIM. (F): Lateral view, (G): dorsal-ventral view. The multiview dataset is a combination of four independently recorded datasets. The image stacks were taken with a Zeiss Fluar 5x/0.25 objective lens.

absorption process can only take place in an extremely small volume. Thus multiphoton absorption yields an inherent 3D effect, making the application of a confocal pinhole unnecessary (see **Figure 11.3B**). A 3D scan of the sample yields the spatial distribution of the fluorophore within the sample via the detected fluorescence. This information can be used to generate high-contrast 3D microscope images (see **Figure 11.3**). Other advantages of multiphoton absorption fluorescence microscopy are:

1. Since the fluorophores are not excited by a single high-energy photon but rather through a simultaneous absorption of many low-energy photons sensitive living biological samples can be investigated in a nondestructive way. Both the emission of fluorescence and possible bleaching effects, i.e., photochemical or thermal degradation of the sample, are limited to a small volume.

2. The excitation wavelength and the fluorescence wavelength are spectrally far apart, resulting in low background noise since the detection of fluorescence light is not disturbed by the broadband excitation laser.

A broad range of synthetic fluorescence labels or natural fluorophores exist such as, for example, NADH, flavones or green fluorescence proteins, and these can be used for the investigation of living biological material. Since its discovery in 1990 [2] two-photon or multiphoton absorption fluorescence spectroscopy has emerged as an indispensable biophysical method, providing valuable information about sub-cellular biochemical processes within living cells (see, e.g., [3]). Multiphoton imaging by means of NIR-femtosecond lasers is characterized by high penetration depths and therefore allows the description of biological tissue with high spatial resolution and good contrast. Multiphoton imaging is therefore particularly suitable for noninvasive tissue diagnosis [4]. The company JenLab recently released the device DermalInspect for multiphoton tomography of skin cancer. This device uses a tunable femtosecond light source to excite multiphoton auto-fluorescence of endogene biomolecules within skin, e.g., NAD(P)H, flavines, elastine, melanin, porphyrin, etc. (see **Figure 11.3C**).

The combination of multiphoton absorption with stimulated emission (another nonlinear light–matter interaction) allows an improvement of the spatial resolution down to a few nanometers. The transition of a photon from an energetically higher lying state into a quantum state of lower energy induced by a photon is called stimulated emission. This effect of stimulated emission can be utilized to further decrease the focal volume which fluorescence upon multiphoton absorption is monitored [5]. In doing so, stimulated emission is used to quench the emission of fluorescence light in such a manner that only fluorescence at the edge of the fluorescence spot is quenched. **Figure 11.4A** displays the excitation scheme applied for this stimulated fluorescence quenching, so-called "Stimulated Emission Depletion" (STED). The molecules are promoted into an electronically excited state by an excitation laser, and would normally relax spontaneously into the electronic ground state

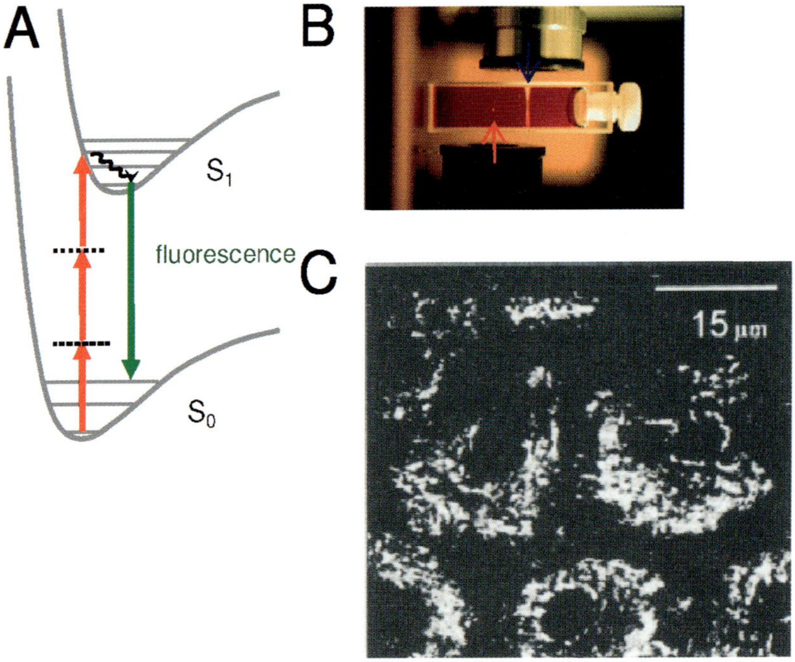

Figure 11.3 (A): Excitation scheme of multiphoton fluorescence spectroscopy. (B): Sample cell containing a fluorescent dye excited with a single photon (excitation from above) and with two photons of a high-intensity femtosecond laser (excitation from below). It is clear that the two-photon fluorescence is reduced to the focal spot of the excitation laser. (C): *In vivo* multiphoton fluorescence image of 7 cells within the *stratum spinosum*. The image was taken in 45 μm on the fore arm of a female volunteer (excitation wavelength 760 nm). The dark spots correspond to nonfluorescent nuclei while the bright spots display the NAD(P)H fluorescence, mainly localized within the mitochondria.

by emission of fluorescence light. However, the application of a high-intensity STED laser puls, red-shifted and slightly shifted in time compared to the excitation laser, leads to a transition of the electronically excited molecules into excited vibrational states of the electronic ground state before they have the change to spontaneously fluoresce. These vibrationally excited ground state molecules cannot be excited by the excitation laser anymore. If one chooses a donut beam profile for the STED laser pulse, i.e., the STED laser profile is almost dark within the focal point of the excitation laser but very intense circularly symmetric to the excitation laser, an extreme reduction of the size of the focal spot occurs, since only molecules inside both laser foci are not affected by the de-excitation process. Thus the fluorescence light originates from an extremely sharp spot only a few nanometers in size. STED microscopy reduces the investigation volume to 0.67 attoliter, being 18 times smaller than what can be obtained by conventional confocal fluorescence microscopy. Another possibility to decrease the spatial dimensions of the focal spot is 4Pi microscopy [6].

Since one can only generate a segment of a perfect spherical wave from the maximum accessible angle of aperture, a microscope objective generates a focal spot stretched along the optical axis (see **Figure 11.4B**). 4Pi microscopy therefore constructively adds the wavefronts of two microscope objectives arranged in opposition to each other in order to approximate a spherical wave exhibiting the full solid angle of 4Pi (see **Figure 11.4C**). This trick of a constructive superposition of two contrary wave fronts generates a 3–4 times smaller focal spot. A symbiosis of STED- and 4Pi-microscopy to a "STED-4Pi" microscope allows resolutions beyond the diffraction limit of light and therefore a transition from "microscopy to nanoscopy" [7]. However, the Abbe diffraction limit of light is still valid but is no longer the limiting resolution factor. In "STED-4Pi" microscopy the sample is excited electronically by a femtosecond laser pulse and subsequently de-excited by two picosecond STED laser pulses travelling in opposite directions, i.e., in a 4Pi arrangement (see **Figure 11.4D**). Such an arrangement generates focal spots of excited fluorophores with a spatial dimension of 33 nm along the optical axis. The application of such spots in microscopy allows the recording of far-field microscopy images with a resolution of a few tens of nanometers (see **Figure 11.4E**). Such a resolution provides completely new insights into living objects. A commercially available "STED-4Pi" microscope was recently released by Leica Microsystems (Leica TCS 4Pi).

Fluorescence technology, however, holds some disadvantages. The fluorescence labels (dyes) can bleach making quantitative studies extremely difficult. Furthermore, sample preparation involves the application of external labels and so is not straightforward, requiring especially skilled personnel. Therefore, the label-free technologies IR absorption and Raman spectroscopy still have a part to play. Both methods yield molecular fingerprint information on the investigated sample. Although Raman scattering is a rather weak effect compared to direct IR absorption or even fluorescence, Raman spectroscopy has proven to be a very flexible method in biophotonics. The renaissance of Raman spectroscopy was mainly triggered by the latest advances in laser technology, the design of very efficient filters to suppress the elastically scattered Rayleigh light and the development of extremely sensitive detectors. The advantages of Raman spectroscopy are its unprecedentedly high specificity and its versatility. Raman spectroscopy is a nondestructive technique and in general requires only minimal or even no sample preparation. Solid, liquid and gaseous samples can be measured, as well as transparent or opaque probes, or samples with different surface textures. Thus, Raman spectroscopy can be applied to any optically accessible sample, where a pretreatment of the sample is not necessary. However, while the specificity of Raman spectroscopy is very high its sensitivity, i.e., the conversion efficiency of the Raman effect, is rather poor. Since only a small amount of the incident photons are inelastically scattered, it takes relatively long to record a complete Raman image. Fur-

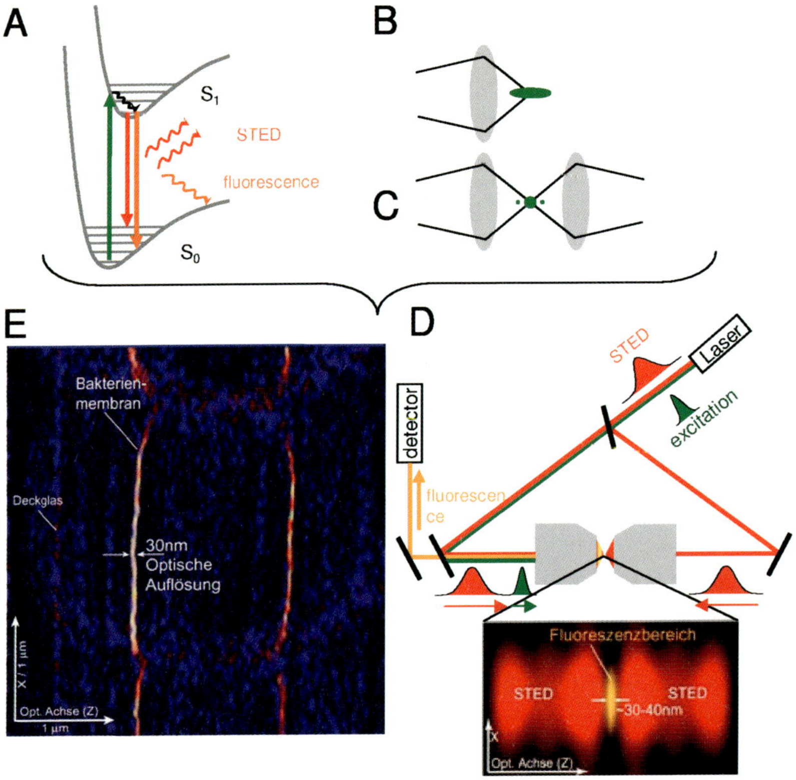

Figure 11.4 (A): Excitation scheme of "STED-spectroscopy". (B): 4Pi microscopy: The Abbe diffraction limit leads to a stretched focus along the optical axis. The size of such a focal spot can be reduced by constructive superposition of two wavefronts generated from two oppositely arranged microscope objectives. This superposition leads to a small central light spot and two side maxima. (C): Beam geometry of a "STED-4-Pi microscope". (D): "STED-4Pi" image of a bacterium (*Bacillus megaterium*) with a resolution of 30 nm.

thermore, for visible excitation wavelengths the Raman spectra are very often masked by fluorescence. Chapter 3 provided an extremely illustrative example of the application of Raman microscopy in Biophotonics. There, single microorganisms were investigated by means of Raman microscopy and identified by means of subtle analysis techniques. A modern nonlinear version of Raman spectroscopy is the coherent anti-Stokes Raman scattering technique abbreviated as CARS. In a CARS process, three laser pulses interact with the sample and generate a coherent spatially directed CARS signal. CARS spectroscopy is based on the fact that two laser pulses (pump (ω_P)- and Stokes (ω_S)-laser) drive the molecules to vibrate coherently, i.e., in phase, if the energy difference between pump- and Stokes-laser corresponds to a Raman transition ω_R. Another pump photon can be subsequently scattered inelastically

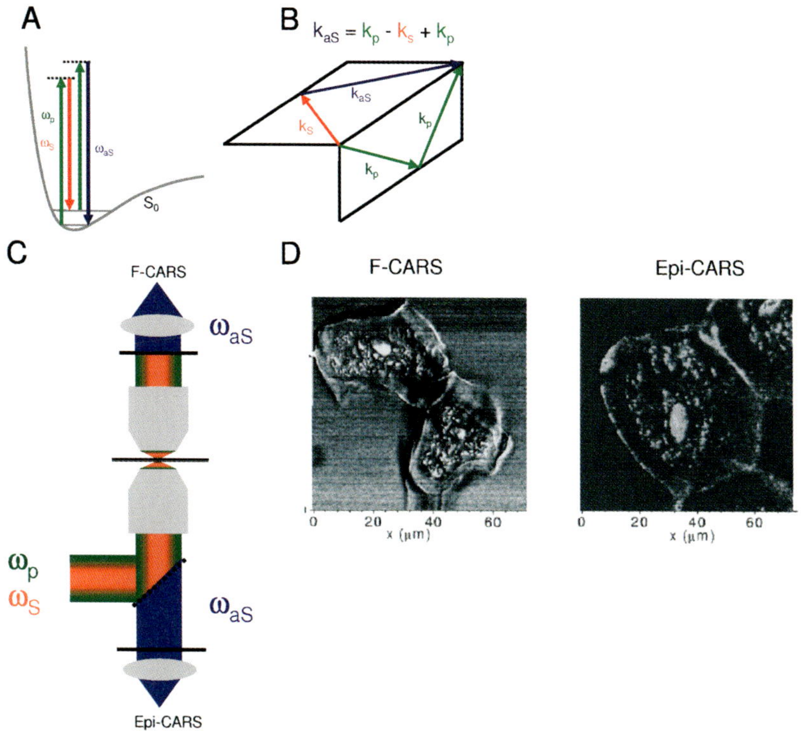

Figure 11.5 Energy level diagram (A) and wave vector diagram (B) to illustrate energy conservation and phase matching within a CARS process. (C): Schematic sketch of an F- and EPI-CARS microscope, respectively. (D): F- and EPI-CARS microscopy images of unstained living epithelian cells recorded for a Raman shift of 1579 cm^{-1} (F-CARS image) and 1570 cm^{-1} (EPI-CARS image), respectively. This wavenumber region is characterized by protein or nucleic acid vibrations.

off this ensemble of coherently excited vibrational modes by emitting an anti-Stokes signal ω_{aS} blue-shifted relative to the excitation lasers. The frequency of the anti-Stokes signal arises from energy conservation: $\omega_{aS} = 2\omega_p - \omega_S$ where $\omega_{aS} > \omega_p > \omega_S$ (see **Figure 11.5A**), while the direction of the CARS signal is determined by wave vector conservation. The sum of the wave vectors of the four participating photons must equal 0 ($\Delta k = 0$; $k_{aS} = k_p - k_S + k_p$, see **Figure 11.5B**) for the CARS signal to become maximal. In a commonly applied beam geometry fulfilling these phase-matching conditions, the three laser beams are arranged three-dimensionally leading to a maximum separation between the incoming lasers and the CARS signal. The main advantage of CARS spectroscopy compared to linear Raman spectroscopy is that the occurrence of fluorescence is not troublesome since the CARS signal is blue-shifted compared to the exciting lasers. Furthermore, the resulting Raman signals in CARS are orders of magnitude more intense than those obtained with the classical Raman effect. Since the CARS signal is a laser-like directed beam it is possible to work without any spectrometers, i.e., a high resolution is obtained. However, CARS spectroscopy requires costly equipment and is experimentally extremely complex. High-intensity (pico or femtosecond) pulses and at least two wavelengths-tunable short-pulse laser sources are needed to generate a CARS signal. In a CARS microscope two collinearly aligned laser pulses (ω_P and ω_S) are focused via a microscope objective with a wide aperture onto a small spot of the sample. The use of a microscope objective, i.e., strongly focused laser pulses, let the phase-matching condition become uncritical since the CARS signal is only generated via a relatively small area. The CARS signal can be detected either in the forward (F-CARS) [8] or the backward (EPI-CARS) [9] direction (see **Figure 11.5C**). For the F-CARS arrangement, a second microscope objective is needed to collect the CARS signal, which is separated from the excitation light by means of a suitable filter. An EPI-CARS signal is only generated if the size of the investigated sample is smaller than the wavelength of the exciting lasers. EPI-CARS offers the advantage that only one microscope objective is required. Thus a conventional fluorescence microscope can be easily modified into a CARS microscope. By selectively scanning the samples, 2D and 3D CARS images of special molecular vibrations can be obtained. These CARS images yield detailed chemical structure information about the investigated sample (see CARS microscopy images of **Figure 11.5**)

These few examples reveal that great progress is being made in changing the above mentioned scientific "daydream" into reality.

References

1 J. Huisken, J. Swoger, F. Del Bene, J. Wittbrodt, E.H. Stelzer, *Science* 305, (**2004**), p. 1007; P.J. Keller, F. Pampaloni and E.H.K. Stelzer, *Current Opinion in Cell Biology*, 18 (2006).

2 W. Denk, J. H. Strickler, W. W. Webb, *Science*, 248 (**1996**), p. 73.

3 K. Svoboda, W. Denk, W. H. Knox, S. Tsuda, *Opt. Lett.*, 21 (**1996**), p. 1411; C. Xu, W. Zipfel, J. B. Shear, R. M. Williams, W. W. Webb, *Proc. Natl. Acad. Sci.*, 93 (**1996**), p. 10763; R. M. Williams, W. R. Zipfel, W. W. Webb, *Curr. Opin. Chem. Biol.*, 5 (**2001**), p. 603.

4 K. König, I. Riemann, *J. Biomedical Optics*, 8 (**2003**), p. 432.

5 T. A. Klar, S. Jakobs, M. Dyba, A. Egner, S. W. Hell, *PNAS*, 97 (**2000**), p. 8206.

6 S. W. Hell, in: *Topics in Fluorescence Spectroscopy*, edited by J. R. Lakowicz, Plenum Press, New York, 1997, Vol. 5, p. 361.

7 M. Dyba, S. W. Hell, *Phys. Rev. Lett.*, 88 (**2002**), p. 163901-1.

8 A. Zumbusch, G. R. Holtom, X.S. Xie, *Phys. Rev. Lett.*, 82 (**1999**), p. 4142.

9 A. Volkmer, J.-X. Cheng, X.S. Xie, *Phys. Rev. Lett.*, 87 (**2001**), p. 023901; A. Volkmer, *J. Phys. D: Appl. Phys.*, 38 (**2005**), p. R59.

Index

Boldfaced page numbers refer to glossary entries.

Biophotonics: Visions for Better Health Care. Jürgen Popp and Marion Strehle (Eds.)

ISBN: 3-527-40622-0

d

e

n

o

t